BUS

**ACPL ITEM
DISCARDED**

AGILE MANUFACTURING:
THE 21ˢᵗ CENTURY COMPETITIVE STRATEGY

Elsevier Science Internet Homepage
>http://www.elsevier.nl (Europe)
>http://www.elsevier.com (America)
>http://www.elsevier.co.jp (Asia)

Consult the Elsevier homepage for full catalogue information on all books, journals and electronic products and services.

Elsevier Titles of Related Interest

BINDER et al.
Management and Control of Production and Logistics.
ISBN: 0-08-043036-8

HASSAN & MEGAHED
Current Advances in Mechanical Design and Production.
ISBN: 0-08-043711-7

KOPACEK & PEREIRA
Intelligent Manufacturing Systems.
ISBN: 0-08-043239-5

KOPACEK & NOE
Intelligent Assembly and Disassembly.
ISBN: 0-08-043042-2

KOPACEK
Multi-Agent Systems in Production.
ISBN: 0-08-043657-9

MOREL & VERNADAT
Information Control in Manufacturing.
ISBN: 0-08-042928-9

UME & SELKE
Mechatronics 2000.
ISBN: 0-08-043703-6

Related Journals

Free specimen copy gladly sent on request: Elsevier Science Ltd, The Boulevard, Langford Lane, Kidlington, Oxford, OX5 1GB, UK

Applied Ergonomics
Artificial Intelligence in Engineering
Automation in Construction
Computers and Industrial Engineering
Computers in Industry
Engineering Applications of Artificial Intelligence
International Journal of Industrial Ergonomics

International Journal of Production Economics
Journal of Manufacturing Processes
Journal of Manufacturing Systems
Journal of Operations Management
Journal of Strategic Information Systems
Mechatronics
Robotics and Computer Integrated Manufacturing

To Contact the Publisher

Elsevier Science welcomes enquiries concerning publishing proposals: books, journal special issues, conference proceedings, etc. All formats and media can be considered. Should you have a publishing proposal you wish to discuss, please contact, without obligation, the publisher responsible for Elsevier's industrial and production engineering publishing programme:

>Dr Martin Ruck
>Publishing Editor
>Elsevier Science Ltd   Phone: +44 1865 843230
>The Boulevard, Langford Lane   Fax: +44 1865 843920
>Kidlington, Oxford, OX5 1GB, UK   E-mail: m.ruck@elsevier.co.uk

General enquiries, including placing orders, should be directed to Elsevier's Regional Sales Offices – please access the Elsevier homepage for full contact details (homepage details at top of this page).

AGILE MANUFACTURING: THE 21st CENTURY COMPETITIVE STRATEGY

A. Gunasekaran

*Department of Management,
University of Massachusetts,
North Dartmouth, USA*

2001

ELSEVIER

AMSTERDAM · LONDON · NEW YORK · OXFORD · PARIS · SHANNON · TOKYO

ELSEVIER SCIENCE Ltd
The Boulevard, Langford Lane
Kidlington, Oxford OX5 1GB, UK

© 2001 Elsevier Science Ltd. All rights reserved.

This work is protected under copyright by Elsevier Science, and the following terms and conditions apply to its use:

Photocopying
Single photocopies of single chapters may be made for personal use as allowed by national copyright laws. Permission of the Publisher and payment of a fee is required for all other photocopying, including multiple or systematic copying, copying for advertising or promotional purposes, resale, and all forms of document delivery. Special rates are available for educational institutions that wish to make photocopies for non-profit educational classroom use.

Permissions may be sought directly from Elsevier Science Global Rights Department, PO Box 800, Oxford OX5 1DX, UK; phone: (+44) 1865 843830, fax: (+44) 1865 853333, e-mail: permissions@elsevier.co.uk. You may also contact Global Rights directly through Elsevier's home page (http://www.elsevier.nl), by selecting 'Obtaining Permissions'.

In the USA, users may clear permissions and make payments through the Copyright Clearance Center, Inc., 222 Rosewood Drive, Danvers, MA 01923, USA; phone: (978) 7508400, fax: (978) 7504744, and in the UK through the Copyright Licensing Agency Rapid Clearance Service (CLARCS), 90 Tottenham Court Road, London W1P 0LP, UK; phone: (+44) 207 631 5555; fax: (+44) 207 631 5500. Other countries may have a local reprographic rights agency for payments.

Derivative Works
Tables of contents may be reproduced for internal circulation, but permission of Elsevier Science is required for external resale or distribution of such material. Permission of the Publisher is required for all other derivative works, including compilations and translations.

Electronic Storage or Usage
Permission of the Publisher is required to store or use electronically any material contained in this work, including any chapter or part of a chapter.

Except as outlined above, no part of this work may be reproduced, stored in a retrieval system or transmitted in any form or by any means, electronic, mechanical, photocopying, recording or otherwise, without prior written permission of the Publisher.
Address permissions requests to: Elsevier Science Global Rights Department, at the mail, fax and e-mail addresses noted above.

Notice
No responsibility is assumed by the Publisher for any injury and/or damage to persons or property as a matter of products liability, negligence or otherwise, or from any use or operation of any methods, products, instructions or ideas contained in the material herein. Because of rapid advances in the medical sciences, in particular, independent verification of diagnoses and drug dosages should be made.

First edition 2001

Library of Congress Cataloging in Publication Data
A catalog record from the Library of Congress has been applied for.

British Library Cataloguing in Publication Data
A catalogue record from the British Library has been applied for.

ISBN: 0-08-043567-X

∞ The paper used in this publication meets the requirements of ANSI/NISO Z39.48-1992 (Permanence of Paper).
Printed in The Netherlands.

PREFACE

Manufacturing has undergone many evolutionary stages and paradigm shifts. The paradigm shifts in going from a craft industry to mass production, then to lean manufacturing, and finally, to agile manufacturing (AM). The concept of agility (flexible and quick responsive manufacturing) will reduce time to reach market with appropriate products/services.

Businesses are restructuring and re-engineering themselves in response to the challenges and demands of the 21st century. The businesses of the 21st century will have to overcome the challenges of demanding customers who will seek high-quality, low-cost products that are relevant to their specific and rapidly changing needs. The time during which many companies competed based primarily on price tag has gone. Now is the time for companies to compete in the global marketplace, and "push the envelope" in delivery-response, product quality, and overall excellence in customer service and customer satisfaction. Agility addresses new ways of running companies to meet these challenges.

Agile manufacturing is defined as the capability of surviving and prospering in a competitive environment of continuous and unpredictable change by reacting quickly and effectively to changing markets, driven by customer-designed products and services. Critical to successfully accomplishing AM are a few enabling technologies such as the standard for the exchange of products (STEP), concurrent engineering, virtual manufacturing, component-based hierarchical shop floor control system, information and communication infrastructure, etc.

The aim of the book is to help the students at the undergraduate and graduate levels, senior managers and researchers in understanding and appreciating the concepts, design and implementation of Agile Manufacturing systems (AMS). One should be able to understand, develop and implement appropriate agile manufacturing strategies after reading this book.

The scope of the book is to present the undergraduate and graduate students, senior managers and researchers in manufacturing systems design and management, industrial engineering and information technology with the conceptual and theoretical basis for the design and implementation of AMS. The book emphasizes on systems methodology approach for the design and implementation of AMS. Also, the book focuses on broad policy directives and plans of agile manufacturing that guide the monitoring and evaluating the manufacturing strategies and their performance. A problem solving approach is taken throughout the book, emphasizing the context of agile manufacturing and the complexities to be addressed.

This book provides a much needed comprehensive coverage of materials required developing and implementing agile manufacturing strategies and systems. The book includes the concept, theory, modelling, and architecture of agile manufacturing system. It covers the state-of-the-art, concepts and methodologies of manufacturing strategy development taking into account the current development in information technologies and the overall trend in agile manufacturing. The book is expected to assist the companies in formulating 21st century manufacturing strategies to flourish in the competitiveness of manufacturing.

The book presents original works and interesting case studies arising from research with the evolving technologies and production concepts of agile manufacturing. The book aims to promote the ideas and technologies that promote agile manufacturing as company wide strategies to reduce the lead times at all stages of manufacturing. The chapters offer ideas on lowering manufacturing costs, increasing market share, satisfying the customer requirements, rapid introduction of new products, eliminating non-value added activities and enhancing manufacturing competitiveness. The book is organized into six parts to cover the introduction, design and development, information technology/systems, supply chain management, operations management and strategies of agile manufacturing. The chapters deal with the following areas as a part of agile manufacturing system development: concepts, strategies and enablers of agile manufacturing, virtual enterprise, managing people in agile organizations, product development in agile environment, application of Information Technology/Systems in agile manufacturing, supply chain management in agile environment, operations planning and control in agile manufacturing enterprise and some strategic approach for the development of agile manufacturing.

The Editor of this book acknowledges Dr. Martin Ruck, Publishing Editor for Industrial Engineering and Control and the staff of Elsevier Science Ltd (UK) for their great support in completing this book.

I am most grateful to my wife, Latha Parameswari and my son Rangarajan for their generous support and permission for staying away during family time in order to complete this edited book. I appreciate the countless hours spent by my wife for proof reading of all the chapters. My heartfelt thanks go to all the authors who have contributed chapter(s) to this book. Without their contribution and overall support, this book should have been hardly realized.

A. Gunasekaran
Editor
June 6, 2000

CONTENTS

Preface   v

PART I - INTRODUCTION TO AGILE MANUFACTURING

Agile Manufacturing as the 21st Century Strategy for Improving Manufacturing Competitiveness   3
Henrique Luiz Corrêa

Agile Manufacturing: Concepts and Framework   25
A. Gunasekaran, R. McGaughey and V. Wolstencroft

PART II - DESIGN AND DEVELOPMENT OF AGILE MANUFACTURING SYSTEMS

A Strategic Approach to Develop Agile Manufacturing   53
Jens O. Riis and John Johansen

BM_Virtual Enterprise Architecture Reference Model   73
G.D. Putnik

Integrated Product/Process Development (IPPD) Through Robust Design Simulation (RDS)   95
Daniel P. Schrage and Dimitri N. Mavris

Developing the Agile Enterprise   113
John Bessant, David Knowles, David Francis and Sandra Meredith

Towards Building of Knowledge-Base in Indian Corporations: Some Strategic Directions   131
R.P. Mohanty

Enhancing Agility in Manufacturing: The Role of QFD   157
David Ginn, Mohamed Zairi and P.K. Ahmed

Product Development Strategies for Agility   175
Sudi Sharifi and Kulwant S. Pawar

Managing People in Agile Organisations   193
David Francis

PART III - INFORMATION TECHNOLOGY/SYSTEMS IN AGILE MANUFACTURING

Application of Information Technology in Agile Manufacturing *Henry C.W. Lau and Eric T.T. Wong*	205
Information Systems for Agile Manufacturing Environment in the Post-Industrial Stage *S. Subba Rao and A. Nahm*	229
Management of Complexity and Information Flow *E. Szczerbicki*	247
An Object-Oriented Optimization-based Software for Agile Manufacturing in Process Industries *M. Draman, I.K. Altinel, N. Bajgoric, A.T. Unal and B. Birgoren*	265
Application of Multimedia in Agile Manufacturing *Ronald E. McGaughey*	279
Computational Intelligence in Agile Manufacturing Engineering *Kesheng Wang*	297
Computer Applications in Agile Manufacturing *M.A. Pego Guerra and W.J. Zhang*	317
Secure Communication in Distributed Manufacturing Systems *István Mezgár and Zoltán Kincses*	337

PART IV - SUPPLY CHAIN MANAGEMENT IN AGILE MANUFACTURING

Agile Supply Chain Management *J. Sarkis and S. Talluri*	359
Engineering the Agile Supply Chain *Denis R. Towill*	377
Information Technologies for Virtual Enterprise and Agile Management *Nijaz Bajgoric*	397
Early Supplier Involvement: A Design-Based Sourcing *Shad Dowlatshahi*	417
Information Technologies in Supply Chain Management *Alexander V. Smirnov and Charu Chandra*	437
Enterprise Integration and Management in Agile Organizations *F.B. Vernadat*	461

Agility, Adaptability and Leanness: A Comparison of Concepts and a Study of Practice   483
Hiroshi Katayama and David Bennett

PART V - OPERATIONS OF AGILE MANUFACTURING SYSTEMS

Computer Control of Agile Manufacturing Systems   499
Robert W. Brennan

Computer Aided Process Planning for Agile Manufacturing Environment   515
Neelesh K. Jain and Vijay K. Jain

Aggregate Capacity Planning and Production Line Design/Redesign in Agile Manufacturing   535
Z. -S. Hua and P. Banerjee

The Control Problems of Agile Manufacturing   559
B. Ilyasov, L. Ismagilova and R. Valeeva

Contingency-Driving Autonomous Cellular Manufacturing - Best Practice in the 21st Century   583
S.-J. Song

Role of IT/IS in Physically Distributed Manufacturing Enterprises   601
Walter W.C. Chung and Michael F.S. Chan

The Method of Successive Assembly System Design Based on Cases Studies within the Swedish Automotive Industry   621
T. Engström, D. Jonsson and L. Medbo

PART VI - STRATEGIC APPROACH FOR AGILE MANUFACTURING

Reengineering and Agile Manufacturing Development   645
G. Doumeingts, H. Kromm, Y. Ducq and S. Kleinhans

Corporate Knowledge Management in Agile Manufacturing   665
Michael Thie and Dragan Stokic

Agile Manufacturing Strategic Options   685
Vicky Manthou and Maro Vlachopoulou

Virtual Enterprise Engineering in Support of Distributed and Agile Manufacture   703
R.H. Weston, R. Harrison and A.A. West

Putting the Pieces Together Using Standards   735
Ricardo Jardim-Gonçalves and Adolfo Steiger-Garção

Enterprise Integration and Management   759
R.H. Weston and A. Hodgson

Gaining Agility Through Supply Chain Management 785
Tareq Suleman and Mohamed Zairi

AUTHOR INDEX 809

Part I

INTRODUCTION TO AGILE MANUFACTURING

Agile Manufacturing as the 21st Century Strategy for Improving Manufacturing Competitiveness

Henrique Luiz Corrêa[a]

[a]Production and Operations Management Department, São Paulo Business School (FGV)
Av 9 de Julho, 2029 / 10º andar 01313-902 São Paulo, Brazil. hcorrea@fgvsp.br

1. DEFINITION OF MANUFACTURING STRATEGY

Manufacturing strategy has increasingly been regarded by academics and practitioners as having an important contribution to make to enhance competitiveness. The growth of the literature in manufacturing strategy has matched the growth of interest in the area. Within the literature three main reasons are identified for this newly found importance.

The first is the increased pressure owing to the growing international manufacturing competitiveness made more intense by the recent movement towards globalisation. The second is the increased potential to be gained from the development of new manufacturing technologies, the potential of which grows much faster than our ability to use it for competitive benefits and, the third is the development of a better understanding of the strategic role of manufacturing. Five characteristics can be listed to help understand the need for a strategic management of the manufacturing function:

- Manufacturing in general involves the bulk of the company's assets and human resources
- Many decisions regarding manufacturing resources require a long time to take effect therefore requiring a long term outlook of the future to support them
- Once made, many of these decisions will normally take a long time and substantial amounts of resources to revert
- Manufacturing decisions directly affect the way companies can compete in the market place because it is increasingly accepted that there is not such a thing as a "best way" to manage manufacturing resources - different configurations of manufacturing resources will result in different levels of manufacturing performance in different aspects (e.g. delivery, flexibility, quality and cost)
- Manufacturing decisions have to support and be supported by other functions in order to properly support the business strategy of the company, therefore requiring strategic orientation

Manufacturing strategy can be defined as a framework whose objective is the increased competitiveness of the organisation: to achieve this it should aim at designing, organising, managing and developing the company's manufacturing resources and shape a consistent pattern of manufacturing decisions in order that they can result in an adequate mix of performance characteristics which will allow the company to compete effectively in the future.

2. THE CHANGING INTERNATIONAL MANUFACTURING COMPETITION IN THE ´60s, ´70s AND ´80s

During the 60s, 70s and 80s, the relative competitive positions occupied by the formerly leading industrial countries changed substantially. Some traditional industrial nations were outperformed by other countries, of which Japan was the most evident example. The United States and the United Kingdom had their leading positions challenged and in many cases lost them, e.g. in the automobile market, long dominated by American companies.

Considering the Japanese manufacturing industry, Buffa (1984) notices that the industries in which they excelled during that period - motor cycles, domestic appliances, automobiles, cameras, hi-fi, and steel production – there were, already developed markets with established market leaders. According to the author, Japanese companies may have succeeded, partially because of their Finance and Marketing related skills, but largely because of the high quality and low cost which they achieved through a sharp manufacturing practice which most of the Western manufacturers initially were not able to match. Japanese companies were using the improvements which they had been achieving in manufacturing as their main competitive advantage, as opposed to the Western companies, which had considered manufacturing as a 'solved problem', focusing their attention on getting competitive advantage through achieving excellence in marketing their products and managing their financial issues.

Not only were Japanese companies on average more cost efficient than most Western companies (though there were many exceptions of Western companies which had maintained or improved their competitive position in the world market during those decades), but they were competing and winning based also on their better quality and reliability performance as well as on their better responsiveness to the market needs and opportunities. In the introduction of new products, for instance, Japanese car manufacturers had cut their product development times (the period between the earliest stages of design and the manufacture of a new model) to an average of less than four years compared to six to eight years in Europe and America of the ´70s.

There is, in general, agreement that (initially, at least) Western companies lacked an effective response to the Japanese challenge. According to the literature, the reasons behind this lack of an effective response are various. Hayes and Wheelwright (1984), in their now classic book, summarize some of them in five main points:

Financial considerations The assessment of companies and their manager's performance based predominantly on short term considerations may have induced managers to avoid long term investments which might have resulted in a more effective manufacturing. Managers may not have decided to invest in improvements whose results would only show in the long term because they needed short term performance.

Technological considerations Western managers would have been less sophisticated, imaginative and even less interested in dealing with technological considerations than the overseas competitors, focusing attention predominantly on financial and marketing issues.

Excessive specialization and/or lack of proper integration Western managers would have tended to separate complicated issues into simpler, specialized ones to a greater degree than their foreign counterparts without having developed proper integration to pull the differentiated responsibilities together and to be able to deal with the total picture.

Lack of focus The separating and specializing mentality would have led many Western firms to diversify away from their core technologies and markets. They would have tended to adopt the *portfolio* approach, used by stocks and bonds investors. This approach considers that diversifying is the best way to hedge against random setbacks. Manufacturing, however, would not be subject only to random setbacks but, more significantly, to carefully orchestrated attacks from competitors who focus their resources and energy on one particular set of activities. Focused manufacturing is based on the idea that simplicity, repetition, experience and homogeneity in manufacturing tasks breed competence (Skinner, 1974).

Inertia Skinner (1985) observed that most factories in the Western world were not managed very differently in the 1970s from the way they were in the 1940s or 1950s. Such practices might have been adequate when production management issues centered largely on efficiency and productivity. However, the problems of operations managers moved far beyond mere physical efficiency. On top of this, managers considered that the production problems were solved, directing attention and resources toward other issues such as distribution, packaging and advertising. According to Hill (1995), there had been a failure, conscious or otherwise, of Western industries and the society at large to recognize the size of the foreign competitive challenge, its impact on their way of life, and consequently to recognize the need for change.

The result of the concurrence of the five factors above is that Western plants and equipment were allowed to age in all senses. What one day had been technological advantage, was eroded by the decline in expenditure and attention to issues such as new products research and development and new process technologies (Hayes and Wheelwright, 1984). Then, Hayes and Wheelwright conclude, "in the beginning of the 1970s, US companies found themselves pitted against companies that did compete on dimensions such as defect-free products, process innovation and delivery dependability. Increasingly, they found themselves displaced first in international markets and then in their home market as well".

2.1. The development of a better understanding of the strategic role of manufacturing

Since the seminal work of Skinner (see e.g. Skinner, 1969), a number of authors have addressed the strategic role of the manufacturing function. Hayes and Wheelwright (1984) and later Hayes et al. (1988) called attention to the need to transform the manufacturing role from being primarily reactive to being *proactive*, where the manufacturing function contributes actively to the achievement of competitive advantage.

Another point which some authors make, e.g. Slack (1991) refers to the fact that the complexity of the manufacturing function calls for strategic management. According to Slack, manufacturing is almost certainly the largest (both in terms of people and capital employed), and probably the most complex and arguably the most difficult of all the functions within the organization to manage.

Hill (1995) argues that the need for a manufacturing strategy to be developed and shared by the business has to do not only with the critical nature of manufacturing within the corporate strategy but also with a realization that many of the decisions in manufacturing are structural in nature. Therefore, unless the issues and consequences are fully appreciated by the business, then it can be locked into a number of manufacturing decisions, which may take years to change. Changing them is costly and time consuming, but even more significantly, the changes will possibly come too late.

More recently some authors (Hayes and Pisano, 1996; Teece and Pisano, 1994; Pisano, 1997; Alher, 1998) have added to the debate by arguing that the recently developed resource-based view of strategic management should play an important role in the development of manufacturing strategy - the resource-based view would help manufacturing strategies to be more difficult to copy resulting in more sustainable competitive advantages. This concept will be further developed later in this chapter.

2.2. Focused manufacturing: a controversial concept

Although the manufacturing function is regarded as one of the most complex to manage within the organization, what creates the complexity is not the technology dimension but the number of aspects and issues involved, the interrelated nature of these and the level of fit between the manufacturing task and its internal capability (Hill, 1995). The level of complexity involved depends largely on corporate and marketing strategy decisions, made within the business, where the competitive priorities are established. These competitive priorities are established because a manufacturing system cannot excel in all aspects of performance at the same time. Trade-offs must be made. Different types of performance demand different manufacturing resources organized in different ways (Slack, 1991; Skinner, 1996). An organization which competes predominantly on cost efficiency, for instance, by manufacturing in high volumes, would need different resources (possibly more dedicated machines) in order to compete effectively if compared to an organization competing on product customization, making products to order (which would possibly call for more general purpose flexible equipment).

This is the rationale behind the concept of *focused manufacturing*. According to this view, for the effective support of competitive business strategy the manufacturing function should focus each part of its manufacturing system on a restricted and manageable set of products, technologies, volumes and markets so as to limit the manufacturing objectives in which it is trying to excel. This means that if an organization has different products or product groups competing in different ways, then its manufacturing function should reflect this in the way it is subdivided and organised so as to maintain focus on what is most important for its competitiveness in the market place.

If a company competes on a broad range of products, the decision to adopt the concept of focused manufacturing can have the disturbing implication of calling for major investments in new plants and new equipment to break down the existing complexity. One alternative approach, which helps to avoid major investments, is a solution that does not involve selling big multipurpose facilities and decentralizing them into small ones. The solution could be the more practical approach of the 'plant-within-a-plant', where the existing facility is divided both organizationally and physically into plants within the original plant. Each of them would have its own facilities. Each plant-within-the-plant can in this way concentrate on its particular manufacturing task, using its own work force management approaches, production control systems, organizational structure and so forth. Each plant-within-the-plant would quickly gain experience by focusing and concentrating every element of its work on those limited essential objectives which constitute its manufacturing task or focus.

The idea of focus should thus permeate all the process of formulation and execution of the business and manufacturing strategies. The establishment of competitive priorities and the decision making process should also take the idea of focus into consideration, in order to make sure that the manufacturing function can really excel in what it is expected to.

Although it is intuitive and appealing, having gained broad support among academics and practitioners, a number of authors (see e.g. Schonberger, 1986) have challenged the idea of focus in manufacturing strategy. Inspired by the Toyota-developed Japanese just-in-time system, the "lean production" system advocates argue that trade-offs do not exist (since at a certain point in time some Japanese companies outperformed western competitor companies in all aspects of performance) and that the principles on which "lean manufacturing" rests:

- broadly trained rather than specialised people;
- people empowered to identify and solve production problems in teams;
- horizontal and informal communication rather than through hierarchical paths;
- emphasis on production throughput flow rather than resource utilisation;
- production flows pulled by demand rather than pushed by centrally defined schedules;
- product based rather than process based layout;
- no acceptable level of defective production;
- inventory is considered as waste and set-ups should be minimised;
- continuous improvement and waste fighting initiatives are central;
- cooperative and long term rather than adversarial supplier relationships; and,
- product development related activities done concurrently by cross functional teams

Would be the "one best way" to organise and manage manufacturing. But is it? Many authors disagree. Hayes and Pisano (1996) for example argue that although many companies experienced improvements by implementing one or more of the "lean manufacturing" principles, "this does not assure that it will be successful financially. For example, the winners of the [American] national Baldridge Award, which recognises American companies that have been unusually successful in improving their quality, productivity and customer satisfaction, have done well on average - however, some of them, entered Chapter 11 soon after receiving the Baldridge, and others (like General Motors, IBM and Westinghouse) soon thereafter began experiencing highly visible problems.

Even more disturbing, a number of Japanese companies are beginning to question many of the same approaches [...] Toyota's newest factory in Japan utilizes neither the JIT system nor mixed model assembly".

Arguments about the trade-offs in manufacturing have sometimes been polarised in two approaches - some of the advocates of "lean manufacturing" argue that trade-offs do not exist. Contrary to this some of the more radical advocates of the trade-off idea sometimes ignore the fact that even considering that trade-offs exist, they are dynamic rather than static in nature and that trade-off relationships can be altered in a number of ways. One of the interesting models to describe trade-offs in manufacturing is Slack's see-saw analogy. According to Slack (1991) manufacturing management is sometimes portrayed as consisting almost entirely of handling trade-offs. Trading off high finished goods inventory against good product availability, trading off expensive preventative maintenance against the reliable provision of capacity are some examples. Improvement in one place should be paid for elsewhere. Schematically this idea can be seen in Figure 1.

When performance objective 2 is improved, performance objective 1 suffers, at least in the short term (B). One example in the field of inventory management would be the trade-off between cost efficiency (associated with lower inventories) and custom service defined as good product availability. If in the short term a company decides to improve service level, one way of doing it is by increasing finished goods inventory. Having done that, it then may be possible to re-gain the lost level of cost efficiency (by reducing inventory levels) without jeopardising the newly acquired improved level of customer service (C) - for example by reducing lead times or improving future demand knowledge (via e.g. improving forecasting systems or better coordinating with the customer and this way, with less uncertainty, less buffer inventory would be needed), represented by the movement of the see-saw pivot.

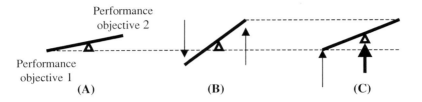

Figure 1. Slack´s (1991) see-saw represents the dynamic nature of trade-offs

None of this pivot moving alternatives however can normally be achieved in the very short term. They are initiatives, which normally take longer than simply increasing inventory levels. This means that managing the trade-offs between performance aspects of manufacturing performance does not mean only managing the position of a static see-saw (which in many situations can be altered in the short term), but it also means managing the movements of the see-saw pivot (which normally takes longer). Hayes and Pisano (1996) add to Slack´s point arguing that trade-offs should be managed considering not only the improvements in each of the performance objectives but also the knowledge and learning that each of different possible dynamic improvement paths will bring to the organisation. The idea of dynamic improvement paths is interesting. Let us use another form of representation for the idea of dynamic trade-offs in relation to the trade-off between service level and inventory level. One of the simplest models used to dimension safety stocks of inventory items (the demand of, which is approximately constant, is):

$$SS = SF \times \sigma_{LT} \tag{1}$$

Where:
SS = safety stock level
SF = safety factor
σ_{LT} = standard deviation of demand (compared to forecast) during replenishment lead time

The *SF* (safety factor) is defined as a function of the service level intended to be offered to the customer (see Chase et al., 1998 for a detailed treatment of safety stocks - the idea here is just to use this simple model as an illustration). Assuming that demand forecast errors behave normally and with some help from statistics, the plotting of a graph relating safety stock level and service level, results in something like what is shown in Figure 2.

This somewhat simplistic model can be used to show the idea of dynamic paths. Movements along the trade-off borders 1 and 2 represent Slack´s "satic" pivot see-saw movements - if one wants to increase service levels one way to do it is surely to increase the levels of inventory (therefore jeopardising the objective of cost efficiency). However, as it can be seen by the formula (1), one can alter the level of service without changing the level of inventory - by changing the other factor of the right hand side term - the standard deviation of the demand forecast during replenishment lead time.

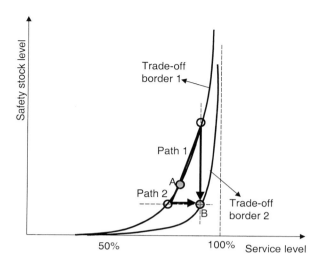

Figure 2. Graph representing Hayes and Pisano (1996) dynamic trade-off paths

If the standard deviation is reduced for example, the trade-off border changes from trade-off border 1 to trade-off border 2. To illustrate Hayes and Pisano (1996) dynamic path idea, one can imagine a manager intending to go from status A to status B, in figure 2, therefore improving both aspects - service level and inventory-related cost efficiency. Several paths of improvement could be chosen. Among them, two are used to illustrate the point: path 1 and path 2. Path 1 would mean first to increase inventory levels (which can be done relatively quickly) to achieve increased service level ("static" pivot see-saw movement) and then to set off efforts to reduce inventory levels without reducing service level through e.g. improving forecasting methods (which takes longer and requires a particular set of capabilities development). The other path which could be used to achieve the same status B (coming from status A) which is path 2, means a different sequence of actions. First, the inventory levels would be reduced and then efforts would be made to increase service levels without increasing inventory levels again, by for example, using JIT-type techniques of identifying production problems by reducing inventory levels and then acting selectively and constantly to tackle such problems (which will result in a rather different set of capabilities being developed). Path 1 is more centred in the traditional methods whereas path 2 is more towards JIT-type management. The final status (B) is the same, but Hayes and Pisano (1996) argue that depending on the path chosen the learning experience which the hypothetical company would go through would differ considerably and therefore the ability of the company to face future competitive challenges would also differ considerably. The conclusion is: trade-off analysis is not as simple as the radical advocates of "lean manufacturing" (one best way) would have liked and they are not as simple as a "static" analysis would have us believe either. Trade-offs exist and will probably always exist but their treatment requires an in depth understanding of the dynamics and dynamic paths involved in each particular situation under analysis.

2.3. Why Manufacturing Strategies for Improving Productivity and Quality

Basically the whole movement set off by Skinner's seminal articles in the beginning of the '70s was basically an attempt of western astonished manufacturing academics and practitioners to understand and respond to the competitive challenged posed by the suddenly successful Japanese companies who had quickly taken a substantial share of the world export market from them. In terms of Slack's see-saw model, Japanese companies had found out how to move the pivots while western companies had been complacently managing the "static" pivot see-saw movements only. One of the formerly accepted trade-offs which the Japanese companies challenged was one between high levels of conformance quality (the ability of the production system to produce outputs according to specifications) and cost efficiency. In the traditional manufacturing systems inspectors would sort good from bad products at the end of the production line - if a company wanted more quality, more inspectors (with the corresponding increased inspection cost) would be needed. Quality used to cost. Japanese companies changed this paradigm (they moved the pivot), by giving operators the responsibility and the ability to detect and solve quality problems, re-directing the attention from product quality to process quality. More conformance quality therefore would not necessarily mean more costs. Quality started to be considered as free (Crosby, 1979).

Western managers started to realise that their role should change: the traditional mass production approach which had reduced the manufacturing strategic contribution to "reducing costs" to something more complex and relevant: the purpose of the strategic management of manufacturing would have to change to specifying the kind of competitive advantage that a company is seeking in the market place and to articulate how that advantage is to be achieved (Hayes and Pisano, 1996). However, the challenge in the '80s was basically one of catching up with the Japanese companies and the most important trade-off involved was the one between cost efficiency and quality: Western companies had to manage better the things which were under their control e.g. levels of defect and wasted manufacturing resources. The '90s brought a different environment. Japanese companies used the lead they had simultaneously achieved in quality and cost and while the western companies spent all their efforts to catch up with them, they had started directing efforts towards moving more pivots - e.g. that between flexibility and cost efficiency, for example, based on set-up time reductions via both technology (flexible automation) and methodology (quick changeover techniques based on different more rational non-technology based methods - see for e.g. Shingo, 1985). At the same time, markets had became increasingly turbulent, globalisation had taken place and technology had reached unprecedented development rates. The challenge then was not only to manage things better which were under control (such as product quality variability) but to manage things better which were not completely under control - to manage better the unexpected change.

3. THE NEW MANUFACTURING ENVIRONMENTS OF THE '90s AND THE 2000s

The manufacturing environments of the '90s and the 2000s have been and will be considerably different from those of previous decades.

Information technology has remarkably changed the patterns of integration and communication within as well as between companies and between companies and consumers. The ERP-type integrated management systems have broadly been adopted and although one could argue that the results were not as spectacular as the consultancy companies and software houses had announced, the levels of integration and communication between customers and suppliers (internal and external) which the companies have achieved so far are superior than the levels they used to work with without the integrated systems. With the integrated systems being connected with the Internet (a reality now) new virtually endless opportunities are available to the companies who are competent enough to use them for competitive benefit. However, one should not make the usual mistake - assuming that it is enough to *possess* the technology to ensure a *good use* of the technology. These are actually different things and anyone who has managed a manufacturing operation knows that. Sometimes some authors tend to neglect this fact considering that operational excellence is easily copied because manufacturing operations are increasingly technology oriented and technology is easily traded. They sometimes mix concepts. Having an integrated software system plugged in, for example, is in fact easy and increasingly cheap. However when one looks at how comparatively well companies use the resources made available by such systems one starts seeing huge differences. So having the same technology is easy; using that technology for the company´s competitive benefit is not - and therefore it is not easily copied.

The rate at which technology has evolved requires, more than ever, that manufacturing managers are *proactive* in anticipating and understanding the newly available technologies and their impact on the company´s competitive performance, both in terms of information, product and process technologies.

Customer requirements are increasingly demanding because competition is increasingly global and fierce and, competitors are increasingly competent. Customers already want it "here, now and customised" (McKeena, 1997) - that means achieving levels of agility never required before from the production systems. That means a level of ability to ensure consistency between manufacturing actions and strategic direction, which was never required before. Required changes in the strategic direction must be quickly mirrored by changes in the pattern of manufacturing decisions. In the same way, changes in the manufacturing resources, newly developed competencies and newly available technologies should also be able to quickly change the strategic direction of the company changing for example the marketing aim to market segments which better value the newly acquired or developed competencies.

4. ROLE OF NEW MANUFACTURING CONCEPTS AND TECHNOLOGIES

4.1. The development of new manufacturing technologies

Manufacturing Technology is regarded as one of the most important decision areas within the manufacturing management function. Traditionally, manufacturing management has influenced manufacturing technology to a much greater extent than the other way round.

Changes in the manufacturing technology were for a long time slow and gradual not calling for profound changes in its management methods and techniques. With the new micro-electronics and information handling technology being quickly incorporated into the process technologies, the resulting changes were not gradual and did not follow the usual pattern. A new paradigm was established. Computer controlled flexible machines challenged the once well established concept of *economies of scale* because they have the potential of making changeover times negligible. The concept of *economies of scope* started to gain importance. Economies of scope (Goldhar and Jelinek, 1983) are said to occur when one production unit can produce a given level of outputs of a variety of products at an unitary cost which is lower than that obtained by a set of separated production units, each producing one product at the same level of output.

The new flexible technology made it possible to produce different products at the same rates, which had only been possible with mass production, with single, or a few products. The strict one-to-one relationship between product and process life cycles would not apply any more (Stecke and Raman, 1986).

In summary, without a clear strategic direction with regard to manufacturing, the new manufacturing technologies can become an expensive 'solution in search of a problem'. In this sense, one of the aims of manufacturing strategy is to give the organization strategic direction with regard to manufacturing issues, technology included, making sure that not only the technologies but also the people and the infrastructure used are consistent with the strategic objectives of the business.

4.2. The resource-based view

The more popular paradigm for approaching competitive strategy has been based on the notion of strategic fit (Hayes & Pisano, 1996). Porter´s (1980) book, "Competitive Strategy" became possibly the most celebrated book in the field. Recognising the existence of trade-offs, Porter argued that the goal of business strategy is to seek sustainable competitive advantage by positioning oneself within industries and businesses that are either structurally attractive or can be made so through deliberate actions. According to Porter, competitive advantage is strongly linked with the idea of good positioning. In the ´90s, Prahalad and Hamel (1990) added to this debate challenging Porter´s ideas by advocating that companies should focus on building "core competencies" that could create competitive advantages in a variety of markets. They argue that only competencies which are difficult to copy actually make a company sustainable competitive and therefore a company who positions itself and then develops the needed competencies will have their recently acquired competencies easily copied and therefore the advantage will not be sustainable. Teece and Pisano (1994) called the attention to the dynamic aspects of the resource-based view, arguing that not only are the capabilities to be developed important but that the mechanisms by which new skills and capabilities are built have an important role to play because they influence the learning processes and knowledge base of the company and these will influence the ability of the company to compete in the future.

The resource-based approach is markedly different from the traditional manufacturing strategy paradigm.

According to most of the early authors, the manufacturing strategy development should follow a predominantly top-down approach. Skinner (1985), Fine and Hax (1985), Gregory and Platts (1990), Slack (1991) and, to a certain extent, Hill (1995), suggest hierarchical models in which the corporate strategy drives the business strategy. This in turn drives the strategies of manufacturing and other functional areas within the business unit. In fact, the manufacturing strategy formulation process has not received as much attention as the manufacturing strategy contents - objectives and decision areas - in the literature (Leong et al., 1990). Among the pioneers in the field, Hill (1995) seems to have been one of the few who actually delved into a more detailed discussion on it, proposing a specific framework to guide the development process on a (also predominantly top-down) step-by-step basis. Rather, the authors in the field tend to focus their work primarily on the manufacturing strategy objectives and decision areas. This approach, according to Leong et al. (1990), seems to consider some sort of implicit process, which depends on breaking manufacturing down into a number of decision areas and making the goals of manufacturing explicit in terms of a number of performance criteria. The steps of identifying these criteria, prioritising them and relating the decision areas to them would form the implicit process. Hayes and Wheelwright (1994), for instance, although describing four stages along a "continuum", which represents the evolution of manufacturing's strategic role, where the key aspect of evolution is the increasing, more proactive involvement of manufacturing in the firm's strategic needs, do not describe how a company should go about reaching the more advanced stages.

The exclusive top-down traditional planning approach does not seem to be adequate for the future - planning is only of use when a good level of stability is present. Otherwise it may easily become a futile exercise. In the future the only certainty companies will face is that changes will be larger, more sudden and quicker than ever before therefore requiring more agile manufacturing strategy development and implementation processes.

5. NEED FOR AGILE MANUFACTURING STRATEGY PROCESSES

The authors in the field of manufacturing strategy are more prolific in prescribing *what* to do than *how* to do it. There is however some authors whose work can help in the difficult task of developing a manufacturing strategy in real situations. Two examples are the worksheets developed by Gregory and Platts (1990), which are interesting tools for helping define the priorities for manufacturing and, the importance-performance matrix proposed by Slack (1991), which is both simple to understand and use and effective in giving managers a clear idea of what performance aspect needs urgent action in manufacturing. Both however are still predominantly top-down planning-based tools. As can be seen, although some very valuable contributions can already be found in the literature, some increasingly important aspects of the manufacturing strategy development process still lack proper operationalizing methods in the literature.

The proactivity of the manufacturing function is an example. Proactivity, particularly in turbulent environments, is not something that simply can give companies an edge. It is the only way to survive. In fact, manufacturing proactivity is suggested by a number of authors (Hayes and Wheelwright, 1994 is possibly the most eloquent example) but few of them actually prescribe how the function should be organized and managed to achieve it. Proactivity relates to the concept of the resource-based view - it is no doubt desirable, but actually how to go about reaching it? You will not find much about this in the current literature.

Breaking functional barriers is a second example. In turbulent environments, where change is not an exception, but the rule, inter-functional communication becomes essential in order to allow for rapid responses to frequent and sudden changes. The authors in the literature generally agree that for an effective manufacturing strategy to be put into practice it is necessary that functional barriers be broken down. Much of the reengineering discussion gravitates around this aspect. However, few authors in the field of manufacturing strategy deal specifically with methods to operationalize ways to break down or at least reduce the negative effects of the inter functional barriers.

The propositions described here aim to contribute to the manufacturing strategy process development debate addressing specifically aspects such as manufacturing proactivity and inter-functional integration, drawing some conclusions that may help companies operate under the turbulent conditions of the future when dealing with unexpected change is central.

5.1. Change is rule, not the exception

Change is a central concept in managing organisations in the future In recent years, the turbulent industrial/economic environment makes long-term planning a difficult task for many companies around the world. The high and unstable levels of inflation and exchange rates, the constantly changing government industrial policies, high interest rates, the political turmoil in which many countries have found themselves in recent years, the globalisation with constant mergers and acquisitions, the break down of trade barriers, the development of communication technologies, e-commerce and e-business changing drastically the way companies relate to each other and the way companies relate to customers, have forced companies to adopt predominantly "fire-fighting" reactive approaches to management.

Responding effectively to change is a dominant part in the manager's activities of the future. Any framework which aims to be effective in supporting the development of manufacturing strategy has to consider *change* and *dealing well with change* as central concepts. By analysing this reality and at the same time bearing in mind the models found in the current literature, some aspects start to emerge as relevant to be taken into consideration for the development of a framework to help the development of more agile manufacturing strategy in the future:

- The internal and external changes affecting the organisation will be so frequent and relevant that *change* should be the main trigger for the strategy reviewing process rather than only time, as the literature generally suggests. Companies cannot afford to wait for, say, 6 months to alter its strategic direction, once a relevant change (such as a drastic change in import taxes affecting the products it makes - favouring competitors or affecting the goods it buys - favouring the company itself) has happened.
- Changes may frequently affect so many functional areas that it is impossible that just one or a few of them keep such changes monitored and under control. Each and every function should adopt a proactive attitude, trying to anticipate changes and thinking contingently about possible future changes with regard to its main field of interest. In the literature, although most of the authors advocate the need for proactive manufacturing, most of the frameworks suggested are, in fact, almost totally top-down planning-based tools. No formal means for the manufacturing function to exercise its contribution proactively seems to be provided. They seem to rely solely on people's attitudes in order to make the manufacturing "proactivity" to happen. It seems to be risky though to assume that managers will assume a proactive attitude in the short term, mainly in environments in which the manufacturing managers historically have had a reactive role.

5.2. Two ways of dealing with unplanned change: control and flexibility

There is an extensive literature under the heading "management of change", generally by researchers on Organisational Behaviour who strongly emphasise the management of planned change rather than unplanned change. The literature on Production Operations Management usually deals with the issue of managing unplanned change under a number of different headings. One of them is "manufacturing flexibility". Although very valuable contributions can be found (mainly in the '80s) in the manufacturing flexibility literature (Browne et. al., 1984; Slack, 1983; Gerwin, 1986; Upton, 1994), since their emphasis is on flexibility, they do not explore sufficiently the fact that unplanned change can also be dealt with by unplanned change *control* - that means avoiding being affected by the changes.

An alternative approach is proposed here, according to which there are two distinct and complementary ways used by managers in order to manage unplanned change in manufacturing systems (Corrêa, 1994):

a. by controlling the unplanned change and therefore by interfering either directly with, or with the way the manufacturing system perceives, the size, novelty, frequency, certainty and/or rate of the changes, before the changes.

b. by dealing with the effects of the unplanned change by being flexible which is the ability to respond to the changes left uncontrolled, after they happen.

5.3. Unplanned changes control

Below are some real examples of unexpected change control mechanisms (Corrêa and Slack, 1996).

Monitoring/forecasting - one company (a first tier supplier in the automotive industry), facing turbulent industrial relations, monitors closely the trends of the Labour Unions' behaviour, in order to avoid being taken by surprise by a possible Labour strike. In doing so, the company is trying to reduce the *uncertainty* of some of its unplanned change.

Co-ordination/integration - one company's (a manufacturer of tractors) engine manufacturing shop reduced its short-term demand changes *uncertainty* by establishing on-line computer links in order to *coordinate* the engine assembly line with the paint shop. With on-line information, the engine assembly line has now accurate and timely information about the car bodies which are coming out from the paint shop and therefore better information about the next few hours' demand for engine derivatives. Now the schedule of the assembly line can be done under less uncertainty.

Focusing/confining - one company's (a manufacturer of off-road vehicles) manufacturing cells are generally *focused* on making a narrow range of parts. The cell, which machines engine blocks, for instance, uses transfer lines to perform only a few slightly different engine block type. This focusing aims at reducing the number of changeovers. Not always however is this possible because there are numerous components, which cannot be made in any one of the product-focused cells. In order to cope with this, one cell exists, which is equipped with expensive computer numerically controlled machines and multi-skilled operators, to perform a multitude of different engine components. This way, the need to be flexible is *confined* to one production cell whilst the others work only on a limited range of parts. With the focused approach, depending on what sort of task the system decides to focus on, the *size, novelty, frequency* and/or *certainty* of the stimuli which is perceived by the system can be altered.

Delegating/contracting - one company (an auto assembler) had always designed its own diesel engines. However, some years ago, they decided to *delegate* this task, by *contracting* a European expert firm to design the engines, mainly because the technology involved with Diesel engines' design was changing substantially and at a very fast *rate* due to new emission control regulations. The company decided not to have to deal themselves with such changes.

Hedging/substituting - one company (a second tier auto assembler supplier), dealing with erratic supplies, decided to run programs on supplier base reduction and supplier development. However, while the suppliers are not sufficiently dependable, the company decided to keep some of the standard components supplied by a number of sources rather than one or a few, *hedging* against their individual *uncertainty*. Another way to limit the stimuli level is by *substituting* the source of the stimuli, replacing it with a less "changeable" one. This applies to either unreliable suppliers, equipment or workers.

Negotiating/advertising/promoting - one company's (shock absorber manufacturer) manufacturing plant is running a program of parts standardisation aiming to reduce the variety of parts they manufacture to avoid unnecessary changeovers. Such an effort involves *negotiation* with the plant's internal customer, the marketing function. *Negotiating* is an attempt to interfere directly with the customer in order to reduce the changes she/he can possibly demand. Another way to interfere with the demand curve shape is by *advertisement* and *promotions*. Promotions and advertisement campaigns are usual ways to stimulate off-peak demand in order to level the overall demand curve, or in other words, to reduce demand change *sizes* and *rates* along the time.

Maintenance/update/training - Many manufacturing managers use preventive *maintenance* as a desirable way to deal with machine breakdowns, which would be one way to reduce possible equipment availability changes with regard to *frequency* and *size*. The idea of maintenance is not only suitable for machines. The maintenance of computer systems' records to ensure data integrity is another way of exercising control over future changes. Managers also emphasised *training* as an appropriate way of reducing the uncertainty and variability of people's behaviour.

5.4. Flexibility - dealing with the effects of the unexpected change

There are several classifications of manufacturing flexibility in the literature. Slack's (1989) classification seems to be one of the most consistent at the manufacturing strategic level. Slack's flexibility 4 types are product, mix, volume and delivery.

- Product flexibility: the ability to develop or modify products and process to the point where regular production can start.
- Mix flexibility: the ability to produce a mix, or change the mix of products within a given time period;
- Volume flexibility: the ability to change the absolute level of aggregate output which the company can achieve for a given product mix; and
- Delivery flexibility: the ability to change delivery dates effectively

We suggest the definition of a 5th and complementary type of system's flexibility:

"System robustness" flexibility: the ability of the system to overcome unplanned changes either in the process (such as machine breakdowns) or in the input side (such as faulty deliveries).

The need for a 5th systems flexibility type comes from the observation that even a system with high levels of performance in the 4 Slack's flexibility types could lack flexibility to deal with some of the changes which may happen to the process or to the supply side.

Each flexibility type can be understood in two dimensions: range and response flexibility, according to Slack (1989):

Range flexibility would be the ability of the system to adopt different states. One production system will be more flexible than another in a particular aspect if it can handle a wider range of states, for instance, to manufacture a greater variety of products or to produce at different aggregate levels of output. However the range of state a manufacturing system can adopt does not totally describe its flexibility. The ease with which it moves from one state to another in terms of costs, time and organizational disruption is also important. A production system, which moves quickly, smoothly and cheaply from one state to another should be considered more flexible than another system which can only cope with the same change at greater cost and/or organizational disruption. The way the system moves from one state to another would define Slack's other flexibility dimension, *response* flexibility.

Agile manufacturing strategies will have to treat flexibility (in its different types and dimensions) as a central concept. That is a fact. We suggest here however that there must be some sort of baseline stability for the manufacturing systems to be adequately flexible to deal with the changes to which it is increasingly subject. This means that in any manufacturing strategy exercise managers should have flexibility as a central concept, however they also should decide what kind and intensity of changes they are willing to deal with flexibly and what kind and intensity of changes they would prefer to "filter" or control via unplanned change control mechanisms. It means that being flexible is desirable, but since it normally comes at some cost, it is important to consider at least as a managerial tool, the possibility to limit the changes with which the company is willing to deal.

6. BASIC ELEMENTS FOR AGILE MANUFACTURING STRATEGY

A general approach is now proposed to the formulation of agile manufacturing strategy. Because of the huge variety of particular situations different manufacturing companies face, we consider that it is impossible to prescribe a step-by-step generic method for companies to develop their manufacturing strategies. However it is possible to outline, based on the previous discussions and concepts, some features, some foundations on which the companies should base the development of their manufacturing strategies in order that they can face the challenges of the future. Some of these features are described below.

6.1. Flexibility is central; and so is change control

Given that change is a central concept in the manufacturing management of the future, manufacturing strategies have to treat change management with the corresponding priority. Change is so broad a concept and can have so many facets that companies will normally prefer to opt to have a certain level of "protection" against some types / levels of change. This is convenient among other reasons because there are environmental changes which affect the whole market (giving an edge to companies who outperform competitors in dealing with them - such as the unexpected requirement of a customer for product customisation) and changes affecting only the company under analysis (such as changes in the availability of human resources because of high or uncertain turnover rates) - therefore only having the potential to hinder competitiveness. Companies who decide to manage manufacturing strategically in the 2000s should seek for the right balance between control and flexibility. Being flexible is no doubt increasingly desirable but it seems that in order to achieve effective flexibility some level of baseline stability should be present. Change control mechanisms may be a valuable resource for companies to achieve this baseline stability.

6.2. Breaking barriers through customer-supplier negotiation

Breaking organizational barriers is absolutely essential for the company to adapt and respond effectively and as a coherent whole to changes.

In order to break down the Organisational barriers, the approach proposed here is based on negotiation between the functions on a "customer-supplier" basis. The basic assumption is that everybody in the organisation has customers (either internal functions or external customers) and should serve them in the best possible way, given the constraints imposed by the availability of resources and also bearing in mind the corporate objectives, policies and strategy. Customer and supplier functions should negotiate and agree on the levels of service or goods, which the supplier is to provide. They have to agree on a specific set of performance criteria which represents the "point of contact" between the two functions. The "negotiation", it is suggested, can be based on "gap analysis" between the required set of performance criteria (by the customer function) and the set which is "offered" by the supplier function. The "point of contact" between marketing and manufacturing, for instance (the one emphasised in figure 3) may be the list of prioritised *order winning* and *qualifying* criteria (levels of delivery, product quality, costs and flexibility) which manufacturing should pursue (borrowing from Hill's (1995) framework). Between other pairs of functions, other "points of contact" are required, although the particular pairs of functions should negotiate and agree on their particular points of contact. Between manufacturing and finance, for instance, the relationship customer-supplier can be defined by the service, which finance supplies manufacturing: availability of capital over a period of the time. Therefore, one aspect, which has to be agreed upon, is the capital cash flow to be made available to manufacturing.

6.3. The time-phased approach

The points of contact or, in other words, the points, which have to be agreed upon between customer and supplier, are not related to a single point in time, either present or future. Instead, they should be "time-phased". This helps the functions agree not only upon objectives on a future point in time but also on the path through which the company will go about reaching some future competitive situation, stage by stage. The list of prioritised competitive criteria is no exception. Competitive criteria and also the other "points of contact" should be considered on a "time-phased" basis. The idea of improvement paths is present here and given the implications of these choices for the knowledge base of the company (Teece and Pisano, 1994), it is suggested that this process is carefully monitored to avoid local optimisations and wrong improvement path choices.

6.4. Proactivity achieved by using scenarios: the role of "contingency models"

In the proposed approach, proactivity is achieved through the explicit consideration of future possible alternative scenarios by all functions. In order to develop these scenarios, the function representatives and analysts have to be aware of current and prospective developments in their fields of interest. In the negotiation process, people from other functions will eventually demand alternatives from them in order that they are able to achieve a better performance in their own functions. Manufacturing people, for instance, will demand from finance people that they are able to offer alternatives for obtaining cheaper capital, in order to make investments. Marketing people will demand alternatives of possible future sets of competitive performance levels with regard to delivery, quality, costs and flexibility in order that they can choose from a broader array of markets to be targeted in the present and in the future. This should motivate the representatives from the different functions to act proactively, in search of new alternatives in their specific fields. For the people within the particular functions to be able to devise scenarios, and also for them to be able to negotiate with other functions, they have to develop what we call "contingency models". Contingency models are defined here as formal conceptual models, which link possible present, and future contingencies (characteristics, actions and decisions) with the various "points of contacts" between the function and other interacting functions. In terms of the manufacturing-marketing interface, manufacturing people should develop contingency models which associated possible future decisions and actions (such as investments in equipment, hiring and training of people, adoption of control systems, developing particular capabilities, among others) with the resulting alternative set of order winning and qualifying criteria. This would require that manufacturing people monitor and acknowledge new developments in production processes in order that they are able to assess the possibility of attending or not to the marketing "time-phased" requirements and also to produce alternative scenarios for them. Marketing people, on the other hand, should develop contingency models which should allow them to associate sets of order winning and qualifying criteria with different market segments, in order that they are able to reformulate marketing plans (target-market, frequency of new product introduction, among others) given that some change occurred in the possible set of "time-phased" competitive criteria which the manufacturing function is able to provide either in the present or in the future.

The contingency model approach is in line with the resource-based approach - capabilities may be proactively developed but what resources and capabilities to develop will be the result of an interactive discussion process to guarantee consistency between resource and capabilities development and strategic directions.

Figure 3 illustrates the negotiation process and Figure 4 is an example of worksheet for the operationalizing of proactivity of the various functions. For an example of an application of this concept in a real situation, see Prochno and Corrêa (1995).

6.5. The consideration of dynamic trade-offs and dynamic paths of improvement

Basically when considering the development of manufacturing strategies one has to be concerned with strategic fit (between the manufacturing task required to win orders in the market place and the manufacturing capabilities) and focus. As already discussed in previous sections of this chapter, the rationale behind these two concerns is the existence of trade-offs between different aspects of manufacturing performance. The concept of trade-offs is actually not a new one. It has been present since the early works published in the field of manufacturing strategy. What has some novelty is the idea that trade-offs are dynamic. What we propose here is that the analysis of strategic fit and focus in the development of agile manufacturing strategies are done carefully considering the dynamics of trade-offs involved and the alternative dynamic paths the company can go through (see section 2.2.).

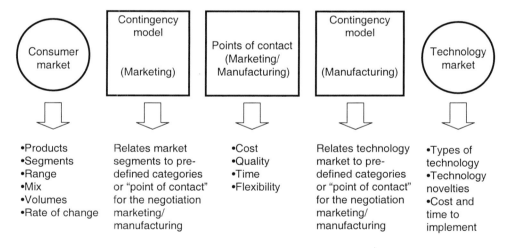

Figure 3. Negotiation Process for the operationalization of proactivity - example for the marketing / manufacturing interface (Prochno and Corrêa, 1995)

Hayes and Pisano (1996) have indicated that these paths may have an important effect in the knowledge base and on the learning experience of the company. The knowledge base is directly linked to the capabilities and competencies of the company and competence-based (or resource-based) approaches seem to be increasingly important to compete in the turbulent environment of the future in which the planning or top-down approaches tend to become more difficult to use because of the difficulties to anticipate changes. On top of that several authors in the literature argue that managerial rationality (planning) on which the traditional manufacturing strategy paradigm is based (the more traditional notions of strategic fit and focus) can not by itself result in a sustainable advantage because it would be too easily copied.

Example of worksheet for building contingency models

Function: _____

Scenario A
Main Characteristics:
Cost & resources required to have it operational:
Time to implement:
Future decisions and actions:
Reflex in criteria:
Criteria A - _____; Criteria B - _____; ... Criteria n - _____
.
.
Scenario X
Main Characteristics:
Cost & resources required to have it operational:
Time to implement:
Future decisions and actions:
Reflex in criteria:
Criteria A - _____; Criteria B - _____; ... Criteria n - _____

Figure 4 - Example of worksheet for building contingency models (Prochno and Corrêa, 1995)

According to this idea, for the company to achieve real distinctive capabilities management it should start paying more attention to the repertoire of organisational routines - which are actually the carriers of knowledge and experience an organisation has (Alher, 1998). Although this may be only partially true (since these repertoires are difficult and time consuming to change and in a turbulent environment, sometimes companies have to perform sharp changes of direction), this is in line with the proposal that dynamic improvement paths are considered and that a time-phased approach is adopted, together with the customer-supplier continuous negotiation because they all have impacts on the organisational routines.

6.6. The replanning process - triggered by relevant events and time

In the proposed approach the replanning process can be triggered by relevant events and time as opposed to that triggered only by time as the main frameworks in the literature suggest. This can prevent the company from delaying to respond to relevant changes, which occur between replanning points in time. The replanning process can also be triggered by any function, which considers that something relevant has changed or may come to change relevantly in his field of interest. A sudden and significant change in import rates is typically a change which can trigger a replanning process in order that the whole of the company realign their efforts in face of the new situation brought about by the change. The worksheet explained in the last section (see Figure 4) helps to formalize the process: the function that wants to trigger the replanning process fills in the worksheet and sends it to the other functions; a meeting is then set to decide the need for a strategy review or just minor adjustments to the new reality.

7. SUMMARY

The manufacturing environments of the present and of the next decade differ substantially from the past. Technology and consumer markets have become extremely difficult posing difficulties for the use of traditional top-down planning-based methods for developing manufacturing strategies. More agile manufacturing strategies are needed in an environment in which dealing with change becomes the central point. In this chapter we discussed some aspects which are increasingly important to be taken into consideration in the development of more agile manufacturing strategies:

It is actually very difficult for companies to be able to develop enough flexibility to deal with the whole number of changes it is subject to. We argue therefore that in order to achieve a basic baseline stability upon which to develop flexibility, companies should direct some efforts to define what types and what magnitude of changes they are willing to be able to deal with. We develop the concept of unexpected change control, which are management mechanisms, which help the company limit the changes with which it intends to cope with flexibly. The types of unexpected change control are monitoring/forecasting, co-ordinating/integrating, focusing/confining, delegating/contracting, hedging/substituting, negotiating/advertising/promoting and, maintaining/updating/training.

Breaking down organisational barriers is another feature, which will have to be taken care of carefully in the development of manufacturing strategies in the next decade. We propose an approach, which is based on internal and external customer-supplier negotiations on levels of service, which the supplier is to provide. This will add to the still prevalent functional organisation of most companies the process orientation needed to react quickly to internal and environmental changes.

Another feature we propose is that this customer-supplier negotiation is done on a time-phased fashion in order that the dynamic paths of improvement are incorporated in the process - not only gap analysis (comparing present status and desired future status) is taken into account, but also the time-phased evolution of the improvements which will take from the current status to the desired future status.

The turbulent environment of the future will require that all functions within the company adopt a proactive stance. Proactivity in the proposed approach is achieved by using scenarios and what we defined as "contingency models" - these are tools which can help companies to achieve the desirable levels of proactivity in a systematic way, rather than by leaving it purely to the initiatives of the individuals involved.

We also propose that any manufacturing strategy in the future should be subject to more frequent reviews. The static model in which it is reviewed periodically, say every year does not seem to be adequate for the turbulence of the future - so we propose that reviews of the manufacturing strategy should be triggered by both - time and possible relevant events which might happen between default review periods.

Finally we also propose that the traditional analyses of strategic fit and focus will still have to be done since there is no such thing as "one best way" for managing manufacturing resources. However, these analyses should be done considering all the dynamics of the trade-off relationships between different aspects of manufacturing performance and also the dynamic paths of improvement, which will directly impact the knowledge base, and therefore the future competencies of the company.

REFERENCES

1. Alher, F. (1998) *A strategic model of operational performance improvement*, Ph.D. Thesis, University of Warwick Business School, U.K.
2. Browne, J. et al. (1984) Classification of Flexible Manufacturing Systems, *The FMS Magazine*, Vol. 2, No. 2
3. Buffa, E.S. (1984) *Meeting the Competitive Challenge*, Irwin
4. Chase, R.B., N.J. Aquilano and F.R. Jacobs (1998) *Production and Operations Management* (8th Ed) Irwin McGraw-Hill
5. Corrêa, H.L. (1994) *Linking flexibility, uncertainty and variability in manufacturing systems*, Avebury, Gower Publishing, U.K.
6. Corrêa, H.L. and N.D.C. Slack (1996) Framework to analyse flexibility and unplanned change in manufacturing systems, *Computer Integrated Manufacturing Systems*, Vol. 9, No. 1, London
7. Crosby, P. (1979) *Quality is Free*, McGraw Hill
8. Fine, C.H. and A.C. Hax (1985) Manufacturing Strategy: A Methodology and an Illustration, Working Paper, Sloane School of Management, MIT, Cambridge, Massachusetts
9. Gerwin, D. (1986) An Agenda for Research on the Flexibility of Manufacturing Processes, *International Journal of Operations and Production Management*, Vol. 7, No. 1
10. Goldhar, J.D. and M. Jelinek (1983) Plan for Economies of Scope, *Harvard Business Review*, Nov-Dec
11. Gregory, M.J. and M.J. Platts (1990) A Manufacturing Audit Approach to Strategy Formulation, *Proceedings of the 5th International Conference of the Operations Management Association U.K.*, University of Warwick, Vol. 2.
12. Hayes, R and G. Pisano (1996) Manufacturing strategy: at the intersection of two paradigm shifts, *Production and Operations Management*. Spring
13. Hayes, R.H. and S.C. Wheelwright (1984) *Restoring Our Competitive Edge*, John Wiley
14. Hayes, R.H., S. Wheelwright and K. Clark (1988) *Dynamic Manufacturing*, The Free Press
15. Hill, T. (1995), *Manufacturing Strategy - Text and Cases*, Irwin
16. Leong, G.K., et al. (1990) Research in The Process and Content of Manufacturing Strategy, *OMEGA, International Journal of Management Sciences*, Vol. 18. No. 2
17. McKeena, R. (1997) *Real Time,* Harvard Business School Press
18. Pisano, G. (1997) *The development factory*, Harvard Business School Press
19. Porter, M (1980) *Competitive strategy*, Free Press
20. Prahalad, CK and G. Hamel (1990) The core competence of the corporation, *Harvard Business Review*, May-June
21. Prochno, P.J.L.C. and H.L. Corrêa (1995) The development of manufacturing strategy in a turbulent environment, *International Journal of Operations and Production Management*. Vol. 15. No. 11
22. Schonberger, R. (1986) *World Class Manufacturing* The Free Press
23. Shingo, S. (1985) *A Revolution in Manufacturing: The SMED System*, Productivity Press
24. Skinner, W. (1969) Manufacturing - Missing Link in Corporate Strategy, *Harvard Business Review*, May-Jun
25. Skinner, W. (1974) The Focused Factory *Harvard Business Review,* May-Jun
26. Skinner, W. (1985) *Manufacturing: The Formidable Competitive Weapon*, John Wiley

27. Skinner, W. (1996) Manufacturing strategy in the "S" curve, *Production and Operations Management*, Spring.
28. Slack, N.D.C. (1983) Flexibility as a Manufacturing Objective, *International Journal of Production and Operations Management*, Vol. 3, No. 3
29. Slack, N.D.C. (1991) *The Manufacturing Advantage*, Mercury, London
30. Stecke, K.E. and N. Raman (1986) Production Flexibilities and Their Impact on Manufacturing Strategy, Working Paper No. 484, Graduate School of Business Administration, University of Michigan
31. Teece, D. and Pisano, G. (1994) The dynamic capabilities firms: an introduction, *Industrial and Corporate Change*, Vol. 3, No. 3
32. Upton, D. (1994) The management of manufacturing flexibility, *California Management Review*, Vol. 36, No. 2

Agile Manufacturing: Concepts and Framework

A. Gunasekaran[a]   R. McGaughey[b]   V. Wolstencroft[c]

[a]Department of Management, University of Massachusetts, North Dartmouth, MA 02747-2300, USA

[b]Department of Business and Economics, Arkansas Technological University, Russellville, Arkansas 72801, USA

[c]Department of Manufacturing & Engineering Systems, Brunel University, Uxbridge, Middlesex UB8 3PH, UK

1. AGILE MANUFACTURING

It is a common misconception that agile manufacturing is synonymous with *lean* and *flexible* manufacturing. Agile manufacturing is a new concept, though not necessarily comprised of new techniques. In agile manufacturing popular approaches and techniques, such as lean and flexible manufacturing, are implemented in a concerted way, thus bringing about enormous improvement in products, responsiveness to customers, innovation, and flexibility. Agility gives companies a commanding competitive edge in the marketplace. This chapter provides insight into the changing role of manufacturing, with emphasis on the latest paradigm, 'agile manufacturing'. The nature of agility and its building blocks are discussed.

The term *agile manufacturing* came into common usage with the publication of *21st Century Manufacturing Enterprise Strategy* (Iacocca Institute, 1991). Some have used the term agility as though it were synonymous with other concepts such as *flexible manufacturing* and *lean production* (Kidd, 1994). It is not! Dictionaries generally define agility as quick moving, nimble, and active. This is not the same as flexibility, which in the manufacturing sense implies adaptability and versatility. Flexibility is a requirement for the competitive markets of today, but on its own, it will not deliver agility. *Lean manufacturing* is so called because it involves doing everything with less. In other words, the excess of wasteful activities ('muda' in Japanese), unnecessary inventory, long lead times, and so on have been cut away through the application of, for example, just-in-time manufacturing, concurrent engineering, overhead cost reduction, improved supplier and customer relations, and total quality management (Womack et al., 1990). Lean manufacturing, however, is not the same as agile manufacturing, because leanness and agility have different meaning and connotations. Computer integrated manufacturing and the computer-integrated enterprise can be considered in the same light. When computers are linked across applications, functions, and enterprises, agile manufacturing is not achieved, yet rapid communications and the exchange and reuse of data are necessary for agile manufacturing.

Agile manufacturing is not flexible manufacturing, or lean manufacturing, or computer integrated manufacturing, but rather it is a useful heading under which to place a series of techniques, methods, and philosophies--these are the building blocks of agility--that excellent companies employ to bring about unprecedented improvements in quality, productivity, and customer service. These building blocks of agility are not new. Many have been available for decades, and others have developed gradually over the last 30 years in innovative companies like Toyota (Schonberger, 1982). These techniques, methods, and philosophies now are employed in a concerted way, bringing about enormous improvements in products, responsiveness to customers, innovation, and flexibility. These improvements can give companies a commanding competitive edge in the marketplace.

Agile manufacturing requires a fundamental change in management philosophy (Maskell, 1994). The aim is to create a manufacturing firm that can produce in volume and simultaneously produce variety for market niches. Agile companies seek to combine the advantages of time compression with techniques to reduce the cost of variety. The goal is to offer almost instant delivery of small quantities of goods that meet individual specifications. To become agile a firm must redesign its processes and products to meet the expectations of customers for both customisation and responsiveness.

Agility is dynamic, context-specific, aggressively change embracing, and growth oriented. Agility is *not* about improving efficiency, cutting costs, or battening down the business hatches to ride out fearsome competitive "storms". It is about winning: about succeeding in emerging competitive arenas, and about winning profits, market share, and customers in the very centre of the competitive storms many companies now fear (Goldman et al., 1995).

1.1 A Change in Paradigm

Manufacturing has undergone many evolutionary stages and paradigm shifts (Esmail, 1996). Figure 1 illustrates the paradigm shifts experienced in moving from a craft industry to mass production, then to lean manufacturing, and finally to agile manufacturing. The two dimensions used in the matrix are "Response to changes in environment" and "Product variety."

The *dinosaur* represents early manufacturing. The dinosaur was heavyweight and slow moving, and thus responded unhurriedly to changes in its external environment. The *craft industry* is a good example of a dinosaur with high product variety and low response times. Trades-people had a one-to-one relationship with their customers from product development to supplying the finished product directly to the customers. There was no need for retailers or distributors. As volumes grew mass markets began to evolve where new methods and principles for planning and organisation were advocated and adopted. *Mass production* and new manufacturing methods made the dinosaur enterprise leaner, resulting in the next stage, the *mule* enterprise. The mule enterprise evolved to meet the needs of the mass market slowly and steadily through dedicated production lines. *Economies of scale* was the driving force, and hence the drive towards *lean manufacturing*, represented by the *winning horse with weights*. The idea behind the winning horse with weights is that removal of the weights from the racehorse enterprise will enable the horse to gallop faster and faster.

The *lizard* enterprise represents the *agile* or *adaptive* enterprise, taking into account factors such as the versatility and adaptability of the enterprise. The prominent feature of the lizard enterprise is its ability to react quickly and accurately to changes in its environment and adapt itself quickly (like a chameleon changes its colour) to the opportunities or threats it faces for its own preservation. Like the flattened body of the lizard, enterprises will need the agility to enter crevices for niche customers and to have the technology and processes to meet the customer needs.

Figure 1: Variety/Response Matrix (Esmail, 1996)

For the lizard enterprise to operate efficiently, every aspect of the enterprise will need to be agile - *S.T.O.P.*(Strategy, Technology, Organisation, and People).

The lizard, or agile competitor, must be a *learning organisation* capable of reinventing itself when the need arises. The focus for the lizard enterprise is not economies of scale driven purely by cost and efficiency, but instead market success based on *economies of scope* (Esmail, 1996). The support of a level of product variety required for market success may necessitate the creation of *virtual companies* that can be disassembled when they outlive their usefulness.

1.2 What is Agility?

Agility is dynamic and open-ended. There is no point at which a company or an individual has completed the journey to agility. Agility demands constant attention to personal and organisational performance, attention to the value of products and services, and attention to the constantly changing *contexts* of customer opportunities. Agility entails a continual readiness to change, sometimes radically, what companies and people do and how they do it. Agile companies are always ready to learn whatever new things they need to know in order to profit from new opportunities.

Agility is context-specific. Markets pull the acquisition of agile business capabilities, and differences among markets limit the generalizability of detailed prescriptions for agility. In the end, the transition to agility is justified by the promise of sharing in highly profitable markets for information-rich and service-rich products configured to the requirements of individual customers.

The profitability of these products rests on marketing and pricing strategies based on customer-perceived value. Successful agile competitors, therefore, must understand not only their *current* markets, product lines, competencies, and customers very well, they must also understand the potential for *future* customers and markets. This understanding leads to strategic plans for acquiring new competencies, developing new product lines, and tapping new markets. The implications of agile competition are highly dependent on the competitive contexts within which individual companies operate.

Agile companies aggressively embrace change. For agile competitors change and uncertainty are self-renewing sources of opportunities from which to fashion continuing success. Thus, to an unprecedented degree, agility is dependent on the initiative of people, their skills, their knowledge, and their access to information. An agile organisation is one whose organisational structures and administrative processes enable fast and fluid translations of this initiative into customer enriching business activities. Agility is aggressive in *creating* opportunities for profit and growth. Agile competitors *precipitate* change, creating new markets and new customers out of their understanding of the directions in which markets and customer requirements are evolving. Although agility allows a company to react much more quickly than in the past, the strength of agile competitors lies in proactively anticipating customer requirements and leading the emergence of new markets through constant innovation. Agility is a comprehensive response to a new competitive environment shaped by forces that have undermined the dominance of the mass-production system (Goldman et al., 1995).

1.3 Market Forces

Agility is a name for the reorganisation of production, adapted to distinctively new market forces that have undermined the mass production organisation of business that dominated in the 20th century. These new market forces are identified in Figure 2. The organisation of a business to make it adaptable to these forces helps it become agile.

The agile enterprise provides solutions to its customers, not just products. It works adaptively, responding to marketplace opportunities by reconfiguring its organisation of work, its exploitation of technology, and its use of alliances. It engages in intensive collaboration within the company, pulling together all of the resources that are necessary to produce profitable products and services regardless of where they may be distributed. The agile enterprise forms alliances with suppliers, and with partnering companies. Finally, the agile enterprise is a knowledge-driven enterprise. It is centred on people and information, not on technologies alone that must be consistent with the manufacturing base of the company (including the supplier network) and, of course, manufacturing success can only occur when there is a *market* for the goods being produced.

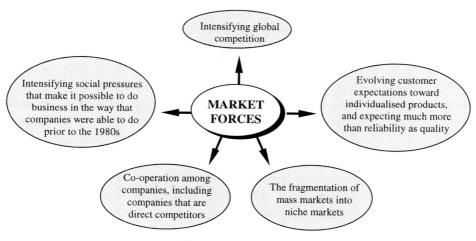

Figure 2: Market Forces

Traditional manufacturing firms are fragmented. The fragmentation of manufacturing is presented in Figure 3 (Booth, 1995). Agile manufacturers must be integrated. Integrated manufacturing is likewise depicted in Figure 3, but it is depicted as a unified whole.

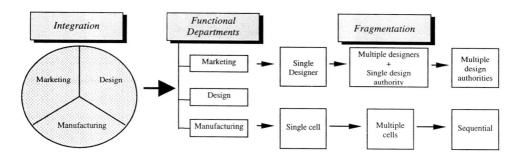

Figure 3: Fragmentation of Manufacturing (Booth, 1995)

Integration requires the concurrent planning of marketing and the manufacturing activities and a concurrent, not sequential, product design process. The objective is to simultaneously design a product--one that meets the needs of a market segment--and the manufacturing processes to produce it, and to do so quickly.

As simple as this may sound, it runs counter to the approach employed by most companies for most of this century. The byword has been fragmentation as evidenced by the following:

- Companies have been fragmented along functional lines with the most serious fissures occurring between marketing, design and production;
- Tasks have been subdivided, sometimes needlessly. This is most evident in the production area, though the fragmentation of the design function is arguably more damaging;
- A departmental hierarchy has grown up to manage the complexity of the divided business.

The agile manufacturing company must eliminate these and other characteristics or sources of fragmentation.

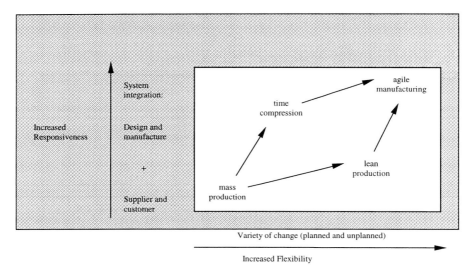

Figure 4: Road to Agile Manufacturing (Lei, 1990)

The road to agility is illustrated in Figure 4, companies approach the goal in different ways (Lei, 1990). Some companies have been pursuing process integration to achieve leadtime compression. Initially, this may simply entail linking order entry with manufacturing. Eventually it involves linking design and operations and extending the links outside the company to customers and suppliers. With these inter-company links a customer could directly enter a design on the company's systems. This would allow production schedules, both internally and among suppliers, to be updated in accordance with the design. Others have been increasing their ability to deal with both planned and unplanned variety. Both the product design process and the production operations are arranged to minimise the costs of complexity associated with variety.

Firms can only benefit from agile manufacturing if they have a broader agility strategy that links agile manufacturing to market performance. Firms must develop a manufacturing strategy that is driven by customer needs and exploits manufacturing agility by shaping market opportunities. Agile manufacturing requires a market driven strategic manufacturing response. With markets becoming increasingly dynamic, and competition increasingly based on improved quality and customer responsiveness, the development of an appropriate

manufacturing strategy is likely to have a major impact on company survival, success, and growth. A strategy for agile manufacturing should focus on designing an enterprise that will lead to (Kidd, 1994):
- Faster response to highly variable customer demand patterns
- Improved productivity;
- Opportunities for system wide innovation, learning and improvement;
- Improved product quality;
- Better utilisation of capital and improved return on investment;
- Improved customer and market focus, better understanding of customer needs, and closer customer relationships;
- Flexibility to cope with a wide range of batch sizes including one-of-a-kind production, and a wide range of products (volume and product flexibility);
- Integration of suppliers into product development and manufacturing processes;
- Short-order capability and the ability to rapidly respond to new windows of opportunity;
- Capability to undertake multi-venturing through virtual organizations;
- Reduced indirect labour and other overhead costs;
- More time and opportunities for management to tackle problems;
- System integrity and robustness.

1.4 The Competitive Landscape

The competitive landscape creates a cornucopia of pressures on the firm (Pine, 1993). These pressures are summarised in Figure 5.

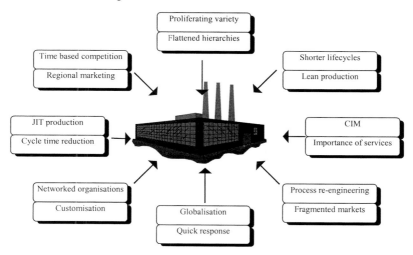

Figure 5: Pressures on the Firm (Pine, 1993)

Companies producing products in mass have traditionally been unable to cope with the lean production challenge. Managers in these companies did not perceive lean production to offer enhancements in proven ways of running their firms. Prior success was a major impediment--the more successful the company, the less likely it was to change by abandoning proven avenues to success. The competitive landscape summarised above portends yet another round of fundamental changes. Demand for standard products has fragmented as can be seen in markets for such things as fast food, breakfast cereals, banking, insurance and IT. The niches are smaller and constantly changing. In fact, the niches *are* the markets!

1.5 Features of Agility

Essentially, there are four basic features of agility (Goldman et al, 1995);
- *Products*: They are solutions to customers' individual problems. Agility is centred on the customer-perceived value of products. It aims to decouple cost of production and lot sizes.
- *Virtual organisation*: Internal and external co-operation are the strategies of choice. The aim is to bring agile products to market in minimum time by leveraging resources through co-operation.
- *Entrepreneurial*: Organizations must "organise to thrive on change and uncertainty." This message is very similar to that found in *Liberation Management* (Peters, 1992). Innovative, flexible organisational structures that promote rapid decision-making are the order of the day. Agile manufacturers must have personnel who can convert change and uncertainty to growth.
- *Knowledge-based*: The key differentiators in tomorrow's world will be people and information. Thus, agility embraces the decentralization of authority and leveraging the value of human and information resources.

These features of agility suggest ideals contrary to the very basis of lean thinking. The lean mindset has been described as follows (Womack, 1996);

"*Our earnest advice to lean firms today is simple: to hell with your competitors; compete against perfection by identifying all activities that are muda and eliminating them. This is an absolute rather than a relative standard which can provide the essential North Star for any organisation.*"

The ideals of leanness, stated this way, encourage companies to become similar to one another. Some would suggest that strategy is actually about making trade-offs and argue that "operations effectiveness is not a strategy" (Porter, 1996);

"*Strategy is about making trade-offs in competing. The essence of strategy is choosing what not to do. Without the trade-offs, there would be no need for choice and thus no need for strategy...performance would depend wholly on operational effectiveness.*"

The mass customisation mindset, in contrast to "lean thinking," demands synthesis of the economies of mass production with individually customised goods and services. Great product variety can be achieved at costs comparable to the costs of mass-production (Harrison, 1997);
- *mass production*: achieved by economies of scale (low cost resulting from high volume and low variety)
- *mass customisation* : achieved by economies of scope (varied products can be produced at a low cost and with little lead time--individual product volumes are low, but total product volume is high).

Mass customisation relies heavily on the concept of manufacturing flexibility--the ability to change what is done.

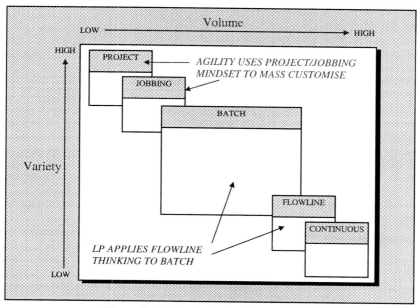

Figure 6: Agility and Process Types (Harrison, 1997)

Agility, in contrast to lean production, is a deliberate, competitive response to constantly changing requirements for competitive success in current and emerging markets. Using the concept of virtual organisation, agility is about rapidly reconfiguring the whole supply chain if necessary. Agility is a competitive strategy that companies with a lean mindset would find difficult to follow. Lean supply chains are characterised by comparative rigidity, achieved over time by the painstaking process of partnership!

A useful way to envisage the difference in emphasis with agility and lean production is to refer to the 'volume/variety grid' (slack et al, 1995) seen in Figure 6. Differences in product characteristics in the market place can be described according to:
- *Volume*: the relative number of units sold in a given range;
- *Variety*: the relative number of product offerings in a given range;

Manufacturing companies typically organising operations in one of the 5 ways depicted in Figure 6.

Lean production has been concerned with applying the advantages of flowline (low cost, low throughput time and low inventories) to batch environments. Thus, concepts like set-up reduction and small batch sizes were used to create harmonisation of flow throughout the supply chain. Agility is more characteristic of the project and jobbing process types. High variety and high uncertainty about what the next specification will be are the operation's orders of the day. Project companies often form consortia to respond to a given bid, pre-qualify and establish terms of trade in advance to develop a blueprint for the virtual organisation (Harrison, 1997).

2. ENABLERS OF AGILE MANUFACTURING

This section provides a conceptual model (See Figure 7) to illustrate the concept and enablers of agile manufacturing. The *virtual enterprise* has received much attention of late. It is a relatively recent development and a fundamental approach for co-operating to compete.

Figure 7: A Conceptual Model to Illustrate the Concept and Enablers of Agile Manufacturing

The idea of the *extended enterprise* is made possible by enabling technologies such as the Internet, Multimedia, and Electronic Data Interchange (EDI). Agility requires a customer-driven approach and integrated product, process, and business information systems.

2.1 Conceptual Model

The conceptual model illustrates the key enablers of agile manufacturing. As with CIM (Computer Integrated Manufacturing), integration of these enablers, and of physically distributed firms, is essential if one wishes to become truly agile. If treated as separate islands, the enablers will not contribute fully to the overall competitiveness of the firm.

2.1.1 Virtual Enterprise

The virtual enterprise (VE) or, more accurately, an organisation with a virtual organisational structure, is only one of many forms of co-operation amongst and within companies (Goldman, 1995). The VE is of special interest because it places the greatest demands on a company to co-operate in achieving collaborative production. If a company is so staffed, equipped, organised, and motivated that the option of creating a virtual organisation structure for a project is routinely available to it, then all of the other elements of agility are likely to be present and functional in that company.

A virtual organisational structure is an opportunistic alliance of core competencies distributed among a number of distinct operating entities within a single large company or among independent companies. The underlying idea is almost trivially obvious and far from new. Organise a group of people with relevant abilities into a team focused on a well-defined problem, create a system that motivates these people, give them access to appropriate resources, reward them based on the value of the solution they create, and then stand back and watch them produce!

Identifying complementary core competencies, and then synthesising them at low cost into a complete production capability designed to satisfy a particular customer opportunity, creates a powerful competitive weapon. It is rapidly becoming available to companies of all sizes as computer networks adapted to the needs of integrated, interactive, electronic commerce grow. Furthermore, while the VE is opportunistic, its objective is to create solution products with lifetimes dictated by the market. As consumer needs change, the "solution" products must evolve, and so will the virtual organisation's resource requirements. Some participants will leave, perhaps to join other ventures, because their competencies no longer add sufficient value to warrant continued participation in the VE. For precisely the same reasons, others will join, because they can add value as the product evolves.

There are six reasons for creating a virtual organization, and all of them are strategic. These six strategic considerations influence the decision to establish a VE instead of a more traditional collaborative approach, such as partnership or joint venture. These considerations are listed below as questions.

- Would forming a virtual organisation to market a new product allow one's company to share infrastructure, resources, R&D, costs, and risks?
- Would a virtual organisation enhance product development opportunities for one's company by linking internal core competencies to core competencies of other companies?

- Would a virtual organization reduce concept to cash flow time by integrating knowledge and skills across company boundaries in concurrent operations?
- Would the VE approach increase the apparent size or scale of the operation at less cost than achieving this scale? First, to the people involved internally in terms of access to expertise and resources; and second, to customers, whose response to a product from a sixty-person company might be very different from their response to the products of what may seem, in effect, a 6000 person company.
- Would a VE give one's company access to new markets through the relationships forged and allow one's organization to share in other companies' customer loyalty base, by adding value with them and for them via the new product?
- Would a virtual organisation accelerate a company's migration from selling products to selling solutions?

Participants in the virtual enterprise share accountability and responsibility for the success of the enterprise as a whole, as well as for their particular portion of its operations. All participants are involved, more or less actively, in all of the virtual enterprise operations.

The benefits of adopting a virtual organisation structure are not only the much greater speed with which new products can be brought to market and the lower cost, but also the better match the structure allows between resources and the specific requirements of a new market opportunity. The inevitability of the virtual organisation structure as a routine option for management follows from the centrality of core competencies to success in the agile competitive environment. Core competencies need to be complemented and supplemented on a dynamic basis for productive resources to be harnessed and effectively employed within the short time lines that determine success in the pursuit of agility.

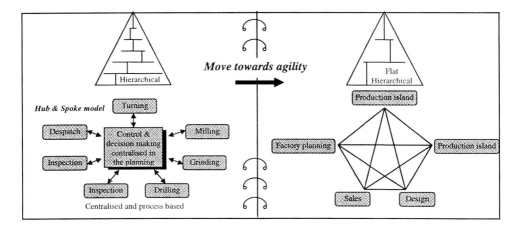

Figure 8: Agility Requires Flatter Hierarchies (DTI)

In the arena of agile competition, the ability to form virtual companies routinely is a formidable competitive weapon. The routine formation of virtual enterprises is possible only if communication and information exchange technologies are in place that are capable of supporting the plug-in compatibility required by the organisational structure of the VE. In effect, virtual organisational structures entail not only the complete integration of the enterprise but also extensive interenterprise integration as well.

Agility success, generally, and VE success, specifically, hinge on changes in managerial values, organisational structure, and prevailing corporate culture paradigms. Work force empowerment, self-organising and self-managing cross-functional teams, performance and skill-based compensation, flatter managerial hierarchies, distributed authority, and point-of-problem decision making are all important to the pursuit of agile business capabilities. The flatter organizational structure required for agility is contrasted with the taller structure more traditional in manufacturing companies in Figure 8. Flatter structures are more conducive to the behaviours and values required for agility and in the VE.

Performance measures that can be used to evaluate the effectiveness of the formation of the virtual enterprise include: (i) Time to identify the core competencies of partner-firms, (ii) New product development time, (iii) Technology levels, (iv) Innovation, (v) Flexibility, (vi) Delivery performance, (vii) Quality, (viii) Inventory, (ix) Virtual enterprise development time, (x) Profitability, and (xi) IT skills and Knowledge.

2.1.2 Physically Distributed Teams and Manufacturing

New types of logical infrastructures supporting agility and quick response reduce the time to reach global markets (Gunasekaran, 1997). Nowadays, companies typically rely on internal shops or a few familiar contractors. Establishing relations with new suppliers can take months because everything from terms and conditions to CAD formats must be negotiated and

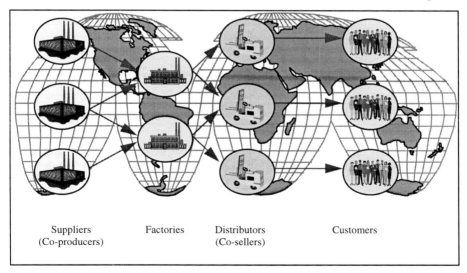

Figure 9: Extended Enterprise Network (Walters, 1997)

the supplier's capabilities must be validated. The physically distributed VE is a temporary alliance of partner enterprises located all over the world where each contributes its core competency to take advantage of a specific business opportunity or fend off competitive threats.

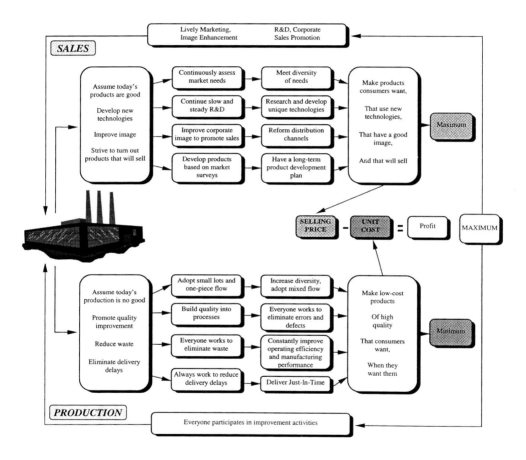

Figure 10: A Customer-Driven Company (Hirano, 1988)

Customer expectations and continually increasing competition on a worldwide scale are forcing industry to rethink business strategies, alliances, partnerships and the mode for day to day operations. In parallel, recent developments in IT and communications offer significant opportunities for all market sectors to operate in fundamentally different ways. The idea of physically distributed teams and manufacturing is captured by the term "Extended Enterprise" and it is illustrated in Figure 9 (Walters, 1997).

In the past, manufacturing was organised primarily on a local level, in towns and cities in relative isolation. With the development of modern transportation facilities, both the market and production widened their sphere of connection to suppliers and customers on a regional level, i.e. district, country and eventually, continent. It is primarily in the last decade that this manufacturing organization has expanded beyond continents. We have witnessed the creation of the globally extended market, liaisons and globally extended manufacturing which spreads across all boundaries, including continents.

Manufacturing expectations are changing. Some of the reasons for those changes are as follows:

- Cheap labour costs in certain parts of the world;
- Increasing flexibility of global transport;
- Continually rising sophistication of manufacturing processes;
- Increasing intelligence of both products and processes;
- Economic conditions.

Apart from the progress in mechanical, electrical and electronic engineering, the most radical technological changes have occurred in communications and information technology. Voice, data and multi-media communications that started with the use of the telephone, faxes and modems have leapt forward in recent years with the growth of the Internet and WWW. At the same time, computers have become smaller and more accessible to users, with available memory well beyond expectations of just a few years ago, with additional storage capacity, calculating power, easy to use word-processing, spreadsheets and graphics facilities, cheap packages, and new ways of presenting results.

The important aspects of the past, such as factory layout, number of lines and machines in the plant, planning and scheduling of production through the factory etc. are still relevant, but there are additional concerns with an extended global network. Matters of concern include locations of the supply, manufacturing, storage and transport facilities, relationships (i.e. contracts) with suppliers and customers, generalised data structures (standards) across the world, common IT systems for easy and fast processing of information, and adequate and reliable support of communications.

2.1.3 Rapid Partnership

Improving the responsiveness of a firm to a changing market, requires a shared partnership between the core parts of the firm. Shared understanding of marketing and manufacturing is the starting point. Necessary, also, is a shared understanding of the market itself. This is key in moving toward the ideal of a customer-driven, knowledge enterprise (Berry et al., 1995).

The concept of rapid partnership is a sub-function of the virtual organisation. Partnership is necessary if one is to remain competitive. It allows a firm to capitalise on the core competencies of others. There has been a significant increase in research carried out on the selection of suitable partners for virtual enterprises. One method of finding an optimal partner for the virtual enterprise is to employ a flexible and interactive decision support system. This

decision support system employs the Fuzzy-AHP (Analytic Hierarchy Process) (Gupta et al, 1997) wherein it (see Figure 11):
- Formally combines concrete quantitative information as well as users' *fuzzy* qualitative information, and
- Employs state-space search algorithms to provide a quick and optimal selection of partners.

As an alternative, Quality Function Deployment (QFD) can be used to benchmark possible partners.

Selecting the right manufacturing partners for a given product is one of the most fundamental decisions in a multi-product, multi-partner scenario as in the case of a VE. The partner selection problem is typical of managerial problems that involve trade-offs among multiple criteria, some of which are qualitative. Important criteria involved in partner selection are the cost, quality and lead time associated with the operations of a manufacturing partner. When the subsequent processes are performed at different plant locations, transport cost becomes an added consideration. Other criteria/considerations such as past performance, market position, and service influence the selection of a partner.

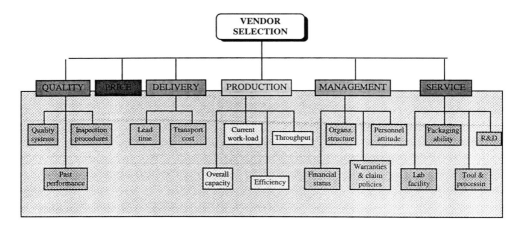

Figure 11: Vendor Selection Attribute Hierarchy (Gupta, 1997)

Rapid partnership formation tools would seem to be quite valuable in helping firms respond quickly to capture markets that are turbulent and uncertain. The advent the Internet, Multimedia, software packages such as Microsoft®Project, and Electronic Data Interchange (EDI), have greatly facilitated the task of forming partnerships quickly.

2.1.4 Concurrent Engineering

The entrepreneurship of empowered teams is necessary to achieve success in agile competition. Uncoordinated entrepreneurship, where individuals and teams worry only about their own projects, leads to chaos (Goldman, 1995). Together with organising for change, the creation of teams and changes to performance measurement systems, the agile competitor must institute systems to co-ordinate actions and deployment of resources across the entire enterprise.

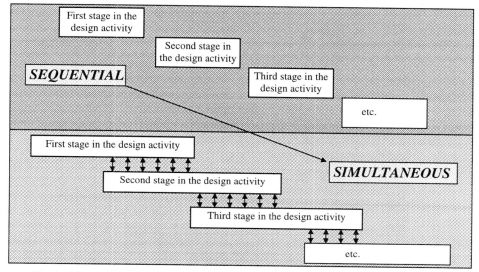

Figure 12: Arrangements of the Stages in the Design Activity (Slack et al, 1995)

Rapid response to change requires that work be completed as rapidly as possible. This makes it necessary for people, functions and processes to work as much as possible in *parallel*. This includes all of the relevant functions of a company, its suppliers and its customers, who together comprise the *extended enterprise*. To achieve agility, every team member in every company must do everything relevant to a project concurrently to the extent possible. This requires suitable communications infrastructure, work practices, and training of workers. It also requires motivation on the part of everyone involved and an ability to see the big picture, as well the details of the task in hand. Concurrent engineering is a term used to describe processes or activities performed in parallel (simultaneously). Concurrent engineering of design activities is depicted in Figure 12. Concurrent engineering is often discussed in the context of product and process design--they proceed in a parallel, coordinated fashion.

Concurrent engineering is synonymous with *simultaneous engineering* (Figure 12, Slack et al, 1995). "*Simultaneous engineering means that people who design or manufacture products work under the same targets and the same sense of values to tackle the same problems enthusiastically from the early phases. The targets here are the reduction of lead time, design for manufacturing (DFM), product development and development of advanced production technologies. The common measure of value is the satisfaction of customers, which is one of the corporate philosophies of the entire company*".

"*Simultaneous engineering attempts to optimise the design of the product and manufacturing process to achieve reduced lead times and improve quality and cost by the integration of design and manufacturing activities and by maximising parallelism in working practices*"

"*Bureaucratic and cumbersome decision processes in product and service development can stifle technical competence. One of the reasons often cited for why IBM failed to dominate the personal computer market in spite of setting the standard for such products is that smaller, more agile and sharper rivals outmanoeuvred them. Often products just arrived on the market too late. Products would arrive a year or so after they would have been a really good idea*"

2.1.5 Integrated Product/Process/Business Information Systems

Communications and information are the technical elements driving the next industrial revolution. They are central to agility. Information is increasingly diverse in form including not only computer text data, but multimedia pictures and voice data as well. It is possible today to have a database of photographs of customers, or employees, or products, and by clicking the picture with the mouse, to get data about the targeted picture item (Goldman, 1995).

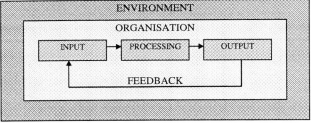

Figure 13: Activities of Information Systems (Laudon, 1995)

An information system (IS) can be defined a set of interrelated components working together to collect, retrieve, process, store, and disseminate information for the purpose of facilitating planning, control, co-ordination, analysis, and decision making in businesses and other organisations. A generic information system is depicted in Figure 13.

Figure 14: An information System: Not Just a Computer

An information system must not be described in terms of the computer alone. An information system is an integral part of an organisation and is a product of three components: technology, organisations, and people (Figure 14).

Table 1 shows some typical manufacturing and production information systems arranged by the organisational level of the problem. Strategic-level manufacturing systems deal with the firm's long-term manufacturing goals such as where to locate new plants, or whether or

MANUFACTURING AND PRODUCTION INFORMATION SYSTEMS

Strategic-Level Systems
Production technology scanning applications
Facilities location applications
Competitor scanning and intelligence

Tactical Systems
Manufacturing resource planning (MRP)
Computer-Integrated Manufacturing (CIM)
Inventory control systems
Cost accounting systems
Capacity planning
Labour-costing systems
Production scheduling

Knowledge Systems
Computer-Aided Design systems (CAD)
Computer-Aided Manufacturing systems (CAM)
Engineering workstations
Numerically controlled machine tools
Robotics

Operational Systems
Purchase/receiving systems
Shipping systems
Labour-costing systems
Materials systems
Equipment maintenance systems
Quality Control systems (QC)

Table 1: Manufacturing and Production Information Systems (Laudon, 1995)

not to invest in new manufacturing technology. Tactical manufacturing and production systems deal with the management and control of manufacturing and production costs and resources. Knowledge manufacturing and production systems create and distribute design knowledge or expertise to drive the production process, and operational manufacturing and production systems deal with the status of production tasks (Laudon, 1995).

Setting up communications and information systems to support agility requires attention to many details. For instance, information flowing across national boundaries must take account of the different laws and regulations regarding privacy and a company's responsibility in regards to these matters.

Manufacturing companies, especially those pursuing agility and those involved in a virtual enterprise, are increasingly aware of the requirement for computer assisted logistics management systems (CALS). The name may imply that a CALS is narrow in scope, when in fact, the use of CALS requires that all technical specifications, drawings, manuals, and other documents relating to products, from all levels of the supplier chain, be computerised. CALS includes a large number of standards and requirements, including the Standard for Product Data Interchange (STEP), which is the developing national standard for the definition of electronic data that describe physical products.

A major challenge for firms striving for agility and for VE is the integration of information systems to support their efforts. Lack of standardization has always been a problem, even within a company or country. When partnering involves multiple companies, perhaps in different countries, the integration of systems becomes more complex. The rapid evolution of E-commerce and standards to support E-commerce has contributed to the development and acceptance of standards that will make system integration less perplexing in the future.

2.1.6 Rapid Prototyping

A prototype is a preliminary model of a system solution that end-users and/or designers can interact with and analyse. The prototype is constructed quickly and cheaply, ideally, within days or weeks. A test group of end users (representative of the customer population), or designers themselves, can try out the experimental model to see how well it performs. In the process, design teams may discover requirements they overlooked, or they may discover other avenues for improvement. The prototype is then modified, re-evaluated, and enhanced over and over until it conforms exactly to what customers or potential customers want (Laudon, 1995).

There are various forms of rapid prototyping such as *three-dimensional printing*, *photochemical machining*, *laminated object manufacturing*, and *ballistic particle manufacturing*, where streams of material such as plastics, ceramics, metals and wax, are injected using an ink-jet mechanism through a small orifice at a surface (target). The mechanism uses a piezoelectric pump, which operates when an electric charge is applied, generating a shock wave that propels 50-μm droplets of wax at a rate of 10,000 per second (Kalpakjian, 1995).

Prototyping is a vehicle for expediting the product design process, thereby reducing time to market. Prototyping helps firms to test proposed designs against customer requirements. This is particularly important in making sure that products meet customer needs--a must for agile manufacturers.

2.1.7 Electronic Commerce

Electronic Data Interchange (EDI) is the computer-to-computer exchange of standard business documents such as purchase orders, invoices, and bills of lading, among organisations. Transmitting these documents electronically, or on-line, saves time and money by cutting down on paperwork and data entry (Laudon, 1995). Figure 15 illustrates an EDI system that transmits a purchase order (P.O.) from the buyer to the seller.

EDI can produce strategic benefits in that it can help firms increase market share by 'locking in' their customers - making it easier for customers or distributors to order from them rather than competitors. EDI can reduce transaction processing costs, cut inventory costs by reducing the amount of time components are in inventory--it can help get them in the pipeline faster--and EDI can reduce errors associated with manual data entry, and re-entry.

Electronic mail, or E-mail, is the computer-to-computer exchange of messages that eliminate telephone tag, costly long-distance telephone charges, and long delays associated with traditional mail service, sometimes called "snail mail," and for good reason. It is possible for someone at that is "on-line" to send notes, letters, and attachments in the form of documents (that can retain their original formats), pictures, videos (if not too large), audio recordings, etc. to one or more recipients. By providing faster and more efficient communication between different functional areas of a firm, and among firms, E-mail can speed up many business processes, including production.

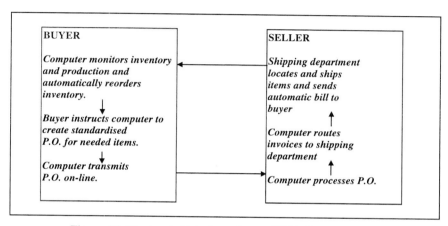

Figure 15: Electronic Data Interchange (EDI) (Laudon, 1995)

Almost faster than they can be evaluated, new information technologies appear that offer promise in improving an organization's competitiveness. Although possible advantages may appear obvious, implementation is another thing. Agile competitors must be up to the challenge of prospering in a rapidly changing environment where every day brings new opportunities and new challenges. The challenge is to know what is available, determine if it has value for the firm, discover how to use it, acquire it, and apply it in a way that produces some advantage for the firm.

3. A FRAMEWORK FOR AGILE MANUFACTURING

The following section provides the reader with a framework for the development of an Agile Manufacturing System (AMS). The section summarises four design strategies and provides a general overview of the implementation of agile manufacturing.

The framework illustrates the four key design strategies (Goldman, 1995):
- Enriching the customer
- Co-operating to enhance competitiveness
- Organising to master change and uncertainty
- Leveraging the impact of people and information

The framework is presented in Figure 16, and suggests that the agile manufacturing system is only complete when all of the pieces have been integrated to produce maximum benefit, thus improving overall competitiveness. The requirements for satisfying the design strategies are summarised in each of the four corner pieces. These, along with the enablers, will pull the pieces closer another until 'locked into' place, and agility is achieved.

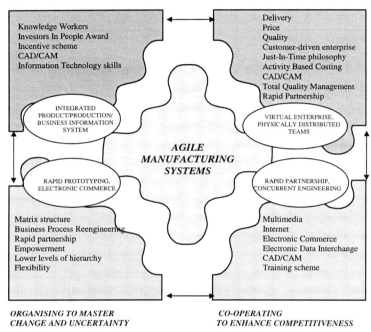

Figure 16: A Framework for the Development of Agile Systems

3.1 Agile Enterprise Design Strategies

All four design strategies suggested should be incorporated to some extent in order to become more flexible, responsive to the needs of the customer, and more importantly, adaptable to the turbulent, unpredictable conditions of dynamic global markets. The requirement to become more flexible and lean, for example, may result in the adoption of the Just-In-Time (JIT) philosophy along with flexible cellular manufacturing to offer make-to-order contracts to customers, thereby enriching them. There can be no single, universal formula for mastering agile competition. Every company must formulate its own, market-specific, dynamic program for becoming an agile competitor. Although there is no real formula, the following design strategies are recommended.

3.1.1 Enrich the Customer

The strategy is simple here, if one's organization cannot enrich the customer, then the customer will find an organization that can. The emphasis is on becoming a customer-driven enterprise. Quality, delivery, and price requirements of customers must be identified, monitored, and understood using approaches such as Quality Function Deployment (QFD). The firm must then satisfy the customers' demands by adopting customer focused philosophies and by using manufacturing methods and technologies such as JIT, Materials Requirements Planning (MRP), and Activity Based Costing (ABC).

3.1.2 Co-operating to Enhance Competitiveness

Co-operation is a key requirement for partnering firms within a physically distributed or virtual manufacturing enterprise. Collaboration between partners is made possible with tools/approaches such as Multimedia, Internet, EDI, CAD/CAM, and electronic commerce. Business Process Reengineering (BPR) may be used to reduce the number of non-value adding activities, resulting in improved information flows and hence, better co-operation between partnering firms. Cross-functional teams, empowerment, reengineering of business processes, virtual companies, and partnerships--possibly with direct competitors-- are all means employed to leverage resources through co-operation (Goldman, 1995).

3.1.3 Organising to Master Change and Uncertainty

An agile company is organised in a way that allows it to thrive on change and uncertainty. Its structure is flexible enough to allow rapid reconfiguration of human and physical resources. *"There is no single 'right' structure for and agile company, and no single 'right' size."* (Goldman et al., 1995).

The goal of very rapid concept-to-cash time implies innovative, flexible organisational structures that make rapid decision making possible by distributing managerial authority. People need to be empowered to convert change and uncertainty into new opportunities for the company.

3.1.4 Leveraging the Impact of People and Information

People--what they know, the skills they possess, the initiative they display--and the information to which they have access, are the differentiators among companies striving for agility. Because knowledge-based products offer the greatest potential for individualisation, continuous work force education and training are integral to agile company operations (Goldman, 1995). Companies must invest in information and manufacturing technologies that can contribute to agility, but they will not get the benefits of the technology without corresponding investments in human capital (education and training as well as appropriate compensation and incentives). Both are necessary to leverage the value of people and information in the pursuit of agility.

4. CONCLUSIONS

Information Technology alone will not guarantee responsiveness, but without it, the road to agility is blocked. The same can be said of people. Strategy is not enough, nor is organization. As stated in the introduction to this chapter, the acronym: **S.T.O.P.** covers every eventuality in an agile manufacturing system. Without suitable *Strategy* (the reader may not be surprised that a large number of companies do not even have a manufacturing strategy written down!) to show the way, suitable *T*echnologies to cope with the latest demands, an *O*rganisation able to deal with change, and *P*eople with the correct training, education and ability to cope with uncertainty, agility will not happen! The key to agility is to put these four elements together in a way that allows an organization, or group of organizations in the case of VE, to not only survive, but to prosper in an environment of continuous and unpredictable change. Though there is no single prescription applicable to all firms, no firm can become agile without successfully blending these four elements in developing its unique recipe' for agility.

REFERENCES

1. Berry, W.L., Hill, T.J., and Klompmaker, J.E., 1995, "Customer-Driven Manufacturing". *International Journal of Operations & Production Management*, **15**, 3, 4-15.
2. Booth, R., 1996, "In the market". *Manufacturing Engineer*, October 1995, p236-239.
3. Esmail, K., and Saggu, J., 1996, "A changing paradigm". *Manufacturing Engineer*, December 1996, p285-288.
4. Goldman, S., Nagel, R., and Preiss, K., 1995, *"Agile Competitors and Virtual Organisations"*. New York: Van Nostrand Reinhold.
5. Gunasekaran, A., 1998, "Agile Manufacturing: Enablers and An Implementation Framework". *International Journal of Production Research*, Vol. 36, No. 5, pp. 1223-1247.
6. Gupta, P., Nagi, R., 1997, "Optimal Partner Selection for Virtual Enterprises in Agile Manufacturing". *Department of Industrial Engineering, 342 Bell Hall, State University of New York at Buffalo, Buffalo, NY 14260*.
7. Harrison, A., 1997, "From Lean to Agile Manufacturing". *IEE Colloquium Digest on Agile Manufacturing*, No. 97/386.
8. Iacocca Institute at Lehigh University, Goldman, S.L., and Preiss, K. (eds.); Nagel, R. N., and Dove, R., principal investigators, with 15 industry executives, 1991, *"21st Century Manufacturing Enterprise Strategy: An Industry-Led View"*. 2 volumes, Bethlehem, PA.
9. Kalpakjian, 1995, *"Manufacturing Engineering and Technology"*. Addison-Wesley Publishing Company, Inc. Third Edition.
10. Kidd, P.T., 1994, *"Agile Manufacturing: Forging New Frontiers"*. Addison-Wesley, Wokingham.
11. Laudon, K.C., and Laudon, J.P., 1995, *"Information Systems"*. The Dryden Press, Harcourt Brace College Publishers, Third/International Edition.
12. Lei, D., and Goldhar, J.D., 1990, "Multiple niche competition: the strategic use of CIM technology". *Manufacturing Review*.
13. Maskell, B.H., 1994, *"Software and the Agile Manufacturer: Computer Systems and World Class Manufacturing"*. Portland, Ore, Productivity Press.
14. Peters, T., 1992, *"Liberation Management: Necessary Disorganisation for the Nanosecond Nineties"*.
15. Pine, B.P., 1993," Mass customisation *the New Frontier in Business Competition"*. HBS Press.

16. Porter, M., 1996, "What is Strategy?" *HBR*, Nov/Dec, pp61-78
17. Schonberger, R.J., 1982, *"Japanese Manufacturing Techniques: Nine hidden lessons in simplicity"*. New York: Macmillan.
18. Slack, N., Chambers, S., Harland, C., Harrison, A. and Johnston, R., 1995, *Operations Management*. London: Pitman publishing.
19. Walters, H., 1997, "Management and Improvement of the Extended Enterprise". *IEE Colloquium Digest on Agile Manufacturing*, No. 97/386.
20. Womack, J.P., and Jones, D.T., *1996 "Lean Thinking: Banish wastes and create wealth in your corporation"*. Simon & Schuster.
21. Womack, J.P., and Jones, D.T., and Roos, D., 1990, *"The Machine That Changed the World"*. New York: Macmillan Publishing.

Part II

DESIGN AND DEVELOPMENT OF AGILE MANUFACTURING SYSTEMS

A Strategic Approach to Develop Agile Manufacturing

Jens O. Riis and John Johansen

Center for Industrial Production, Aalborg University, Fibigerstraede 16, DK-9220 Aalborg, Denmark

1. NEW CHALLENGES TO INDUSTRIAL ENTERPRISES

Industrial conditions have changed radically over the last 15 - 20 years. In this period of time technology, market conditions, and customer demands have changed at a speed and in directions barely seen before. This for instance includes dynamic market fragmentation, shrinking time-to-market, increasing product variety and production to customer specification, reduced product lifetimes, globalization of production, etc.

At the same time, competition is becoming global, as the global economy is rapidly replacing local markets. The emergence of the open markets, reductions in trade barriers and improvements in transportation and communications links have led to a situation where local competition and markets operate in the context of global standards. As a consequence, today's industrial enterprises face new challenges and competitive pressures. The Next-Generation Manufacturing Project at MIT emphasizes agility and customer responsiveness, networking in a global market, employee participation, integration in an extended enterprise, knowledge management and competence development.

The changed industrial context calls for new capabilities. The ability of industrial enterprises to adjust quickly and accurately to changing conditions will be an important key to success in the future. Within this process enterprises must be able to integrate a multitude of technological, organizational and managerial viewpoints.

Especially within the last two decades literature has brought forward new manufacturing philosophies each offering a solution as to how a company should be managed and organized to be competitive. The list includes concepts like Just-In-Time (JIT) and Total Quality Management (TQM); but also concepts like Continuous Flow Manufacturing, Integrated Logistic/Fast Cycle Time, Time Based Manufacturing and Supply Chain Management may be seen in this connection. Some of the recent manufacturing philosophies that have appeared on the industrial scene are Lean and Agile Manufacturing focusing on leanness and agility, respectively.

Within several of the manufacturing philosophies a tendency can be traced from an enterprise view to an extended enterprise view; and from an intra- to an interorganizational company perspective, where the role and the development of the company are discussed more explicitly in connection with its markets, customers, distribution and supply networks. The company's cooperation with other companies and suppliers is especially important. Furthermore the discussion is closely attached to concepts like core competence and key technology and the ongoing discussion about externalization and outsourcing to specialized suppliers, (Kragh-Schmidt & Johansen, 1998).

1.1 Agility and Lean Manufacturing

Lean Manufacturing concept integrates the most essential Japanese production management principles from the 1970s and 1980s, e.g. by emphasizing the elimination of waste from Kaizen, Just-In-Time principles applied to the entire supply chain, TQM principles, and the empowerment of employees to take on the responsibility for their own work. The Lean Manufacturing concept was first coined in The International Motor Vehicle Program – IMVP in which a collection of automotive assembly plants in North America, Japan and Europe were described and analyzed in a number of measurement and performance areas, e.g. quality performance, flexibility and productivity, (Womack et al, 1990). The Lean Manufacturing was launched as a concept describing Best Practice within the automotive industry, but has gradually evolved, and today the concept is widely used in industry.

Whereas Lean Manufacturing presents a well-structured set of methods and enablers, Agile Manufacturing is rather a broad philosophy originated from the work of the Agile Manufacturing Enterprise Forum (AMEF), which is affiliated with the Lehigh University and was initiated in 1991. The work is documented in the report 21st Century Manufacturing Enterprise Strategy. The point of departure of Agile Manufacturing is the increased dynamics and unpredictability of industrial enterprises' environment. Accordingly, agility can be defined as the capability of operating profitably in a competitive environment of continually and unpredictably changing customer opportunities, (Goldman et al, 1995).

Agility is more than the traditional interpretation of organizational flexibility. Organizational mastery of uncertainty and changes is in focus in the agile organization; therefore people and knowledge are regarded as the most important organizational assets. Also organizational learning and the capability to reconfigure the business on a continuing basis are important characteristics of an agile enterprise - often associated with the ability to intelligently innovate and invent new responses, e.g. to new markets demands and business processes.

Agile Manufacturing accepts the significant trends for industrial enterprises towards working in networks, and consequently seeks enablers to facilitate appropriate responses to the dynamics imposed on a network of companies. Often a total product life cycle design philosophy is implemented, in such a way that design is integrated into a holistic production process incorporating the company's business processes from supplier relationships to product disposal.

Accordingly, the organization should be capable of performing well in cooperative relationships, in internal and inter-company teams that are cross-functional and require multi-skilled members. Agility also embodies such social concepts as self-directed business cells and virtual partnerships for the rapid formation of multi-company alliances to introduce new products to the markets in ways previously considered impossible, (Gunneson, 1996).

In the literature one can observe a divergence in the perception of the concepts of lean and agility. Some authors see the concepts as two opposite or orthogonal philosophies, others claim that Agile Manufacturing has an enterprise view whereas Lean Manufacturing is usually associated with the efficient use of resources on the operations floor. Others again regard the concepts as complementary philosophies mutually supporting each other.

Industrial companies often view the two manufacturing philosophies as modern buzzwords competing for the attention of industrial managers to the extent that one is superior to the other. Furthermore, some companies hold the position that there is no hurry to decide which one of the two to adopt, because in one or two years' time a third manufacturing philosophy probably will appear.

We want to adopt a more constructive view by relating the two manufacturing philosophies to two opposing criteria that any industrial enterprise somehow must reconcile. The first criterion is productivity, i.e. the ability to utilize and optimize the resources of the company. The other criterion is effectiveness, pointing to the ability to select and implement strategies and market opportunities with interesting future perspectives for the enterprise. Also the capability of the organization to reconfigure and change the business on a continuing basis are important characteristics.

With the risk of simplifying the issue, we see Lean Manufacturing as very much related to productivity, and Agile Manufacturing to effectiveness. This implies that any company actually needs to address both manufacturing philosophies and to seek a company-specific reconciliation of the two criteria. Manufacturing strategy development is very much about finding such a balance.

However, they are strongly interconnected. For instance, an effort to reduce time will contribute to both leanness and agility, as compressed time in business processes often results in increased productivity and quality which pertains to productivity. At the same time, speed and fast response time in business processes are often a pre-requisite for an effective reorganization and reconfiguration of the enterprise that relates to effectiveness. Similar examples could be found in relation to learning and organizational competence.

In this chapter we shall present a process for developing a manufacturing strategy. An essential underlying theme is to analyze and synthesize the balance between leanness and agility. This includes a phase aimed at achieving a broad acceptance of the need to change and to initiate a strategic manufacturing development process and a phase for the development of an overall manufacturing vision to depict the future production system and its mode of operation. Three case examples will be presented illustrating how a strategic manufacturing development process may be staged and which type of manufacturing system has emerged. Finally we shall draw implications of the process and case examples for a company-specific strategic response to the quest for agility.

2. A STRATEGIC MANUFACTURING DEVELOPMENT PROCESS

In this section a process for developing a manufacturing strategy will be presented. It aims at supporting enterprises in their strategic effort to develop innovative, agile and competitive manufacturing systems. An essential underlying idea is that managers and employees have unreleased ideas and capabilities to develop new solutions. However, they are seldomly voiced explicitly, discussed jointly or brought into a unified context. Accordingly, the framework is grounded on a collaborative dialogue designed to capture managers' and employees' innovative ideas and knowledge about the present situation in their company. Placed in the right strategic perspective, such ideas and knowledge, in our experiences, often have great potential for contributing significantly to the survival of the company.

The process is designed to develop the knowledge, attitudes and motivation of managers and employees in such a way that they are willing to take responsibility for the implementation of a production strategy with speed and efficiency. Part of the strategic manufacturing development process is therefore organized as intensive seminars involving both managers and employees. Between the events project groups or appointed taskforces will detail and consolidate results and inputs from a seminar, and prepare the agenda for the next seminar. The process can be characterized as a gradual refinement process where resources in a continuous assessment process are canalized to clarify critical points in the strategy.

One of the corner stones in the process is a so-called manufacturing vision. With its strategic aim it adopts a holistic and extended enterprise point of departure focusing on creating innovative and visionary, forward pointing solutions at the expense of details and detailed analysis. A manufacturing vision is different from existing solutions, which is often a prerequisite for creativity and innovation. Furthermore, a manufacturing vision should represent a coherent picture of the future production system illustrating and integrating many complementary perspectives. This includes technological, organizational, and managerial perspectives as well as a business process perspective.

In our experience a manufacturing vision contributes to form an overview of the production system and its mutual interplay of systems and organizational units in the enterprise, and therefore contributes to secure coherence between the various subsystems. Consequently, design and implementation of production systems are not only concerned with technical specifications and system design; but also with creating an organizational consensus among the many interested parties with their different background and individual opinion about criteria for a good and a poor solution. A production system interacts with different internal and external sections and organizational units. Perhaps, this mutual interplay may be the most crucial element for developing an appropriate production system.

Phase	Content
1. Initiation	Staging and organizing the process, plus clarifying the starting point and the ambition and scope of the process.
2. External trends and strategic challenges	Creating an organizational shared picture as regards the need for change, external trends and the strategic challenges of the enterprise.
3. Development of a manufacturing vision	A collaborative dialogue based process designed to capture managers and employees' innovative ideas and knowledge.
4. Evaluation of the manufacturing vision	Evaluation of ideas and elements of a manufacturing vision with respect to strategic challenges defined, and an examination of the risks and resources associated with implementing the developed manufacturing vision.
5. Application and planning of the next steps	Planning how to proceed by making use of the organizational momentum created, the potential strategic contribution of the manufacturing vision and critical areas for designing a production system

Figure 1: A five-phase process of developing a manufacturing vision.

We shall propose a five-phased process for developing a manufacturing vision, as the first step in design and implementation of a production system. The process will include creative elements that encourage a mood of dreaming and playfulness. This will stimulate generation of new ideas even if they are not well-thought out, because there is no risk of loosing face. In this way experimenting with new ideas is encouraged. It is our experience that such an experimental mood will enable a company to develop a manufacturing vision within a short period of time and with a relative limited effort.

The phases will be carried out in a spiralling process that includes both sequential and iterative elements, as illustrated in figure 2 and discussed below.

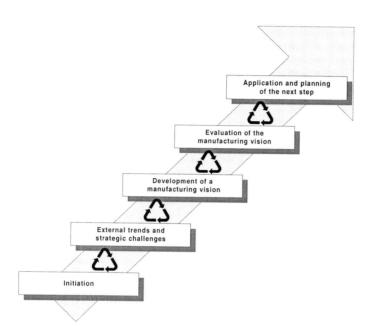

Figure 2: Phases of strategic manufacturing development are part of a spiraling process.

2.1 Initiation

Before starting the actually work of developing a manufacturing vision, the stage must be set and the process carefully prepared, organized and planned. In particular this includes a discussion of the scope of the process – does it embrace the entire extended enterprise or is it delimited to a production plant, or a production cell. Also it is important to harmonize the expectations of the organization and the interested parties. This may also influence whom to involve in the process.

The starting point and the ambition of the process should also be carefully discussed in this phase. Clearly it is not sensible to develop a manufacturing vision in isolation; it needs to be done in the context of an overall strategy process, as it should be consistent with corporate strategy.

In our work in industry we often have seen different degrees of clarification of an enterprise's strategy. For enterprises with a well-defined corporate strategy a manufacturing vision can be deduced with a clear focus. In this situation the manufacturing vision helps formulate the role of the production more precisely in accordance with the corporate strategy.

If a corporate strategic situation is characterized by uncertainty and no clear direction, e.g. due to dynamic and unknown surroundings, then it is difficult to develop a clear manufacturing vision as a response to distinct strategic challenges. In this situation the manufacturing development process will play quite a different role. Here the process can be used to develop several alternative holistic manufacturing visions and thus contribute to a strategic clarification process in the enterprise.

The defined scope of the process naturally will influence the resources needed, as it will influence who should attend and be involved in the process, which is not an easy question to answer. On the one hand, developing a manufacturing vision within a short time frame may call for professionalism, which points to a smaller group of specialists. On the other hand, it is important to involve managers and employees who are expected to implement the developed solutions in order to get their acceptance. At least the key players of the process should be involved.

To balance the point of views presented here, we have successfully held seminars with up to 15 – 20 participants combined with appointed task forces or project groups working out the details in between the seminars. In this way ideas, opinions, and expectations are harmonized during the seminars by developing a shared picture of the need to change and of future directions. Details are worked out in between the seminars with participation of experts, if necessary. Furthermore, one can add discussion meetings and other meetings to inform people not participating in the project.

2.2 External trends and strategic challenges

One of the most important prerequisites for accomplishing a successful strategic development process is that members of the organization share a common picture of the need to change, of external trends and of strategic challenges of the enterprise. For several years we have been concerned with, through a participative process, to find ways of developing such a shared picture which would form a sound basis for developing an overall vision of the future role and functioning of production.

To capture future external and internal trends we have developed a so-called "world-picture". The picture is divided into three spheres representing different views of the challenges of the enterprise. The first sphere concerns what we name the production task of the enterprise covering internal conditions, as for instance current planning and control systems and production technology. The second sphere relates to the close surroundings of the enterprise including markets, the community, and other functions of the company. The third sphere is the one furthest away from the enterprise and deals with the world society including general trends in economic, demography, technology, environmental issues, ethics, and welfare etc. An example of a "world-picture" can be seen in figure 3.

Most often the "world-picture" is used at a seminar where managers and employees are asked to identify future trends related to the three spheres that somehow may affect the enterprise. The time horizon is typically elected to be about 4 – 6 years. The seminar is usually conducted over a one-day seminar.

The trends identified are then used to derive a commonly shared picture of future challenges and conditions for the company, thus serving as a platform for seeking appropriate future solutions. Based on our experiences with these methods it seems fair to conclude that it is in fact possible to develop a broadly shared appreciation of the need to change, even when a seminar leaves a number of questions that need further investigation and explanation.

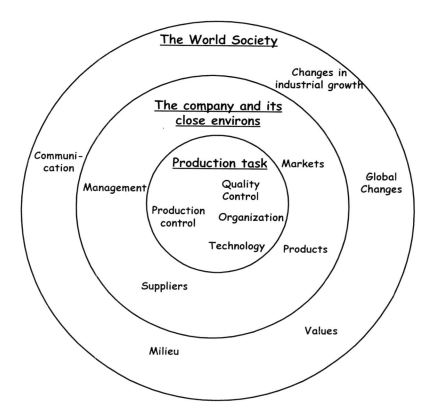

Figure 3: An example of a "world picture".

Some times it is appropriate to assess the present mode of operation. In this context we have developed a rather simple method, called Problem Matrix. The tool is useful for a group of persons from different sections and functions in a company to achieve a comprehensive understanding of the mutual interaction between departments and sections, (Riis, 1994). Also this method is used at a one-day seminar.

Management has a key role in prioritizing the future challenges and to select focal areas for the improvement effort. This will include an indication of the desired combination of agility and leanness derived from future challenges.

2.3 Development of a manufacturing vision

While the previous phase was concentrated on forming a common organizational understanding of the strategic challenges of the enterprise, this phase will focus on the development of a manufacturing vision. It seeks to combine creative thinking and professional production management and engineering knowledge and uses a combination of seminars and task forces.

In our terminology a manufacturing vision is a holistic picture of the way in which future production will be carried out, expressed by (i) structural elements and subsystems of a production system and (ii) processes in which they interact. Emphasis is placed on including only the essential features in order to provide an overall view. Furthermore, the holistic nature is underlined by focusing on the mutual interplay between different systems.

In our work we have identified three major challenges associated with the development of a manufacturing vision:

- The vision should integrate different elements of a production system and integrate complementary perspectives. A technical, organizational, economical and managerial perspective needs to be employed at the same organizational unit. This requires a bridging language to allow different disciplines and professionals to communicate constructively about the same object of study, e.g. industrial company, plant or workshop, (Riis et al., 1996).

- The vision should be developed through a participative process involving managers and employees from various functions and sections. In this way the manufacturing vision will represent a commonly shared opinion about the future production, as also discussed by Senge, (Senge, 1990).

- The vision should be innovative and address important future manufacturing capabilities. By addressing future expected situations the vision should capture novel ideas and break new grounds. Hence, the process of developing a manufacturing vision needs to be experimental and allow for opportunities to play with new ideas without fear of being committed to a yet unknown solution.

In this creative process of developing a vision there is not just one single procedure to be followed. In fact we have tried several different paths; for example (i) to let the group be inspired by one or more of the existing manufacturing philosophies, e.g. Lean Manufacturing, and Agile Manufacturing; (ii) to adopt ideas from Best Practice companies; and (iii) to record all ideas among managers and employees related to either the overall operations or to a specific subsystem. We have learned to include both structural and process oriented elements, and in particular to focus on their interaction. The structure, e.g. plant-layout, organizational structure, the structure of the product program, will provide the stage, whereas the processes will describe the daily life in the company, e.g. handling of a customer order, procurement, and development of a new product through to its launching on the market. We have noticed how the process point of view holds a key to visualize how future production may look by capturing sequences of events. It becomes a story telling about a day in a plant of the future.

Along the same lines, Maslen and Platts (1997) suggest that ideas be clustered into manufacturing decision areas according to (i) structure (facilities, process technology, capacity, vertical integration, products), (ii) infrastructural (production control, quality, new product introduction, suppliers, performance measurement), and (iii) human aspects (culture, organization, skills and training, rewards and incentives, communications).

In many respects the process of developing a manufacturing vision is an organizational learning process that prepares the mind-set for new options and may generate an organizational momentum for change pointing to both short-term and long-term initiatives. The process is highly experimental and offers opportunities for playing with ideas without being forced to make firm commitments to specific details. The process holds many elements of divergent thinking, as contrasted with the typical convergent thinking of daily operations in which decisions are to be made quickly.

In view of the behavioral patterns pointed out in the introduction, the experimental climate established during the process of developing one or more visions may hold a key to overcoming resistance to divert from a working mode suitable for the daily operations, partly because the risk associated with exploring new grounds is significantly reduced. Ideas and proposed solutions are deliberately kept floating and undecided for a long time and a playful climate is stimulated. Visualization is important, because ideas, visions and concepts are often better expressed in terms of pictures, images and metaphors.

Few managers and employees are accustomed to the abstract thinking implied in developing a manufacturing vision. However, when supported by facilitators, we have seen how they are able to engage themselves very actively in the process, and how it is possible to develop an idealized manufacturing vision through a mixture of seminars and task forces. On the basis of future challenges to production we have asked participants at seminars to propose ideas and solution elements. Usually a task force, with assistance of external consultants, has been asked to combine these ideas into one or two manufacturing visions, which were then presented at a seminar for discussion, refinement and approval.

A major task of an industrial enterprise is to continually aim at developing new capabilities that are valued by the customers better than or different from those of competitors. This often implies a shift of paradigm. For example, a new production technology that may require new competencies and working modes; a new way of managing customer orders may imply assignment of new roles; or the demand for shorter delivery times may require a closer cooperation between engineering design, sales, purchasing and production.

If a manufacturing vision implies a drastic shift of paradigm, the process brings to the open this drastic future change. It is not an easy managerial task to handle such a shift of paradigm that often requires a shift of corporate culture, working modes and qualifications. However, the process of developing a manufacturing vision holds potential for a shift in mentality, especially if it is combined with experimental activities, such as role-playing games and simulated demonstrations. A consequence may be that it becomes obvious to both management and employees that some persons may not be part of the future organization and better be asked to seek other jobs.

In our industrial research we have proposed to adopt a dialectic planning approach in which opposing ideas or visions are developed. In some cases the visions represent rather extreme cases to be combined at a later stage into a new vision. The process will generate a sense of robust direction with stable conceptual elements that are expected to be applicable for future solutions. Also, the discussion has led to identification of areas of future attention.

2.4 Evaluation of the manufacturing vision

Before a manufacturing vision is approved by management as a basis for negotiating new modes of cooperation with sales, purchase and product development and for further detailing a new production system it is useful to evaluate the vision. Due to lack of details, this will not constitute a rigorous assessment, but an early evaluation of the potential of the proposed vision with respect to the strategic challenges defined. Traditionally it is unusual that such an evaluation is carried out at this early stage. But the existence of a holistic picture of how the future production will look provides a better basis.

Instead of asking a taskforce to do the evaluation and submit the result to management, we have proposed and tested a mixed process. After a taskforce has completed a proposal for a new manufacturing visions and evaluated its potential, it is presented at a seminar to management and representatives from production, sales, purchase and product engineering for an evaluation. Participants will look at the vision from their own point of view and identify advantages and limitations and in this way provide a broad evaluation of the potential of the vision and identification of areas of attention.

Partial evaluations may take place during the process of developing a vision when a proposed guiding principle is held against strategic challenges. The more well defined these strategic challenges are, the more likely it is that they will guide the process of selecting principle solutions.

A manufacturing vision is comprehensive and embraces many dimensions as the vision relates to corporate strategy of the enterprise. Consequently, the evaluation process should also be multi-dimensional. Inspired by Maslen and Platts (1997) we suggest the following dimensions as a checklist to be elaborated in view of the specific company situation:

- Corporate strategy
  1. Does the vision support the corporate strategy and the strategic challenges identified?
  2. Does the vision provide a clear focus for change?
  3. Does the vision express manufacturing contribution to corporate strategy?
- Economic and business opportunities
  1. Does the vision open up new business opportunities?
  2. Does the vision have a satisfactory return on investment?
- Market and competitors
  1. Does the vision provide capabilities that are valued by customers?
  2. Does the vision contribute to provide a competitive edge?
- Manufacturing
  1. Do the vision and chosen solutions appear consistent?
  2. Does the vision support leanness and agility?
  3. Are the chosen manufacturing solutions mutually coherent, sufficiently innovative and visionary?
- Organization
  1. Does the vision support and build on the core competencies of the organization?
  2. Does the vision hold new possibilities for management and employees?
  3. Does the vision imply a drastic change in working mode and corporate culture?

If the evaluation is not convincing, the development process should return to a preceding phase. Relatively little effort has been spent so far, and no commitments have been made. The significance of this early comprehensive evaluation is that it is rather easy to repeat earlier phases. It is important for the next phase that the potential of the new manufacturing vision has been subjected to a broad discussion and that there is a broad organizational support for pursuing its realization.

2.5 Utilization of the manufacturing vision and planning its implementation

The manufacturing vision represents an idealized solution not associated with any time horizon. To exploit its potential we need to transform it to a number of holistic solutions each related to a specific time horizon, e.g. a short-term solution, a medium-long-term solution, and a long-term solution. This transformation should be carried out with due regard to the current situation of the enterprise, e.g. its manufacturing system, management systems, corporate culture, and capabilities for carrying out major organizational changes.

Because industrial enterprises may have different strategic situations with respect to turbulent environment, a manufacturing vision may play different roles:

- A manufacturing vision merely may serve as a coordinating role to align current improvement activities to secure maximum synergy, thus serving as a vehicle for a constructive dialogue with other functions of an industrial enterprise.
- A manufacturing vision may be used to point to a new paradigm of operations. Few investments may be necessary, but a shift in mentality is needed. In this way the vision may help orchestrate a concerted effort in different functional areas.
- A manufacturing vision may serve as a blue print for future investments and as a master plan for an innovative, concerted effort in which production engineering, management systems and organizational initiatives are integrated into a new, coherent solution. The vision thus provides a basis for a coordinated detailed development and design of the production system.

It is our experience from several action research projects that the high degree of involvement of middle management and employees in the process will create an organizational momentum that may support implementation of more drastic changes that may represent paradigm shifts and cultural changes. For example, the process may have convinced many persons in the organization that it is indeed necessary to improve the performance and that there exist ways of realizing it. If an organizational momentum is not utilized by initiating a change process that within a short period of time may show visible results, the motivation and morale in the company may fall far below the original level. Thus, in a sense, the process of developing a manufacturing vision has irreversible features.

In this chapter we shall not discuss how an implementation plan may be prepared taking into account both business, technical and organizational aspects; but refer to some of the extensive literature Kotter (1996), Riis & Mikkelsen (1997).

3. CASE EXAMPLES

In this section three cases will to illustrate various aspects of the proposed process. The main emphasis will be on how a strategic manufacturing development process was staged and how a manufacturing vision was developed. The cases cover a spectrum of different types of industrial enterprises, as well as manufacturing and market situations.

3.1 Alpha – a supplier manufacturing vision

Alpha is a medium-sized supplier of welded parts and equipment for a number of different industries (chassis parts for buses, shovels for front loaders for construction, ventilators for cement factories, hydraulic lifts for trucks, etc.). The enterprise has been in a continuous development, and many ideas for improvement are waiting for the needed capacity and financial capability. Yet, management was not clear of the direction to move.

As a supplier with great variations in customers and their demands, the company needs to be flexible. However, as a first step it was possible to divide the portfolio of tasks into three distinct categories: (i) continuous flow manufacturing; (ii) Flexible welding; and (iii) Small batch production of equipment. The volume and flexibility requirements vary from category to category and support an idea of three mini factories in a factory.

The decision to divide the factory into three sections according to the three distinct groups of tasks allowed a clear focus for developing a manufacturing vision for each category with specific ideas for production planning, production processes, work organization and management, etc. Not only did the vision hold significant potential improvements in terms of reduced delivery time and costs, but it also provided a clear picture of desirable customers and the company's competitive strength for different types of customers. The three manufacturing visions inspired management to implement a number of changes, e.g.

- Adjusting the plant-layout towards three separate production units and in particular to improve flow in the first mini factory
- Introducing production groups that could be given a clear role in the respective production unit
- Developing a pallet with parts for a specific customer order organized in the sequence in which the welder was to use them (this method is known from assembly as kitting).

In this way, the manufacturing visions were able to bridge short-term initiatives and long-term thinking and action. Some of the changes implied in the manufacturing vision were seen as a continuation of the current thinking and working mode, but they could now be placed in a larger, strategic context. Other initiatives represented a mental shift in working mode that implied a new paradigm. For example, the new production groups suggested a change in production planning whereby the foremen were given new roles.

The division of the plant into three mini factories provided an opportunity to discuss different aspects of agility and leanness, and their mutual interaction. For example, the continuous flow factory encouraged focus on modular welding equipment to support the capability to configure and produce a large variety of different shovels. In the case of the production of equipment agility should be achieved by versatile and skilled workers.

3.2 Beta – a process industry manufacturing vision

Beta is a small plant affiliated with an international group producing among other thing enzymes and natural colors for the food industry. Dairies for manufacturing cheese use the enzymes. For several years the plant has not maintained its technology and management. The former plant manager was inefficient and managed the plant in an old fashioned way. Consequently, middle managers and many of the employees lost their motivation and interest in developing the business.

To revitalize the plant and to reestablish the responsibility of middle managers for running the production a new manufacturing vision was developed. In fact, headquarter had considered moving production abroad and closing down the plant, unless it was able to demonstrate considerable enhancements in productivity and flexibility within one or two years. However, if the plant was able to produce a powerful and visionary plan for developing the business, the headquarter was willing to invest considerable resources and capital in renewing the company. To support the process a new plant manager was recruited.

The process of developing a manufacturing vision was planned in cooperation with the new plant manager and middle management. The process followed the suggested phases of the presented framework and was organized as a mixture of seminars, project groups and task forces. It soon became clear, however, that it was necessary to include educational activities to prepare and upgrade middle managers' business knowledge, as their understanding of the internal interplay and of the plant's interaction with external supply chain and customers was inadequate.

It was decided to involve as many employees as possible in the development process. For all practical purposes, two persons from each department were chosen to represent the employees and were made responsible for informing the rest of the group about the progress of the process. In addition, seminars and meetings were organized with participation of the entire company. Altogether about 40 people were involved in the process.

The resulting manufacturing vision of the plant is illustrated in figure 4. It consists of three autonomous "production" units and a support center offering services, e.g. administrative and maintenance tasks, bookkeeping and IT.

Besides the support center, manufacturing was divided into three focused production units, each with specific tasks and features.

- An extraction unit to produce concentrated quality enzymes, high volume at low costs. Its bulk products are sold to internal and external customers. Production is organized in a continuous flow layout supported by specially developed technology.
- A flexible packing unit to manufacture high-tech quality enzymes customized to individual customer orders. The unit is using new advanced flexible packing equipment, e.g. new innovative principles for tapping liquid into plastic bags and cans.
- A powder-manufacturing unit to produce mass customized products, in terms of unique powder mixes and strengths, but packed in standard packaging sizes.

Figure 5 illustrates the vision for one of the autonomous production units. Besides an illustrative drawing, the capabilities and functionality of each production unit are described. To validate some of the advanced technological solutions several pilot experiments have been carried out.

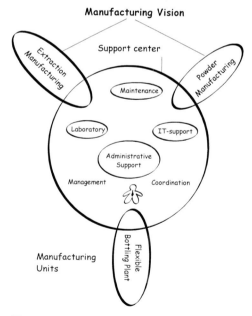

Figure 4: The manufacturing vision of Beta.

The case has shown that it is possible within a short period of time to create an overall, visionary and coherent picture of how a production unit could function and be organized in the future. Besides convincing the headquarter about the viability of the business, the vision has pointed to new innovative solutions and has re-established the optimism of management. In addition, through the process employees also have realized that they can contribute to develop the plant.

Compared to the situation at the outset of the development process, the manufacturing vision that is now being implemented represents a combination of improved agility and leanness, and thus demonstrates that they may support one another.

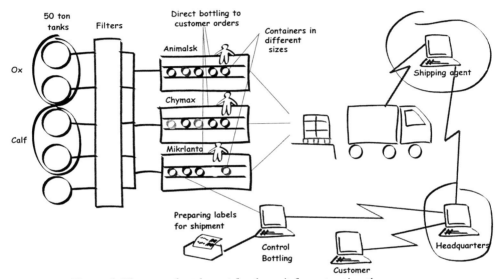

Figure 5: The new plant-layout for the unit for extracting the enzymes.

3.3 Gamma – an international manufacturing vision

Gamma produces exhaustion systems for various industries such as the wood industry and industries using welding processes. Gamma is positioned among the 10-15 largest manufacturers of industrial ventilation systems in the world with an ambition to become among the top three. In the beginning of 1998 the company merged with two other factories, one in Germany and one in England, each with own product program and production facilities. Management wanted that production activities should function as one production center and was interested in developing an overall production strategy[1].

The work force was anxious, because of the change in ownership. They were worried about their future jobs due to fear of rationalizations and transfer of part of the production to other countries. Also white-collar employees were nervous, and several key personnel had left for other companies.

The production manager at Gamma was faced with the dual challenge to develop a manufacturing strategy for his production facility and an international production strategy that would indicate how the three distributed facilities would function as an integrated unit. He

---

[1] The case is adopted from a Ph.D. project carried out by Brian Møller, Aalborg University.

would present the strategy to the manager in charge of international production and the top-management of group.

Several people were invited to participate in the strategy development process. Head of production, head of orders, head of purchasing, and an export salesman represented management. Correspondingly, shop stewards and selected members of the workforce represented employees of the company.

The method applied in this case was concentrated around four seminars, figure 6. In between the seminars a task force worked on the results from each seminar. Typically, a seminar was conducted over one day. External persons facilitated the entire process.

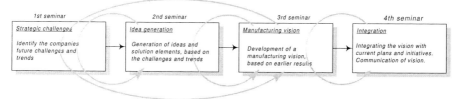

Figure 6: A series of four seminars represented corner stones of the development process.

At the first seminar where the "world-picture" was presented, the participants were asked to think about 3-4 trends or tendencies that they thought would affect the company in the future. The objective of the first part of the seminar was to generate as many statements as possible and in the second part to cluster the statements into overall themes. Most of the statements dealt with trends in conditions close to production, such as employees, products, markets, environmental issues, supplier-relationships and communication.

The objective of the second seminar was to generate solution elements concerning the challenges identified at the challenge seminar. The third seminar aimed to integrate the solutions developed at the second seminar into one integrated holistic vision. In view of the dual strategy task it was decided to develop two manufacturing visions in parallel, one for the Danish factory and one for the international manufacturing including the joint operation of the three factories. The two visions were mutually interdependent; how would it be possible to develop a vision for the Danish factory without knowing its role and production tasks; and how would it be possible to develop a vision for international manufacturing without knowing the capabilities of the Danish factory.

The employee representative and shop stewards together with the head of production were asked to address the Danish factory. They developed the notion of a self-propelled factory indicating that employees would themselves solve problems, plan production, organize work, initiate changes etc. Two distinct production tasks were identified that suggested a division of the factory into two separate plants, a streamline and a flexible plant. The former should focus on costs and process optimization, whereas the focus of the latter should be on flexibility and multi-skills. The resulting manufacturing vision held potential for significant improvements in terms of increased productivity and faster response time for customer orders.

The head of purchasing and the head of orders were asked to lead the development of an international manufacturing vision with focus on the overall logistics. The group agreed on the following headlines as a basis for the idea generation: Gamma's role as one production center out of three; Order flow; Material flow/layout; Supplier collaboration in Supply Chain Management context; Planning/capacity control; and Operation economy.

Each headline was written on large sheets of paper and attached to the wall. Participants were then given 10 minutes to write down or draw ideas regarding different areas directly on the sheets. Afterwards the ideas were presented and discussed succinctly, which spurred even more ideas. After a while it was suggested to draw a picture illustrating the flow of orders from the first customer contact to the final delivery of a ventilation system. This method was used to create an overview of the entire logistic chain and to join the different ideas generated previously into one coherent picture.

During the period of sketching the order process, several questions were raised as to how various tasks should be solved in this new situation. Trying to make many of the ideas concrete and integrate them through a sketch made it clearer which concrete solutions had to be developed for the entire system to function. Drawing of sketches worked as a way to externalize and make explicit many aspects of the various ideas and suggestions, see figure 7.

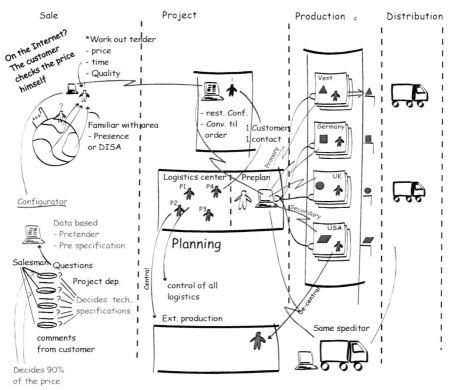

Figure 7: A sketch illustrating the international ordering process at Gamma.

The sketch describes the process from the configuration of a ventilation system through a dialogue between the customer and the salesman until the customer has his system installed. In the sales phase the salesman will be supported by a system configurator, that through a series of questions will be able to provide a price estimate that is 70-90% accurate. If the customer decides for the system, the salesman can confirm the order through the IT-system.

Then the project department will prepare specifications of the remaining 10-30% of an order and submit a tender to the customer. If the customer accepts, the order is confirmed and broken down and spread out to the different production facilities.

Each production facility will be an expert on a set of primary products and will be capable of also manufacturing secondary products in periods of over-capacity. The primary products are usually secondary products for another production facility and vice versa, which means that ideally the production facilities only manufacture primary products.

The project department will be the only contact to the customer.

A global shipping company will handle all transportation of goods from the production facilities to the customer. It will be hooked up to the computer system of the project/logistics centers such that they are automatically notified when they have to carry out a transport.

The vision is still very crude and many things need to be worked out for the concept to be feasible. However this is a preliminary sketch of an emerging vision, where many of the ideas are beginning to appear, although unpolished and relatively abstract.

The Delta case examples shows how two manufacturing visions are developed in parallel, representing two distinct, but interdependent areas, respectively a national factory and the international level including all production facilities. The sketches of the two visions helped the international production manager initiate a strategic process through a dialogue with the three international production factories. The sketches demonstrate, by providing a holistic picture, the possibility to increase both agility and leanness.

4. DISCUSSION AND IMPLICATIONS

The management literature is rich on proposals for development of a vision, e.g. Kotter (1996) and Womack & Jones (1996). But very few authors discuss in detail how a vision may be developed, how it may look, and which roles it may play.

From practice we have learned that the development process may itself be just as valuable to the company as the emerging vision, because of the broad discussions of essential issues. In this way an organizational momentum is generated. However, further studies are needed to understand the role of a manufacturing vision and how the development process may be staged and supported. However, based on more than ten case studies that we have been working with, we have derived a set of both descriptive and normative propositions, which we offer as postulates for further dialogue and research:

- *Agile and Lean Manufacturing philosophies may be combined into a company-specific manufacturing vision.* Although a manufacturing philosophy often represents a comprehensive framework, its proposed principles and methods need to be transformed into a company-specific context. As Agile and Lean Manufacturing may be viewed as representing two opposing criteria relevant to any company, respectively efficiency and productivity, they may be combined into a manufacturing strategy and visualized through a manufacturing vision.

- *Significant results with limited time and effort.* The proposed method for collaborative conceptual design for manufacturing has several advantages in comparison with traditional methods for production systems design. By challenging individual experiences and ideas and by bringing them into a common picture it has proven possible within a short period of time to develop a shared vision of the company's future production.

- *Complexity syndrome.* In a turbulent environment with different interests, it is more likely that the organization implements an improvement that is merely an adjustment or an amendment to the existing system. This is saving much energy and time. However, this behavior is likely to increase the complexity of the system. A manufacturing vision holds great potential for overcoming the complexity syndrome by providing a coherent picture of the way in which the organization functions, especially the intricate mutual interaction processes between individuals, sections and departments.
- *Experiential working mode.* Experimentation is encouraged in such a way that it can be carried out without fear of loosing face, e.g. by addressing an idealized future situation like a manufacturing vision, which does not directly relate to their current operations and position. Examples of methods are role-playing games, simulation, virtual solutions, Technology of Foolishness, and creation of diversity (appointing new staff with alternative background). The dialogue is important.
- *Connecting different time horizons.* As mentioned in the previous proposition, it is important to focus on connecting activities of different type. But it is equally essential to connect activities with different time horizons.

    Relating short-term to long-term activities. Look for long-term aspects of a short-term activity to see, if it could be "a step in the right direction".

    Relating long-term to short-term activities. From long-term solutions identify implications for the organizational and cultural dimensions, and seek to identify specific short-term activities that may support these future dimensions and which may be seen as part of organizational learning processes.

    The development of a manufacturing vision may provide a solid basis for identifying both types of relationships.
- *Early anticipation of a drastic change.* The sooner future drastic changes are anticipated and accepted broadly in the organization, the easier it is to implement the change in the organization, even if the change-over takes place during a short period of time. This supports involvement of a large part of the organization in appreciation of a need to change and in discussing and experimenting with future solutions. In the predominantly departmentalized organizations very few persons have a comprehensive picture of what is going on. This emphasized the benefit of initiating a collaborative process whereby commonly shared pictures are established of the current situation, future challenges and, later on, of future solutions.
- *Orchestrating implementation.* An improvement activity is often carried out as a self-contained project in a section or function with its specific focus, approach and performance measurement. If they could be better orchestrated, a more significant overall effect could be achieved. We have observed that the number of ideas and proposed solutions at any time may far exceed an organization's capability to implement them, although, at the same time, many organizations have much unused 'slack' change capacity. This also calls for a careful orchestration of the implementation process.

    A manufacturing vision will provide a coherent picture that may serve as a vehicle for relating implementation initiatives. Furthermore, it may help managers and employees involved in implementation to define the role of their own activities in a larger context which may release more energy in the organization.

- *Involve people in both exploitation and exploration.* In most companies exploitation and exploration are organized into separate functions. Would it not be desirable to loosening this to allow operations people to be involved in recognition of a need to change and creative development of new solutions, and to engage development people in the mechanisms of daily operations?

5. CONCLUSION

In this chapter we have presented a process for developing a company-specific manufacturing vision to serve as a bridge between corporate strategy and production systems design. A manufacturing vision may include a combination of lean and agile strategic elements suited to the specific situation of the industrial enterprise.

The five-step process aims at involving managers and employees in a constructive dialogue that leads to a commonly shared appreciation of the need to change and a common vision of the structure and operation of future production.

We identified three challenges associated with developing a manufacturing vision: (i) to integrate different elements and perspectives of a production system; (ii) to ensure a collaborative process with a high degree of participation; and (iii) to include innovative features.

A manufacturing vision may play different roles. If the corporate strategy is clear and robust, a manufacturing vision may constitute a blue print for (re)design of a production system. If, on the other hand, the environment is turbulent, a dialectic planning approach may be applied. Two opposing manufacturing visions may serve as a means for generating a constructive dialogue and may contribute to clarification of directions and potentials for the entire company.

Three case examples demonstrated different manufacturing visions and how elements of agile and lean manufacturing could be combined.

ACKNOWLEDGEMENTS

The authors wish to thank members of the P2000 project team: Steen Hildebrandt, Mogens Myrup Andreasen, Kristian Stokbro, Kresten Kragh-Schmidt, Jesper Olesen, Niels Rytter, Brian Møller and Irene Odgaard for inspiration through a collaborative process of developing ideas and testing them in practice.

REFERENCES

1. Goldman, S. L., Preiss, K., Nagel, R. N., and Dove, R. (Eds.): 21st Century Manufacturing Enterprise Strategy: An Industry-Led View, 2 volumes, Iacocca Institute at Lehigh University, Bethlehem, PA, 1991.
2. Goldman, Steven L., Roger N. Nagel & Kenneth Preiss (1995): Agile Competitors and Virtual Organizations, Strategies for Enriching the Customers, Van Nostrand Reinhold, USA.

3. Gunneson, A. O. (1996): Transition to AGILITY – Creating the 21st century enterprise. Addison-Wesley Publishing Company.
4. Kotter, John P. (1996): Leading Change, Harvard Business School Press.
5. Kragh-Schmidt, K. & John Johansen (1998): An outline of production philosophies, Department of International Marketing and Management, Southern Denmark School of Business and Engineering.
6. Maslen, Roy & Ken W. Platts (1997): Manufacturing vision and competitiveness, Integrated Manufacturing Systems 8/5 p. 313 – 322.
7. Riis, Jens O. & Hans Mikkelsen (1997): Capturing the nature of a project in the initial phases: Early identification of focal areas, Project Management Vol. 3 No. 1, pp. 18 – 22.
8. Riis, Jens O. (1994): Situational production management: a practical theory for the development and application of production management, International Journal of Production Planning & Control, Vol. 5, No. 3, p. 240 – 252.
9. Riis, Jens O., de Haas, Henning, Drejer, Anders and Møller, Brian (1996): An Experimental Production System Design Lab for Increasing Creativity in Manufacturing System Education, in Proceedings of the International Conference on Education in Manufacturing, San Diego, Society of Manufacturing Engineers.
10. Senge, Peter M. (1990): The Fifth Discipline, Random House.
11. Womack, James P. & Daniel T. Jones (1990): The Machine that Changed the World, Rawson Macmillan.
12. Womack, James P. & Daniel T. Jones (1996): Lean Thinking, Simon & Schuster.

BM_Virtual Enterprise Architecture Reference Model

G. D. Putnik[*]

University of Minho, Department of Production and Systems Engineering
4800 Guimarães, Portugal

The virtual enterprise (VE) reference model, named *BM_Virtual Enterprise Architecture Reference Model (BM_VEARM)*, is proposed. The BM_VEARM is defined as a hierarchical multilevel model of the enterprise/manufacturing system control and satisfies the requirements for integrability, distributivity, agility and virtuality. It is conceived to cover all processes in an enterprise, from the macro to the micro level, and for any type of production. A formalization of the BM_VEARM is presented, as well as a laboratory installation for the VE enterprise, BM_VEARM based, demonstration and validation.

1. INTRODUCTION

The combination of the shorter life span of new products, rapid technological developments, frequent changes in demand as well as "social and political environment changes" increase an enterprise's need for a new organisational model to keep competitiveness.

In (Kim, 1990) is given a very illustrative specification of the performances required for a new manufacturing system, or enterprise, organisational model. An "ideal" (target, future) manufacturing system, or enterprise, should be able to:
1. manufacture from 1 to 1.000 products simultaneously;
2. accommodate lot sizes from 1 to 1.000.000;
3. the system should reconfigure for a new product within 1 second (in order to satisfy 1 and 2).

The requisites 1) and 2) express the need for the highest level of the manufacturing system, or enterprise, adaptability. But, the adaptability is not sufficient by itself. If the enterprise takes too much time to adapt the opportunity may be lost, and therefore the competitiveness is not achieved. So, one of the most important factors is the manufacturing system's, or enterprise's, capability of *fast adaptability* or *fast reconfigurability*, i.e. *flexibility*, in order to satisfy the new circumstances (the new market opportunity, new demand, new tasks, optimisation of old tasks, "deadlocks", etc.).

As an answer to the above-mentioned requisites for the enterprise organisation, it is conceived the concept of *virtual enterprise (VE)*.

There is not a universally accepted definition of the virtual enterprise concept (depending on application domain there are also referred terms, or concepts, as *virtual company, virtual corporation, virtual organisation, virtual factory, virtual manufacturing*, etc.). Our analysis shows at least two (main) approaches in virtual enterprise concept definition, or specification.

---

[*] Prof. Goran D. Putnik, Department of Production and Systems Engineering, University of Minho, 4800 Guimaraes, PORTUGAL, fax: +351-253-510268, e-mail: putnikgd@eng.uminho.pt

By the first approach the most important characteristic of the virtual enterprise concept is dynamic networking of enterprises, see e.g. (Goldman et al., 1995), (Kidd, 1994), (Hormozi, 1994), (NIIIP, 1996), (Browne, 1995).

The second approach emphasises the "virtuality" of the system as something "not physically existing as such but made by software to appear to do so" (Oxford Dictionary). For the second approach enterprise networking is irrelevant. The VE is reduced to the simulation program, see e.g. (Kim, 1990), (Onosata and Iwata, 1993), (Fujii et al.; 1999), (ISR, 1995).

Regarding virtual enterprises reference models we didn't find too much of them, e.g. (NIIIP, 1996), (Camarinha-Matos et al., 1999). Usually VE definitions are presented but VE reference models are not presented.

In this text we present a reference model for virtual enterprise architecture, named *BM_Virtual Enterprise Architecture Reference Model*, under development at the University of Minho, concerning primarily the production or manufacturing enterprises.

The text is organised as follows. In the first part of the text, we introduce basic concepts, which make a framework for the virtual enterprise reference model development (Chapter 2). The second part of the text is dedicated to the presentation of the *BM_Virtual Enterprise Architecture Reference Model* derivation and structure (Chapters 3 and 4). Finally, the third part of the text presents shortly a project on development of the demonstrator of distributed and virtual manufacturing systems that follows the reference model conceived (Chapter 5). At the end of the text the conclusions and references are provided.

2. HIERARCHICAL SYSTEM MODEL OF THE MANUFACTURING SYSTEM

BM_Virtual Enterprise Architecture Reference Model is based on a hierarchical system model as a global view of the enterprise/manufacturing system. The underlying starting formalisation is a theory of hierarchical multilevel systems (Mesarovic et al., 1970).

The hierarchical, multilevel, systems are specified in a following way:

The system S is specified as

$$S : X \to Y \qquad (1)$$

where X is the set of outside stimuli and Y is the set of responses. Both, X and Y, are representable as Cartesian products, i.e. X and Y are assumed as a families of sets such that:

$$X = X_1 \times \ldots \times X_n \quad \text{and} \quad Y = Y_1 \times \ldots \times Y_n \qquad (2)$$

representing ability to partition the input stimuli and responses onto components.

Each pair of (X_i, Y_i), $1 \le i \le n$, is assigned to a particular level of a system S_i, represented as a mapping, Figure 1:

$$\begin{aligned}
&(i) \quad S_i: X_i \times W_i \to Y_i, \quad \text{if} \quad i = 1, \\
&(ii) \quad S_i: X_i \times C_i \times W_i \to Y_i, \quad \text{if} \quad 1 < i < n, \\
&(iii) \quad S_i: X_i \times C_i \to Y_i, \quad \text{if} \quad i = n.
\end{aligned} \qquad (3)$$

A family of systems S_i, $1 \le i \le n$, represents a hierarchical system S if exist two family of mappings $h_i: Y_i \to W_{i-1}$, $1 < i \le n$, and $g_i: Y_i \to C_{i+1}$, $1 \le i < n$, such that for each x in X and y=S(x):

(i) $y_1 = S_1(x_1, h_2(y_2))$,
(ii) $y_i = S_i(x_i, g_{i-1}(y_{i-1}), h_{i+1}(y_{i+1}))$, if $1 < i < n$, (4)
(iii) $y_n = S_n(x_n, g_{n-1}(y_{n-1}))$.

The mappings h_i and c_i are referred as the i_{th} *information function* and *decision function* respectively.

The hierarchy of the system means that there are no influences between C_i, and C_{i+1}, and W_i and W_{i-1}. In other words there is a complete decomposition between two levels in a hierarchical system (from practical point of view a complete decomposition is a strong condition and represents an idealised case).

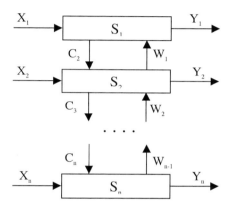

Figure 1. A hierarchical multilevel system.

By application of sequential ("AND"), parallel ("OR") and feedback ("$\mathcal{F}$") operators for system composition/decomposition it is possible to represent (or model) different engineering systems and, especially, manufacturing system and its components. It is important to notice that the composition/decomposition operators enable not only development of a rigorous hierarchical control processes structure, equal to a tree structure, Figure 2a, but also enable representation of sequential and parallel processes with, or without, feedback, Figure 2b.

Although some advance concepts advocate "heterarchical" architectures, we strongly believe that hierarchy can't be avoided, especially in manufacturing enterprises. One simple reason is that in practise, as well as in theory, we have to recognise the structural complexity of processes.

By analogy, and necessary abstractions, we have extended the model to the higher level processes of an enterprise. Consequently, the enterprise system is modelled as a hierarchical process based system. The model is further specialised in order to build the specific (reference) model of a virtual enterprise.

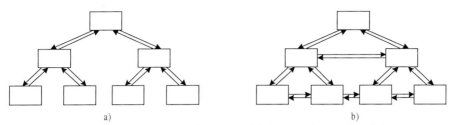

Figure 2. a) Model of the multilevel system with rigorous hierarchy of the processes; b) Model of the multilevel system with hierarchy and sequences of processes.

3. INTEGRATED, DISTRIBUTED, AGILE AND VIRTUAL SYSTEM OR ENTERPRISE

There are four global properties that the virtual enterprise architecture reference model should provide to a particular virtual enterprise model:
1) *integrability*,
2) *distributivity*,
3) *agility*, and
4) *virtuality*.

The specialisation of the general model of a hierarchical, multilevel, system, building the specific (reference) model of an virtual enterprise and implementing the above-mentioned four properties is presented over two control levels (S_i, S_{i+1}). The pair of two control levels (S_i, S_{i+1}) is *an elementary structure* for specification of different functional systems each one of two hierarchical levels having different terms in different areas. Some terms used are e.g., "controller-object of control", in area of production control or devices control, "client-server", in area of communication and object-oriented programming and modelling, "principal-agent" (Tirole, 1986) in economics and organisational sciences, "control-resource", etc.

Another term that we will use frequently is *"resource"*.

A *resource* is (a view of) an enterprise object which is used to realise, or "to support the execution of", one or more processes and it is the subject of control (or management). In the context of open distributed systems, a *service* is a kind of resource. A "service" is "an object which can perform one or more specific operations" (Vernadat, 1996). In terms of implementation the "resource" is a physical support for the service realisation or execution, e.g. material, machine tool, computer, human operator, time, money, software. The resource is a recursive construct, i.e. resources can be made of resources. So, we recognise *primitive and complex resources*. But, a process is not a resource.

An *enterprise or a company is a resource* (primitive or complex) when the enterprise (*server*), or its part, is used (contracted) by other enterprise (*client*) to carry on some process (*service*) required by that enterprise.

We will use interchangeably the terms "resource" and "enterprise".

3.1. Integrability

One of the most important requirements for the virtual enterprise is the capability for efficient access to heterogeneous candidate resources (enterprises) to be integrated in the enterprise, efficient negotiation between them and their efficient integration in the virtual enterprise. By "heterogeneous" resources here we mean that the resources work internally in their own specific, proprietary language, e.g. in the case of software or "business" application, or they don't conform to the same standard(s), e.g. in the case of mechanical or electrical/electronic devices.

For our purposes we would say that the integration is primarily the task of improving interactions among the system's components using computer based (information and communication) technologies with the following goals (Vernadat, 1996):
1) to *hide underlying heterogeneity and distribution* of functions, data, knowledge, and functional entities to business applications and users, therefore ensuring *portability*;
2) to *facilitate information exchange and/or sharing* among applications, and
3) to provide an open environment, i.e. an *interoperable 'plug and play' environment* in which new components can be easily added or connected, updated, or removed, for integrated enterprise operations.

A portability and interoperability among heterogeneous applications and devices (platforms), as well as extendibility, reconfigurability, longevity, are the characteristics of the so-called *open system architecture*.

An open system is defined as: "Open systems are those that conform to internationally agreed standards defining computer environments that allow users to develop, run and interconnect applications and the hardware they run on, from whatever source, without significant conversion cost" (Hugo, 1991). The condition "without significant conversion cost" in fact makes the difference from the other systems. Phenomenologically it is possible to make conversion between any two systems but the problem is the conversion cost.

The open system architecture uses some integration mechanism.

One way to support open system architecture is based on the well-known "*neutral file data transfer*" principle. This principle is applied as a standard approach in Computer Aided Design (CAD) systems. This approach is standardised through the ISO STandard for Exchange of Product model data (STEP). The product model data are interchanged between two heterogeneous CAD systems transferring the corresponded file which contains the product data presented by the STEP formal language EXPRESS. The STEP data format is called "neutral format" and the STEP file is called "neutral (format) file". The process of data interchange is called "neutral file data transfer".

From CAD systems has came another important example. For the problem of the product model visualisation in the CAD environment the computer graphics committee of ISO has developed a basic *reference model for computer graphics*. Graphical output and input are provided in a device- and language-independent manner. As an interface it uses the concept of a *workstation*, or *normalised device*, which is *an abstraction from physical devices*. Transformations to the coordinate system of the physical display device, from the user application, is accomplished in two stages (Encarnação et al., 1990):
1. normalisation transformation, maps from world (user) coordinates (WC) to *normalised device* coordinates (NDC) and
2. workstation transformation, maps from NDC to device coordinates (DC).

The same logic was followed in different domains of manufacturing systems. There is developed a number of international standards with the objective to provide integrability of the systems.

Another way the open systems are applied are "distributed computing system", or distributed software applications, or simply "distributed systems". By (Wu, 1999): "A distributed system is one that looks like an ordinary system to its users, but runs on a set of autonomous processing elements (PEs) where each PE has a separate physical memory space and the message transmission delay is negligible. There is close cooperation among these PEs. The system should support an arbitrary number of processes and dynamic extensions of

PEs". Obviously, there should exist some integration mechanism for the distributed systems (as well as for the enterprise integration). An example of the integration mechanism for the distributed systems, conceived for integration of the object-oriented software components, is the well-known *Common Object Request Broker Architecture (CORBA)*. CORBA is "the object bus" architecture which "lets objects transparently make requests to - and receive responses from - other objects located locally or remotely. The client is not aware of the mechanisms used to communicate with, activate, or store the server object. ... (it) lets objects discover each other at run time and invoke each other's services" (Orfali et al., 1997).

The "distributed system" concept is of the greatest importance for the VE concept and we would say that it is *a model of the VE*. Virtually, there is a homomorphic relation between these two concepts primarily on an abstract level.

In *BM_Virtual Enterprise Architecture Reference Model* the integration mechanism is presented through the "*Integration Mechanism*" (IM) "level", Figure 3. Conceptually, both a translator as the integration mechanism (e.g. the file transfer mode, STEP) and a distributed systems integration mechanism (e.g. CORBA) are supported. Furthermore, the IM "level" will play the role as a component of the *Normalised Virtual Enterprise (NVE) Model*.

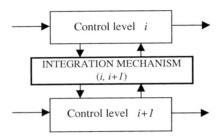

Figure 3. Elementary structure for an integrated and open hierarchical multilevel system control.

3.2. Distributivity

Distributivity has different views. One view of distributed systems, i.e. distributed software applications, is already discussed shortly in the previous section.

Another views of distributivity, especially for the manufacturing system or enterprise are related to the *distributed control* of the (manufacturing) enterprise, based on *multi-agent system* model, and to the *spatial (or geographical) distribution* of the (manufacturing) enterprise functions and physical components.

In the context of the VE reference model the distributivity will be considered from the view of the VE components spatial distribution.

The spatial distribution of the VE components is important from the following reasons.

The VE requirement for reconfigurability, as a part of flexibility, implies the new resources search, to be allocated to the task to be performed.

However, the traditional organisational model, for the problem of reconfigurability, uses the own resources existing within the organisation, i.e. "*within the company boundaries*".

The set of the own resources of the company represents the resources selection domain. As the selection domain is of a relatively limited, small, size in general it can't provide the desired performances neither for actual products nor for new products. To solve the problem of the lack of resources that could bring to the company a competitive advantage, the company searches for co-operation with other companies simply buying components, subcontracting other companies or creating strategic or joint-venture associations. In other words, the company tends to integrate independent resources "*across the company boundaries*". This requisite implies that the resources candidates to integrate an association, to fulfil a specific market opportunity, i.e. to integrate VE, are, in the best case, globally distributed and inter-connected using Wide Area Networks (WAN) communication and telematics technologies, with the objective to enable the negotiation capability (to integrate the association or virtual enterprise) and operation (of the virtual enterprise, as they got into it) in effective and efficient (real time) way.

The effective and efficient access and operation of spatially distributed objects is the main idea under the concept of Distributed Manufacturing Systems (DMS) or Distributed Enterprise (DE). We define a "distributed manufacturing system or enterprise" as a manufacturing system or enterprise which *performances does not depend on the physical distance between the enterprise elements* (Putnik et al., 1998).

In *BM_Virtual Enterprise Architecture Reference Model* the distributivity of the VE is provided through the use of Wide Area Network (WAN) communication and telematics technologies that enables efficient access to remote resources distributed geographically (all over the world), Figure 4.

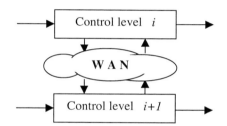

Figure 4. Elementary structure for a distributed hierarchical multilevel system control.

3.3. Agility

To be *agile* means to be "quick-moving, nimble, active" (Oxford Dictionary). "The competitive foundations (of the agile manufacturing or enterprise) are continuous change, rapid response, quality improvement, social responsibility and total customer focus" (Kidd, 1994). We could say that the *agility* is a capability for *fast adaptability* or *fast reconfigurability* in order to respond rapidly to the market (or customer demand) changes (see Chapter 1). We also state that the *flexibility* is either equal (synonym) to *agility* or is a part of it (flexibility $\subseteq$ agility). Regarding development of the VE reference model the virtual difference between "flexibility" and "agility" is irrelevant.

In any case, the reconfigurability, as a part of agility or flexibility, implies the new resources search, which we would allocate to the task to be performed. If the enterprise searches for resources "within the company boundaries" then we talk about *intracompany agility*. On contrary, if the enterprise searches for resources "across the company boundaries" then we talk about *intercompany agility*. The concept of virtual enterprise concerns intercompany agility (by the "first approach", Chapter 1).

As the virtual enterprise, or agile enterprise, implies interactions between various independent companies, i.e. enterprises, there will need to be controlled inter-company organisational configuration management to permit and manage these interactions. It is essential to be able to define domains of responsibility for configuration management, which

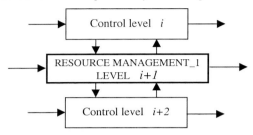

Figure 5. Elementary structure for an agile hierarchical multilevel system control.

reflect organisational policy and permit limited configuration management facilities to be offered, or to be contracted, across company boundaries. A domain, i.e. an environment, for configuration management could represent a set of enterprises, or companies, being managed by a particular manager, or a set of enterprises, or companies, to which a particular access control policy applies. We designated the domain for configuration management by the *Market of Resources* (Cunha et al., 1999). The management structuring needs to be flexible to reflect a wide range of organisational policies. The enterprises may be members of multiple domains to represent the fact that an enterprise is subject to multiple different management policies in different contexts. For example, an enterprise may be a member of a trading domain indicating it is offering a particular service while at the same time it is a member of the domain which is the responsibility of a particular manager. Subdomains are domains containing groups of enterprises, which are members of other domains and provide the means of structuring management and partitioning responsibility. Some special subdomains for configuration management are designated as the *Focused Market of Resources* (Cunha et al., 1999).

In the section 3.1 we have said that virtually there is a homomorphic relation between the "distributed system" concept and the VE concept, i.e. the "distributed system" concept is virtually *a model of the VE*. In fact, the above paragraph is the paraphrased paragraph about "domains for open (distributed) systems" from (Twidle et al., 1992). We did it changing the terms "object" by the term "enterprise" and similar, as we have adapted and added some specific information.

Based on the previous discussion, the VE's agility must be carried on by some "organisation configuration manager". For the "organisation configuration manager" we will use the term *resource manager* or *broker*.

In *BM_Virtual Enterprise Architecture Reference Model* the "organisation configuration management", i.e. the agility function is presented through the *"Resource Management_1"* level, Figure 5.

From the implementation point of view, the "Resource Management_1" level can be owned by the control level i or it can be independent. There are arguments that the "Resource Management_1" level should be a part of, or owned by, a control level i. This model is the classical "two-layer hierarchy" organisation model. Another expressions used for the model are "principal/agent" or "manager/worker" hierarchy. The "principal" is the owner of the vertical structure and the "agent" is responsible for production and affects the principal. But

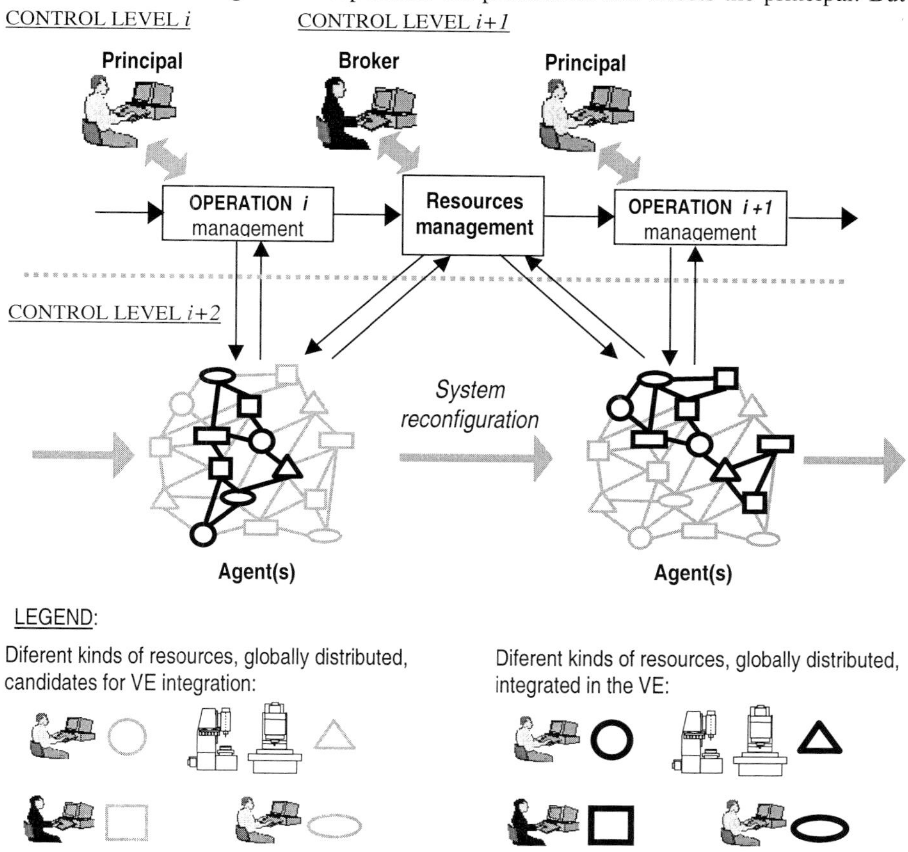

Figure 6. Agile enterprise operation scheme (elementary structure).

in organisational theory are known higher-order vertical structures as well. The independence of the resource management function in VE corresponds to the "three-layer hierarchy" organisation model or, in other words, "principal/supervisor/agent" or "manager/foreman/worker" hierarchy. The main motivation for application of the "principal/supervisor/agent" model is that "the principal, who is the owner of the vertical structure or the buyer of the good produced by the agent, or, more generally, the person who is affected by the agent's activity, lacks either the time or the knowledge required to supervise the agent" (Tirole, 1986). The direct implication of this approach is that the "resource management" function is carried on by an independent agent *resource manager* or *broker*.

In the Figure 6 is presented a scheme of the agile enterprise elementary structure operation. It is important to notice that the structure proposed provides the enterprise reconfigurability between two operations, assuming that during one operation there is not changes of the organisational structure. When the operation is finished the resource manager, or broker, can reconsider the organisation structure and act with the objective to adapt it (to reconfigure it). The resource manager, or broker, is the principal agent of agility.

The model could be described as a model by the *operation off-line reconfigurability* of the enterprise. The consequence of the "operation off-line reconfigurability" model is that the underlying physical structure of the enterprise is not hidden to the manager, i.e. to the "principal" as the broker acts only between the operations. During the operation the manager, (the "principal") has direct contact with the worker (the operator or the "agent"), who provides the service (or production).

Although the model is represented as three-level hierarchical system, in practise the model can work as a *rigorous hierarchy* as well as a *non-rigorous hierarchy*.

3.4. Virtuality

Our critic to the definition of the VE as "dynamic (agile) networking of enterprises" only ("the first approach", see Chapter 1) is that there is not present the original meaning of the word "virtual". "Virtual" means something "not physically existing as such but made by software to appear to do so". So, although the VE is interpreted as an agile enterprise integrated over "intercompany" domain we would say that these enterprises are still only agile enterprise as they exists as real. The "virtuality" is only in the design phase. Another argument to keep the term "virtual" for the agile networked enterprise, for which we think it is a better argument, could be that although we work in a real enterprise at one moment the actual real organisational structure will be virtually changed in some future. Thus, the actual organisational structure is a virtual one. However, we would keep only the attribute "agile" for the networked enterprises (the attribute "extended" seems to be good as well, as it points to the "intercompany" integration domain. The attribute "flexible" also fits very well).

We critic also "the second approach" where the VE is reduced to the simulation program. First, we don't find justification to substitute the designation (enterprise) "simulation" or (enterprise) "simulator", although the simulation could be now much more advanced (including multimedia and virtual reality). Second, the real enterprise in fact doesn't exist.

In the conclusion, no one approach applied as a pure concept is acceptable by our requirements. And our requirements are that we need the real, physical enterprise, which will produce real products (not simulated), and in the same time to keep the meaning of the word "virtual", i.e. to keep some part of the enterprise that doesn't exist in reality. How it is possible to conciliate these two requirements? Another question could be, if we need the real enterprise why we need some virtual part.

We will introduce the virtuality in the similar way as it is introduced in CAD systems and in distributed (software) systems.

To implement the "virtuality" in the enterprise we think on introduction of an interface layer between the "Control level i" (principal, manager) and the "Control level i+1" (agent, "worker"), which passes now to be the "Control level i+2". The role of this level is management of underlying physical structure, i.e. management of resources, which will carry on the process ordered by the upper level, or by itself based on the delegated responsibility. Therefore, the VE's agility must be carried on by some "organisation configuration manager", i.e. *resource manager* or *broker*, similarly as for the concept of agility.

In *BM_Virtual Enterprise Architecture Reference Model* the "organisation configuration management", i.e. the function which provides virtuality is presented through the "*Resource Management_2*" level, Figure 7.

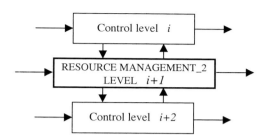

Figure 7. Elementary structure for a virtual hierarchical multilevel system control.

The model is represented as a three-level hierarchical system with a *rigorous hierarchy*. During the operation the "Control level i" (principal, manager) and the "Control level i+2" (agent, "worker") communicate through the "Resource management level i+1", i.e. through the resource manager. During the operation the manager, (the "principal") does not have the direct contact with the worker (the operator or the "agent"), who provides the service (or production).

In the Figure 8 is presented a scheme of the virtual enterprise elementary structure operation (includes agility as well). It is important to notice that the structure proposed provides the enterprise reconfigurability during the (single) operation, i.e. the organisational structure changes during the operation, *at the run time*. The resource manager, or broker, is the principal agent of virtuality (and agility).

The model could be described as a model by the *operation on-line reconfigurability* of the enterprise. As a consequence of the "operation on-line reconfigurability" model, *the underlying physical structure of the enterprise is hidden to the manager*, i.e. to the "principal". The broker must provide the transition from one physical structure to the another in a way that the "principal" can't be affected by the system reconfiguration, in which case the operation would be interrupted and split in two implying immediately some lost time. The lost time can have two components: by interruption of the operation itself (e.g. set-up time for restarting the operation), and the principle's adaptation time to the new specific organisational (hardware) structure. Additionally, the three-level hierarchical model, i.e. the "principal/supervisor/agent" organisation model, as it is conceived here, is in fact an application of the principle of "simultaneity" of the processes. In the previous Chapter we referred that the main motivation for application of the three-level hierarchical model, i.e. the "principal/supervisor/agent" model is that "the principal lacks either the time or the knowledge required to supervise the agent". But, even in the case the principal has "either the time or the knowledge required to supervise the agent", in order to cut further processing time of the production operation and the enterprise reconfiguration it is necessary to perform them in parallel. This is the main principle of the concurrent or simultaneous engineering. In the "agility" scheme, as it is defined in the previous Chapter, the production operation and the enterprise reconfiguration are still performed in a sequence.

These are the reasons why we need the virtuality. The virtuality in this sense (the hidden underlying hardware structure) is actually present in the (open) CAD systems and distributed (software) applications. All these systems are virtually the models of a VE. In other words, the "Resources Management Level" together with the "Integration Mechanism Level", e.g.

the translator, *emulate* the underlying organisational (hardware) structure in a format that is

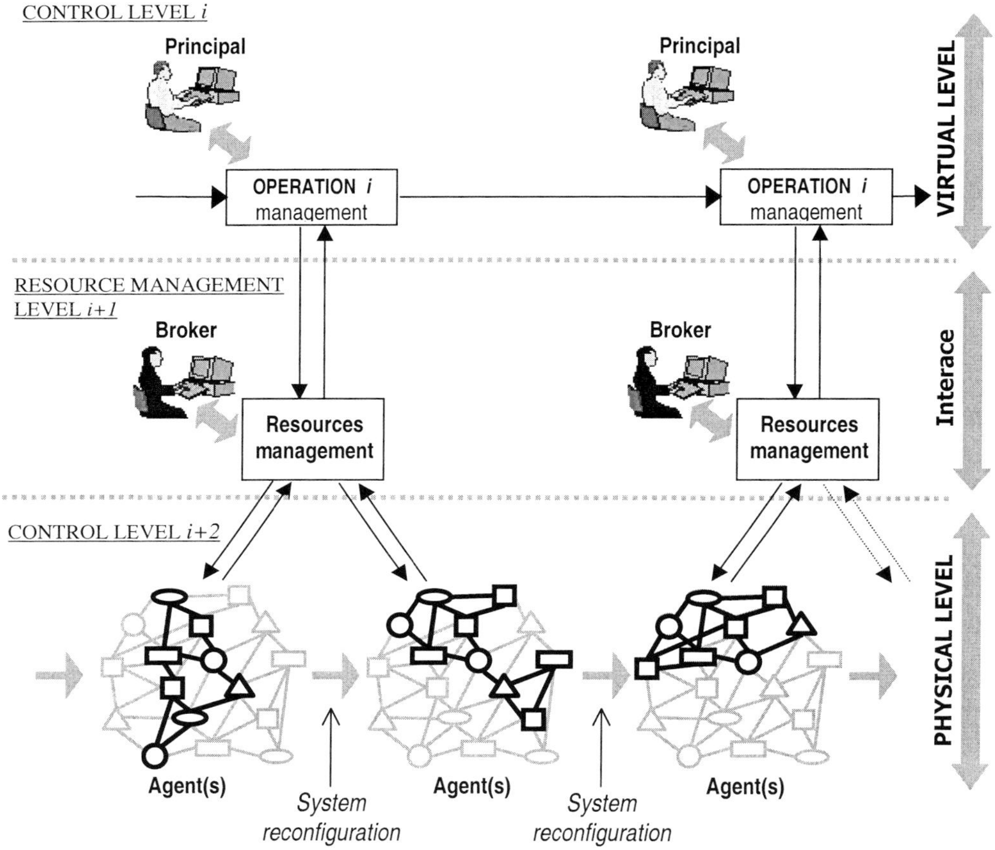

Figure 8. Virtual enterprise operation scheme (elementary structure).

understandable by the manager, or "principal". The "principal" doesn't see the real structure, he sees some "virtual" structure that doesn't exist.

3.5. Virtual enterprise model space - Integrated, distributed, agile and virtual enterprise

The integrability (I), distributivity (D), agility (A) and virtuality (V) are four *independent ("uncoupled") functional requirements* to be satisfied by an enterprise model, and in the same time we will say that they represents four "*design parameters*" for the enterprise model synthesis. An enterprise could be integrated by only one of the "design parameters" or by any combination of them.

We would say that integrability, distributivity, agility and virtuality, as an enterprise design parameters, form a kind of a 4-dimensional enterprise model *(meta) space*, which is presented on the Figure 9 (each dimension is further defined by some model subspace, meaning additional dimensions of an enterprise characterisation. For other important enterprise model space dimensions see (Petrie, 1992). Naturally, it is necessary further research in order to clarify relationship among different enterprise model dimensions).

In this space we can recognise 2^4, i.e. 16 different enterprise model classes. We will use the following notations: I-class will be the class of "integrated" (only) enterprise models, IA-class will be the class of "integrated and agile" (only) enterprise models, etc. IDAV-class will be the class of *"integrated, distributed, agile and virtual"* enterprise models. The set of 16 different enterprise model classes is {∅, I, D, A, V, ID, IA, IV, DA, DV, AV, IDA, IDV, IAV, DAV, IDAV} (model of the type ∅ means that the enterprise is neither I nor D, nor A, nor V).

The question is: *which model maximises the enterprise performances.*

To answer this question we could use an algorithm which will compare models and select the best one. The requirements for competitiveness to be satisfied are 1)-3) from the Chapter 1, over global market.

Our hypothesis is that the "IDAV" model is the best one. Why?

One simple argument could be that the IDAV-model is the most complex and all other models are its special cases. Another argument is based on the following reasoning:

1) An enterprise should be integrated (by computer based technologies as an open system). This feature implies an I-model. We will call this model Integrated Enterprise (IE);

2) An enterprise must be integrated and distributed. Capability to perform globally distributed tasks (remote design and control of the system) brings to the enterprise independence of the physical distances among enterprise's elements and an additional spatial and temporal flexibility. The distributed system could be conceived as a proprietary system but on the cost of flexibility. Therefore, the system should be integrated and distributed. These features imply an ID-model. We will call this model Distributed Enterprise (DE);

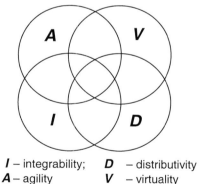

I – integrability;   *D* – distributivity
A – agility   *V* – virtuality

Figure 9. Enterprise model (meta) space.

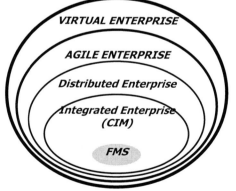

Figure 10. Relationship among Integrated, Distributed, Agile and Virtual Enterprise.

3) An enterprise must be agile. The agility could be implemented over intracompany resources. It is expected that use of computer based open systems as a tool for agility would contribute to improve it (the agility). But to improve further the enterprise performances it is required to integrate the most competitive resources "across the company boundaries". To access any candidate resource at any point of the globe, it is necessary to apply technologies inherent to distributed enterprise. These features imply an IDA-model. We will call this model Agile Enterprise (AE);

4) An enterprise must be virtual. To improve further the agile enterprise performances it is introduced the feature of "virtuality". *The virtuality provides the system with the capability of the system on-line reconfigurability without interruprtion of any process.* The virtuality, combined with the agility, distributivity and integrability, brings to the enterprise the highest level of flexibility. These features imply an IDAV-model. We will call this model Virtual Enterprise (VE).

Based on the above considerations, we would establish the following relationship among Integrated, Distributed, Agile and Virtual Enterprise, see also Figure 10:

Integrated Enterprise ⊂ Distributed Enterprise ⊂ Agile Enterprise ⊂ Virtual Enterprise

We could give now an informal definition of the VE:

A Virtual Enterprise (VE) is an optimised enterprise synthesised over universal set of resources with the real-time substitutable physical structure. The design (synthesis) and control of the system is performed in an abstract, or virtual, environment.

The "universal set of resources" means that the VE integration can consider any kind of resources, primitive or complex, that they can be distributed globally and that they can belong to some domain "within the company boundaries" or to some domain "across company boundaries". The combination of "the real-time substitutable physical structure" and the "virtual environment" for the enterprise design and control gives the highest level of the enterprise flexibility or agility, i.e. gives the enterprise the ability to "reconfigure within 1 second". Additionally, the "virtual environment" justifies the attribute "virtual" of the VE.

4. BM_VIRTUAL ENTERPRISE ARCHITECTURE REFERENCE MODEL

BM_Virtual Enterprise Architecture Reference Model is defined as a hierarchical multilevel model of the enterprise/manufacturing system control and satisfies the requirements for the integrability (I), distributivity (D), agility (A) and virtuality (V).

The *BM_Virtual Enterprise Architecture Reference Model* is build up of the *BM_Virtual Enterprise Architecture Reference Model Elementary Structures*. The *BM_Virtual Enterprise Architecture Reference Model* elementary structure is synthesised over elementary structures of the VE architecture, which provide I, D, A, and V (described informally in the previous chapter, Chapters 6.). Thus, I, D, A, and V, are the design parameters of the *BM_Virtual Enterprise Architecture Reference Model* elementary structure and of the model as a whole.

Recalling the general definition of the multilevel hierarchical system (1) - (4) (Chapter 2) and specialising it, the *BM_Virtual Enterprise Architecture Reference Model* is specified as follows, Figure 11 and Figure 12.

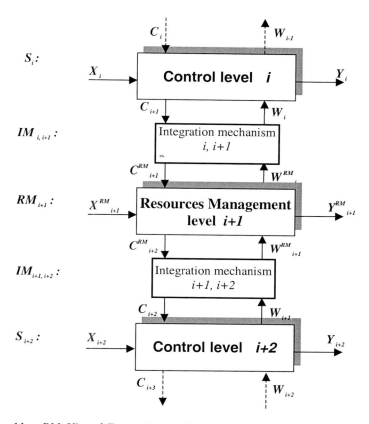

Figure 11. *BM_Virtual Enterprise Architecture Reference Model Elementary Structure.*

Each 'input-output' pair, i.e. each pair (X_i, Y_i), $i = 1, 3, 5,..., n$ (n odd), is assigned to a particular *Control Level* represented by:

(i) $S_i: X_i \times W_i \to Y_i$,   if $i = 1$,
(ii) $S_i: X_i \times C_i \times W_i \to Y_i$,   if $1 < i < n$,   (5)
(iii) $S_i: X_i \times C_i \to Y_i$,   if $i = n$,

and each pair (X^{RM}_i, Y^{RM}_i), $i = 2, 4, 6, ..., n-1$, is assigned to a particular *Resource Management Level* represented by:

(iv) $RM_i: X^{RM}_i \times C^{RM}_i \times W^{RM}_i \to Y^{RM}_i$ (6)

There are defined the folowing families of mappings, i.e. there are defined decision functions, information functions and translation functions:

$h_i: Y_i \to C_{i+1}$ $h^{RM}_{i+1}: Y^{RM}_{i+1} \to C^{RM}_{i+2}$ (decision functions)

$$g_i: Y_i \rightarrow W_{i-1} \qquad\qquad g^{RM}_{i+1}: Y^{RM}_{i+1} \rightarrow W^{RM}_i \qquad \text{(information functions)}$$

$$tg^{RM}_{i+1}: C_{i+1} \rightarrow C^{RM}_{i+1} \qquad tg_{i+1}: C^{RM}_{i+1} \rightarrow C_{i+1} \qquad \text{(translation functions)}$$

$$tw_i: W^{RM}_i \rightarrow W_i \qquad\qquad tw^{RM}_i: W_i \rightarrow W^{RM}_i \qquad \text{(translation functions)}$$

such that for each x in X and y=S(x), with i = 1, 3, 5,..., n (n odd):

$$
\begin{array}{lll}
\text{(i)} & y_i = S_i(x_i, tw_i(W^{RM}_i)), & \text{if } i = 1, \\
\text{(ii)} & y_i = S_i(x_i, tg_i(C^{RM}_i), tw_i(W^{RM}_i)), & \text{if } 1 < i < n, \qquad\qquad (7)\\
\text{(iii)} & y_i = S_i(x_i, tg_i(C^{RM}_i)) & \text{if } i = n,
\end{array}
$$

and for each x^{RM} in X^{RM} and $y^{RM}=RM(x^{RM})$, with i = 2, 4, 6, ..., n-1:

$$\text{(iv)} \qquad y^{RM}_i = RM_i(x^{RM}_i, tw^{RM}_i(W_i), tg^{RM}_i(C_i)) \qquad\qquad (8)$$

The integration mechanism functions, i.e. the integration mechanism blocks from the Figure 11. are not levels of the model. They only represent the interface (translation functions) between control levels and resources management levels.

Additionally, we propose a concept of the *Normalised Virtual Enterprise (NVE) Model*, Figure 12, similarly with the CAD systems. The *NVE model* is "an abstraction" from the physical VE and it serves as an interface, i.e. together with the translation functions serves as an integration mechanism between two control levels. Transformations, communications or integration between two particular (heterogeneous) enterprises, or resources, on two Control levels i, i+1, is accomplished in two stages (by analogy with the CAD systems):

1) "normalisation transformation", maps information (orders) from the "principal" (enterprise, or resource) or manager on the Control level i, to the *normalised VE model*, and
2) NVE transformation, maps from NVE to or "agent" enterprise, or resource on the Control level i+1.

(conceptually it is not important whether the transformations are performed between levels i and i+1, i.e. between "principal" and "resource manager", or between levels i+1 and i+2, i.e. between "resource manager" and "agent").

The expected advantage of the NVE definition is independence of the VE components, i.e. tools and technologies development, as well as VE formal theory development. Also, an independent (of VE tools and technologies producers) organisation or institution, for example ISO, could provide the specification of the NVE model.

5. DEVELOPMENT OF THE VE DEMONSTRATOR BASED ON BM_VEARM

The validation of the VE reference model proposed will be carried on along with the number of research projects under development at the University of Minho on VE theory and VE design and control tools and technologies. However, in the same time, the VE reference model proposed already serves as a framework for cooperation and coordination of the group of research projects referred.

In order to fulfil the requirements of the project(s) validation, including the VE reference model, it is implemented a laboratory installation which will serve as a demonstrator for the VE design and control. The laboratory installation is conceived as a Distributed/Virtual

Manufacturing System (D/V MS) Cell, named *AURORA 98* (Putnik et al., 1998). In the first

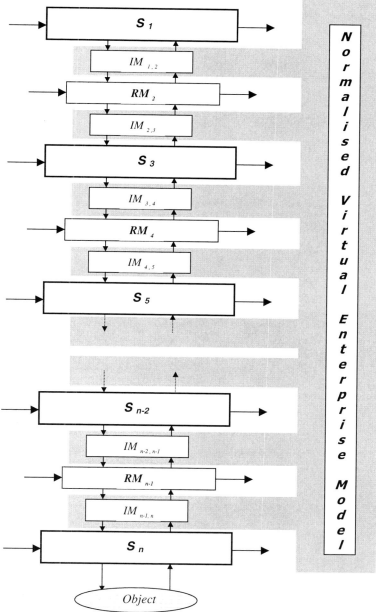

Figure 12. *BM_Virtual Enterprise Architecture Reference Model and the corresponded Normalised Virtual Enterprise (NVE) Model.*

period the laboratory was used for research of distributed manufacturing system. In the second (present) period the laboratory is extended with the components which are expected to provide the full demonstration of the VE concept based on the *BM_Virtual Enterprise Architecture Reference Model*.

The components of the D/V MS Cell structure which will be used in the first phase for the VE model validation, based on the *BM_Virtual Enterprise Architecture Reference Model* and therefore for its validation as well, is composed of, Figure 13:

1) *Machine cell*: Two machine simulators, PLC, external sensors and actuators, Robot SCORBOT ER-VII, vision system, conveyor system, computer based local controller;
2) *Broker*: Computer based remote resource manager;
3) *Control center_1*: Computer based remote machine cell controller;
4) *Control center_2*: Computer based remote machine cell controller;

The cell formal specification (ESTELLE based) is given in the Figure 14.

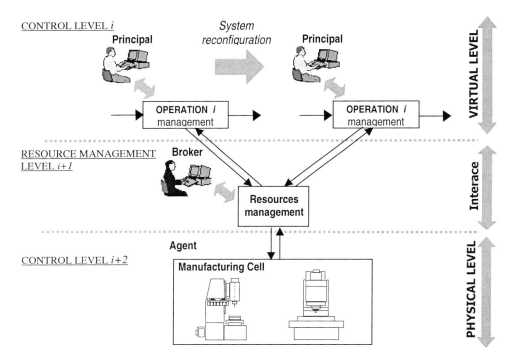

Figure 13. An informal scheme of the virtual enterprise demonstrator based on the *BM_Virtual Enterprise Architecture Reference Model*.

The reconfiguration of the system consists of switching between two manufacturing cell

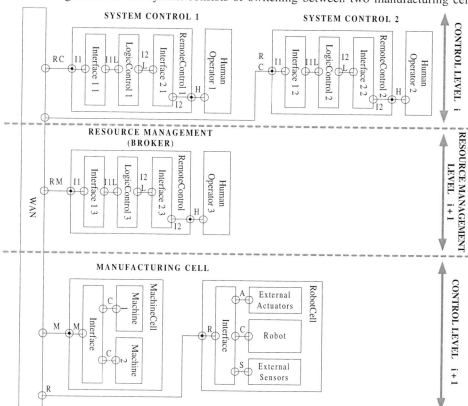

Figure 14. A formal scheme (ESTELLE based) of the virtual enterprise demonstrator based on the *BM_Virtual Enterprise Architecture Reference Model*.

controllers in accordance with their availability, service cost and quality. The broker performs the function of the system configuration management. Manufacturing cell controllers, as well as broker, could be located at any point in the world as the system communication is based on WAN, in particular in this phase it is used Internet as a communication protocol and JAVA based applications for resources management and the manufacturing cell control.

The creation of the network of the system controllers, brokers and other manufacturing cells or services (e.g., CAD, CAPP, CAM, business processes, etc.) is planned for the near future. The network will include in principle laboratory installations and researchers from academia as well as it will be open for other participants from academia and especially for participants from the industry.

6. CONCLUSION

Besides the integrability, distributivity, agility and virtuality, the VE reference model presented is characterised by the following additional attribute values (of an enterprise integration (EI) model space dimensions referred in (Petrie, 1992)):
1) Semantics oriented – the reference model provides translation of the syntax between particular models as well as a part of semantics. The part of semantics is implemented through the *Normalised Virtual Enterprise (NVE) Model* proposed;
2) Global – it is conceived to link any particular model to any other model;
3) Translator – it is chosen a "translator" as an intermediate mechanism as it doesn't require modifications of the actual applications;
4) Unification – the reference model is conceived to serve as a unification model, i.e. all other models must translate to it, but its internal structure should support a "federation" approach as it would surely integrate international standards already specified, e.g. STEP. In the future a "federation" with other VE reference models is considered;
5) Dynamic – it is conceived to support the highest dynamics of the enterprise reconfiguration ("within 1 second") for any type of production. This is provided by specification of the "three-levels" hierarchical organisation model and by introduction of the principle of virtuality;
6) Intercorporate – it is conceived to support the enterprise integration over (domain of) intracompany resources.

Additionally, it is intended to cover all processes in an enterprise "from business management to end effector", from macro to the micro processes level, for any type of production.

By the *BM_Virtual Enterprise Architecture Reference Model (BM_VEARM)* the VE is seen as a general enterprise model from whom *all other enterprise models are special cases*. For example, the agile, distributed, integrated and other enterprise models are special cases and can be derived from the *BM_VEARM*.

It is a proprietary model, as it is not developed within some standardisation organisation.

At the end we would mention some important research topics related with the VE reference models development (for some topics see e.g. (Gielingh, 1992), (Petrie, 1992)):
1) Algebraic specification of the VE reference models;
2) Representational classes for VE models;
3) Reference model(s) integration;
4) Reference model modifications;
5) Reference model extensions for other views (e.g. information system, implementation);
6) Reference model extensions for domain-specific models;
7) Metrics and certification criteria for EI models and software developed under the particular reference model, etc.

(Note: All figures in the text, except the Figure 1, are originals published by the first time)

REFERENCES

1. Browne J. (1995) The Extended Enterprise – Manufacturing and The Value Chain, in Camarinha-Matos L. M., Afsarmanesh H. (Eds.) *Balanced Automation Systems – Architectures and design methodologies*, Chapman & Hall;

2. Camarinha-Matos L. M. et al. (1999) Partners search and quality-related information ezchange in a virtual enterprise, in Mertins K. et al. (Eds.) *Global Production Management*, Kluwer Academic Publishers;
3. Cunha M., Putnik G., Avila P. (1999) Towards Focused Markets of Resources for Agile/ Virtual Enterprise Integration, (submitted for publishing);
4. Encarnação J. L., Lindner R., Schlechtendahl E. G. (1990) *Computer Aided Design - Fundamentals and System Design*, Springer-Verlag;
5. Fujii S., Kaihara T., Morita H., Tanaka M. (1999) A distributed virtual factory in agile manufacturing environment, in *Proceedings of ICPR '99*;
6. Gielingh W. (1992) Requirements for the Development of Layered Information Models, in Petrie C. (Ed.) *Enterprise Integration Modeling*, The MIT Press;
7. Goldman S. L., Nagel R.. N., Preiss K. (1995) *Agile competitors and virtual organizations*, Van Nostrand Reinhold;
8. Hormozi A. M. (1994) Agile Manufacturing, in *Proceedings of the 37th International Conference*, American Production and Inventory Control Society, San Diego;
9. Hugo I. (1991) *Practical Open Systems – A Guide for Managers*, Data General Ltd.;
10. ISR (1995) What Virtual Manufacturing is, in *Virtual Manufacturing User Workshop Report*, Lawrence Associates Inc., URL: http://www.isr.umd.edu/Labs/CIM/vm/vmdesc.html;
11. Kidd P. T. (1994) *Agile Manufacturing – Forging New Fron*tiers, Addison-Wesley;
12. Kim S. H. (1990) *Designing Intelligence*, Oxford University Press;
13. Mesarovic M. D., Macko D., Takahara Y. (1970) *Theory of Hierarchical, Multilevel, Systems*, Academic Press;
14. NIIIP (1996) The NIIIP Reference Architecture, http:/www.niiip.org;
15. Onosata M., Iwata K. (1993) Development of a Virtual Manufacturing System by Integrating Product Models and Factory Models, *Annals of the CIRP*, Vol.42/1/1993, pp 475-478;
16. Orfali R., Harkey D., Edwards J. (1997) *Instant CORBA*, John Wiley & Sons;
17. Petrie C. (Ed.) (1992) *Enterprise Integration Modeling*, The MIT Press;
18. Putnik G. D., Sousa R. M., Moreira J. F., Carvalho J. D., Spasic Z., Babic B. (1998) Distributed/Virtual Manufacturing Cell: An Experimental Installation, in *Proceedings of 4th International Seminar on Intelligent Manufacturing Systems*, Belgrade;
19. Tirole J. (1986) Hierarchies and bureaucracies: On the role of collusion in organization, in *Journal of Law, Economics and Organization*, **2** (2), Autumn, 181-214;
20. Twidle K., Sloman M., Magee J., Kramer J., Dulay N., Crane S., Cheung S. C. (1992) Configuring Heterogeneous Open systems, in Petrie C. (Ed.) *Enterprise Integration Modeling*, The MIT Press;
21. Vernadat F. (1996) *Enterprise Modeling and Integration*, Chapman & Hall.
22. Wu J. (1999) *Distributed System Design*, CRC Press.

Integrated Product/Process Development (IPPD) Through Robust Design Simulation (RDS)

Daniel P. Schrage and Dimitri N. Mavris
School of Aerospace Engineering, Georgia Institute of Technology, USA

1. INTRODUCTION

The accepted systems approach for complex manufacturing systems for most of the world during the 1970s and 1980s was based on the systems engineering methodology that was initiated and developed for U.S. defense and space systems during the early 1960's. It provided a top down, system decomposition approach so that complex systems could be broken down into subsystems, components, and parts that could be developed and manufactured by subcontractors, suppliers, and vendors around the U.S., as well as the world. While this systems engineering approach has been successful in the U.S., as indicated by the high performance weapons systems developed and by putting the first man on the Moon, it is not sufficient by itself, for development of both complex defense and commercial systems in today's competitive marketplace.

On-line manufacturing was an important downstream element in the hierarchical and sequential systems engineering methodology; however, it was the recipient of a design that had to be transitioned into manufacturing processes, often with substantial, costly re-design changes. While the systems engineering approach recognized the need to address "design for's", e.g. manufacturing, supportability, etc., the configuration design was usually synthesized by a small advanced design team responsible for conceptual design. This team usually emphasized performance based on their experiences and the design tools that were available. The impact of this approach was to lock-in the Life Cycle Cost (LCC) of the complex system early in the life cycle process, as illustrated in the "now famous curve", Figure 1 (Ref.1). Two figures are included in Figure 1. The small one in the upper half is the generic curve often referenced. The larger one illustrates the actual data, from a Boeing ballistic missile system, upon which the generic curve is based.

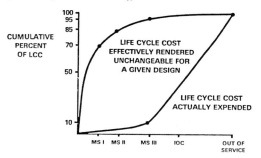

Figure 1. cont'd..../

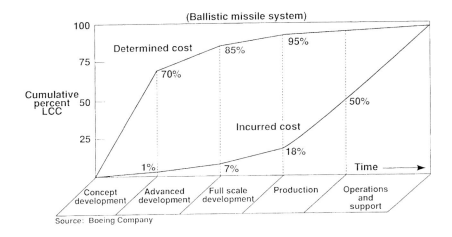

Figure 1. Life Cycle Cost Gets Locked in Early for Complex Systems Using Only a Top Down Systems Engineering Approach

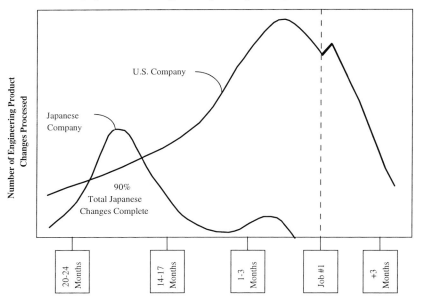

Figure 2. Japanese/U.S. Automotive Engineering Change Comparison

Around 1980 a *manufacturing enterprise flexibility* was sought in the U.S., particularly in the automotive and electronics sectors, in order to insure U.S. products would remain competitive with foreign equivalents. This was particularly true of Japan, which was redefining what was meant by a quality product. As an example, a timeline comparison of where design changes were taking place during the development of a Japanese automobile compared with those for a U.S. automobile is illustrated in Figure 2 (Ref.2). As can be seen, the Japanese automobile company made design changes earlier; thus, they could produce a car with higher quality in a shorter period of time. Direct comparisons, such as this, served as a "wake-up call" for the U.S. automotive and electronics industry and the race for improved quality and reduced cycle times continues today. Subsequently, it served to wake-up other manufacturing sectors, such as aerospace, as well as the U.S. government which embraced Total Quality management (TQM) and Concurrent Engineering as its preferred modes of operation. (Ref. 3)

The quality revolution, which has been ongoing for approximately the last twenty years, has determined *where competition is today*, and is illustrated in Figure 3 (Ref.4). The manufacturing *Cost Advantage* in the 1960s was *Cheap Labor, High Volume, and Low Mix Production*, i.e. mass production. The emphasis on *Quality* in the 1970's, initiated in Japan and emulated elsewhere, contained the three essential elements of *Statistical Process Control, Variability Reduction, and Customer Satisfaction*. The main emphasis with *Manufacturing Enterprise Flexibility* was to move on-line, quality methods such as Statistical Process Control (SPC), off-line for a more robust design. This change, in essence, constituted a quality engineering *recomposition* effort to complement the traditional systems engineering *decomposition* effort. Robust design approaches, such as Taguchi's Robust Design and Six Sigma process capability, were introduced. The quality revolution in the late 1980's moved to a *Time-to-Market* strategy with the emphasis on *Cycle Time Comparison, such as Just-In-Time (JIT) manufacturing, Integrated Product/Process Development (IPPD), Product/Process Simulation, and High Skill Adaptable Workforce*. This has been followed in the 1990's by *Product Variety* and an emphasis on *Cost Independent of Volume, Agility, Commercial/Military Integration, and Virtual Companies*. For the next millenium the progression is to *Company Goodness* with an emphasis on *Enterprise Integration* (EI).

Ideally, EI connects and combines people, processes, systems, and technologies to ensure that the right people and the right processes have the right information and the right resources at the right time. EI should enable successful operation, in a world of continuous and largely unpredictable change, of a single manufacturing company or an ever-changing set of extended (or "virtual") enterprises – by enabling quick and accurate decisions and adaptation of operations to respond to emerging threats and opportunities. However, in the most advanced manufacturing enterprises today, many technologies exist to integrate elements of the product realization and business systems. Enterprise Resource Planning (ERP) systems integrate a number of the business functions, but lack the detail and fidelity needed for the product realization side of the house. CAD systems have expanded to include product and process simulation and planning capabilities, but typically lack the horsepower and tools to perform specialized analyses and integrate the information into the business and manufacturing execution functions. (Ref. 5)

Company goodness can imply a number of things, including how the employees view the company, as well as external perceptions of the company, such as its customers, suppliers, and shareholders. Organizational structures based on "Communities of Practice" (Ref.6), self-organization and complexity science principles (Ref.7), rather than trying to emulate other successful company organizational models seem to be the new norm that is emerging. By the same token, the *Environment* also has several perspectives, including both internal and external. This could include advanced engineering environments (AEE's) (Ref. 8), as well as societal constraints and concerns, such as noise, pollution, emissions, safety, and disposal.

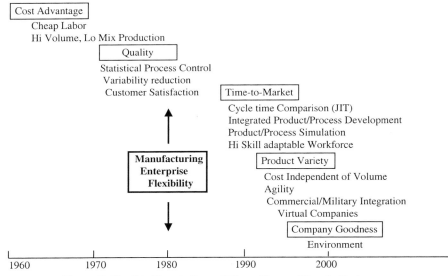

Figure 3. Quality Revolution – Where Competition is Today

Over the past five years there has been an industry driven collaboration effort to define the Next Generation Manufacturing (NGM) environment (Ref.9). The NGM project, completed in 1997, was a groundbreaking effort to examine long-term trends in U.S. manufacturing in light of unprecedented changes taking place in the global business environment, and to identify actions required to respond to these new challenges. More than 500 technologists and business leaders from industry, government, and academia participated in the project. Following the NGM project an Integrated Manufacturing Technology Roadmapping (IMTR) initiative (Ref.5) was undertaken to provide a comprehensive plan to:

- Define key technology goals that cut across all manufacturing sectors
- Provide focus for concentrated effort to achieve the goals
- Promote collaborative R&D in support of critical needs
- Move these developments from the laboratory to industrial use

IMTR has been a focused effort, sponsored by the National Institute of Standards and Technology (NIST), U.S. Department of Energy (DOE), National Science Foundation (NSF), and Defense Advanced Research Projects Agency (DARPA), to develop a manufacturing R&D agenda that cross-cuts the diverse needs of government and industry across all major manufacturing sectors. Leveraging the work done by NGM, IMTR conducted a structured process to define future manufacturing enterprise technology requirements and outline solution paths to meet these requirements in four interrelated areas:

- Information Systems for Manufacturing
- Modeling & Simulation
- Manufacturing Processes & Equipment
- Enterprise Integration

The Technologies for Enterprise Integration (TEI) roadmap (Ref.10) has been approached from five different perspectives, or levels:

- *Sub-enterprise level*: the functionally of the integrated application or system is limited to a relatively homogeneous area, typically at a single local site under a single ownership, e.g. flexible manufacturing systems at the integrated sub-enterprise level.
- *Single-site enterprise*: complete functional integration assures that business processes, manufacturing processes and product realization are united using a common architecture to fulfill a common goal. This is most likely found at a single plant under a single owner, such as an automated factory.
- *Multi-site, extended, and virtual*: these three levels occur over multiple geographic settings. Multi-Site enterprise integration is generally an issue faced by large enterprises (e.g., Boeing, Lockheed Martin, IBM, General Motors, Ford, and Caterpillar) in integrating heterogeneous systems through out their enterprises. An extended enterprise, which generally involves complex supply chains, concerns the integration of all members of the supplier and distribution chain to the common goal of market share capture through product realization. Virtual enterprises are very similar to extended enterprise, but they have the feature of being created and dissolved dynamically on an as-needed basis, and integration of member entities is largely electronic.

All levels, to varying degrees, influence and are influenced by integrated product realization, integrated business systems, and tools enabling integration. While the objective is to support creation and operation of extremely efficient, flexible, and responsive extended manufacturing enterprises, the path to reach this will require capturing the wisdom achieved at each of the enterprise integration levels. As stated in the IMTR TEI roadmap (Ref.10) the path to EI has already started. Several sub-enterprise elements have already been integrated, as illustrated in Figure 4 from Ref. 10, with powerful new tools in different domains. The goals of Integrated Product Realization are being supported with integration of CAD, CAM, and computer-aided manufacturing planning systems, coupled with the use of integrated product teams (IPTs), leveraging the emerging disciplines of *Integrated Product/Process Development (IPPD)*. Similarly, ERP systems have integrated the business functions of finance, accounting, human resources, and material requirements planning. Electronic data interchange is a beginning toward integration of the extended enterprise. Finally, inter-, intra- and extranets are starting to provide the infrastructure required for integrated distributed enterprise operations. (Ref.10)

The remaining portions of this chapter will consecrate on the *IPPD* and *Product/Process Simulation* effort that has been initiated and developed at Georgia Tech in the 1990s (Ref.11). This is done for several reasons. *First*, as can be seen in Figure 3 these are two of the key 1990's elements in the *quality revolution* and *where competition is occurring today*. Second, the IMTR TEI plan identified some 70 top-level goals and more than 265 supporting requirements and tasks to achieve the IMTR vision. However, out of these goals and requirements there were 10 "nuggets" – critical capabilities or attributes – that underpin the IMTR vision and which offer the greatest return on investment by virtue of their broad applicability to industry. Under the *Manufacturing as an Integrated System (Integrated Product Realization)* "nugget" it is assumed that the *concepts of Concurrent Engineering and*

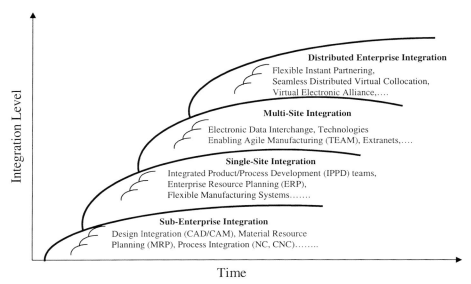

Figure 4. Technology Paths to Integrated Enterprises

IPPD will mature with creation of enabling tools that support complete integration of all functions and disciplines involved in converting product from initial concept to completed units ready for delivery. Finally, for IPPD to be properly implemented a "new systems methodology", that couples *systems engineering decomposition* methods/tools (mostly deterministic) with *quality engineering recomposition* methods/tools (mostly probabilistic) must be formulated. It must then be demonstrated for early system level design tradeoffs in a realistic, robust design simulation environment. This chapter will discuss how a generic IPPD methodology (a new systems methodology) has evolved into a Robust Design Simulation (RDS) environment.(Ref.11) This environment is now being transferred to industry and government to provide more affordable complex systems for today and tomorrow.

2. EVOLUTION OF IPPD THROUGH RDS

The cultural change-taking place in industry and government due to the *quality revolution*, Figure 3, has also identified the need for education, research and training as well as *new systems approach methodologies* and *computer integrated advanced engineering environments*. These elements are necessary to capture the essence of IPPD and Product/Process Simulation. What is needed is much like the Systems Engineering methodology that was developed in the late 1950s and early 1960s for designing and building large scale complex systems, such as ballistic missiles and manned space flight systems.

The generic IPPD Methodology that has been taught formally, and used as the education and training approach for the Navy's Acquisition Reform effort is illustrated in Figure 5 (Ref.11). This "new systems approach methodology" consists of four key elements, illustrated at the top in "umbrella" form. These four elements are *Systems Engineering (SE) methods/tools*, *Quality Engineering (QE) methods/tools*, a *Top Down Design Decision Support (TD3S) process*, and a *Computer Integrated Environment (CIE)*. Below the "umbrella" are the sub-elements of each key element. As illustrated by the downward arrow, the SE methods/tools flow are *product design and decomposition driven*, while the QE methods and tools flow are *process design and recomposition driven*. The arrows from the SE and QE methods/tools feeding into the TD3S process, *the heart of the methodology for tradeoff assessment*, represent the information flow, which for timely integration, cycle time reduction and decision making requires a CIE. *The primary design/synthesis iteration is illustrated in gray boxes in Figure 5, i.e. between the SE method: System Synthesis through Multidisciplinary Design Optimization (MDO), to "Generate Feasible Alternatives" and the QE method, Robust Design Assessment & Optimization, to "Evaluate Alternatives" and finally to update and provide a robust System Synthesis*. It will be shown later how the iterative process is exercised in a RDS environment

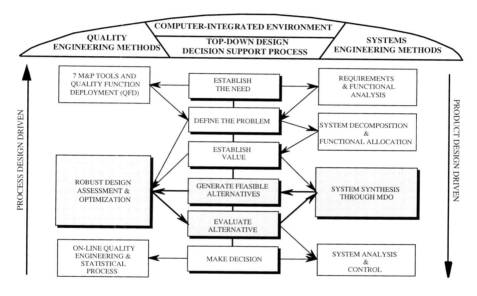

Figure 5. Georgia Tech Generic IPPD Methodology

The Methodology illustrated in Figure 5 is considered as a *procedural* approach to design, but also encompasses an *analytical* approach in the SE method, "Systems Synthesis through MDO", and an *experimental* approach in the QE method, "Robust Design Assessment and Optimization" (Ref. 12). The *procedural* approach is a trade-off process where the objective is modified as the design proceeds. The solution that results is the solution that satisfies all the design objectives in the *best manner*. The *analytical* approach is a function of the problem attributes that are precisely defined - much of engineering optimization, especially in academia, has followed an *analytical* approach. The *experimental* approach to design relies on a matching of design attributes to the objective of the design process - use of Design of Experiment methods characterize the *experimental* approach. The procedural approach illustrated in Figure 5 has also been called a *Design Justification* approach. Design Justification is a term used to describe a design process where the economic ramifications of design decisions are considered concurrently with design development and are used to guide the design process so as to result in the most economical criteria satisfying design (Ref. 12). This is the basis for the IPPD through RDS that will be discussed later.

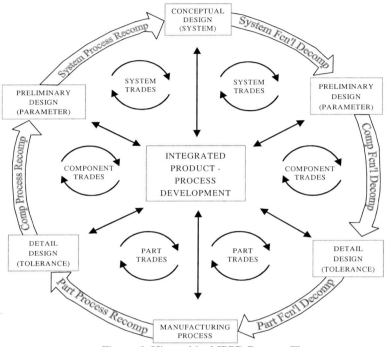

Figure 6. Hierarchical IPPD Process Flow

For large-scale integration of complex systems, the IPPD methodology provides the centerpiece in the hierarchical tradeoff process flow illustrated in Figure 6. The right half of the figure represents the SE decomposition from system (conceptual design) to component (preliminary design) to part (detail design) to the on-line manufacturing process; while the left half represents the QE recomposition from the manufacturing process back to the system design, i.e. to tolerance, to parameter, to system design. Inside the circle are parallel trades at the system, component, and part levels. An IPPD methodology (the center box), such as that in Figure 5, is necessary if true IPPD is to be exercised. The hierarchical process flow in Figure 6 is also useful in understanding why the Japanese were able to make design changes earlier (Figure 2) and shorten the development cycle time. This is further illustrated, in a more generic way, in Figure 7, which illustrates a traditional serial approach versus a CE approach. As can be seen, the IPPD focus should be at the front end, i.e. in design and development. The traditional serial approach is illustrated in Figure 7 and is based on SE decomposition. Also shown is the wall that has often separated design and manufacturing in many companies. While SE decomposition has served its purpose in producing high performance large-scale systems, such as aerospace, it has also served to lock in life cycle cost early as was illustrated in Figure 1. Therefore, for IPPD SE methods/tools are considered *necessary*, but *not sufficient*.

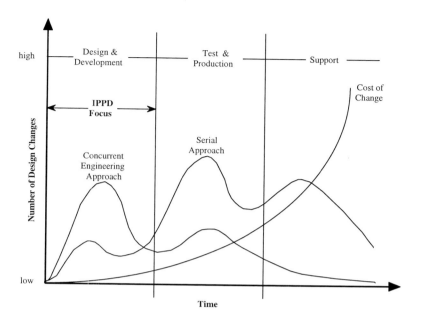

Figure 7. Traditional Serial Approach versus CE Approach

Quality Engineering methods/tools have evolved mostly from Japan and have become part of the quality revolution. They provide the means of bringing downstream manufacturing process information back into the design process, *thus emphasizing a recomposition rather than a decomposition approach* (Ref. 13). They basically consist of the flow illustrated in Figure 9. The Seven Management and Planning Tools and Quality Function Deployment (QFD) are used to transform the "Voice of the Customer" and prioritize where improvements are needed. Robust Design Assessment and Optimization methods, such as Taguchi, then provide the mechanism for identifying the process improvements. Statistical Process Control (SPC), an on-line manufacturing process, provides the means to hold these gains as well as to insure continuing quality improvement, through variability reduction. In the U.S., the emphasis on achieving a "Six Sigma" process capability has been evident in the electronics and propulsion sectors for at least the past five years and is now being emphasized for large scale complex systems, such as aerospace. (Ref. 14)

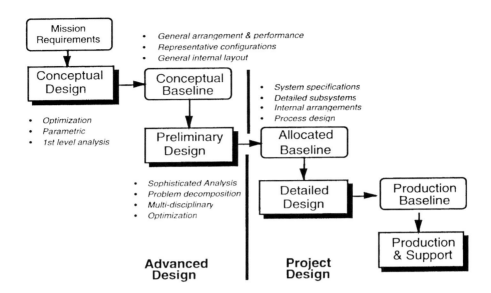

Figure 8. Traditional Development Process (Using Systems Engineering Only)

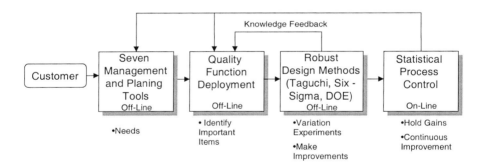

Figure 9. Quality Engineering Flow

To better understand the Hierarchical IPPD Process Flow illustrated in Figure 6, the identification of product/process *metrics* for design trade-offs at various levels of decomposition and recomposition is provided in Figure 10. The right half product metrics, such as speed, power, weight, range, volume, productivity are familiar to most engineers, while the left half process metrics, such as life cycle cost, return on investment, etc. are not as familiar.

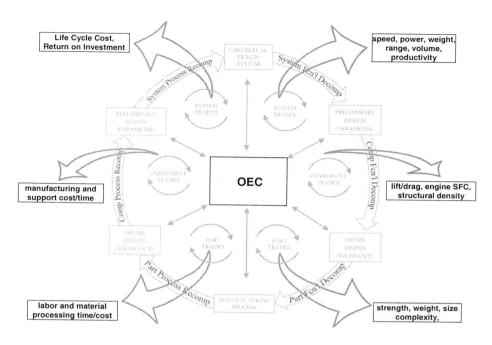

Figure 10. Product/Process Metrics for Design Trade-Offs using an Overall Evaluation Criterion (OEC)

All of the process metrics involve cost/time relationships, as illustrated in Figure 11 for theoretical production (Ref.15). The relationships in Figure 11 can be used to discuss some of the recent manufacturing initiatives, such as *lean manufacturing* and *just-in-time (JIT) manufacturing*. As can be seen there is a split in the cost/time relationship depending on whether it is the "largest" or "smallest" production run. The intersection with the Cost/Time curve, in essence the learning curve from Theoretical First Unit Cost (TFUC or T1), shows that the "largest run" takes more time but has the lowest cost/unit while the "smallest run" takes an opposite path. Reducing the TFUC and flattening out the learning curve are the essence of "lean manufacturing". By the same token the relationship between "Setup time" and "Setup cost" is what Toyota Production Systems attacked with JIT (Ref.16). In many manufacturing industries "Setup time" has been considered relatively fixed to handle cyclic variations in orders and to achieve Economic Batch Quantities (EBQs). Along with this assumption is that inventory is considered an asset, in order to be able to ramp up when necessary. Under the Toyota system, with its suppliers as an integral part of the production process, "Setup times" and the related "Setup costs" are driven toward zero and inventory becomes a *liability*, rather than an *asset*. Finally, Figure 12 illustrates how the Cost/Time curve can become a constraint curve for candidate manufacturing processes for use in design tradeoffs. As can be seen Process E lies outside the constraint curve, while Processes A - D fall within the constraint curve. Thus, if the product technology warrants the benefits in reduced weight, volume, etc. and can only be used with Process E, then a parallel manufacturing technology development program must be initiated to bring the manufacturing process into the feasible design space.

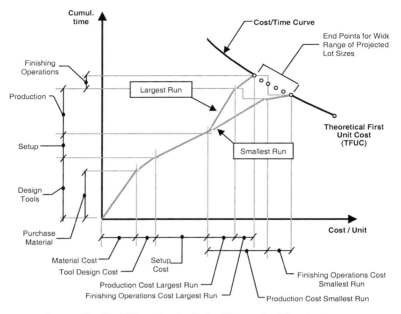

Figure 11. Cost/Time Analysis for Theoretical Production

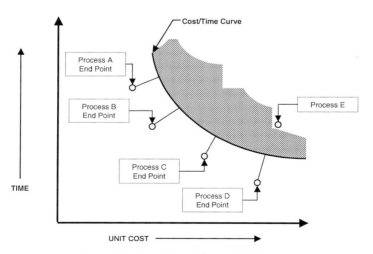

Figure 12. Cost/Time Constraint Curve

An approach to modeling the Hierarchical IPPD Process Flow in Figure 6 and to evaluate the metrics and the Overall Evaluation Criterion (OEC) shown in Figure 10 is depicted in Figure 13 for an aircraft system. Illustrated are typical decomposition models, i.e. Aircraft Synthesis (Sizing), Finite Element Analysis (FEA), and recomposition process models, i.e. knowledge based system (KBS), component cost models, and a Top-Down Aircraft LCC model. These models were developed and exercised in a Ph.D. thesis, *Integration of Design and Manufacturing for a HSCT,* and highlighted in several journal papers and conference proceedings (References 17, 18, 19, and 20). A key element of this research was to convert the NASA/Georgia Tech Aircraft Life Cycle Cost Analysis (ALCCA), Figure 14, into a more process-based cost model, as illustrated in Figure 15. As can be seen in Figure 15 the weight-based Aircraft Manufacturing Costs module was replaced with a New Wing Production Module which included the capability to establish the cost/time relationships illustrated in Figure 11, using the NASA Knowledge Based System (KBS), CLIPS, to generate manufacturing heuristic input.

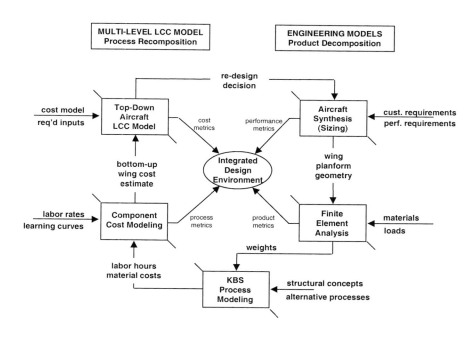

Figure 13. Typical Models Used for Decomposition and Recomposition

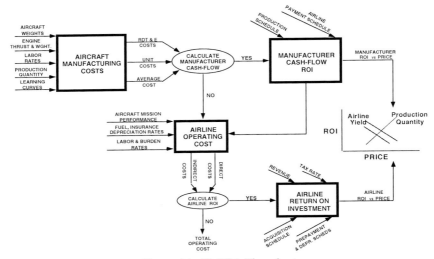

Figure 14. ALCCA Flowchart

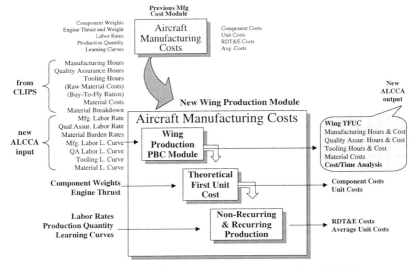

Figure 15. New ALCCA Wing Manufacturing Cost Module

It is noted that the ALCCA model in Figure 14 is more than a LCC model, i.e. an economic analysis model, as it includes the capability to assess cash flow analysis and the ability to assess price, as well as cost.

Before beginning further discussion on IPPD through RDS, the terms "affordability" and "robust design" should be defined. *Affordability*, as used here, is associated with a *benefit-cost ratio* (BCR), which is used in economic analysis when economic resources are constrained and relates the desired benefits to the capital investment required to produce the benefits. This method of selecting alternatives is most commonly used by governmental agencies for determining the desirability of public works projects (Ref.1). A project is considered viable when the net benefits associated with its implementation exceed its associated costs. For the assessment and the selection of new aircraft or technologies for insertion into existing aircraft, the ratio may be more appropriately considered the *system effectiveness to the system cost ratio* or *operational effectiveness to operational cost ratio*, which has often been used in the military for Cost & Operational Effectiveness Analyses (COEA's). The term system effectiveness can be considered a function of the capability, dependability, and availability of the system; while the system cost should be the life cycle cost of the system (Ref.21). *Robust design* is defined in Reference 1, as the *systematic approach to finding optimum values of design factors which result in economical designs with low variability*. A slightly modified version of this definition is being used for the Office Naval Research (ONR) Affordability Science research program and has been defined as the *systematic approach to finding optimum values of design factors which results in economical designs which maximize the probability of success*. (Ref. 22)

As a result of this research effort a "Roadmap for Affordability" has been defined and is being implemented through the use of Robust Design Simulation (RDS), as illustrated in Figure 16. In the center box is the linkage between Synthesis & Sizing and Economic Life-Cycle Analysis, which has evolved from the primary iteration in the generic IPPD Methodology illustrated in Figure 5 and discussed earlier. This approach thus provides the IPPD through Robust Design Simulation described in this chapter.

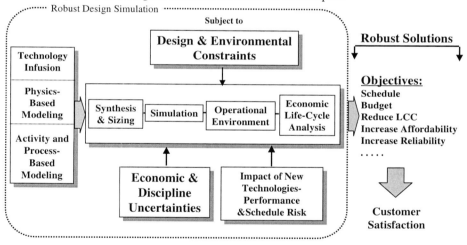

Figure 16. Roadmap to Affordability through RDS

As this linkage between *Synthesis & Sizing* and *Economic Life-Cycle Analysis* has been developed, *Simulation of the Operational Environment* has been included to address additional life cycle issues and constraints in the affordability assessment, as well as operational effectiveness. Inputs into this center RDS box come from three areas that will now be addressed.

From the left side comes *Technology Infusion* which must be handled through the use of improved modeling since most current *Synthesis & Sizing* models and *Economic Life-Cycle Cost Analysis* models are based on historical data and linear regression of this data, i.e. weight equations, drag polars, cost estimating relationships, etc. If the new aircraft or system is to be similar to the existing database then the current models are sufficient for synthesis and economic analysis. However, if new technologies, either product or process are required for innovative or out-of-the-box designs then these historical databases must be replaced with more relevant data. Physics-Based Modeling is a way of bringing higher fidelity analysis (CFD, FEA, etc.), or experimental results, into the synthesis & sizing models and directly links disciplinary analysis and the science and technology (S&T) program into advanced design. Physics-Based Modeling is more applicable to product technologies in today's environment than it is to process technologies. Therefore, Activity and Process-Based Modeling based on heuristic type models, such as knowledge based systems (KBSs), must be developed to establish and provide the cost/time analysis discussed earlier and depicted in Figure 11.

There are two inputs from the bottom into the center box in Figure 16. The left one is *Economic and Discipline Uncertainties*, which indicate the need for probabilistic approaches. The right input illustrates how the RDS can be used to address the *Impact of New Technologies*, to include *Performance and Schedule Risk*. Finally, the top inputs to the center box in Figure 16 illustrate that Design and *Environmental Constraints* must be addressed. The output from the use of RDS are *Robust Solutions* that result in attainment of multiple objectives to achieve *Customer Satisfaction*. A more detailed description of RDS is provided in Reference 23.

3. SUMMARY AND CONCLUSIONS

Integrated Product/Process Development (IPPD) has been identified as a key element in future manufacturing systems. While some use of IPPD is being applied in industry and government; in reality, it can not be fully executed without a "new systems approach" methodology and the creation of a robust design environment for its implementation. This "new systems approach" methodology must capture, especially for complex systems, the key elements of the quality revolution, as well as the traditional systems engineering methods/tools. It must reflect the complex system design trade-offs that have to be addressed early in the design process, as well as take advantage of the information technologies that are creating the necessary computer integrated environment. This computer-integrated environment must provide for robust design simulation, where probabilistic approaches are used for both product and process design. Development of the "new systems approach" methodology, along with the creation of the accompanying robust design environment, is being prototyped as IPPD through RDS. A number of companies and government agencies are beginning to use and evaluate this prototype, which should help it mature and provide a foundation for the next generation agile manufacturing systems.

REFERENCES

1. Dieter, G.E., Engineering Design – A Materials and Processing Approach, Third Edition, McGraw Hill, Boston, MA (2000).
2. Hauser, J.R., and Clausing, D, "The House of Quality", The Product Development Challenge, A Harvard Business Review Book (1994).
3. Schrage, D.P., "Concurrent Design: A Case Study," Chapter 21, Concurrent Engineering – Automation, Tools, and Techniques, Edited by Kusiak, A., John Wiley & Sons Inc., New York (1993).

4. Technology for Affordability: *A Report on the Activities of the Working Groups – Integrated Product/Process Development (IPPD), Simplified Contracting, and Dual-Use Manufacturing*, The National Center for Advanced Technologies (NCAT), January (1994).
5. Integrated Manufacturing Technology Roadmapping (IMTR) Project, IMTR Project Office, URL: http://imtr.ornl.gov (1999)
6. Wenger, E.C., and Snyder, W.M., "Communities of Practice: The Organizational Frontier," Harvard Business Review, January-February (2000).
7. Complexity Science, Santa Fe Institute (1999).
8. Advanced Engineering Environment – Achieving the Vision, Phase I Report, National Research Council Committee on Advanced Engineering Environments, June (1999).
9. Next Generation Manufacturing (NGM): An Industry Driven Coalition, Appendix A, Technologies for Enterprise Integration Roadmap, URL: http://imtr.ornl.gov (1999)
10. IMTR Technologies for Enterprise Integration Roadmap, URL: http://imtr.ornl.gov (1999)
11. Schrage, D.P., "Technology for Rotorcraft Affordability Through IPPD," Nikolsky Lecture, 1999 American Helicopter Society (AHS) Forum Proceedings, Montreal, CA, May 25 (1999).
12. Noble, J.S., and Tanchoco, J.M.A., "Design for Economics", Chapter 16, Concurrent Engineering: Automation, Tools, and Techniques, Edited by A. Kusiak, John Wiley & Sons, Inc. (1993).
13. Schrage, D.P., and Mavris, D.N., "Recomposition: The Other Half of the MDO Equation," Multidisciplinary Design Optimization – The State Of The Art, Edited by N.M. Alexandrov and M.Y. Hussaini, Society for Industial and Applied Mathematics (SIAM), (1997).
14. Aviation Week and Space Technology, January 1, 1999.
15. MIL-HDBK –727, Military Handbook: Design Guidance for Producibility, April 1984.
16. Harrison, A., Just-In-Time Manufacturing In Perspective, The Manufacturing Practitioner Series, Prentice Hall, 1992.
17. Marx, W.J., "Integrating Design and Manufacturing for the High Speed Civil Transport," Ph.D. Dissertation, Georgia Institute of Technology, Atlanta, GA, 1996.
18. Marx, W.J., Mavris, D.N., and Schrage, D.P., "Cost/Time Analysis for Theoretical Aircraft Production", AIAA Journal of Aircraft, Vol 35, No.4, July-August, 1998.
19. Marx, W.J., Mavris, D.N., and Schrage, D.P., "A Hierarchical Aircraft Life Cycle Cost Analysis Model," AIAA Paper 95-3861, Sept. 1995.
20. Marx., W.J., Schrage, D.P., and Mavris, D.N., "A knowledge-based system integrated with numerical analysis tools for aircraft life-cycle design", Journal for Artificial Intelligence for Engineering Design, Analysis and Manufacturing, I2, 211-229, Cambridge University Press (1998).
21. Defense Systems Management College (DSMC) System Effectiveness Definition.
22. A Comprehensive Robust Design Simulation (RDS) Approach to the Integrated Product/Process Development (IPPD) of Affordable Systems, Grant from the Office of Naval Research, Georgia Institute of Technology (1997-1999).
23. Mavris, D.N., Bandte, O., and DeLaurentis, D.A., "Robust Design Simulation: A Probabilistic Approach To Multidisciplinary Design," Journal of Aircraft, Volume 36, No. 1, pp. 298-307 (1999).

Developing the Agile Enterprise

John Bessant, David Knowles, David Francis and Sandra Meredith

Centre for Research in Innovation Management, University of Brighton, Falmer, Brighton, BN19PH, United Kingdom

The problem facing enterprises in the late twentieth century can be seen as the latest version of two long-standing puzzles to do with responding to demanding internal and external environments. Searching for solutions to these puzzles leads to investment in innovation - via R&D, technology transfer, etc. But evidence suggests that the key requirement is not solving the puzzle for one set of circumstances but in continually solving the problems as the puzzles mutate. This places emphasis on organisational capability - it is not what you know or what you can buy but how well you learn and adapt which is the key. We term this ' agility' - and this chapter explores the definition in terms of the 'dynamic capability' view of strategic management. It draws on case study research being carried out as part of a major UK programme of work looking at the development of agility in small and medium-sized manufacturing enterprises. It presents a reference model which seeks to explain and guide the development of agility within organisations.

1. INTRODUCTION

In the turbulent conditions characterising the new century it is clear that successful manufacturing firms will need to innovate. This is neither a new nor a surprising observation; the history of manufacturing is about creating new ways of producing and new things to produce. But much of the emphasis has been on seeing innovation as an *occasional* response – either deploying new technology in a way which confers competitive advantage as a 'first-mover' or by responding quickly and effectively to demand signals – for lower prices, higher quality, greater choice, etc. Arguably the challenge of the current environment is one in which the nature of innovation needs re-examining; firms need not just to innovate but to do so *continuously*.

Dealing with turbulent and shifting environments requires a combination of strategic assessment of the nature and direction of change required and the ability to deploy innovative capacities to deal with it. Innovation in this sense does not have to be dramatic and radical in nature – although sometimes this is necessary. Most innovation is more concerned with incremental problem-solving – continuously improving things within an existing framework rather than rewriting the rules of the game. But the capacity for both is needed to survive; the key skill is one of constant re-configuration of internal knowledge resources to seize and defend competitive a quality which several writers call 'dynamic capability' .[1]

For this reason we see agility as the core strategic capability within the manufacturing organisation associated with being able to configure innovation (in product and process) on a continuing and pro-active basis. Such capability is increasingly needed in an environment where – as de Geus points out, the survival rate of firms which fail to learn and develop is worryingly low. [2] Much of the thinking about manufacturing strategy in the late 20th century was dominated by assumptions about relatively stable conditions in which the main challenge was identifying clear market segments and focusing the manufacturing operations on meeting the needs of those segments as well as possible [3, 4]. But an emergent theme – highlighted by the regular 'Manufacturing Futures' survey amongst others – has been the shifting parameters and the decline in the ability of firms to 'trade-off' different competitive priorities [5]. Today's markets require responsiveness across a range of dimensions, and their requirements are likely to change again with increasing frequency. Consequently the challenge is less about maintaining or returning to equilibrium conditions than about managing in what is an essentially chaotic and unpredictable environment. This places considerable emphasis on learning and innovating capabilities within the organisation – what Hayes et al call 'dynamic manufacturing' [6].

2. THE NATURE OF AGILITY

Although used with increasing frequency, agility is a difficult concept to operationalise because it admits of many interpretations. The dictionary defines it as:

'... *having the facility of moving quickly; quick, nimble, active...*'
(New Elizabethan Reference Dictionary, Newnes, London.)
Agility is a theme much discussed in a variety of literature sources; examples include:
- operations strategy and management where it appears in the context of organisations trying to cope with uncertain and turbulent [7]. Here it is associated with flexibility in responding to market conditions and with the ability to change (or not) the manufacturing systems involved. This brings back into focus the discussion associated with the concept of 'flexible manufacturing systems'. These typically offer some combination of different kinds of flexibility – in offering choice to customers, in seasonality, in speed of response, etc. Studies of FMS from the 1980s highlight two themes which are still relevant; first is the recognition of the multiple nature of solutions to the 'flexibility problem'. Different kinds of flexibility can be delivered via many configurations of equipment and organisation [8-10]. Second, solutions which are relevant at a particular time may become inappropriate at a later stage. (The example here of 'economic batch quantity theory is relevant here; although originally an appropriate response to the problems of flexibility in batch production it gave way in the face of new developments in set-up time reduction which opened up the possibility of significant batch size reduction). [11, 12]
- a theme of interest is that of 'mass customisation' – a concept which argues that firms should exploit new technologies and organisational forms to move towards a much more customer-specific offering [13, 14]. There are clear cases where this kind of solution is relevant, but we need to be careful to match the application of this approach with operating contingencies. Mass customisation represents one solution for a particular kind of firm, but others – for example, prototype producers or subcontractors – have always operated with a low volume/high customisation orientation and have developed alternative and appropriate solutions for their needs. Different firms operate at different points on the 'volume/variety' spectrum originally outlined by Hayes and Wheelwright [15].

- a different perspective comes from the literature associated with small firms and clustering into sectoral or regional groupings which convey some form of collective efficiency. the argument here suggests that being quick on ones organisational feet may represent an alternative strategy to traditional views of scale economy [16, 17]. New forms of inter-firm co-operation are critical to this view and there is an emerging line of argument which sees agility as an emergent property of networks and clusters – rather than necessarily being a property of individual firms [18, 19]. This has particular significance for smaller firms; as one commentator put it, 'the problem for small firms is not that they are small, but that they are isolated'. Agile networks may represent a viable alternative model for dealing with this weakness; certainly the experience of clusters in highly competitive sectors like textiles, furniture and ceramics gives some support to this view [20, 21].
- much of the current discussion of agility centres on technological developments, especially those concerned with information and communications technologies. Here the strong influence of programmes like the DARPA and related US military projects in creating the concept of 'agility' can be seen; a good review of the 'technology enabled' route to agility can be found in the work of the US Agility Forum [22].
- another strand of relevant literature concerns organisational arrangements which support greater agility to help cope with uncertain and unpredictable environments. Discussion here ranges from strongly structural – for example, the work on 'fractals' and cellular organisation – to behavioural, where the role of teamworking, of learning and of employee involvement is particularly stressed [23-26]. In particular there is considerable discussion of the ways in which high involvement in 'kaizen' and similar programmes can increase flexibility and innovation in response [27-29].
- it is also important to mention the growing literature on 'lean thinking' in both enterprises and value streams [30]. A criticism of much of this is that the focus is often on the constant search for ways of eliminating waste from operations – in other words, 'doing what we always do but better'. The difficulty with this is that it neglects the proactive aspect of agility. It is not just a matter of getting lean and fit to respond to a challenging environment but also the ability to configure new and unexpected offerings for that environment.

Viewed in this way agile manufacturing is not a single solution to a particular set of problems associated with current conditions, but a capability to adapt and innovate on a continuing basis. Whilst advanced information and communications technologies can radically extend the range of innovative options open to firms, they are not in themselves agile. Similarly whilst lean production and its associated organisational arrangements represent powerful aids to improving performance, they too do not represent manufacturing agility. Rather it is the ability to configure and select from these and other options which offers potential strategic advantage.

3. DEVELOPING MANUFACTURING AGILITY

If manufacturing agility is essentially an example of what Pisano and others call 'dynamic capability' then a key question becomes that of how such capability can be developed within the firm [31]. The approach taken within our research has been to try and identify key behavioural routines – 'practices' – associated with agility and to develop an organisational development approach to auditing and extending capabilities in these [32].

One approach of value in this connection is that of reference models. In similar fashion to benchmarking, this approach involves providing a framework against which firms can position themselves and from this comparison identify directions and options for future development work. Examples of reference models include the Capability Maturity model of software development or the European Business Excellence Model [33, 34].

An important component of reference models is the use of two distinct dimensions - one concerning performance references and the other concerning what are sometimes called 'practice' references [35]. The former are concerned with result outputs - how does the organization perform (in terms of responsiveness, flexibility, high quality, etc.) with reference to an absolute or to industry/sector 'best standard'? This is often the focus of benchmarking research and provides external indicators of the extent to which agile behaviour is meeting the strategic challenges of the enterprise.

The second dimension, of practice, is essentially concerned with how that performance is arrived at. What are the particular organisational behaviours (and their supporting structures, systems and procedures) which contribute to good performance? Agility, as we have defined it above, involves various aspects of organisational behaviour and these can be scored with reference to real 'best in class' examples or to a notional 'best practice' model.

Using reference models to enable organisational development is essentially a process of audit and review against the structured framework, followed by introduction of relevant changes and review of their impact. (We can take the analogy of an athlete in training as an illustration of this process. The strategic targets have to do with the highly specific objective of, say, a distance of 30m in the long jump – and by definition, not aiming for the high jump or the marathon – by 2000. The performance dimension assesses current performance – distance achieved – and the gap to be closed. The practice dimension looks at the particular abilities contributing to overall fitness for long jumping (acceleration, aerodynamics, co-ordination, muscle tone, mental ability, etc.) – and the gap to be closed. From analysis of the gaps a development programme – for example, working on diet, weight training, positive visualisation, etc.) – can be created).

4. THE RESEARCH APPROACH

The model described here has emerged from an 'action research' programme of work with a network of 10 small/medium-sized enterprises (SMEs) with an interest in developing agility. Extensive interaction with these firms has led to the development of a set of longitudinal case studies, the design and implementation of several interventions and the establishment of an experience-sharing 'learning network' which meets on a monthly basis.

The model 'best practice' framework has gone through several phases of development over the past two years and is still being tested and refined. In particular the validity of the model is at present based on 'face validity' – does it make sense of their experiences to the companies involved – and, via secondary sources, does it contain relevant insights from other research? Its use as an OD tool is not compromised by this WIP status but we recognise that further work needs to be done in the area of testing and validation. Table 1 indicates the key stages of development:

Table 1: Key stages in the development of the agile reference model

Stage	Key elements	Commentary
Version 1 – summer 1998	Emphasis on agile 'practice' – what are the key behavioural routines and accompanying structures/processes which are associated with agility? First version based on 18 core factors, grouped roughly into 4 quadrants – strategy, processes, people and linkages	Initial model used to capture insights from literature and begin testing ideas with companies in network.
Version 2 – autumn 1998	Based on feedback with companies the model was refined to 16 basic clusters of behaviours. Some attempt made to add performance dimensions, particularly those associated with 'agile' behaviour – for example, frequency of product innovation, speed of response, etc.	Development was assisted by comparison with an existing benchmarking tool (Microscope) and identifying where changes would be required to reflect agility issues
Version 3 – spring 1999	Model further refined following pilot testing with 6 new companies as part of a London-based manufacturing improvement programme ('Made in London'). This involved using the tool as an OD resource and enabled the development of scaling system to facilitate scoring on the 16 core factors	
Version 4 – the 'Manufacturing Agility Quotient' tool	Based on further review of literature on relevant agility performance measures and on the pilot company responses. This version includes a 5 point scale on each of the 16 key areas, together with (for OD purposes) a self-assessment section identifying current and desired scores and priority rating for change.	This version is still targeted primarily as an OD aid but contains an attempt at a scaling framework on the 16 key dimensions. This scale is derived from literature and case examples and provides more extensive reference framework for thinking about agility within a particular organisation.

5. 'THE AGILE WHEEL'

Development of the model involved identifying groups or clusters of behavioural routines – the semi-automatic set of behaviour patterns which define 'the way we do things around here' -associated with agility These routines can be observed directly, and can be inferred from the structures and procedures operating in the organisation. Four key groups of routines emerged as relevant in the literature and in the experience of the case study firms; these are concerned with The four major dimensions of the reference model are:

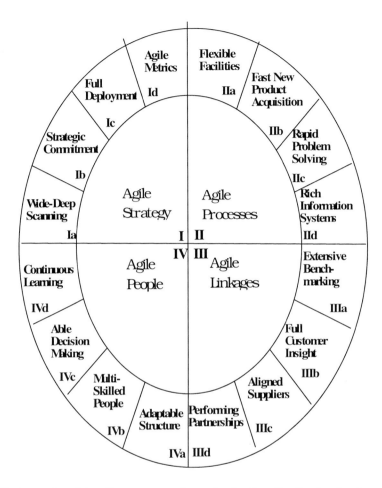

- Agile Strategy - involving the processes for understanding the firm's situation within its sector, committing to agile strategy, aligning it to a fast moving market, and communicating and deploying it effectively.
- Agile Processes - the provision of the actual facilities and processes to allow agile functioning of the organization
- Agile Linkages - intensively working with and learning from others outside the company, especially customers and suppliers.
- Agile People - developing a flexible and multi-skilled workforce, creating a culture which allows initiative, creativity and supportiveness to thrive throughout the organization.

Each has four constituent routines associated with it, and for ease of presentation they are grouped as segments in a wheel.

In outline the 16 key routines identified in the model are given in table 2, together with relevant supporting references to literature in the field.

Table 2: Dimensions of the 'Agile wheel'

Key ability	Supporting references	Underlying questions
1a. Wide/deep scanning – active search behaviour	[2, 36-39]	Do you extensively study and evaluate state-of-the-art technologies / processes that could be useful to you? Do you track market and competitor developments and consider their possible impact on your manufacturing facilities?
1b. Strategic thinking – agility built into the planning process and content	[2, 3, 40-42]	Is there a strategic planning process? Are manufacturing managers dedicated to agile strategies? Do decisions that affect the development of manufacturing increase agility (and reduce rigidity)?
1c. Strategy deployment – communication, commitment-building and alignment	[43-46]	Are all teams and individuals in the manufacturing facility striving to adopt agile principles?
1d. Measuring – monitoring and measurement of relevant parameters (performance and practice) to drive improvement in agility	[33, 47-49]	Do you understand what measurable goals you need to achieve in order to be a truly agile manufacturing organisation? Are you monitoring all relevant critical factors (for example, do you accurately measure the time taken to complete all key processes) for achieving agility? Are the results of these measures instantly available?
2a. Configuring flexibly – using technologies and practices which allow rapid and frequent reconfiguring	[9, 30, 50-53]	Are your production facilities (assets, equipment, systems etc.) inherently flexible and capable of rapid reconfiguration so that you can always complete any production task that may be required? Do you rapidly adopt technologies that help you to become more agile.
2b. Creating/ acquiring and implementing new products – renewing rapidly and frequently the product/service offer	[38, 54-58]	Is your speed of implementation of new product development / improvement clearly superior to your rivals?
2c. Problem-solving- continuously and systematically and rapidly finding and solving problems	[59-63]	Do problems get solved in hours not weeks?
2d. Informing and communicating – building awareness and sharing of key knowledge content in the organisation through information and communication management	[13, 22, 24, 64]	Do your information systems allow all staff to know immediately the status of processes and work flows?
3a. Benchmarking – constant and creative comparison with others across a range of relevant measures	[33, 35, 65, 66]	Do you extensively benchmark your production products, processes, and services against those of rivals in terms of speed, flexibility and capacity to meet customers' needs?
3b. Understanding customers – behaviours which bring deep insight of customer needs and reactions	[67-71] [72].	Do key production staff have direct contact with customers? Do all production staff know exactly what their customer wants?

3c. Aligning and developing suppliers – replicating and enabling learning around relevant agile capability throughout the value stream	[22, 63, 72-74]	Are your suppliers capable of delivering exactly what you want when you need it?
3d. Networking – building and sustaining relevant complementary alliances outside the firm	[75-80]	Do you have strong partnerships with other firms and organisations that provide support, opportunities and additional capabilities at all levels of the company as and when you need it?
4a. Structuring – creating/recreating appropriate organisational structures to enable agility	[25, 81-86]	Does the way you structure your production organisation enable things to get done quickly and flexibly, as well as supporting people taking initiatives to seize opportunities?
4b. Developing flexible human resources – multi-skilling and competence developing the people within the organisation	[26, 87, 88]	Do you have extremely extensive and flexible competencies in all staff in the production function?
4c. Rapid decision-making	[28, 84, 89, 90]	Are decisions taken quickly but carefully?
4d. Continuous learning – managing the knowledge creation, capture and sharing process at all levels	[6, 63, 90-95]	Is everyone, both individually and in teams, actively developing skills and learning continuously?

Since we are concerned with behavioural routines, each of these is expressed as a verb, implying a set of active behaviours which together create the capability to be agile in an uncertain environment. These define the 'practices' of agility and are the things which an organisation can change – through training, structural interventions, capital investment, etc. The model has a number of levels and each of the 16 major abilities can be subdivided further – for example, 'structuring' contains themes like teamworking which represent particular ways of achieving agility. (In the interview schedules and questionnaire instruments based on the model which we use in companies we have increasingly tried to reflect this kind of sub-division). A brief description of each area is given below.

5.1 Agile Strategy

Ia Wide-Deep Scanning - Tracking the external environment

A key principle in agility is the development of awareness of external signals requiring a response, and this places emphasis on mechanisms for tracking the external environment. It involves not only market and competitor analysis but also technology scanning and development of an understanding of relevant social and political trends. Another key element is the ability to explore the future dimension of these variables through various forecasting and planning techniques.

Research has consistently shown that innovative firms are those which have a strong external orientation and which make use of multiple methods for scanning their environment and picking up even weak signals about relevant changes. [2, 36-39]

Ib Strategic Commitment

Agility is not simply reactivity, nor is it blind technological push. It is a strategic posture, an approach which sees continuous innovation as a way of securing and maintaining competitive advantage. It involves processing the signals available from its scanning activities (what could we do?) and analyzing appropriate courses of action (what will we do?) based on the commitment of limited resources. Making this happen requires a clear understanding of the core resource base which can be deployed, and a deep understanding of the dynamics of the marketplace in which the firm is operating. It also requires a flexible approach to planning, such that possible futures can be translated into specific actions and resource commitments – but in such a way that these plans can be adapted or revised in the face of new information – what de Geus calls 'planning as learning'. [2, 3, 40-42]

Ic. Full Deployment

Having a clear (and flexible) manufacturing strategy is critical but it will only succeed if it is communicated and understood deep within the organization. Central to the development of agility is the ability to deploy strategy quickly and effectively, breaking down key strategic goals into manageable and focused projects around which the resources of the firm can be aligned. In part this is a planning activity setting clear and shared objectives – but in part it is also a learning process, building an understanding within the firm and generating commitment to its strategic objectives. It is of particular relevance in underpinning high involvement of the workforce in innovative behaviour. [43-46]

Id. Agile Metrics

Closely associated with the above is the ability to measure and direct activities through the use of suitable metrics. Whilst many firms have strong financial measurement tools and systems, fewer have the necessary capabilities to monitor and manage other elements. Performance metrics are about measuring the operation of the whole system, which includes people, machinery, technology, logistics, marketing and financial and market environments - not in isolation, but in the way that each area interacts and which explores ways to maximize the synergy that occurs between the different elements. The underlying philosophy is one of measurement not for control but as an input to a process of continuous learning and organisational development. [33, 47-49]

5.2. Agile Processes

IIa. Flexible Facilities

Agility is concerned with the capability for rapid configuration and reconfiguration to suit changing environmental conditions. For this reason there can be no single 'best' way for organizing manufacturing or for laying out the factory or for any other aspect of operations. Instead the aim is to develop flexibility and rapid changeover. At the physical level this is likely to involve an equipment philosophy which suits rapid rearrangement – either through physical relocation (e.g. putting machines on wheels) or reprogrammability – an option which current information technologies makes available.

But a more fundamental challenge lies in ensuring that ways of thinking – mental models – about manufacturing operations are equally flexible. The principles of lean manufacturing are powerful and have had a major impact – yet in many ways they are simple. The nature of the 'revolution' in manufacturing which their application has brought about in recent years is more concerned with challenging and 'unlearning' principles laid down in the early days of mass production than with radically new concepts. Firms need to develop the capacity to challenge and adjust their mental models and the physical and operational systems which follow from them on a continuing basis. [9, 30, 50-53]

IIb. Fast New Product Acquisition

Shorter life cycles, demand for greater product variety and narrowing windows of market opportunity, mean that fast new product development or acquisition, is an important element of competitive advantage. Within the agile organization emphasis is not placed on internal new product development alone, but includes the skill of recognizing what new products externally developed, are important and acquiring them as appropriate to the organization's needs. (This issue is particularly important for SMEs, who may not have the resources for in-house product development and may have to rely upon external acquisition to develop their product line or services).

Inevitably this poses challenges for the behavioural routines, systems and structures within the firm. Agility in this area needs characteristics like empowered cross-functional teams, devolved responsibility, parallel working, early involvement of multiple functions, clear project management structures, systems and responsibilities, phased risk management systems and the capacity to review and learn from projects. [38, 54-58]

IIc. Rapid Problem Solving

In relatively stable environments manufacturing becomes something which can be standardized and managed by procedures – 'doing it by the book'. This was a strong feature of the original mass production philosophy characterized by Ford's early factories and in the ideas of Frederick Taylor. But in uncertain and turbulent environments the requirement shifts from being able to predict or forecast to being able to respond quickly and to solve problems rapidly and creatively. This ability requires an embedded approach to finding and solving problems systematically and the capacity to learn and capture knowledge such that the same problems are not repeatedly solved. [59-63]

IId. Rich Information Systems

Free flows of information throughout the organization allow for more effective devolution of decision-making – essentially passing control to 'the sharp end' and enabling flat and responsive organisational structures. Within an agile organization the role of middle managers becomes more important in line with the increasing number of decisions they are required to make. But a reliable decision can only be made when the context within which the decision is being taken is known to the manager/decision maker. This places considerable emphasis on the design of information systems which enable rapid communication (within and increasingly between firms) but which also allow for effective knowledge capture and sharing; it is why information technology remains one of the key enabling resources in developing agility. [13, 22, 24, 64]

5.3. Agile Linkages

IIIa Extensive Benchmarking

By its nature agility is about being in a constant state of preparedness for change – but to enable this requires a continuing flow of information about positioning and about the availability of potential external linkages. Being aware of competitors, or markets, of technological developments, etc. is critical to recognizing opportunities for change (see 1a) and being able to identify potentially fruitful linkages and alliances is an important complement to this. Agile organisations make extensive use of external linkages but this implies the capability to recognise with whom such linkages can be built and from whom learning can take place. For this reason a continuing search and monitoring process is needed to build and maintain such environmental maps. [33, 35, 65, 66]

IIIb. Full Customer Insight

Innovation research consistently shows the importance of getting close to the customer and identifying and understanding their needs. At the limit this may involve active participation of users in the design and development process. Agile organisations make extensive use of multiple approaches to get close to key customers, work with lead users and also to pick up emerging but not yet critical trends – something to which Christenson's work has drawn attention. US research views as a generic agility issue the developing and sustaining of loyal relationship with customers to a deep level [71]. This relationship transcends product technology cycles, ascertaining unarticulated needs, using emerging electronic commerce effectively, integrating intra-enterprise information systems, developing and employing a customer knowledge base and developing more responsive and more robust logistic and distribution system.

Importantly agility is not simply about response however fast and effective that is. It is also about proactive deployment of innovative capabilities – at the limit providing customers with solutions even before the customers know what their needs are [67-71] [72].

IIIc. Aligned Suppliers

One of the key areas of development over the past twenty years has been in the area of supply chain management. Originally born out of a recognition that traditional models were often confrontational and based on 'win-lose' outcomes, the emerging literature began to argue for closer and more co-operative relationships. Experience with supply chain management and development suggests that much more can be gained through working towards integrated development of agility within the entire value stream and through developing a learning and continuous improvement capability at the inter-firm level. The principles of the 'extended', 'virtual' or 'boundaryless' enterprise depend critically on making the transition towards such new relationships. [22, 63, 72-74]

IIId. Performing Partnerships

Perhaps the most distinctive characteristic of the 'new' model for manufacturing is the recognition of the importance of seeing the firm as part of a system rather than in isolation. Networking and co-operation offer even small firms significant opportunities for improving competitive performance, and the experience of a number of industrial clusters supports this view. Italian furniture makers, for example, have dominated the export league table for many years, yet the average firm size is less than 20 employees. Their success – which is mirrored in many other examples – comes through developing co-operative relationships into what economists have come to term 'collective efficiency'. Agile organisations make extensive use of external linkages of this kind – in resource sharing, in access to technology or markets, in shared risk ventures, etc. [75-80]

5.4. Agile People

IVa. Adaptable Structures

In keeping with the emphasis on reconfigurability (2a) it follows that there is unlikely to be a single 'best' organization design for firms pursuing this approach. Instead there is a need to develop flexibility in the ways in which manufacturing can be organized and co-ordinated. Whilst there may be underlying archetypes around which such configuration can take place there is growing recognition of the need to explore new forms. In particular some of the principles of what Morgan calls 'holographic design' (in which the functions of the whole organization are replicated in its parts) are becoming a key focus for experiment – giving rise to concepts like 'fractal' organisations. Ideally the agile organization will have the "organisational flexibility to adopt for each project the managerial vehicle that will yield the great competitive advantage." [75]. The form this takes will be contingent on the circumstances in a moment in time. For example, a specific project may well require the participation of an internal cross-functional project team supported by both suppliers and customers. For new product development, it may require a collaborative venture engaging with other interested companies, or even the formation of virtual enterprises in order to pool people skills and competencies. [25, 81-86]

IVb. Multi-skilled People - agile workforce

Pfeffer and others have argued persuasively for the importance of human resource practices in contributing competitive edge. High involvement and commitment on the part of the people within an organization coupled with the development of their skills and capabilities offers a powerful problem-finding and solving 'engine' with which to tackle the challenges of an uncertain environment and through which to secure rapid and effective implementation. There is increasing recognition that new forms of organization at both intra- and inter-firm level and new technologies have shifted the emphasis from capital intensity to knowledge intensity as a source of competitive advantage. But implicit in this is a new role for human resources as important assets – not as pairs of hands within the factory but as creative and flexible knowledge-workers. This places considerable emphasis on developing and retaining such resources. [26, 87, 88]

IVc. Able Decision Making

The integrated organization where rich information systems allow information to flow throughout the company, enables employees to be able and willing to take decisions. Continuous workforce education and continuous growth in the quality of the workforce must be considered as a long-term investment as part of developing a decision support system. It is becoming apparent from research that able and frequent decision making is a core characteristic of the agile organization in line with a rapidly changing business environment. Sharing knowledge across a broader base of employees will facilitate better decision making. [28, 84, 89, 90]

IVd. Continuous Learning

With the increasing emphasis on knowledge as the basis for competitiveness comes a recognition that learning is probably one of the core processes which a firm has to manage. According to Senge [69] "the rate at which organisations learn may become the only sustainable source of competitive advantage". Development of learning capability is not simple – it requires a combination of training and development at the level of individuals (developing learning skills), supporting structures and processes (for example, for problem finding and solving, for strategic direction of learning activities through policy deployment and for knowledge management) and the ability to maintain this pattern of behaviour across and between organisations. [6, 63, 90-95]

6. USING THE MODEL

Each of the above elements represents behaviours which the organisation must carry out if it is to be agile; the question is the extent to which they are well or weakly developed. Agile capability is not so much a function of being good at any one of these as an all-round capability. For example, an organisation which has weak capability in terms of deploying its strategy throughout the organisation is unlikely to be able to achieve its objectives, even if it can define them clearly. Equally one which has rapid product development capabilities will only succeed if it can link these to deep understanding of customer needs – otherwise it will rapidly develop the wrong products.

As we have shown it is possible to map the dominant themes in the literature on to this model, and to show how their relative emphasis is linked to one or more quadrants. For example, the IT-driven views of agility relate strongly to quadrant 2 and 4 but less to the organisational quadrant 3. By contrast the socio-technical literature emphasizes quadrant 3 behaviours whilst the networks/clusters dominates in quadrant 4.

In order to use such a model to help enable the development of agility we need to have some dimensions along each of these characteristics related to strong or weak performance. Our research work at present is focused on trying to introduce such measures – and also to identify enablers of progress along each dimension. For example, in the case of 'continuous improvement' the ideal is clearly one in which every member of the organisation is involved in actively seeking out problems, solving them and sharing the resulting learning with others. The reality – for most firms – is much patchier, with varying levels of development and performance. Research has enabled the development of a detailed 'road-map;' for the evolution of CI capability and identified a range of specific resources which can enable progress in particular stages of the journey [Bessant, 1997 #570]. Similar approaches have been used in the development of project management capabilities around the difficult theme of software development [34].

7. PERFORMANCE DIMENSIONS

So far the discussion has been around those practices which are associated with agility. But we also need to consider the performance dimension in our reference model. The extent to which agile practices actually contribute to competitive advantage will depend on how well they re aligned to strategic objectives. For example, efforts to develop agile practice which particularly emphasize speed of response may not be helpful if the particular environment is one in which quality or variety are the key 'order-winning' factors. For this reason it is difficult to specify a generic set of performance indicators but instead to identify a set of relevant measures which are configured in the light of strategic factors. These include:
- Speed of response/time compression
- Volume flexibility
- Variety

- Frequency of product innovation
- Time to market for new products – concept to cash cycle

8. SUMMARY

Agility is not a new idea – but it is essential for survival in the emerging global competitive environment. Possession of resources will matter far less in determining strategic advantage than the ability to configure and reconfigure resources rapidly. As we move towards more network-based models so ownership of resources becomes less important than knowing where to access them and how to manage the relationships within the network.

There is no doubt that new tools and techniques have much to offer in helping firms develop agility – for example, the potential of information and communication technologies for facilitating global networking and 'virtual companies' has only begun to be exploited. But such tools, no matter how sophisticated, will not be sufficient to create agility; experience with earlier generations of flexible technologies teaches us that they only work when integrated into the rest of the business and used within a clear strategic framework [Voss, 1986 #66; Bessant, 1993 #284].

Agility does offer a route to strategic competitive advantage, but this comes from not only reacting quickly and appropriately to demands from the environment but also in pro-active behaviour, trying to change and shape the rules of the game. Capturing strategic advantage depends on having a particular firm-specific 'edge' which others find hard to emulate – rather than simply following the fashion. The big advantage in agility lies not so much in any particular solution at any time (since these can always be licensed, copied or stolen) but rather in the internal capabilities which make the creation of such solutions possible. This dynamic capability will be at the heart of the 'knowledge-based' learning organisations of the future.

One last point concerns firm size. Whilst much of the twentieth century has been dominated by the experience of large organisations, and the models which they used to achieve strategic advantage the pattern is changing. Agility depends less on size or resources than on the ability to move and change quickly and continuously. For these requirements big may not be beautiful – and it may even represent a positive barrier. In a world where networking enables access to knowledge and resources, the potential for agile smaller firms is considerable.

But developing such capability is not a simple process, as the above model suggests. There are multiple dimensions to agile capability and all need to be addressed through a process of sustained organisational learning. Whilst frameworks and reference models can help, the internal commitment to and management of the learning process should be high on the strategic agenda for any organisation concerned with becoming more agile. This is challenging since one of the lessons about the agile approach is that it involves considerable 'unlearning' – letting go many of the old ways in which the organisation worked in the past.

REFERENCES

1. Teece, D., G. Pisano, and A. Shuen, *Dynamic capabilities and strategic management*, 1992, University of Berkeley.
2. de Geus, A., *The living company*. 1996, Boston, Mass: Harvard Business School Press.
3. Hill, T., *Manufacturing strategy*. 2nd ed. 1993, London: Macmillan. 230.

4. Skinner, W., *Manufacturing in the corporate strategy*. 1978, New York: John Wiley.
5. De Meyer, A., *Report on the Global manufacturing Futures survey*, . 1998, INSEAD: Fontainbleu.
6. Hayes, R., S. Wheelwright, and K. Clark, *Dynamic manufacturing: Creating the learning organisation*. 1988, New York: Free Press.
7. Slack, N., *The manufacturing advantage: Achieving competitive manufacturing operations*. 1992, London: Mercury.
8. Ettlie, J., *Taking charge of manufacturing*. 1988, San Francisco: Jossey-Bass.
9. Tidd, J., *Flexible automation*. 1989, London: Frances Pinter.
10. Bessant, J., *Managing advanced manufacturing technology: The challenge of the fifth wave*. 1991, Oxford/ Manchester: NCC-Blackwell.
11. Suzaki, K., *The new manufacturing challenge*. 1988, New York: Free Press.
12. Monden, Y., *The Toyota Production System*. 1983, Cambridge, Mass.: Productivity Press.
13. Boynton, A., B. Victor, and B. Pine, *New competitive strategies: Challenges to organisations and information technology*. IBM Systems Journal, 1993. 32(1): p. 40-64.
14. Pine, B.J., *Mass customisation: The new frontier in business competition*. 1993, Cambridge, Mass.: Harvard University Press. 333.
15. Hayes, R. and S. Wheelwright, *Restoring our competitive edge: Competing through manufacturing*. 1984, New York: John Wiley.
16. Best, M., *The new competition*. 1990, Oxford: Polity Press.
17. Piore, M. and C. Sabel, *The second industrial divide*. 1982, New York: Basic Books.
18. Grandori, A. and G. Soda, *Inter-firm networks: Antecedents, mechanisms and forms*. Organization Studies, 1995. 16(2): p. 183-214.
19. Nohria, N. and R. Eccles, *Networks and organisations:Structure, form and action*. 1992, Boston: Harvard Business School Press.
20. Nadvi, K., *The cutting edge: Collective efficiency and international competitiveness in Pakistan*, . 1997, Institute of Development Studies.
21. Martinussen, J., *Elements of success in cluster policies : from a practitioner's point of view*. 1995, Worcester: Business Net Ltd.
22. Preiss, K., S. Goldman, and R. Nagel, *Co-operate to compete: Building agile business relationships*. 1996, New York: Van Nostrand Rheinhold.
23. Holti, R., J. Neumann, and H. Standing, *Change everything at once: The Tavistock Institute's guide to developing teamwork in manufcaturing*. 1995, London: Management Books 2000.
24. Tranfield, D. and e. al., *Teamworked organisational engineering: Getting the most out of teamworking*. Management Decision, 1998. 36(6).
25. Warnecke, H.-J., *The fractal company*. 1992, Berlin: Springer-Verlag.
26. Pfeffer, J. and J. Veiga, *Putting people first for organizational success*. Academy of Management Executive, 1999. 13(2): p. 37-48.
27. Imai, K., *Kaizen*. 1987, New York: Random House.
28. Boer, H., et al., *CI changes: From suggestion box to the learning organisation*. 1999, Aldershot: Ashgate.
29. Bessant, J., *Developing continuous improvement capability*. International Journal of Innovation Management, 1999. 2(4): p. 409-429.
30. Womack, J. and D. Jones, *Lean thinking*. 1997, New York: Simon and Schuster.

31. Teece, D. and G. Pisano, *The dynamic capabilities of firms: an introduction.* Industrial and Corporate Change, 1994. 3(3): p. 537-555.
32. Pentland, B. and H. Rueter, *Organisational routines as grammars of action.* Administrative Science Quarterly, 1994. 39: p. 484-510.
33. Chiesa, V., P. Coughlan, and C. Voss, *Development of a technical innovation audit.* Journal of Product Innovation Management, 1996. 13(2): p. 105-136.
34. Paulk, M., et al., *Capability maturity model for software,* . 1993, Software Engineering Institute, Carnegie-Mellon University.
35. Voss, C., *Made in Britain,* . 1994, London Business School.
36. Carter, C. and B. Williams, *Industry and technical Progress.* 1957, Oxford: Oxford University Press.
37. Jantsch, E., *Technological forecasting in perspective.* 1980, Paris: OECD.
38. Thomas, R., *New product development: Managing and forecasting for strategis success.* 1993, New York: John Wiley.
39. Van de Ven, A. and G. Raghu, *Innovation and industry development: The case of cochlear implants*, in *Research on technological innovation, management and policy*, R. Burgelman and R. Rosenbloom, Editors. 1993, Jai Press: Greenwich, Conn.
40. Prahalad, C. and G. Hamel, *Competing for the future.* 1994, Boston, Mass.: Harvard University Press.
41. Francis, D., *Step by step competitive strategy.* 1994, London: Routledge.
42. Voss, C., *Manufacturing strategy.* 1992, London: Chapman and Hall.
43. Shiba, S., A. Graham, and D. Walden, *A new American TQM; Four practical revolutions in management.* 1993, Portland, Oregon: Productivity Press. 565.
44. Akao, Y., ed. *Quality function deployment - Integrating customer requirements into product design.* . 1990, Productivity Press: Cambridge, Mass.
45. Smith, S. and D. Tranfield, *Managing change.* 1990, Kempston: IFS Publications.
46. Bessant, J. and D. Francis, *Developing strategic continuous improvement capability.* International Journal of Operations and Production Management, 1999. 19(11).
47. Garvin, D., *How the Baldrige award really works.* Harvard Business Review, 1991(November/December): p. 80-93.
48. Kaplan, R. and D. Norton, *Using the balanced scorecard as a strategic management system.* Harvard Business Review, 1996. January-February.
49. Deming, W.E., *Out of the crisis.* 1986, Cambridge, Mass.: MIT Press.
50. Shingo, S., *A revolution in manufacturing: the SMED system.* 1983, Cambridge, Mass.: Productivity Press.
51. Senge, P., *The fifth discipline.* 1990, New York: Doubleday.
52. Bessant, J. and J. Buckingham. *Beyond substitution: organisational implications for successful use of integrated technologies.* in *A flexible future? Prospoects for employment and organisation in the 1990s.* 1989. Cardiff Business School.
53. Wickens, P., *The road to Nissan: Flexibility, quality, teamwork.* 1987, London: Macmillan.
54. Wheelwright, S. and K. Clark, *Revolutionising product development.* 1992, New York: Free Press.
55. Cooper, R., *Third-generation new product processes.* Journal of Product Innovation Management, 1994. 11(1): p. 3-14.
56. Bessant, J. and D. Francis, *Implementing the new product development process.* Technovation, 1997. 17(4): p. 189-197.

57. Smith, P. and D. Reinertsen, *Developing products in half the time*. 1991, New York: Van Nostrand Reinhold.
58. Stalk, G. and T. Hout, *Competing against time: How time-based competition is reshaping global markets*. 1990, New York: Free Press.
59. Francis, D., *Effective problem solving*. 1990, London: Routledge.
60. Sirkin, H. and G. Stalk, *Fix the process, not the problem*. Harvard Business Review, 1990. July/August: p. 26-33.
61. Taylor, F., *The principles of scientific management*. 1947: Harper and Row (original published in 1911).
62. Duguay, C., S. Landry, and F. Pasin, *From mass production to flexible/agile production*. International Journal of Operations and Production Management, 1997. 17(12): p. 1183-1195.
63. Leonard-Barton, D., *Wellsprings of knowledge: Building and sustaining the sources of innovation*. 1995, Boston, Mass.: Harvard Business School Press. 335.
64. Goldman, S. and R. Nagel, *Management technology and agility; the emergence of a new era in manufacturing*. International Journal of Technology Management, 1993. 8(1/2).
65. Miller, J. and et.al., *Benchmarking global operations*. 1992, Homewood, Ill.: Irwin.
66. Oliver, N., *Benchmarking product development*, . 1996, University of Cambridge.
67. Shillito, M., *Advanced QFD: Linking technology to market and company needs*. 1994, New York: John Wiley.
68. Von Hippel, E., *The sources of innovation*. 1988, Cambrdige, mass.: MIT Press.
69. Rothwell, R., *Successful industrial innovation: Critical success factors for the 1990s*. R&D Management, 1992. 22(3): p. 221-239.
70. Van de Ven, A., H. Angle, and M. Poole, *Research on the management of innovation*. 1989, New York: Harper and Row.
71. Christenson, C., *The innovator's dilemma*. 1997, Cambridge, Mass.: Harvard Business School Press.
72. Lamming, R., *Beyond partnership*. 1993, London: Prentice-Hall.
73. Kaplinsky, R., J. Bessant, and R. Lamming, *Using supply chains to diffuse 'best practice'*, . 1999, Centre for Research in Innovation Management: Brighton.
74. Harland, C., *Supply chain management: Relationships, chains and networks*. British Journal of management, 1996. 7(March): p. 863-880.
75. Schmitz, H., *Collective efficiency:Growth path for small-scale industry*. Journal of Development Studies, 1995. 31(4): p. 529-566.
76. Schmitz, H., *Collective efficiency and increasing returns*, . 1997, Institute of Development Studies, University of Sussex.
77. Humphrey, J. and H. Schmitz, *The Triple C approach to local industrial policy*. World Development, 1996. 24(12): p. 1859-1877.
78. Meade, L., D. Liles, and J. Sarkis , *Justifying strategic alliances and partnering: a prerequisite for virtual enterprising*. Omega, 1997. 25(1).
79. Baden-Fuller, C. and M. Pitt, *Strategic innovation*. 1996, London: Routledge.
80. Arnold, E., *et al.*, *Strategic planning in Research and Technology Institutes*. R&D Management, 1998. 28(2): p. 89-100.
81. Mintzberg, H., *The structuring of organisations*. 1979, Englewood Cliffs, N.J.: Prentice-Hall.
82. Morgan, G., *Images of organisation*. 1986, London: Sage.

83. Coyne, W. *Building the innovative organisation*. in *The UK Innovation Lecture*. 1996. London: Innovation Unit, Department of Trade and Industry.
84. Leonard-Barton, D., *The organisation as learning laboratory*. Sloan Management Review, 1992. 34(1): p. 23-38.
85. McCloughlin, I. and M. Harris, *Innovation, organisational change and technology*. Management of Technology and Innovation, ed. J. Bessant and D. Preece. 1997, London: International Thomson Business Press.
86. Smith, S., et al., *Factory 2000: Organization design for the factory of the future*. International Studies of Management and Organisation, 1992. 22(4): p. 61-68.
87. DTI, *Competitiveness through partnerships with people*, . 1997, Department of Trade and Industry: London.
88. Teece, D., *Capturing value from knowledge assets: The new economy, markets for know-how, and intangible assets*. California Management Review, 1998. 40(3): p. 55-79.
89. Berger, A., *Continuous improvement and kaizen: standardisation and organisational designs*. Integrated Manufacturing Systems, 1997. 8(2): p. 110-117.
90. Garvin, D., *Building a learning organisation*. Harvard Business Review, 1993. July/August: p. 78-91.
91. Bessant, J. and S. Caffyn, *Continuous improvement and organisational learning*, in *Knowledge, Technology and Innovative Organisations*, J. Butler and A. Piccaluga, Editors. 1997, Edizione Angelo Guerini i Associati SpA: Milan.
92. Bessant, J. *Developing learning networks*. in *2nd IPSERA conference on strategic purchasing and supply*. 1998. London.
93. Pedler, M., T. Boydell, and J. Burgoyne, *The learning company: A strategy for sustainable development*. 1991, Maidenhead: McGraw-Hill.
94. Pisano, G., *Knowledge, integration and the locus of learning: An empirical analysis of process development*. Strategic Management Journal, 1994. 15: p. 85.
95. Prokesch, S., *Unleashing the power of learning*. Harvard Business Review, 1997. September/October: p. 147-168.

Towards Building of Knowledge-Base in Indian Corporations: Some Strategic Directions

R. P. Mohanty

Vice President, Human Resources Division
The Associated Cement Companies Ltd., Mumbai-400020
FAX: 0091-22-2080076; E-mail: rpmohanty@acccement.com

1. INTRODUCTION

India as a sovereign democratic republic during the last half-century has made very significant progress in political front. Democracy has matured. Economic development/progress has taken place, but the progression curve is not similar to any other developed democratic nation. There are myriad of reasons for slow economic progression. Indian corporate sector has already entered into the process of large-scale liberalization and globalization. This process is in inexorable and irreversible. It is not the purpose of this paper to discuss the historical background of Indian slow growth. The intention here is primarily to elaborate only *"why, what and how"* of building the knowledge-based corporate sectors in the ever expanding competitive landscape in a boundaryless world. The growing prominence of economic liberalization and accelerating change in the environment of Indian organizations are compelling to renew the existing knowledge base and acquire new knowledge to greatest strategic effect of profound growth; because the knowledge as well as the knowledge systems are important in shaping social outcomes. The new millenium demands the coexistence of creativity and productivity i.e. value creation as well as maintenance. In post capitalistic economy, which is termed as knowledge based economy, wealth flows to those who can develop, direct and acquire knowledge. Adams [1] pointed out that India needs to redirect its attentions towards education-knowledge building i.e. enhancing intellectual capital. Such a view according to this author is very critical, since, the pursuit of knowledge building requires that corporations allocate adequate financial resources and faceup to the relevant changes. The truth of organizational knowledge building and thereby formation of intellectual capital depends on a very complex proposition that how do we envision the 21st century corporations and the organic changes in the associated work systems and co-evolving participative style of human resource development and management. Given the current socio-economic-political climate, many corporations have expressed the view that the existing knowledge base is entirely inadequate for innovation and improvement in sustaining their competitive potential [2]. Since innovation and improvement require sustained investment in human resources [3], corporations and educational institutions are required to play a central role in the process of knowledge building. Our intentions in this paper is three-fold:

- *To understand the universal background and prepare the Indian foreground for knowledge building*

- *To discuss the various institutional modalities of knowledge building, and*
- *To show how one can develop a strategic perspective to critical role profiling, which may lead to better and more practical insights into human capital formation and upgradation.*

2. ENVISIONING THE FUTURE OF ORGANIZATIONS

The most exciting view in any day is the view of tomorrow – *the view of the future*. What will it be like? Will it be a return to the stability of earlier years? The author has put these questions to a very large number of Indian corporate executives in the recent past, while interacting with them in several management development programs. The answer is definitely no. Each one of us, who are engaged in the affairs of the world has realized that tomorrow will be a world of *greater complexity, fiercer competition, rapidly accelerating change*. We will encounter dramatic changes in our work place and work force as well as in the demands we will place on them. There will be radical changes: *in the number of changes, the content of changes, and the speed of changes*. As we move into the next century, all organizations will confront a completely new set of changes that will represent major discontinuities. These discontinuities will not be autonomous. They will take place both in human and organizational contexts. Through them we can anticipate improvements to our quality of life and further advances to the level of modern civilization. They will manifest as cultural artifacts.

Organizations of the future will make a quantum paradigm shift:

- *from manual work to knowledge work*
- *from the efficient motion of work to value innovation*
- *from closed system to more permeable and flexible boundaries*
- *from fat to lean: the new staffing principle*
- *from vertical command to horizontal processes: the new organization*
- *from homogeneity to diversity: the new work force*
- *from status and command rights to competencies and relationships: the new power source*
- *from authoritarianism to empowerment: the new pattern of decision making*
- *from ritualistic performance assessment to relativistic benchmarking*
- *from organizational capital to reputation capital: the career asset*
- *from single career path to multiple career path*
- *from single loop reactive learning to double loop proactive and interactive learning*
- *from experience based mundane actions to knowledge based innovations and contributions*
- *from compliance to commitment, vulnerability, and accountability*
- *from stand-alone competing to simultaneous strategic collaborating and competing*
- *from the relatively stable hegemony of financial factor-ruled to the dominance of knowledge as the driving force*

These shifts will be all permeating in a new competitive landscape configured by technological, economic, managerial, political, social, and ecological sectors etc. The new organizations will be more complex places than they have been. They will certainly be less predictable, less measurable, less amenable to the traditional disciplines of knowledge. According to Handy [4], we cannot reject the future because it is uncomfortable.

The world competitiveness report [5] establishes that:

Competitiveness = (competitive assets) x (competitive processes) . . .. (1)
From such an identity, we derive some propositions:
- *Competitive success can only be harnessed from exceptional human capital, possession of excludable knowledge assets, and the creation of a generative environment for innovation.*
- *The competitiveness of an organization rests on human resource as the most important asset (both tangible and intangible).*
- *The market economy will die in its infancy, if it is not grounded in strong human resource development processes/systems.*
- *Those organizations, which will recognize the need for investment in human resource and gaining access to and absorbing new knowledge and simultaneously act with a fast pace, will be the winners in the corporate business Olympic*
- *Those organizations, which can rethink fundamentally their human resources management models to meet the new requirements, will establish early breakpoints in business parameters.*

According to Sternberg et al [6], there are basically six resources required for the total development of human resource. They are:
- *Knowledge*
- *Intellectual abilities*
- *Thinking styles*
- *Motivation*
- *Personality*
- *Environment*

The organizations of the future will require renewed investment in human resources and formulating new policies, new modalities of learning, and innovative motivational tools. Bill Gates, head of the 21st century organization-Microsoft, has said that the only security his employees have is their knowledge and he supports their education to maintain and improve the knowledge base. President Bill Clinton has also expressed a similar view for the American people. The increasingly dynamic nature of competition during the last two decades has made the improvements of organizational learning and the developments of more effective methods for managing *knowledge* a crucial but predominant issue of contemporary organizations. Mascitelli [7] mentions that traditional competitiveness factors cannot provide a sustainable advantage in a highly dynamic, knowledge-driven global marketplace. The scarcest resources in any organization are the performing people endowed with knowledge. Knowledge within the organization is living, developmental, and synergistic, if it is applied and internalized in organizational activities. Barney [8] is of the opinion that the most fundamental criterion for sustainable competitive advantage is the building of economically valuable knowledge base of a company: both tacit and explicit. Morgan [9] stated that in the new economy-knowledge based economy managers should 'find ways of developing and mobilizing the intelligence, knowledge and creative potentials of human beings at every level of the organization." Knowledge is the only resource, which can only guarantee long-term sustainable advantage. Knowledge is at the heart of an organization for creating value. Knowledge originates in human beings. It is insight, judgment, and innovation, based on experiences, heuristics, passions, and neural connections. It provides the intellectual frameworks, conceptual models, governing ideas and ideals that allow a company's human resources to identify opportunities, to make strategic and tactical decisions and generate values for the stakeholders. According to Handy [4], knowledge is the principal assets that reside in the heads and hands of the people in the organizations. Knowledge comprises strategy, practice, method, or approach. It has become the most important factor of production in contemporary social and economic life. The knowledge-based view of the firm "can yield insights beyond the production function and resource-based theories of the firm by creating a new view of the firm as dynamic, evolving, quasi-autonomous system of knowledge production and application"[10]. Today, knowledge per se is not the power but the ability to deploy and use knowledge for the welfare of the human system is recognized as power [11]. Bontis [12] is of the similar opinion that knowledge and power are correlated very strongly in a modern and change-embracing organization. The scarcest resources in any organization are performing people endowed with knowledge. Knowledge within the organization is living, developmental and synergistic, if it is applied and internalized in organizational activity [11]. Knowledge management caters to the critical issues of organizational adaptation, survival and competence in face of increasingly discontinuous environmental change. Essentially, it embodies organizational processes that seek synergistic combination of data and information processing capacity of information technologies, and the creative and innovative capacity of human beings. Sarvary [13] is of the opinion that knowledge management is a business process through which firms create and use their institutional or collective knowledge. Knowledge has its greatest value when it is transparent and transferable: powerful assets to amplify our very latent capacity to learn, create, and innovate.

3 REVIEWING THE WORK SYSTEMS

Work systems are the bedrock of any human endeavor. As the essence of a corporation's philosophy for achieving success they provide a sense of unifying direction for all employees and common guidelines for their day-to-day behavior. They are the building blocks exist for the sake of the results and have to organize resources to attain the results. They are the primary organs capable of producing results outside of themselves. The emergence of this new form of work organization is fully displaying humanity in our work as an important requisite for achieving the mission of work. This new form is displaying a set of new dimensions which are basically be interpreted as activators of human energy (physical, mental, spiritual) and collective momentum. Ettorre [14] with an interview with Charles Handy has elaborated on the future of work and an end to the century of the organization. Along with the form of work systems, there is also a shift in the thinking process of the workers [15]. Nishibori [16] mentioned that human work should always include the following three elements:
- *Creativity*
- *Sociality*
- *Physical activity*

More fundamentally, a few companies across the globe have reviewed the basics of work. Thus, following its experiment at its Kalmar, Sweden plant in the early 1970s, Volvo now is returning automobile production to the handicraft age at its plant in Uddevalla-much to the skepticism of other Swedish enterprises. A few examples of the fundamental revisionism; Work is in teams which instead of working under foremen, choose rotating coordinators. Teams are a microcosm of local society, with respect to both age and sex. Team workers participate in an initial 16-week training course, the first stage in a 16-months program. Special hand tools, ergonomically designed for women, have been introduced, and each worker has his/her own personal set of tools. There is a very low decibel noise, sophisticated ventilation, lack of dirt and smells, use of natural light and stress-free color designs on the walls.

The Norwegian industrial democracy experiments of the 1960s and the Volvo Kalmar plant of the 1970s indicated the limitations of using unique experiments as a means of generally disseminating new concepts or organizational development. The Germans also started humanization of work program and were unable to replicate "kalmar" in Volkswagen, Moreover, sifting out the key features of best performance or, more generally, of excellence, inevitably causes at least mild derision when the mighty stumble or fall. On the other hand, building up discrete clubs of companies with similar concerns and motivations to change appears to be an effective means for disseminating such experimental innovations. Examples are many, such as the Belgian "Laboratoire" and the Klubs of the Danish Management Center. Furthermore, the importance of stable sub-contracting relationships (i.e. not just based on the cheapest short-term prices) has become more important under pressure of Japanese examples.

As we approach the beginning of the new millennium, we need to make a very fundamental review of our contemporary work systems and to pay as much attention to *the means of work as to its end.* This calls for a shift in emphasis from:
➢ *Outer resources to inner resources i.e. knowledge and intuitive wisdom.*
➢ *Maximizing profit to maximizing organizational and human capital*

Traditionally, many companies take their business conditions and their associated work systems as given and set their strategies accordingly. Over the last few years, companies around the world are looking for phenomenal growth in performance through redesign of work systems and strategic value innovation. Competitiveness of an economy today rests on its business premises such as; *geographical spread, scope of operations, ownership and control, factor conditions and demand conditions etc.,* which can be shaped by restructuring and reorganization. Industrial age principles articulated by Karl Max-resting on the control of the means of production and scientific management principles developed by Taylor-centering on industrial efficiency are now undergoing drastic revisions. Business today is a social practice calls for collective concerns and undertaking. Factor conditions around the work systems are being shaped in pursuit of multiple objectives aiming at transforming environmental circumstances in which the human system is engaged not only in the production of goods and services but also in the search for new knowledge and design and development of knowledge based work systems [17]. Traditional concepts of ownership and control, which creates managerial hegemony are moving towards social cohesion-creating corporate community; a new form of corporate governance. Ecological interface of business is shifting from unbounded growth to intelligent growth that is both productive and ecologically benign.

4. CO-EVOLVING STYLES OF MANAGING HUMAN RESOURCE

Industrial society has traditionally drawn a distinction between the workers-on the one hand, and first craftsmen and artisans, and then the professionals, on the other. The work system/enterprise is the key unit of industrial society. Within enterprises, the "doing" masses-the unskilled workers implementing orders-were separated from the "thinking" decision-takers; managers and professionals and their surrogates, foremen, and supervisors. However, since the middle of this century this binary division and distinction between thinkers and doers have shown signs of wear. Now it has shown the trend of disappearance under social pressures (for example, greater education, a more democratic society, and higher quality of life) and economic realities-knowledge-intensive industry with the pressure of the shortening life cycles of goods and services compounded by the process of corporate globalization. And one of the most significant aspects of this internationalization is the arrival and absorption of information technology and system. Summarily, the era of managing by dictate is being replaced by an emerging era of managing by knowledge and inspirational learning and the work system is now a space filled by many ideas generic from the pluralistic stakeholders.

The conflict of the binary "thinkers-doers" has also been expressed in the past. Organizations have attempted to resolve such historic conflicts through *"participative"* approaches to management. But participative approaches have been severely objected because of [18]:

➤ *Control will be lost*
➤ *Decision making will take too long*
➤ *Group-think will reduce quality and efficiency*
➤ *Individuality will be lost*
➤ *Rights and responsibilities will not be in balance*
➤ *Focus on performance will be lost*
➤ *Managerial authority will be lost*

However, participative approaches are differing from those of paternalistic companies in the past in that they will be based essentially on economic and technological necessities, and the scale of their introduction will be larger and more rapid in the 21st century. Only time will tell if the current participative approaches are really the manifestation of changing attitudes or only of adaptive behaviors, which, under different circumstances, could revert to previous patterns. But it is an imperative to understand that participation to become successful requires the coordinated development and deployment of *knowledge base* within and between enterprises. Participation is a cognitive process by which, individuals form, structure, and articulate their ideas and cooperate to inscribe those ideas into a body of knowledge. It has a richer social and psychological dimension. It is an organic transformational (value adding) process, wherein the knowledge as a sustainable resource is shared between individuals and collective minds and evolves over time. Participation in the context of knowledge building has three sub-processes: organizational learning, knowledge production, and distribution. Nonaka [19] termed this as people-embodied knowledge. Participation is a human construct that dictates the design, direction and outcome of a worksystem.Lane [20] mentioned that the strengths of German manufacturing enterprises are widely seen to emanate from two core institutional complexes- the system of education and training and the system of participative human relations style of co-determination. Miyai [21] expressed that innovative ability is the sole natural wealth of Japanese human resources developed through group consciousness. Donegan [22] in his study reported that, British Petroleum realized an incredible $260 million benefit by focusing on the building of a climate of learning and participation- taught people how to bring difficulties to the table and ask for collaboration. The organization supported people by helping them to move across the boundaries, but helping them to talk to anybody they felt comfortable with. Lowendahl and Haanes [23] presented the case of Alcatel Telecom Norway, which is a very relevant example for competence leveraging through *the unit of activity framework;* wherein people participate in problem solving and continuous progression in knowledge building. Bernstein [24] mentions that to create an environment propitious for change and learning requires actionable information. This process requires visibility, transparency, and universal access to information to bind corporate resources together. He terms this type of company as *Visual Organization*. Maccoby [25] refers to AT&T's

workplace of the future. Here, the doers and the operational leaders together attend courses where they learn about stakeholder needs and values. In 1998, Novamad used *multi business unit team* concept of knowledge sharing to build the critical mass of doctors and patients to increase the health care business [26]. Price Waterhouse Cooper has created knowledge centers that act as focal point for knowledge exchange and provide a repository of best practices for different business processes.

Every company is vitally concerned with its performance; this implies that to improve performance, participation should be inherent in its corporate culture. In the competitive era, performance improvement means creation of value. The process of creating value from resources is based on the interactions of people and depends upon the level and kind of knowledge base. This is because participation in bringing together the existing competence and creating new knowledge provides a strategic focus so that everyone associated with the company understands and works towards the same objective. Norman and Ramirez [27] are of the opinion that one of the chief strategic challenges of the new economy is to integrate knowledge and relationships. Participation in reality is a value creating chain within and between organizations, which connects collective knowledge to deployable actions and finally to collective value creation [28]. Thus, participation is the foundation of best practices in knowledge building. This must surely be the real challenge to all organizations in the new millenium. Wenger and Snyder [29] reemphasize that a new organizational form is emerging in companies that run on knowledge and term it as *the communities of practice*. They are groups of people such as; cross-functional teams customer-or product focused business units, and work groups - which capture and spread ideas and know-how, galvanize learning to develop members' capabilities; to build and exchange knowledge and change. They emphasize that communities of practice add value to organization in several important ways:

➢ *They help drive strategy*
➢ *They start new lines of business*
➢ *They solve problems quickly*
➢ *They transfer best practices*
➢ *They develop professional skills*
➢ *They help companies recruit and retain talents,*

For example, Storck and Hill [30] have elaborated on knowledge diffusion though startegic communities in case of Xerox. Tiessen [31] has narrated an *epistemic community perspective* for developing intellectual capital globally. What is observed is that clusters of organizations of all kinds are building knowledge base through strategic collaboration and communities of practice on global and national fronts, but the modalities of knowledge building are differing from organization to organization and from nation to nation.

In the following section, we intend to discuss the larger perspective relating to institutional vehicles for knowledge building.

5. INSTITUTIONAL MODALITIES OF BUILDING KNOWLEDGE BASE

Corporation all over the globe are becoming increasingly explicit and conscious of the process and modalities by which knowledge is created, identified developed, accumulated, shared and above all, applied. However, management is aware of the fact that the knowledge building is inextricably linked to the organizations internal context characterized by core competency, skills inventory, training and development practices, learning curve etc. Knowledge building in a corporation is driven by the strategic aspirations. A first step to that end involves creating a deeply and widely shared knowledge vision within and throughout the extended corporation. Knowledge building i.e. creation, development, maintenance, and

deployment is a systemic process. It involves three kinds of understanding i.e. know-how (the state of knowledge), know-why (the process of knowledge), know-what (the purpose of knowledge). Know-how is learning by doing. Know-why is theoretically directed learning by doing. Know-what is strategic understanding i.e. learning both top down and bottom up. Buckler [32] explains these three kinds of understanding in the context of a learning organization. Knowledge building is the foundation of the concepts of skills, capabilities, and competence [33]. Knowledge base has 3 dimension:
- Residing knowledge either in individualistic terms or collective social terms
- Knowledge sourcing either internally focussed or externally focussed
- Knowledge dissemination either through informal encounters or in more structures ways.

According to Nonaka and Takeuchi [34] the five phases of knowledge creation process include:
- *Sharing of tacit knowledge* – correlated closely to the socialization mode of knowledge conversion
- *Creating concepts* – involves the conversion of the shared tacit knowledge into explicit knowledge
- *Justifying concepts* – is an internal verification mechanism
- *Building an archetype* – is a form of rapid prototyping, this can either be a 'hard' product development or a 'soft' organizational entity; various forms of explicit knowledge are combined in this phase
- *Cross-leveling knowledge* – ensures a wide exchange of knowledge both within the organization and in the exchange with its external environment.

Knowledge transformation process has four distinct phases [34]:
- *Socialization*
- *Externalization*
- *Internalization*
- *Combination*

To launch and bolster the drive for more knowledge, dynamic enterprises do engage in the process of learning, which changes the state of knowledge of individuals or organizations (the knowledge base); and do adopt one of the following three modalities, which we may term here as participative styles of knowledge building:

5.1 Training

Chronologically, the first strategy for developing skills, competence and capability has been to train. Training is envisaged as the primary approach for competence progression and takes place at three distinct levels: (1) individual; (2) groups; and (3) organizations. Training is a process of transferring skills and enabling trainees to act, of building on their strengths. It is reinforcing on other people to make easy the way of carrying out tasks by giving them enough time and room to practice. Developed countries have perpetually accorded top priority to this strategy and have developed national training systems and institutes of national importance. These countries have all the times worked on a single mission to make the succeeding generations more competent than the present generation. The focus of training is-not just knowing the analytical idiosyncrancies of the trade or vocation or profession, but also involves strategic understanding of processes and values. Training helps in acquiring explicit knowledge, codified knowledge, and experiential knowledge. Because of the limitations of Indian training systems, and lack of concern by the Governments-both national and states, companies are often constrained first, to develop their own training systems at all levels of corporate competence and, second, to develop joint ventures with parts of the national educational system. Even if some educational institutes exist at the national and state levels, the competence of such institutes needs to be audited and evaluated whether they are really capable of competence progression in the corporations of the 21st century.

In the case of Indian companies developing their own training systems, one can possibly follow the German, "dual system", which provides the classical approach. This system is being adapted for "apprenticeship" training by other community countries, for it combines off-the job teaching with on-the-job practice and learning. Pre-requisites for its successful implementation are; industrial attitudes must be positive, and the training must be relevant for future employment. (In contrast, the British Youth Training Scheme has been criticized for being a "make-training" scheme along the lines of "make-work" schemes).

Training within enterprises needs to be both economical as well as knowledge driven. Belgian government provides a checklist of aspects, which have to be carefully studied to ascertain the economic return on investment in training in the short, medium and long term. Included on the list; the costs of the training service (personnel, overhead, development); production of off-the-job training; participation in external courses; production of on-the-job training; and cost of the time of trainees.

To be really effective, top management must demonstrate that it truly considers training is important for the future of the company. For example, GE's Crontonville institute is a staging ground for corporate revolution [35]. Particularly companies such as international airlines, which have mounted broad "quality campaigns", have showed this intended for the whole of the work force. A few companies in situations of having to restructure rapidly because of changing markets and technologies also have undertaken very large-scale retraining activities. Recognition of the importance of training has increased over the recent years too, because it is included in the educational baggage of top management, particularly of business school graduates. Some corporations in USA have initiated corporate universities. Corporate university differs from a training department in several ways. A training department tends to be decentralized, reactive, and targeted primarily to instructing internal employees in job skills. A corporate university is the centralized strategic umbrella for the education and development of employees and value chain members such as customers, suppliers, and dealers. Most importantly, a corporate university is the chief vehicle for disseminating an organization's culture and fostering the development of not only job skills, but also such core workplace skills as learning-to-learn, leadership, creative thinking, and problem solving.

Outside the individual enterprise, mass training can also be an effective tool for raising the general level of competence. In Ireland and Sweden, for example, the trade unions and employers' organization have broad-based "economic and financial awareness" programs. These are typically 40-hours packages designed jointly, but usually implemented separately by trade unions; they are intended to give the entire work force an understanding of the balance sheet, profit and loss accounts, etc. Once run, such programs can only show positive results if enterprises really do adopt participative styles and disclose figures and trends honestly and before news is leaked to the media.

A third aspect of broad based training program as a means of raising corporate competence concerns distance learning. Several countries have launched Open University training in management. Though intended for persons and geographical areas outside the normal university catchment areas, it is used to a very large extent by the better-educated in metropolitan areas. Indira Gandhi National Open University is an example in India making an attempt to provide developmental opportunities to many professionals. Professional societies and associations have also entered into imparting such training. The issue is again the *quality* of training and *intensiveness* of the scope and dimensions of training above all the *means* adopted to impart meaningful training to cope up with the changing demands of the competitive landscape of the new millenium.

The development of a set of new competencies through training must explicitly recognize the holistic role of individuals, groups, and the organizations and above all the processes adopted by which, scientific knowledge, creative and innovative behaviors of individuals are transformed into collective learning and shared across organizational boundaries. It should be noted here that according to a recent survey conducted by the University of Pennsylvania, companies that invest 10% more in education see an 8% increase in productivity. However, 10% increase in capital expenditures boosts productivity only by 3% [36]. A similar view has been expressed earlier by Stewart [37] with reference to a study carried out by National Center on the Educational Quality of the Workforce. Therefore, there is a greater need for influencing the broader system.

5.2 Influencing the broader system

Training only within the corporate system as a means of raising corporate competence as well as human resource capability is indeed insufficient; wider systems are necessary. But enterprises have become increasingly ill at ease with national educational systems overall and the general schooling system in particular. There are three manifestations of this trend:

➢ *Industry and enterprise representatives are assuming key governing positions in an increasingly privatized vocational training system.*

➢ *Enterprises in developed countries are entering into new working relationships with schools, particularly Secondary schools. Under specified circumstances, they guarantee future jobs. Although they usually focus on the average and more gifted students, because of demographic trends they also are starting to pay attention to under-privileged school children. Thus in the UK, the Foundation for Educational Business Partnerships was set up in mid-1989 to stimulate change in enterprise-schools relationships. It particularly aims to instill more drive in the lower achievers by supplanting traditional teaching approaches of traditional subjects by more active learning around more directly relevant themes. Although, Indian political parties speak of social justice, our social justice is confined to job reservations for backward communities' only-not for developing their competencies and capabilities and thus empowering through knowledge base. Political-administrative systems in India are fundamentally responsible for the current chaotic situations.*

➢ *An enterprises learns to prize the formally taught competence of their professionals and managers by providing time-off and financial incentives for studies, managers in turn have increasingly demanded academic recognition of their new knowledge. This has led to a spate of enterprise-university links in many countries. Do we have such plans? Our enterprises are short-term oriented, purely commercial and inward looking. Our universities and other institutions both at state level and national level do not have futuristic planning rather have already diluted the quality of education. Today, many institutions have lost their credible status and are just sustaining by mere government subsidies and some sponsorship. There exist no remarkable or profound knowledge base in the real sense. Research is negligible and almost absent to create knowledge base for the future generation. Many institutions even do not have infrastructure and faculty competence. In the final analysis, it is the quality of institutions that determine the future of the nation. In the age of competition, the present and future will belong to those who can shape broader educational systems.*

5.3 Corporate human resource management

Human resources management of late has become an "in" concept in some Indian organizations. Its use is totally cosmetic and merely change of name in the structure. Vary

rarely, the strategy follows the structure. Alignment is a very long cherished intent. But where HRM is seriously practiced it consists of bundles of critical elements, in particular:
➢ **Human resource planning**
➢ **Human resource development**
➢ **Human resource relationship enhancement**
➢ **Alignment of vision and values**
➢ **Human resource performance feedback**

At the total enterprise level, HRM has to become an integral part of corporate strategy. It should identify all strategic pathways for the optimal use of human resources. Of particular significance in such integration into corporate strategy is the real and visible involvement of top and middle management. Information and communication are becoming indispensable for motivation and commitment. To this end, many progressive companies, for instance, have adopted a multi-pronged policy of limiting the size of manufacturing plants to 250-300 persons; reducing the number of supervisory levels and staff functions; rotating managers and limiting their stays in particular jobs to five years; encouraging all managers to spend part of their career abroad; and making foreign experience mandatory for all vice presidents. Developing competence throughout the total enterprise is particularly fruitful for multi-national corporations. Honeywell Europe, for example, has set up, under its Single Market Coordination Council, international task forces in 11 potential areas of activity important for human resource planning and development. Philips, too, has established international policy councils for planning and development within its product divisions. McKinsey & Co., which provides management consulting services to companies, has a unique process for training its management consultants. Procter and Gamble has a reputation as an academy producing high quality brand managers. These companies have recognized that their profitability critically depends on their internal human resource management processes that have been difficult for the competitors to imitate. All over the globe, maximizing human capital formation is the real time strategic intent. For example, World Bank in the year 1999 made significant investments in retooling the Bank's knowledge base and revamping institutional capabilities through human resource management. HRM in the 21st century will be engaged in creating an energizing culture through continuous investment in:
➢ **Human resource planning and development,**
➢ **Recognizing contributions with fairness and equity, and**
➢ **Providing high performers both dignity and security.**

The author believes that the human resource function in India has a very specific mandate in order to build the knowledge-based corporations of the future. Therefore, HRM function must gain in stature and significance, move from presently undertaking transactional role to a more value-adding business partner and as the necessary catalyst.

The impression should not be gained that these three modalities of knowledge building are mutually exclusive. Rather, they represent continuums, with certain ones being preferable in some types of enterprises and industries. What is important everywhere is that there is a dynamics in each enterprise, which requires critically reviewing, and seeking to improve its own system. But, the basics of work must be related to people's knowledge and skills. There are differences in basic skills, in motivation, in expected reward systems, in culture, and in experiences, which must be taken into account. Reskilling the corporations is to be undertaken for chasing dreams and to be driven by the energy of aspirations and therefore must be proactive. The alternative to reskilling is regression, obsolescence and finally decay and extinction. Corporations of the new millenium will become self-defeating inhibiting their performance-which will be the most dangerous consequence, if they destroy the system of knowledge base that gives the power to compete and excel.

6. SOME STRATEGIC DIRECTIONS FOR INDIAN CORPORATIONS

The importance of 21st century for Indian corporations is necessarily to be over-stated here. Generally, the pressures of and for globalization have been apparent for a decade. Meeting the challenges of globalization automatically requires meeting the demands of the next millenium, which might be more demanding because the major competitive pressures are from those multinationals with headquarters outside India. On the other hand, it is of major significance to the vast majority of enterprise, which will be experiencing more competition in their traditional markets. These firms are primarily small-to-medium size enterprises typically serving local markets with local (often-family) management without specific management training and development. Indian organizations are currently faced with a paradox: that the knowledge developed is the generic source of competitive advantage, but according to Covey [38], the value of knowledge is susceptible to decline over time. Furthermore, although the maintenance and the enhancement of knowledge base are a condition to success in the short term, but the continuation is a real threat in the long term. The capability of Indian corporations to integrate knowledge will play a fundamental role in the transformation of work systems into competitive advantages. However, this integrative capability must also play a role in a changing environment by improving the knowledge base, which is vital for the development and growth. In view of this, it is necessary here to make a strategic role profiling for Indian corporations.

6.1 Implications for human resource management

The implications for HRM are both profound and far-reaching. For instance, although there will continue to be places for all types of career models, in most industries only those career systems emphasizing continuous development and adaptation will survive the dawn of the flexible, process-orientated organization. Moreover, in a turbulent and ever-changing business world, replacing the recruitment and training systems that have provided organizations with qualified people for many decades is quickly becoming a necessity, not an option. As we have realized that more than ever the world is in a flux. And therefore, organizations and their managers must recognize the necessity of developing the mindsets, skills and abilities that will allow them to cope with the flux.

It is essential to note here that *the success of an organization is a multiplicative function of human resources' competence and commitment.* Competence is manifested in our *abilities to do work* and commitment is our intrinsic *willingness to do work.* Increasingly, organizations will need to find systems and practices that promote entrepreneurship and learning. Managers and professionals, in turn, will have to learn how to use the vast amounts of data which have come available with the information technology revolution, and learn to live with the complexity and ambiguity created by the competitive forces buffeting organizations today. So how can Indian organizations help people prepare for the future? What are the implications for the management of human talent and resources? Once, again, we may not have all the answers, but a number of suggestions can be made to accelerate some of the changes needed in organizations. Therefore, we suggest some contemporary imperatives for HRM, They are:

- *enriching people*
- *cooperating to enhance competitiveness*
- *organizing to master change and uncertainty*
- *leveraging the impact of people and information in the business processes*

All these call for a new theory incorporating developmental principles.

6.2 Initiating mentorship development and making it effective

Mentoring is an important role for organizations for professional development in many countries. Its underlying principle is that a more knowledgeable colleague can facilitate the professional development of a new employee. Bush and Coleman [39] describe mentorship is a relationship building mechanism and has the potential to enhance the knowledge base of both individuals. Mentoring has always been present in the business environment, usually to help all employees to learn new skills [40,41]. This is especially true in the new millenium, because one can expect the skills one has to be obsolete in three to five years [42]. These programs are even more necessary when our contemporary work systems are undergoing organic transformation. Many research studies [43,44,45&46] in the recent years have revealed the following benefits of mentorship development programs:

➢ *Helping newly hired employees or promoted employees become fully productive and understand the organization's future in a compressed time frame*
➢ *Creation of future entrepreneurial leaders*
➢ *Low cost transfer of skills*
➢ *Increased ability to manage participative relationship*
➢ *Increased learning potentials*
➢ *Positive affirmative action results*
➢ *Strengthened link between business strategy and developmental needs*

The author has designed and intervened in some mentorship development programs for a number of Indian companies. Of particular significance here to mention about The Associated Cement Companies Ltd., which is the largest cement producer and the market leader in India. The company believes that human resource development is a key to building knowledge base. Mentoring is the creation of a formal relationship between two people of different business processes and status in the company's cement manufacturing units. Some of the advantages that the program may claim are as follows:

➢ *Better adoption of the organizational values (this company is most respected in the society for it's high corporate ethics and values)*
➢ *Effective transfer and absorption of circumstantial and experiential knowledge*
➢ *Low cost but highly relevant learning and better cross-functional knowledge*
➢ *Cooperative development of knowledge*
➢ *Increased job satisfaction*
➢ *Low turnover of employees*
➢ *Meaningful career guidance*

The above findings are not subjective. The author has monitored the performance objectively for the last three years. It is worth mentioning here that the company's approach is in line with both the scientific evidence and with recent proponents of achieving competitive advantage through people [47]. However, it may be noted here that there are a number of difficulties faced in starting of a mentorship program, but its benefits are many provided that the focus is made on the areas of learning -informally.

6.3 Building competencies through innovative 'Practice Fields'

Many progressive organizations have emphasized on the need to give managers and employees more opportunities to practice the skills that are needed to perform well in the emerging business environment. It has been argued by many that classroom teaching and role-playing are necessary but not sufficient. Therefore, many researchers and practicing managers suggest that organizations create 'practice fields; that let managers and employees hone their skills and gain experience under realistic but risk-free conditions. The Productivity Enhancement Program at Bell Labs is a useful example. According to Cannon, the company asked a number of its star engineers to develop an expert model. The result was a set of nine prioritized work strategies the engineers believed other employees could master. Training sessions to pass on these strategies occur in the normal workday. Productivity increases in both star and average performers have been striking, from a 10 percent increase immediately after the sessions to 25 percent after a full year. A number of companies across the globe have adopted this approach. However, the most important *'product'* of this approach is managers who understand how to create a *learning environment* for those around them. *Action learning* has been a very successful approach in U.K. British Petroleum [22] calls it as *Learning Engine*-an elegant system that meant:

> *People and systems demonstrate learning before, during and after tasks.*
> *Communities of practice access, apply, validate and renew existing knowledge through performance histories and real time observation, both within and without their own organization.*

This author has been a pioneer in initiating *action research and exploratory projects* in some Indian companies. The experiences are very encouraging in terms of knowledge acquisition, deployment, and utilization for different companies. These projects have helped the attainment of mastery of some knowledge, and building a better and better fit between relationships and skills transferring by reconfiguring roles and structures. An organization's processes for articulating, codifying, and transferring knowledge within are important determinants of its ability to leverage its existing knowledge effectively- and thus of its ability to leverage its competence to greatest strategic effect. The ability of some companies to survive and thrive in the future hinges more on an optimal management of skills through participation than on the implementation of new technologies and manufacturing processes. Moreover, these companies saw that the new technological breakthroughs could not be integrated unless their staffs were able to adapt to ever-quicker cycles of change and their organizations able to cut the cost brought about by this unceasing need for human resource adaptation.

Companies that have enjoyed enduring success during the last several years have created learning organizations around people who have transformed business strategies and practices endlessly adapting to a changing world. If the core purpose of an organization is to remain in business in a competitive world, the organizational members collectively accomplish certain tasks, which ultimately should result in making a product, or service, which is of value to the human system. The basic dynamics of successful companies in the recent years has been in terms of decisions to build the strength of the organization and its people or in other words creating and nurturing a learning organization where people are capable and competent enough to make effective decisions perpetually. Without knowledgeable workers, a corporation will be at a competitive disadvantage.

6.4 Strengthening the articulation mechanisms

In any learning process, the following three phases can be identified [11]:

❖ *A priori articulations*: These articulations may be based on prior experience, historic data base and prior knowledge of the processes, as well as future projections about environmental and business trends. These articulations are oriented towards proposing conjectures about diversity.

❖ *Articulations during the process*: These are the articulations expressed during the process and they are the reflections mirrored to enhance interactions and promote interdependence.

❖ *Posteriori articulations*: These articulations are expressed based on additional knowledge gained and on performance feedback and intrinsic desire of an individual to discover potentialities and limitations.

At each phase, the individual or group may attempt to learn more and more. The process of learning is iterative and evolves dynamically. Commitment to learning may manifest in many ways: change in behavior, change in attitude, adapting to new values etc. The postulation here is that there are basically six generic and interactive forces that influence any business corporation to evolve into a learning organization. These are:

➢ *Customer power*
➢ *Information power*
➢ *Global investors power*
➢ *Global market power*
➢ *Power of simplicity*
➢ *Power of the organization*

The customer power, the predominant one stems from the fact that an organization has to perpetually learn/unlearn and relearns as dictated by the customer's choice, and his/her requirement. This power will compel an organization to move from bureaucratic mode to responsive mode and will necessitate it to be flexible, lean and yet be able to meet the customer demand to stay in the market.

The information power, (with advances in Communication and Information Technology) will help an organization to continuously update and upgrade on technology, information systems and be thus able to learn at a faster pace. Now it is possible to transfer volumes of data globally from one organization to another. The information power will enable to promote knowledge networking.

The power of global investors will affect the learning mode of the organization. Because of liberalization that is globally evident, an organization is stimulated to learn because now there are no boundaries for investments. The organization will continue to invest in its development by fostering global search for all resources.

The power of market place will generate fierce time based competition which will motivate an organization to learn faster to provide quality and value.

By power of simplicity, we mean streamlining of systems and procedures within the organization and moving away from bureaucratic culture to more towards autonomous structure. Because of this, the organization can quickly undertake reengineering/redesign of business processes and has to forge organic partnership with the multiple stakeholders to eliminate delays and bottlenecks.

The power of organization itself will rest in its ability to quickly transforming market opportunities into tangible bottom line results. Such a force will lead to recreate lean and agile organization structure and high performance teams. The power of organization will be manifested in making profits and growth.

In summary, these forces will compel an organization in the 21st century to transform through reskilling. Reskilling the corporations are not possible by irrational and adhoc principles; rather, very systematic and concerted efforts are necessary. In this connection, we would like to stress the importance of identifying different types of out side –in forces for the development of knowledge base and competence progression of the organization. The entire world class organizations are strengthening their knowledge articulation mechanisms by adhering to double -loop learning [48]. Double-loop learning refers to the theories of action that are to be governed by a set of imperative generic from the outside-in forces. Knowledge building in the organizations becomes successful when double-loop learning involves the individuals/groups to articulate and reconsider and revise their governing values to find out an improved solution to the situation being faced. Therefore, to strengthen the articulation mechanisms *making the realities visible and visibility of information* are most vital. We suggest here some propositions:

- *The organization is to be viewed as a human community capable of providing diverse meanings to information outputs generated by the various business processes, instead of the traditional emphasis on command and control.*
- *Business as usual approach is to be de-emphasized, so that such prevailing practices may be continuously assessed from multiple perspectives for their alignment with the dynamically changing generic and interactive forces.*
- *Diverse viewpoints have to be encouraged by avoiding premature consensus on issues that need deeper analysis of underlying assumptions. Often, viewpoints of persons with differing backgrounds and expertise can provide a much broader focus that is essential for completely grasping the essence of the core issues, particularly when the changing context demands a fresh look at what was yesterday defined as a "benchmark" or a "best practice."*
- *Greater proactive involvement of human imagination and creativity be encouraged to facilitate greater internal diversity to match the variety and complexity of the forces.*
- *More explicit recognition to tacit knowledge and related human aspects be given such as ideals, values, or emotions, for developing a richer conceptualization of knowledge articulation.*
- *Attempt should be made to implement new, flexible technologies and systems that support and enable communities of practice, informal and semi-informal networks of internal employees and external individuals based on shared concerns and interests.*
- *Organizational information base be made accessible to organization members who are closer to the action, while simultaneously ensuring that they have the skills and authority to execute decisive responses to changing conditions.*

The author has been able to derive these propositions from his close observations and learning from the best practices of many successful global consulting firms. These firms view the implementation of these issues in terms of the shift from the traditional emphasis on transaction processing, integrated logistics, and work flows to systems that support competencies for communication building, people networks, and on-the-job learning. For example, McKinsey &Co adopts a three level architecture needed for enabling such competencies:

- *A new information architecture that includes new languages, categories, and metaphors for identifying and accounting for skills and competencies.*
- *A new technical architecture that is more social, transparent, open, flexible, and respectful of the individual users.*
- *A new application architecture oriented toward problem-solving and representation, rather than output and transactions.*

On a similar note, Bob Hiebeler, Arthur Andersen's managing director of Knowledge pace intranet observed at a recent panel discussion of knowledge management experts: "To me, this is the essence of knowledge sharing. It's all about contribution, it's all about the respect for others' opinions and views, it's all about a good facilitation and synthesis process, it's all about the distribution of lessons learned from this knowledge process, and it's all about access to packaged knowledge and key insights that become the starting points for individual learning."

Managers need to develop a greater appreciation for their intangible human assets, captive in the minds and experiences of their knowledge workers. Without these assets, companies are simply not equipped with a vision to foresee or to imagine the future and able to articulate strategic directions.

6.5 Identifying and activating the ageing workforce

With the projection of middle-aged employees comprising a larger part of the workforce in most industrialized countries, and with work becoming more unstable and demanding, organizations have good reason to be interested in preparing people for the future. Indian organizations should prepare now for the inevitable frustrations of career stagnation in the middle years. Already there are signs that the number of individuals who experience career entrenchment is increasing dramatically. The author conducted a survey of over one thousand middle-aged men in managerial and professional positions and found that five out of every six respondents endured a period of severe frustration and trauma that began in their early 40s. Work performance, emotional stability and physical health were seriously affected. Some people would like to change careers, but wanting to change, of course is not the same as doing it. Many organizations are burdened with workers who do not want to learn and at the same time aspire to jump ship, but who stay firmly on board grasping for long-term security in the face of widespread job cuts. Out of desperation, many employees stay with the organizations in which their careers have unfolded, but do not stay committed to them in the way management would like. Candidates for second careers tend to be in their mid 40s and report a perceived discrepancy between personal aspirations and current opportunities for achievement and promotion. This group is likely to become larger as the opportunity for advancement decreases, resulting in more career frustration and entrenchment among middle managers. There is a need for a *serious dialogue and resolution* on this critical issue. More we globalize, this issue will be more predominant. There is as such a vast number of unemployed youths.

6.6 Facilitating career mobility and change

Some MNCs are already attempting innovative solutions to increase job mobility between and within their organizations. Cable and Wireless, for example, has set up what it calls career action centers to help people make inter-company moves and to encourage a *'contract mentality'*, where employees think of their work in terms of a series of projects rather than as a life-long career. Furthermore, to minimize some of the potentially adverse consequences of career entrenchment, organizations can take a number of approaches, including the following:

> *Providing on-going career counseling, mentoring and outplacement assistance to all employees, not just those who are made redundant.*
> *Offering training, time-off and financial help to those who want to attend to improve their skills and competence, even if these skills are not highly organizational or career specific.*
> *Allowing employees to make career changes within the organization, rather than forcing them to stay within their (functional) career ladder.*
> *Encouraging employees to think about career planning issues and not making them feel guilty or disloyal as they explore new career options.*
> *Allowing employees who wish to change careers to leave in good standing and to return if they do not succeed.*
> *Providing portability of benefits such as pension plans, health plans and other accumulated forms of compensation.*

Each of these methods is aimed at reducing the progression of career entrenchment and at encouraging people to take more responsibility for their self - development and career planning. They may result in some turnover, as some individuals may recognize the need to abandon their current occupational paths and explore new ones. However, as we enter into the 21st century, the cost of turnover is likely to be less concerning than the issues of career stagnation and entrenchment. The author has designed multiple careers profiling system for some companies, and the results are yet to be seen.

6.7 Promoting individual growth and enabling development through real participative teamwork

Increasingly, managers and professionals face complex situations in which they must rely on others to get the work done. Organizational restructuring and delayering has also produced ill-defined roles requiring people to seek the support of others in the hope of attaining enhanced levels of productivity and performance. Further workplaces may require managers to rely on their peers and subordinates for their rewards, recognition, appraisal and training. Moreover, employees in the 21st century may periodically have to backtrack their own careers, moving from expert back to novice, as they required developing new competencies. Many researchers suggest that adaptation to these changes and movement into unfamiliar roles may take place more smoothly within a supportive team environment. To summarize some of the research findings: 'in a team model, the responsibility for career development is shared among the individual employees, the team and the organization. Individuals continue to assume primary responsibility for career planning, career goal setting, education and training. Organizations provide job-related training, an environment in which growth and development are valued, and human resource systems supportive of career development. Teams acquire the roles of supervisors, and help individuals by providing feedback on skills, identifying opportunities for growth and development, coaching and mentoring, and serving as training grounds for the acquisition of new skills and knowledge areas'. Hence, as organizations evolve to become more flexible, a compelling case can be made for team-oriented career development systems. In any case, the challenges of the 21st century call for innovative solutions that can complement the existing methods of performance evaluation, compensation, training, and life long learning.

6.8 Role of HRM in implementing the new mandates

Preparing people for the future requires systems and procedures to align individual and organizational objectives, communicate and consult with managers and professionals, develop them effectively, assess their potential and performance, give them feedback and help them plan and manage their careers. Each of these activities has traditionally not been the responsibility of the HR department in most companies in India. Despite all the talk about strategic human resource management, however, most HR managers remain stuck in their administrative roles processing complaints and paperwork. Indeed, one of the most worrying findings in studies of HRM is how few HR managers put themselves at the forefront of developing human talent in their companies. For instance, this author has asked many HR managers about their priorities in the 21st century, only a very small number of such managers, gave top priority to improving the quality of their company's workforce or rated employee training and management development as central thrusts of their function. What is more, 50per cent of the HR managers reported that either they have no major responsibility for meeting their company strategic objectives or they simply are not sure where they fit in. Conclusively, therefore, that very few human resource departments are true partners with line managers in running the business. In other words, there is no alignment between corporate mission and human resource development objectives. The author's experience as a top management professional is that top managers in some good professionally managed corporations are becoming more seriously interested and attentive to the issues and challenges of HRM. At the same time, current economic conditions, along with downsizing, restructuring, globalization and international competition mean that most organizations are preoccupied with cost reductions and increasing the financial performance of the organization. It remains to be seen, therefore, what form the increased interest in human resource development will take in the coming years.

From where do we start?
➢ *Developing both the understanding of the need for change and the willingness to do*
➢ *Acquiring the minimum capability required to learn new skills, behaviors and relationships*
➢ *Preparing some action plans however rudimentary they may be*
➢ *Taking action*
➢ *Responding to the reinforcement that follows action*

All these efforts require much dedication and a strong motivating force. HRM Professionals can perform an important role by becoming more aware of opportunities for common action, taking initiatives to bring that action about, and developing the critical skills to do so effectively. We may call these professionals *as the critical mass of collective interest and an institutional representation.*

6.9 Role of higher education and research institutions

It is a known proposition that the experience and scope of approaches to linking the broader learning systems with enterprises are rich indeed, particularly since so much real mutual learning can be developed. Necessity rather than fashion have brought about the emerging pattern of closer cooperation between the worlds of "action" and "reflection". Universities and independent research establishments had their budgets pruned throughout the 1990s and are constantly searching for other sources of income, notably from enterprises. But increasingly universities' staff are finding that not only can they sell their knowledge to the outside world but, in so doing, they have to increase their knowledge; and increasingly at a faster pace such that enterprises will be finding it profitable to utilize them. This is true in case of most developed countries. For example, Jouan S.A.A small hi-tech enterprise in the west of France has achieved 80% of its market niche of blood centrifuges by tapping the local technological institute's local, national and European networks. Two recent examples of such cooperative initiatives are, at the undergraduate level, the four-year courses of the Middlesex Business School and Cesem-Mediterranee at Marseille and, at post-experience level, the joint venture of Ashridge Management College (UK), CPA (France) USW (Germany). Jointness can also be in the form of a single institution with trunks (not branches from a main trunk) in several countries: For example, EAP, the European Business School, has its establishments in France, Germany, and the UK. Thailand 's Asian Institute of Technology (where the author has worked as a visiting Professor) has provided collaborative research and continuing education support to all South East Asian countries. The *"shared learning"* approach (pioneered by the Irish Productivity Center) in essence twins similar types of enterprises in dissimilar industries in different countries. Participants generate cross-sectional teams to present issues from their own enterprise and try to help solve those of the twinned enterprise.

It may be pointed out here with an emphasis that role of the institutes of higher learnings should not be merely the knowledge communication, rather the knowledge building involves giving people the chance to explore their subjective positions i.e. creating a trajectory for continuous cultural change. What change-makers need to do is to increase their awareness about the six generic forces identified earlier in respect of their own institutions, to find out more elaborately where they position themselves and search for ideas, meanings, standards, practices structures and above all resources.

We suggest here a much broader role for institutes of higher learning and the vital responsibility towards undertaking *"innovation capability audit"* for organizations. And based on such continual audits the academic curricula should be restructured and faculty resources be developed. The philosophy of such an audit is not manipulation but education for progress. Powerful global forces for change- such as liberalization of markets, communication technology, and the integration of the world's economies- have inescapable implications for business as well as academia, and demand rapid and innovative responses. The new imperatives of business world compel managerial/academic practices to shift from *business-as-usual to value innovations* [2]. It must be mentioned here that value is subject to constant pressure for change. It is no longer sufficient to fix it once and expect the value dimensions to remain stable over time. The challenge in the real time is to create performance break points in the value metrics geared to expecting the unexpected – one that can take speedy advantage of new patterns of demand, new markets, and new ways of service while still coping with the expectations of various stakeholders. The intent and purpose of an educational institute must be to provide a *cultural leadership* i.e. to upgrade the whole intellectual potentials of society, and to pattern socially active individuals in order to create, express and communicate new ideas and ideals and undertake design, operation and maintenance of systems of value innovations, which can help corporations to deploy such innovations and recreate new demands. The tasks before the institutions are designing curricula that can allow participants to utilize their intellectual abilities, expertise and experience more effectively. But, do our institutes need rejuvenation? We propose here the following:

➢ *Educational Institutions require a coaliational, multiple stakeholder change model.*
➢ *Institutions must adopt the new philosophy of competence building and create a missionary approach to implement the new philosophy with urgency.*

6.10 Role of top management

Unions are striving for legislation prescribing how participation in enterprises should be organized. But the top management has to become the driving force in *corporate social innovation*. For instance, in The Netherlands, a "management-labor new style" grouping of half a dozen leading companies was founded in 1983 on the participative management concepts of Juran and Deming. In northern Europe; labor-management agreements are putting equal stress on "participation" and "productivity". Their ultimate purpose is to make better use of *"person-power"* by encouraging local level developments, rather then designing national models. The success of such private agreements contrasts with the relative failure of national agreements in the early 1980s on the introduction of new technologies. The lack of impact of the latter can be explained by their striving to lay down general rules rather than broad guidelines, which are adaptable to local situations. Olivetti management has introduced a plethora of approaches to HRM, particularly in the tapping of corporate person-power. Requalification and acquisition of talent and expertise have become essential. The new demands under which Olivetti's "competence drive" is conducted are:

➢ *Rigorous selectiveness of investment in both new personnel and existing staff.*
➢ *Introduction of sophisticated techniques, such as matrix management and multi-dimensional career plans.*
➢ *Radical transformation of management style with delegation, participation, intrapreneurship, etc.*

➤ *Changes in the structures and organization, particularly in the reduction in the number of hierarchies.*

In itself, this change in the shape of organizations from tall pyramids to interconnected processes vastly enhances learning, particularly when it is coupled with a constrained proliferation of responsible and responsive units-small businesses within large businesses. The example of Asea Brown Boveri (ABB) is very significant. To ensure that the benefits of constructively working together in smallness are combined with the advantages of scale, most management have paid particular attention of the development of *"identity"*, expressed in terms of *corporate culture and discipline*. Though its association with the development of identity and use of competence might seem somewhat difficult, it is an imperative for Indian organization to developing the team spirit that encourages the individual to do his/her all for the enterprise and thereby for the national development [49].

With the advent of Japanese practice of management, India has become fascinated with Japanese styles of management. The first such approaches borrowed were *"quality circles"* - workplace groups set up to determine and resolve workplace problems. Quality circles set the way for other developments because they combined the knowledge and insights of workers of a specific task for a specific product with training, basic problem analysis and follow-up. Quality circles have not prospered in India as a set of moral and material incentives have developed around them. Subsequently, there has been a strong drive to complete the quality circles technique with broader approaches focusing on *"total quality management."* For quality as a concept needs to stretch from the boardroom to the broom cupboard, without missing any single corporate function or member. So far, TQM lacks any comprehensive guide or description [50]. Yet, this does not stop some companies from demonstrating and promoting its significance. For instance, top companies have to launch a national campaign to publicize what they are doing to enhance their quality and what they would like to do better. To have meaning and to be successful, top management needs a sense of purpose-long term dominant logic and a culture of reskilling the corporations' [51]. For that quality has to be *a permeating ethic*. What are needed are the *Corporate Joint Ventures*. With the movement from training and teaching to "learning and particularly "action learning", enterprises have to become more interested in participating in reskilling by setting up joint learning ventures. Such joint ventures can be on the national levels. For example, in 1988, 14 major European multinational enterprises jointly created the "European Federation of Quality Management". Its objective is to create conditions to enhance the position of European products and services in the world market by strengthening the role of management in quality strategies. These strategies are characterized by, excellence in all managerial, operational and administrative processes; an understanding that quality improvement results in cost advantages and better profit potential; creation of more intensive relationships with customers and suppliers; involvement of all personnel; and market-oriented organizational practices. EuroPACE is an enterprise-supported satellite distribution system providing top-quality, high level; pre-recorded courses to support the continuing educational needs of technology based enterprises, universities, and research centers. In early 1989, programs covered six fields; microelectronics; software engineering; Tele-communications; artificial intelligence; advanced manufacturing technologies; and technology management. Students are provided feedback, especially through electronic mail.

These types of approaches have led to increasing talk of the *"learning organization"* [48]- one in which there is some sort of harmony between the growing competence of the individual and the smooth adaptation of the enterprise to other internal and external change. What this author has observed and realized in Indian corporate sectors that there is no "*performance support*" provided by top management to HRD functions, and many are unfamiliar to such a role. Even, many business leaders express that they do not have the time to devote for slow and tedious tasks of teaching and coaching people and helping to grow in the high-pressure economy. However, it is this role of the leaders, which is essential for making sure that corporations have the training, skills, information, systems, tools, resources and support to execute knowledge building process and undertake continuous improvement in the knowledge base. This role has a remarkable positive impact on building a high performance climate in the organization. A very significant attempt has been made during the past few years by a large public limited company – Crompton Greaves Ltd., to adapt and diffuse Japanese manufacturing practices across the company. This has been possible by the high level of commitment and extra-ordinary personal involvement of the CEO. The CEO has been instrumental to create a techno-managerial infrastructure, whose role has been to facilitate engagement, collaboration, exploration and experimentation of Japanese best practice [52]. The CEO is willing to put up with short-term failure if it furthers long-term learning for individuals and building knowledge base for the company. This is what may be viewed as the "*expanded leadership*" role in creating a learning organization. Further to this development, the CEO has established a corporate university. The intentions are three folds:

➢ *Learning focus – challenging and testing the assumptions what is being done*
➢ *Experimentation – encouraging and supporting exploration of innovative ideas and concepts*
➢ *Leadership – seeking to engage and integrate leadership at every level.*

The role of the trop management in enhancing intellectual capital formation in Indian corporate sector is fundamentally to recognize that all knowledge has the equal importance to the corporation's core competencies. Knowledge must be prioritized for socio-economic relevance. We suggest here that the modalities of knowledge building discussed earlier must be supported by an infrastructure within the corporation. This infrastructure is required to have two basic elements:

➢ *Appropriate information systems, which can provide widely, distributed access to the knowledge base of the corporation*
➢ *Fostering knowledge – sharing culture to encourage employee to disseminate their tacit individual knowledge throughout the organization.*

We may not be totally able to predict/plan/control the future accurately, but certainly we can influence the future by building quality in/of people through learning organizations across the nation. In order to make the construct of a learning organization usable, an articulation of reskilling the corporations has to emerge. If India and other south Asian countries have to enjoy the status of developed nations in the 21st century; it becomes mandatory that they prudently collaborate to build learning organizations, which will accelerate knowledge building. But, it is submitted here that the attempt must be with urgency, because whether we like it or not time waits for none and competition is relativistic, but not absolutistic. It will require not simply changing our action priorities, but changing our management pattern and more importantly our relationship within and between corporations. Top management, being the strategic apex has the profound responsibility towards building learning organizations and the exclusive accountability towards competence progression. Therefore, there is a need for a strategic mindset and evolving a road map. Swami Vivekananda [52] propounded a century ago: "*if we could get rid of the belief in our limitations, it would be possible for us to do everything just now. It is only a question of time. If that is so, add power, and so diminish time*".

7. CONCLUSIONS

The process of globalization consists of much more than simply seeing an organization integrated in the economic environment. Looking to the emerging competitive landscape in the new millenium, Indian corporate sectors are viewing transfrontier operations as a logical and even unavoidable step in developing their own competitive potential. Managing the complexity associated with such a move require not only a sound corporate strategy, but also the knowledge base of people and organizations that can implement it. This paper is an attempt towards understanding the consequences of rapid globalization process for building knowledge-intensive Indian corporations. Summarily, to build knowledge base and enhance intellectual capital, Indian corporations are required to build momentum for:
➢ *Corporate vitality*
➢ *Strategic organizational connectivity within and between organizations.*

Therefore, it is argued here that we the management professionals, as the custodian of 21st century organizations must provide the real time strategic leadership to build learning organizations with a mission to meet the aspirations of our future generation i.e. to make them more competent than the present generation. This shift in organizational philosophy has profound implications for management practice. Change is where the action is.
➢ *Do we have that commitment?*
➢ *We are paid to create wealth in the enterprises, but what is the form and content of this wealth?*

The author suggests serious research on these fundamental issues both at theoretical and empirical level. These underlying issues, as well as the strategic context of change dynamics are seldom explicit. They can be brought into focus by viewing the participative process of knowledge building as the most fundamental concern. The process of knowledge building is the most daring and difficult for corporation to attempt, articulate and codify. This process is essentially an educational and empowering task, which requires ongoing instruction, motivation and regulation to assure continuance in the corporate environment. Traditionally, knowledge building occurs in crisis situations when the lack of knowledge threatens survival. Global competition and the change dynamics are evolving toward the capturing of the vast cognitive, experiential, and creative potentials of human system. Accepting corporations as exclusive economic entity is an oversimplification and according to the author is unethical. The common good is that structure of relationships in which the life of all is enhanced by the actions of each one of us. Its common name is knowledge building. The knowledge building process will represent the glue that can bind individuals/groups/clusters/organizations /institutions, enabling the creation and rapid diffusion of new knowledge. The cumulative and continuous directing of an organization's efforts towards reviewing the basics of work and strengthening of those processes in the enterprise-wide participation framework, which are most vital to its perpetual well being. To effectively implement the organizational changes, including changes in the modalities of knowledge building, the participation process has to be understood, engineered, and configured to be change-embracing, strategy-enabling and value-enhancing. The feudal emphasis with hierarchical structure and non-people systems and above all apathy towards human development of Indian corporations as the leverage points for eventuating change in the competitive landscape must be heavily supplemented with an intense focus on ensuring that enterprise-wide participation supports the change and invests in an ever-broadening range of knowledge resources. Increasingly, all of us belong to complex configurations of groups and organizations in the competitive landscape of many aspects of our lives. We can perform and fulfil an important role by becoming more conscious of opportunities for common action, taking initiatives to transform those actions into fruition, and developing the knowledge base to do so effectively. It is hoped that this will be a critical success factor in attaining our competitive advantages.

REFERENCES

1. Adams, J. (1996) *Current History*, April, Current History Inc. USA.
2. Mohanty, R.P (1999) 'Value innovation perspective in Indian organizations', *Participation & Empowerment – An International Journal*, Vol. 7, No.4.
3. Porter, M.E. (1990) *The competitive advantages of nations*, Basingstoke, McMillan.
4. Handy, C. (1997), 'Unimagined futures: The organization of the future, edited by Frances Hesselbein et al, Jossey-Bass Publishers, SanFransisco.
5. *"The World Competitiveness Report"* (1993) IMD and World Economic Forum.
6. Sternberg, R. J., Linda, A., Hara, O. and Lubert, T.I. (1997) 'Creativity as investment', *California Management Review*, Vol.40, No.1.
7. Mascitelli, R. (1999) 'A framework for sustainable advantage in Global high-tech markets', *International journal of technology management*, Vol. 17, No. 3.
8. Barney, J.B. (1997) *Gaining and sustaining competitive advantage*, Addison-Wesley Pub. Co.
9. Morgan, G. (1998) riding the waves of change: developing managerial competencies for a turbulent world', San Fransisco: Jossey-Bass, Vol.7.
10. Grant, R.M. (1996) 'Prospecting in dynamically competitive environments: organizational capability as knowledge integration', *Organization Science*, Vol. 7, No.4.
11. Mohanty, R.P. and Deshmukh, S.G. (1999) 'Evaluating manufacturing strategy for a learning organization: A case' *International Journal of Operations and Production Management*, Vol.19, No.3.
12. Bontis, N. (1999), 'Managing organization knowledge by diagnosing intellectual capital: framing and advancing the state', *International Journal of Technology management*, Vol. 18, Nos. 5/6/7/8.
13. Savary, M.(1999) 'Knowledge management and competition in the consulting industry", *California Management Review* ,Vol. 41, No.2, Winter
14. Ettorre, B. (1996) 'A conversation with Charles Handy on the future of work and an end to the century of the organization', *Organizational Dynamics*, summer.
15. Hammer, M. and Stanton, S. (1999) 'How process enterprises really work'*, Harvard Business Review*, November – December 1999.
16. Nishibori, E.E. (1971) *The development of humanity and creativity*, Japan Productivity Center, Tokyo.
17. Mohanty, R.P. and Deshmukh, S.G. (2000), 'BPR: the value innovations in IE practices', *International Journal of Technology Management* (Forthcoming).
18. McLagan, P. and Nel, C. (1997) *The age of participation in executive guide to everyday management*, World Executive Digest Ltd.
19. Nonaka, I. (1994) 'A dynamic theory of organizational knowledge creation, *Organizational Science*, Vol. 5, No. 1.
20. Lane, C. (1989) *Management and labor in Europe: the industrial enterprise in Germany, Britain and France*, Aldershot, Grower.
21. Miyai, J. (1990) 'Human resources: Japan sole natural wealth', *International productivity journal*, Spring.

22. Doregan, J. (1993) *The Learning Organization: Lessons from British Petroleum* in Enterprise School of Management. MCB University Press.
23. Lowendahl, B.R. and Haames, K. (1997) 'The Unit of Activity: A new way to understand competence building and leveraging; in strategic learning and knowledge management', Edited by Sanchez, R. and Heene, A., John Wiley & Sons Ltd., England.
24. Bernstein, P.L. (1998), 'Are networks driving the new economy', *Harvard Business Review*, November-December.
25. Maccoby, M. (1996), 'The Human side: Knowledge workers need new structures', *Industrial research institute.*
26. Eisenhardt, M. and Galunic, D.C., [2000] 'Co-evolving at last, a way to make synergies work', Harvard Business Review, Jan-Feb.
27. Normann, R. and Ramfrez, R. (1993) 'From value chain to value constellation: Designing interactive strategy', *Harvard Business Review*, July -August.
28. Mohanty, R.P. and Deshmukh, S.G. (1999) *'Advanced manufacturing technology selection: A strategic model for learning and evaluation', International Journal of Production economics, Vol.55.*
29. Wenger, E.C. and Snyder, W.M. (2000) *'Communities of Practice: The organizational frontier' Harvard Business Review, January-February.*
30. Sanchez, R., Heene, A. and Thomas, H. (1996) *Dynamics of competencies and competition: Theory and practice in the new strategic management*, Oxford, Elsevier.
31. Tiessen, J.H. (1999), 'Developing intellectual capital globally: an epistemic community perspective', *International Journal of Technology Management,* Vol. 18, No. 5/6/7/8.
32. Buckler, B. (1998) *'Practical steps towards a learning organization: applying academic knowledge to improvement and innovation in business processes', The learning organization*, Vol.5, No.1.
33. Sanchez, R., Heene, A. and Thomas, H. (1996), *Dynamics of competencies and competition: Theory and practice in the new strategic management*, Oxford, Elsevier.
34. Nonaka, I. And Takeuchi H. (1994) *The Knowledge Creating Company*, Oxford University Press, New York.
35. Tichy, N.M. (1993) *GE's Crontonville: A staging ground for corporate revolution in enterprise school of management*, MCB University Press.
36. Bennis, W. (1998) *Rethinking leadership, Executive Excellence, April.*
37. Stewart, T.A (1995)., *How a little company won big by betting on brainpower, Fortune, September 4.*
38. Covey, S.R., (1998) *Constant renewal, Executive excellence, April*
39. Bush, T. and Coleman, M. (1995) 'Professional development for heads: the role of mentoring', *Journal of Educational Administration*, Vol.33, No.5.
40. Gunn, E. (1995) 'Mentoring: The democratic version'. Training, Vol.32, No.6.
41. Smith, M. L. (1994) 'Creating business development talent through mentoring', *Journal of management engineering*, Vol.10, No.2.
42. Davenport, S., Grimes, C. and Davies, J. (1999) 'Collaboration and organizational learning: a study of a New Zealand collaborative research program', International Journal of technology management, Vol.18, Nos.3/4.
43. Whiteley, W., Dougherty, T.W. and Dreher, G.F. (1992), 'Correlates of career mentoring for early career managers and professionals', *Journal of organizational behaviour*, Vol.10.
44. Loeb, M. (1995) 'The new mentoring', *Fortune*, Vol.X, No.11.

45. Orpen, C. (1997) 'The effects of formal mentoring on employee work motivation, organizational commitment and job performance', *The learning organization*, Vol.4, No.2.
46. Tabbron, A., Macaulay, S. and Cook, S. (1997) 'Making mentoring work', *Training for quality*, Vol.5, No.1.
47. Pfeffer, J. (1994) *Competitive advantage through people: Unleashing the power of the work force*, Harvard Business School Press, Boston.
48. Senge, P.M. (1990), *The fifth discipline – the art and practice of the learning organization*, Century business, New York.
49. Mohanty, R.P. (1998) 'Understanding the interconnection between productivity and quality', *Journal of TQM*, Vol.9, No8.
50. Mohanty, R.P. (1997) 'TQM: some issues for deliberation', *Production Planning and Control*, Vol. 8, No. 1.
51. Mohanty, R.P. and Lakhe, R.R (1998) 'Factors affecting TQM Implementation: an Empirical study in Indian industry', *Production Planning and Control*, Vol.9, No.5.
52. Iyer, K. and Mohanty, R.P. (1995), 'Adaptation of Japanese manufacturing management practices: A case study of an Indian organization', *Proceeding of the 13th International conference in production research*, Israel, August.
53. Swami Vivekananda (1900), 'Work and its secret', *The lectures delivered at Los Angeles, California*, 4th January (Ref. Advaita Ashrama, Publication department, Calcutta).

Enhancing agility in manufacturing: The role of QFD

David Ginn, Mohamed Zairi and P.K. Ahmed

European Centre for TQM, University of Bradford, Management Centre, UK

1. THE COMPONENTS OF THE QFD SYSTEM

A Roadmap to Understanding QFD

To begin this roadmap it is necessary start with at least one fundamental definition of QFD, that can be understood and accepted by all levels. To choose just one author or definition as a baseline for understanding is courting a prejudgement, or bias to the discussion. However, the definition given here is sufficiently broad enough to allow latitude as discussion proceeds and specific enough to retain focus on what will become key elements of the research that follows. In a simple sentence, Karabatsos (1988) quotes Larry Sullivan (chairman of the American Suppliers Institute), as stating in 1986 that QFD is the *'mechanism to deploy customer desires vertically and horizontally throughout the company'*. At a fundamental quality process level QFD can also be seen as a 'positive' quality improvement approach as opposed to a (traditional) 'negative' quality improvement approach to deliver customer satisfaction (Ford Motor Co. 1983). For a more detailed baseline definition, Sullivan (1986) proposes that there are six key terms associated with QFD, which are as follows;

i) *'Quality Function Deployment'* (an overall concept that translates customer requirements into appropriate technical requirements for each stage of product development and production).

ii) *'Voice of the Customer'* (the customers' requirements as expressed in their own terms).

iii) *'Counterpart Characteristics'* (the voice of the customer expressed in technical language).

iv) *'Product Quality Deployment'* (the activity required to translate the voice of the customer into technical requirements).

v) *'Deployment of the Quality Function'* (the activity required to assure that customer required quality is achieved).

vi) *'Quality Tables'* (the series of matrices used to translate the voice of the customer into final product characteristics)

The above six key terms of QFD described by Sullivan (1986) can be further simplified as follows; i) a 'concept' for translating customer wants into the product, ii) a requirement to understand 'what' the customer 'wants', iii) the requirement to identify 'how' to technically deliver the 'what' the customer wants, iv) the requirement for a 'team' to carry out the 'translation' of 'whats' into 'hows', v) the requirement for a 'team' required to 'deliver' the hows into the product, vi) the requirement for 'charts' that facilitate the translation of whats and hows into the product. In even simpler terms, this can be distilled down to just one 'concept' of QFD with four key 'requirements' of; customer 'whats' (or wants), technical 'hows', 'team(s)' and 'matrices'. This can be taken a step further by proposing that the first requirement of 'customer whats' needs the second requirement of 'technical hows' to translate itself into the product, this second requirement in turn needs the third requirement of 'teams' to translate itself into the product, and finally this third requirement needs 'matrices' to translate its decisions into the product. This systematic trace from customer subjectiveness, to technical objectiveness, to team decision making with the aid of matrices into product characteristics is a fundamental basis for QFD.

This fundamental step by step process of QFD also lays down the foundation of the concept of the customer to customer process through teamwork. Furthermore, this step by step process is a fundamental description to the mechanics of the matrices and how they are used within a QFD process to translate 'customer whats' into 'technical hows' throughout the product development cycle. It is important to understand the mechanics of the QFD matrix charts early in the discussions, but before this is done, it is necessary to broaden the baseline definition as stated by Sullivan (1986) to include the term 'quality tools'. As a next step, it is beneficial to understand the two terms of; quality tool, and matrix diagram, as this adds the dimension of 'why' use QFD in the first place, and they are important signposts to understanding the QFD system.

The Quality Tool of QFD

QFD is often referred to as a *'tool'* in broad terms (Reynolds 1992), and in more specific terms; a *'competitive tool'* (Kathawala & Motwani 1994), a *'communication tool'* (Fowler 1991), a *'marketing tool'* (Potter 1994), a *'design tool'* (Slinger 1992), a *'planning tool'* (Sullivan 1988, McElroy 1989, Ford Motor Co. 1989, 1992, 1983, 1983), and a *'quality tool'* (Ealey 1987, Barlow 1995, American Supplier Inst. Inc. 1992). This last reference of, 'quality tool', perhaps best summarises all the tool references, and needs a definition in itself to better understand the basic roots of QFD. Straker (1995) describes quality tools as *'structured activities that contribute towards increasing or maintaining business quality'*.

By 'structured activities', Straker (1995) means repeatable and using a defined set of rules, by 'contribute', he means add value, by 'increasing or maintaining' it is meant for use in all areas of quality improvement, and for 'business quality' it means that the company benefits from the quality tool use. In simple terms, Straker (1995) suggests that quality tools are both serious and valuable ways of doing business. Straker (1995) also proposes that tools can be used at either the organisational level or (structuring the way people work together), or at an individual level (helping people and groups solve problems and tasks in their everyday business). Straker (1995) finally suggests three areas where tools can be used, which are;

i) *'collecting various levels of numeric and non-numeric information.'*
ii) *'structuring the information in order to understand aspects of process and problems.'*
iii) *'using the information to identify and select information and plan for specific actions.'*

The definition of quality tools and the three areas of use as described above by Straker (1995) helps outline the fundamental basis of any quality tool including QFD as defined already by Sullivan (1988), Barlow (1995) and Clausing (1994). However, according to Straker (1995), who lists some 33 individual tools in a relationship diagram with their information uses, it is apparent that not all tools are suitable for all three areas of use, or are of equal use. Asaka and Ozeki (1990) list some 15 individual quality tool types, while Nickols (1996) lists just three suites of tool types. It is clear then that the interpretation of what constitutes an tool, a tool type, or a suite of tools is largely dependent on the perspective the various authors and the application of the tool(s) in question. Nickols (1996) considers the question of tools in terms of its 'problem solving' capability, and proposes his three tool types in terms of;

'Repair Tools' for technical trouble shooting
'Improvement Tools' such as Kaizen, continuous improvement, TQM and re-engineering
'Engineering Tools' for design or solution engineering from scratch

This approach by Nickols (1996) is based on the premise that tasks are best performed using the proper tool. Although Nickols (1996) does go into more detail as to what tools fit into the above groups, it is clearly based on findings later on in the thesis that QFD could fit into either of the second two groups and a tool such as FMEA would fit into the first of these groups.

The Matrix Diagram of QFD

Asaka and Ozeki (1990) describe matrix diagrams as a method to *'show the relationships between results and causes, or between objectives and methods, when each of these consists of two or more elements or factors'*. Asaka and Ozeki (1990) continue by stating that *'various symbols are used to indicate the presence and degree of strength of a relationship between two sets of essential items'*. Asaka and Ozeki (1990) propose some four key benefits of using matrix diagrams with symbols as follows;

i) The use of symbols makes it visually clear whether or not a problem is localised (symbols appear isolated) or more broad ranging (symbols in rows or columns).
ii) It possible to show the problem as a whole, and view all the various relationships between the various at once
iii) By testing and evaluating each relationship intersection of the essential factors it becomes easier to discuss the problem at finer levels of detail.
iv) A matrix makes it possible to look at specific combinations, determine essential factors and develop an effective strategy for solving the problem.

Some Basic Mechanics of the QFD Process

It may benefit the reader, at whatever level of understanding of the topic of QFD, to begin with a baseline assumption of the way a QFD matrix chart or *'house of quality'* is constructed. It is also essential to explain the way in which the *'customer whats'* and *'technical hows'* that make up the basis of any QFD project are incorporated into the matrix and analysed.

The House of Quality Mechanics Within QFD

To begin explaining the mechanics, Kim and Ooi (1991) remind the reader that *'QFD is a set of planning and scheduling routines that has proven effective in producing high quality as well as low cost products'* Kim and Ooi (1991). Burton (1995) proposes that the QFD chart, often referred to as a *'house of quality'* due to its' so called construction of *'rooms'* and a *'roof'* is essentially a chart comprising nothing more complicated then a series of *'lists'* and *'relationship matrices'* Clausing (1994) agrees with the term rooms, but adds they can also be referred to as *'cells'* and adds that the QFD matrix diagram comprises of 8 such rooms (or cells) which in turn contains 20 steps in completing the *'Basic QFD'* matrix. The American Suppliers Institute (ASI) (1992) also refer to 10 *'analytical steps' for* studying the completed house of quality at the product planning level. ASI (1992) suggest the same principles apply to all of the QFD matrix charts used at each phase of the process, and add that these steps can take anywhere from a few minutes to several days to complete. In Burton's (1995) description of lists and relationship matrices he is also referring to any of the QFD phases. However, in Clausing's (1994) description of 8 rooms and 20 steps for the Basic QFD he is referring specifically to the first phase of the QFD process. Clausing (1994) proposes, however, that to complete an *'Enhanced QFD'* matrix a total of 43 steps (another 23 steps beyond the first 20) are required for a successful concept phase. The initial 20 steps described by Clausing (1994) are divided among the eight rooms of the QFD planning matrix, while steps 21 to 43 are product planning enhancements that include selecting a *'winning concept'* (Clausing 1994) via Pugh's concept selection chart which in turn leads to the subsequent deployment to a subsystem level.

The eight rooms Clausing (1994) describes are effectively the same basic rooms Ford Motor Company use for their House of Quality charts at a planning level, but Ford (1994) go further by adding a ninth' *'Quality Plan*' room', (excluding the Relationship Matrix) which is a key strategic aspect of the QFD process within the Company.

Burton (1995) adds to his description of the House of Quality chart comprised of lists and relationships matrices by stating that they are aligned along two axes, where the x-axis is called the customer axis, and the y-axis is called the technical axis. This twin axis description is supported by Asaka and Ozeki (1990), who suggests QFD is generally charted using a *'two dimensional diagram'*, with customer quality requirements on the vertical axis and the quality requirements needed to satisfy the customer requirements on the horizontal axis. The sum of these two axis of customer and quality requirements Asaka and Ozeki (1990) refer to as quality information that consists of the problems and desires of the market and workplace. Sullivan (1988) emphasises that the use of matrix charts is key to this process as the correspondence or interaction between heterogeneous elements cannot be viewed in one dimensional space. Sullivan (1988) adds that this process requires two or three dimensional space to evaluate interactive relationships effectively, and confirms what most authors suggest, that symbols (in whatever shape or form) are the ideal way to identify strong, medium or weak relationships between the vertical and horizontal axis of communication. The symbols are also usually assigned numeric values, often weighted in favour of the strongest relationships, 9 = strong, 3 = medium, and weak = 1 as used by ASI (1992) and ITI Burton (1995) or in a linear fashion 3 = strong, 2 = medium and 1 = weak as described by Aska and Ozeki (1990). In most cases however, the company standard default for the strengths can be altered and customised as required, depending on whether the process is carried out on paper as traditionally done by Japanese companies (Akao 1988), or with specifically designed in-house QFD software which is typical of companies such as Lucas Engineering. In all cases however, if no relationship (or correlation) is apparent then the 'cell' or 'value' in the relationships matrix remains blank (or zero). Akao (1988) refers to these symbols within the quality charts used for QFD as indicators of correlation between the customers *'demanded qualities'* and the technical *'quality elements'*. Akao (1988) also refers to the traditionally used symbols depicting; strong, medium and weak as the; double circle, circle and triangle respectively, which is corroborated by Asaka and Ozeki (1990). These traditional QFD relationship symbols originally came from the Kobe shipyard employees who first used QFD, as they represented the horse racing symbols of win, place or show (strong medium or weak) (Ford Motor Co. 1989).

Akao (1988) also differentiates between two types of quality charts within QFD. the first is called the *'Demanded Quality Deployment Chart'* and the *'Quality Elements Deployment Chart'*. The first of these, the demanded quality chart includes information provided by the customers about the qualities they want from the product. These demanded qualities can also be arranged in first, second and third level order Akao (1988), ie first level is 'easy to manoeuvre', second level is 'easy to hold' and third level is 'easy to hold because it is light'. However it must be noted here that what Akao (1988) refers to as a demanded quality chart is what Clausing (1988), (1994), Burton (1995), The ASI (1992) have collectively referred to as; customer requirements, customer wants, customer attributes and whats, and invariably as a room, list, cell, field or list, and not as a chart. This mismatch between what Akao (1988) refers to as a chart, and many Western practitioners refer to as a room continues with Akao's

(1988) quality elements deployment chart. The Akao (1988) quality element deployment chart is the technical translation of the customers' demanded qualities. An example of a quality element for the customer demanded quality of 'easy to hold because it is small' would be 'weight'. These quality elements, have been referred to over time by the various Western QFD practitioners Clausing (1988) (1994), Burton (1995), ASI (1992) as engineering characteristics, substitute quality characteristics, product expectations, design characteristics, how's and technical system expectations. Also within the West, this quality element chart Akao (1988) tends to be referred to as a room. What is clear, however, is that the first building blocks to any QFD chart are these two components of the initial customer requirements (regardless whether its called a want, requirement, demanded quality) and their subsequent interpretation into the product or service technical measurable (regardless of whether they are called (quality elements, engineering requirements, hows, substitute quality characteristics, or technical systems expectations) is the starting point for any QFD chart (regardless of whether it is called quality deployment, quality tool, house of quality or quality matrix).

Cascading Phase to Phase Mechanics of QFD

Sullivan defines four levels of QFD matrices that reflect different stages of application in the product development cycle. The first of these is the *'Planning Matrix'* that culminates with selected control characteristics (based on customer importance, selling points and competitive evaluations). The second is the *'Component Deployment Matrix'* which culminates in defining the finished component characteristics (based the planning matrix targets). The third stage is the *'Process Plan Chart'*, which culminates in the production process monitoring plan required by the operators. Finally the fourth stage is the *'Control Plan'* which culminates in defining quality controls that would typically include control points, control methods, sampling size frequency and checking methods. In each case Sullivan outlines that the previous charts' key outputs feed into the next chart as key inputs, and represent the transition from the development phase to the execution of the production phase within the product development cycle. This four phase process is consistent with most authors. This four stage, step or phase approach is also typically taught by the American Suppliers Institute (Verduyn & Wu 1995) even though flexibility, customisation and overlap with other quality tools (such as FMEA, Taguchi Methods and TIPS (theory of inventive problem solving) is becoming more typical. In common with Ford Motor Company reference to the 'process clock', IBM reference to 'dynamic QFD' (Claxton 1995, Hochman & O'Connell 1993) support the argument for a flexible approach to the QFD process with an emphasis to the cyclical nature to the customer input and feedback loop.

Phase 1 Prioritisation Mechanics of QFD

Four key areas of this prioritisation process will now be discussed. These four areas are crucial components to the Phase 1 'House of Quality' (HOQ). The first is benchmarking, and the second is, in Ford Motor Company language 'Customer Desirability Index' (or CDI) (Ford 1994). The CDI has also been typically been referred to as Customer Importance Rating (CIR) (Ford 1987, 1989, 1992, 1983). The third prioritisation process is ultimately an end product of the first two, and relates to the technical importance rating of the technical systems expectations (TSE's) which represent the company measurables. These measureables typically take the form of a test or metric that can be assigned a target with technical data to support an actionable follow up by the system or component engineer who are the next 'internal' customers of this data. The fourth, and often least used form of prioritisation within the QFD House of Quality is the 'Roof Correlation matrix' (Ford 1987, 1989, 1992, 1983, 1994). This room is effectively where conflicting technical system expectations (or company measureables) can be identified. It is least used partly due to the extra time it takes to complete, and partly because it is often difficult to resolve the technical conflicts that ensue from setting optimised targets for all of the key technical measureables. The term 'roof" is due the triangular nature of this technical relationship matrix on top of the main wants and hows relationship matrix. As a result of the roof is rarely completed or viewed with fear and suspicion.

The Mechanics of the Benchmarking Process Within Prioritisation

Benchmarking within the Phase 1 HOQ comes in two forms, the first is the Customer Competitive Assessment (or Evaluation), (CCA or CCE) (Ford 1994). As the title suggests this is the qualitative benchmarking that the customer participates in within the horizontal customer axis Ford (1994). Customers evaluate the products by comparing the relative 'perceived' performance according to the key customer requirements (using customer language) as identified by prior market research with the support of the QFD team. This exercise will involve the company product (or service) amongst its key competitive products (or services). The second benchmarking activity is the quantitative Engineering Competitive Assessment (or Evaluation) (ECA or ECE) (Ford 1994). This technical benchmarking exercise will compare the same products (or services) through conducting tests that are 'global and measurable' (Ford 1983, 1994) and have been correlated objectively or subjectively to best represent the technical function of the subjective customer wants. These tests have been typically referred to as Substitute Quality Characteristics (Akao 1988), or Design Requirements (Ford 1987, 1989), Technical System Expectations (Ford 1994), or Hows (ASI 1996). These are the technical Company Measures (Verduyn & Wu 1995). These make up the key element to the technical axis (Ford 1994). The benefit of conducting both benchmarking exercises within the same HOQ matrix is that it is then possible to compare subjective customer ratings to objective engineering ratings. The first benefit is to show the company where improvements are required the most, and where there is already high satisfaction relative to competition. The second key benefit is that it is possible to compare discrepancies between customer perception and technical reality. Where discrepancy occurs it is either due to the wrong technical measure being in place, there are more 'hidden' customer wants that require further research, or quite simply as occurred with a Ford Driveability QFD benchmarking exercise in Germany and Britain in July and September 1990 (Ginn 1995), a complexity of 'brand image' (despite efforts to 'debadge') and other complex secondary factors play a part in customer perception. The specific example involved the performance feel of two vehicles, the first was a BMW, and the second was a Citroen. The customer perception was that the BMW was faster, while the technical reality was that the Citroen was faster. The findings showed a complex web of secondary factors that included; brand image, sound quality, interior and exterior styling, the accelerator pedal ergonomics and throttle progression and torque curve rise. The basic element of vehicle acceleration, peak power and velocity over time where in the Citroen's favour on paper. However it proved to be a powerful lesson to the Driveability QFD. This prompted a later Performance Feel QFD research with outside suppliers such as Lotus, Braunschweig University in Germany and Loughborough University to study these secondary factors, that were outside of the time resources of the powertrain engineering community supporting the QFD exercise. These lessons learnt are both a feature and the power of benchmarking within QFD.

Competitive benchmarking to set goals is a powerful tool and is supported by Vaziri (1992), who adds that it assists companies to anticipate customer needs. This ability to anticipate customer wants is a critical measure of success within any QFD exercise, and in the absence of any other form of futuring provides the engineer a key tool in setting so called 'stretch' targets (Ginn 1995). Vaziri (1992) adds that it is important to obtain this benchmarking data in a timely fashion to be effective. Vaziri (1992) also argues that QFD derived customer requirements are a precursor to benchmarking, but not a pre-requisite, although he does reinforce the argument that the combination of QFD and benchmarking culminates in feeding information to quality improvement teams. Ohinata (1994) supports the idea that benchmarking was originally a Japanese invention (rather then an American invention, typically attributed to Xerox) used by small companies who used this tool for modelling best practice from other larger Japanese and American companies. Ohinata (1994) cites some five areas for benchmarking of; product, function, process, management and strategy. Ohinata (1994) adds to this the five steps for successful benchmarking as; clarifying goals, organising a team, selecting target organisations (products or services), collecting and analysing information and devising an action plan. These five areas and steps are arguably a mirror image of the basic key areas and steps required to set up and run a QFD exercise. It is therefore perhaps no coincidence that the synergy of the QFD process with benchmarking is complete when it is recognised that the two key axis of QFD include a benchmarking exercise to support the target setting and prioritisation of both axis. Finally when considering benchmarking, as with all tools, De Toro (1995) warns of 10 pitfalls that confront the benchmarking team which De Toro (1995) refers to as 'miscues'. These ten miscues, or pitfalls, support the argument that QFD and benchmarking are from the same mould of teamwork and process.

The Mechanics of the Quality Strategy Plan Within Prioritisation

The quality strategy or plan is the area or room within the QFD HOQ where consideration of the customer importance rating (CIR) or customer delight index (CDI) Bergeon (1996) for the key customer wants is effectively weighted using a combination of techniques. First it is important to emphasise the subtle difference between CIR and CDI. Typically CIR's were individually rated by the customer during drive surveys (within Ford Motor Company) (Ginn 1995), although this practice still exists a more recent practice initiated by the Quick QFD process is based on the Thurstone methodology (Ford 1994, Guilford 1954, Bergeon 1996) of triplicate comparisons. The CDI method as based on Thurstone is only one of many methods that can be used to compare customer wants. In simple terms the CDI is a customer-assigned rating of desirability for each customer want relative to every other want. From this process a pareto list of customer wants is developed, where typically only the top 25% of wants are taken and put into the QFD House of Quality matrix (Ford 1994). Effectively this is a form of prioritisation before the QFD HOQ is constructed in an effort to keep the total matrix size containable. The more traditional form of QFD also still practised within Ford Motor Company will take all of the identified customer wants, and rely on the prioritisation of resources and the end of Phase 1 by taking only the top 25% of Technical Importance Ratings of the Technical System Expectations (How's, or Company Measureables) into Phase 2, the Component Design level (or Phase 1A the System or Phase 1B sub-System level as appropriate) (1994). With either route the basic mechanics for the Quality Plan (or Strategy) remains the same. The CIR or CDI will then be weighted by a combination of strategic pointers such as Sales Points, Product Attribute, Leadership Strategy, Customer Satisfaction Data, Marketing Brand Strategies and, as already described benchmarking. Sales points are directly influenced by benchmarking results and support weightings to customer wants CIR's or CDI's by assigning pre agreed weighting factors such as 1.5 for strong sales point or 1.2 for moderate sales point.

The Mechanics of Technical Importance Ratings Within Prioritisation

Although the software algorithms and strategies for determining weightings of customer wants CIR's and CDI's are often a closely guarded secret with most companies using QFD, the basic QFD HOQ maths for determining the final technical axis TIR's remains universal. Each TIR is the sum of the 'final' weighted CIR multiplied with each respective relationship value (typically 9, 3 or 1) across the horizontal axis, and then the summed down the vertical axis. Typically the CIR's are also normalised between 1 to 5, although the Strategic CDI (which is the weighted CDI as a result of the Quality Strategy maths and algorithms to produce a futuring effect) may vary, and even include decimal points (Ford 1994, Bergeon 1996).

The Mechanics of the Roof Correlation Matrix Within Prioritisation

The mechanics of QFD is perhaps the least utilised part of many QFD teams. The full function of the roof correlation is to assign weak and strong positive and negative relationship symbols between the technical measureables of the QFD HOQ. As a result it has become the practice to just assign strong negatives that highlight the critical conflicts between optimised technical measureables. A common set of attribute level conflicts within an automobile are; sound quality (or noise vibration and harshness), vehicle weight, safety packaging, emissions packaging, performance and fuel economy. The list could easily be expanded to include sub level conflicts such as idle quality, air conditioning, smoothness and styling. The key issue here is that to resolve these conflicts in a rational approach, a structured data driven process is required. Such a process already exists with QFD, with the support of other quality tools to assign optimised target values. There is also a formula for weighting the key TSE's from different QFD Attribute or Systems. This customised 'extended' version of the QFD HOQ 'roof' correlation matrix is known within Ford Motor Company as the 'Super Roof'. Where companies suffer the most, particularly when developing a complex product over a protracted product development cycle, is that conflicting targets set early in the process become increasingly more difficult to rectify by the time the final product leaves the factory floor. It is these conflicts that can be identified and resolved early in the product development process through the use of QFD, particularly within the least used 'room', the roof correlation matrix.

2. THE PHILOSOPHY, COMPONENTS AND DEVELOPMENTS THAT HAVE LED TO CURRENT QFD USE

It is now appropriate to understand the basic philosophy that led to the development of QFD and the key components that make up the House of Quality that is typically associated with QFD. A critical part of this understanding will be a series of discussions that considers the development of QFD, initially in Japan, and later how it was translated and applied in USA, Europe and the 'Western' world in general. This discussion on the developments will highlight both the cultural difference on interpretation and application. It will also identify why QFD usage today is still in a state of development within the West in particular. First, however, it is essential to identify from a broad base of literature, East and West, what are the formal definitions of QFD, and what are the fundamental agreements or disagreements within the worldwide QFD as a whole.

Some Working Definitions and Descriptions of Quality Function Deployment

Kathawala & Motwani (1994) simply state *'QFD can reduce the risk of misinterpreting customer requirements'*. Kathawala & Motwani (1994) further quote from the work of Maddux, Amos & Wyskis (1991), that *'QFD's objectives are to: identify the customer,* determine what the customer wants, and provide a way to meet the customer's 'desires'. Asaka and Ozeki (1990) place great emphasis on the word 'planning' in their descriptions of QFD as do Sullivan (1988), McElroy (1989) and Ford Motor Company (1983, 1989, 1992). Asaka and Ozeki (1990), however, prefer to shorten the term 'quality function deployment' to just *'quality deployment'*, and state that quality deployment (or QFD) *'defines the functions of planning, development, design and manufacturing of a product to satisfy the quality requirements of customers'*. This shortening of QFD to just quality deployment is consistent with Akao (1988). Quality deployment refers to the charts, tables and descriptive matrices used to design in the quality (or *'goodness'*) required by the customer in the product Akao (1988). Akao (1988) has two definitions for QFD, one narrow, and one broad;

i) narrow QFD definition: *'The business or task functions responsible for quality (design, manufacturing, production).'*

ii) broad QFD definition: *'A combination of these business or task functions responsible for quality (design, manufacturing, production etc.) and the quality deployment charts.'*

    Akao (1988) adds that *'function deployment is often a later step in QFD where the basic functions of the product or service are identified by experienced people at the production company.'* Akao (1988) likens function deployment to the *'voice of the engineer'* who has the task of identifying the *'must be'* attributes of the product, where Akao (1988) gives the example of 'must be' as an unspoken customer requirement, an attribute that must be there, otherwise it is a source of dissatisfaction to the customer (such as a bed, and bathroom in a hotel, that the customer must have). However Akao (1988) asserts that to have these 'must be' attributes, or functions, does not guarantee customer satisfaction, it only ensures no strong dissatisfaction. Akao (1988) summarises this argument by stating that when customer's spoken quality demand opposes these 'must be' attributes or functions, then the producer of the product or service must balance the spoken demands with practical functional requirements of the product or service. Akao (1988) ties in the purpose of the quality charts or quality tables (which have already been referred to as houses of quality or QFD matrices by the previously referenced authors) as a *'means to..'* not *'an end in themselves'* , that is to say they are there to provide insight into the nature of the product or service and what is necessary to improve it with relation to the spoken quality demands of the customer.

    Asaka and Ozeki (1990) further develop what they mean by quality requirements of the customer by stating the product or service must meet or fulfil customer standards, needs, expectations and future unanticipated needs and aspirations, 100% of the time. This total product development cycle definition of QFD driven by an extreme level of customer expectation by Asaka and Ozeki (1990) proposes a very stringent test for QFD success.

    Slinger (1992) neatly proposes that *'Quality Function Deployment is a design tool which is a powerful support to 'encouraging' engineering design teams to take a structured, thorough approach to product design'*. Slinger (1992) and Metherell (1991) further describe a four stage (phase) QFD process as part of an integrated engineering process, which they illustrate as linked into Simultaneous Engineering using teamwork, training and planning. Metherell (1991) adds to the setting of QFD and Simultaneous Engineering in context with Integrated Engineering by emphasising the focus for team effort. Metherell (1991) also intimates that QFD as part of this Integrated Engineering process, is consistent with the highest 'opportunity for change' at the concept levels, and offers traceability throughout the product cycle.

    Consistent with the previous two authors (Metherell 1991 and Slinger 1992), Hauser and Clausing (1988) propose a definition of QFD through reference to its classic House of Quality matrix that reads 'the house of quality is a kind of conceptual map that provides the means for inter-functional planning and communications'. They further suggest that people with different problems and responsibilities can thrash out design priorities by referring to patterns of evidence from the house of quality. This interpretation adds to argument for QFD being more than just a planning tool scenario, but also a tool for interdisciplinary communications within any company. Hauser and Clausing's (1988) definition proposes that QFD is both a planning and communications tool that helps focus and coordinate skills within an organisation from design to manufacture into a product customers want and will continue to buy. This definition is concurred by McElroy McElroy (1989) who refers to QFD as a 'powerful planning tool', and quotes Dana Cound (a VP within GenCorp Automotive) as saying it is 'a typical Japanese take-nothing for granted procedure that makes you write everything down' (as opposed to the traditional approach that leaves too much to chance). McElroy (1989) also quotes Bill Eureka (ASI president) who states that 'QFD is a process that will bring out the 'hidden knowledge' in your organization'. Bob Porter (Texas Instruments) also suggests that the QFD process is an 'exercise in culture change' and that anytime a group a group of people sit a room discussing the customer there will be conflict but from this conflict comes creativity (McElroy 1989). Fowler (1991) states that the QFD 'matrix is a communication tool for members of a broad based, cross-functional design team that serves three key functions; i) develops within the team members and the organization a better understanding of how the customer needs relate to design requirements, ii) focuses design effort on areas where effort is justified, and iii) identifies problems during the design phase to minimise later redesign effort. From all the quotes in the above section it is clear that QFD is more then a customer satisfaction delivery tool but is also; an improver of communications, a prompter for creativity, a discoverer of latent knowledge, a documentation of process, an identifier of problems, and perhaps most importantly a changer of culture.

Sullivan (1988) corroborates th view that QFD is a both a planning tool and aid to communication, and observes that several U.S. companies are being very successful in applying the QFD matrix charts, which in turn has helped integrate the various diverse activities within that company. Sullivan develops this argument, however, by suggesting that QFD can be used as the 'hardware' through which 'policy management' which he refers to as the , 'software' can be integrated. The difference with policy management to 'objective management', the more typical style of management, is that the latter is based on measuring performance by results, while the former focuses on developing the means of achieving results through methods, systems, or resources. The foundation of policy management (Sullivan 1988) suggests is 'business planning'. Business planning in turn is based on employee ownership or entrepreneurship to set goals through a comprehensive planning process across the whole organization, by reducing the void between departments. The results from this level of detail then become the results of the policy means and a measure of policy management success. In summary, Sullivan (1988) proposes that 'soft technologies' such as policy management are important to achieve the business plan, and that this must be integrated through congruent objectives with the use of 'hard technologies' such as QFD, Taguchi Methods, SPC, to deploy product requirements. All these elements combined deliver the key goal of meeting customer expectations. This argument for QFD being an integral part of business planning is corroborated by Barlow (1995). He refers to 'policy deployment' in the same context. Greenall (1995) describes policy deployment as process focused, rather then management by objectives, which is reiterated by Barlow (1995) who uses Kawneer UK Ltd as an example of policy deployment in action. Barlow's (1995) description of policy deployment mirrors the key elements of a QFD in that both ensure a clear understanding of the company objectives, goals and direction, both are diagnostic tools that set targets through focusing on the *'vital few objectives'* , both place emphasis on team building and good communications, and both focus on the interaction of all tools (including QFD) to achieve an integrated business plan. Greenall (1995) adds to this by suggesting that the policy deployment process is formalised and measurable, with goals and targets and negotiated and set by the employees which often tend to be stiffer, than had they been set by management. Benefit of policy deployment is that improvements are continuous and everyone ends up pulling in the same direction (1995). These three arguments by Sullivan (1988), Greenall (1995) and Barlow (1995) strongly suggest that a suite of quality tools including QFD must be used as part of a process oriented business plan and that the full benefits of any one tool cannot be realised without such an approach. Ealey (1987) adds to this line of argument for using quality tools in support of one another by suggesting that QFD can be used to identify where a company should use such powerful tools as Taguchi Methods without waste, which is an example he concludes of QFD's often unmentioned benefit of being able to tell manufacturers where NOT to invest time and money.

The idea of using QFD within an organization as an aid to business planning becomes clear when placed in the context of its numerous and varied benefits which will now be discussed. Zairi (1993) summaries four key benefits as being; higher quality, lower cost, shorter timing and marketing advantage. Akao's (1988) survey of QFD benefits within Japanese industry quotes five key process benefits of; decreased start-up problems, competitive analysis became possible, control points clarified, effective communications between divisions, design intent carried through to manufacturing. Hideaki Aoki, Yukio Kawasaki and Takao Taniguchi (1990) relate the benefits of QFD as being in conjunction with *'quality charts, related procedures of new product development and quality assurance activities'* and summarises these into two broad benefits that lead to;

i) the development of new products that both meets the customers' demands and wins their trust as well as being developed in a timely manner to lead the market.
ii) the improvement of interdepartmental communication on product development, by identifying problems from early predesign stage to ensure development and process time reductions.

Finally Aoki, Kawasaki and Taniguchi (1990) add that from planning to preproduction QFD enables the relationships between systems to be clearly understood thus benefiting the development of more diversified projects. This argument implies that QFD although complicated in itself can help clarify complex inter-system relationships. This line of thought is captured by Sullivan (1986) who describes the overall QFD system based on four key documents that trace a continuous flow of information from customer requirements to plant operating instructions. This Sullivan (1986) considers is in line with what W.Edwards Deming calls a 'clear operational definition'.

The argument from Aoki, Kawasaki and Taniguchi (1990) regarding the ability of QFD to assist more diversified projects from planning to pre-production to is seen by Hiroshi Takamura and Tadayoshi Ohoka (1990) as a key aim of QFD. They add to this by stating that the goal of QFD is to 'achieve mass production of a product with assured quality, with ease of manufacturing, and at minimum cost'. They then goto develop the argument for production participation at the product development stage to allow for greatest efficiency. A key process within QFD, Takamura and Ohoka (1990) continue, particularly in today's competitive market, is more focused prioritisation. Methods that can assist with prioritisation is reviewed by Nabuo Takezawa and Masyuki Takahashi (1990) who suggest using 'fault tree analysis (analysing the system)' to accurately deploy high priority quality items, as relationships alone cannot do this. Later they also suggest using a 'concept deployment chart' based on the component feature values that are to be deployed to establish the best design policy. The details of these component feature values identified within the QFD are also examined and refined using such tools as FMEA (failure mode and effects analysis). Takezawa and

Takahshi (1990) suggest by using this method key feature values are determined and engineering bottlenecks are identified for further research and deployment. Concept selection is also seen as valuable part of QFD by De Vera, Glennon, Kenny, Khan and Mayer (1988) for four reasons; i) a much needed focus on satisfying customer needs as a quality issue, ii) as a reasonable technique to consider alternatives, iii) it represents a paradigm shift by defining quality by design concept rather then quality control, and iv) it is a simple process that invites more interaction from management. This last point perhaps suggests that concept selection charts may act as an intermediary between the QFD team and management. It is proposed that when QFD is integrated with concept selection, the team's learning curve increases and it's level of efficiency climbs with the increased range of experience bought to the forum when conducting such an exercise. This in turn benefits new product development and start-up times (1988).

It would appear, however, that QFD is only one of many techniques available to companies wishing to improve product development times. Reinertsen (1991) reviews how companies can overcome 15 common barriers to timing product-development cycles, and refers to QFD and CE (concurrent engineering) as valuable to trim development cycles down, but from the 15 common barriers QFD is only completely successful in just two areas while CE is successful in only four areas. These are; 'hitting moving targets' (QFD), 'lack of concurrency' (CE), 'moving locus of control' (CE), 'phased development systems' (CE), 'focus on communication' (QFD/CE). Reinertsen's (1991) list of remaining barriers are that QFD specifically does not adequately address include; taking giant steps, ignoring market clocks, overloading capacity, ignoring queue time, burn rate management, lack of concurrency, inattention to architecture, moving locus of control, phased development systems, inappropriate testing strategies, failure to quantify the problem, make/buy decisions, and when efforts pay off. Reinertsen (1991) does, however, acknowledge the crucial role of communications in developing products rapidly.

Finally a neat description of QFD is given by Reynolds (1992). He proposes that the planning process of Quality Function Deployment is the major development in the Quality sector of business. Reynolds (1992) describes QFD as a tool that uses 'a sophisticated subjective analysis to design an "optimal" product with maximum customer satisfaction assured.' This definition places emphasises on the subjective approach QFD offers and that optimisation of the product is the route to maximum customer satisfaction.

3. SOME KEY BENEFITS OF QFD PHILOSOPHY AND PROCESS

Already many of the inherent benefits of QFD have been alluded to in the previous sections, but the following discussion will review the types of benefits already mentioned and establish other key benefits so that the effectiveness of QFD can be better understood.

Kim and Ooi (1991) neatly summarise some of the benefits of Quality Function Deployment as a useful tool for *'integrating the human expertise of marketing, design, production, and service personnel to address all relevant issues and to achieve the single goal of customer satisfaction'*. Kim and Ooi (1991) place this benefit in the context of an organization's need to be effective by making full use of its disposable knowledge. It does they suggest by using QFD to focus and coordinate the skills of the organisation from the design stage through to manufacturing (Kim and Ooi 1991). In agreement with QFD providing a knowledge base De Melo Cavalcanti (1993), in his summary of key features that QFD offers, suggests most of the typical benefits such as already identified but specifically quotes the benefit of computer files of company generic information as a source of knowledge. De Melo Cavalcanti (1993) also notes QFD gave the process of benchmarking to

ICL, another great source of knowledge. In further agreement with Kim and Ooi (1991) De Melo Cavalcanti (1993) quotes several QFD case study benefits from, ICL, Elida Gibbs and Milliken (with whom he benchmarked) and observed the reiterated benefit of teamwork which helped ICL (1993) in particular to develop improved customer/supplier chains within the organization that developed into partnerships rather then barriers.

Hauser and Clausing (1988)] reiterate the emphasis on coordination and focus suggested by Kim and Ooi (1991) but place the word *'focus'* in relation to the customer by stating a Toyota example as follows, *'Toyota improved its rust prevention record from one of the worst in the world to one of the best by coordinating design and production decisions to focus on this customer concern'*. Hauser and Clausing (1988)] also quote three other benefits that include;

i) Reduced pre-production and start-up costs by using QFD,
ii) The emphasis QFD places on moving the design changes upstream to avoid more costly and time consuming downstream changes.
iii) The help QFD brings, through using the house of quality technique, in breaking down functional barriers and encouraging teamwork.

Cristiano, Liker and White III reinforces the second benefit of QFD shortening the overall product development cycle (with time and cost savings) by shortening both the product design stage and the product redesign stage, despite the fact that the product definition stage is extended. The third benefit of breaking down barriers is has already been noted by De Melo Cavalcanti (1993) within ICL. It is also noted as a fundamental benefit that Hunter & Van Landingham (1994) propose that even the detractors of QFD concede. It is suggested that a key barrier QFD help break down is that between marketeers and engineers who often feel that QFD does not provide anything they would not have guessed with a little more effort Hunter & Van Landingham (1994). Potter (1994) also emphasises the importance of *'breaking down organisational barriers'* and proposes that in instances of truly cross functional teams, such as the Digital Equipment Corporation example he cites from Van Treek and Thackery (1991). This example cited by Potter (1994) states that *'the product development QFD team gets closer to the customer than to the marketeers,...destroying the traditional barriers between marketing and engineering'*. These arguments would suggest that it is precisely the fact the barriers are broken down that QFD succeeds, where perhaps isolated more entrenched processes would not. The benefit that QFD offers in terms of teamwork in breaking down organisational barriers and QFD's ability to solving quality problems in key market segments as a marketing tool is a key argument for Potter (1994), however, it is the potential QFD offers as a marketing tool for finding new customers, and not just satisfying or exceeding the expectations of current customers that Potter (1994) explores more deeply. Potter (1994) describes this as *'finding customers to implement the strategy of obtaining broader horizontal market position'*. Potter supports this claim by proposing two frameworks, the first being *'cross fertilisation across market segments'* and second being *'co-suppliers as matchmakers'*. In the first framework Potter (1994) suggests that by compiling the customer requirements from several QFD charts selected for their fit with the company's skills, this would add a valuable tool for the sales force when prospecting new business. the second framework Potter (1994) suggests that by selectively finding contacts with needs that would fit the company's skills a broader customer base could be established. A key benefit of QFD, states Potter (1994), is that QFD helps to *'visualise the information on organisational fit'* (to the customer's application process)

4. SOME FAILINGS AND ROADBLOCKS WITHIN THE QFD PHILOSOPHY AND PROCESS

One key failing of the method Sullivan (1988) describes when using QFD with policy management is the tendency to sub-optimise within the company on internal objectives which inhibits the business plan. Zairi (1993) agrees with this in particular where activities driving QFD efforts are driven by individuals within a functionally oriented companies. Zairi (1993) and Sullivan (1986) identify this as a cultural incompatibility with using QFD.

Other failings are simply related to the lack of understanding of the basic mechanics of the tool itself (1993). In this context probably one of its most common criticisms is that it is not a simple tool to use or introduce Zairi (1993), Hauser & Clausing (1988), De Melo Cavalcanti (1993), and is also an intellectually very demanding (1993). It can also be difficult to translate QFD into requirements, specifications or design in a concurrent manner, with more traditional processes and tools if executed robustly having a better chance of success Goldense (1993), De Melo Cavalcanti (1993).

Another common criticism is that QFD can and invariably does, take a long time to implement, with only a few companies having completed all four Houses of Quality, which can be frustrating to those wishing to see instant results Goldense (1993), Reinertsen (1991), De Melo Cavalcanti (1993), Kathawala & Motwani (1994). Often the protracted time it takes to complete a QFD House of Quality (which is an all too frequently experienced aspect of QFD), is a direct a function of the size of the QFD chart Liner (1992), Kathawala & Motwani (1994). When it comes to chart size, it really is a case where biggest is most certainly is not best and Liner (1992) suggests anything above a 20 item matrix becomes impractical time wise. Kathawala & Motwani (1994) also note another common chart process failing seen mostly in large organisations, is to mix engineering and customers demands together. The average time it takes QFD to pay off is estimated by experts to be two to five years (1991), and Reinertsen suggests that companies who only use QFD (and, or CE) may find themselves behind competitors who use other more immediate techniques for product development cycle improvements. Linked to this is the problem that many large companies in particular complete their QFD's too late to be implemented into their intended programmes, or more frustratingly still, complete the QFD on time but fail to get the QFD recommendations recognised Kathawala & Motwani (1994).

5. AGILITY THROUGH QFD: INTEGRATION WITH OTHER TOOLS

Ford Motor Company Engineering Quality Improvements Programme (EQUIP) within Europe has been the linking of some seven quality tool techniques including QFD as the core link Henshall (1995), Herrick (1995). Another key element of EQUIP is the training of people skills to enhance the teamwork processes required to bond the usage and communication of the technical tools. The seven technical tools within EQUIP are taught within a Systems Engineering framework and are also integrated into the 'Ford Customer Satisfaction Process'. Ford (1994) Henshall (1995). The seven EQUIP tools are; TOPS 8D, Process Management, FMEA, Experimentation, Taguchi, Reliability Engineering and QFD (referred to as Customer Focused Engineering) Ford (1994), Henshall (1995), Herrick (1995).

Clausing (1994) also strongly recommends the integration of QFD with Taguchi's quality engineering tool and also recommends their integration into the Systems Engineering process. To apply either of these tools in isolation is seen by Clausing (1994) as ineffective at best. Clausing (1994) considers that when developing complex and dynamic products, such as cars, two key enhancements of QFD must include Pugh concept selection and sub-system development between design requirements and component development. Clausing further adds that other quality tools such as FTA, (Fault Tree Analysis), FMEA (Failure Mode Effects Analysis), FAST (Functional Analysis System Techniques), VA/VE (Value Analysis/Value Engineering), and QE (Quality Engineering).

This EQFD (Enhanced QFD) Clausing (1994) considers as an important corporate capability that integrates the corporation holistically with a concentrated focus on customer satisfaction. Clausing (1994) proposes total quality development has three major elements;

1..Basic improvements in clarity and unity (basic quality engineering)
2..Enhanced quality function deployment (EQFD)
3..Quality engineering using robust design.

Each of these Clausing (1994) sub divides into further subsections. For basic quality engineering, Clausing (1994) adds; concurrent process, focus on quality, cost and delivery, emphasis on customer satisfaction, competitive benchmarking, and better teamwork which includes; multi-functional teams, employee involvement and strategic supplier relations. The point Clausing (1994) makes is that to attain a truly *'world class concurrent engineering'* process, all of the above criteria need to be met, otherwise the results in attaining a few of these elements, will be both only promising and disappointing at the same time. Clausing (1994) provides a series of enhancements to basic QFD, that include the integration of various quality tools including; FTA (Fault Tree Analysis), FMEA (Failure Mode Effects Analysis), FAST (Functional Systems Techniques), VA/VE (Value Analysis/Value Engineering), as well as the key elements of EQFD of concept selection using Dr Stewart Pugh's concept chart analysis and Taguchi's quality engineering techniques for robustness. Clausing (1994) describes robustness as *'small variation in performance'* and suggests *'only robust products provide consistent customer satisfaction'*.

REFERENCES

1. Aiba, K., *'What is Quality Design'*, Quality Control, Union of Japanese Scientists and Engineers (JUSE), Vol.17, No.1, 1966, pp.88-89.
2. Akao, Y., *Quality Function Deployment: Integrating Customer Requirements into Product Design,* English translation: Productivity Press, 1990, (originally published a, *'Hinshitutenkai katsuyo no jissai'*, Japan Standards Association, 1988).
3. American Suppliers Institute Incorporated, *Phase 1 - Product Planning, analyzing and diagnosing the product planning matrix,* ASI Quality Systems QFD, 1992, pp.51-54.
4. Aoki, H., Kawasaki, Y., and Taniguchi,T., *'Using Quality Deployment Charts:Subsystems, Parts Deployment, Quality Assurance Charts'*, Chapter 4 in *Quality Function Deployment, Integrating Customer Requirements into Product Design,* Akao, Y., (ed.), Productivity Press, English translation, 1990, pp.83-111.
5. Asaka, T., Ozeki, K., *Handbook of Quality Tools, The Japanese Approach,* English translation Published by Productivity Press, Inc., 1990. (Originally published as *'Genbacho no tameno QC Hikkei'*, by Japanese Standards Association, Tokyo, 1988).
6. Barlow, K., *'Policy Deployment in Action at Kawneer'* , ASI Quality Systems 6th European Symposium for Taguchi Methods & QFD, Kenilworth, England, May 16-18, 1995.
7. Bergeon, S., ' *Strategic CDI and Parent Process with Quick QFD'*, A Presentation by SSO (Strategic Standards Office), QFD/MRO (Market Research Office) Conference, FAO, Fairline Training & Development Centres, Dearborn, Michigan, USA, March 11-15 1996.

8. Burton, D., *'The Ideal Lunch, building the heart of quality, the complete 'how to'', QFD'*, QFD Workshop Conference, Bradford Management Centre, University of Bradford, June 27, 1995.
9. Clausing, D. *Total Quality Development, A Step-by-Step Guide to World-Class Concurrent Engineering*, ASME Press, 1994.
10. Claxton, T., *'Using Dynamic QFD to Improve Manufacturing Competitiveness'*, IBM Consulting Group Presentation, QFD Workshop, Bradford University, Heaton Mount, 27 May 1995.
11. Cristiano, J.J., Liker, J.K., White III, C.C., *'An Investigation into Quality Function Deployment (QFD) Usage in the US*
12. De Melo Cavalcanti, L.M., *An Investigation into the Use of QFD in ICL, The Establishment of Best Practice*, MBA Thesis, Management Centre, University of Bradford, 1993.
13. De Toro, II., *'The 10 Pitfalls of Benchmarking'*, Quality Progress, Vol. 28 No.7, January 1995, pp 61-63.
14. De Vera, D., Glennon, T., Kenny, A.A., Khan, M.A.H., Mayer, M., *'An Automotive Case Study'*, Quality Progress, June, 1988, pp.35-38.
15. Ealey, L., *'QFD-Bad Name For A Great System'*, Automotive Industries, Vol.167, July 1987, pp.21.
16. Ford Motor Company *QFD Reference Manual*, Car Product Development, Technical Training and Educational Planning, Ford Motor Company, Dearborn, Michigan, USA, May 1992.
17. Ford Motor Company, *Module 7, Customer Focused Engineering, Level 1, QFD Manual*, EQUIP (Engineering Quality Improvement Programme), Ford Motor Company Ltd, Published by Education and Training, EQUIP Centre, 26/500, Boreham Airfield, Essex, England, 1983.
18. Ford Motor Company Limited, *Module 7, Customer Focused Engineering, Level 2, QFD Manual*, EQUIP (Engineering Quality Improvement Programme), Ford Motor Company Ltd, Published by Education and Training, EQUIP Centre, GB-26/500, Boreham Airfield, Essex, England, 1983.
19. Ford Motor Company Limited, *Quick QFD, The Marketing – Engineering Interface*, Automotive Safety & Engineering Standards Office, Ford Motor Company Limited, Fairlane Plaza, Dearborn, USA, (Restricted access).Version 3.0, 1994.
20. Ford Motor Company Limited, *Ford Customer Satisfaction Process*, European Automotive Operations Powertrain QFD Steering Team, Issued by the Customer Focused Engineering Group, Ford Motor Company, Vehicle Centre 1, Dunton Research & Engineering Centre, Essex, SS15 6EE, England, Version One, (Restricted access), December 1994.
21. Ford Motor Company *Quality Function Deployment, Executive Briefing*, American Suppliers Institute Incorporated, Dearborn, Michigan, A.S.I. Press, QFD00250, 1987.
22. Ford Motor Company *QFD Awareness Seminar*, QETC (Quality Education and Training Centre), Ford Motor Company, Dearborn, Michigan, USA, First Issue: January 1989, Second Issue: May 1989.
23. Fowler, C.T., *'QFD-Easy As 1-2-3'*, 1991 SAVE (Society of American Engineers) Proceedings, Kansas City, MO, USA, SAVE National Business Office, Vol.26, 1991, pp.177-182.
24. Ginn, D., *'Worldwide Powertrain QFD Library'*, Ford Motor Company QFD/CFE Department, 814/220, Research & Engineering Centre, Dunton, Essex. England.

25. Goldense, B.L., *'QFD: Applying 'The 80-20 Rule'*, Design News, December 20th, 1993, pp.150.
26. Greenall, R., *'Policy Deployment'*, ASI Quality Systems 6th European Symposium for Taguchi Methods & QFD, Kenilworth, England, May 16-18, 1995.
27. Guilford, J.P., *'Psychometric Methods'*, Second Edition, published by McGraw-Hill Book Company 1954.
28. Hauser, J.R., Clausing, D., *'The House of Quality'*, Harvard Business Review, May-June 1988, pp.63-73.
29. Henshall, E., *'EQUIP (Engineering Quality Improvement Programme) at Ford Motor Company'*, ASI Quality System 6th European Symposium for Taguchi Methods and QFD, Kenilworth, Warwickshire, England, May 16-18, 1995.
30. Herrick, R., *'Quality Methodology Application - The Application of Technical & Behavioural Methodologies by Companies involved in Industrial Design, Development and Manufacturing Processes'*, MSc by research, University of Bradford, Department of Mechanical Engineering and Management, December, 1995
31. Hochman, S.D and O'Connell, P.A., *'Quality Function Deployment : Using the Customer to Outperform the Competition on Environmental Design'*, IEE International Symposium on Electronics and Environment, Arlington NA, USA, 1993, pp 165-172.
32. Hunter, M.R., and Van Landingham, R.D., *'Listening to the Customer (Using QFD)'*, Quality Progress, April 1994, pp.55-59.
33. Kano, N., *'A Perspective on Quality Activities in American Firms'*, Quarterly California Management Review, Spring 1993, pp 12-31.
34. Karabatsos, N., *'Listening to the Voice of the Customer'*, editorial, Quality Progress, June 1988, p5.
35. Kathawala, Y., & Motwani, J., *'Implementing Quality Function Deployment, A Systems Approach'*, The TQM Magazine, MCB University Press, Vol.6, No.6, 1994, pp.31-37,
36. Kim, S.H., and Ooi J.A., *'Product Performance as a Unifying Theme in Concurrent Design-II. Software'*, Robotics & Computer-Integrated Manufacturing, Vol.8, No..2, 1991, pp127-134,
37. Liner, M., *'First Experiences using QFD in Product Development'*, Design for Manufacture, American Society Mechanical Engineers (ASME), Anaheim, CA. USA, DE-Vol. 51, Nov. 8-13, 1992, pp.57-63.
38. Maddux, G.A., Amos, R.W. and Wyskidcy, A.R., *'Organisations Can Apply Quality Function Deployment As a Strategic Planning Tool'*, Industrial Engineering, Vol.23, September 1991, pp. 33-37.
39. McElroy, J., *'QFD, Building The House of Quality'*, Automotive Industries, January, 1989, pp.30-32.
40. Metherell, S.M, 'Quality Function Deployment, Less Firefighting and More Forward Planning' , IFS Conference Proceedings, 1991.
41. Morrell, N.E., *'Quality Function Deployment, disciplined quality control'*, Automotive Engineering, Vol.96, February 1988, pp.122-164.
42. Nickols, R.W., *'Yes, It Makes a Difference'* Quality Progress, January 1996, pp 83-87.
43. Nishimura, K., *'Designing of Ships and Quality Charts'*, Quality Control, Union of Japanese Scientists and Engineers (JUSE), Vol.23, May 1972, pp.16-20
44. Ohinata, Y., *'Benchmarking : The Japanese Experience'*, International Journal of Strategic Management & Long Range Planning, Vol. 27 Issue 4, August 1994 pp 48-53.
45. Potter, M., *QFD as a Marketing Tool*, MBA Thesis, Management Centre, University of Bradford, December 1994.

46. Reinertsen, D., *'Outrunning the Pack in Faster Product Development'*, Electronic Design, 1991, January 10th, pp.111-124.
47. Reynolds, M., *'Quality Assertive Companies to Benefit From Recovery'*, Elastometrics, February, 1992, pp.19.
48. Slinger, M., *'To Practice QFD With Success Requires a New Approach to Product Design'*, Kontinuert Forbedring, Copenhagen, 20-21 February, 1992.
49. Straker, D., *'The Tools Of The Trade'*, Quality World, Vol.21, Issue 1, January 1995, pp.28-29.
50. Sullivan, L.P., *'Quality Function Deployment, A system to assure that customer needs drive the product design and production process.'*, Quality Progress, June 1986, pp.39-50.
51. Sullivan, L.P., *'Policy Management Through Quality Function Deployment'*, Quality Progress, June 1988, pp.18-20.
52. Takamura, H., and Ohoka, T., '*Using Quality Control Process Charts: Quality Function Deployment at the Pre-production Stage*', Chapter 5 in *Quality Function Deployment, Integrating Customer Requirements into Product Design*, Aqua, Y., (ed.), Productivity Press, English translation, 1990, pp.112-146.
53. Takayanagi, A., *'QC in Order Taking/Production in Our Company (No.1) QC Activities for Ordered Products: Concepts of the Quality Chart'*, Quality Control, Union of Japanese Scientists and Engineers (JUSE), Vol.24, May 1973, pp.63-67.
54. Takezawa, N., and Takahashi, M., *'Quality Deployment and Reliability Deployment'*, Chapter 7 in Quality *Function Deployment, Integrating Customer Requirements into Product Design*, Akao, Y., (ed.), Productivity Press, English translation, 1990, pp.180-210.
55. Vaziri, K., ' *Using Competitive Benchmarking to Set Goals'* Quality Progress, Vol. 25, October 1992, pp 81-83
56. Verduyn, D.M. and Wu, A., *'Integration of QFD, TRIZ & Robust Design Overview & "Mountain Bike" Case Study'*, ASI Total Product Development Symposium, Novi, Michigan, USA, November 1-3, 1995.
57. Zairi, M., *Quality Function Deployment: A Modern Competitive Tool*, TQM Practitioner Series, European Foundation For Quality Management in association with Technical Communications (Publishing) Ltd., 1993.

Product Development Strategies for Agility

Sudi Sharifi[a] and Kulwant S Pawar[b]

[a]The Graduate School of Management, University of Salford, Salford, M5 4WT, UK
E-mail: S.Sharifi@man-school.salford.ac.uk

[b]School of Mechanical, Materials, Manufacturing Engineering and Management
University of Nottingham, University Park, Nottingham NG7 2RD, UK
E-mail: Kul.Pawar@Nottingham.ac.UK

In recent years the long term success of some manufacturing organizations has been enhanced by their ability to bring new products onto the market at regular and shorter intervals. Although, innovation and new product development (NPD) may be pivotal to success, growth and survival of manufacturing organizations, a number of issues and in the organisational and managerial processes need to be addressed. Indeed, the criteria for competitiveness in the market have been changing continuously. For instance, levels of product complexity, market demands, extent of globalisation of markets and degree of consumer awareness have varied over the last few. This has meant that organisations have to become more agile and responsive to the changing needs of customers and consumers. Within this context the role of design as a function within organisations has changed significantly. Intra-organisational as well as inter-organisational levels of involvement of the design function has significantly increased which matches the desired level of integration. The evolving NPD context thus calls for innovation and innovative approaches to design and manufacturing. In this evolving context not only there has to be uniqueness and novelty in the product but also in the design process itself. An innovative design process implies involvement incorporation of other collaborators at inter-and intra-organisational levels. The nature and type of interactions between different actors during the product design phase often influences the nature of the final product and its subsequent manufacture. The iterative nature of the product design process highlights the point that a great deal of the design activity includes re-defining and redesigning of the conceived ideas.

Organizations, including manufacturing firms, have been seeking strategies and organisational frameworks, which would allow for /facilitate the integration of differentiated technologies, methods and organisational process. Such an integrative framework will thus be responsive to the requirements of the environments. The framework and context of NPD have evolved over the years from traditional engineering to Concurrent Engineering. Indeed the uncertainty and unpredictability of the environment and needs of customers and consumers calls for an agile enterprise. The product design and development within agile environment is viewed as a series of transaction of knowledge, information between people and groups they represent. multi-functional teams and teaming become enablers in such processes. The 'agility' framework offers an alternative way of looking at NPD in relation to a network of organisations and organisational relationships. Here customer and consumer are put at the forefront of the design process.

1. NEW PRODUCT DEVELOPMENT IN ADVANCED MANUFACTURING ENVIRONMENT

In the early 1990's Peters argued that 'uncertainty is the only certainty' for all organisations irrespective of sector or industry. And such a situation requires a shift from rigid and mechanistic structures and processes to flexible and organic ones. Indeed the changing economic, social, technological and political contexts in the last few decades have been translated into forms of organising which embrace intra and inter organisational collaboration, partnering and thus closer coupling of roles, tasks and knowledge.

In the last few decades, some notable changes have occurred in the manufacturing sector. Price traditionally has been the most competitive factor for most organisations. Many organisations tried to achieve this objective through mass production and thus economies of scale, product rationalisation, division of labour, improved machine utilisation, automated production and de-skilling and elimination of direct labour and so on. However, in the last few decades markets have become more global, dynamic and customer driven and competition on price alone is no longer a viable business strategy. Other aspects such as quality, flexibility, delivery and responsiveness have become equally important. Product price is still a relevant competitive factor but customers are demanding a wide range of high quality products, which may be configured to meet their specific requirements. In essence rising customer demand combined with intensified international competition has resulted in a transition from sellers to a buyers market. Thus, many firms on the one hand have to be lean from manufacturing point of view and whilst on the other hand agile manufacturing has to cater for diverse markets and tastes (Figure 1). In contrast to lean manufacturing which is a reinforcement of mass production systems the agile paradigm forces the organisation to challenge its basic business assumptions'…it cannot be an add-on, it requires rethinking of what the organisation is doing and changing the nature of business relationships both internally and externally.

New product development is often pivotal to the success and survival of most manufacturing organisations. Kidd (1994) in his book debates this point extensively and has summarised the main characteristics of the design paradigm underlying the traditional manufacturing as:

- A reductionist approach
- Monodisiciplinary professions and views
- Serial engineering
- Optimisation at the task level implies optimisation at the enterprise level
- Concentration on technology
- Emphasis on cost reduction

In contrast, the products demanded are becoming increasingly varied and complex. The fierce competition and globalisation of the last decade combined with increasingly environmental awareness have led to even higher demands. Shorter innovation cycle and recycling oriented product design represent additional requirements. In the future changes in both demands and market structures will accelerate even further. This development will intensify the globalisation of firms' activities and will make it necessary to produce market-oriented variations of any new product. Notably, product life cycles are becoming shorter and there is increased emphasise on bringing products to market quicker than competitors (Pawar, Riedel & Menon 1994). Thus, the opportunity to recoup product development costs over a long period is becoming outdated.

In this context, serial approaches to product development are becoming obsolete. Concurrent engineering (CE) principles and practices have replaced this outdated way of designing and developing new products. By adopting CE principles (parallelism, continuous improvement, goal sharing, co-operation, and integration) many organisations aspire to reduce product development lead-time, improve quality, reduce costs, and enhance customer focus. The benefits of bringing products to market quicker than competitors are: extra sales revenue and earlier breakeven, extended sales life, premium price giving bonus profits from being first, early introduction means "hooking" customers before competition and thus developing their loyalty, leading to increased market share, produces a technological edge, which improves firm's innovative image, and increases its product range.

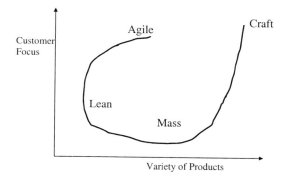

Figure 1: Shifts in Manufacturing Practices

2. GLOBAL CUSTOMER

To be competitive most manufacturing firms face some paradoxes. There is growing competition, changing customer demands (e.g. increased customer requirements such as better products in shorter times), new regulations, rising environmental concerns, changing working environments (e.g. increased use of computer-aided technologies) shorter product life cycles and other factors which are complicating the introduction of new products. Yet, manufacturing firms are forced to carry this out faster, better and at a lower cost. The markets for products and associated services are becoming fragmented and geographically dispersed at national and international level. Affordable alternative strategies to help firms adapt to these new challenges, especially with respect to time-to-market performance, are not readily available.

Most notable example can be found in Britain, where the market place has also undergone many changes over the last two decades due to the political moves to privatise major sectors these, amongst others, include electricity, gas, water, coal, transport, telecommunications, aerospace, defence and automotive etc. This has meant that these organisations have had to undergo substantial reorganisation in terms of products, services, structure, management style, processes, policies and procedures. Accordingly, a complete cultural shift from not for profit to that of competitiveness and accountability to shareholders has become necessary. This coupled with increased level of globalisation and setting up of trading blocs such as the

European single market and within the US region and the Pacific rim, has meant that goal posts are constantly changing. Thus, most firms are required to redesign their businesses on the strategic, tactical and operational levels in order to remain competitive. The emergence of globalised customer base has meant that there are newer threats as well as opportunities.

In the current highly dynamic information and communication age, the customer is likely to be more aware and better informed about the range of products and services available in the market place. Hence the suppliers/designers of products have to be competitive on a number of fronts such as variety, quality, cost, flexibility, complexity, responsiveness, delivery and service (see Figure 2).

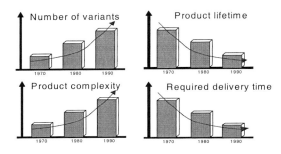

Figure 2: Changes in different aspects of a product over the last few decades

To be competitive in any market place requires continuous improvement in quality whilst at the same time reducing manufacturing costs and purchased material costs. This often requires radical changes to designs to reduce component count and variation along with significant new component investigations. This has increased the requirement on the design and development of new products and changes to existing products. Moreover, shorter timescale for design of products has increased the demand on designers and design department and other functions involved in New Product Introduction (NPI) process. In parallel, there is pressure to reduce costs, time, and staffing levels. This means that there is constant push for efficiency and effectiveness in utilisation of resources and thus to achieve much more with fewer staff. This has necessitated a rethink of organisational structures and method of operation within the design process and the establishment of NPI procedures to achieve a step change in this process. These changes have placed increased pressure on design departments and all those involved in the NPI process. Changes to the detailed processes are required to speed up the total process and meet the new challenges. This has necessitated many firms to seek new and innovative ways of designing and developing product. It also implies that manufacturing organisations need to be agile and responsive.

An 'agile' mode of organising is thus seen as one which allows for the required collaboration, facilitates rapid response ability and exploits the advances in technology.

'Agility' here involves a set of competencies; learning to learn and thus developing and managing 'organisational knowledge' which in itself highlight the organisation's ability to be proactive, analytical and responding.

Dove (1999) suggests that an agile organisation/enterprise will be engaged in 'collaborative learning'. That the knowledge created during such a process will easily flow through the organisation. The diffused knowledge can be presented in the form of a template (Sharifi 1988) or as Dove calls it 'plug knowledge', a kind of knowledge which will be shared and transferred if different units within the enterprise perceive it as relevant and compatible. Furthermore, it is argued and shown that information technology in the form of advanced communication systems is the 'necessary' medium for such transformation (Campbell, 1998).

3. NEW PRODUCT DEVELOPMENT PROCESSES IN A CHANGING CONTEXT

Increasing levels of sophistication in customers' behaviour, rapid developments in technology, formation of strategic alliance between customers and suppliers, the changing span of control of designers and influence over a given product (including design for disassembly, disposability and recylability etc.), have been some of the features of the business environment.

The changing global customers also have provided firms with opportunities for new outlets to increase exports. These new markets require particular products which may vary in a number of ways from those being sold on the home market and thus need to be designed and developed.

Indeed industrial arena has become immensely complex and the possible alternative strategies adopted by companies in terms of improved engineering design and manufacturing processes, especially with respect to the time-to-market performance factors, are crucial. Within the current highly complex and dynamic environment design managers have to ensure that their designs are compatible from a number of points of view (see Figure 3).

In recent years the long term success of some manufacturing organizations has been enhanced by their ability to bring new products onto the market at regular and shorter intervals. It seems that a number of issues in the organisational and managerial processes need to be considered. Indeed, the criteria for competitiveness in the market have been changing continuously. For instance, levels of product complexity, market demands, extent of globalisation of markets and degree of consumer awareness have varied over the last few decades. This has meant that organisations have to become more flexible and responsive to the changing needs of customers. Within this context the role of design as a function within organisations has changed significantly (Pawar, 1994). Intra-organisational as well as inter-organisational levels of consideration of the design function has significantly increased which matches the desired level of integration. Thus, the nature and type of interactions between different actors during the product design phase often influences the nature of the final product and its subsequent manufacture (Sharifi & Pawar, 1993). The iterative nature of the product design process highlights the point that a great deal of the design activity becomes re-defining and redesigning of the conceived ideas.

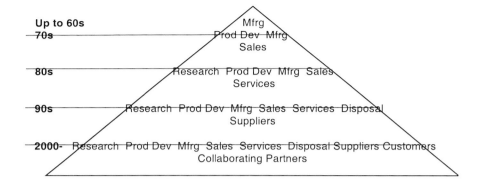

Figure 3 - Expanding span of design and development

Redesigning implies that customer requirements and their scrutiny of the design outcome, their experiences are integral to the frame of reference of the enterprise. Furthermore, the role of design function is re-defined here, and becomes one which will maintain the continuity of the process and is to do with co-ordination and communication which are the essence of concurrent engineering. In essence CE has a set of principles including parallelism, continuous improvement, goal sharing, co-operation and so on (Walker, 1997). Traditional organisational structures and approaches are restrictive regarding the changing environments. They break down the product development process across the specialist functional departments which renders poor communication, convoluted problem ownership, excessive complexity and considerable non-value adding activities, and thus prolonging time-to-market (Pawar et al, 1994).

However, CE has been described and defined in different ways with various implications. The variation in descriptions has arisen from different emphasis put by theoreticians and practitioners. For instance, Garrett (1990) describes CE in terms of "simultaneously designing of products and defining the best way to make it in order to reduce costs and cycle times..". However, Hurst (1993) reminds us that the purpose of CE is not just to reduce time to market. It can, as a method, improve the performance of the organization as a whole. Most descriptions share the assumption that CE is a means to improve the quality of product design process which is reflected in their suggestions for the development of procedures for implementation. A widely accepted definition of CE, is developed by the Institute for Defence Analyses (Pennel, & Winner, 1989): "Concurrent engineering is a systematic approach to the integrated, concurrent design of products and their related processes including manufacture and support. This approach is intended to cause the developers from the outset to consider all elements of the product life cycle from conception through to disposal, including quality, cost schedule and user requirements."[1]

A further point, here, is about the emphasis which is put on the 'integration' function and roles in the form of 'multifunctional teams' or task forces. CE calls for changing actors' mindsets or ways of thinking about problems or their personal and organizational needs or what is frequently and loosely referred to as 'culture'. The propositions are about collaboration within the organization and with stakeholders who operate outside the assumed boundaries of the organization. In Bryson and Eden's terms (1995) "collaboration is the process intended to foster sharing that is necessary among involved or affected groups or organizations in order to achieve the collective gains or minimise the losses". This process requires definition of goals which will be the desired outcomes that the organization intends to pursue.

4. A CONCEPTUAL FRAMEWORK FOR INTEGRATING CAD/CAE/CAM FOR NEW PRODUCT DEVELOPMENT IN AN AGILE ENTERPRISE

The main purpose of this framework is to show how the main elements of an agile and responsive enterprise relate to each other. The conceptual framework presented in Figure 4 provides an overview of interrelations between three dimensions namely: NPD process; support infrastructure; and tools, techniques and methodologies. All these are encapsulated within the outer 'shell' or overall context of agile manufacturing environment and enterprise.

The NPD process covers, in principle, all activities that are relevant from the conception of a product idea to the satisfactory delivery to the customer. The next dimension, support infrastructure, entails further four sub categories. These are organisational structure, processes, people and technology. The third dimension, providing support to the previous two, are supporting tools, techniques and methodologies which are required to perform various tasks or activities. From a functional perspective this layer can be envisioned as technical applications/ support service.

5 CROSS-FUNCTIONAL TEAM DEVELOPMENT IN AGILE MANUFACTURING

The most obvious and prevalent feature of NPD is the use of interdisciplinary teams. More commonly referred to as "multidisciplinary teams", the primary mechanism for implementing CE philosophy centres around the assemblage of people with specialised skills, experience and perspectives on the product development process. Such teams are expected to comprise capable individuals representing the relevant departments and/or functions in the extended enterprise and to continue for all phases of product life cycle. Further to reducing costs and time to market, which are expected primary outcomes of establishing multidisciplinary teams, there are some other macro and micro organizational outcomes identified, such as: pooling of knowledge; opportunity to leverage skills; conducive work environment; increased self motivation; increased inter-departmental communication and improved product quality. However, these outcomes are more likely to be achieved if the continuity of the membership of the team, addressing similar development efforts, is maintained over time. Moreover,

---

[1] *The PACE project set out to provide some tools for manufacturers to implement or improve CE methods and principles. CE is defined by the PACE consortium (Walker, 1997) as: 'A structured and controlled way of managing product or service development with respect to integrating resources and calendar time, sharing common goals and accurate information throughout.'*

duration and continuity are focal to getting the diverse understandings, different ways of doing things of different members closer to each other and thus in developing 'a common sense of purpose' and concordance (Pawar et al, 1999).

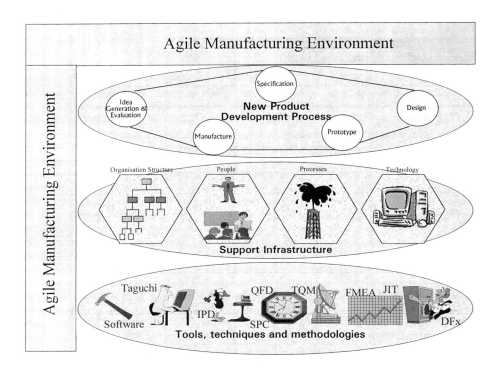

Figure 4: A Conceptual Framework showing interrelations between different dimensions of the agile enterprise

Successful use of teams is not new, nor is it unique to CE. Lockheed's use of highly motivated teams, called "Skunk Works", was used effectively in the 1940's. "Tiger Teams", "Process Action Teams" and "Integrated Product Development Teams" constitute only a small portion of representative examples where teamwork has been valued and practised. Indeed, 'organisation' by definition implies collaboration amongst its members. Team working is neither new, nor is it unique to CE. For instance Chrysler Corp. established 'platform' teams in 1989 to develop a family of large cars which were launched (Kisiel, 1998) and this is still the case with a number of car manufacturers. Similarly, over the past few decades the success of Japanese style of organising has frequently been associated with the establishment of interdisciplinary and cross-functional groupings. These teams/groupings, acting as entrepreneurial units can create the elusive collaboration between different functional disciplines by means of collocation. The following figure illustrates possible patterns in the formation of NPD Teams in a manufacturing context over the last few decades.

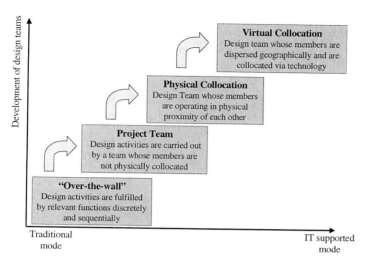

Figure 5: Patterns in the formation of design teams

The NPD team ideally should comprise product design, process engineering, manufacturing, marketing, services, purchasing, and if possible customers and selected vendors. During the selection process, team members are often selected for their specific skill and ability to deliver. This implies that teams will comprise experienced and expert individuals whose commitment can be maintained by empowering them and endowing them with autonomy and discretion over their task related decisions.

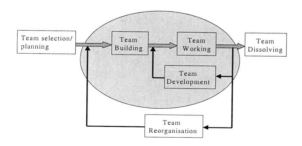

Figure 6: A Typical Team Life Cycle for New Product Development Projects

Each team member needs to be experienced enough and proficient at their individual functional tasks to ensure success. If team members are lacking in expertise or knowledge, training need to be provided. During the selection process, the importance of compatibility of members should be recognised. A willingness to take part to share information and commitment to the team is inherent part of the selection process. The overall aim is to develop a well-rounded and balanced team with the right blend of skills and experience. The size of the team must be manageable. This may mean breaking a product down into several teams, each being responsible for major component. For instance in the case of automotive industry these major components include; chasis, power train, braking system, body, heating/cooling system etc. Here, inter-team communication would become an issue for careful management. A team philosophy is the one to adopt for achieving integration and experience shows that this has to be developed and nurtured through team based training. A Typical NPD team life cycle may be as shown in Figure 6.

The way in which teams are assembled and developed can go a long way towards establishing such an environment in which team spirit and creativity flourishes. The managers, in particular, those at the locus of decision making play an important role in this and must be committed to the integration philosophy and team working. They must give their backing to the project and allow team members the time to build up relationships within the team (Bower & Hout, 1988).

The effectiveness of teams largely depends upon individual team members' qualifications, experience, technical competence and experience, the "teamwork skills" and personal characteristics of individual team members, overall team efficiency and the existence of an environment for meaningful and productive interaction (often characterised as breaking down cultural and disciplinary boundaries). Furthermore, Katzenbach and Smith (1992) have identified four 'team basics' that need to be present for teams to perform well. The team must:
- Have complementary skills;
- Establish goals and individual and collective accountability;
- Agree a common approach to getting the work done;
- Have a common purpose

The first three 'basics' can be attained irrespective of the extent of proximity or collocation (Benson-Armer & Hsieh, 1997). The fourth 'basic' is a lot harder to achieve and, unlike the others, not easy to quantify.

5.1 Collocation of NPD Teams: some issues and analysis

Collocation of teams, implies that functions contributing to the design and development of a new product are located in close proximity of each other. It has been considered as one of the main tools for enabling CE (Bergring & Andersin, 1994). Rafii (1995) has defined collocation as, "...physical proximity of various individuals, teams, functional areas and organizational sub-units involved in the development of particular product or process..." (p. 78). Major benefits of 'collocation' are described in terms of increased interactions and ease of informal communication and increase in efficiency of resource use. However, Rafii sees these gains as illusory and argues that in a globalised manufacturing and trade world 'centralised collocated product development' activities will become inefficient. That, there are also 'differences in outlook, beliefs and goals' of those involved in the process which are 'reinforced by the structure of the organization'.

However, in a globalised environment, physical collocation may not always be an efficient use of resources or an appropriate course of action to follow (Rafii, 1995). Electronic mail and engineering databases are often used to facilitate a kind of virtual collocation (Pawar & Sharifi, 1997). Virtual integration, accordingly, is achieved via the use of information technology which can provide a design data base to be shared and a disciplined development process. It may also neutralize, to an extent, the hierarchical positions of team members in the development process. It needs to be noted here that 'teaming'- virtual and physical – presents the context for collaboration and co-operation and thus 'collaborative learning'. However, the case examples in section 9.6 will highlight the point that the diffusion of knowledge is not a 'natural' progression, expected outcome or occurrence and thus cannot be taken-for-granted. Therefore the features of the 'plug knowledge' (Dove, 1999) becomes essential to the collaborating teams including the new product development team and whether or not it is diffused, shared and used.

Table1: A comparison of the typical characteristics of physically or virtually collocated design teams

	Physical Collocation	Virtual Collocation
Physical Proximity	Close	Remote
Typical Use	In small and medium sized companies with one or few sites	Multi-national and Inter-national organisations with different sites
Cultures	Limited variety of cultures, since the team members may come from the same company or site	Different people from different countries or sites, with a variety of experiences
Information Exchange	Opportunity for sharing formal and informal information (ideas, dilemmas) between team members	Limited opportunity to share informal information because of the dispersed location
Relationships	Ample opportunity for face-to-face interactions	Limited opportunity to interact and build relationships
Purpose	An evolving common sense of purpose	A directed common sense of purpose
Resources	Ample opportunity for sharing of resources (technical, human, financial)	Limited access to similar technical and non-technical resources
Technology	Fewer hiccups due to possible sharing of technical systems	Possible problems in terms of hardware, software and resources, due to variations in technical systems
Working Environment	A higher sense of belonging within the team	Feelings of isolation, and frustration, and possible absence of a sense of belonging
Accessing Information	Availability of information at anytime to every team member	Limitation in time and space for accessing information
Transparency of design activities	Greater visibility of the design work	Lack of visibility of the work being carried on by the group
Education/Training background	Similarity of work method and employment	difference in education, language, training, time orientation and expertise
Empowerment and Management of the team	A lower degree of empowerment and closer supervision	A higher degree of empowerment and delegated authority and looser control

In virtually collocated teams, integration of design /development activities is often achieved via the use of such communication means as post, telephone, fax, video conferencing, E-mail and sharing electronic data bases. These are meant to simulate face-to-face communication. Tables 1 and 2 provide a set of variable on the basis of which the two forms of teaming may be compared.

Table2: Pre-requisites for operationalizing and maintaining collocated teams

	Physical Collocation	Virtual Collocation
Monitoring of Design Activities	The possibility of immediate access to the details of the activities and the progress of the project	An ongoing monitoring of the activities for short term projects in particular
Negotiation and Compromise	A high degree of interaction and bargaining in collocated teams irrespective their being Physical or Virtual	Ditto
Commitment	Commitment and Proficiency must characterise every team member irrespective the nature of Collocation	Ditto
Participation, Cooperation, Collaboration	These characteristics are very important for the team performance	Ditto
Integration	Integration is very important in a CE environment	Ditto
Conflicting Goals	Each team member representing a different department/function may support different goals. Moreover there are Personal Conflicts as well	Ditto
Establishing Goals	The importance of setting and establishing clear goals is the same in Physical and Virtual Collocation	Ditto
Collective Accountability	Every team member wherever is located must be characterised by willingness and accountability	Ditto

6. CASE STUDIES

The following section provides a brief account of the authors' observation of new product development (NPD) processes in three organisations namely, Fast Cook Ltd., Flow Well Ltd. and Easy Chair Ltd.

Fast Cook Ltd.

Fast Cook Ltd. designs and manufactures various domestic appliances in the UK. It is a medium sized company with over 2000 employees including a design team of approximately 100 staff with various functional specialisms. Fast Cook is part of a larger group of companies with a turnover of £600 million per annum.

Fast Cook's major home market for domestic appliances has been affected by the opening up of European and World markets with greater competition from many other countries. The nature of their market has radically changed over the last 2 decades due to the political move in the UK to privatise the major electricity and gas utilities. This has increased the requirement for the design and development of new products and for changes in the design of the existing products. The time scales for developing these designs has become increasingly shorter due to competitive environment. All of this has increased the workload of the design team plus that of the departments involved in the NPD process. These design changes are required at a time when there has been a pressure to reduce the number of staff. This combination of reduced time to market and reduced human resources has necessitated a rethink of the structure and method of operation within the NPD process. Many of the established NPD procedures have had to be rewritten to achieve an incremental change.

Fast Cook Ltd. has a functional organizational structure. It used to set up teams on an *adhoc* basis whereby contribution from other functional disciplines to the design process were sought as and when required. A pilot project was introduced by the top management and lasted for 18 months and aimed at developing 20 products in a family. The specific aims of the ***pilot project*** were to:
- minimise parts and fixings
- maximise features and minimise costs
- maximise usage of standard components
- standardise basic core design
- offer the flexibility for customisation at the later stages of design process

These aims, therefore, It required the design team to be physically collocated which was seen as a way of introducing 'concurrent engineering' principles and practices within the firm. The collocated team comprised 5 full time and 4 part time members from different functions. The main objectives for the establishment of the ***team*** were to:
- shorten product design and development lead time
- break down functional barriers
- improve communication
- eliminate reiteration in the design process
- empower engineers

The designers designed according to given guidelines and specifications drawn up by marketing and pass it onto other members to be costed, and considered for scheduling. Team set targets taking into account the complexity of specifications. A prototype would become the object of discussions for improvements and possible redesign.

The performance of the team was monitored monthly by the Technical Director of the firm. The physical collocation, in the first instance, reduced the design and development lead-time and reiterations by 17%. There was a significant reduction in the number of components used in the end product (Team managed to reduce the complexity of the design yet maintaining its functionality and characteristics). There was also a 35% reduction in total investment in tooling. The ownership of design problem was more widely shared. For instance one team member commented 'we have shortened time scales and broken down communication barriers to a certain extent by physical collocation of the design team'. Whereas, project teams set up previously did not render the desired outcomes regarding innovation in the design of products. The team did not gel-together over the set time period. There was also a tendency among members to deal with design issues on casual basis and designers did not share ideas with other actors.

At Fast Cook Ltd. the implementation of physical collocation itself produced some 'tangible' and measurable benefits. These were seen at a macro level in the project performance. The main benefits were a reduction in the development time by 40% and reductions in parts, fixings and labour costs, resulting in an overall reduction of product cost.

There were 'intangible', yet noticeable, benefits too. These were found mainly at a micro level or operations level and one could argue were the drivers behind the success achieved at a macro level. These were generation of a better understanding, less conflict and improved morale amongst team members. This was achieved through re-design of the process using process modelling and analysis, which encouraged early involvement and a more controlled product development. The collaboration between members of the core team, i.e., design, development and production engineering, who produced the process models and were totally committed to collocation, was considered "a major improvement" by the senior project manager in charge of projects.

Some structural problems were encountered during the lifetime of the *collocation project*. For instance, there seemed to be an on-going delay in the communication of market research information. This led to the design freeze being postponed and dealt with later in the development process. Subsequently, some mistakes in design were identified at pre-production trial runs. Whilst others were identified during assembly operations. It may be noted here, that physical collocation of design team was part of an experiment to improve time to market in the design process, and was parachuted on the existing structural arrangements and thus roles and responsibilities of those involved were not redefined. Moreover the firm's approach to collocation was fragmented and the need to improve quality meant increase in the amount and forms of information available which complicated the process further. Indeed, lack of involvement from some other departments caused severe problems and overshadowed the successes. According to the senior project manager though collocated projects had their benefits there were some major issues and problems which had to be resolved at the outset before further collocated projects could continue. These issues were:

Selection of Team Leader/ Project Manager: The team leader/project manager needed to be committed full time to that project or at least would have plenty of time allocated to the task of project management and be a strong leader. In one of the collocated projects a design engineer was chosen to manage the project, but was also still responsible for the design of the product. This meant that not enough time could be allocated to leading and managing the team, especially in the most crucial early phases of the project, when design activities were at a peak.

Participation and commitment of non-core team members: Department representatives, not part of the core team and managers needed to be **committed** to the team and the concept of collocation. Managers would be expected to participate in meetings so that they would understand the problems of their delegated staff and hence appreciate more the benefits of collocation and CE in general.

Flow Well Ltd

Flow Well Ltd is a domestic heating and cooking appliance manufacturer. As medium sized firm with 930 employees decided to introduce two new models of gas fires, basing its design on an existing model. It was keen to reduce it time to market by overlapping its design product design and development activities with manufacturing. The case provides a brief outline of sequence of events

In Flow Well, the Production Engineering (PE) Manager plays an important role in the design process. There is close collaboration between the design and production engineering departments and their activities are coordinated by the PE manager. Here, he co-ordinates closely with the design department to ensure that problems are resolved at the drawing board stage. The firm also has a Production Co-ordinator located in the design department who is responsible for ensuring that designs are drawn up in collaboration with the production departments.

A board meeting of the company agreed to introduce two new models of gas fires to replace the ageing ones. The Technical Director discussed the basic design parameters with the Design Manager and passed the details to the industrial and engineering designers. This was all done in an informal manner, with no written design brief. It was, though, specified that the two new fires were to have differing external appearances with a functional design based on an existing gas fire.

In the first month of the design process two operations proceeded in parallel. Firstly, a working model was constructed using the functional elements from an existing gas fire and the body shell of another. Second, the designer produced detail drawings of the two new fires.

In Month 2 meetings were held almost every week which variously included the Production and Technical Directors, Design Manager, Designer, Purchasing Manager, Production Co-ordinator, Production Engineering Manager and Development Engineer. These meetings had the following outcomes. Foreseeable production queries were discussed; Timescales for buying materials and tools, for component production and for commencement of production were estimated. The possible degree of standardisation was discussed; Parts lists for both fires were produced by the Production Co-ordinator; Preliminary drawings were issued to the Production Engineering Manager by the Designer.

Month 3 saw a number of modifications made to both fires with the resulting models being sent for statutory testing. All the tooling was ordered, which would have been a risk if modifications would have been required after statutory testing. The pre-production run was scheduled for Month 6. In the following Month, 5, it was determined that the advertising campaign would begin in Month 8 and, therefore, the fires would have to be in production and in stock in the warehouses. Also specifications for the assembly shop and a spare parts list were compiled. While the prototypes were still under test, in Month 6, a final meeting on colours and design of the fires was held. In Month 7 the pre-production run was started, one month behind schedule, using components produced by production tooling. Some statutory modifications were also carried out. Month 8 saw the production of both models commence, as scheduled in Month 5. Also the delayed production tooling arrived. A few small modifications were made in Month 10. Production was then running at 85 fires per week, which it was hoped to increase to 120. This low volume was put down to the inexperience of the workers in producing the new fires.

The official launch and sale of both gas fires was completed in Month 11. In the following month full production of 120 fires/ week was achieved. In Month 16 a number of major and minor modifications were carried out which necessitated fires being recalled from the warehouse. These modifications were carried out using a change note procedure.

Overall, this was an excellent attempt to pursue a concurrent engineering approach to the introduction of new products. Two new gas fires were successfully launched 11 months after the initiation of the design process. The approach and its effective management gave the firm the edge in achieving a short lead-time to launch. The presence of two of the staff with responsibility for co-ordination between design and production - located in each department was also crucial.

Further, the approach also meant that the new fires had a high degree of compatibility with existing products and component commonality between them. A high degree of standardisation was achieved, 41 and 45% of components of the new fires were carried forward from the old designs. Secondly, 43 components were shared between the two fires (total number of components for both fires was 229). No special problems were encountered in transferring the designs to production. The somewhat risky strategy of ordering tooling before statutory approval for the designs had been given paid off in shortening lead-time, but led to a number of modifications being made and thus some extra tooling cost. This, in combination with the other design modifications carried out, resulted in an extra 21% of tooling costs being incurred.

However, regarding the quality aspect, fewer modifications were made during the transfer of the design to production, also fewer adjustments to the design for manufacture were made during production. In the sequential engineering case the designer was unaware of the available production facilities resulting in a large number of modifications being made. This and the lack of time allocated to transferring the design to production meant no attention was given to economic manufacture meaning quality was not optimised. The lead of concurrent over sequential engineering for quality is explained by the attention paid in the simultaneous approach to manufacture of the product before it reaches production.

Virtual Collocation: a mechanism for integrating design activities
Easy Chair Ltd. is a recently acquired subsidiary of an American manufacturer of steel components and one of the suppliers of automotive systems. It manufactures a range of products including electric and manual integrated seating mechanisms to control all motions of an advanced automotive seat, to fully trimmed ones. Easy Chair Ltd. has been innovative in both product and manufacturing process design and believes in the concept of 'design for manufacture'. There have been improvements in productivity through process design led by a team of manufacturing engineers.

Manufacturing in Easy Chair Ltd. is organized on a JIT basis with cells dedicated to each customer's volume product. The company has adopted Kaizen and TQM throughout its plants. It set up a cross-functional 'virtual' team to develop a seat design. The team had two geographical bases namely, UK and US. They provided complimentary technical expertise required for the seat design which was for the US market. The UK team was multidisciplinary as it included product design manager and engineers, product and manufacturing engineering directors, quality and production engineers, tool engineers, commercial director, management accountant and a project manager. Different aspects of the project [product] were developed in different sites in the UK. The members of the US team included production and tool engineers, a design and a quality manager and tool engineers. The final product was hosted in the US base.

Suppliers were located in the UK and US. Initial meetings were face to face which provided the grounds for familiarising and motivating members. Further into the project, the teams used telephone, tele-conferencing, e-mail and ISDN link as the means of communication between the teams and their customers. The effectiveness of the communication means was eroded due to time zone differences. The differences in the size of paper used in fax machines on each side of the Atlantic generated other snags prolonging the design attempts. Furthermore, the difference in CAD language and other technical languages and terminology led to certain confusions as to characteristics of the product. The systems were neither standardised nor compatible at some points. This prolonged the design process and was detrimental to the end product and time to market. Video-conferencing reduced some of the discrepancies and enhanced convergence between the team members. An exchange visit for team members on either side intended to enable them to establish common means for technical communication.

7. CONCLUSIONS

This chapter has provided a detailed account of various facets to consider when organising NPD activities with an agile enterprise. The chapter at the outset explored the relationship between the NPD and the advanced manufacturing context. It has been argued that NPD plays a central and crucial role in underpinning the highly competitive, global, distributed, and dynamic market in which we have to do business in. Customers are increasingly seeking better value-for-money high quality products and services. The authors have developed a conceptual framework showing the linkages between the NPD process; support infrastructure and tools, techniques & methodologies within the overall agile context of agile manufacturing environment.

The second half of the chapter explores the issues relating to the formation, development and operationalisation of multi-disciplinary NPD teams. Here the concept of collocation of teams (physical or virtual collocation) are compared and contrasted in detail. The chapter ends by providing practical examples in the form of three case studies from authors' own research and consulting experience to elaborate and illustrate new product development in operation.

REFERENCES

1. Benson-Armer R. and Hsieh T., (1997), Team Work Across Time and Space, *The McKinsey Quaterly*, Number 4
2. Bergring J. & Andersin H., (1994). Designing Performance Measurement Systems for Improving the Visibility of the Concurrent Engineering Process. Concurrent Engineering Research and Applications Conference.
3. Bower JL & Hout TM., (1988) 'Fast-cycle capability for competitive power, Harvard Business Review, 66 (6) Nov-Dec. pp 110-118.
4. Bryson J M & Eden C., (1995) 'Addressing public problems through collaboration: the role of 'not-goals' and the problem of assessing accountability for their achievement'. Paper presented at the 2nd International Workshop on Multi-organisational Partnerships: working together across organizational boundaries, Glasgow, June.
5. Campbell A., (1998) 'The agile enterprise: assessing the technology management issues', International journal of technology Management, vol 15, nos ½ pp82-96
6. Coutu DL (1998) 'Trust in virtual teams', Harvard Business Review, May-June, Vol. 76, No3, pp20-22
7. Dove R., (1999) 'Knowledge Management, response ability, and the agile enterprise, Journal of Knowledge Management, vol 3, no 1, pp18-36
8. Garrett R., (1990) 'Eight steps to simultaneous engineering'. Manufacturing Engineering. 105, 5. pp 41-47.
9. Hurst D., (1993) 'Concurrent engineering-a management challenge'. In Nolan P Innovation in Product Design. IMC-10 Proceedings. pp 27-44. September.
10. Katzenbach J. and Smith D., 1992, *The Wisdom of Teams*, Harvard Business School Press, Boston, Mass.
11. Kidd PT (1994) Agile Manufacturing: Forging new Frontiers, Addison Wesley London.
12. Kisiel R (1998) 'Chrysler taking platform teams to the next level', Automotive News, July 27, Vol. 72, No 5777.
13. Lipnack J & Stamp J (1997) Virtual Teams: reaching across space, time and organizations with technology. New York: John Wiley & Sons
14. Pawar, KS; Menon, U & Riedel, JCKH (1994) Time to Market. Journal of Integrated Manufacturing Systems, Volume 5, Issue 1, pp 14-23
15. Pawar KS, (1994) Implementation framework for Concurrent Engineering in the European context, Proceedings of the first conference on Concurrent Engineering, Research and Application (CERA), West Virginia, USA, August pp 111-118.

16. Pawar, KS & Sharifi S, (1997), Physical or Virtual Team Collocation: Does It Matter? International Journal of Production Economics, Volume 52, No3, December, pp283-290
17. Pawar KS, Sharifi S & Weber F (1999) 'Managing Concordance and Knowledge in Virtually Collocated Design Teams'. International Conference on Concurrent Enterprising (ICE'99), edited by Wognum N, Thoben K-D & Pawar KS, The Hague, The Netherlands, 15-17th March, pp 433-443.
18. Pennel, J.P. and Winner, R.I, (1989), Concurrent Engineering: Practices and Prospects. Institute for Defense Analyses. IEEE Global Telecommunications Conference and Exhibition Part 1. 27-30 Nov, pp 647-655.
19. Rafii F., (1995) 'How important is physical collocation to product development success?' Business Horizons Jan-Feb. pp 78-84.
20. Sharifi S (1988) 'Managerial work: the diagnostic model in AM, Pettigrew (ed), Competitiveness and the management process, Oxford: Blackwell
21. Sharifi S & Pawar KS, (1993) 'Product design: an interdisciplinary approach'. In Nolan P Innovation in Product Development. IMC-10 Proceedings. September. pp 15-26.
22. Smith PR & Reinertsen DG, (1991) Developing Products in Half the Time. New York: Van Nostrand Reinhold.
23. Walker R (1997) The need for industry to improve their concurrent engineering practices, Proceedings of the PACE'97 workshop, (Walker R & Weber F), SET, Marinha Grande, Portugal, 15-16th May, pp 1-11.

Managing People in Agile Organisations

Dave Francis

Center for Research in Innovation Management
University of Brighton, Brighton, England

1. INTRODUCTION

The annual convention of the American Society for Training and Development attracts more than 10,000 delegates. The largest attendance of any conference in Europe is at a huge event sponsored by the UK Institute for Personnel and Development each October. Approximately one person out of 200 employed people is a full-time personnel specialist. The management of people is big business.

It is a changing business too. Over the last 50 years Human Resource Management (HRM) ceased to inhabit a managerial cul-de-sac preoccupied with welfare and industrial relations. Now, for many firms, HRM is a core discipline represented on the top team and responsible for overseeing efforts to develop the human potential of the organisation as a whole.

The 'technology' of HRM has developed enormously. Today there are elaborated methodologies for defining jobs, assessing skills, developing competencies, building teams, defining rewards and so on. HRM is so pervasive that a sociologist might observe that HRM is the dominant force for intellectual standardisation across the enterprise (Mintzberg, 1983).

Let's unpack the phrase 'intellectual standardisation' for a few moments as it provides an insight into the historic contribution of HRM to the effective functioning of an enterprise. 'Intellectual standardisation' refers to how people define what can and can't be done. Many agree that standardised ways of thinking and behaviour are important, even critical. Sometimes it is called 'alignment' or 'deployment'. Intellectual standardisation provides an organisation with the capacity to be organised - it is a force for coherence.

Intellectual standardisation also reduces conflicts. After all, what happens if the managers of two employees doing the same job offer them different pay scales? Dissatisfaction can grow in a trice, industrial relations become a minefield and the lower paid person has a powerful argument for upgrading. Only when managers think alike can policies be applied uniformly.

Coherence, consistency and conformity bring benefits. People know where they are, decisions can be taken confidently and balance can be achieved between interest groups. HRM helps to achieve this. In fact, one of the major contributions of HRM has been to provide an organisation with a methodology for developing an integrated framework of policies that all stakeholders accept as being, largely, 'fair'.

This is not easy to achieve. Webs of integrated policies are frequently interdependent. For example, a firm's recruitment policy affect rewards which affects motivation and so on. For this reason HRM policies can become monolithic and slow to change.

But coherence, consistency and conformity can be dysfunctional. Especially if an organisation is seeking to be flexible, dynamic, opportunistic and re-configurable - in a word, 'agile'. Many conventional principles and practices of HRM become negative for a firm that is pursuing an agile strategy.

Consider this example. A human resource (HR) policy is introduced in a large firm to avoid excessive use of consultants. The policy says that anyone working for more than five days on an assignment must have a temporary contract signed by the Head of HR. One day the marketing department receive an fax asking for a proposal to be developed within two weeks to supply a vast military facility in the Middle East. It is decided that two specialist consultants will be needed for 10 days each to prepare the bid. The Marketing Director phones the HR Director for permission and discovers that she is on a two-week vacation. Her deputy agrees to consider the matter and asks for background papers. A welter of e-mails is generated. Eventually, the deputy HR director is unwilling to take a decision and discusses the matter with the Chief Executive - the request is finally signed. However, three days have passed, one of the consultants (the best) is no longer available and the marketing director walked around her office threatening to strangle HR "with their own red tape". The contract for the Middle East opportunity is ill prepared, does not please the client and goes to a rival firm. The next time the marketing director needed more than five days consultancy she signed two contracts (without telling HR), each of four days, so not to appear to offend the HR policy.

We see that it is not just HR policies themselves that can create rigidities. The time and effort needed to negotiate exceptions and make ad hoc arrangements inhibits all but the most tenacious managers. HRM can become a sheet anchor inhibiting flexibility. For this reason some managers spend a considerable amount of effort trying to circumvent HR policies. In one research project, I videoed a senior management group plotting for 11 minutes how to evade a current requirement of the personnel department. At least in this case, HR was seen as the enemy of effectiveness.

HRM can be a force for conservatism. It can inhibit or destroy organisational agility. But is this always true? Perhaps a different kind of HRM may promote agility. I will argue that HR managers can lead an organisation towards agility - if their policies and practices are in tune with the over-arching mission to be strategically differentiated through superior agile capacity.

2. CHANGE DRIVERS

Elsewhere in this book (chapter xx) my colleague John Bessant describes the characteristics of agile organisations. His description of the 'agile wheel reference model' can be seen as a particular configuration of the famous 7Ss framework (Structure, Strategy, Skills, Staff, Style, Systems and Superordinate Goals) (Waterman Jr., Peters, & Phillips, 1980). Work undertaken in the Agile Manufacturing Research Group at the Centre for Research in Innovation Management (CENTRIM) into three of these areas - structure, staff and style - provides specific insight into people management issues. In describing this work I shall draw on material published in (Woodcock & Francis, 1999).

There can be few organisations that have not gone through a significant redefinition in recent years. This is inevitable. New strategies, technologies, processes and markets drive changes so frequently that they can be said to be continuous. Managing a river of change requires building new competencies but, also, destroying part of what has gone before. It is impossible to re-construct an organisation without some element of destruction.

Change, inevitably, causes disturbance to people - a central concern of HR. Some argue that people do not like change and are naturally resistant. The reality is more complex. Often, change is desired, accepted or acknowledged as being unpleasant but inevitable. After all, if all change were resisted people would unwilling to get married, move house, go on holiday or visit a new restaurant. The scenes of joy shown on the world's television screens as the Berlin Wall was demolished provided a vivid image of people relishing the prospect of change.

Agile firms cannot manage change as turbulent episodes. The famous change model of (Lewin, 1947) suggests that change occurs in the movement from one frozen state to another via a period of unfreezing. This is inappropriate in an agile world. There is no frozen state - no unfrozen state - but a constant blend of both. An agile firm is always in a process of inventing itself. This requires frequent adaptation, sometimes to new or unpredicted challenges. The turbulence of opportunity in external environments must be matched by the speed of a firm's adaptive capability.

Increased turbulence renders the management challenge more demanding and increases the significance of the contribution of managers. Static firms focus managerial attention on perfecting current systems. Agile firms adapt systems, methodologies and processes all the time. There is more managerial work to be done.

Prudent opportunism is one of the most pervasive characteristics of an effective agile enterprise. Managers need to become intrapreneurs (Pinchot, 1985) as well as guardians. Driven by intrapreneurship agile organisations seem to be intelligent and alive, not bureaucratic machines mindlessly implementing the bidding of a small cadre of top managers. The language needed to describe such an organisation, and the metaphors that we use to make sense of its intelligent wholeness, is organic, dynamic and fluid (Morgan, 1986).

In those firms in which the pursuit of organisational 'agility' has become a deeply embedded objective the dominant question has changed from "How should we organise for this task?" to "How should we organise to achieve any task we might need to perform?" An agile organisation is more extensively integrated than previous organisational forms. In an ideal case, elements of the firm work together with a near seamless effectiveness. Decisions are taken throughout the organisation with an awareness of the impact of the commitment on the whole firm.

This requires the rethinking of many underlying assumptions. Flexibility needs to be built-in to facilities, equipment, systems, people and organisational forms. The cost of maintaining agility must be brought down so that the principles of lean organisation are not prejudiced by an escalation of costs incurred in the search for market-focused flexibility.

Agility helps managers rediscover their true role. In an agile organisation intelligence is distributed widely and decisions are taken much more frequently than in a rigid organisation. This increases the need for hands-on management and affirms the critical role of managers. Their role has changed dramatically in fifty years; from administrators to intrepreneurs.

No longer is the achievement of an objective the sole criterion; each decision can become a stepping stone to a more capable and agile organisation. This concern to build dynamic flexibility and capability fits well with contemporary theories that see organisations as portfolios of competencies. This approach has become known as 'the resource theory of the firm' (Hamel & Prahalid, 1994).

3. THINKING AND DOING

Until recently many employers valued, above all, skilful obedience: they liked people to do what they were told. The need for organisational agility has changed our definition of 'a good employee' to a person who can use his or her skills intelligently, depending on the needs of the moment.

Sometimes we overlook the magnitude of this change. For almost a century the principles developed by F W Taylor have dominated thinking about the best way to organise (Taylor, 1911). Taylor recommended the separation of thinking and doing - so that specialists designed work processes and operators operated. In many cases, operators were actually punished for using their intelligence and told "that is not your job".

Agile organisations cannot afford a separation of thinking and doing - they need to enlist the intellect of most, if not all, employees. This is doubly important because those at the 'coal face' are often best able to find opportunities and improvement possibilities, since they deal with them all the time.

There is a second, perhaps more important, reason for seeking ways to channel the intelligence of people into their work - involvement increases motivation. Its opposite, treating people as machine substitutes, provokes resentment, alienation and enmity.

4. A CHALLENGE TO HUMAN RESOURCE MANAGEMENT

Agility presents a challenge to HR managers as conventional HRM terminology, indeed the underlying structure of thought that guide the development of HR policy and practice, is largely administrative rather than organic. An example makes the point. In the journal People Management the following sentence introduces a news item, "D-Day for the UK's implementation of the parental leave directive is 15 December. On that date, changes to family rights envisaged by the Employment Relations Act 1999 come into force" (28/19/99 - page 27). The news item summarised enhanced rights for new parents and the conditions in which they apply. All over the UK personnel managers pondered the impact of this new legislation, amended policies, briefed managers, devised forms and so on. This was essential work but administrative in nature, no doubt mirrored by similar legislation in other countries.

If agile firms need to be well-functioning organisms rather than perfectible bureaucracies is there a place for standardised administrative practices like the implementation of the Employment Relations Act 1999? To begin to answer this question it is helpful to consider one of the most agile organisations of all - the military.

The armies, navies and air forces of the world must be agile - one week they may be helping victims of an earthquake, then they may be curbing a terrorist threat and, a few days later, forming part of a U.N. task force in the Balkans. Military leaders have a long history of developing agile competencies. The agility of military units is hard won and systemised. Much effort is invested in building flexibility into people and systems. Military managers place great emphasis on selection, training, development, career progression, reward systems, team building, communication, discipline and so on. It seems that, at least in this case, formalised policies and practices are not eschewed but embraced. An agile army marches on a road of dense administration.

This insight is helpful. We see that policies, practices and disciplines are not automatic hindrances to the development of agility. Indeed, if the military example is valid, formality and systemisation can be enablers of agility - if their purpose is defined and dysfunctions avoided.

How then do the military achieve such a high degree of agility? I will draw an example from the US Army (Lengy, 1996) who describe the procedures being used to create tomorrow's military capability (Force XXI) in the following way:

"Using Joint Venture (JV), the Army is executing a series of Advance Warfighting Experiments (AWE) and Advanced Warfighting Demonstrations (AWD) to define the force of tomorrow: FORCE XXI. As the Army creates FORCE XXI, we must concurrently develop the means and methods to train and sustain the force. To achieve the maximum potential of FORCE XXI, the Army must use a spiral development process allowing early decisions based upon projected requirements and emerging concepts. By using the spiral development process, the Army can leverage technological improvements to continually integrate changes as tomorrow's force is developed" (1).

This quotation emphasises the importance of experiment, definition, training and "a spiral development process allowing early decisions based upon projected requirements and emerging concepts". The US army's deliberate management of constant innovation requires agile structures, staff resources and styles - it is the job of military HR managers to provide them.

If HR policies and practices are to facilitate such agility then HR managers need to understand the concept. There is some evidence that they are doing this. (Ulrich, 1997) writes:

"HR professionals... make sure that initiatives are defined, developed and delivered in a timely manner; that processes are stopped, started, and simplified; and that fundamental values within the organization are debated and appropriately adapted to changing conditions... Successful HR change agents replace resistance with resolve, planning with results, and fear of change with excitement about its possibilities" (152).

(Stredwick & Ellis, 1998) sees the primary requirement in terms of flexible working practices. They write "flexibility is required both *vertically* and *horizontally*. In the vertical sense, employees carry out work which could be regarded as above their job... in the horizontal sense, employee's skills and knowledge are stretched to cover a variety of activities in their area" (9-10) (authors' italics).

Both Ulrich and Stredwick & Ellis argue that HR managers can enable firms to be agile. But is this enough? Is a facilitative contribution from HR sufficient? I suggest that the answer is 'no'. HRM, I believe, has a major task to *lead* organisations towards agility.

In a single chapter it is not possible to explore all aspects of the new HR task. There are many questions, for some of which of which there are no good answers. My top ten questions (in random order) are:

- How do you motivate great people who you want to use (just) as a temporary resource?
- Is the task of constantly inventing and reinventing the organisation excessively stressful for individuals?
- How do you reward people whose jobs are always changing?
- What is the role of teams when work groups are constantly forming and reforming?
- What is the function of strategy when it is devised in real time?
- If initiatives come from the bottom of the organisation, how is organisational alignment maintained?
- What is the role of communication when the top of the organisation does not know what is going on?
- How is learning to be managed when it is a core process for driving the organisation forward (rather than a means of bringing everyone up to a standard)?
- What is the role of social status in an organic organisation?
- How are managers to be developed for an agile organisation?

These are demanding questions and I shall explore the last, concerned with management development, in more detail as an example of the scale and scope of new thinking required.

5. DEVELOPING AGILE MANAGERS

Becoming an agile organisation requires a nexus of changes in responsibilities, cognition and behaviour of managers. This affects all who take decisions including, sometimes, front-line staff - their decisions may commit the company and can involve making complex trade-offs. The definition of who is a manager becomes fuzzy and the task of management becomes diffused.

Managers need to prioritise and re-prioritise with a frequency that can be breathtaking. Previously, for example, it was considered possible for a manager to sit down with his/her boss and agree a raft of annual objectives. In an agile organisation objectives could change weekly, even daily. In such an environment I argue that a manager needs to:

- become more outgoing and decisive;
- develop business skills so that s/he becomes an internal entrepreneur;
- manage learning since this, more than any other activity, provides agile potential to individuals;
- gain skills in the management of innovation - both incremental and radical;
- be close to potential customers, not just existing customers.

These points provide clues about the emerging role of management in an agile organisation. It is both exciting and daunting. We expect more from the manager of the future, but need to give more too. Tomorrow's managers need specific rewards, training, recognition and business-building skills. Their development agenda is substantial but provides managers with dignity and purpose in an increasingly central role.

More generally, the need for organisational agility has reaffirmed the vital significance of people at the heart of an enterprise. It is possible to automate most routine systems but difficult to automate systems which are constantly changing. The increase in uncertainty in an agile enterprise, which is inevitable, requires that the most adaptable animal of all - the human being - to become central. Numerous options, choices, tradeoffs, changes of plan and resource allocation decisions must be made. Many of these require creativity, problem analysis and rapid problem solving - a task for skilled, motivated and flexible people.

6. SEVEN AGILE COMPETENCIES

There has been little systematic analysis of the competencies required for an agile organisation. Doubtless more work on this topic will be undertaken in the next few years. In order to initiate a discussion, I will return to the example of the military, in particular examining ways that army personnel are socialised and developed. From my studies, seven competencies can be identified

- All army personnel are taught to analyse situations carefully and act decisively (they show perceptive decisiveness).
- All manner of problems are simulated so that army personnel become expert in rational problem solving but are encouraged, also, to use their intuition (they show two-brained problem solving skills).
- Those in command are trained in team leadership and form effective teams quickly (they show multiple team leading skills).
- When new situations occur a strategy, tactics and an implementation plan are devised quickly (they show change leader skills).

- Military units and individuals work together for a common cause (they show partnering skills).
- Opportunities are identified and seized (they show intrapreneuring skills).
- There is a continuous emphasis on training and development. Military training has gone beneath the level of skills to develop the 'character' of each solder and officer (they show dynamic learning skills).

Interestingly, military units have found that effectiveness in turbulent times requires more attention to structure, not less. Fluid structures serve to prevent rigidity, not promote it. It seems that only with intensive attention to structure and process can short-cycle re-invention of organisational paradigms occur. Agile organisations require integrity, reliability and predictable quality of performance - otherwise they fall off the edge and into chaos. Agility, in this case, is not an alternative to disciplined management but a consequence of it.

Key to the success of military units has been an extraordinary emphasis on training. It has been recognised for many centuries (certainly since Roman times) that abilities, skills and knowledge need to be deeply embedded in each officer and soldier. The breadth of competencies developed in individuals provides a behavioural foundation for agility.

In non-military firms a similar pattern can be observed. Objectives and priorities are revised frequently. Indeed, the rapid cycle-time of review and reformulation provides a dynamic organisational structure (Eisenhardt & Brown, 1999). For many firms this is a process innovation - leading to the conclusion that HR managers need to become dynamic process designers.

Seven key competencies can be identified (see (Woodcock & Francis, 1999) for a more extensive discussion). These are shown on figure 1 and discussed below.

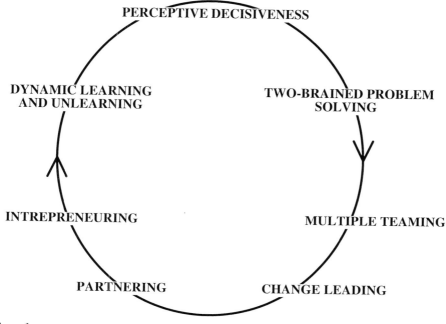

Figure 1

Competency One - Perceptive Decisiveness

In agile firms more people take more decisions. This is inevitable in responsive, fast-moving organisations as those at the top are unable to take the quantity of decisions needed. People low in the organisational hierarchy may take decisions, or fail to take decisions, that shape the firm's future. Some of these decisions are 'judgement calls' rather than puzzles with a correct answer. Assessment of risks, options and likely consequences is key. Decision-making requires the ability to understand when a decision is necessary, find and weigh options and carry a decision through to action. The decision-maker needs to be both perceptive and decisive.

Competency Two - Two-Brained Problem Solving

Finding and solving problems has been an element of the managerial role for many years. More recently we have realised that rational techniques (which are located in the left-hand side of the brain) need to be enriched with intuition and emotion (located in the right-hand side of the brain) (Ornstein, 1972). 'Two brained' problem solving combines both the rational and the intuitive approaches.

Competency Three - Multiple Teaming

When team building developed in the space industry of the 1960s the main requirement was to construct high-performing groups that shared objectives - for example, to launch an Apollo rocket. Once they had been developed, team-building techniques proved effective in a wide variety of settings, from factory workers making dog food to the executive corridor of a global business (Francis & Young, 1994). More recently, we have realised that many teams form and re-form frequently and team building can become counterproductive if it encourages a silo-mentality. Today's managers need to have the skills to construct and de-struct multiple teams. The skills include joining, facilitating, contracting, leading and exiting.

Competency Four - Change Leading

It is often said that 'change is the only constant'. This statement is only partly true, as change is episodic with periods of relative stability being followed by times of fundamental upheaval. Distinct skills are needed in periods of incremental and transformational change. In periods of incremental change the manager acts as a catalyst, avoiding complacency and finding a myriad of (often small) opportunities to make improvements. Leading transformational change requires visioning, extensive planning, coordinated initiatives and, frequently, managed culture change.

Competency Five - Partnering

Agile managers become partners with people inside and outside of the organisation. The partnering style of management extends to suppliers, contractors, customers and other affiliations. Partnering is based on the belief that many relationships work best when there is equality and responsibility - not dominance, formality and subordination. Easy, open communication is needed. Partnering requires managing by negotiated agreements. Managers need to acquire skills of developing authentic win-win relationships and of giving recognition to contributions.

Competency Six - Intrepreneuring

For many generations, perhaps even in the embryonic trading businesses which developed in the Paleolithic period, certain men and women have become entrepreneurs - they made things happen, created businesses and acquired wealth. Until the late 20th century such people thrived largely outside of corporate life. Established organisations were uncongenial places for entrepreneurs as opportunism was stifled - except at the very top of the firm. In the final decades of the 20th century this began to change, as just just a few people at the apex of the firm could not fill the need for marketing-led agility. Employees needed to think like business-builders and champion innovative ideas. A new term, 'intrepreneur', was coined for people who created intrepreneurial opportunities within firms. Their skills include innovation management, building businesses, advocacy and risk assessment.

Competency Seven - Dynamic Learning and Unlearning

Management agility requires an agile mind. Many of the impediments to agility are psychological - they exist within the mind of man. Becoming an agile organisation requires that people learn to perceive their existing way of seeing things as open to question. No-one could operate without habits, practices and routines. In the past such unconscious standardized behaviour was often constructive - if good habits were learned they provided a set of inner guidelines which assured capable performance. In an agile world the habitualisation of thought and practice is undesirable. People need new ways of thinking and doing. This requires learning: not the simple acquisition of new skills but a fundamental re-appraisal of the way things are done. We call this 'dynamic learning' - since the intention is not to acquire new knowledge and skills and, also, to unlearn at the same time.

7. DEVELOPING AGILE COMPETENCIES

A key question is "How can these competencies be developed?" Each competence makes a distinctive contribution to agility, but agile organisations require more than agile people - systems, processes, infrastructure, strategy and measures need to be agile as well.

Intellectual appreciation will be insufficient. There are many examples of people who know the rhetoric of management but cannot fulfil the role requirements in the real world. Learning must take place at a deep level. The intellect must be developed but also perceptions and habits. It is not until we perceive things differently that we act differently. Since management is a science of action, it is necessary that learning should be applied in the real world, not a classroom or counselor's office.

The challenge for management developers is simple in concept but complex in execution. Management developers must become agile themselves! They need to create waves of innovation in learning, employ state-of-the-art methodologies, work on just-in-time principles, supply a market size of one and develop the seven competencies described above.

The same message applies to other parts of HR - recruitment, benefits, motivation, employment policy and so on. HRM needs to become and remain agile - otherwise it will inhibit rather than lead organisation development.

Whilst agility is an increasing requirement it is not everything. People will always save for their pensions, need to abide by employment laws and be required to adopt standardised practices for dealing with racial harassment, discipline and the like. HR managers need to provide both coherence and agile capability.

Firms will differ, perhaps more than they have in the past. We can all think of an example of a firm, perhaps a favourite restaurant, which hasn't changed in years and is "doing very nicely thank you". Some industries are inherently more agile than others. If you build computers then agility is a prized competence: if you manufacture 20 year old malt scotch whisky then agility is less likely to be a concern (at least for the time being). Also, not all parts of an organisation are required to be equally agile. For example, a television company's news team need to be highly agile but their video library department less so - their work requires the painstaking cataloguing of each day's output. This differentiation has consequences for human resource managers who have hitherto sought to unify and standardise whereas, in the future, they may need to devise local solutions - to become agile themselves.

Human resource management specialists must weave administrative excellence and relevant development of agile capability – helping their firms become intelligent, proactive, fast and innovative.

In the introduction to this chapter I mentioned that approximately one employed person out of 200 is a full-time personnel specialist. All need to consider whether their activities are enhancing their organisation's agility and whether their own departments have adopted agile principles and practices.

REFERENCES

1. Eisenhardt, K. M., & Brown, S. L. (1999). Patching: Restitching Business Portfolios in Dynamic Markets. HBR (May-June), 72-82.
2. Francis, D., & Young, D. (1994). Improving Work Groups (2nd ed.). San Diego: Pfeiffer & Company.
3. Hamel, G., & Prahald, C. K. (1994). Competing for the Future. Boston: Harvard Business School Press.
4. Lengy, R. (1996). Army Training XXI. Vol. 1999, US Army.
5. Lewin, K. (1947). Frontiers in Group Dynamics: Concept, Method and Reality in Social Sciences. Social Equilibria and Social Change. Human Relations, 1(1), 5-41.
6. Mintzberg, H. (1983). Structure in Fives: Designing Effective Organisations. Eaglewood Cliffs, N.J., USA: Prentice-Hall.
7. Morgan, G. (1986). Images of Organization. Sage.
8. Ornstein, R. F. (1972). The Psychology of Consciousness. New York: Freeman.
9. Pinchot, G. (1985). Intrapreneuring. New York: Harper & Row.
10. Stredwick, J., & Ellis, S. (1998). Flexible Working Practices. London: IPD.
11. Taylor, F. W. (1911). The Principles of Scientific Management. New York: Harper.
12. Ulrich, D. (1997). Human Resource Champions. Boston: Harvard Business School Press.
13. Waterman Jr., R. H., Peters, T. J., & Phillips, J. R. (1980). Structure is not Organization. Business Horizons (June).
14. Woodcock, M., & Francis, D. (1999). Developing Agile Organizations: Theory and Interventions. Aldershot: Gower.

Part III

INFORMATION TECHNOLOGY/SYSTEMS IN AGILE MANUFACTURING

Application of Information Technology in Agile Manufacturing

Henry C. W. Lau[a] and Eric T.T. Wong[b]

[a]Department of Manufacturing Engineering, The Hong Kong Polytechnic University, Hong Kong.
[b]Department of Mechanical Engineering, The Hong Kong Polytechnic University, Hong Kong.

1. INTRODUCTION

The era of competing primarily on the basis of quality, price, or reliability is passing quickly for many manufacturers. It is generally accepted that the current competitive priority for a world-class firm is agility. Agility refers to the ability to produce and market successfully a broad range of low cost, high quality products with short lead times in varying lot sizes, which provide enhanced value to individual customers through customization. It can be seen that the emphasis is now on adaptability to change in the business environment and a proactive way of approaching to market and customer needs.

Common features of agile manufacturing include flexible production technologies, rapid product and process development, as well as partnerships with suppliers and competitors. While flexible manufacturing is focused on rapidly setting up the plant for producing different types of components, agile manufacturing focuses on rapidly setting up the entire organization for manufacturing different products. This includes monitoring customer demands closely, identifying market opportunities for new products, rapidly developing products jointly with other companies, developing the production operations simultaneously, and using a flexible and responsive physical distribution system for product delivery.

These characteristics have triggered structural changes in how companies are organized to create, produce, and distribute goods and services. At the same time, the economics of doing business have shifted. The strength of the mass-production system -- its ability to function well in less developed societies -- makes it increasingly difficult to realize adequate profits to justify locating such a facility in the more developed economies. It is no longer sufficient just to choose between functionally or divisionally organizing. Operations that produce market-segmenting, knowledge-based, service-oriented products customized to the requirements of individual customers, however, are likely to succeed in those societies. The current technological environment enables businesses to virtually organize. As information technology(IT) has advanced, coordination costs have declined significantly, and companies are now able to form partnerships where separate firms specialize and activities are coordinated through decentralized information systems.

The information-processing capability to treat large numbers of customers as individuals is permitting more and more companies to offer customized products while maintaining high volumes of production. The most prominent feature of the information world of tomorrow is the expansion of the cyber space. The emergence of a cyber counterpart to our physical reality is becoming an all-encompassing experience. As the volume and variety of on-line transactions grow, manufacturing activities in cyber space will be as influential on most organizations as those conducted in the physical world. The cyber world, however, could operate in ways quite different from its physical counterpart. Its use and exploitation would therefore require somewhat different approaches in many aspects. The emergence of information technology is making it possible for groups of companies to coordinate geographically dispersed facilities into a single virtual company and to achieve powerful competitive advantages in the manufacturing industry. The authors wish to present the case that data mining and distributed object technology can be significant tools to achieve manufacturing agility, especially where rapid distribution of information is a feature, as in virtual organizations. In the following sections the potential benefits of IT applications are first described, followed by a consideration of the information needs of a manufacturing system. Then the role of data mining and distributed object technology in supporting the information infrastructure of virtual organizations are discussed. Finally, an illustration of IT application in agile manufacturing is shown.

2. POTENTIAL BENEFITS OF IT APPLICATIONS

2.1. Marketing and Retailing

The development of IT may help to improve product marketing and retailing in several ways. Bar-coding, product recognition systems, Internet Protocol(IP) international toll-free telephone service and Electronic Point Of Sales (EPOS) are in wide use and help to increase the accuracy and speed of sales which lead to improved customer service. The Internet certainly enables organizations to reach customers who is geographically remote (Quelch and Klein, 1996), providing opportunities, in particular, for small and medium-sized enterprises. Gilmore and Pine (1997) outlined how the Internet also facilitates customisation. Rathnam et al., (1995) have examined the usefulness of IT to increase the co-ordination among customer support teams and Kauffman and Lally (1994) have developed an approach to evaluate the advantages of customer access information technologies.

More and more companies understand the speed and extent of the shift to electronic commerce conducted between businesses, homes and countries and start to put into place the means of controlling such invisible processes. For example, companies can offer their products through the Internet, including photos and product information, and clients can procure products and services through this means and return opinions about the features of the products/services through the e-mail system (Chandler, 1998).

Companies can also use IT to analyse customer surveys, which information can be stored as a database and used for mail shots, for targeting specific products or services, and for facilitating the decision making processes on new products and services. IT can also lead to efficiency in market research since the statistical techniques required for this are too complex to apply without computer aids. Although there are obstacles to be overcome prior to a full-scale implementation of e-commerce, e.g. software interfacing problems, transaction security, etc., it is anticipated that an increasing number of companies will use IT as marketing and retailing mechanisms.

2.2. Product Development

The capacity to innovate increases with the use of IT (Schein, 1994). Computer-aided design (CAD) and artificial intelligence (AI) technologies are a fundamental aid in the design process because, through CAD and expert systems, the design of products according to consumers' needs can be undertaken at a faster rate and the innovation can be greater. Moreover, an effective new product design and development process requires information from different departments (production, marketing and R&D) and IT should aid the effective and speedy transmission of this information and design decision making, e.g. through IP video conferencing and IP telephone messaging facilities. Hameri and Nihtila (1997) report a case study in which the design projects involved numerous teams from various locations. These Web-based applications in new-product development efforts provide the effective media for communicating and disseminating information.

2.3. Production Operation

IT has been found useful in the task of process flow management. Automation helps to reduce deviations in the production process, because machines usually demonstrate less variability than workers and increases the speed of production processes with a significant quality enhancement (Freund et al., 1997). Both electronic detection and signaling devices also help to reduce process variance. These types of applications have led to the reduction and eventual elimination of a number of periodic inspections (Litsikas, 1997).

IT also assists the maintenance function through the use of remote sensors and telemetering systems to detect the need for machine maintenance and diagnose what needs to be done; this can be carried out at a location far away from the main plant (Dilger, 1997).

Statistical Process Control (SPC) may be facilitated, through the automated measurement of product and process parameters and the registration and processing of data (Kendrick, 1995; Papadakis, 1990). IT can also be useful in bridging the gap between SPC and statistical problem solving as described by Layden and Pearson (1992). For instance, Gong et al. (1997) have proposed a procedure for combining an on-line sensor and a control chart to improve statistical process control decisions. Such types of application are a clear indication of the way in which IT can help to achieve agility at a shop floor level.

The design of processes to ensure that products conform to quality requirements is a key issue along with the control of processes in which transactions are conducted on-line. Patterson et al. (1997) provide an example of the need of new quality control and improvement tools created as a consequence of the use of CNC machinery. In relation to this there is a need to develop appropriate algorithms and software interfaces to evaluate the effects of process interfaces and changes to processes and systems, prior to their implementation.

IT also enables companies to run all their global businesses on one system. They can operate the same manufacturing processes in every country and have real time data such that the management is well informed of the current state of the business anywhere in the world (Palvia et al., 1996; Paxton, 1997).

2.4. Supplier Relations

As in the case of customers, systems of electronic data interchange (EDI) can help to develop improved communication links with suppliers. The electronic transmission of data can be used to place orders, send product specifications, design details, etc., along with confirmation of invoices and paying for suppliers (Jonscher, 1994). Indeed suppliers can be involved earlier in the design process by the use of IT (Teague ET al., 1997). In some cases, companies can access the inventory systems of their suppliers and place orders automatically and there can also be access to production scheduling systems. Mukhopadhyay et al. (1995) report the considerable savings achieved by an automobile manufacturer using EDI systems between itself and suppliers. The study of Banerjee and Sriram (1995) shows that those organizations that have encouraged their vendors to use EDI appear to have significantly improved organizational efficiencies. The research of Srinivasan et al. (1994) concluded that investments in IT to support both the sharing of Just-in-time (JIT) schedules and the establishment of integrated information links are related to significant reduction in the level of shipment discrepancies.

Currently, quite a number of authors postulate that electronic transactions and their accompanying systems will re-configure how manufacturing organizations function and this, in turn, will influence the advancement of manufacturing agility. Consequently, the necessary IT interactions between a company and its supplier should be considered in terms of how they can enhance the communication process between the concerned parties, including access to databases, systems and the necessary integration and software interfaces.

3. INFORMATION INFRASTRUCTURE OF AGILE MANUFACTURING SYSTEMS

To construct an information infrastructure to support the management of virtual organizations, companies could streamline the functions of order processing and distribution, product development, supply network, and production operation - the essence of agile manufacturing - through information technologies. In this section the information needs for a manufacturing system are discussed.

3.1. Order Processing and Distribution

A manufacturer that provides customers with a tangible product must perform the activities of order processing, product distribution and invoicing. In recent years the effort to shorten time has focused on the manufacturing phase. However, even if manufacturing lead-time is reduced to a short duration, the total revenue opportunity interval could still take a long time if the other functions are not improved. To achieve agility manufacturing systems must be streamlined to support minimum activity lead-time. This will require new methods to rapidly assimilate information from the customer-shared databases, high speed networks for transmitting information, and paperless transactions (Maital, 1994; Pandiarajan and Patun, 1994). Customer orders can be placed directly to the factory using electronic data interchange (EDI) which immediately updates planning and execution systems while, at the same time, issuing material requirements to suppliers. At the end of the activity cycle, invoices and financial transactions are handled electronically. For such an online order process, planning and execution system, software companies have responded with a variety of offerings, often referred to as enterprise resource planning (ERP) systems.

3.2. Product Development

Shortening product developing cycle time is a fundamental concern as the next millennium approaches. According to Meyer and Purser (1993), manufacturing companies must be able to identify, satisfy, and be paid for meeting specific customer requirements for products faster than the competition. Various approaches have been proposed to achieve faster product development. For instance, it has been suggested that most product development strategies should be based on a team approach (Wolff, 1992; Ranney and Deck, 1995), so as to enhance horizontal communications and cross-functional co-operation (Kelsey, 1995). Others suggested that the development process should be viewed as one in which projects move through the knowledge-work equivalent of a job shop (Adler et al., 1996). Again, this requires a cross-functional project team approach, which is usually referred to as concurrent engineering (Swink et al., 1996).

As fast product development involves input from interest groups other than engineering, the information needs of such groups should also be coordinated. Ideally, operational requirements of engineering, marketing, production, design, R & D, finance, purchase, quality, suppliers and customer representatives should be made known to each other, perhaps under the co-ordination of engineering.

3.3. Production Operation

It is widely accepted that production process capability is the main attribute to achieve speed, flexibility, and better response time/service to customer demands. In order to produce highly customized products in very short lead times many firms will require a dynamic reconfiguration of their manufacturing processes to meet customer demands. For instance, through internet-based technology a direct link to the factory is provided. This allows product design, programming of tool paths for equipment and production to be carried out simultaneously as soon as orders are confirmed. The shop floor then uses electronic customer instructions to fabricate and assemble the product with automated equipment.

Indeed, when orders enter the system, production plans, manufacturing schedules, material flows and vendor requirement databases are updated at once and monitored electronically. This requires an enterprise-wide view that takes advantage of forming virtual alliances with other support organizations, such as suppliers and freight forwarding agents (Moskal, 1995). In such case, the Intranet can be used to support most of the communication necessary for partnership formation since it provides a system that is more flexible than EDI and GroupWare, is not geographically constrained like a local area network (LAN), and is also more secure than simply using the Internet.

3.4. Supplier Needs

Partnering with suppliers is becoming recognized as a potential trend. Companies adopting supplier partnership programme have reported that the speed they have gained from being able to leverage the combination of their own resources with those of their suppliers have outweighed the risk of leaking proprietary information (Willis, 1998). In order to achieve agility, improved communication links (both internal and external) and enhanced information systems are needed to create a more cooperative customer-supplier environment.

3.5. Comparison of IT

The information technologies mentioned above include EDI, GroupWare, the Internet and Intranets. EDI involves the direct routing of information from one computer to another without interpretation or transcription by people, and to achieve this the information must be structured according to predefined formats and rules, which a computer can use directly. Although EDI has been shown to facilitate accurate, frequent, and timely exchange of information to coordinate material movements between trading partners, and suppliers receiving JIT schedule information achieved better shipping performance, the problem with EDI is the lack of a globally recognized standard format for data storage and transfer. Because of this, organizations must agree upon the translation software and data format on a project by project basis. Consequently EDI may not be a good choice for supporting quick response in agile manufacturing.

GroupWare applications can help coordinate work through: (1) making available to project members a common body of information, (2) tracking work flows so that project members can collaborate from a distance, and (3) provision of a platform for communication and interactive discussion. However, it can be expensive and it cannot be used to gain access to remote computers that are not GroupWare servers. Although GroupWare provides more flexibility than EDI, it is still not flexible enough to enable firms to quickly form a partnership to react to a market opportunity. One solution to the inflexibility problems of EDI and to a lesser extent GroupWare is using the Internet. Several technologies have been integrated successfully into the Internet and these include the WWW, Telnet, FTP, Email, and Videoconferencing. Specifically the Internet provides a mechanism for global access to both external data and customers. The Web would allow collaborators in remote sites to share their ideas and all aspects of a common manufacturing project.

An Intranet is essentially any site based on Internet technology but placed on private servers and designed not to allow unauthorized users. Hence it merges the advantages of the Internet (global access) with those of LANs (security). It appears that Intranets, utilizing the WWW and any Net browser, can be used to support the communication required for agile manufacturing, viz. external access and inter-organizational coordination.

4. IT DEVELOPMENT FOR AGILE MANUFACTURING

With the advances in information technology, the core of manufacturing activities has shifted from the physical production of goods to the systematic processing of knowledge to create value for customers, capitalizing on the utilization of innovative information-based tools in the global marketplace. In general, manufacturing-related business services that exploit the advances of information technology are expanding fast. The competitiveness of tomorrow's manufacturing enterprise lies in the development and adoption of a wide range of digital technologies and tools to meet the challenge of the next decade. The formation of agile manufacturing networks, taking advantage of the latest development in information technology, is taking up momentum to meet this challenge. In particular, data mining and distributed object technology have received significant attention for achieving agility of manufacturing systems. This section will discuss the development of these two technologies, which have played an important role in transforming the business operations of manufacturing companies.

4.1 On Line Analytical Processing

In this new business model of agile manufacturing, the partnership synergy, which is built upon the mutual effort, collaboration and trust among business partners with possibly dissimilar core competencies, is an important issue to be addressed. In particular, a model to evaluate the performance of partners in "real time" based on the data mining technique is to be discussed here. A case example has been covered in this Chapter to evaluate the feasibility of this approach.

Data mining is a technology that provides sophisticated analysis based on a set of complex data. Data mining tools enable the management of different data formats in relational and multi-dimensional database systems. The shared data access interface of data mining tools will enable exchange of data as well as results among various computer systems. The typical example of data mining tool is On-line Analytical Processing (OLAP), which provides a service for accessing, viewing and analyzing on large volumes of data with high flexibility and performance. The essential characteristic of OLAP is that it can perform numerical and statistical analyses of data.

OLAP data model consists of descriptive data (dimensions) and quantitative value (measures), both of which build up the OLAP data cube. Typical dimension includes location, company and time whereas typical measure includes price, sales and profit. In multi-dimensional data model, data is organized in a hierarchy that represents different levels of details. It allows users to compute a complex query, arrange data to appear on reports and switch the view of data in different dimensions more easily. In the OLAP data cube, it comprises two elements: fact table and dimension table. In the fact table, user-defined measures (calculated members) are used for data analysis. In the dimension table, different dimension levels are used for different views of the OLAP data cube.

In the traditional approach, when a user needs to retrieve information across a multi-table, the tables used for finding the specific information must be clearly defined. For example, when a user needs to know how many sales was achieved for the year 1997 in USA and uses Internet as promotion media, the tables and their relationship must be clearly defined. Then, the Structured Query Language (SQL) is used to retrieve information from the tables. An extract of the SQL script is shown below.

"Select sum (sales_fact_1997.store_sales) from sales_fact_1997, promotion, region, store where sales_fact_1997.store_id = store.store_id and region.region_id = store.region_id and promotion.promotion_id = sales_fact_1997.promotion_id and region.sales_country = USA and promotion.media_type = Internet"

In the above example, a complex statement for retrieving simple result across a number of tables (in this case 4) must be determined accurately. When a user needs to get the result from more tables, the statement will be more complex. Moreover, SQL statement cannot perform a decision support function such as If-Then statement and cannot use the intermediate result to perform calculation. So, in the traditional method, the SQL statement is not appropriate for decision support on complex calculation.

In the OLAP approach, the table needed for the query and the data used to perform calculation is defined in the dimension and measures separately. And then, a user can build up a complex calculation on a calculated member to meet his specific requirements.

Because all the calculation and analysis are pre-computed in the OLAP server, only a simple multi-dimension expression is needed to construct for retrieving identical result as follows.

"Select [Measures].[Store Sales] on columns, [Store].[Sales_country] on rows from sales where ([Promotion].[Media_Type].[Internet], [Region].[Sales_country].[USA])"

It can be seen that the above expression is simpler and clearer than the SQL statement. When the user requirements are changed, only minor part of the OLAP data cube and simple multi-dimension expression are required to be modified to fulfil the user requirements. In the traditional approach, the SQL statement may need to be rewritten to satisfy a new requirement.

In addition, OLAP approach can define some decision support measures on the data cube so OLAP approach may be more suitable for decision support system. A typical example of decision support system is shown below.

4.2 Case example of OLAP

Before the implementation of OLAP approach on partner selection, the data cube must be built in the OLAP server. In the data cube, the dimension, measures and calculated member are defined to find suitable partners for the customers.

- Dimension

  In "Jobs Details" dimension, the "Quotation" and "Company_General" table is used to find the historical job for the service provided and retrieve their performance such as delivery date, price and quality. By the same token, the "Competence", "Requirement" and "Requirement_competence" tables are used to find the core competence of the customer's requirements.

- Measures

  On the partner selection, the "Requirement Price", "Requirement Delivery Date", "Quotation Price", "Quotation Delivery Date", "Actual Delivery Date", "Start Date of the Job" and the "Survey Result about the Product Quality" are defined as measures to determine a calculated member for finding suitable partners. After the dimensions and measures are created in a data cube, the relationship of these two dimensions is built through the fact table.

- Calculated Member

  The calculated member of the partner selection is constructed by the measures. The method of finding suitable partners for customer is described below.

  Calculation Method:

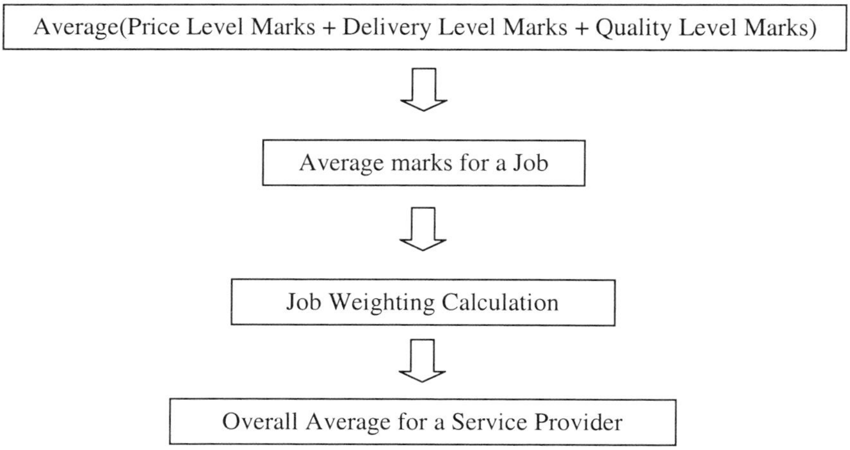

  | Average(Price Level Marks + Delivery Level Marks + Quality Level Marks) |

  ⇩

  | Average marks for a Job |

  ⇩

  | Job Weighting Calculation |

  ⇩

  | Overall Average for a Service Provider |

In the above diagram, the marks of delivery level, quality level and price level of each job are defined by the measures. Then, the average mark of a job is the average of the delivery level mark, price level mark and quality level mark. Finally, the overall average of the service provider is determined by summing the different weightings of the latest 4 jobs. The table below shows one way of assigning the job weightings.

Job Available	Weighting Method
1	Average Marks of Job x 100%
2	Average Marks of Job1 x 75% + Average Marks of Job2 x 25%
3	Average Marks of Job1 x 50% + Average Marks of Job2 x 30% + Average Marks of Job3 x 20%
4 or above	Average Marks of Job1 x 40% + Average Marks of Job2 x 30% + Average Marks of Job3 x 20% + Average Marks of Job4 x 10%

Price Level Marks Calculation:-
Formula used:

$$Deviation\ Percentage = \frac{Quoted\ Price - Required\ Price}{Required\ Price} \times 100\%$$

In the above formula, the "Deviation Percentage" is determined by the required price and quoted price. When "Deviation Percentage" gets a negative value, it means that the service providers can provide a lower price level for their job. Otherwise, they can provide a higher price level for their job. In the following table, the marks for different price levels are determined by the "Deviation Percentage" of price level.

Deviation Percentage	Level	Score
Below –25%	Extremely Low	7 marks
Between –25% and –15%	Very Low	6 marks
Between –15% and –5%	Fairly Low	5 marks
Between –5% and 5%	Average	4 marks
Between 5% and 15%	Fairly High	3 marks
Between 15% and 25%	Very High	2 marks
Above 25%	Extremely High	1 marks

Delivery Level Marks Calculation:-
Formula used:

$$Deviation\ Percentage = \frac{Actual\ Delivery\ Date - Quoted\ Delivery\ Date}{Days\ needed\ for\ the\ job} \times 100\%$$

In the above formula, the "Deviation Percentage" is determined by the actual delivery date, the proposed delivery date and the days needed for the job. When "Deviation Percentage" gets a negative value, it means that the service providers deliver the product/service on or before the proposed date. Otherwise, the product/service may be delivered after the proposed date. In the following table, the marks for delivery levels are determined by the "Deviation Percentage" of delivery level.

Deviation Percentage	Level	Score
Below –25%	Extremely Low	7 marks
Between –25% and –15%	Very Low	6 marks
Between –15% and –5%	Fairly Low	5 marks
Between –5% and 5%	Average	4 marks
Between 5% and 15%	Fairly High	3 marks
Between 15% and 25%	Very High	2 marks
Above 25%	Extremely High	1 marks

Quality Level Marks Calculation:-

In the quality level marks calculation, the questionnaire is used to collect the quality level from the customers. Therefore, the quality level marks will be determined by summing up each positive answer from the questions below.

1. Does the service provider have a proper customer support procedures for their product?
2. Does the service provider deliver a product in your expected quality?
3. Does the service provider afford a good communication within the processing procedure?
4. Does the service provider give sufficient information for you to monitor the job in progress?

The scoring method of the quality level marks is show in the following table.

Question Number	Calculation Method	Marks
Question 1	7marks/4	1.75marks
Question 2	7marks/4	1.75marks
Question 3	7marks/4	1.75marks
Question 4	7marks/4	1.75marks

This OLAP approach enables the timely supply of analyzed information to the user for supporting decision making based on a number of data tables and pre-determined calculation procedures. This is difficult to achieve by using the traditional SQL approach.

4.3 Distributed Object Technology

Effective information interchange among various computer systems is a prerequisite for achieving agile manufacturing which requires a data flow system to enable the access of corporate and relevant manufacturing data in a distributed way. In general, distributed systems go beyond the client/server models in the sense that networks of platforms can alternatively function as clients or servers depending on the situation. Moreover, these systems allow applications to send messages to other data objects that may reside in databases or on other client machines in other applications – anywhere in an organization. At the moment, the effective approach to develop distributed systems relies on object technology (Harmon & Morrissey, 1996) – distributed object technology.

The example here is based on the Common Object Request Broker Architecture (CORBA) specification, which defines a software bus -- the Object Request Broker (ORB) -- that provides the infrastructure necessary to enable cross-platform

communication among distributed objects and client programs (Vogel 1997). CORBA, a well-accepted standard, supports different computing languages and runs on different machines in heterogeneous distributed environments. There are a number of specifications and standards associated with CORBA. The core ones is:
- Object Request Broker (ORB) is a middleware with which the developers are able to access data from objects over remote systems
- Internet Inter-ORB Protocol (IIOP) is the protocol that ORBs use to communicate over TCP/IP networks.
- Interface Description Language (IDL) is used to specify the interface between the client ORB and the server ORB.
- The applications developers to encompass data related to business operations, such as inventory and sales information create business objects. OMG defines business objects as high-level representations of things that exist in a business domain.

The benefits of distributed object model in enhancing the agility of manufacturing systems can be demonstrated by the object-based modeling of value chain activities from product development to logistics distribution, featuring the "dynamic and interactive linkage" of people, parts and machinery within an organization. CORBA and the Distributed Component Object Model (DCOM) technologies make it possible to distribute information across virtually any number of physical servers located on a local and/or wide area network forming the infrastructure of the value chain (Brown and Kindel, 1996). By providing a seamless architecture for distributed services, CORBA, and DCOM-based systems avoid the single server bottlenecks that can plague traditional client-server systems (Flynn, 1998). To achieve efficient data flow over the value chain network, an information flow system, which leverages the latest technological development, is of the utmost importance.

This section outlines the component modules which are required in the formation of the infrastructure of an object-based manufacturing information system that, in turn, is characterized by its ability to provide accurate and relevant information to enhance the performance of the value chain network. In addition, this agile manufacturing system will support the distributed information interchange between the customer and the entire supply chain, so that companies can use the information that already exists to support diverse strategies for design, manufacturing and distribution.

Regarding the software tools to implement an object-based system, Microsoft has released its DCOM (Distributed Component Object Model) architecture, based on which services to build distributed applications can be developed within the Windows environment. In brief, DCOM is an architecture that enables components (processes) to communicate across a network in a distributed way. This architecture includes two types of services, one is to be provided at runtime and the other is to be used to develop distributed applications. DCOM provides distributed messaging services, object request broker services, distributed transaction services, and data connectivity services all layered over its own Remote Processing Control (RPC) mechanism (Rock-Evans, 1998).

The main advantages of the DCOM are:
- DCOM is able to facilitate a robust transaction processing which can be accomplished through Microsoft's transaction server, which runs on Windows NT platform.
- DCOM is claimed by Microsoft that it is "free" with Windows NT though the cost of DCOM is "implicitly" included in the Windows NT. Nevertheless, it is comparatively cheaper than other similar products.

- Some of the built-in capabilities of DCOM are well thought out and able to work favorably with other middleware products such as Microsoft Message Queue Server (MSMQ).
- The wide customer base of Microsoft products is favorable for the growing acceptance of DCOM in the object technology market, which may be further accelerated by the potential support of third party developers.

DCOM services are closely associated with the Windows NT platform. Though Microsoft has extended the support of DCOM to other operating system, such as Unix, by cooperating with third-party companies, it is impossible to implement all the DCOM functions on other platforms because of their inherent differences. For heavy users of Unix and the mainframe or other non-Windows platforms, the Distributed Computer Environment (DCE) protocol may be the better alternative.

The purpose of using distributed object technology (CORBA and DCOM) is to ensure inter-operability between applications on different machines in a heterogeneous distributed environment. This technology can simplify the communication between the heterogeneous objects and each business objects in can remain unique with shared data and logic elements. This approach is able to improve the manageability of the company (more effective information handling on business activities and relationships) and enhance the speed on application development via better communication between functional departments. In brief, distributed object technology is characterized by its provision of a transparent information communication platform. This feature allows a wide range of organizations to have transparent access to information and data. In effect, the boundaries between applications disappear and each object in an enterprise-wide environment can locate any other object without having to know where the object is located. Once this distributed platform is established, a company can change any one application without having to worry that any other application will be affected (Harmon & Morrissey, 1996).

5. APPLICATIONS OF IT IN AGILE MANUFACTURING

There are various techniques that can be applied to facilitate the realization of an agile manufacturing system. In this Section, the techniques to be discussed include the use of multi-agent modeling and virtual manufacturing enterprise. These techniques will capitalize on the application of information technology to support the infrastructure of agile manufacturing.

5.1 Multi-agent Modeling

A Multi-Agent Model(MAM) can be formulated to achieve basic task decomposition using an inference mechanism and to facilitate the subsequent execution of these tasks by responsible agent(s). With the autonomous and collaborative nature of agents, a manufacturing system, which is responsive to external influences, can be developed, enabling greater agility to cope with the changing environment.

The principle of MAM can be embraced to develop a Task Management Scheme (TMS) for the task allocation and monitoring. In general, a TMS consists of three modules, namely, the Rule-based Inference Mechanism (RIM), Object-Oriented Virtual Agent (OOVA) Module, and Task Control Subsystem (TCS). An inference mechanism can be regarded, in short, as a searching process, which ploughs through the knowledge base, containing facts, rules and templates, to arrive at decisions (goals). The inference process operates by selection of rules, matching the symbols of facts and then "firing" the rules to establish new facts (Krishnamoorthy & Rajeev, 1996). The process is continued like a "chain" until a specified goal is arrived at.

A template in the RIM module is analogous to a structure definition of a "user-defined variable" in programming languages such as Pascal and C. For example, the template "goal-is-to" contains two "symbols", namely, *action* and *argument*. The templates are used in writing rules, the patterns of which have a well-defined structure. A template contains slots (attributes), single slot or multi-slot. A single-slot (or simply slot) contains exactly one field while a multi-slot contains one or more fields.

Basically, templates are designed to be used in the building of rules. In the process of decomposition of a client request, templates should be designed to suit the overall requirement particularly taking into consideration the operations of the inference process. It should be noted that although the example templates shown in the following context are designed in compliance with the specific operational process of a particular company, the same principle can be applied to organizational processes of other companies.

- On-duty-agent – The *attributes* of the on-duty-agent include *location* (where is the agent ?), *at* (the exact office-room or floor number), and *holding* (Is he/she holding something or just doing something ?). The pseudo-code for the *on-duty-agent* template is as follows.

    Template name : on-duty-agent
    Includes 3 attributes :
    location with default value "general-building"
    at with default value "common-room"
    holding with default value "nothing"

This template has the meaning that the on-duty-agent has the three attributes, namely, *location*, *at* and *holding* and when the inference process of decomposition starts, the on-duty-agent is in the "common-room" of the "general-building" without "holding" anything.

- Thing – This refers to an object, which can be a dossier or an office-room. There are three attributes for the *thing* template, namely, *name* (the name of the *thing* object), *location* (where is the *thing* object ?) and *at* (the exact location of it). The pseudo-code for the *thing* template is as follows.

    Template name : thing
    Includes 3 attributes :
    name with default value "none"
    location with default value "general-building"
    at with default value "common-room"

This template has the meaning that the *thing* object has three attributes, namely, *name*, *location*, and *at* and when the inference process of decomposition starts, the *thing* object has no designated name and is located in the "common-room" of the "general-building".

- File – This refers to a document. This template is characterized by the *unlocked-by* attribute, which means that the file has to be opened with a permit or a password. There are three attributes for the *file* template, namely, *name* (the name of the thing object), *contents* (what does it contain ?) and *unlocked-by* (the permit or password required to open the file). The pseudo-code for the *file* template is as follows.

    Template name : file
       Includes 3 attributes :
          name with default value "none"
          contents with default value "none"
          unlocked-by with default value "none"

This template has the meaning that the *file* has three attributes, namely, *name*, *contents*, and *unlocked-by* and when the inference process of decomposition starts, the *file* has no designated name and there is nothing inside and it does not need to be unlocked by any key.

- Goal-is-to – This refers to the goal to be satisfied. This template of *goal-is-to* includes attributes, namely *action* (the *verb* involved in the goal) and *arguments* (the *object* related to the *verb* of the goal as specified)
  The pseudo-code for the *goal-is-to* template is as follows.
      Template name : goal-is-to
        Includes 2 attributes :
          action with default values "none" which only allows one of the following
            actions : hold, unlock, change, move, on, walk-to
          arguments with default value "none"

This template has the meaning that the *goal-is-to* has two attributes, namely, *action* and *arguments* and when the inference process of decomposition starts, the *goal-is-to* has no designated action and argument. Notice that the attribute *arguments* is multi-slot meaning that it can contain more than one field.

Facts are normally asserted during the start of the inference process, which operates, by selection of rules, matching the symbols of facts and then "firing" the rules to establish new facts. The assertion of facts is analogous to the initialization of a structured program, where the variables (whether user-defined variables or system variables) are assigned with certain values. In this rule-based program, the structure of the templates is used for the generation of the facts. For easy understanding, a practical example with realistic manufacturing data is adopted to illustrate the design of the facts. The facts making use of the four templates including *on-duty-agent*, *thing*, *file* and *goal-is-to* are shown as below.

(on-duty-agent (location general-room) (at general-building) (holding nothing))
(thing (name general-building) (location general-room))
(thing (name doc-storage-room) (location manuf-mgr-office))
(thing (name filebox) (location manuf-mgr-office) (at doc-storage-room))
(thing (name gen-request-form) (location manuf-mgr-office) (at filebox))
(file (name gen-request-form) (contents manuf-dept-approval) (unlocked-by endorsement-document))
(thing (name File-Target) (location master-schedule-office) (at restricted-area))
(file (name File-Target) (contents form-for-changing-prod-schedules) (unlocked-by Permit-Target))
(thing (name document-room) (location gen-admin-office))
(thing (name permit-target-appl-doc) (location gen-admin-office) (at restricted-area))

(file (name permit-target-appl-doc) (contents Permit-Target) (unlocked-by endorsement-document))
(thing (name endorsement-document) (location prod-supervisor))
(goal-is-to (action change) (arguments form-for-changing-production-schedules)))

The facts shown in the above context are self-explanatory. It should be noted that during the inference process, the *fields* of the *attributes* of the *facts* are changing continuously depending on which rules are fired. For example, the first fact, i.e. (on-duty-agent (location general-room) (at general-building) (holding nothing)), indicates that at the beginning, the agent is in the general-room of the general building without holding anything. As it will be shown in the following context, the agent will move from one place to another, holding documents and files to be authorized by relevant departments. Another point that needs to explained here is the fact "(file (name File-Target) (contents form-for-changing-prod-schedules) (unlocked-by Permit-Target))" contains the *unlocked-by* attribute. This fact means that the file-target (the "ultimate" document to be accessed for meeting the goal) contains the form for changing production schedules and it needs to be unlocked (approved for making change) by a special permit (the permit-target). The last fact is the goal of the inference process, which is to change the production schedule of a certain production line.

Generally speaking, a rule is a collection of conditions and the actions to be taken if the conditions are met. A rule is made up of two parts; the Left Hand Side (LHS) or antecedents consisting of a series of conditional elements to be matched against the given facts, and the Right Hand Side (RHS) or consequents containing a list of actions to be performed when the LHS of the rule is satisfied. Facts are "asserted" and modified during the inference process. In most cases and also in this example, the facts are asserted when the first rule is "fired" during the inference process. In this example, the first rule to be fired contains the facts to be asserted including the fact containing the goal of the inference process, which is to "change the production schedule of production line 3".

The practical example here is taken from a manufacturing firm and the service request, in this case, is to change the production schedule of a certain production line. Before that starts, the procedures required to meet this objective need to be clearly understood. To have the job accomplished, the involved departments must agree to the change. In fact, the change of production schedule affects several relevant departments. Other related issues also need to be addressed, such as the possible ramifications in case the goods cannot be delivered on time, resource problems in terms of equipment/manpower availability if schedule is to be shortened. Firstly, the procedures required to accomplish a certain task have to be worked out among various departments. The procedures in this case include :

- An endorsement document has to be obtained from the Production Supervisor's Office about the request and then this form is attached with a general request form obtainable from the Manufacturing Manager's Office.
- The Manufacturing Manager's Office will issue the manufacturing department approval if the request is granted.
- The Manufacturing Manager's Office approves the relevant document to be sent to the General Administration Office which will check the request based on the administrative perspective and taking into consideration the reasons stated on the endorsement document (from the Production Supervisor's Office) in order to decide if a special permit for this request is to be issued.

- The Master Scheduling Office considers the change approval (from the Manufacturing Manager's Office) and the scheduling situation to issue the file, which together with the special permit (from the General Administration Office) will officially approve the change of production schedule as requested.

All these procedures which may somewhat differ from company to company are taken into consideration to build the rules. However, the important point is that the rules have to be "generalized" which means that they are not just designed for this particular request, as other requests of different natures should also be able to use these rules without any program rewriting.

In the rule-based expert system, inference can be done primarily in two ways, namely, forward chaining and backward chaining. Backward chaining is a goal-driven process, whereas forward chaining is data driven. As the details of these basic inference mechanisms are covered in a number of publications (Giarratano & Riley, 1993; Krishnamoorthy & Rajeev, 1996), they are not described any further here. In short, the rules are grouped in compliance with the action field of the goal template. As illustrated in the goal-is-to template, the fields of attribute action include *unlock*, *hold*, *change*, *move*, *on*, *walk-to*, which will form the different categories of the rules.

In the group of *unlock* rules, the rules are all built based on the *unlock* field of the *action* attribute. A typical rule in this group is called "get-key-to-unlock" with pseudo-code as shown below.

The Rule with name "get-key-to-unlock"
   IF   The goal is to unlock a certain document for access
   AND   The document is stored in the common room
   AND   The document has to be unlocked by a special key
   AND   The on-duty-agent is <u>not</u> holding that special key
   AND   The fact states that "the goal is to hold that special key" does not exist
   THEN   Assert "the goal is to hold that special key" as a fact in the knowledge base

It can be shown here that the rule can deal with any document or file. The conditions are that if the document is in common-room (the default location of on-duty-agent and thing), and the document requires to be unlocked by a key (a password or any sort of authorization), which is not possessed by the agent and the goal to hold that key is not existent in the knowledge base, then the rule will be fired resulting in the assertion of a new fact which is to hold the special key. A number of documents have to be unlocked by special "keys". The general request form has to be "unlocked" by the endorsement document in order to obtain the approval from the Manufacturing Manager's Office.
For the group of *holds* Rules, the typical example in pseudo-code is as follows.

   The Rule with name "unlock-file-to-hold-object"
   IF   The goal is to hold a certain document
   AND   A certain file (say, file-A) contains that document
   AND   The fact states that "the goal is to unlock file-A" does not exist
   THEN   Assert "the goal is unlock file-A" as a fact in the knowledge base

Most of the rules in this example are designed with this "goal-action-field" methodology which categorizes the rules based on the field of the action attribute of the goal template and the consequent is another goal with probably a new action field such as *unlock* in the above example. It should be emphasized here that the rules should be "generalized", which means that it is not designed for only one type of goals. They should be able to cope with various goals as those rules are basically designed in accordance with the action field of the goal-is-to template as well as the actual operational process of the company.

With the basic tasks available, responsible agents concern the next step with the execution of these tasks. The OOVA module unit contains the details of the virtual agents, which are objects created by the object-oriented programming tool. A number of tools can be used to develop these objects. The Window-based ones include Visual Basic, Delphi, PowerBuilder, Visual C++ and others. Visual Basic (VB) has turned out to be a comprehensive object-oriented programming tool and easy to use with its syntax similar to Basic language . VB is hence used to develop the sample code for the design of this module. In object-oriented programming, most of the objects contain elements such as attributes, object methods and interface with the outside world. The detailed functions of these elements are described in most of the programming books for object-oriented programming and therefore not to be covered here.

Generally speaking, each object is responsible for performing some duty depending on the *methods* and *attributes* encompassed within the object. For example, the security agent (object) is responsible for checking the access level of users so that it can determine what sort of information the individual users can access. Each method is responsible for a certain task. For example, the method "entry_security_check" is responsible for checking the access status of users when they log in the system.

The methods of an object (agent) indicate the sort of tasks it can carry out. For example, the project agent has a number of methods such as "get_est_time" (for acquiring information related to the estimated completion time of certain component or product).

Based on the features of objects in object-oriented programming, one object can access the methods of another object by creating an instance of the other object using the command

Set instance_of _objectA = New objectA

The object *instance_of_objectA* is now an instance of *objectA* and can access some methods of *objectA* as long as these methods are declared to be publicly accessible. Objects can communicate and exchange information by virtue of this feature.

It should be noted that like the human agents in companies, virtual agents may also face the situation that they may be phased out or modified and in some cases other new agents may be added to the system as well. Various agents (objects) contain their own relevant methods for performing duties but the next immediate question is how to coordinate the agents to carry out the separate tasks. This issue will be dealt with in the following section.

The TCS plays the role as a coordinator as well as administrator for the RIM and OOVA modules. It performs two important functions (a) monitoring the status of the basic tasks deduced from the RIM and (b) coordination of the tasks to be carried out by the relevant VAs in compliance with the type and nature of tasks to be completed.

The basic tasks produced after the decomposition process have to be monitored and assigned to the relevant VA for processing. The TCS will firstly check through the recommended actions deduced from the inference engine to ensure that the agents in the OOVA module are able to carry out the tasks. If any one of the tasks cannot be processed by any of the included agents, the user has to be informed of this so that an alternative solution should be worked out. When the TCS is satisfied that the included agents can do the tasks, commands will be sent to the relevant agents for task execution. It is important that the TCS should follow closely every process carried out by the responsible agents and to ensure that individual agents will be assigned with the tasks deduced and the whole job is not considered completed until the goal, in this case the "change of schedule form", is achieved.

In order to ensure smooth and efficient exchange of information between the RIM and OOVA modules, it is important that they are working under the same operating environment. For example, if the RIM is developed with an expert system shell called CLIPS while the OOVA module is developed with an object-oriented programming tool VB5, these two development tools cannot "naturally" talk to each other. In this respect, it is important that these two modules should be "integrated" in order to achieve efficient bi-directional data transfer. Fortunately in Microsoft Windows, some Dynamic Link Library (DLL) programs can be developed to link Window-based products to achieve information exchange among the software applications. As a matter of fact, there are DLL programs available for integrating CLIPS to VB5. These programs include "clips.dll" and "clipshll.dll' which can be downloaded from the Internet.[1] With these DLL programs added to VB5, the inference mechanism of CLIPS becomes a part of VB5, thus enabling free and automatic data exchange between these two modules.

As the inference mechanism becomes part of the object-oriented programming environment (in this example VB5), the list of tasks generated is directly sent to the TCS, which is a program within VB5. The task items are treated as the *list items* inside a *ListBox*. The task items will be collected one by one and the content is checked to decide which agent is responsible to carry out which task. A command in VB5 called *Instr()* can be used to check the keywords within the "string". A function of TCS called Extract_Keywords() is invoked to extract the keywords of the tasks. Notice that for every task, there are "pairs" of words; one is the *movement* word and the other is the *destination* or *object* word. For example, the statement can be the task is "on-duty-agent takes the Gen-request-form off with the Filebox onto the Common-room". There are three "pairs" of movement-object keywords (i) *takes* and *Gen-request-form*, (ii) *off with* and *Filebox*, and (iii) *onto* and *Common-room*. These three pairs of keywords in this case can sufficiently suggest which agent is responsible for the relevant task.

The next step is to assign the task to appropriate agents based on the keyword-string. TCS invokes a specially-written subroutine called *assign_based_on_keywords()* which is designed for the assignment of tasks to appropriate agents. It should be noted that the *assign_based_on_keywords()* subroutine only suggests the agents who are considered suitable for the relevant tasks. The final decision of task assignment lies on the outcome from the TADS which takes into consideration other factors related to the assignment of tasks to appropriate agents. The consideration for the acquisition and selection of agents typically includes the following factors :

- Resource levels – In a manufacturing environment, the proper control of resource levels in terms of manpower and manufacturing material is essential. Virtual agents, like human agents, are also considered as the resource of the company. It is important to plan beforehand the resource levels of virtual agents, e.g., what type of agents is required. In manufacturing, the type of virtual agents required may include manufacturing progress agent, production planning agent, cost estimation agent, product status agent, quality control agent, manufacturing research agent. In addition, the capability and performance record of individual agents are also taken into

---

[1] The web site address is : http://ourworld.compuserve.com/homepages/marktoml/clipstuf.htm

consideration related to whether a particular agent is suitable for a task suggested by the ITMS.
- Cost involved - Apart from the capabilities and the types of agents required, the consideration related to the cost involved for "acquiring" the virtual agents should also be on the agenda. Basically software routines need to be developed and therefore the cost involved in employing software engineers in doing this job should be considered. Generally speaking, a higher cost is required for the generation of agents with higher capability and also with more varieties. TADS should consider whether the capability of existing agents should be upgraded to cope with the future requirement if it is found that the existing agents are not likely able to carry out the future tasks effectively and efficiently.
- Priorities – Just like the management of human agents, there may exist some "grey-area" regarding which agent is responsible for which task. For example, in manufacturing, the checking of the reject percentage of a certain product may also be the job of a quality control agent or the product status agent. This can be resolved by taking into account the record of performance of these two agents in carrying out this task. The other consideration is that if the following task(s) is/are supposed to be undertaken by one of the candidate agents, it is more appropriate to assign the "examined" task to the same agent as it would be more cost-effective to have more tasks to be done by one agent continuously.
- Time constraints – In real situations, human agents have their working time related to the nature of their job. For example, security agents should be there all day while other agents, such as production status agent and quality control agent are confined to office working hours e.g. from 9.00 a.m. to 5.00 p.m. The benefit of using virtual agents is that there will be no time constraints. As long as the network server is on, the agents should be available. The other point related to time constraints is that some tasks have to be completed before the other can start or some tasks can be undertaken concurrently. In some systems such as the multi-agent system by Findler and Elder (1995), this time constraint factor can be handled during the assignment of tasks.

5.2 Formation of VME to achieve agility

Virtual Manufacturing Enterprise (VME) is an internet-based "dynamical" organization consisting of dispersed enterprises with various core-competencies. There are a huge number of enterprises with different organization and topological structure, and for the efficient operation of all kinds of enterprises the two essential supporting systems are information system and material handing system. Every enterprise has an Intra-enterprise Information System and an Intra-enterprise Material Handing System. In an era of information society and knowledge-based manufacturing, the enterprise must "plug into" the Global Information Transmission Superhighway (GITS), such as Internet, and the Global Material Transportation Superhighway (GMTS), such as FedEx[1], TransPark[2], so as to be connected with other enterprises and to be possible to cooperate with each other.

Considering impacts of Internet on enterprises, it offers the opportunity to go into virtual cyber-organization. In order to make great usage of the opportunities for dynamic cooperation among numberless enterprises over the world, the potential enterprises

---

[1] http://www.FedEx.com/
[2] http://www.gtp.net/

should join together to form an organization, called "Virtual Enterprise Organization (VEO)", which is a group of loosely connected enterprises or service entities that are potential partners for future cooperation and/or joint ventures for a particular sector of industry and/or from a particular economy. The process of searching and identification of partnership is the most crucial as well as time-consuming part in the formation of a VME. To facilitate the temporary alliance of different enterprises with a high successful rate, the VEO is needed to perform the negotiation, communication and coordination among interested clients to form a VME for a particular project. While the VEO members have a relatively loose connections and join together for the long term opportunities of cooperation, a VME is a sub-set of the VEO members with a certain carrying-on cooperation project and have closer relationships than other members. The VEO members that form a VME are called partners of the VME. Out of the members of VEO, various VMEs will be formed under mutual-benefits rule. The necessities of VEO to VME are mainly because:

- Firstly, it is for the security of the cooperation among VME partners. The VEO members are recommended by the old members and qualified by certain assessment system, so as to ensure the legality and quality of the potential VME partner.
- Secondly, it is for the efficiency of operation. An originator of a VME can look for the right partners from within the well-organized VEO efficiently, and need not search randomly form millions of the enterprises all over the world.
- Thirdly, it is for the exploitation of the high-performance infrastructures. A VEO can greatly improve the common infrastructures of global information system and global material transportation system, which are crucial for the efficient negotiation among the VME partners.

The VEO has four main tasks. Firstly, the VEO provides an enterprise model, which describes the core-competencies of the constituent enterprises, which is available to be selected as an internet-oriented partner. The competencies can be classified in a hierarchical model. In general, the competencies are modeled at enterprise level, functional level and machine level. Different kinds of project may require different cooperation level and need to search the partners at different competence levels. At the enterprise level, the model may describe the capabilities of new product, the quality assurance, the marketing. At the functional level, it provides the information about product design, manufacturing, testing, etc. At the machine level, it provides the information about dimensions, precision, efficiency, etc. This multi-layer model provides a more completed and flexible description for the various kinds of members of the VEO, and so provides a more flexible mechanism for VME formation and control. Secondly, the VEO provide an Intelligent Coordinator (IC) for each VEO members. The IC is a multi-agent system that performs the communication, coordination and negotiation with other members through the Internet, and is described later. Thirdly, the VEO provides and/or improves the infrastructure of global information system and the global material transportation system. Fourthly, the VEO sets up the regulations and rules for the cooperation among members. An enterprise with some competencies and qualification and with expectations to cooperate with others can apply to join the VEO. For the membership establishment, it must have general VEO protocols.

For each VEO member, there is an opportunity to be invited by other VEO members to become a partner of a new VME. When there is a chance for exploiting, and none can finish it alone, the VEO members with similar competence may compete with each other to become the TO-BE VME partners. The VME is an optimal restructuring of the various kinds of enterprises with the appropriate core competencies. For an enterprise to join the VEO and to become a potential VME partner, it must follow a stipulated registration assessment and monitoring procedures, so as to ensure the qualities and reputations of the VEO members are maintained. The enterprise must have some fundamental qualifications and "cooperative capabilities" or core competencies. It is obvious that the sharing of resources and expertise of geographically isolated firms with dissimilar core competencies to achieve the common purpose of producing a product based on the virtual enterprise model can significantly enhance the agility of companies to meet the ever-changing demands of customers in the global marketplace.

6. REMARKS

IT has a key role to play in achieving manufacturing agility and can affect all the major functions of a manufacturing system. On balance, IT facilitates the operation of an agile manufacturing system and in the main acts as an enabler. However, the support of senior management is necessary for the achievement of manufacturing agility. It has been shown that the introduction of a new IT intervention may generate uncertainty within the workforce and the support of senior management is vital in maintaining the manufacturing process. On some occasions, the introduction of IT has created problems with the workforce and other members of staff(Wilson, 1994), so top management must be cautious in this task. If IT increases management control by top management, this needs to be applied without creating undue stress and concerns.

REFERENCES

1. Adler, P.S. Mandelbaum, A., Nguyen, V. and Schwerer, E.(1996) "Getting the most out of your product development process", Harvard Business Review, March-April, pp.134-152.
2. Banerjee, S. and Sriram, V. (1995), "The impact of electronic data interchange on purchasing: an empirical investigation", International Journal of Operations and Production Management, Vol. 15 No. 3, pp. 29-38.
3. Brown, N. and Kindel, C. (1996), Distributed Component Object Model Protocol - DCOM/1.0, Microsoft Corporation, Network Working Group, http://www.microsoft.com/
4. Chandler, K. (1998), "Quality in the age of the networked society", Quality Progress, Vol.31 No. 2, pp. 49-52.
5. Dilger, K. (1997), "To protect and preserve", Manufacturing Systems, Vol. 15 No. 6, pp. 22-8.
6. Flynn, J. (1998), " The Marriage of Document Management and Electronic Commerce", Inform Magazine, November/December 1998, p16.
7. Freund, B., Konig, H. and Roth, N. (1997), "Impact of information technologies on manufacturing", International Journal of Technology Management, Vol. 13 No. 3, pp.215-28.
8. Giarratano, J.C. & Riley, G.D. (1993) Expert Systems: Principles and Programming, International Thompson Publishing.
9. Gilmore, J.H. and Pine, B. J. II (1997), "The four faces of mass customization", Harvard Business Review, Vol. 75 No. 1, pp. 91-101.

10. Gong, L., Jwo, W. and Tang, K. (1997), "Using on-line sensors in statistical process control", Management Science, Vol. 43 No. 7, pp. 1017-28.
11. Hameri, A. and Nihtila, J. (1997), "Distributed new product development project based on Internet and WorldWide Web: a case study", Journal of Product Innovation Management, Vol. 1 No. 2, pp. 77-87.
12. Harmon, P. and Morrissey, W. (1996), The Object Technology Casebook, OMG, John Wiley & Sons Inc., Canada.
13. Jonscher, C. (1994), "An economic study of the information technology", in Allen, T.J. and Scott Morton, M.S. (Eds), Information Technology and the Corporation of the 1990s, Oxford University Press, New York, NY, pp. 5-42.
14. Kelsey, G.S.(1995) "Flatten the pyramid and speed product development", Research-Technology Management, March-April, pp.12-13.
15. Kendrick, J.J. (1995), "SPC on the line", Quality, Vol. 34 No. 1, pp. 35-9.
16. Krishnamoorthy, C.S. and Rajeev, S. (1996) Artificial intelligence and expert systems for engineers, CRC Press.
17. Layden, J.E. and Pearson, T.A. (1992), "A missing link in Total Quality", Controls and Systems, Vol. 39 No. 3, pp. 42-4.
18. Litsikas, M. (1997), "Electronic downloads eliminate inspection audits", Quality, Vol. 36, No.1, pp. 50-4.
19. Maital, S. (1994),"A 'made in America' system", Across the Board, April, pp.45-6.
20. Meyer, C. and Purser, R.E. (1993), "Six steps to becoming a fast-cycle-time competitor", Research-Technology Management, September-October, pp.41-8.
21. Moskal, B.S.(1995) "Son of agile", Industry Week, 15 May, pp.12-16.
22. Mukhopadhyay, T., Kekre, S. and Kalathur, S. (1995), "Business value of information technology: a study of electronic data interchange", Management Information Systems Quarterly, Vol. 19 No. 2, pp. 137-56.
23. Palvia, P., Kuma, A., Kumar N. and Hendon, R. (1996), "Information requirements of a global EIS: An exploratory macro assessment", Decision Support Systems, Vol. 16 No. 2, pp. 169-79.
24. Pandiarajan, V. and Patun, R. (1994), "Agile manufacturing initiatives at concurrent technologies corp.", Industrial Engineering, February, pp.46-9.
25. Papadakis, E.P. (1990), "A computer-automated statistical process control method with timely response", Engineering Costs and Production Economics, Vol. 18 No. 3, pp. 301-10.
26. Patterson, D.W., Anderson, R.B. and Rockwell, H.E. (1997), "Increased use of automated machinery requires changes in quality control procedures", Forest Products Journal, Vol. 47, No. 1, pp. 33-6.
27. Paxton, C. (1997), "Putting your best (ADC) foot forward", Automatic I.D. News, Vol. 13, No. 6, pp. 36-7.
28. Quelch, J.A. and Klein, L.R. (1996), "The Internet and international marketing", Sloan Management Review, Vol. 37 No. 3, pp. 60-75.
29. Ranney, J. and Deck, M. (1995), "Making teams work: lessons from the leaders in new product development", Planning Review, July-August, pp.6-12.
30. Rathnam, S., Mahajan, V. and Whinston, A.B. (1995), "Facilitating coordination in customer support teams: a framework and its implications for the design of information technology'', Management Science, Vol. 41 No. 12, pp. 1900-21.
31. Rock-Evans, Rosemary (1998), DCOM Explained, Digital Press, Boston

32. Schein, E.H. (1994), "Innovative cultures and organizations", in Allen, T.J. and Scott Morton, M.S. (Eds), Information Technology and the Corporation of the 1990s, Oxford University Press, New York, NY, pp. 125-46.
33. Stone, M., Woodcock, N. and Wilson, M. (1996) "Managing the change from marketing planning to customer relationship management", Long Range Planning, Vol. 29 No. 5, pp.675-83.
34. Srinivasan, K., Kekre, S. and Mukhopadhyay, T. (1994), "Impact of electronic data interchange technology on JIT shipments", Management Science, Vol. 40 No. 10, pp. 1291-304.
35. Swink, M.L., Sandvig, J.C. and Nabert, V.A.(1996), "Adding zip to product development: concurrent engineering methods and tools", Business Horizons, March-April, pp.41-9.
36. Teague, P.E., Bak, D.J., Puttre, M., Fitzgerald, K.R. (1997), "Suppliers: the competitiveedge in design", Purchasing, Vol. 122 No. 7, pp. 32S5-23.
37. Vogel A, and Duddy, K. (1997), Java programming with CORBA. New York: J. Wiley.
38. Willis, T. H. (1998) "Operational competitive requirements for the twenty-first century", Jr. of Industrial Management of Data Systems, Vol.2, pp83-86.
39. Wolff, M.F.(1992),"Working faster", Research-Technology Management, Nov-Dec, pp.10-12.

Information Systems for Agile Manufacturing Environment in the Post-Industrial Stage

S. Subba Rao[*] and A. Nahm

Department of Information Systems and Operations Management, College of Business Administration, The University of Toledo, 2801 W. Bancroft, Toledo, Ohio 43606, USA

1. INDUSTRIAL TO POST-INDUSTRIAL ENVIRONMENT

The environment in which business and organizations operate is changing. While Huber [1] has described this change in a broad, societal terms, Skinner [2], Doll and Vonderembse [3] and Vonderembse et al. [4] have described it in terms that are more manufacturing specific. It is a change from industrial to post-industrial environment.

Huber [1] has described how the society in general is moving from an industrial to a post-industrial stage. The post-industrial society is "characterized by *more and increasing knowledge, more and increasing complexity*, and *more and increasing turbulence*" ([1]; emphasis from the original text). Not only there is "Knowledge Explosion," but the development of communications and computer technologies has greatly increased the *availability* of whatever knowledge is produced. Increased knowledge will cause many technologies to be more effective. As a result of these heightened levels of effectiveness, individual events (such as developments in R&D, market development, or carrying out of military campaigns) will be shorter in duration. The eventual effect is increased turbulence.

As the rate of changes in life style and technology is accelerating, it imposes a greater challenge to organizations in the post-industrial age to cope with such changes. "The greater turbulence of the post-industrial environment will demand that *organizational decision making* be *more frequent* and *faster*. The greater complexity of this environment will also cause decision making to be *more complex*.... The heightened turbulence of post-industrial environment will require that these *organizational innovations* be *more frequent* and *faster*" ([1]; emphasis from original). "In the post-industrial society, the central problem is not how to organize to produce efficiently (although this will always remain an important consideration), but how to organize to make decisions – that is, to process information" [5]. Decision makers in the post-industrial era need to create a structure, through which they could deal with the equivocality in the environment and consistently make appropriate decisions for their organization [6].

The overall change of society from industrial to post-industrial is affecting manufacturing firms as well. Doll and Vonderembse [3] have identified the driving forces for this change, which is illustrated in Figure 1. When manufacturing was evolving from craft to industrial, the driving force was technology, particularly mass production technology, such as conveyer systems and the managerial practice of division of labor. This was also accompanied by growth in domestic market, which enabled firms to produce standardized products in mass volume and

---

[*] Corresponding author. Tel: 419 530 2421; e-mail: srao5@uoft02.utoledo.edu.

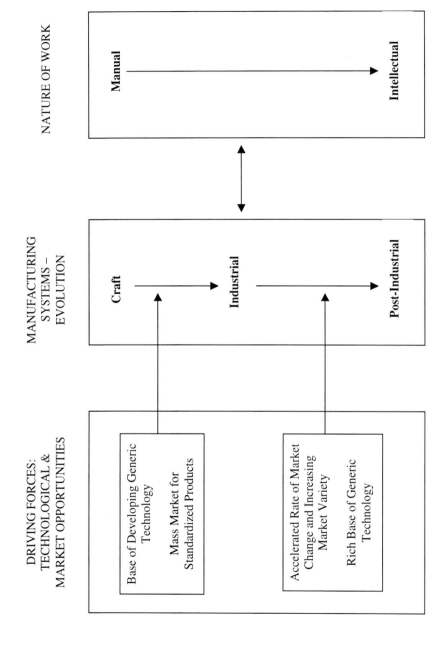

Figure 1. The evolution of manufacturing systems (adopted from Doll and Vonderembse [3])

still sell them to customers, who did not require much except for low price and acceptable quality. However, as the society moved from industrial to post-industrial, the scope of the market became global, and the customer requirements became much more diversified. With the increase in competition, firms need to deliver products/ services to customers who now require satisfaction across multiple criteria, such as cost, quality, delivery, flexibility, time, and service, all at the same time. Also occurring is the development and sophistication of flexible and agile manufacturing technologies, which are enabling firms to satisfy those multiple customer requirements simultaneously and in greater speed. The industrial revolution was technology driven and market enabled, but the post-industrial revolution is market driven and technology enabled [3].

Some examples of market-driven and technology-enabled changes in the post-industrial environment are electronic commerce, virtual enterprise, and increased use of web technology. Electronic commerce has a potential to revolutionize the way customers shop in the future, for customers are demanding greater convenience and economy while acquiring the skills and technology necessary to shop electronically [7]. On the supplier side, several medium-to-small firms can collaborate to form a virtual enterprise, which pools the expertise of member firms together to deliver the product or service the customer requires. The Agile Web, a group of 19 small companies in eastern Pennsylvania, is one example of such a virtual enterprise [8]. Member firms in such virtual enterprise participate in the bidding and production/delivery process while fully maintaining their autonomy as independent entities. The web technology, which supports the electronic commerce and the formation of virtual enterprise, provides a channel for information flow and collaborative work. By supporting the rapid flow of information throughout the supply chain, web technology enhances the integration among chain members and reduces the cycle time for product design, product/service delivery, and order fulfillment. The material flows and the information flows in the supply chain are depicted in Figure 2. The links on the material flow side represent physical movement (e.g., transportation, carriers), storage (inventories, warehousing) and material handling. It should be noted that while Figure 2 conveys a sequential configuration, real supply chains are essentially network configurations.

Figure 2. The material flow and the information flow throughout the supply chain

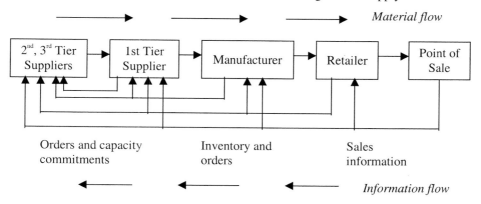

2. AGILE MANUFACTURING

As one of the many market-driven and technology-enabled changes in manufacturing, agile manufacturing is gaining recognition as a necessary condition for competitiveness. The concept is still being refined to further its understanding, but at its most basic level, agility, according to Oleson [9], is understood as "the ability to respond effectively to unexpected or rapidly changing events." Yusuf et al. [10] compiled a list of "main points in the definitions of agile manufacturing," which is shown in Table 1.

Table 1. Definitions of agile manufacturing (adopted from Yusuf et al. [10])

Main points of definitions	Authors
High quality and highly customized products	Goldman and Nagel [11]; Kidd [12]; Booth [13]; Hilton and Gill [14]
Products and services with high information and value-adding content	Goldman et al. [15]
Mobilization of core competencies	Goldman and Nagel [11]; Kidd [12]
Responsiveness to social and environmental issues	Kidd [12]; Goldman et al. [15]
Synthesis of diverse technologies	Kidd [12]; Burgess [16]
Response to change and uncertainty	Goldman et al. [15]; Pandiarajan and Patun [17]
Intra-enterprise and inter-enterprise integration	Kidd [12]; Vastag et al. [18]; Youssef [19]; Yusuf [20]

Yusuf et al. [10] also proposed a comprehensive definition of agility as being "the successful exploration of competitive bases (speed, flexibility, innovation proactivity, quality and profitability) through the integration of reconfigurable resources and best practices in a knowledge-rich environment to provide customer-driven products and services in a fast changing market environment." Imbedded in the above definition, a hierarchy of agility of three levels are to be noticed – elemental agility, micro-agility and macro-agility. A truly agile organization, to achieve the hierarchy of agility has to focus on individual resources (people, equipment and machine, and management) at the elemental level, and on enterprise-wide functions and decisions at the micro-agility level. To bring about macro-agility, the focus will be at the inter-enterprise level where core competencies of partners have to be brought into joint venturing to maximize the gains of cooperation [10].

The concept of agility was originally coined by a group of researchers at Iaccoca Institute at Lehigh University [21]. It was recommended as holding potential for the USA to resume a leading role in manufacturing [10]. However, in a constantly changing and ever-intensifying post-industrial environment, "the ability to respond effectively to unexpected or rapidly changing events" not only holds potential for any firm or country to gain competitive advantage, but would increasingly become a requirement for a firm to sustain its leadership position. In fact, agility is becoming a requirement for gaining and maintaining leadership position in post-industrial manufacturing because agility confers on its practitioners the competitive capabilities of speed, flexibility, quality and innovation proactivity to stay ahead of the pack.

The literature suggests that agility is not a single attribute but a bundle of many related attributes. Building upon the collective insights provided by researchers in the field of agile manufacturing, Yusuf et al. [10] have summarized the agile practices and attributes of an agile organization in Table 2. The table presents 32 attributes, in 10 decision domains, of an agile manufacturing enterprise. The pathways, facilitators and barriers to achieving these attributes are important issues for consideration if progress is to be achieved in moving towards agility. The metrics for the processes required for achieving agility have to be carefully thought of and chosen. The attributes, pathways and metrics have to be integrated to achieve the competitive edge conferred by the agile organization.

Table 2. The attributes of an agile organization (adopted from Yusuf et al. [10])

Decision domain	Related attributes
Integration	Concurrent execution of activities
	Enterprise integration
	Information accessible to employees
Competence	Multi-venturing capabilities
	Developed business practice difficult to copy
Team building	Empowered individuals working in teams
	Cross functional teams
	Teams across company borders
	Decentralized decision making
Technology	Technology awareness
	Leadership in the use of current technology
	Skill and knowledge enhancing technologies
	Flexible production technology
Quality	Quality over product life
	Products with substantial value-addition
	First-time right design
	Short development cycle times
Change	Continuous improvement
	Culture of change
Partnership	Rapid partnership formation
	Strategic relationship with customers
	Close relationship with suppliers
	Trust-based relationship with customers/suppliers
Market	New product introduction
	Customer-driven innovations
	Customer satisfaction
	Response to changing market requirements
Education	Learning organization
	Multi-skilled and flexible people
	Workforce skill upgrade
	Continuous training and development
Welfare	Employee satisfaction

3. ROLE OF IS/IT

IS/IT is a general term that describes the application of computers, communications and electronic engineering to the specification, design and construction of information-rich systems. To achieve agility at micro, intra-enterprise and inter-enterprise levels and to derive the full benefits of agility along the value/supply chain, appropriate design, architecture, and implementation of the IS/IT are critical.

The point of interest is the enabling role of technology in gaining agility. Agility is thought of as a synthesized use of already developed and well-known technologies and methods of manufacturing, such as lean manufacturing, CIM, TQM, MRP II, BPR, employee empowerment, and OPT [10], [11], [12]. When one looks at this wide array of technologies that support agility, one can easily realize the enormous magnitude of the role IS/IT plays in gaining agility in manufacturing. A proper and clear understanding as well as assessment of IS/IT needs in one's road map towards agility is essential to build up the capability of a firm to compete effectively.

The two most critical aspects of IS/IT to achieve agile manufacturing environment would be integration and flexibility. Vonderembse et al. [4] have argued that in post-industrial manufacturing, companies have to focus on integration issues because information exchange becomes a critical factor when flexibility is a competitive requirement, and that integration should be given higher priority in technology deployment than automation. Field studies of four different firms verify their notion that higher level of integration among business functions and technologies contributes for better business performance, including short lead time within wide variety of products, than higher level of automation [4]. Further evidence of the need to integrate enterprise-wide can be seen in Yusuf [20] where it is reported that integrated organizations were superior to their non-integrated counterparts.

Agile manufacturing is possible only when different business functions and members of the supply chain move together as parts of an organic whole. This requires a high level of integration in all information flows and material flows between and among business functions and members of the supply chain. Information systems requirement in the agile manufacturing environment basically calls for integrated information systems encompassing the entire value chain. Accurate, timely and reliable provision of critical and appropriate information to all internal and external constituents of the value/supply chain will be the ultimate goal of the information systems in an agile environment.

However, pursuing high level of integration by all means may result in systems characterized by lack of robustness and flexibility [22]. Popular enterprise resource planning (ERP) systems (such as SAP) or custom-made manufacturing control systems (MCSs) can be good examples of this. Such a system may exhibit a very high level of integration within itself, but when it comes to flexibility, i.e., reconfiguring some or all of its components, it could become extremely difficult, if not impossible to change [22], [23]. As such, one should be careful, in designing and implementing IS/IT in an agile manufacturing environment, to equally emphasize integration and flexibility.

4. INFORMATION SYSTEMS ARCHITECTURE FOR AGILE MANUFACTURING

The architecture of such a system would be quite different from that in a traditional industrial environment characterized by mass production. Jung et al. [24], Aguirre et al. [22], and Song and Nagi [25] have each proposed such architecture for agile manufacturing. To get a feel of what is required in an architecture for agile manufacturing, let us look into these examples more closely.

Jung et al. [24] state that agile manufacturing can be successfully accomplished using a well-defined system architecture. There are several aspects of system architecture: control, function, process, information, communication, distribution, development, and implementation. Control architecture specifies how each system component is arranged and interacts. Process architecture specifies how queries, synchronization, and other status reports flow and interact. Function architecture specifies a structured representation of the functions of a manufacturing system. Information architecture specifies the requirements for what information is managed by manufacturing systems. Communication architecture specifies the communications paradigm to enhance the timeliness of message delivery for real-time control systems. Distribution architecture specifies the appropriate location of physical and logical control entities. Development architecture specifies what information is used to instantiate an instance of manufacturing systems and how it happens. Implementation architecture specifies the actual system software subroutines.

Jung et al. [24] contend that there should be a "Reference Architecture" for each of these aspects of system architecture. Reference architecture can be defined as the scope and purpose, domain analysis, components and their integration methods, and the development procedure of various system architectures. They argue that various problems, which occurred in the past in the course of rapidly developing computer integrated manufacturing (CIM) systems, were caused by lack of such reference architectures. The reference architecture provides a transparent way to the users when they establish the automated CIM systems for agile manufacturing. The system architecture for IS/IT can be centralized or decentralized, as well as hierarchical, heterarchical or hybrid. Schematically, the architectures are shown in Figure 3. For agile manufacturing enterprises, heterarchical and hybrid architectures are preferred. In what follows, we present two examples, MCSARCH and AMIS, which are hybrid architectures.

Figure 3. Control architectures (adopted from Jung et al. [24])

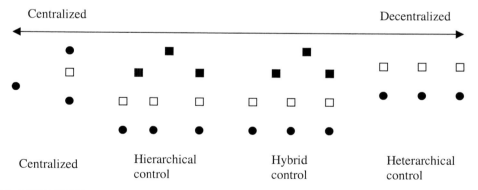

4.1. MCSARCH

While Jung et al. [24] highlighted the various aspects of system architecture for agile manufacturing in a broad, conceptual level, Aguirre et al. [22] attempted to define such architecture in greater detail. As proposed by them, Manufacturing Control Systems Architecture (MCSARCH) is perceived as structured in four different levels (Figure 4). At the center lies the object-oriented software development level, where techniques such as object-oriented analysis and design, distributed computing, parallel computing, and object persistence are deployed. Above this level lies the standards level. In order to provide a common development computational platform, which promotes open systems development in an incremental and scaleable way, MCSARCH uses standards as much as possible. Some examples are:

- For application data interchange: STEP part AP213 (ISO/DIS 10303-213)
- For message protocol between manufacturing objects: MMS standard (ISO 9506)
- For communication protocols: DCE, CORBA, DCOM, TCP/IP, ISO/OSI, Java, Ada, C++, Smalltalk, RS-232/422, fieldbus, LANs, etc.
- For orders sequence synchronization: Petri Nets

The service modules level, the third level of MCSARCH, is composed of libraries, pre-compilers/translators and applications which ease and encapsulate standards usage. The main function of this level is that, when modification or substitution of any standard occurs, it could be accomplished without effecting higher level applications built with this architecture. The fourth level is the manufacturing control systems level. This level comprises manufacturing object servers and manufacturing object client applications, which communicate with each other by using MMS messages within the CORBA environment. Each manufacturing object server and associated client applications is distributed in a computer network, so that the system behavior is indifferent to the place of execution of a given object.

4.2. AMIS

While Jung et al. [24] and Aguirre et al. [22] focused on IS/IT architecture issues for agile manufacturing *within* a manufacturing company, Song and Nagi [25] address the issues related to designing and implementing an agile manufacturing information system (AMIS) that integrates manufacturing databases dispersed at various partner sites. Typically an agile manufacturing, in this view, is a virtual enterprise along the value chain consisting of agile partners who collaborate with each other. The environment in which such an enterprise operates will have the following characteristics:

(1) Partner information interoperability across companies: Product data and knowledge are segmented and distributed across the distributed sites of partners who need to access product information frequently and dynamically.
(2) Information consistency across partners in the virtual enterprise: The data is own by multiple owners, and mutual agreements or protocols are needed to resolve data ownership. It is up to the AMIS to maintain system information consistency in this environment. Limiting the rights to update primitive data while providing multiple information views from the same primitive data storage becomes necessary. This goal is accomplished by having three layer for data hierarchy: local primitive data/information layer, atomic objects layer, and composite objects layer.
(3) Partner policy independence and autonomy maintenance: Agile partners have full autonomies, in the sense that there are no hierarchical organizations in an agile enterprise. AMIS must reflect the partner's own policy set regarding the information flow in/out of that partner.
(4) Partner heterogeneity accommodation: Heterogeneity in partner computer hardwares, database systems and applications should be accommodated.
(5) Open and dynamic system architecture: AMIS must provide a flexible and open architecture to host partners dynamically, while ensuring individual partner information security.

The system that Song and Nagi [25] propose is illustrated in Figure 5. In this illustration, the virtual enterprise is composed of three autonomous partners: Company A, which has expertise in computer aided process planning (CAPP); company B, specialized in design and is responsible for product computer aided design (CAD); and company C, which is to construct enterprise manufacturing requirements planning (MRP). Looking closely to a single partner, there exist seven modules in a typical agile partner model. Local primitive information storage stores local primitive information and forms a part of the enterprise

Figure 4. MCSARCH architecture (adopted from Aguirre [20])

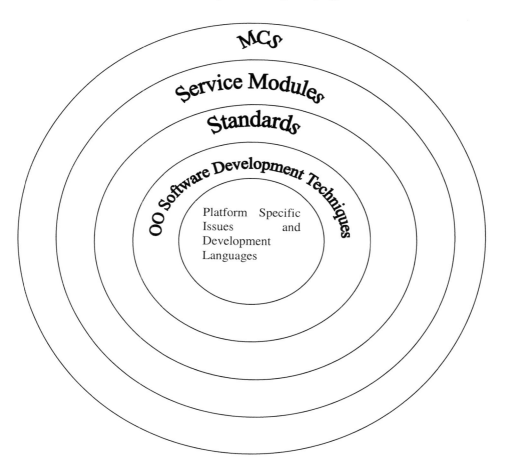

primitive information storage. DBMS manages local primitive information accessing, updating and constraint maintenance. By interacting with the DBMS, the server is responsible for local atomic object generation and instantiation. Application programs access information storage through the workflow manager (WM), the knowledge base stores partner policies and inter-partner protocols supporting partner transaction workflow specifications.

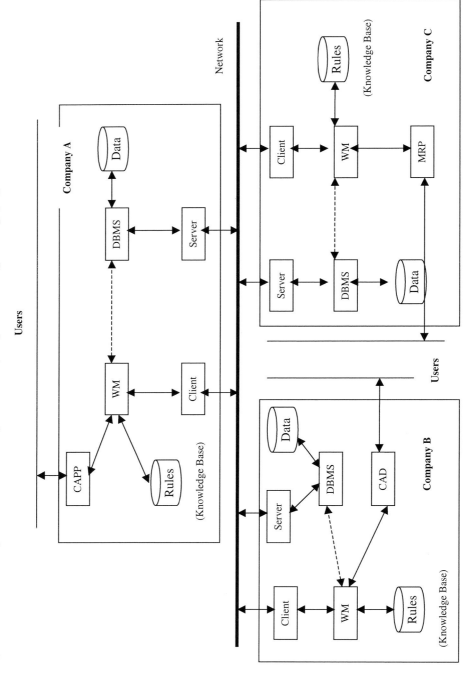

Figure 5. A virtual enterprise and information system model (Adopted from Song and Nagi [23])

The client and server are responsible for connecting to the network and transmitting information back and forth. The overall information system has the following features:
(1) Each partner is connected to the network through a client and a server. The server is responsible for serving other partners. It receives queries, interacts with the DBMS to execute the queries and sends the results back to the inquirer. The client is responsible for coordinating the workflow execution. It sends subqueries to other partners and receives results.
(2) The local workflow manager controls the information flow. Based on the query from the user, the workflow manager has to retrieve relevant rules to analyze the query, and form workflows to be executed by the client to keep system data consistency.
(3) The correct workflow execution is ensured by the cooperation of partner servers and clients. The client at the global workflow initialization site becomes the temporary workflow coordinator. It uses dynamic two-phase-commit protocol to ensure the workflow execution atomicity [25].

5. A FRAMEWORK FOR IS/IT REQUIREMENT

As seen in these examples, researchers are trying to address the issues of IS/IT for agile manufacturing by carefully defining the system architecture. The two primary objectives in IS/IT architecture for agile manufacturing are integration and flexibility. Some common features that can be found in the above examples are:
(1) Use of object oriented programming and data retrieval to reach a high level of integration throughout the value chain/supply chain, maintaining primitive data consistency while allowing multiple information views.
(2) Use of open architecture, hybrid and heterarchical control architecture, and distributed computing architecture, which enables a system to exhibit flexibility (in reconfiguring manufacturing objects) and dynamic collaboration (among agile partners).

These IS/IT requirements for agile manufacturing are illustrated in Figure 6.

Figure 6. Framework for IS/IT requirements for agile manufacturing

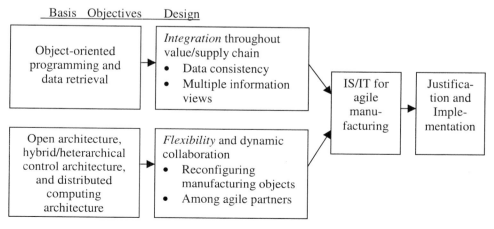

As we have seen, the primary objectives of information systems architecture in agile manufacturing environment are quite different from those in industrial manufacturing

environment characterized by mass production. Information systems architecture in industrial manufacturing traditionally focused on providing high level of efficiency through specialization, which was a suitable strategy for a stable environment. Information systems architecture in an agile manufacturing environment emphasizes high level of data integration and the ability to change quickly its configuration and components to adapt to fast changing post-industrial environment. In the course of pursuing this new set of goals, IS/IT in an agile manufacturing environment may seem to loose some degree of efficiency in its execution, compared to the previously specialized, proprietary systems. However, ever-increasing speed of technology development would enable us to overcome this hindrance in the near future. By then, those who have the ability to adapt quickly through agile manufacturing information system in the midst of volatile business environment would not only continue to survive but rise as leaders in the industry.

6. JUSTIFICATION AND IMPLEMENTATION

What can be the issues involved in justification of investments for IS/IT in agile manufacturing environment? We can find a similar set of issues discussed in the literature in relation to advanced manufacturing technology (AMT) justification. Chen and Small [26] explain that while the costs (hardware, software, planning, training, operations, etc.) are generally easily quantifiable, the benefits are often very difficult to quantify. In particular, major strategic benefits such as early entry to market, perceived market leadership, the ability to offer a continuous stream of customized products and improved flexibility, although extremely important for the growth and survival of the firm, are not readily convertible into cash values [26]. But these *strategic benefits* are what agile manufacturing is all about! Ramamurthy and King [27] assert that serious problems may be faced when top management tries to employ traditional evaluation and justification methods focusing largely on savings, e.g., direct labor costs. Even quantifiable benefits such as improved quality through lower defectives, reduced inventory and reduced floor space are often ignored. How much more would it be for intangible benefits such as flexibility to changes in demand or design, and short lead-time in new product development! [27]

Since strategic planning approach takes a long-term, comprehensive view of both business and technology issues, there is a greater possibility of adoption success if the decision to implement an IS/IT for agile manufacturing is based on strategic consideration [26]. Vonderembse et al. [4] state that investments in technologies in the post-industrial environment can be justified not only through highlighting the potentials for gaining operational benefits, but also certain strategic and marketing advantages which can be created with the new technologies [4], [28]. Vrakking [29] also suggest that projects might have to be justified on the basis of strategic arguments. Arguments based on comparison with competitors, the retention, attainment or perception of industry leadership and expected future developments in the industry might serve as alternative factors for decision makers to approve IS/IT projects for agile manufacturing ([29]; quoted from [26]).

To alleviate the problems inherent in using purely financial or purely strategic appraisals, hybrid financial and strategic appraisal techniques are being promoted [30], [31], [26]. Kaplan [30], for example, suggests that managers should adopt a two-step process when evaluating technology investments using discounted cash flow (DCF) technique.

First, estimate the annual cash flows about which there is the greatest confidence: the cost of the new process equipment and the benefits expected from labor, inventory, floor space, and cost-of-quality savings. If at this point a discounted cash flow analysis - done with a sensible discount rate and a consideration of all relevant alternatives – shows the investment to have a positive net present value, well and good.... If the DCF is negative, however, then it becomes necessary to estimate how much the annual cash flows must increase before the investment does have a positive net present value.... Rather than attempting to put a dollar tag on benefits that by their nature are difficult to quantify, managers should reverse the process and estimate first how large these benefits must be in order to justify the proposed investment. Senior executives can be expected to judge that improved flexibility, rapid customer service, market adaptability, and options on new process technology may be worth $300,000 to $500,000 per year but not, say, $1 million. This may not be exact mathematics, but it does help put a meaningful price in [the project's] intangible benefits [30].

The emergence of several weighted scoring techniques allows management to assign weights to each tangible and intangible economic or strategic factor under consideration. More sophisticated scoring methods, such as the analytic hierarchy process which can correct for managerial inconsistencies, have also been proposed ([32]; quoted from [26]).

In implementing IS/IT for agile manufacturing, it is important to adopt a *total systems approach*. There are two aspects to this: One is to maintain holistic view of the value chain/supply chain, with eyes continually focused on the customer [4], and second is to be aware of the many attributes of an agile organization. As mentioned before, the information flows and material flows between members of the supply chain are essentially network configurations, even though typical illustration (such as Figure 2) might convey a sequential configuration. Since agile manufacturing environment seeks to achieve high level of integration among all members of the value chain/supply chain that are connected to each other like a web, it is critical for managers to keep their eyes on the whole system in the process of IS/IT implementation. Failing to do so could result in undesirable situations such as 'enterprise islands' and not an integrated enterprise network, or *agile webs*, in which case the whole agile IS system breaks down. Equally important is to keep in mind the many attributes of an agile organization as summarized in Table 2. Among the 10 decision domains listed, the following are closely related to IS/IT implementation for agile manufacturing, and thus require careful consideration and consistent effort to build up their respective attributes:

- Integration (concurrent execution of activities, enterprise integration, value/supply chain integration, and information accessibility to employees/partners/customers)
- Team building (empowered individuals working in teams, cross functional teams, teams across company borders, and decentralized decision making)
- Technology (technology awareness, leadership in the use of current technology, skill and knowledge enhancing technologies, and flexible production technology)
- Partnership (rapid partnership formation, strategic relationship with customers, close relationship with suppliers/carriers, and trust-based relationship with customers/suppliers)

7. STRATEGIC ROLE OF INFORMATION SYSTEMS FOR AGILITY

Then what are the strategic roles that information systems can play in enhancing agility? There are many, such as in strategic planning, new product development, time to market, quality assurance, mass customization, relationship building, and cost leadership, to name a few. By having an information system that is highly integrated and flexible, firms could explore niche markets that require faster service and dynamic collaboration among partners. Pooling core competencies of member firms, virtual enterprise can quickly come up with a design for new product, manufacture a prototype product in a short lead time, test customer responses, and rapidly improve the product based upon customer feedback. Repeated collaboration would allow firms to test the ability of cooperating members. If critical deficiency were to be found in any of them, then the flexibility feature in their information system would allow them to quickly replace member firms, thus allowing the virtual enterprise to ensure quality product/service to their customers. By rapidly distributing customer orders throughout the supply chain and responding to them as an integrated whole, agile firm can provide customized products and services. By having IS/IT that greatly reduces direct and indirect costs associated with material flows and information flows (Figure 2), agile firms can realize cost leadership. Through repeated supply of high quality, customized products/services at low cost, agile firms can build lasting relationship with their customers. The relationship between IS for agile manufacturing, agile capabilities, and competitive capabilities can be depicted in the following diagram.

Figure 7. Strategic role of IS

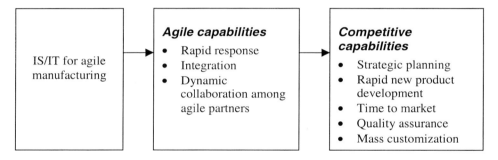

8. SUMMARY

In conclusion, we have looked into the rapidly changing environment of manufacturing characterized as a change from industrial to post-industrial environment. Agile manufacturing is proposed to be an effective paradigm in dealing with such changes in the external environment. We discussed the enabling role of IS/IT in support of agile manufacturing. We have seen that the system architecture for IS/IT in agile manufacturing environment is quite different from that in an industrial manufacturing environment. Information systems in agile manufacturing environment should exhibit high level of integration and flexibility simultaneously. To achieve these goals, it is suggested that object-oriented programming and data retrieval, open architecture, heterarchical and hybrid control architectures, and distributed computing should be adopted. MCSARCH and AMIS, as examples of architectures, have been discussed. We also provided a conceptual framework for IS/IT requirements for agile manufacturing. We furthermore discussed the justification and implementation issues in IS/IT for agile manufacturing. Just like in AMT justification, strategic and intangible benefits should be incorporated in making IS/IT investment decisions for agile manufacturing. In implementation, a total systems approach should be adopted. When properly planned and implemented, IS/IT for agile manufacturing would lead to agile capabilities, such as rapid response, integration and dynamic collaboration among agile partners, which would yield multiple competitive capabilities.

REFERENCES

1. G.P. Huber, The Nature and Design of Post-Industrial Organizations, Management Science, 30 (8), August (1984), 928-951.
2. W. Skinner, The Taming of Lions: How Manufacturing Leadership Evolved, 1780-1984, In The Uneasy Alliance: Managing the Productivity-Technology Dilemma, K. B. Clark, R. Hayes, and C. Lorenz (eds.), Harvard Business School Press, Boston, MA (1985), 63-114.
3. W.J. Doll and M.A. Vonderembse, The Evolution of Manufacturing Systems: Towards the Post-Industrial Enterprise, OMEGA, 19 (5) (1991) 401-411.
4. M.A. Vonderembse, T.S. Ragunathan, and S.S. Rao, A Post-Industrial Paradigm: To Integrate and Automate Manufacturing, International Journal of Production Research, 35 (9) (1997) 2579-2599.
5. H.A. Simon, Applying Information Technology to Organization Design, Public Administration Review, May/June (1973) 268-278.
6. R.L. Daft and R.H. Lengel, Organizational Information Requirements, Media Richness and Structural Design, Management Science, 32 (5), May (1986), 554-571.
7. R.R. Burke, Do you see what I see? The future of virtual shopping, Journal of the Academy of Marketing Science, 25 (4) (1997) 352-360.
8. J. Sheridan, The Agile Web: A model for the future? Industry Week, 245 (5) (1996) 31-35.
9. J.D. Oleson, Pathways to Agility, John Wiley & Sons, Inc., New York, NY, 1998.
10. Y.Y. Yusuf, M. Sarhadi, and A. Gunasekaran, Agile Manufacturing: The Drivers, Concepts and Attributes, International Journal of Production Economics, 62 (1/2), May (1999), 33-43.
11. S.L. Goldman and R.N. Nagel, Management, Technology and Agility: The Emergence of a New Era in Manufacturing, International Journal of Technology Management, 8 (1/2) (1993) 18-38.
12. P.T. Kidd, Agile Manufacturing: Forging New Frontiers, Addison-Wesley, Reading, MA, 1994.
13. R. Booth, More Agile than Lean, Proceedings of the British Production and Inventory Control Society Conference (1995) 191-207.
14. P.D. Hilton and G.K. Gill, Achieving Agility: Lessons from the Leaders, Manufacturing Review, 7 (2) (1994).
15. S.L. Goldman, R.N. Nagel, and K. Preiss, Agile Competitors and Virtual Organizations: Strategies for Enriching the Customer, Van Nostrand Reinhold, New York, NY, 1995.
16. T.F. Burgess, Making the Leap to Agility: Defining and Achieving Agile Manufacturing through Business Process Redesign and Business Network Redesign, International Journal of Operations and Production Management, 14 (11) (1994) 23-34.
17. V. Pandiarajan and R. Patun, Agile Manufacturing Initiatives at Concurrent Technologies Corporation, Industrial Engineering (1994) 46-49.
18. G. Vastag, J.D. Kasrda, and T. Boone, Logistics Support for Manufacturing Agility in Global Market, International Journal of Operations and Production Management, 14 (11) (1994) 73-85.
19. M.A. Youssef, Agile Manufacturing: A Necessary Condition for Competing in Global Markets, Industrial Engineering, December (1992), 18-20.
20. Y.Y. Yusuf, The Extension of MRP-II in Support of Integrated Manufacture, Unpublished Ph.D. Thesis, University of Liverpool, 1996.

21. Iacocca Institute, 21st Century Manufacturing Enterprise Strategy, Lehigh University, Bethlehem, PA, 1991.
22. O. Aguirre, R. Weston, F. Martin, and J.L Ajuria, MCSARCH: An Architecture for the Development of Manufacturing Control Systems, International Journal of Production Economics, 62 (1/2) (1999) 45-59.
23. M.I. Barber, S. Jennis, R.H. Weston, and J.D. Gascoigne, A Study of Business Process Re-engieering Practice in the UK, MSI Publication, Loughborough University, UK, 1996.
24. M. Jung, M.K. Chung, and H. Cho, Architectural Requirements for Rapid Development of Agile Manufacturing Systems, Computers and Industrial Engineering, 31 (3/4), December (1996), 551-554.
25. L. Song and R. Nagi, Design and Implementation of a Virtual Information System for Agile Manufacturing, IIE Transactions, 29 (1997) 839-857.
26. I.J. Chen and M.H. Small, Implementing Advanced Manufacturing Technology: An Integrated Planning Model, Omega, International Journal of Management Science, 22 (1) (1994) 91-103.
27. K. Ramamurthy and W.R. King, Computer Integrated Manufacturing: An Exploratory Study of Key Organizational Barriers, Omega, International Journal of Management Science, 20 (4) (1992) 475-491.
28. V. Kumar, S.A. Murphy, and S.C.K. Loo, An Investment Decision Process: The Case of Advanced Manufacturing Technology in Canadian Manufacturing Firms, International Journal of Production Research, 34 (1996) 947-958.
29. W.J. Vrakking, Consultants' Role in Technological Process Innovation, Journal of Management Consulting, 5 (3) (1989) 17-24.
30. R. Kaplan, Must CIM be Justified by Faith Alone? Harvard Business Review, 64 (2) (1986) 87-95.
31. P.L. Primrose, Investment in Manufacturing Technology, Chapman & Hall, London, 1991.
32. T.L. Saaty, The Analytic Hierarchy Process, McGraw-Hill, New York, NY, 1980.

APPENDIX: ACRONYMS

AMIS: Agile manufacturing information system. Proposed by Song and Nagi [25].
AMT: Advanced manufacturing technology. Includes a variety of computerized technologies, such as computer-aided manufacturing (CAM) and computer-aided process planning (CAPP). These technologies can be combined into various types of integrated systems, such as flexible manufacturing systems (FMS) and computer-integrated manufacturing systems (CIM).
BPR: Business process re-engineering. A radical or breakthrough change in a business process, often making use of modern information technology.
CAD: Computer aided design. Product design using computer graphics.
CAPP: Computer aided process planning. A computer system that helps in selecting sequences of operations and machining conditions.
CIM: Computer integrated manufacturing. A system for linking a broad range of manufacturing activities through an integrating computer system.
CORBA: Common object request broker architecture. An evolving framework being developed by the Object Management Group to provide a common approach to systems interworking.

DBMS: Database management system. Groups of software used to set up and maintain a database that will allow users to call up the records they require. In some cases, DBMS also offer report and application-generating facilities.

DCE: Distributed computing environment. A set of definitions and components for distributed computing developed by the Open Software Foundation, an industry-led consortium.

DCF: Discounted cash flow. A technique to obtain the present value of a future project by multiplying the sum of the future cash flows by discount factors.

DCOM: Distributed component object model. Microsoft's system for spreading an application across more than one computer on a network. Built upon Windows NT 4.0 and Windows 98.

ERP: Enterprise resource planning. An MRP II system that ties customers with suppliers.

IP: Internet protocol. A connectionless (i.e. each packet looks after its own delivery), switching protocol. It provides packet routing, fragmentation and reassembly to support TCP. IP is defined in RFC 791.

IS/IT: Information systems/Information technology. IT is a very general term coined in the 1970s to describe the application of computer science and electronics engineering to the specification, design and construction of information-rich systems.

ISO: International organization for standardization. Specialized international organization founded in Geneva in 1947 and concerned with standardization in all technical and non-technical fields except electrical and electronic engineering.

ISO/DIS: Draft international standards of ISO. Revision of five ISO 9000 series.

LAN: Local area network. A communication network consisting of many computers (mostly personal computers and workstations) that is placed within a local area, such as a single building or company.

MCS: Manufacturing control systems. A group of information systems designed to control shop floor activities on a manufacturing plant.

MCSARCH: Manufacturing control systems architecture. Proposed by Aguirre et al. [22].

MMS: Manufacturing message specification. ISO approved international communication standard for messaging between various manufacturing machines.

MRP: Manufacturing requirements planning. Computer-based information system for ordering and scheduling of dependent-demand inventories.

MRP II: Manufacturing resource planning. Expanded approach to production resource planning, involving other areas of a firm in the planning process, such as marketing and finance.

OPT: Optimized production technology. A scheduling software package that emphasizes identifying bottleneck operations and optimizing their use.

OSI: Open systems interconnection. The ISO reference model consisting of seven protocol layers. These are the application, presentation, session, transport, network, link and physical layers. The concept of the protocols is to provide manufacturers and suppliers of communications equipment with a standard that will provide reliable communications across a broad range of equipment types.

SAP: An ERP system, provided by the company with the same name. SAP stands for Systems, applications, and products.

STEP: Standard for the exchange of product model data. A neutral format for product definition data, which aims to eliminate the cost, complexity and time related to multiple CAD systems without pulling the plug on any of those systems. With STEP, companies can move data between dissimilar systems.

TCP: Transmission control protocol. The most common transport layer protocol used on Ethernet and the Internet. It was developed by DARPA. TCP is built on top of Internet protocol (IP) and the two are nearly always seen in combination as TCP/IP (which implies TCP running on top of IP). The TCP element adds reliable communication, flow control, multiplexing and connection-oriented communication to the basic IP transport. It is defined in RFC 793.
TCP/IP: Transmission control protocol/Internet protocol. The set of data communication standards adopted, initially on the Internet, for interconnection of dissimilar networks and computing systems.
TQM: Total quality management. Management of an entire organization so that it excels in all aspects of products and services that are important to the customer.
WM: Workflow manager. An automated system that directs application programs' access to information storage. A part of AMIS, proposed by Song and Nagi [25].

Management of Complexity and Information Flow

E. Szczerbicki

The University of Newcastle, Department of Mechanical Engineering, Newcastle, NSW 2308, Australia

Agile manufacturing offers to industry the promise of surviving in changing, uncertain, and imprecise market environments of the new millenium. The implementation of agile manufacturing strategy might increase the complexity of the manufacturing system itself as well as the complexity of information flow in both internal and external environments in which the system operates. To adequately manage Agile Manufacturing Systems, it is suggested that a crucial focal point is the development of tools and approaches that support management of complexity of such systems and that further the understanding of the nature and role of information flow within such environments.

1. INTRODUCTION

Faced with fierce, world-wide competition, organizations struggle to improve quality while reducing costs. Agile manufacturing is a strategy developed to lead this quest. Agile Manufacturing Systems (AMS) to be flexible and responsive have to deal with complexity of their own operational structure and that of information flow.

Managing complex AMS that function in changing and uncertain information-rich environments requires greater understanding and knowledge about the role of information in systems operation. To gain this understanding, an approach is needed that could be used to model and evaluate information flow in different situations. Such an approach is presented in this Chapter.

In fact, this Chapter goes well beyond the above in proposing an approach considering important practical issues of information flow in AMS, i.e. delays, incompleteness, imprecision and loss in value. The current practice of dealing with such issues are mostly when problems are detected and reactively. This situation may not be desirable and definitely be a major drawback for complex AMS that more and more rely on the timeliness and quality of information for their operation. The proposed approach, in this respect, would greatly enhance the understanding of the various factors that influence the quality of information to the benefit of better decisions in adequate time, which in fact is the core of the philosophy behind AMS concept.

AMS become increasingly complex. Their decomposition into smaller units is the usual way to overcome the problem of complexity. This has historically led to the development of atomised AMS structures consisting of a limited number of *autonomous subsystems* that decide about their own information input and output requirements, i.e. can be characterised by what is called an *information closure*. Autonomous subsystems can still be interrelated and embedded in larger AMS, as autonomy and independence are not equivalent concepts. These ideas are recently gaining very strong interest in both academia and industry, and the atomised approach to AMS modelling, design and development is an idea whose time has certainly come [1, 2, 3, 4]. The issues discussed in this Chapter will focus on information flow for autonomous subsystems.

In a real-world context of AMS, autonomous subsystems consist of groups of people and/or machines tied by the flow of information both within a given subsystem and between this subsystem and its external environment [5]. We will briefly present a modelling approach that could be used to evaluate such an information flow. The suggested approach allows for the evaluation of an information flow to be performed for different types of external and internal environments of a given AMS. It takes into account two basic cases, i.e., static and dynamic processes describing the external environment. Such issues as the role of correlation and interaction, and the losses caused by incomplete and delayed information are considered. The approach also accommodates the question of uncertain and imprecise information flow modelling.

The proposed modelling approach will provide the basis for a wide variety of real-world AMS applications which require advanced methods, and have thus hitherto been unachievable. In particular it will address the frequent situations in which the following should be answered:

"How to structure an exchange of information between an autonomous subsystem within AMS and its environment?"
&
"What is better, complete information but heavily delayed, or incomplete information less delayed?"

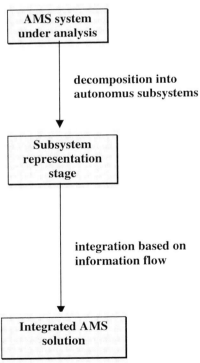

Figure 1. Overview of a three stage approach to AMS analysis

A three stage approach is proposed for the analysis of complex AMS. The involved stages are system's decomposition, modelling representation, and integration at the level of information flow. Figure 1 depicts the general underlying idea behind this approach.

Decomposition stage can easily be modelled by logical AND/OR trees. This stage occurs at the highest level of abstraction and it must provide enough information to begin the representation and integration stages. After the overall AMS is decomposed into a number of autonomous subsystems, the stage of subsystem modelling representation is entered. Representation stage aims at providing models of the components generated in the first stage. In Figure 1 an integrated system is a system that consists of components that efficiently contribute to the tasks, functional behaviour, and performance of a given system as a whole. It is believed that one can achieve such an integration through the flow of information [6].

This Chapter focuses on modelling approaches used to depict information flow at the representation stage of AMS analysis. Both analytical and soft modelling platforms are discussed.

2. ANALYTICAL APPROACH

In the analytical modelling platform at the representation stage of AMS analysis information is defined as knowledge about the realisation of random variables $X_i(T)$ describing the external environment of an autonomous subsystem. Characteristic features of these variables are modelled by the linear autoregressive process given as [7]:

$$X(t)=(w_1 + w_2)X(t-1) - w_1 w_2 X(t-2) + p(t) \qquad (1)$$

where $p(t)$ are uncorrelated random variables with zero mean and constant variance and w_1 and w_2 are the equation coefficients. Coefficients w_1 and w_2 define the dynamics of the external environment.

The transformation process of any given AMS subsystem describes the actions of its elements designed in order to achieve the goals. As introduced in [5], such actions can be expressed by:

$$A_{opt} = \min E[f(A,X)|C] \qquad (2)$$

where A represents the set of all possible actions, A_{opt} an optimal one, X describes the state of the environment, and C is an information structure. Information structure defines the flow of information for a given subsystem and is modelled by matrix C in which $c_{ij}=1$ if the ith functional element of subsystem has obtained information about the jth variable X realisation (if $c_{ij}=0$, no information has been obtained).

The control process of a subsystem, or AMS as a whole, uses the value of information structure VC given as:

$$VC = \min E[f(A,X)| \text{ no information}] - \min E[f(A,X) | C] \qquad (3)$$

The value of information (3) will be different for different information structures and different environments. It can be considerably affected by two major attributes of information: incompleteness and delay. The highest VC value will be possessed by a full information structure C ($c_{ij} = 1$ for each i and j). On the other hand, however, gathering information in dynamic environment causes its delay. Both delay and incompleteness can be represented by losses in the value of the information structure denoted as LVC.

Delay and incompleteness represent two contrary information attributes. Namely, one type of loss rises with increases in the amount of information, the other type falls, and total losses have a minimum which represents the required balance. Mathematical models can be developed for assessing the values of incomplete and delayed information. The basic notion of these models are the best decision functions of the subsystem elements (β_i) and the value of the information structure VC expressed by the following systems of equations:

$$\beta_i(h_i) + \sum_{j \neq i} qE[\beta_j(h_j) \mid h_i] = E[b_i \mid h_i], \quad i = 1, 2, \ldots, n \qquad (4)$$

$$VC = E[b^T \beta] \qquad (5)$$

where h_i stands for information of the ith element of a subsystem, b_i stands for realisation of the ith variable describing the external environment in a form in which it may reach the AMS, and q stands for interaction coefficient describing interaction between AMS elements. Equations (4) and (5) are the tools for the quantitative evaluation of an information flow. Such an evaluation depends on the following state parameters: delay of information (d), amount of information (a), dynamics in the external environment (w), correlation in the external environment (r), and interaction in the internal environment (q).

With the modelling tools given by (4) and (5) it is possible to extract knowledge about autonomous AMS subsystems functioning in some decision situations. This knowledge is easily codified as production rules and can be used in control, command, and management of a given AMS. A simple example of such rules related to the role of correlation and interaction in structuring and analysis of information flow for AMS is presented next.

2.1. The role of correlation and interaction

In static case external environment of a given AMS subsystem is represented by correlation between random variables denoted as r, and internal environment by interaction between AMS elements denoted as q. If q=0 the actions within AMS are independent. They are dependent for q≠0. The value of information structure VC (5) is influenced by r and q, and is different for different C representing different information structures. Each information structure (C) is a collection of single pieces of information about realisation of random variables X. Each single random variable is described by its variance s^2.

For illustrative purposes let us consider the following simple information structures for a two-element AMS subsystem:

$$C1 = \begin{bmatrix} 1 & 0 \\ 0 & 0 \end{bmatrix} \quad C2 = \begin{bmatrix} 1 & 0 \\ 0 & 1 \end{bmatrix} \quad C3 = \begin{bmatrix} 1 & 0 \\ 1 & 1 \end{bmatrix} \quad C4 = \begin{bmatrix} 1 & 1 \\ 1 & 1 \end{bmatrix}$$

Information structures C1 and C2 are created only by observation. In C3 and C4 both observation and communication are involved. Before further analysis is carried on, a simple introductory example of a two-element AMS subsystem is presented to illustrate the relation of the above information structures to real situations.

Two parts (PA and PB) are manufactured by two AMS elements (A and B). The external environment of the AMS is described by the amount of material available for this production (MA and MB). The assembly process of PA and PB determines the character of the internal environment of the AMS. For example, if PA and PB take part in the assembly of two different end products, or are the end products themselves, the AMS elements actions are independent. They are dependent if PA and PB are assembled into the same end product. For example for information structure C2 the external environment of the AMS is described by MA and MB (X_1=MA and X_2=MB). Information that is available for the AMS element A is given by d_1=[MA] (c_{11}=1). For the AMS element B we have d_2=[MB] (c_{22}=1). As there is no communication between the elements A and B the values of c_{12} and c_{21} are equal to zero. The AMS element A decides about the amount of material that is ordered for production of part PA. The element B decides the same about the amount of material for production of part PB.

The above interpretation is easily translated into analytical representation using (4) and (5). As it can be seen from such an interpretation for information structure C1, the value of a single piece of information depends on the variance of X; the bigger the variance the more valuable is the information about random variable X realisation. Thus we have:

RULE 1
    IF   an external environment of AMS is static,
    AND it is described by a random variable,
    THEN the value of information about this variable realisation decreases with decreasing value of its variance.

For information structure C2 there is no exchange of information between AMS elements and we can formulate the following rule for decentralised information flow:

RULE 2
    IF   an external environment of AMS is static,
    AND it is described by random variables,
    AND there is an interaction between AMS elements and correlation between random variables,
    THEN the value of decentralised information structure (created only by observation) increases with increasing correlation and interaction

Information structures C3 and C4 involve both observation and communication inside AMS and we can formulate the following general rule:

RULE 3
    IF   an external environment of AMS is static,
    AND it is described by random variables,
    AND there is an interaction between AMS elements and correlation between random variables,
    THEN the value of information structure increases with increasing amount of information

Using analytical representation it is also easy to notice that the value of VC can increase only up to the point where the information structure reaches C4 (or more generally the full information structure). This special case can be coded in the following production rule:

RULE 4
    IF   an external environment of AMS is static,
    AND it is described by random variables,

> **AND** there is an interaction between AMS elements and correlation between random variables,
> **THEN** full information has the value that is always greater than any other information structure

2.2. Special cases of q=0 and r=1

Special case of q=0 represents the decision situation of a given AMS in which there is no interaction between its elements. For this case the analytical model reduces to zero the value of any piece of information resulting from communication (AMS element seeks only this information that describes its own part of the external environment). For q=0 there is no interaction in the internal environment and the actions of the elements of AMS are independent. For such a case information exchange inside AMS does not affect the value of the information structure. For example, although information structure C4 contains a larger amount of information than C2, their VC values are the same for q=0. Thus we have:

RULE 5
> **IF** an external environment of AMS is static,
> **AND** it is described by random variables,
> **AND** there is no interaction in the internal environment,
> **THEN** it is enough to restrict the information flow only to observation; organising an information exchange does not improve the value of resulting information structure

Special case of r=1 represents the decision situation in which the relationship between the variables describing the external environment is of functional character (not statistical). It is similar to the case of q=0 as the value of any piece of information resulting from communication is reduced to zero. Thus we have:

RULE 6
> **IF** an external environment of AMS is static,
> **AND** it is described by random variables,
> **AND** the relationship between variables describing the external environment is given by function dependence,
> **THEN** communication between agent elements does not affect the value of information structure; information flow should be restricted to observation

The rules 1 through 6 form the knowledge base capturing the evaluation of information flow for an AMS functioning in static environment. They can be easily delivered using rigorous mathematical solution and tools described by (4) and (5). However, the application of these tools in the knowledge acquisition process for larger AMS structures is very limited because:

- the computation involved in information flow analysis and evaluation may become complex and prohibitive,
- complete and certain information that the analytical models require is often not available.

Simplified modelling approaches are needed that can be used to describe and understand the flow of information in AMS. Such approaches are presented next.

3. SOFT MODELLING AND SIMULATION

The underlying idea in soft (qualitative) modeling is that the complexity of a given system can often be reduced by taking into account only certain abstractions of its behaviour [8]. In quantitative modeling the idea is to represent the system as a set of parameters that can assume real values and mimic the functioning of the system by the set of changes of these parameters in time. Soft modeling deals with an abstraction of the above values.

In a broader context, we can formulate the following motivations for developing qualitative models of real-life AMS systems:

(a) to provide simpler computational mechanisms than those already existing,
(b) to provide description for systems where traditional methods are ineffective
(c) to provide modelling approach that mimics more closely our common sense and intuition
(d) to provide modelling methods for reasoning with partial, uncertain or incomplete information, and for effective explanation facilities.

Soft modelling tools that can be useful in information flow representation and complexity modelling for AMS include decision trees, connectionist systems, digraphs, and simulation. Their possible applications are outlined next.

3.1 Decision tree classifiers

Decision tree classifiers are used successfully in many diverse areas. Their most important feature is the capability of capturing descriptive decisionmaking knowledge from the supplied data. Decision tree can be generated from training sets. The procedure for such generation based on the set of objects (**S**), each belonging to one of the classes $C_1, C_2, ..., C_k$ is as follows [9]:

Step 1. If all the objects in **S** belong to the same class, for example C_j, the decision tree for **S** consists of a leaf labelled with this class.

Step 2. Otherwise, let T be some test with possible outcomes $O_1, O_2, ..., O_n$. Each object in **S** has one outcome for T so the test partitions **S** into subsets $S_1, S_2, ... S_n$ where each object in S_i has outcome O_i for T. T becomes the root of the decision tree and for each outcome O_i we build a subsidiary decision tree by invoking the same procedure recursively on the set S_i.

The above procedure is applied to the training set of objects related to the flow of information for a given AMS subsystem in Table 1. The training sets are delivered from the initial analysis based on the quantitative model of AMS functioning as presented in Section 2. Each object is described by the relating attributes and belongs to one of the agent decision classes exchange_information ("yes" in the last column) or do_not_exchange_information ("no" in the last column).

Table 1
Training set for agent functioning

external environment	internal environment	type of dynamics	correlation	delay of information	decision
static	independent_actions	0	0	0	no
static	independent_actions	0	0.7	1	no
dynamic	dependent_actions	1.5	-0.5	1	yes
static	dependent_actions	0	0	0	yes
static	independent_actions	0	1	2	no
static	dependent_actions	0	0.5	2	yes
static	dependent_actions	0	-1	3	no
static	independent_actions	0	-1	0	no
static	dependent_actions	0	0.9	1	yes
static	dependent_actions	0	1	1	no

Suppose, as it was done in Section 2, that we are interested in decision making situations involving static environment only. When for this case the set is partitioned by testing on internal_environment and then on correlation, the resulting structure is equivalent to the decision tree shown in Figure 2.

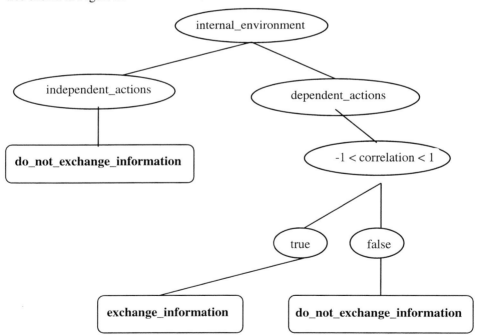

Figure 2. Decision tree classifier for AMS information flow related decisionmaking

The following rules can be delivered from Figure 2.

RULE 1
IF an external environment of an agent is static
AND it is described by random variables
AND there is no interaction in the internal environment
THEN communication (exchange of information) between agent elements is not necessary

RULE 2
IF an external environment of an agent is static
AND it is described by random variables
AND there is interaction in the internal environment
AND the relationship between variables describing the external environment is of statistical character
THEN exchange of information between agent elements should be organised

RULE 3
IF an external environment of an agent is static
AND it is described by random variables
AND there is interaction in the internal environment
AND the relationship between variables describing the external environment is given by function dependence
THEN exchange of information between agent elements is not necessary

The above shows how the simple decision tree in Figure 2 can be used to retrieve some of the knowledge concerning the functioning of an AMS subsystem in static environment. The resulting rules are exactly the same as those that were developed using the analytical model presented in Section 2. The decision tree, once developed, can support decision situations that are not covered by the training set. That is why the production rules can be formulated as a generalised statements. As it has been illustrated the use of decision trees is simple and as effective as the analysis based on a rigorous mathematical model. Another tool that can be of much help in the process of knowledge retrieval for AMS information flow and evaluation is the technique based on connectionist systems.

3.2. Connectionist systems

Neural networks which learn mappings between sets of patterns are called mapping neural networks [10, 11]. A key property of mapping networks is their ability to produce reasonable output vectors for input patterns outside of the set of training examples (please note the similarity to the decision tree classifiers). The above is especially important in areas such as discussed in this paper, i.e. areas for which it is possible to develop only a very limited number of IF...THEN rules and thus also to make inferences only for a very limited number of decision situations.

Problem solving tasks, such as information structure development for AMS, may be considered pattern classification tasks. The system analyst learns mappings between input patterns, consisting of characteristics of agent's external and internal environment, and output patterns, consisting of information structures to apply to these characteristics. Thus, neural networks (neural-based expert systems) offer a promising solution for automating the learning process of the analyst.

As we already know, systems analyst, while developing an information structure for an AMS or its subsystem, transforms certain characteristics of a given AMS into recommendations concerning the flow of information. These characteristics represent the input for the system and their full description (for both static and dynamic environments) includes 5 parameters:

correlation in the external environment (r), dynamics (t), interaction in the internal environment (q), delay (d), and type of the process describing the external environment (w). Output consists of the following decisions (recommendations): (i) observation (or sensing) should be present, and (ii) exchange of information should be present. Please note that the decision concerning sensoring of information is added for this case (it was not considered in Section 3.1). An input portion together with an output portion of the data represents a training pair. The training pairs were used to train a 5-10-2 neural network.

The target values for each output node were normalised in such a way that the maximum target for each node received a value of 0.75 and the minimum target for each node received a value of 0.25. The training values for each input node were identically normalised. The learning rate and momentum term of 0.9 were used in the network. The network was trained using error back propagation procedure with a training tolerance of 5%. The network was considered trained if, for all training pairs and output nodes, |(desired output - actual output)/(desired output)| < tolerance.

Table 2
The use of the trained network

no	value	description	observation	exchange
1	$r=0.95$	strong relationship between variables describing external environment		
	$t=0$	external environment is static		
	$q=0.01$	there is no interaction in internal environment	yes	no
	$d=0$	information is not delayed		
	$w=0$	process is independent		
2	$r=0.2$	weak relationship between variables describing external environment		
	$t=0$	external environment is static		
	$q=0.90$	there is interaction in internal environment	yes	yes
	$d=0$	information is not delayed		
	$w=0$	process is independent		

After training, additional characteristics of AMS were generated for use of the network. Five sets of characteristics were submitted to the network. In response, the network suggested five information flow recommendations. As an example, Table 2 presents two sets of characteristics submitted and the obtained recommendations after the trained network has been used. For the first input set (no. 1 in Table 2) the network recommends decentralised information structure (only observation, no exchange of information). For the second, full information structure is recommended (observation and exchange). In both cases the recommendations agree with the IF ... AND ... THEN rules discussed in Section 2 which was based on analytical model.

3.3 Signed directed graphs

Signed directed graphs can be used to build simple qualitative models of complex AMS, and to analyse those conclusions attainable based on a minimal amount of information.

A directed graph, or digraph, is a graph in which all edges are directed [12]. A signed digraph is a digraph with either + or - associated with each edge. SDG nodes are chosen as variables relevant to or representative of the problem that is studied. There is an edge from variable A to variable B if a change in A has a significant direct effect on B. The sign of the edge is + if an increase in A leads to an increase in B, and a decrease in A leads to a decrease in B. The sign is - if the effect is opposite; an increase in A leads to a decrease in B, and a decrease in A leads to an increase in B.

According to the mathematical model presented in Section 2, information flow depends on the following state parameters: delay of information (d), amount of information (a), dynamics in the external environment (w), variance in the external environment (s), and interaction in the internal environment (q). The above parameters influence the loss in the value of information caused by delay (L1), the loss in the value of information caused by incompleteness (L2), and total loss (LV). Based on relationships and dependencies described by mathematical model, the SDG can be developed for this case as depicted in Figure 3.

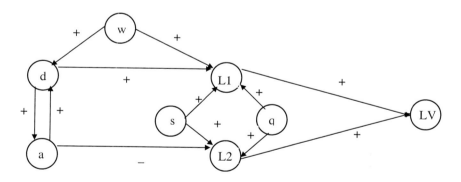

Figure 3. SDG model of informational flow for AMS subsystem

SDG models can be simplified. Two principles are used for the simplification process. The first one is the principle of removal of intermediate nodes and the other one is the simplification of positive feedback loop. There are two simplification steps that can be applied to the SDG model shown in Figure 3. First, the evaluation of information flow is represented by total losses denoted as LV. The parameters L1 and L2 are not of interest in our reasoning process and can be removed as intermediate notes.

Next, the positive feedback between (a) and (d) that generates spurious solutions as well as parameters (s) and (q) that are constant for a given decision situation of an AMS can also be eliminated from the model. Figure 4 shows the SDG model after the overall simplification.

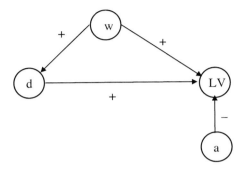

Figure 4. SDG model after two simplification steps

The following logic rules can be developed for the model in Figure 4.

SDG Rule 1:
    IF [d=+] .and. [pdLV]
    THEN it is a possible solution pattern for a positive change in d

SDG Rule 2:
    IF [a=+] .and. n[aLV]
    THEN it is a possible solution pattern for a positive change in a

SDG Rule 3:
    IF[w=+] .and. p[wLV]
      .and. p[wd]
      .and. p[dLV]
    THEN it is a possible solution pattern for a positive change in w

The above logic rules describe qualitative behaviour of the SDG model and thus the modelled AMS as well. They translate easily into the corresponding IF... AND.... THEN production rules. They point into exactly the same behaviour of information as analysis based on mathematical model. For example, they depict the adverse character of two contrary information attributes, i.e. delay and incompleteness. They also show clearly the effects of increasing dynamics in the external environment of an AMS. More generally, the results show that as far as the analysis of overall directions of a given AMS behaviour is concerned the simple qualitative model can be sufficient at a minimum level of complexity.

3.4. Simulation

Simulation can be defined as a numerical technique for conducting experiments on a digital computer, which involves logical and mathematical relationships that interact to describe the behavior and structure of a complex real-world system over extended periods of time [13, 14]. It is important for us to model and simulate processes by which tasks are performed in AMS and through which we evaluate alternative designs to improve system performance.

For the modelling purpose in the case studies presented in this Section, AweSim and Visual SLAM simulation environment has been used [15]. SLAM provides a simulation language that allows alternative approaches to modelling by easily altering parameters so that many variations of a given AMS can be analysed. It permits network, discrete event, and continuous modelling perspectives, or any combination of the three, to be used in developing a single simulation model.

AweSim and Visual SLAM supports both graphical and textural modes of programming. The graphical mode makes program design and debugging easier and quicker, whereas the textural mode is used for specific instructions. The software can easily convert between either mode as it is written. It also has the ability to allow the modeller to insert C language sub-routines where required.

SLAM creates entities and sends them through networks (or paths) which consists of activities and branching that allows the modeller to represent both the physical and decision making processes of the system that is modelled. The entities can have attributes assigned to them to allow for conditional branching. Resources can be used to control the entity utilisations and the program to control specific details can use system variables. Also, a number of probability distribution functions that allow reality to be represented more closely are supported.

For both cases described in this Section, after SLAM models have been developed and implemented, they have been tested for correctness and accuracy. Well established verification scheme was used [16]. All simulation functions that were included have been tested using simple entity flow tests to determine if they work properly. Then simulation models were executed under different straightforward conditions to determine if the computer program and its implementations are correct. The bottom-up dynamic testing strategy was used. First, all program modules representing different parts of modelled systems were tested. Then, overall models were run with a number of TRACE options to include in the testing process the values obtained during the program execution. Models were also validated and their satisfactory accuracy with the study objectives was determined. The techniques presented in [16], i.e. event validity, face validity, and historical data validation have been used for validation of all modules as well as overall models.

3.4.1. Case study: steel processing

The focus of this case study is on the system represented by further processing area of a bar mill in steel manufacturing process. Further processing utilises various resources (equipment, labour, energy) to process bar length and coil products and represents one of the last stages in steel manufacturing sequence.

Further processing area for a company that was studied includes the following activities (Figure 5):
- WH (Warehousing),
- MR2 (Inspection),
- RE2 (Reinspection),
- BCL (Bar Classify),
- ST3 (Straight Press),
- CPR (Coil Press).

SLAM network model was developed to depict the logic of the steel flow in the area depicted in Figure 5. The model includes all activities that influence the functioning of the system under study.

The particular real world steel manufacturing case studied in this Section faced the following two problems:
- The lead time for products that need to be further processed is considered to be too long.
- The dispatching performance is too low for a targeted shorter lead time.

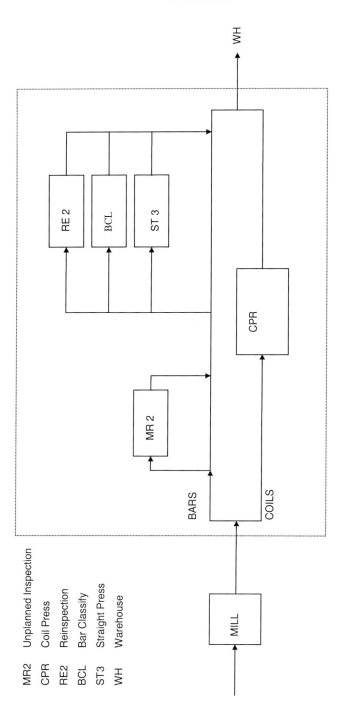

Figure 5 Steel processing area flow chart

MR2 Unplanned Inspection
CPR Coil Press
RE2 Reinspection
BCL Bar Classify
ST3 Straight Press
WH Warehouse

The above are very typical problems faced by AMS. The manufacturer needed a modelling and decision support tool to analyse and solve the above problems and in particular to:
- Determine ways to reduce the lead times for the further processing in order to reach a dispatching performance of 95% within 1 week.
- Improve the current rolling sequence as to reach the targeted lead time and dispatching performance.
- Find the optimum number of operators working at the further processing stations.

Computer simulation and modelling platform based on AweSim and Visual SLAM was proposed as the required decision support tool. The modelling formulations were developed to represent a platform that can be easily customised and applicable to almost any particular case involving scheduling and resource allocation for AMS in steel industry.

The model was run using data supplied by the Planning Department of the manufacturer for whom the study was conducted. The main results were provided in histograms describing dispatching performance for all activities areas included in further processing, i.e. warehousing, inspection, classification and pressing. In Histogram No. 2 below, the results for BCL (Bar Classify) are shown as an example. A 7 day lead time is highlighted.

```
                    **HISTOGRAM NUMBER 2**
                          TIME AT BCL
 OBS   RELA    UPPER
 FREQ  FREQ    CELL LIM  0        20        40        60        80       100
                         +    +    +    +    +    +    +    +    +    +    +
   0   .000    .000E+00  +                                                 +
   8   .195    .100E+01  +**********                                       +
   8   .195    .200E+01  +**********        C                              +
  11   .268    .300E+01  +*************              C                     +
   8   .195    .400E+01  +**********                              C        +
   5   .122    .500E+01  +******                                           +
   1   .024    .600E+01  +*                                               C+
   0   .000    .700E+01  +                                                 C
   0   .000    INF       +                                                 C
  ---                    +    +    +    +    +    +    +    +    +    +    +
  41                     0        20        40        60        80       100

         **STATISTICS  FOR  VARIABLES  BASED  ON  OBSERVATION**
                MEAN      STANDARD   COEFF. OF   MINIMUM   MAXIMUM   NO.OF
                VALUE     DEVIATION  VARIATION    VALUE     VALUE     OBS

TIME AT BCL     .245E+01  .137E+01   .559E+00    .178E+00  .508E+01    41
```

The summary of simulation results included in similar histograms for all activities shown in Figure 5 is presented in Table 1.

Table 3
Summary of simulation results with a 7 day lead time.

Activity area	Dispatching Performance
MR2	85%
CPR	93%
BCL	100%
ST3	93%
WH	100%

The results in Table 3 clearly show the bottleneck (station MR2) which performs well below the targeted 95%. A number of "what if" scenarios were run and conditions for increased performance for this particular area were found. Also, the overall performance of the system was studied and the recommendations to improve it included:
1) changes in the rolling sequence,
2) changes in the allocation of operators.

The above example shows how simulation can help in adding agility to a manufacturing system in meeting market requirements. Probably one of the most important advantages of simulation modelling is its adaptability to various stages of AMS analysis and design. It can be easily applied concurrently to all project stages as the AMS that we design evolves. The stages usually involve (i) concept design, (ii) detailed design, (iii) implementation, and (iv) operation. Using simulation models developed concurrently with each stage we can:
- understand basic AMS operation at the concept design level,
- select the best concept to proceed with to the detailed design,
- test all proposed operating and control procedures,
- test the impact of all design changes made during the implementation stage,
- test the impact of all proposed changes during the operation stage,
- predict necessary changes in the AMS operation to follow the envisaged changes in the external and internal environments of the system.

A question that is often overlooked but should be asked is as follows: Is simulation modelling the right tool for the problem? The following may be good guidelines to consider before selecting simulation as a tool for AMS analysis [17].

Do not simulate AMS or any other system when:
- the problem can be solved using common sense analysis,
- the problem can be solved analytically,
- it's easier to change or perform direct experiment on the real system,
- the cost of the simulation exceeds possible savings,
- there are not proper resources available for the project,
- there is not enough time for the model results to be useful,
- there is no data - not even estimates,
- the model can not be verified or validated,
- project expectations can not be met,
- the system's behaviour is too complex or can't be defined.

4. CONCLUSION

Information flow, resource allocation, and scheduling are usually the key areas in the operation of complex AMS. This Chapter proposes a three stage approach to deal with the complexities of the above issues. One can argue that manufacturing systems have grown in complexity over the years mainly due to the increased striving for performance enhancing combined with a greater degree of uncertainty and imprecision in system's external and internal environments. This complexity, always present in real life systems, makes the application of quantitative modelling tools as problem solvers questionable in many instances. We propose a soft (quantitative) modelling approach to add analytical modelling for such cases. Soft modelling can be especially useful in the area of evaluation of information flow for AMS.

Simulation technique, being one of soft modelling tools, can be used as a decision support platform for enhancing performance of any real life AMS. To illustrate the above, steel manufacturing is presented as the case study in this Chapter. The problems faced by the manufacturer were of sequencing and resource allocation nature. The complexity of the real life system made the application of analytical tools as problem solvers impossible. Simulation technique proved to be an adequate, effective and economically efficient problem solver in this case.

The formulations presented in the Chapter are general enough to relate to any AMS for which information flow needs to be modelled and evaluated. The presented modelling approach can be easily tailored to a specific domain by providing external and internal environment attributes that are characteristic for this domain.

REFERENCES

1. D.M. Anderson, Agile Product Development for Mass Customization, Irwin, Boston, 1997.
2. M.C. Zhou (ed), Petri Nets in Flexible and Agile Automation, Kluwer, London, 1995.
3. R. G. Askin and C.R. Standridge, Modelling and Analysis of Manufacturing Systems, Wiley, New York, 1993.
4. A. K. kamrani and P. R. Sferro (eds), Direct Engineering: Toward Intelligent Manufacturing, Kluwer, London, 1999.
5. E. Szczerbicki, Acquisition of knowledge for autonomous cooperating agents, IEEE Transactions on Systems Man and Cybernetics, 23 (1993) 1302-1315.
6. J. Wyzalek (ed), Systems Integration Success, Auerbach, New York, 1999.
7. H. Theil, Principles of econometrics, Wiley, New York, 1971.
8. B. Kuipers, Qualitative reasoning: modelling and simulation with incomplete knowledge, Automatica, 25 (1989) 571-585.
9. J. R. Quinlan, Decision trees and decisionmaking, IEEE Transactions on Systems Man and Cybernetics, 20 (1990) 339-346.
10. C. H. Dagli (ed), Artificial Neural Networks for Intelligent Manufacturing, Chapman and Hall, London, 1994.
11. R. Rojas, Neural Networks, Springer, Berlin, 1996.
12. G. Chartrand and L. Lesniak, Graphs and Digraphs, Chapman and Hall, London, 1996.
13. T. H. Naylor, Balintfy J.L., Burdick, D.S., and Kong Chu, Computer Simulation Techniques, Wiley, New York, 1996.
14. B. S. Bennett, 1995, Simulation fundamentals, Prentice Hall, London, 1995.
15. A. A. B. Pritsker and J. J. O'reilly, Simulation with Visual SLAM and AweSim, Wiley, New York, 1999.
16. R. G. Sargent, A Tutorial on Validation and Verification of Simulation Models. Proceedings of the 1988 Winter Simulation Conference, New Jersey, 1998, 33-39.
17. J. Banks and Gibson R. Don't Simulate When...., IIE Solutions, September 1997, 30-32.

An Object-Oriented Optimization-based Software for Agile Manufacturing in Process Industries[1]

Draman, M., Altinel, I.K., Bajgoric, N., Unal, A.T., and Birgoren, B.

Bogazici University, Department of Industrial Engineering
80815 - Bebek, Istanbul, Turkey

In addition to most commonly used information technology facilities that enable agile manufacturing, MS/OR techniques can also be used as a way of enhancing informational agility in decision making. In agile management, decision-makers need GUI-based tools for using these techniques, without any necessity for knowing complex skills from decision analysis techniques. Therefore, MS/OR scientists, working together with IT specialists, look for better representations of optimization models, for solving them and for analyzing optimum results. The paper outlines a visually interactive graphical modeling approach for optimization of production planning in process oriented production systems. It promotes combining optimization techniques with object-oriented paradigm and visual interactive modeling through an integrated approach and implementing it in agile manufacturing. The concept is demonstrated with a prototype linear programming modeler - AREMOS, which is developed to generate optimal production plan in a petroleum refinery.

1. INTRODUCTION

Modern management is characterised by using MS/OR techniques that are used by decision makers in order to make better decisions. In agile management, decision makers need effective means for accessing these techniques, without necessarily being experts in mathematical programming, simulation and other methodologies. Managerial needs for agile management and manufacturing in complex process industries such as refineries for making sound and optimal production planning decisions constitute the main motivation for this work. Operations researchers have developed a variety of mathematical modeling approaches for representing these systems and algorithms to solve these models with optimum results. Unfortunately, managers and production engineers do not have sufficient mathematical programming expertise to formulate complex objective functions and constraints of mathematical models directly. Even when custom built by experts in the field and used in the production environment, mathematical models need to be modified to reflect new additions or modifications that take place in real production processes. Such modifications are only possible with the intervention of software and mathematical programming experts that have a thorough understanding of the earlier model. Design for easy maintenance and update is bound to solve these problems, but it is questionable whether the design of current systems is flexible enough for effective software maintenance. Use and maintenance of such complex

---

[1] This research is supported by Bogazici University Research Fund, Scientific and Technical Research Council of Turkey TUBITAK-MISAG, and hardware and software donation by Hewlett-Packard

optimization programs are therefore a major issue. Moreover, lacking such a background, production engineers are generally unable to interpret the solutions of a mathematical programming based modeling system correctly. They have difficulty in communicating with the system, since it restricts the interventions of the user, and require data in a strict format. Lack of user friendliness is also a big problem, since typically management argues that some "hard to use" modules of the system are not utilized. These typical issues were exactly the ones encountered in a case study, the TUPRAS Izmit refinery in Turkey.

Problems and requirements formulated above lead to the necessity of modeling environments that let users interact only with simple models of their "real" system and that hides all mathematical modeling aspects, while assuring model formation and automatic maintenance. This paper proposes a "*clone-based*" modeling approach where users build a graphical model of their processing system using clones of real processing units. The user specifies process parameter values and numerical data for these clones, so that their real processes are represented properly by the graphical model. These clones in turn build and maintain a hidden mathematical model. This mathematical model is generated, solved, and its optimum results are reported, without the user's knowledge of its existence. The user will interact only with a graphical model of the real processes and with real parameter values to run scenarios for optimum production control. The crucial point is in object-oriented design principles that enable easy reconfiguration and reuse of common network units within a process industry system. The main objective is defined as highly adaptable (agile) production or manufacturing system, which will be able to respond to the specific requirements of customers, while still respecting and leveraging mass production paradigm.

Proposed modeling concepts are demonstrated with AREMOS, (A REfinery MOdeling System) a prototype decision support system designed specifically for a real complex process industry, a refinery environment. The structure of AREMOS is object-oriented; its graphical modeling objects represent a refinery environment and corresponding behaviors of refinery units exclusively. As it is a prototype, clone behaviors in AREMOS are built-in through coding (not through interactive template editing). AREMOS demonstrates clone modeling concepts in a refinery, but as mentioned problems are similar for all process industries, current research for a generic modeling environment seems to be greatly justified.

2. VIRTUAL ENTERPRISE, AGILE MANUFACTURING AND AGILITY

By definition, a Virtual Enterprise is such a system which has Web-enabled information architecture that connects information systems of companies participating in virtual business. Information agility or informational efficiency in virtual enterprises represents the major prerequisite for agile management and means eliminating inefficiencies in accessing, exchanging and disseminating of all kinds of information.

Virtual enterprises can be formed in different ways. Five most common approaches are the following:
- Group of people working together separated geographically can form a virtual enterprise. Doing such a kind of business is possible thanks to contemporary communication technology (phone, fax, GSM) and Internet technology (e-mail, talk, videoconferencing). This is very simple type of VE and its information architecture is based on two or more computers connected through Internet.
- Group of small or large companies that specialize in some specific business activities can establish a new company on temporary or permanent basis.

- Business based on outsourcing some manufacturing operations can be done on a virtual basis. Actually, this is a new approach within agile manufacturing philosophy, called virtual manufacturing. Virtual enterprise framework enables such a business.
- Virtual enterprise can also be the case in which a company decides to form special relationships with its customers and/or suppliers through an extranet infrastructure.
- We can also consider a case in which distributed objects are used between two or more companies usually connected through Internet, as some sort of virtual business. These are applications created using distributed computing standards.

Agile manufacturing represents a conceptual framework for more efficient and effective manufacturing. It does this mainly by taking advantage of information technology with key IT facilities such as object-oriented technology, c/s technology, internet technology, distributed application standards (DCOM, CORBA) and enterprise application integration technologies. As Dove [1] points out "... Adaptability (Agility) actually became a reasoned focus with the advent of object-oriented software interests in the early '80s". All these efforts resulted in integrated application suites such as: enterprise resource planning (ERP), supply-chain management (SCM), customer relationship management (CRM), data warehousing (DW), business intelligence-related systems (BI), Intranets and extranets. Norrish [2] refers to ERP system as an "agile software for agile manufacturing".

Agile manufacturing makes use of modern information technology to form virtual enterprises, which agilely respond to the changing market demands. The operational unit in agile manufacturing industry is a dynamic virtual enterprise consisting of agile partners who collaborate with each other. The overall objective of an agile manufacturing industry is to achieve quick response to marketing demands, with compatible product quality and lower manufacturing costs. Song and Nagi [3] propose a framework for an agile manufacturing information system integrating manufacturing databases using Object Oriented Methodology in information modeling. According to them Agile Manufacturing Information Systems (AMIS) must posses the following functionality:
- Provide consistent information to distributed partners
- Resolve multiple data ownership and form multiple data views
- Reflect partner individual policies and retain partner autonomy.

Even though the agile manufacturing concept is considered today dominantly within the concept of virtual enterprise, the term "agility" emerged before the Internet-Web boom. Originally it was defined as "the ability of an organization to thrive in a continuously changing, unpredictable business environment" [4]. It was a workshop at Lehigh University 1991 that gave birth of the concept of the agile manufacturing enterprise. Therefore, it should not be correct to consider agile manufacturing only within the concept of virtual enterprises. In such a context it is better to use a term "virtual agile manufacturing", or simply "virtual manufacturing". This means that companies that do not participate in any virtual enterprise framework still can have agile manufacturing capabilities. "Being Agile means being a master of change" says Rick Dove from Paradigm Shift International [5]. The crucial point is in some design principles that enable reconfiguration facilities and reuse of common modules across a framework. The result should be highly adaptable (agile) production or manufacturing system, which is defined as being able to respond to specific requirements of customers, while still respecting and leveraging the mass production paradigm.

Dove defined ten Agile RRS (Reusable, Reconfigurable, and Scalable) design principles that are based on object-oriented concepts [6, 7]:
- Self Contained Units
- Plug Compatibility

- Facilitated Re-Use
- Non-Hierarchical Interaction
- Deferred Commitment
- Distributed Control & Information
- Self Organizing Relationships
- Flexible Capacity
- Unit Redundancy
- Evolving Standards.

In order to promote agility, companies can use a variety of strategies that include both internal and external initiatives. Examples of prevalent internal initiatives include business process reengineering, adoption of new technology and management planning tools for cycle time and order response time reduction, teamwork, employee empowerment, and employee education and training. Examples of external initiatives, which focus upon supply channel performance improvements, include new forms of partnerships, outsourcing, schedule (information) sharing, postponement and technology adoption [8]. In addition to traditional improving programs such as BPR (Business Process Reengineering) or TQM (Total Quality Management), application of standard IT facilities that enable agile manufacturing, another key point in improving informational agility is the application of MS/OR techniques, supported again by IT, in making more efficient and more effective decisions. In agile management, decision makers need effective means for using these techniques, without necessarily being experts in mathematical programming, simulation and other techniques.

3. AGILE MANUFACTURING IN A REFINERY PRODUCTION SYSTEM

Refineries are typical examples for complex process industries that would considerably benefit from optimization techniques, given their importance within local economies. Computer-based mathematical programming techniques have been used and proved to be very successful in the petroleum industry applications for more than 30 years [9-21]. It is a worldwide practice in petroleum refineries to use such techniques for production and process planning, strategic planning, and evaluating capital investments.

Refinery processes can be perceived as a network of processing units connected by material flows [22]. Primary inputs of this network are different crude oils, with costs dictated by market supply conditions and with procurements subject to market constraints. Crude oils are first processed in distillation units (of HP and Vacuum type) where components are produced.

The primary inputs of a refinery are different crude oils. Crude oils are supplied by purchase on the market, and hence have a cost. In addition, their procurement is subject to market constraints. Crudes are first processed in distillation units, where components are produced. Two different types of distillation units are used in TUPRAS Izmit Refinery; these are HP distillation units and Vacuum distillation units. The residual remaining after crude distillation in an HP unit is fed into a Vacuum unit, which refines the residual into new components.

The components produced by distillation units can be used as feedstock for other processes, or can be blended into end products. Heavy components tend to be reprocessed whereas light ones are more directly blended into end products. The processes for heavy products primarily include chemical reactions with catalysts under high temperatures or pressures. For these processes, yield tables give the nature of component input, nature of component output, and the relationship between inputs and outputs. These processes are performed in FCC (catalytic cracking unit), HCC (Hydro-cracking unit), CCR (Unleaded Gasoline unit), Platformer, Unifier, Desulphurizer, Hysomer and LPG (liquefied petroleum gas processing) units in TUPRAS Izmit Refinery. Apart from the own production of the refinery, components can be imported for processing in these units or for blending.

The final step in oil production is the blending of components into end products. Most of the end products should conform to some specification limits, such as viscosity, sulphur content, density and octane number limitations. The end products within these limitations are obtained by blending different products with different specification levels.

In order to have a healthy object oriented design, the Model class is designed in such a way that among all unit and flow classes, only Unit and Flow are known to it. In other words, a Model object cannot distinguish between, for example, an HPDist and an FCC object, or between a Flow and a CrudeFlow object. Therefore application specific unit classes or flow classes can be derived from Unit and Flow classes respectively (or from any descendant), and can be easily plugged into a system model. Consequently, unit and flow classes are designed according to this design consideration. They make use of the principle of polymorphism, and provide functions that respond differently to the same messages. For example, an abstract 'Receive-input-through-dialog' function is defined in Unit class, and each child unit class redefines it for its specific dialog communication.

The Blender and the GasoBlender classes model the blending and gasoline blending operations. These operations are mostly performed as the last refinery operations to obtain the final products with required quality limitations. A blending unit takes the quality specifications of input product flows, ie. density, viscosity index, etc., and the quality requirements of its output flows (as upper and lower bounds). It uses these values to generate the mathematical model of the input-output relationships in the form of mathematical constraints. Gasoline blending is a special type of blending operation in which extra operations such as Tetraethyllead addition are performed. As a result, the GasoBlender class defined as a child of the Blend, so that it inherits all general blending data and methods.

Process industries such as refinery usually customize their outputs according to customer preferences. To meet these customization requirements, production procedures may require adding new nodes and flows or excluding them from the production process, changing internal parameters within the units, changing parameters in final outputs, etc.

3.1. Clone-based Modeling Approach

Better representations for solving problems are the main concern of Management Science/Operations Research scientists, since end users, in other words decision makers, want to have effective means for accessing techniques from these disciplines, without necessarily being experts in mathematical programming. An obvious solution to this problem is to provide decision support tools in the form of advanced computer programs for easy modeling, because easy-to-obtain, easy-to-use computer programs are the most effective communication link between high-powered research teams and the average professionals hoping to apply the results of research to real life situations [23, 24].

Application environments such as AREMOS will answer to these requirements. The proposed *generic* modeling environment will allow the creation of such systems generically: clone templates representing all types of process-specific units of an application system will be built using non-programming interfaces. Customizing a generic clone template that accepts any number of input/output variables, parameters and mathematical expressions does the creation of an application template. This template is customized integrating its typical process variables (integer, 0/1, real, positive etc.), parameters, and complex mathematical expressions of any form (linear, non-linear, indexed etc.). These expressions represent constraints and partial objective functions to be contributed by instances (*clones*) of the clone template to the mathematical model of the application system.

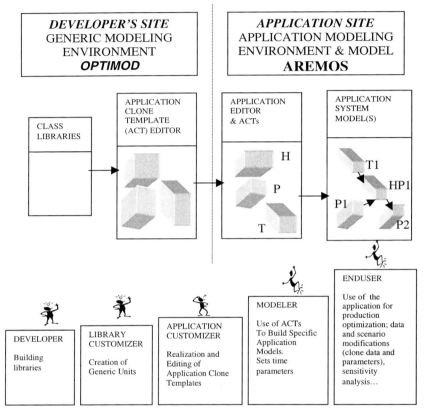

Figure 1. Generic Modeling Environment and Application Modeling Environment.

Validation and verification of clone templates is done by their creator, an "*application customizer*", the only person in the application domain that needs to be knowledgeable about internal processes and their mathematical expressions. As seen in Figure 1, an *application modeling environment* (such as AREMOS) is then delivered to the application site with all its own clone templates integrated. A *modeler* builds a working model of the system by generating clones from these templates, and integrates them in the model through their links and individualized data/parameter settings. The modeler has major modification privileges, such as unit (clone) inclusion, exclusion, and analysis period definition. Such a model is given to an *end-user* in charge of generating and interpreting optimum production plans, and running scenarios. An end user has only editing privileges for parameter and numerical data within clones, and as the modeler, has no knowledge of an underlying mathematical model.

3.2. Methodological Framework

Object-Oriented approach has been adopted for the design and implementation of the prototype AREMOS and its proposed generic successor. The main reason for this selection is the consideration of object-oriented programming (OOP) as an obvious vehicle for the development of complex systems such as visual interactive system packages. Object-oriented programming manages increasing complexity through a number of interrelated principles such as data abstraction, data encapsulation, and inheritance [25]. In addition, OOP offers a solution for the software maintenance problem by providing upgradable modular code, through reuse of hierarchically organized modular data and functionalities, and availability of compact maintenance-free system code activating distinct functionalities in an identical manner with the use of polymorphism. Functionalities and data that may need modifications, such as platform-dependent visual interface requirements and appearances of objects in the visual model, are properly isolated in this approach. New additions required for expanding the applicability of the system are guided by modular format and functionalities, and are expected to fit within existing code without system update.

The major benefit expected of the object-oriented approach is seamless transition among software development processes, starting from the analysis of the real system, through the design of the software and its implementation, therefore facilitating traceability [26, 27]. In addition to these benefits, today's on-line, interactive systems devote a great attention to the user interface, and object-oriented approach provides a natural way of dealing with such user-oriented systems by allowing a direct, obvious mapping between "real" and "software" objects. With these characteristics, object-oriented VIM has found numerous successful applications in simulation, such as MODSIM [28] and PROMODEL [29].

Object-oriented methodology has been applied in a few mathematical programming systems. ASCEND [30] is such a model building environment for complex models comprising large sets of simultaneous nonlinear algebraic equations, which uses and extends object-oriented concepts. In this respect, integration of the concepts from VIM and object-oriented programming with mathematical programming is a fairly new research area.

The principle behind the proposed *clone-based* modeling approach is to visualize a production model uniquely by one-to-one isomorphic abstractions of real elements of the corresponding production system. These abstractions, *clones*, expose only data and parameters required by the user in daily operations of modeled system units. Interaction with clones requires only system expertise. The mathematical model of the system (i.e. the input for a commercial solver), to whom these clones contribute without any external visualization, remains hidden from the end-user.

This approach differs from classical algebraic and structured modeling approaches where visualizations are based on elements of mathematical model itself. Algebraic modeling framework maps visualizations strictly on the algebraic expression of the mathematical model: as seen within GAMS [31] and AMPL [32], mathematical models are compactly "formulated" with reserved keywords in a specialized format, without a possibility for granularity. Structured Modeling framework [33, 34] allows users to build models as a collection of visually and semantically interrelated objects.

Application of object-oriented paradigm in agile manufacturing and especially in virtual manufacturing will be boosted by new enhancements in distributed object computing technology. Both CORBA and DCOM, represent an application specification framework for creating, distributing, and managing distributed objects within a network. They allow programs at different locations and developed by different vendors to communicate within a network.

AREMOS presents itself as a prototype interactive visual modeling tool for building a visual clone representation of a complex process industry -a refinery-, with all relevant unit-specific production data and parameters. Without any intervention by the user, this visual model will automatically create a hidden linear programming model for optimizing production decision variables. Its purpose is to provide a means for the optimum management of such complex systems with a user-oriented, maintenance-free and easily upgradable software. The merits of this approach are:
- One-to-one mapping of visual model clones on real process units, allowing a trivial understanding of the visual model by production professionals.
- The elimination of the need for mathematical programming modeling and associated computer language/package expertise from the optimum production planning process, therefore allowing access to such techniques by production professionals that are not experts in those areas.
- A design that solves major concerns of management for the use, maintenance, update and upgrade of such a system, and that does so without requiring any mathematical programming knowledge for these activities.
- A design that is reusable and adaptable to a variety process industries.

As a summary, the proposed approach integrates two well developed techniques, mathematical programming and object-oriented design, with an emerging technology, visual modeling, for automated and hidden generation of optimization mathematical models of production systems by a visual model made of clones of units of the production system.

3.3. Application of Clone-based Modeling to a Refinery: AREMOS

AREMOS is a prototype system built to demonstrate mathematical modeling through clones in the refinery environment. It lets the user model the refinery visually on the screen through clones of its units and flows. Only refinery engineering knowledge is required to construct the visual refinery model and enter the associated refinery operations data and parameters. AREMOS maintains an internal object-oriented model of the visual refinery model, and internally generates mathematical constraints and objective function of a linear program through the objects of this model, without any need for the end-user to possess general or refinery-related mathematical programming knowledge.

AREMOS provides means for building a visual representation of the refinery in the form of a network of clones, on which refinery process units are represented as iconic nodes and product flows as arcs. It enables the user to integrate all the necessary data through dialog boxes. When asked, this graphical model automatically generates the corresponding mathematical programming model in the background. This "hidden" linear programming model is internally solved by a commercial optimization package, and obtained optimal production results are presented to the user with their interpretations. In this approach, the need for mathematical programming modeling and language expertise is virtually eliminated from the process for building optimization models, as the clones that are inserted by the user into the graphical model possess the capability of internally generating their portion of the mathematical model. AREMOS also helps by giving comments on errors and inconsistencies in the refinery building process and on the results.

Functions within AREMOS can be grouped under four major activities: refinery modeling interface providing users with a visual refinery model development platform, underlying object-oriented model manager responsible for the coordination of all system components, object-oriented refinery network model, and optimization module that can be considered as a distinctly separate tools base. The system is designed using an object-oriented approach: the design of the management system doesn't include any distinct components; functions to be performed by the system are designed as behaviours (i.e. "methods" in object-oriented design) of objects encapsulated with their distinct data. These three object types are central to the design of the management system and are reusable for any process industry implementation. As a prototype, there are other object types usable only for refinery modeling: these will be replaced in the final system by clone-templates customized interactively.

3.3.1. User Interface

A typical Windows application, AREMOS appears on the screen as a window frame with multiple overlapping refinery modeling windows, enabling the user to review multiple visual refinery models simultaneously. Each window lets the user build and operate on a model through commands listed in menus. Data is entered and reviewed by clicking on visual screen objects by a mouse, resulting in the display of multilevel dialog boxes.

A scrollable refinery modeling window with zooming capabilities supports the construction of a network-based representation of a refinery production scheme. This representation consists of graphic icons and graphic arrows standing for refinery process units and product flows respectively. Figure 2 and Figure 3 illustrate a sample model generated for the case study. This visual model is almost identical to and visually more manageable than the usual process flow sheets used by refinery engineers, as most of the entered information stays hidden within the clones to be accessed if needed.

AREMOS interface guides the user throughout the modeling process, checking the validity of the construction at any step, and communicating feedback in case of any infeasible construction attempt, such as when creating an invalid flow connection or entering invalid data in a dialog box. As seen in Figure 4, active dialog boxes offer help buttons to provide the user with on-line help for data entry.

The interface doesn't permit the user to access neither the hidden mathematical programming model for optimization nor the mathematical relations that define petrochemical operations within any unit: each clone possess all information and behaviours of a production unit but hides all information unnecessary to the user.

3.3.2. Model Manager

Model Manager maintains an internal object-oriented representation of the visual refinery production model, consisting of internal representations for refinery unit and flow objects, with graphical interface, database and optimization behaviours. Model Manager performs its user interface management, database management and optimization management duties by organizing these behaviours.

Model Manager retrieves user commands from the interface, such as menu selections and mouse clicks, and communicates them to objects in the refinery network model for performing required operations. Database management is responsible for maintaining multiple refinery models simultaneously, and for performing refinery model storage and retrieval operations. Optimization management has the responsibility of coordinating the generation of the mathematical model by directing clones to prepare their contribution to the linear programming model matrix. If errors are encountered in this process, interface management is invoked for the display of the generated error messages. Once the validation and construction of the mathematical model is successfully completed, a text file containing the linear model is sent to the optimization module for obtaining an optimum production plan. After getting and refining the results, a report to presented to the user in an organized format.

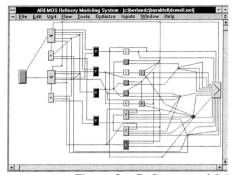
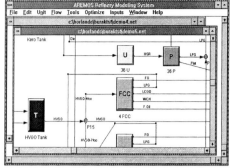

Figure 2. Refinery model in AREMOS

Figure 3. Clones in a refinery model

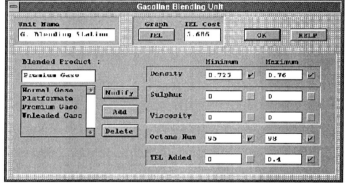

Figure 4. Parameter Input Interface -Dialog Box- for a Gasoline Blending Unit

3.3.3. Optimization

Optimization activities within AREMOS integrate an optimization software package for solving the linear programming model internally generated by the visual interactive refinery model. The selection of a specific software package is not important, what is needed is merely a linear programming solver for solving the refinery linear programming model and to obtain an optimal production plan. A step by step optimization is not the aim in AREMOS, rather each execution should be regarded as a step in the modeling process. Refinery modeling and optimization may include multiple runs of the system, each successive run aiming at better set of control parameters, and at better objective function values.

Formulation of a Refinery Linear Programming Model by Clones of Process Units

All processes mentioned in the refinery environment are localized in specific units, so behaviors of these units can be isolated in their visual computer clones represented by their mathematical formulations Clone optimization functions basically comprise a validation function, a variable generation function, an objective function derivation function, and a constraint derivation function typical of each type of refinery unit in the model. The activation of those functions are managed by the optimization management function of AREMOS.

All unit clones validate their flow connections and their data before the generation of the linear programming model. Once the model is validated, they generate their variable names, followed by the activation of their objective function derivation and constraint derivation functions. Each unit clone is responsible for internally generating its contribution to the model with its own typical list of constraints and partial objective function and individualized parameters. Unit clones make use of well-established refinery linear programming techniques in their objective function and constraint derivation functions. Contributions of each and every one to the optimization model and their functionalities are internally modeled in their object-oriented class description, as mathematical formulations of well defined relations between their inputs and outputs.

Validation

In the model building phase, the validation behavior of a unit clone is responsible for checking its inflow and outflow connections (i.e. units that must have a required quantity of typical connections), and its internal data and parameters with respect to the linear program generation (i.e. positive costs, valid upper and lower quality specification limits for refinery units based on the quality of their inputs). If an invalidity is detected, the clone displays a visual alarm, and an associated error message is fed back to the modeler for the correction of the error.

Variable Generation

Each unit clone is responsible for creating its own distinct variables: these are internally generated and used by the unit (i.e. in a blending unit, amount of inflow i blended into its outflow j is expressed as such a variable), or refer to variables corresponding to the unit's outflows (quantity of product flowing out of outflow j, later inflowing into another clone).

Objective Function Derivation

The entire objective function of the linear programming model of a refinery is produced by internally requesting each unit clone to write its objective function terms in a text file in function of its variables and related cost or revenue coefficients. The objective of the model is to maximize the daily net profit of the refinery and can be expressed as crude oil costs, import product costs and unit utility consumption costs subtracted from sales revenues.

Constraint Derivation

Each unit clone is requested to write its own constraints in a text file. Constraints of a unit express its process description (i.e. are material balance equations, yield equations, and unit capacity constraints), and quality descriptions for outflows of blending units and other restrictions. Constraints about market conditions and requirements are integrated by general purpose "refinery-source" and "refinery-sink" nodes.

Reporting Optimal Results

AREMOS prototype has a simple report utility which states whether the mathematical programming model results in an infeasible solution, an unbounded solution or a feasible solution. If the solution is feasible, it reports the objective function value, and optimal values and descriptions of important variables. These amounts of product flows between refinery units, used crude oils, catalysts, imported components and output final products. Optimal results are reported with a dialog box appearing after optimizing a sample refinery model.

4. CONCLUSIONS

Contemporary corporations today, more than ever, are faced with tremendous competition in a rapidly changing business environment. Markets are becoming more dynamic and customer oriented. In order to respond to these market demands with competitive quality and lower costs, agile manufacturing concept makes use of information technology to improve efficiency and effectiveness of both manufacturing and decision making. Agility or adaptability in manufacturing is of crucial importance and can be boosted by information technology in several ways. With the emergence of object-oriented paradigm in software engineering, new approaches in developing effective manufacturing applications have become possible. The main design principles of agile systems, namely reusability, reconfigurability and scalability, are based on object-oriented concepts, applied in production systems.

AREMOS was developed to demonstrate the "clone-based optimization modeling" concept to management of a refinery and of other complex process industries. It aims at providing an effective decision support tool for obtaining optimal production policies, and to do so without requiring neither any mathematical programming background from its users, nor any adaptation to a new representation: users work with a visual environment that virtually duplicates real processes. Illustrating its concepts in a refinery environment, it brings together ideas from visual interactive modeling methodology and object-oriented design and programming, and uses well-established mathematical programming techniques as the underlying optimization paradigm. AREMOS implements results of a case study in TUPRAS-Izmit Refinery, but it is designed in such a way to cover the general characteristics of a typical petroleum refinery, hence it is also applicable to other petroleum refineries using the same technology.

REFERENCES

[1] Dove, R., Design Principles for Agile Production,
http://www.parshift.com/Essays/essay014.htm
[2] Norrish, D., Agile Software for Agile Manufacturing, APICS - Agile Manufacturing, Volume 6, Number 12, December 1996
[3] Song, L., Nagi, R., Design and Implementation of a Virtual Information System for Agile Manufacturing, IIE Transactions on Design and Manufacturing, special issue on Agile Manufacturing, 1997, Vol. 29(10), pp. 839-857
[4] Dove, R., Agility = Knowledge Management + Response Ability,
http://www.parshift.com/Essays/essay051.htm
[5] Dove, R., Introducing Principles for Agile Systems, Production Magazine 8/95, Gardner Publications
http://www.parshift.com/Essays/essay010.htm
[6] Dove, R., Design Principles for Highly Adaptable Business Systems, With Tangible Manufacturing Examples
http://www.parshift.com/docs/RrsPrinciplesMIH.htm
[7] Dove, R., Knowledge Management, Response Ability, and the Agile Enterprise

http://www.parshift.com/docs/KmRaAeX.htm

[8] Fliedner, G., Vokurka, R., Agility: The Next Competitive Weapon, APICS - Agile Manufacturing, Volume 7, Number 1, January 1997

[9] Baker, T. E., Lasdon, L. S. (1985). "Successive Linear Programming a Exxon", Management Science, Vol. 31, No. 3, pp. 264-274.

[10] Buchanan, J. E., Garven, S. C., Genis, O., Shapiro, J. F., Singhal, V., Thomas, J. M., Torpis, S. (1990). "A Multi-Refinery, Multi-Period Modeling System for the Turkish Petroleum Refining Industry", Interfaces, Vol. 20, No. 4, pp. 48-60.

[11] Charnes, A., Cooper W. W., and Mellon, B. (1952). "Blending aviation gasoline-A study in programming interdependent activities in an integrated oil company", Econometrica, Vol. 20, No. 2, pp. 135-139.

[12] Dewitt, C. W., Lasdon, L. S., Waren, A. D., Brenner, D. A., Melhem, S. A. (1989). "OMEGA: An Improved Gasoline Blending System for Texaco", Interfaces, Vol. 19, pp. 85-101.

[13] Gurkan, T., Kartal, N. (1989). "Model for the Development of the Turkish Petrochemical Industry", Engineering Costs and Production Economics, Vol. 18, pp. 145-157.

[14] Symonds, G. H. (1956). "Linear Programming Solves Refining and Blending Problems", Industrial and Engineering Chemistry, Vol. 48, No. 3, pp. 394-401.

[15] Manne, A. (1958). "A linear programming model of the US petroleum refining industry", Econometrica, Vol. 26, No. 1, pp. 67-106.

[16] Kavrakoglu, I., Or, I., Eyler, M. A., Kaylan, A. R., Dogu, G. (1986). "TUPRAS Izmit Refinery: Final Report for Production Planning Development Project", School of Engineering, Bogazici University, Istanbul (in Turkish).

[17] Kawaratani, T. K., Ullman, R. J., Dantzig, G. B. (1960). "Computing Tetraethyllead Requirements in a Linear Programming Format", Operations Research, Vol. 8, pp. 24-29.

[18] Klingman, D., Phillips, N., Steiger, D., Wirth, R., Padman, R., Krishnan, R. (1987). "An Optimization Based Integrated Short-Term Refined Petroleum Product Planning System", Management Science, Vol. 33, No. 7, pp. 813-829.

[19] Klingman, D., Padman, R., Phillips, N. (1988). "Intelligent Decision Support Systems: A Unique Application in the Petroleum Industry", Annals of Operations Research, Vol. 12, pp. 277-283.

[20] Kosal, H. (1981). "A methodology for Production Planning in a Refinery", Unpublished M. Sc. Thesis, Department of Industrial Engineering, M.E.T.U., Ankara.

[21] Uhlmann, A. (1988). "Linear Programming on a MicroComputer: An Application in Refinery Modeling", European Journal of Operations Research, Vol. 35, pp. 321-327.

[22] Chinneck, J. W. (1992). "Viability Analysis: A Formulation Aid for All Classes of Network Models", Naval Research Logistics, Vol. 39, pp. 531-543.

[23] Jones, C. V. (1994). "Visualization and Optimization", ORSA Journal on Computing, Vol. 6, No. 3, pp. 221-257.

[24] Jones, C.V. (1996) "Visualization and Optimization", Kluwer Academic Publishers.

[25] Thomas, D. (1989) "What is an Object", Byte, March 1989, pp. 231-240.

[26] Coad, P., Yourdon, E. (1991a). "Object-oriented Analysis", Yourdon Press.

[27] Coad, P., Yourdon, E. (1991b). "Object-oriented Design", Yourdon Press.

[28] Marti, J. (1999). "Object-Oriented Modeling and Simulation with MODSIM III", CACI Products Company Publications.

[29] Harrell, C. and Tumay, K. (1995). "Simulation Made Easy: A Manager's Guide". Promodel Publications.

[30] Piela P.C., Epperly, T. G., Westerberg, K. M., Westerberg, A. W. (1991). "ASCEND: An Object-Oriented Computer Environment for Modeling and Analysis: The Modeling Language", Computers Chemical Engineering, Vol. 15, No. 1, pp. 53-72.

[31] Brooke, A., Kendrick, D., Meeraus, A. (1992). GAMS, Release 2.25, The Scientific Press.

[32] Fourer, R., Gay, D.M., Kernigham, B.W. (1993). "AMPL: A Modeling Language for Mathematical Programming", Student Edition, The Scientific Press, Redwood City, CA, USA.

[33] Geoffrion, A.M. (1987). "An Introduction to Structured Modeling", Management Science, 33, 547-589.

[34] Geoffrion, A.M (1989). "The Formal Aspects of Structured Modeling", Operations Research, Vol. 37, 30-51.

Application of Multimedia in Agile Manufacturing

Ronald E. McGaughey

School of Business, Arkansas Tech University, Corley Bldg, 1811 North Boulder Avenue, Russellville, AR 72801, USA

1. INTRODUCTION

The economic, political, social, cultural, and legal landscape is changing rapidly producing a competitive environment characterized by continuous and unpredictable change [1]. Agile manufacturers strive to profit from change by developing organizational structures and cultures that make them fast and effective in recognizing the need for change, designing and redesigning products, creating or modifying manufacturing processes, developing or modifying supply chains to assure a timely supply of all manufacturing inputs, manufacturing high quality, low cost products, and consistently delivering products to customers at the right place, at the right time, and in the right quantity. Agile manufacturers must be capable of sustaining and improving performance in the aforementioned areas to develop and sustain competitive advantage.

Agile manufacturers rely heavily on Information Technology (IT). IT provides an infrastructure to support fast and accurate communications, without which many of activities that make manufacturers agile would not be possible. Many tools, capabilities, activities and methodologies fall under the IT umbrella. One particular area of IT that is making an impact in agile manufacturing is multimedia. In this chapter we will examine the supportive role of multimedia in agile manufacturing.

2. MULTIMEDIA

Collectively, IT developments make it possible to capture, process, store, access, share, transfer, or exchange data and information in the media form of audio, text, graphics, and video. The term multimedia describes the use of two or more of these media in combination. A multimedia system is one that "allows end users to share, communicate and process a variety of forms of information in an integrated manner" [2, p. 441]. Multimedia is a powerful way to package information in a form and context appropriate for recipients. Multimedia technology is a means to the very important end of better communications [3].

The discussion of multimedia herein is not limited to multimedia presentations. The term multimedia presentation has a rather specific meaning–a presentation prepared using multimedia. Multimedia presentations have become commonplace in classrooms, in corporate meetings, in training sessions, and even in political speeches. One can find multimedia presentations on the Web, on CD-ROMs prepared for marketing or training purposes, and elsewhere. A multimedia presentation can in some ways be much like watching television; however, a multimedia presentation can support greater user interaction with the presentation. Interaction via a mouse, voice command, or other approach can give the presentation viewer some control over the direction, pace, repetitions, and duration of a multimedia presentation [4].

Multiple media used in combination can reinforce the meaning of a message. Combining an image with sound, for instance, allows the sender to convey the message to the recipient(s) through more than one of the body's sensory devices; the eyes and the ears in this case. Reinforcing a message with integrated media can promote a clearer understanding of its meaning. Since some people tend to better understand what they read and others better understand what they hear, multimedia provides a way to prepare a message that may be easily understood by numerous recipients. This is particularly important in creating messages that must be broadcast to multiple recipients.

In this examination of the role of multimedia in agility, the definition of multimedia is not limited to multimedia presentations. Multimedia is any combination of media used for the purpose of communicating with an audience of one or many. Multimedia communications convey a message/deliver the information by employing at two or more of the following media in combination:

Text (numbers, letters, words, symbols, etc.)
Audio (Sound–voice, music, etc.)
Graphics (animated or still images in the form of pictures, charts, graphs, diagrams, etc.)
Video (slow of full motion)

Multimedia communications may involve the use of multimedia presentations, but the less restrictive definition of multimedia used herein is any communications, electronic or otherwise, that employs multiple media to communicate a message/transfer information or data.

3. MULTIMEDIA'S ROLE IN AGILITY

Information Technology (IT) has in the last 12 to 15 years advanced at an astounding pace. Improved IT has greatly expanded an organization's capacity to integrate various media, and it has expanded the organization's choice of delivery mediums. The information technology explosion has made available, at a reasonable price, technology with tremendous potential to enhance communications efficiency and effectiveness within and among organizations in support of agile business practices. The technology exists to use whatever media or combination of media (multimedia) is appropriate to convey information in a very rich, meaningful, and easily understood format.

What is the role of multimedia in agile manufacturing? Where is it being used? How is it being used? With what effect? In what areas could it be used? These are some of the questions that will be addressed in this chapter on the application of multimedia in agile manufacturing.

4. EXAMINING MULTIMEDIA'S CONTRIBUTIONS TO AGILITY

Preiss, Goldman and Nagel [5] identified four strategic dimensions of agile manufacturing: 1) Enriching the customer, 2) Cooperating to enhance competitiveness, 3) Organizing to master change and uncertainty and 4) Leveraging the impact of people and information. These strategic dimensions of agility have been used by the Agile Manufacturing Research Institute to consider the tactical and technological dimensions of agility [6] and they provided a basis for examining the Internet's role in Agility [7]. Because the strategic dimensions provide a meaningful way of organizing competencies and capabilities that contribute to agility, they are employed herein to explore the role of MM in agile manufacturing.

The framework below portrays multimedia's contribution to agile manufacturing. The use of multimedia makes communications more efficient and effective, thereby helping agile manufacturers to more quickly and accurately discern the need for change and to formulate and implement an appropriate response. Strategically, multimedia communications enhance an agile manufacturer's ability to enrich the customer, cooperate to enhance competitiveness, organize to master change and uncertainty, and leverage the impact of people and information. Multimedia, thus contributes to the strategic thrusts that facilitate agility and long term success. This framework provides a foundation for the discussion of multimedia's contribution to agile manufacturing and is the basis for organizing this chapter. Major sections address the strategic thrusts and examine how multimedia, by promoting more efficient and effective communications, enables agile manufacturers to do what they must to achieve success.

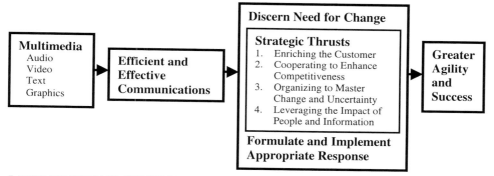

5. MULTIMEDIA'S CONTRIBUTIONS TO CUSTOMER ENRICHMENT

Enriching the customer involves supporting the customer's processes in a way that is perceived by the customer to be enriching [7]. Enrichment of the customer is an ongoing activity that involves much more than simply making a sale to the customer. Enriching the customer involves establishing a long term mutually beneficial relationship between the customer and the manufacturer in which the customer gets what he or she wants/needs, and the manufacturer gets the customer's loyalty and repeat business, and of course, profits. A manufacturer that enriches its customers is not just a source of supply, but instead a permanent part of its customers' processes–a partner in problem solving and key contributor to its customers' success.

Multimedia can be a valuable tool in promoting products. Agile manufacturers must sell their products and multimedia can assist in the task. Perhaps more importantly from the standpoint of achieving agility, multimedia can help a manufacturer to quickly and effectively identify and even anticipate changing customer needs, and to design or redesign products to meet changing needs. Furthermore multimedia can help a firm to build the capability to do these things on an ongoing basis. Private (Intranets and Extranets) and public (Internet) networks and CD-ROM are common mediums for multimedia communications. The Web, a term used to describe a mega-network of public and private networks navigated with the aid of a user friendly graphical user interface (GUI), and CD-ROM are both used to promote products, but the Web has two major advantages over CD-ROM in this context. First, Web pages can be updated frequently at a comparatively low cost. Second, with E-mail the Web can support efficient two-way communications. These advantages give the Web greater potential in promoting products and ascertaining and quickly responding to changing customer needs. Using Web based Multimedia communications, manufacturers can include customers in the design process by allowing them to participate in the customization of products to meet their individual needs, and through soliciting customer input to assist in the design or redesign of a product(s) intended for broader market segments. A well designed, dynamic Web page–multimedia is essential in making it dynamic--can be instrumental in establishing a ongoing relationship with customers, thus helping a firm to become part of the customers' process.

5.1. Using Multimedia in Promoting Products to the Customer

One prominent use of multimedia in retailing and wholesaling, and in manufacturing to a lesser extent, is the promotion of products to the customer. The most popular medium for delivery of multimedia promotions is currently the Web. Manufacturers and other types of operations can use web pages employing a rich combination of media in promotions. Web pages can be endowed with pictures to show products, text that describe products, graphs and charts that further highlight desirable performance characteristics of products, and more. Videos can be included that allow the consumer to see the product in action, or perhaps to deliver a customer testimonial that actually shows a satisfied customer using the product. Though not yet common, multimedia can be used to provide customers with a virtual reality experience. Virtual reality can allow the consumer to, at least in a limited sense, "experience" the product or specific product attributes. The effective use of multimedia on web pages can be a powerful tool of persuasion.

Some of the multimedia "effects" described above can be provided in the form of CD-ROM. The use of CD-ROM requires that potential customers, or current customers, have an adequate computer system to run the CD, and the CD must be delivered to the customer, perhaps by a delivery service (UPS, Post Office, etc.) or hand delivered by a company representative. This can be more cumbersome and costly than Web delivery.

An advantage of Web promotions is that they can be updated frequently, giving customers, or potential customers, a reason to revisit the site and further experience multimedia promotion of company products. It is not unusual for companies to deliver something of value on their web pages, such as a free service (banks provide online calculators that visitors can use to compute their monthly payments based on a variety of loan options and terms), coupons, contests, or a "fun" or "useful" experience such as a product demonstration (downloadable demos are common in the computer software industry). These enticements are intended to not only encourage the purchase of company products, but to encourage visitors to return for future visits. Nearly all of these "added values" involve the use of multimedia in some way. These practices implicitly assume that at least some of the visitors, or repeat visitors will become or continue to be customers.

5.2. Using Multimedia to Learn about Customers or Potential Customers

A company Web page that effectively uses multimedia content to attract visitors is of value in gathering information about customers. Information can be gathered in a number of ways. A company can use cookies to gather information. Cookies, small programs that are transferred to and stored on a site visitor's computer, can record information about the visitor's actions and preferences. Cookies, thus, are a rather subtle vehicle for learning about customers and potential customers visiting a web site. Web based surveys and other tools can be used to gather information on customers or potential customers. A fairly common Web convention for collecting data on visitors is "membership." Visitors can be required to become a member to gain access to a site. To join, visitors must fill out forms designed to collect relevant personal data. Some sites charge a fee for membership. Other sites charge no fee, opting instead to use membership as way to gather data on visitors. Prospective members may be required to provide basic demographic data when filling out membership forms, and they may be required to answer other questions that will provide site owners or sponsors with useful data. A well designed membership form can be quite helpful in building customer profiles. Beyond indirectly assisting in the gathering data about customers and potential customers, multimedia can directly contribute to clarifying the meaning of data gathered about customers, or potential customers. Graphs and charts can, when augmented by descriptive text or narration in the form of voice, go far in transforming data (loose facts) into information (relevant facts aggregated and/or analyzed and explained to give them meaning in a specific context). Multimedia can improve the quality of information used by managers and others by making it easier to understand.

5.3. Using Multimedia to Involve Customers in The Design Process

Multimedia can be instrumental in involving consumers in the product design process. Multimedia presentations or virtual reality experiences can be employed to showcase a new product, or new product concept. A company can create a virtual reality experience or less sophisticated multimedia presentation to describe a product or proposed product. Multimedia, because it can combine text, video, audio, graphic images, and support some level of participant interaction, is a very powerful vehicle for delivering to recipients a very realistic "experience" with product attributes. Volvo has used virtual reality for this purpose, and reports good results [8]. After exposure of customers or potential customers to the virtual reality experience, or the multimedia presentation, surveys or other techniques can be employed to gauge their response. Multimedia helps get customers or potential customers involved in the product design process by allowing them to test designs, critique designs, and offer suggestions for modification and improvement.

Groupware tools such as Lotus Notes can assist in the design process as they allow multiple participants to examine ideas described/depicted using multiple media such as drawings in combination with texts, and to interact in ways that would be useful in examining and modifying the design of a new or existing product. With audio conferencing, and perhaps desktop video conferencing, the design activity can be very similar to having the participants in one place for a joint design session. Three-dimensional modeling tools are already being used to assist manufacturers in involving suppliers and customers in the design of certain types of parts and products [9]. As these technologies mature, these types of activities are likely to become commonplace and the Web will likely be the medium of choice for such activities. These tools and techniques can help involve, not just customers, but all relevant parties in the product design process to make sure that customers get products that meet their needs.

5.4. Using Multimedia to Assist in Taking Customer Orders

Multimedia can assist in taking customer orders to make sure that the customer gets what the customer wants [10]. This capability is not really new, but it has in recent years become more automated, thanks to IT. For years customers, both individuals and institutional customers, have ordered products from catalogs. Catalogs have long contained pictures of products, narrative descriptions of products, general specifications, and product prices. Catalogs used multiple media to convey descriptions of products to make sure that customers received what they wanted.

Modern electronic multimedia supports the experience more richly as online shoppers search out desired products, purchase, arrange shipment, make payment, and monitor the delivery process via the Web. This electronic shopping experience, which falls within the domain of what is now called Electronic Commerce, offers many advantages over traditional catalog shopping. First, multimedia can be used to create a virtual shopping environment familiar to and comfortable for online shoppers. Shoppers can visit a virtual mall or virtual store that closely resembles the "real thing." The on-line presentation of the products with multimedia is much richer than catalog presentation, more closely emulating the "real life" physical presentation of products. Some sites use advanced storage and processing technologies that allow shoppers to view products from various angles, to expand images to show more detail, or contract them for a more panoramic view. In some instances, where product sound is an important characteristic, customers might be given the option of listening to the product perform. Shoppers can place products in their virtual shopping cart, review and modify the cart if necessary, check out, and pay with a credit card if so desired. Multimedia makes a Web based shopping experience much more like the experience of shopping at a "real" place. The Internet and Extranets can support similar experiences for institutional customers who might not require the full "shopping" experience, but who would benefit from the use of multimedia to increase the accuracy of orders and simplification of the ordering process.

5.5. Using Multimedia to allow Customers to Tailor Products to their Needs

Some companies allow the customer to design, within limits, their own product. Dell Computers, for instance, allows one to make a choice of product category, then if one is not satisfied with the standard configuration of the product, one can create one's preferred configuration by selecting from among a variety of standard components, to design a product to better meets one's individual needs. Upon completing one's "custom" system, one can re-price the product to determine the system's cost. If not satisfied with the price, one can reconfigure and re-price as many times as are necessary to design a system that meets one's needs. The apparel industry is exploring the use of IT in the design, ordering, and production of custom clothing [11]. Multimedia will play a supporting role in their efforts, also, because various media must be used in describing product designs completely.

5.6. Using Multimedia to Provide Value After the Sale

Multimedia can be used to provide customers with added value after the sale. Multimedia can be used to provide assembly and operating instructions, it can be used to deliver updated information on product use, and multimedia can be used to deliver information on other products that might, based on customer profiles, be of value to customers.

It is not uncommon today for a customer to receive with a product purchase a video tape, in addition to or as a substitute for the traditional paper assembly and operating instructions manuals. The video will walk the user through the assembly process, or provide instructions for product use. Other mediums, more specifically CD-ROM and the Web, will be used increasingly for delivery of multimedia, after-sale customer support. For assembly instructions, operating manuals, and other after sale support, CD-ROM and the Web have advantages over video tapes. Among their more important advantages are the capacity to support greater customer interaction and delivery of a richer combination of media. CD-ROM and Web based multimedia manuals will allow users to have more choice about the nature of assistance they receive. This will avoid some of the problems associated with a "one size fits all" approach to creating support materials. CD-ROM or Web based multimedia manuals can be designed to give customers options from which to choose a form of assistance appropriate for them (voice instructions, diagrams, texts, video clips, or some combination thereof). The customer will decide what media to view and in what order. The Web has the added advantage of making it possible for a company to provide updates as necessary at a low delivery cost. Updates can take the form of new ways to use the product, safety alerts, product add-ons or upgrades as they become available, and more. Because CD-ROM and the Web have the advantage of supporting video, audio, text and graphics, and to some extent interaction, they offer great promise in this application of multimedia to assist the customer after the sale, thereby adding value and enriching the customer.

6. MULTIMEDIA'S CONTRIBUTION TO FACILITATING COOPERATION TO ENHANCE COMPETITIVENESS

Cooperation is an essential element of agility [12]. Cooperation is necessary within an organization as it means of synchronizing people, departments and functions that must interact in responding to changing internal conditions, changing customer needs, or other changes in the external environment. Cooperation among firms, customers, suppliers, stockholders, government, and even competitors may be necessary in formulating a quick response to internal or external change.

Inter and intra-organizational cooperation requires the interaction of individuals and groups. Verbal and non-verbal communications are the primary vehicles for meaningful human interaction in and among organizations, and thus, are essential in bringing about cooperative effort [7]. Multimedia, because of its capacity to deliver information in a rich variety of formats, offers great potential for improving communications. Furthermore, multimedia can support not just human to human interaction, but also machine to machine, human to machine, and machine to human interaction, providing a solid foundation for the inter- and intra-organizational cooperation necessary to respond to change and carry out the tasks and activities associated with agile manufacturing.

Adjusting to change must be a way of life for an agile competitor and its internal and external constituents. In order to encourage a cooperative response to change, within or among organizations, it is helpful to get agreement on the need for change and the direction of change. When internal or external environmental scanning suggests a need for change, it becomes necessary to communicate that need for change to relevant internal and external organizational constituents. The likelihood of a cooperative response to change is increased if there is agreement among those who must cooperate about the need for change and direction of change. Furthermore, once a direction for change is determined, cooperation is necessary to bring about action. Multimedia can be useful in all of these important areas.

6.1. Communicating the Need for Change and Direction of Change

Because multimedia is a powerful tool for conveying information in a rich variety of formats, it can be instrumental in conveying a persuasive message about the need for change. For many years the written and spoken word (text and voice) were the primary media used in conveying the need for change, but that is changing. Written and spoken words are still important, but they can now be augmented by other media that offer great potential in presenting a persuasive argument for change.

As an illustration of the potential of multimedia in conveying a powerful message about the need for change, take a case where a company is experiencing a loss of market share and declining profits due to problems with product quality. The aim of managers in preparing the message might be to build a case for adopting total quality management. A multimedia message could be prepared for release to targeted groups within and outside of the company. The message could make use of graphs and charts to demonstrate the declining market share and profit margins as well as the nature and magnitude of quality problems. Graphs could compare relevant company performance data to that of competitors. Narration in voice and text could be used explain and interpret the graphs and help convince the recipients of the seriousness of the problem and of the company's position. If delivered via a computer connected to the Web, a look at the company's stock performance in "real time" could be included. Video clips documenting the problems in the manufacturing setting and documenting consumer responses to the problems could help create powerful images of the nature and magnitude of the problems. Some level of interactivity could allow recipients to view select parts of the presentation (graphs, tables of data used in graphs, relevant facts and computations) in more detail. At some point in the message the focus could shift to an examination of potential responses, again using multimedia. Ultimately, the message might present a strong case for the adoption of TQM, and lay the groundwork for the cooperation that will be required to implement TQM as a means of addressing the company's quality problems. A message or a series of messages over time, like the one described above could be delivered to a group or groups using video- conferencing, desktop video-conferencing, or by a person with a laptop computer and projector system. The message could be delivered to individuals via Web pages, E-mail (depending on the size of the presentation and capabilities of the E-mail system), or CD-ROM. Advantages of using multimedia in the manner described above include: 1) reduced travel time; 2) possibility of conveying the message about the need for change without spending time in meetings; 3) capability to create a well conceived, powerful message (can be modified and rehearsed to maximize effectiveness); and 4) possibility of supporting some level of interaction by recipients that would allow them to better understand the case for change and direction of change. Affected parties could voice their reaction and provide input if they so chose, perhaps by E-mail or desktop video-conference. The same multimedia tools used to deliver the message could be used to produce a follow-up message, or messages as necessary.

Messages could be similarly prepared for distribution via CD-ROM, the Web (Internet or Extranet), or E-mail, to convey the need for change and direction of change to external constituents. Suppliers could be provided with much detail about the need for change, the direction of change, and the company's expectations of them as the same or similar rich combination of media could be used to convey the message to them. Similarly, a message for distribution to customers or potential customers, about the need for change, the direction for change, and the expected benefits to customers, could be used to prepare customers for change. In short, multimedia provides companies with a powerful tool for conveying persuasive and informative messages about the need for change, direction of change, and consequences of change. Agile manufacturers must have the ability to respond quickly to change. Effective and efficient communications is essential in laying the groundwork for the cooperative inter- and intra-organizational effort that will bring about successful change. Multimedia can help!

6.2. Using Multimedia to Communicate in Effecting Change

Before anything is changed in an organization, information that describes the change must flow to those involved. The more accurate, thorough, and understandable the description of the what, when, where, how, and by whom pertaining to change, the greater the likelihood that the desired change will be take place. The motivation to cooperate in making change successful has much to do with the effectiveness of the above described justification for change and direction for change. Even if people and groups are convinced of the desirability of change, and understand the direction of change, they must still know in detail what part they must play and have the resources necessary to implement the change. Multimedia offers significant advantages in conveying instructions and plans to people and machines. Firms have recognized the value of multimedia in conveying the details of change. Boeing, for instance, used what it called an "architecture book" to communicate the details of product design and changes therein throughout the development phase of its Delta III Launch Vehicle program. The architecture book has been described as a "living document" because it was changed as necessary and promoted better communication of changes to the many project leaders scattered across the country [8].

To illustrate the value of multimedia in helping to bring about change, take the example of a change in product design. Project management tools provide the capability to describe, using multimedia, plans for the project of redesigning the product. Diagrams such as PERT diagrams, charts such as Gantt charts, text, and even audio and video clips can be used to describe the plans. Some of the same or similar tools can then assist managers in controlling the project once it is underway. Even relatively simple and inexpensive project management tools use multiple media in depicting plans and progress in implementing plans. Design tools like Computer Aided Design (CAD) systems allow those individuals and groups involved in the design process to communicate ideas in more that one media format. Most CAD systems support at least the use of pictures (drawings) and text (words and numbers) to describe a product design in detail. Depending on the CAD system, participants may or may not be able to work together on the design itself, but copies of the design can be distributed using E-mail, or on Web pages via Intranet, Extranet, or Internet [13]. If all participants can view the design simultaneously, desktop video conferencing can be used to allow relevant parties, even if scattered across the globe, to analyze and discuss the design without the need to travel [14]. Other advantages of CAD systems include expediting the design process, rapid modification of design, thorough and easily maintainable documentation of design, and design sharing or transfer [14]. CAD generated designs can be transferred to or shared by computers that will use the design details in the form of a Bill of Materials as the basis for procuring materials and components with MRP, MRPII, and ERP systems [10]. CAD designs can likewise be shared by or transferred to computers and machines (as in CIM, FMS, CAM, NC machines) anywhere in the world that will use the design directly to manufacture the product.

A change in product design may necessitate substantial changes in process design and more. Machines may have to be modified or replaced, supply chains adjusted, work methods changed, and customer knowledge updated. Because of the rich media combinations possible with contemporary multimedia, just about any change in the above areas can be depicted in a way that makes it understandable to those who must cooperate in making changes. If maintenance workers must make changes in equipment or install new equipment, well prepared multimedia instructions that incorporate video and pictures to show, and voice and text to explain, can go far to teach maintenance personnel what they must know to better complete their tasks [2]. Similarly, workers can receive multimedia training, on the job or elsewhere, to prepare them to cope effectively with the changes in their jobs and their work environment. Multimedia can be used to convey to suppliers the requirements for new components or materials. The use of multimedia, perhaps in the form of specifications generated by a CAD system to show and describe requirements in detail, offers great promise in making component and material requirements clear and precise. If necessary, multimedia can be used to explain to customers how they should use the modified product, and apprize them of benefits they might expect from the new design.

6.3. Multimedia's Role in Tasks and Activities Associated with Agile Manufacturing

Multimedia makes contributions in the normal day to day operations of an agile manufacturer, largely by improving the speed, efficiency and effectiveness of necessary communications. Among the manufacturing areas where multimedia communications can and does make valuable contributions to cooperative effort are quality management, maintenance, production planning and control, inventory management, and education and training. Routine tasks and activities requiring cooperation within and among the areas of engineering, marketing, finance, and human resources management can likewise benefit from improved communications made possible by multimedia.

Quality management is an important part of agile manufacturing [15]. Multimedia provides a rich choice of media used in the tools of quality management. The house of quality is an important visual tool associated with quality function deployment. This communications tool combines multiple media in an effort to translate information on customer needs and wants into goals and specific actions for marketing personnel and engineers attempting to create or modify products to better meet customer wants and needs. The tool promotes cooperation among marketing and engineering as they attempt to meet customer needs. Multimedia is used in process control in the form of text and graphics used to convey information about the state of manufacturing processes, that is whether they are in or out of statistical control. The combination of text and graphics employed in process control charts make it easy for production workers and managers to quickly assess the state of critical processes. Pareto analysis employs charts (frequency histograms), diagrams (cause and effect diagrams), and text (describing in narrative and numbers the symptoms, problems, and possible causes) in a collaborative effort to identify, understand, and solve quality problems. Multimedia is being used in ISO9000 for document control and for worker training [16]. These are among the many areas where multimedia can contribute to quality management by promoting better communications within and among internal and external groups that must cooperate to promote agility.

Production planning and control is an area where multimedia can and is being used to improve operations efficiency and effectiveness [17] resulting in reduced production time, efficient work flows, better quality and better worker morale. As noted, multimedia offers many choices for the presentation of plans in a format suitable for use by people and machines. Various combinations of charts, graphs, diagrams, pictures, text, video and audio can be used to create very complete and understandable instructions and schedules for use by workers and machines. The format for those used by workers would typically be quite different from the format required for machines. Boeing used multimedia to deliver to assembly workers and departments the instructions they needed to perform their assigned tasks on the C-17 Globemaster [18]. Boeing's success with this application of multimedia demonstrated that multimedia communications can improve the efficiency and effectiveness of manufacturing. Multimedia tools can help knowledge workers communicate effectively and efficiently with one another, with management, customers, suppliers, or others as necessary to clarify instructions, request permission for modifications or substitutions, request needed resources, or make other inquiries that might help them coordinate their efforts to get their work completed in a timely and accurate fashion. Multimedia offers many options for conveying information regarding the status of machines, processes, and more. Graphics and text can be combined to inform workers about the state of a machine. Colors can be helpful in manufacturing systems. Red is often used to denote problems, yellow to represent unstable conditions, and green conveys the message that all is well. Text can augment pictures with added detail to guide interpretation, and text instructions can suggest an appropriate response. Sound can be used like colors to signal various conditions and voice can guide a response. Tools that make use of the conventions described above are becoming more an more common in the manufacturing environment. The growth in use of these tools would seem to be proof of the value of multimedia in manufacturing planning and control.

Maintenance is critical part of a manufacturer's day to day activities. Multimedia can be used in documenting and describing procedures and tasks to assist technicians in their work and it can be used in the on-going training of maintenance personnel. If a new machine is purchased, or an old one is upgraded, it may necessitate upgrading the skills of the personnel that maintain the machine. CD-ROM or the Web can be used to deliver a rich multimedia training experience very similar to the experience of attending a training workshop. An advantage of the multimedia approach is that the worker can review the materials at work or at home, as often as is necessary to understand the new requirements of his/her job. Another area where multimedia can assist in maintenance is in the presentation of maintenance records and schedules. Graphs can be used to show what has been completed as well as what is needed at present and in the future. Sound, text and pictures can be used to provide daily reminders for individuals or groups. Machine images, supplemented with text and or sound, can be displayed on computer screens to assist technicians in trouble-shooting and in the ordering of parts. Corrective maintenance can be facilitated by multimedia as it allows maintenance technicians to see and hear, even from a remote location, the equipment in question in order to diagnose problems, prescribe/take corrective action, and evaluate results. Collaborative maintenance work involving internal and external personnel can be facilitated with multimedia communications. Remote preventive maintenance is likewise possible with multimedia. Multimedia has the potential to increase maintenance efficiency and effectiveness. Where a timely response to problems and equipment reliability are critical, as they are for agile manufacturers, multimedia contributions to maintenance can be quite important.

Multimedia can be valuable in inventory management. Multimedia can be used to describe components, materials, work in progress, and finished goods with a rich combination of media. Part numbers and text descriptions of inventory items might not be completely effective in describing an item to a worker, or a supplier. The addition of a picture, and sound if relevant to the description, can go far in preventing mistakes such as one product being mistaken for another. With a part number one might need only change one digit when typing the number to make a mistake. If, however, a person or a vision system is familiar with the appearance of an item, a picture might well be "worth a thousand words," as it could provide the visual clue that would help the person or system recognize the item. Expanded computer storage and processing capacities along with compression techniques, and a broader range image storage formats are making the use of multimedia in this area more practical and cost effective. The evolution and growth of E-commerce has contributed greatly to the speed of adoption of multimedia in this important area. Multimedia provides, also, those responsible for inventory management and purchasing with useful tools for monitoring inventory levels and inventory activity. Graphs and texts can be used to portray inventory positions. Graphs can highlight inventory positions much more quickly than numbers as actual inventory levels can be graphed against desired targets with deficiencies or projected deficiencies highlighted with the use of color, text and sound. Graphics can depict inventory departures (shipments), receipts, scheduled receipts and more in a way that is easy for managers and workers to recognize, respond to, and monitor. The same kinds of conditions and actions can be conveyed from one computer to another in a machine-readable format. Orders and shipments can be executed, verified, modified, and canceled with the assistance of systems that use multimedia.

Earlier illustrations demonstrated some of the ways in which multimedia could be of value to marketing and engineering. It has value, also, for finance and human resource management. The benefits of multimedia can be realized in these areas in the form of better communications within the areas, and in supporting interaction and cooperation with other areas of the organization or with external constituents.

7. MULTIMEDIA CONTRIBUTIONS TO ORGANIZING TO MASTER CHANGE AND UNCERTAINTY

An Agile enterprise must be appropriately organized to thrive on change and uncertainty [6]. An organization must attain a level of flexibility that will allow it to change products, processes, structures and relationships in whatever manner is necessary to facilitate success. As any part of its environment changes, a manufacturer must be sufficiently flexible to reorganize its human systems and technical systems to not just adapt to change, but to take advantage of change [7]. In order to be in a position to thrive on change and uncertainty, an agile competitor must be capable of discerning the need for change. The need for effective and efficient internal and environmental scanning is paramount and multimedia has value in this important area. Multimedia has value, also, in facilitating and documenting organizational change.

7.1. Multimedia's Contribution to Environmental Scanning

Environmental scanning involves what might best be described as intelligence gathering. It involves examining data on the external environment and on the internal environment for the purpose of identifying forces for change, often in the form of problems, potential problems, or opportunities. External scanning focuses on intelligence gathering beyond the focal organization, whereas internal scanning focuses on examining the organization's performance in key areas. Internal scanning might focus on the Critical Success Factors (CSFs) for the organization, gathering data for presentation in whatever format or formats that might be most useful to those who will use the information. Executive Information Systems are tools that provide assistance in internal and external scanning and they typically present the results using multimedia. EIS presentation of the results of internal and external scanning may include tables, charts, figures, narrative interpretations of findings, news stories (in the form of paper, audio or video recording) relevant to the company, and more [19]. Although EIS were originally designed for executives, the capabilities that they possess are now embodied in many other systems.

Environmental scanning helps organizations detect relevant change inside or outside the organization. Visual tools like charts and graphs can assist greatly in detecting changes in internal performance, and even in anticipating change. A relatively new phenomenon called data mining–looking to "discover" something significant about internal or external data–is aided by the use of charts and graphs, for the picture may help analysts see important trends or relationships that might not be obvious when viewing the data itself, or even from computations performed on the data.

Graphical components, supplemented with relevant and useful text (text in this context includes numbers), audio, or video to further describe internal or external conditions, situations, actions, or events, assist managers and others in understanding the results of environmental scanning. A better understanding of the forces for change can help managers formulate more timely and more effective responses to change. The value of multimedia in this context is evidenced by its use in EIS as well as in other planning and control systems that serve managers and others throughout modern organizations.

7.2. Multimedia Role in Organizational Change

An organization that is agile must have a organizational culture that is accepting of change and capable of adapting quickly to change. People's knowledge and skills, their organizational roles, and their relationships with others within and outside the boundaries of the organization must and will change over time. Structural and behavioral changes are essential for ongoing success in a dynamic environment of continuous and unpredictable change as an organization continually reinvents itself and its relationships to take advantage of opportunities and solve problems. Effective and efficient communications are essential in bringing about structural and behavioral changes. Multimedia can assist by contributing to better communications.

7.2.1. Multimedia's role in Documenting Change in Organizational Structure

It has been said that organizational structures that focus attention on hierarchy, rigid reporting relationships, division of labor, and accountability, must be abandoned and replaced by a new organizational designs that emphasize results [20]. In firms striving for agility, this pronouncement seems particularly fitting. New organizational designs associated with agility include self-directed, cross-functional, and intra-organizational teams, as alternatives to traditional organizational structures. The composition and organization of a group must reflect the group's mission, goals, activities, and task as well as group member characteristics, motives, and potential contributions. It is noteworthy, also, that in agile manufacturing these workgroups may be temporal in nature. Capturing and communicating the information necessary to describe these organizational groupings in terms of their composition, their purpose, their tasks and their internal and external relationships presents a challenge when using traditional tools such as organizational charts and job descriptions. Although the traditional tools are appropriate for more stable elements of an organization's structure, they seem too rigid and restrictive for describing some of the new structures described above. Multimedia offers great promise is depicting these structures in a manner that describes them accurately, as any media that has value in communicating a more fluid vision of the structure, purpose and relationships embodied in these new organizational structures can be used. As noted, project management toolkits use multimedia to describe projects, but they also include tools for rather thoroughly describing project teams. Most have tools for documenting, with graphics and text (multiple media), the composition and responsibilities of teams, as well as means of relating the teams to project activities, tasks, schedules and more. There is no reason why audio and video could not be incorporated, also, if they would assist in conveying the purpose and function of organizational units. Using project management tools or other tools, multimedia representation of dynamic organizational units offers promise in thoroughly documenting the structures, relationships, and roles in a way that can be understood by those who comprise the organizational units or those who must interact with them.

7.2.2. Multimedia's Role in Upgrading Knowledge and Skills

The fluid and changing structures of agile competitors will necessitate the acquisition of new knowledge or skills. Multimedia can be quite effective in the communication of new knowledge and in upgrading skills for people throughout an organization–management, staff and workers. The capacity of multimedia to support some level of interaction and self testing, as well as learning at the recipients pace and convenience, is significant. It is noteworthy, also, that multimedia education and training may be more entertaining and satisfying for recipients, especially those who may be made uncomfortable by group settings like a classroom or group training session, or by those who do not have the time or inclination to travel in order to acquire new knowledge and skills.

7.2.3. Multimedia's Role in Documenting the Success of Change

Multimedia can be valuable in documenting progress in the direction of desired change. Multimedia offers promise in improving the effectiveness of control by providing a rich presentation of information that makes deviations from plan easier to detect, and multimedia can be useful in formulating the communications necessary to bring about corrective action–

describing the plan for corrective action. The benefits of using multimedia in the context of clarifying plans and facilitating control are substantial. Project management toolkits demonstrate how multimedia can be used in describing plans, monitoring performance, and relating performance to plans for the purpose of exercising control over outcomes.

8. MULTIMEDIA CONTRIBUTIONS TO LEVERAGING THE IMPACT OF PEOPLE AND INFORMATION

Agility requires an entrepreneurial company culture that "leverages the impact of people and information on operations" [6, p. 814]. A culture supportive of agility should encourage individual creativity, initiative and responsibility. Values consistent with agility encourage human interaction to promote the exchange of ideas, cooperation, and collaborative intra and inter-organizational work. A culture conducive to agility is one in which people and groups are empowered to act quickly and decisively in response to internal or external change.

8.1. Leveraging the Impact of People through Empowerment

People and groups are empowered when they have the knowledge, information, resources, and power to act. Information needs include information on what to do, when to do it, how to do it, who to do it with, and how to evaluate results. People and groups require tools, materials, supplies, and often, other people, to act in response to change. Last, people and groups must delegate authority or be recipients of that delegation. Authority is the legitimate power to make a decision. In addition to expanding knowledge through its contribution to education and training, multimedia is a powerful tool for disseminating information, managing resources, and for clearly delegating authority.

8.1.1. Empowerment with Information

Multimedia is a powerful vehicle for delivering information to people or groups in a form that they can understand [10]. Relevant information must be presented in whatever form is appropriate for the recipient be they a manager, technician, worker, or someone in another organization or country. The richness of multimedia communications makes it possible to send and receive information in a form(s) that are comprehensible by even those who may be illiterate, speak a different language, or who may be restricted in the use of their senses by a work context (like a noisy work-place) or physical disability. Pictures, text, video and audio used in some appropriate combination offer the potential to package a message in a way that can be understood by practically anyone–assuming of course that the message is well developed by the sender and that the sender understands the recipient, or recipients. Given the global nature of business relationships and increased complexity of inter-, or intra-group communications, multimedia's value in supporting effective and efficient communications can hardly be overstated.

8.1.2. Empowerment with Resources

Multimedia offers significant advantages in conveying information about the supply of various resources (from materials to human resources) and needed adjustments in resources. Graphic images in the form of charts, pictures and graphs can convey powerful images of resource levels available, as compared to needed resource levels. Supplementary text or sound can further assist in conveying relevant information about resource positions and the need for adjustment. These tools can help workers ascertain whether or not needed resources are available, and can be used to convey to others the need for resource adjustments. A video, or still picture of a machine standing idle can paint a powerful image of a need for materials, supplies, parts or maintenance. It was noted in an earlier section that visual tools such as these are becoming more commonplace in areas like inventory management and production planning and control. Their potential to enrich the communications capabilities of ERP are clear [10]. Pictures augmented with text can be very helpful in conveying a message about what is needed in ways that make misinterpretation less likely. In short, multimedia offers rich choices for conveying the status of resources, the need for resources, and the types of resources required.

8.1.3. Empowerment with Authority

Multimedia cannot give people or groups the authority to act, but it offers potential in conveying request for authority to act, and it can help make the lines of organizational authority and the situational delegation of authority more clear. In illustrating the latter, take the case of a production worker who just completed a machine setup in preparation for the production of a custom order. Having produced a small sample of the product, the worker is unsure of whether the output will be acceptable to the customer. The worker could convey the image of the product, using a picture and descriptive text relating critical measurements to specs, to a manager, on or off site, and request authorization to produce the product. Conversely, the worker could convey the same information to the customer to obtain the customer's approval. Given the manager's and/or customer's approval, perhaps in the form of an electronic signature or voice, the worker could produce the product.

As an example of multimedia's contribution in making clear the delegation of authority, it can be used to convey to people details about actions they can, or cannot take in the form of pictures and text. Text instructions often leave gaps in understanding. Those gaps can be bridged with the addition of pictures to supplement text instructions. Although not commonly used in job descriptions at present, there are no longer significant technological barriers to the use of pictures, video, and audio to augment text in making job descriptions more clear. Additionally, multimedia has value, assuming the existence of a delivery system, in clearly communicating routine daily instructions to workers and departments without the need for face to face interaction [18]. Greater storage capacities, processing capabilities and bandwidth will likely make this commonplace in the future.

In a broader context the authority that accompanies specific organizational roles, or groupings has long been portrayed in documents like organizational charts which have a picture component and a text component to show who has authority over who and for what. Contemporary multimedia provides a means of explaining organizational structure with a richer combination of media. In addition to the chart and texts, pictures can be added to place faces with names and positions. Sound or video could be used to further describe details that cannot be conveyed with a static picture. Augmented with additional media, these traditional organizational devices can prove much more powerful in clarifying the lines of authority, responsibility, and communications relationships. As noted, project management tools provide multiple media for clarifying individual and group authority for tasks, activities and projects.

People and groups are empowered when they have what they need to perform assigned tasks and activities well. People require knowledge, information, resources and the power to act. Multimedia can contribute in improving the communications necessary to empower people, thereby leveraging the power of individuals and groups (inter- and intra-organizational groups) to do what they must to respond successfully to change.

8.2. Leveraging the Value of Information

This entire chapter has been about the application of multimedia to leverage the value of information in one context or another. Leveraging the value of information means getting the most out of the information, or conversely, applying it the greatest benefit of the organization and its constituents (partners, suppliers, customers, owners, etc.). The value of information is leveraged to the extent that it is gathered for, disseminated to, understood by, and used by personnel at all levels to make decisions and carry out activities and tasks that maximize the attainment of stated goals while minimizing the use of resources and occurrence of undesirable consequences. Since a primal goal of agile competitors is to profit in an environment of continuous and unpredictable change, information is leveraged to the extent that it contributes to the attainment of that important goal.

Multimedia contributes to agility broadly by helping agile competitors to discover change in the internal environment and in formulating efficient and effective responses to that change. Many instances have been described herein where multimedia contributes to an organization's efforts on both of these important fronts. Multimedia really cannot make incorrect information better, but the use of multiple media to convey a message might help highlight incorrect information–picture shows that the text is incorrect or vice-versa. Multimedia can definitely help to move information from individual to individual, group to group, organization to organization, from machines to people, from people to machines, and from machines to machines in ways that are efficient and effective. Information has value when it is formatted and conveyed in a way that can be understood by the recipient, or recipients–human or machine. The richness of forms and formats available for the transfer and presentation of information using multimedia communications gives is great value in making information more understandable and more useful, thus more valuable to an agile competitor.

9. SUMMARY AND CONCLUSION

Agile manufacturers strive to operate profitably in an environment of continuous and unpredictable change. Agile manufacturers must, as a consequence of the dynamics of their competitive environment, be fast and effective in discovering change and in responding profitably to change. Multimedia has much potential to assist agile manufacturers. From a broad perspective, multimedia can make agile manufacturers more efficient and effective in discovering change and in formulating appropriate responses to change. Strategically, multimedia can assist agile manufacturers in understanding customers and in taking actions that enrich customers; multimedia can improve the inter- and intra-organizational communications that facilitate cooperative effort among relevant parties; multimedia can help change and thoroughly document organizational culture and structure to help agile manufacturers master change and uncertainty; and multimedia can help leverage the impact of people and information, making both more valuable to the organization. Multimedia contributes by improving communications among people, groups, organizations, systems and machines.

Numerous examples have been provided herein to demonstrate the contributions, or potential contributions of multimedia. At present, it seems that the potential benefits of multimedia are only partially realized in manufacturing. One would conclude, from the attention multimedia receives in the literature, that education and training is the fastest growing application of multimedia in the manufacturing environment. The value and potential of multimedia in the area of education and training is significant, but so is its value and potential in many other areas where its use appears to be much more limited. The technology is available, although not necessarily in widespread use, to support more extensive use of multimedia in many other areas of manufacturing. It seems reasonable that agile manufacturers, or those hoping to become agile, should take the lead in applying multimedia creatively to support their ongoing efforts to achieve/sustain profitable operations. It will be interesting to see to what extent this will be the case as we move into the first decade of the 21st century. Multimedia has much potential, but we must wait and see to what extent that potential is realized in manufacturing, and by whom.

REFERENCES

1. S. Goldman, R. Nagel, and K. Preiss, Agile Competitors and Virtual Organizations, van Nostrand Reinhold, New York, 1995.
2. A. Gunasekaran, R. Bignall, and S. Rahman, Multimedia in Manufacturing, Production Planning and Control, Vol. 7, No. 5 (1996) 440.
3. J. Burger, Multimedia for Decision Makers, Addison-Wesley Publishing, New York, 1995.
4. T. Kuster, Training is Big Business and High-Technology, Metal/Center News, Vol. 37, No. 7 (1997) 76.
5. K. Preiss, S. L. Goldman and R. Nagel, Cooperate to Compete: Building Agile Business Relationships, van Nostrand Reinhold, New York, 1996.
6. R. DeVor, R. Graves and J. Mills, Agile Manufacturing Research: Accomplishments and Opportunities, IIE Transactions, Vol. 29 (1997) 813.
7. R. McGaughey, Internet Technology: Contributing to Agility in the 21st Century, International Journal of Agile Management Systems, Vol. 1, No. 1 (1999) 7.
8. M. Larson, Knowing More, Earlier: Profiles in Agility, Quality, Vol. 37, Issue 4 (1998) 48.
9. R. Webb, Modeler Packs in Tools for Assemblies, Surfaces, and 2D Drafting, Machine Design, Vol. 71, Issue 14 (1999) 84.
10. J. Benadretti, Using Pictures, Sound, Video adds Human Element to Computing, Manufacturing Systems, Vol. 15, Issue 10 (1997) R12.
11. 9th SAMAB Offers a Glimpse of the Industry's Real and Virtual Future. Apparel Industry Magazine, Vol. 59, Issue 2 (February 1998), 8.
12. R. Nagel and R. Dove, 21st Century Manufacturing Enterprise Strategy, Iacocca Institute, Lehigh University, Bethlehem, PA, 1993.
13. S. Greenberg, Computer-supported Cooperative Work and Groupware: An Introduction to the Special Issue, International Journal of Man-Machine Studies, Vol. 34 (1991) 133.
14. S. Rahman, R. Sarker and B. Bignall, Application of Multimedia Technology in Manufacturing: A Review, Vol. 38 (January 1999) 43.
15. R. Narasimhan, Manufacturing Agility and Supply Chain Practices, Production and Inventory Management Journal, Vol. 40, No. 1 (1999) 4.
16. J. Edwards and P.R. Gibson, Integrated Multi-Media Computers in Execution of ISO9000 Quality System Requirements for Document Control and Training, Computers and Industrial Engineering, Vol. 32, No. 3 (1997) 529.
17. J. Frook, Linking the Supply Chain with the Cash Register, Internetweek, Issue 709 (April 6, 1998) PGS07.
18. M. Mecham, 'Paper Lite' Instructions Benefit C-17 Assembly, Aviation Week & Space Technology, Vol. 148, Issue 23 (1998) 53.
19. L. Volonino, H.J. Watson and S. Robinson, Using EIS to Respond to Dynamic Business Conditions, Decision Support Systems, Vol. 14, Issue 2, (1995)105.
20. B. Dess, M. McLaughlin and R. Priem, The New Corporate Architecture, Academy of Management Executive, Vol. 9, No. 3 (1995) 7.

Computational Intelligence in Agile Manufacturing Engineering

Kesheng Wang

Department of Production and Quality Engineering, Norwegian University of Science and Technology, N-7491 Trondheim, Norway

Computational Intelligence or Soft Computing, which mainly includes Artificial Neural Networks, Fuzzy Logic Systems and Genetic Algorithms, is a new advanced information processing technique that exhibits characteristics closely associated with human intelligence. Many approaches have been proposed to implement Computational Intelligence to solve problems in agile manufacturing engineering. The topics of this chapter focus on how to implement Computational Intelligence for improving learning ability and agility of manufacturing systems from three basic functional aspects: agile design, agile planning and agile production. The relevant Computational Intelligence techniques and their uses in agile manufacturing were surveyed, and a number of significant research issues and applications were described.

1. INTRODUCTION

Recently Computational Intelligence (CI) (Bezdek, 1992) or soft Computing (SC) (Jang, 1997) is a rapidly growing area of fundamental and applied research in advanced information processing technologies. The main components of CI encompass Artificial Neural Networks (ANN), Fuzzy Logic Systems (FLS) and Genetic Algorithms (GA). The ANN simulates physiological features of the human brain, and has been applied for non-linear mapping by numerical approach. The FLS simulates psychological features of the human brain, and has been applied for linguistic translating by membership functions. The GA simulates evolution on the computer, and has been applied for solving combinatorial optimization problems. These techniques play the important role in the development of the Intelligent Agile Manufacturing System (IAMS).

This paper describes the theories and applications of CI in the field of Agile Manufacturing (AM). Because it is very difficult to cover all aspects of agile manufacturing, it was decided to confine the topics of the paper to three main issues of agile manufacturing systems, namely: agile design, agile planning, and agile production.

A discussion about Intelligent Agile Manufacturing System from these three basic functional points of view is introduced. Then, the relevant CI components are described, their applications in agile manufacturing systems are surveyed, and a number of significant applications and research are presented.

2. AGILE MANUFACTURING

2.1 Manufacturing strategy

Agile manufacturing is an emerging concept in industry that aims at achieving manufacturing flexibility and responsiveness to the changing market needs. Three important characteristics of agile manufacturing are as follows:
- Great product customization
- Dynamic reconfiguration of system to accommodate swift change in product design
- Prompt adaptability of manufacturing process and decision-making.

Companies have to face with innovative approaches in order to remain competitive in the fast-changing manufacturing environment. The activities in an agile manufacturing, for example, design, planning, production, market are affected by these changes. The global competition has been based on three main factors: price, quality and time. Successful firms dominated all the three factors. Figure 1 shows the company-related features: innovation rate, agility and ability to learn that appear increasingly decisive points on the global markets in addition to the process-related ones mentioned above. Companies that are capable of developing innovative products and adapting their production structure rapidly according to fast-changing market requirements and technologies will assert themselves in the competitive global market.

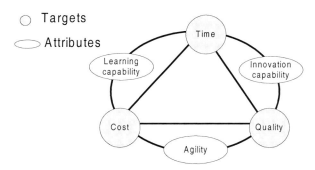

Figure 1: Important factors of the global competition

2.2 Functional view of agile manufacturing systems

Agile manufacturing systems can be conceptually thought of as being an integrated whole of complex interacting sub-systems, organized in such a way as to endeavor towards a common set of goals (Merchant, 1984). Due to the inherent complexities and agility associated with the modern manufacturing systems, modeling these complex interacting sub-systems using common analytical and mathematical approaches has proved to be very difficult.

The functional scheme of typical agile manufacturing systems may be represented as shown in Figure 2. An agile manufacturing system is a multi-objective seeking system. At the top level, an agile manufacturing system takes in the customer needs, feedback (responses), and part of society's total energy information (in raw materials, human power, resources, etc.), then transforms them in such a way as to produce the outputs (products) more efficiently. From the life cycle point of view, it has to deal with scraps, waste disposal, personnel issues, and environmental issues.

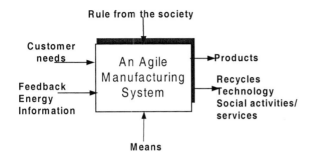

Figure 2: A functional scheme of an agile manufacturing system

Functional view of agile manufacturing systems (shown in Figure 2) and all the issues related to these sub-systems have been combined into three major issues: *agile design*, *agile planning*, and *agile production*.

2.2.1 Agile design
In agile manufacturing systems, according to the dynamic customer requests, along with the feed-back from the system and/or surroundings, the product conceptual design is carried out and normally there is a great need for product innovation. Subsequently, the product is configured and parameterized. Product manufacturability is considered and tested using simulation tools. The corrections are made iteratively and finally the product is manufactured. CI approaches have been implemented for conceptual, configuration, and parametric design activities.

2.2.2 Agile planning
Process Planning, scheduling and manufacturing resource planning (MRP) are considered as three major items in the planning sub-system. Process planning transforms information of a product into sequences of operations with a schedule. Scheduling and MRP are carried out as service functions. CI techniques appear to be the perfect approach in dynamic planning, scheduling and manufacturing resource planning.

2.2.3 Agile production
In most advanced agile manufacturing environments, completely and semi-autonomous systems are used. In order to ensure that the agile manufacturing process is under good control, it is necessary to monitor the process, obtain process and product information, diagnose the problems, and control the process. The main functionality in production is process modeling, monitoring, diagnosis, control, inspection and assembly. ANN and FLS techniques are frequently applied to the production functionality for improving the learning ability and adaptability of systems.

3. COMPUTATIONAL INTELLIGENCE TECHNIQUES

Artificial Intelligence (AI) is a generic term used to describe computerized approaches, which employ knowledge, reasoning, self-learning, and decision making to make machines act smarter. In general, AI might be divided into two categories: the first is Symbolic Intelligence, which includes Expert systems, Knowledge-based systems, Case-based reasoning, etc. and the second is Computational (Numerical) Intelligence, which includes ANN, FLS, GA etc. In this chapter, the focus will be put on Computational Intelligence rather than Symbolic Intelligence. There is no clear definition of Computational Intelligence. Simply, CI techniques consist of Artificial Neural Networks, Fuzzy Logical systems and Genetic Algorithms as shown in Figure 3.

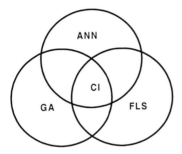

Figure 3: Categories of Computational Intelligence

4. ARTIFICIAL NEURAL NETWORK (ANN)

4.1 Introduction to ANN

ANN offers a powerful and paralleled computing architecture equipped with significant learning abilities. Because the models of ANN are based on biological neural networks, they are able to analyze complex and complicated problems. Today, ANN has undergone a significant metamorphosis becoming an important reservoir of various learning methods and learning architecture. ANN learns from experience and previous examples. They modify their behavior in response to the environment, are ideal in cases where the required mapping algorithm is not known and tolerance to faulty input information is required. They have been successfully used in many system modeling, pattern recognition, classification, predication, novelty detection, robotics, and process control applications. ANN algorithms are most frequently used in agile manufacturing systems particularly to increase the learning ability of the systems.

4.1.1. Structure of ANN

ANN contains electronic process elements (PEs) connected in a special way. A PE (shown in Figure 4) is a simple device that approximates the function of a biological neuron.

- *Inputs, bias and outputs*

Each PE can receive many input signals simultaneously, but there is only one output signal that depends on the input signals, bias, weights, threshold, and transfer function for that PE.

The input signals of a PE comes from either the outside environment or outputs of other PEs and form an input vector **X**, given by:
$$X = (x_1, x_2, ..., x_n)$$

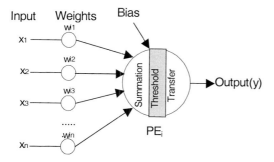

Figure 4: A Processing Element

Some PEs have an extra input called bias which represents other influences from outside the network. Some input signals may be more important than others, so there is a weight vector corresponding to the input vector:
$$W = (w_{i1}, w_{i2}, ..., w_{in})$$
These weights express the relative strength (or mathematical value) of the initial input data or the various connections that transfer data from layer to layer. After a PE receives all its inputs, it computes the total input (summation) being received from its input paths according to these weights. The commonly used method is to use the summation function to find the weighted average of all the input elements to each process element. A summation function multiplies each input value by its weight and totals them together for a weighted sum:
summation = WX^T
If bias exists, another term should be presented when computing **summation:**
summation = WX^T+Bias

- *Transfer (Transformation) function and threshold*

The summation function computes the internal stimulation or activation level of the neuron. Based on this level, the neuron may or may not produce an output (the output of the network is the solution to a problem). The relationship between the internal activation level and the output may be linear or nonlinear. Such relationships are expressed by a transfer function. Some networks use the threshold value in determining the output of the PE:

$$\text{output}(y) = \begin{cases} f \text{ (summation) if summation} > \text{threshold} \\ 0 \quad \text{otherwise} \end{cases}$$

where f is the transfer function.

The four commonly used transfer functions are linear, ramp, step and sigmoid functions where the sigmoid function is the most commonly used one. A number of interconnected PEs constitute a neural network structure (shown in Figure 5) that can be categorized into three types:

- *Input PEs*: are those that receive input from external sources to the system.
- *Output PEs*: are those that send the signals out of the system.

- *Hidden PEs*: have their inputs and outputs within the system.

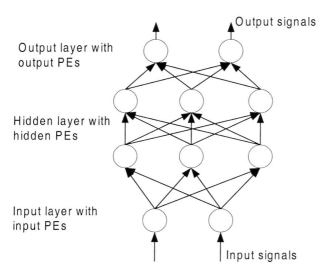

Figure 5: A Multiple-Layered Feed Forward ANN

4.1.2 Network learning strategies, classifications and characteristics

A general definition of learning process can be described as:

Learning is a process by which the free parameters of a neural network are adapted through a continuing process of stimulation by the environment in which the network is embedded. The type of learning is determined by the manner in which the parameter changes take place (Simon Haykin, 1994).

Different ANNs learn in different ways (Figure 6) depending on learning rules (a prescribed set of rules for the solution of a learning problem) and learning scheme (a model of the environment in which the ANN operates).

Hebbian learning comes from the biological world, where a neural pathway is strengthened each time it is used. Error correction learning takes place when the error (i.e., the difference between the desired output and the actual output) is minimized, usually by a least square process. Competitive learning, on the other hand, occurs when the artificial neurons compete among themselves, and only the one that yields the largest response to a given input modifies its weight to become more like the input. Boltzmann learning is a stochastic algorithm derived from information theory and statistical thermodynamics that is known as simulated annealing (Van Laarhoven and Aarts, 1987).

Besides learning rules, different learning processes are also often mentioned including the following two major learning schemes:

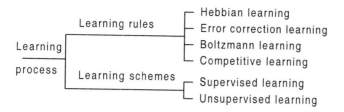

Figure 6: Learning Process

- *Supervised learning scheme*

   Supervised learning uses a set of inputs for which the appropriate (desired) outputs are known. External teacher exists in this scheme, and the weights are changed according to the desired outputs. Some famous supervised learning schemes are perceptron, back-propagation, Boltzmann machine etc.

- *Unsupervised learning scheme*

   In unsupervised learning scheme, the network will change its weight matrix according to the local information and/or internal control mechanism without the presence of external teacher. That is, only input stimuli are shown to the network. The network organizes itself internally so that each hidden processing element responds strategically to a different set of input stimuli (or groups of stimuli). No knowledge is supplied about what classifications (outputs) are correct, and those that the network derives may or may not be meaningful to the person training the network. However, the number of categories into which the network classifies the inputs can be controlled by varying certain parameters in the model. In any case, the final categories should be examined to assign meaning and to determine the usefulness of the results. Some famous unsupervised learning schemes are Hopfield learning method, adaptive resonance theory (ART1 and ART2) (Carpenter and Grossberg, 1987), Kohonen self-organized learning, etc.

   Usually, ANNs can be classified according to:
   1. Structure: such as one-layered structure (or even structure with only one PE) or multiple-layered structure (shown in Figure 2).
   2. Input and/or output signals: the type of input/output signals to/from an ANN might be binary or real valued.
   3. Interconnection degree: such as fully connected or partially connected among different PEs.
   4. Direction of information flow: such as forward-feed or backward-feed (recurrent).

   Generally, there are four types of ANNs that are mostly used in Agile Manufacturing:

- *Standard Back Propagation (SBP, supervised);*

   This is a fast feed forward network using back propagation for training. It is one of the most versatile and consistent of ANNs available, and is used to model patterns within data and therefore has wide and extensive application across many areas, including decision support, process modeling and diagnosis.

- *Radial Based Function (RBF, supervised and unsupervised);*
   The optional RBF supervised ANN offers an advanced fast classification alternative to SBP. For example, training time is fast with large number of variables and it can train with contradictory samples in the data set.
- *Self-Organizing Maps (SOM, unsupervised) or Kohonen Networks (KOH).*
   This tool provides one of the most popular forms of self-organizing maps. It is used in classification and grouping of data sets, and may be used for data filtering and cluster visulization. Again, various properties of the network can be altered as required. A wide range of views of the classification during and after training provides an understandable feedback of progress.

4.2 Applications of ANN

4.2.1. Agile design

ANN systems were applied for recognizing complicated intersecting features for computer aided fixture design (Gu *et al.*, 1995); for associating functions and structures for conceptual design (Kumara *et al.*, 1992); an automated construction method of press brake tools for Metal sheet bending (Frank *et al.*, 1996); Selecting the type of spindle bearing sets based on sample solutions quoted in catalogues of machine tool and bearings manufactures (Kowal *et al.*, 1996).

4.2.2. Agile planning

ANN architectures were developed for selection of grinding wheel (Rowe *et al.*, 1994); bending sequence determination (Geiger *et al.*, 1995a); modeling for planning for agile manufacturing cells (Westkamper *et al.*, 1997); computer aided process planning in multiple-blow cold forging (Alberti, 1997a) and in machining process (Wang *et al.*, 1997); and automated selection of tools in turning (Dini, 1995).

4.2.3 Agile production

ANN paradigms were employed for noise analysis of machine tools (Pfeifer, 1993b); identification of tool wear state (Tansel *et al.*, 1998a; Barschdorff *et al.*, 1993 and Monostori, 1993); identification of cutting conditions (Warnecke *et al.*, 1994); prediction of tool wear development (Wang *et al.*, 1992); monitoring progressive tool wear in single point turning operations (Dornfeld, 1990); estimating the remaining cutting tool life under given cutting conditions (Teshima *et al.*, 1993); tool wear sensing in quasi-orthogonal cutting (Teti *et al.*, 1995b); monitoring of agile manufacturing processes (Grabec *et al.*, 1994); monitoring of turning processes (Etxeberria *et al.*, 1996); diagnosing of machine failures (Wang *et al.*, 1998b); developing thermal actuators to actively compensate for thermal deformation (Hatamura *et al.*, 1993); developing a decision making process model for grinding operations with multi-stage structure consisting of feed forward and brain-state-in-a-box NNs (Sakakura *et al.*, 1992); detecting the onset of chatter vibrations in grinding by removing the subjectivity of operator decision making (Chen *et al.*, 1996); compensating the deformations due to temperature variations in a large milling machine (Revilla, 1996); and mapping the cutting force to the hole quality (Bahr *et al.*, 1998).

5. FUZZY LOGIC SYSTEMS (FLS)

5.1 Introduction to FLS

The philosophy of Fuzzy Logic (FL) may be traced back to the diagram of Taiji that is created by Chinese people before 4600 B. C. But the study of Fuzzy Logic Systems began as early as the 1960s. In the 1970s, fuzzy logic was combined with expert systems to become a FLS, which with imprecise information mimics a human-like reasoning process. Fuzzy Logic Systems make it possible to cope with uncertain and complex agile manufacturing systems that are difficult to model mathematically. A fuzzy logic system basically consists of three main blocks: fuzzfication, fuzzy inference mechanism and difuzzfication.

5.1.1 Fuzzfication

Fuzzfication is a mapping from the observed crisp (numerical) input space to fuzzy sets defined in the corresponding universes of discourse. The fuzzfier maps a numerical value denoted X into fuzzy sets represented by membership functions (MBF) in U.

1. Fuzzy sets

A fuzzy set A is a collection of elements defined in a universe of discourse labeled X. It generalizes the concept of a classical set by allowing its elements to have partial *membership* (usually $\in [0,1]$), and the degree to which the generic element x belongs to A is characterized by a *membership function* $\mu_A(x)$, which associates with each element $x \in X$ a number $\mu_A(x)$ representing the grade of membership of x in A, and is designated as:

$$A = \{(x, \mu_A(x)) \mid x \in X\}$$

Associated with a classical binary, crisp set is a characteristic function, which returns 1 if the element is a member of that set and 0 if not. The fuzzy membership function generalizes this concept by allowing elements to be partial members of a set, reflecting degrees of uncertainty about the information.

2. Membership function

Each linguistic term, such as *cool*, *medium* or *hot*, is represented by a membership function and the set of all these terms determines how an input variable is represented within the fuzzy input (shown in Figure 7).

The *support* a fuzzy set A is the set of inputs that have a non-zero membership function value, i.e. the support for fuzzy set "cool" is $\{temperature: \mu_{cool}(temperatue) > 0\}$.

For example, a temperature x with 22.5 degrees can be regarded as belonging to fuzzy set "cool" with a membership grade of 0.5 and at the mean time it can also be regarded as belonging to fuzzy set "medium" with a membership grade of 0.5.

Besides the triangle and trapezoid functions used in Figure 7, other commonly used membership functions include Gaussian functions, S-curve function and B-spline functions of different orders.

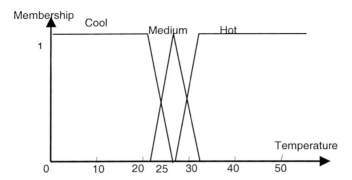

Figure 7: Membership Functions for Temperature

5.1.2 Fuzzy inference mechanism
Fuzzy inference mechanism is the fuzzy logic reasoning process that determines the outputs corresponding to fuzzified inputs. The fuzzy rule-base is composed by IF-THEN rules like

> **IF** Process time is low
> **AND** Queue length is long
> **AND** Slack Time is zero
> **AND** Machine breakdown is very small
> **THEN** Selectibility factor is Medium

Each fuzzy rule defines a fuzzy implication between condition and conclusion rule parts. Using fuzzy sets, the behavior about the object can be represented as the form of fuzzy relations. These relations are composed of fuzzy expressions that are connected by fuzzy logical operators. Three important logical operators are commonly applied in a fuzzy relation: *intersection (AND), union (OR) and complement (NOT)*.

- **Intersection** (AND)

The intersection of m univariate fuzzy sets is calculated as:

$$membership\left[(x_1 \ is \ A_1^i) AND...AND(x_m \ is \ A_m^i)\right] = t(\mu_{A_1^i}(x_1),...,\mu_{A_m^i}(x_m))$$

where t is a class of functions called *triangular norms*. Triangular norms provide a wide range of functions to implement the intersection operator of which the two most popular are the *min* and the *product* operators. The shape of the multivariate fuzzy membership function $t(\mu_{A_1^i}(x_1),...,\mu_{A_m^i}(x_m))$ is influenced both by the shapes of the univariate membership functions and by the operator used to represent the triangular norm.

With *min* operator the minimum membership value in the expression is returned as the membership value of the whole expression:

$$membership\left[(x_1 \ is \ A_1^i) AND...AND(x_m \ is \ A_m^i)\right] = \min(\mu_{A_1^i}(x_1),...,\mu_{A_m^i}(x_m))$$

With the *product* operator, the membership value of the whole expression equals to the product of individual membership value:

$$membership\left[(x_1 \ is \ A_1^i) AND...AND(x_m \ is \ A_m^i)\right] = \mu_{A_1^i}(x_1) \bullet ... \bullet \mu_{A_m^i}(x_m)$$

The multivariate membership functions formed using the *product* operator retain more information than when the *min* operator is used to implement the fuzzy *AND* because the latter scheme only retains one piece information whereas the *product* operator combines *m*-pieces. This fact has led some people to claim that the *min* operator is more robust than the *product*, as noise in any of the *(m-1)* inputs which do not affect the output is not propagated through the network. However, this does not happen in practice because when the fuzzy sets are distributed evenly throughout the input space, at least one fuzzy multivariate membership function depends on each input variable and so the information propagated to the next layer of the fuzzy system is sensitive to any measurement noise. Besides, using the *product* operator generally gives a smoother output surface, which is a desirable attribute in system modeling.

- **Union** (OR)

The union of *m* fuzzy sets is calculated as:

$$membership\left[(x_1 \; is \; A_1^i)OR...OR(x_m \; is \; A_m^i)\right] = s(\mu_{A_1^i}(x_1),...,\mu_{A_m^i}(x_m))$$

where *s* is a class of functions called *triangular co-norms*. Triangular co-norms also provide a wide range of suitable functions but the most popular one is the *max* operator.

Using the *max* operator, the membership value of the whole expression equals to the maximum membership value in the expression:

$$membership\left[(x_1 \; is \; A_1^i)OR...OR(x_m \; is \; A_m^i)\right] = \max(\mu_{A_1^i}(x_1),...,\mu_{A_m^i}(x_m))$$

- **Complement** (NOT)

The complement membership value of a fuzzy expression:

$$x \; is \; A \; \text{equals to} \; 1-\mu_A(x).$$

5.1.3 Defuzzification

Defuzzification is the process of representing a fuzzy set with a crisp number. Internal representations of data in a fuzzy system are usually fuzzy sets. But the output frequently needs to be a crisp number that can be used to perform a function such as commanding a valve to a desired position in a control application or indicate a problem risk index as discussed in next section.

The most commonly used defuzzification method is the center of area method (COA), also commonly referred to as the *centroid* method. This method determines the center of area of fuzzy set and returns the corresponding crisp value. The center of sums (COS) method and the mean of maximum method are two alternative methods in defuzzification.

Figure 8 shows the complete structure of a Fuzzy Logic System. Once all input variable values are translated into respective linguistic variable values, the fuzzy inference step evaluates the set of fuzzy rules that define the evaluation. The result of this is again a linguistic value. The defuzzification step translates this linguistic result into a numerical value.

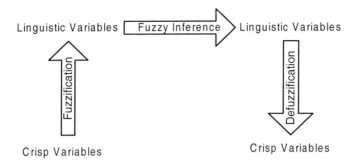

Figure 8: Structure of a Fuzzy Logic System

5.2. Applications

5.2.1. Agile design
FLS were used (Dupinet *et al.*, 1996) in order to deal with tolerance allocation problems encountered in design of mechanical parts.

5.2.2. Agile planning
FLS were designed for allocating jobs to the most suitable machines and assigning load priorities (Balazinsky *et al.*, 1995b); selecting optimal cutting tool (Balazinsky *et al.*, 1994b); selecting object surfaces to be grasped by a gripper (Dini, 1993); automating the tool path planning (Balazinsky *et al.*, 1994b); implementing a chipform classification technique and mapping methodology for use in CAPP systems (Fei *et al.*, 1993); assessing feasibility of cold forged products in single blow operations (Alberti *et al.*, 1997b); and scheduling in industry (Jurgen, 1999).

5.2.3. Agile production
FLS were used for evaluation and selection of cutting conditions (Balasinsky *et al.*, 1993a); monitoring work material heat treatment conditions during machining (Teti, 1995a); pattern recognition of metal cutting states (Wang *et al.*, 1985); control of machine tools to obtain the desired surface finish by controlling the machining process (Zhu *et al.*, 1982); nondestructive inspection of composite material electrical insulators (Teti *et al.*, 1997); and selecting machining procedures and cutting tool size in manufacturing of steel molds through milling (Balazinsky *et al.*, 1994a). FLS systems were also applied to the monitoring of turning processes (Balazinsky *et al.*, 1993b); automatic control of EDM (Kruth, 1995); tool wear sensing in turning (Monostori *et al.*, 1996a); tool wear detection in milling (Monostori *et al.*, 1994); provision of recommended grind wheel specifications for particular grinding problem (Li, 1994); and decision-making for the identification of vibration onset in grinding by removing the subjectivity of human operators (Chen *et al.*, 1996).

6. GENETIC ALGORITHMS

6.1. Introduction to GAs
Genetic Algorithms (GAs) were first introduced by John H. Holland in the 1960s and were developed by Holland and his students in the 1960s and 1970s (Holland, 1975). The basic

idea behind GAs is to evolve a group (also called *generation*) of possible candidate solutions (also called *chromosomes*) to a problem at hand, using several operators (such as *crossover, mutation* and/or *inversion*) which are inspired by natural selection and evolution theory proposed by Charles Darwin. Besides GAs, there are two other similar strategies available: evolutionary programming and evolution strategies. The distinctive trait of evolution strategies and evolutionary programming with respect to genetic algorithms is that in the latter the simulated evolution takes place at the genotype level, that is at the level of coding sequences, whereas the former put the emphasis on phenotype adaptation.

A GA is an iterative procedure that maintains a constant population size and works as follows. An initial population of a few tens to a few hundreds individuals are generated at random or heuristically. During each iteration step (generation), the individuals in the current population are evaluated and given a fitness value. To form a new population, individuals are selected with a probability proportional to their relative fitness. This ensures that the expected number of times an individual is chosen is approximately proportional to its relative performance in the population, so that good individuals have more chances of being reproduced. This selection procedure alone cannot generate any new point in the search space. GAs traditionally use two genetic operators (crossover and mutation) for generating new individuals i.e., new search points. Crossover is the most important re-combination operator, which takes two individuals called parents and produces two new individuals called the offspring by swapping parts of the parents. Through crossover the search is biased towards promising regions of the search space. The second operator (mutation) is essentially background noise that is introduced to prevent premature convergence to local optima by randomly sampling new points in the search space.

GAs are stochastic iterative algorithms without converge guarantee. Termination may be triggered by reaching a maximum number of generations or by finding an acceptable solution. The following general schema summarizes a standard genetic algorithm:

Produce an initial population of individuals
While termination condition not met *do*
- *Evaluate the fitness of all individuals*
- *Select fitter individuals for reproduction*
- *Generate a new population by inserting some new good individuals and by discarding some old bad individuals*
- *Mutate some individuals*

End while

Holland's GA is basically a global searching and optimization strategy. Compared with other traditional searching or optimization techniques such as hill-climbing methods, which depend solely on local information to decide the best direction along which the next step should move, GAs use global information, perform parallel search and do not require local gradient information, which enable it to find globally optimal or near globally optimal solutions. Some major characteristics of GAs which distinguish them from other conventional methods are listed as follows:

- **Direct manipulation of a coding**

Genetic algorithms manipulate decision or control variable representations at a string level to exploit similarities among high-performance strings. Other methods usually deal with functions and their control variables directly.

GAs deal with parameters of finite length, which are coded using a finite alphabet, rather than directly manipulating the parameters themselves. This means that the search is unconstrained neither by the continuity of the function under investigation, nor the existence of a derivative function. Moreover, by exploring similarities in coding, GAs can deal effectively with a broader class of functions than many other procedures.

Evaluation of the performance of candidate solutions is found using objective, payoff information. While this makes the search domain transparent to the algorithm and frees it from the constraint of having to use auxiliary or derivative information, it also means that there is an upper bound to its performance potential.

- **Search from a population, not a single point**

By searching from a population, GAs find safety in numbers and by maintaining a population of well-adapted sample points, the probability of reaching a false peak is reduced.

The search starts from a population of many points, rather than starting from just one point. This parallelism means that the search will not become trapped on a local maxima - especially if a measure of diversity - maintenance is incorporated into the algorithm, for then, one candidate may become trapped on a local maxima, but the need to maintain diversity in the search population means that other candidates will therefore avoid that particular area of the search space.

- **Search via sampling, a blind search**

GAs achieve much of their breadth by ignoring information except that concerning payoff. Other methods rely heavily on such information, and in problems where the necessary information is not available or difficult to obtain, these other techniques break down. GAs remain general by exploiting information available in any search problem. GAs process similarities in the underlying coding together with information ranking the structures according to their survival capability in the current environment. By exploiting such widely-available information, GAs may be applied to virtually any problem.

- **Search using stochastic operators, not deterministic rules**

The transition rules used by genetic algorithms are probabilistic, not deterministic. A distinction that exists between the randomized operators of GAs and other methods is simple random walks. GAs use random choice to guide a highly exploitative search.

In the past thirty years, GAs have been successfully applied to many areas such as, rules learning, feature partitioning, fuzzy controller design, function estimation, software testing and many others (Goldberg, 1989). These areas share a common characteristic. Although these problems can be different in terms of form and concrete objective, they usually can be transformed into a search or optimization problem in a huge searching space or with too many possible solutions that conventional techniques usually cannot find satisfactory results in an effective way.

6.2. Applications

6.2.1. Agile design

GA were employed for improving the efficiency of vehicle designs (Osman, 1998) and balancing assembly lines (Xu, 1998).

6.2.2. Agile planning

A GA based system was used for selecting sequences of machining operations (Wang *et al.*, 1998); process planning of job shop machining and multicriteria production process planning (Zhang *et al.*, 1997); optimizing the position of production resources, positions of robots and movement sequences of cooperative handling tasks (Rossgoderer, 1996); and job shop scheduling (Moldskred, 1999).

6.2.3. Agile production

GAs were used for identification of cutting conditions by using an analytical model and genetic algorithms for micro-end-milling operations (Tansel et al., 1998b).

7. HYBRID CI SYSTEMS (HCIS)

7.1 Introduction

All of these methodologies above stem from the essential cognitive aspect of Fuzzy Logic Systems, underlying evolutionary mechanisms of genetic algorithms and biological foundations of Artificial Neural Networks which provide essential foundations when dealing with engineering problems. With their increasing complexity, it becomes apparent that all of the technologies should be used concurrently rather than separately. A hybrid intelligent system, which merges ANN, FLS and GA into one system, will be an important methodology in solving complex manufacturing problems. Some of these techniques are fused as:
- Artificial Neural networks for designing Fuzzy Logic Systems
- Fuzzy Logic Systems for designing Artificial Neural Networks
- Genetic Algorithms for the design of Fuzzy Logic Systems
- Genetic Algorithms for design of Artificial Neural Networks

7.2 Applications

Hybrid CI systems were used for tool state classification in milling on the basis of measured cutting force and vibration signals (Egretits, 1996); and integration of expert knowledge into ANN process models by means of FLS in grinding process (Weatkamper et al., 1996).

8. CONCLUSIONS

Computational Intelligence (Soft Computing) is a new concept for advanced information processing. The objective of CI approaches is to realize a new approach for analyzing and create flexible information processing of humans such as sensing, understanding, learning, recognizing and thinking. The ANN simulates physiological features of the human brain, and has been applied for non-linear mapping by numerical approach. The FLS simulates psychological features of the human brain, and has been applied for linguistic translating by membership functions. The GA simulates evolution on computer, and has been applied for solving combinatorial optimization problems. These techniques play important roles in increasing the agility and learning ability of manufacturing systems. By reviewing a great number of applications and researches in CI techniques regarding the main issues in design, planning and production in agile manufacturing systems, this chapter has demonstrated that CI techniques make manufacturing systems more flexible, effective, robust, adaptive and productive.

The main developments of agile manufacturing systems comprise the integration of CI methods into computer-aided systems, such as CAD, CAPP, CAM, etc., and the improvements of the performance of present CI paradigms. As a matter of fact, CI systems in the future are expected to be integrated, modular and hybrid in nature.

REFERENCES

1. Alberti, N. et al, (1997b), Computer aided process planning of cold forging based on a fuzzy logic approach, III AITEM Conf. Sept. Salerno, pp. 18-20.
2. Alberti, N. et al., (1997a), Intelligent computation technique for process planning of cold forging, 2nd World Conf. On Intelligent Manufacturing Process and Systems, June 10-13, Budapest, pp. 229-235.
3. Bahr, B., et al. (1998), Mapping the cutting force to the hole quality using neural networks, Proceedings of ANNIE '98, Smart Engineering System Design, St. Louis, Missouri, ASME Press, pp.773-778.
4. Balasinsky, M. et al, (1993a), Application of fuzzy logic techniques to selection of cutting parameters in machining processes, *Int. J. for Fuzzy Sets and Systems*, 61, pp. 307-317.
5. Balazinsky M, et al, (1995b) Intelligent job allocation on machine tools using fuzzy logic, The First World Congress on Intelligent Manufacturing Processes and Systems, Feb., Mayagues, San Juan, pp. 950-961.
6. Balazinsky M. et al, (1994b), Automatic process planning using a fuzzy logic decision system, 10th ISPE/IFAC Int. Conf. On CAD/CAM, Robotics and Factories of the Future CARs & FOF '94, Ottawa, pp325-330.
7. Balazinsky, M. et al, (1994a), Fuzzy logic techniques in the automatic manufacturing of molds, Int. Fuzzy Systems and Intelligent Control Conference '94, Louisville, KY, pp. 316-325.
8. Barschdorff, D., et al, (1993) Wear estimation and state classification on cutting tools in cutting via artificial neural networks, Tooldiag'93, Int. Conf. On Fault Diagnosis, April, Toulouse.
9. Bezdek, J. C., (1992), On the relationship between neural networks, pattern recognition and intelligence, *Int. J. Approximate Reasoning*, 6, pp. 85-107.
10. Chen, X., et al, (1996) Grinding vibration detection using a neural network, *J. of Engineering Manufacture*, 210, pp. 349-352.
11. Dini, G., (1993), A module for the automated planning of grasps in robotized assembly operations, *Annals of CIRP*, 42, 1, pp. 1-4.
12. Dini, G., (1995), A neural approach to the automated selection of tools in turning, II AITEM Conf., Sept. Padua, pp. 1-10.
13. Dornfeld, D., (1990), Neural network sensor fusion for tool condition monitoring, *Annals of CIRP*, 39, 1, pp. 101-105.
14. Dupinet E, et al, (1996), Tolerance allocation based on Fuzzy and simulated annealing, *J. of Intelligent Manufacturing*, 7, 6, pp. 1-11.
15. Egretits, C., (1996), 7th Int. DAAAM Symp., Oct., Vienna, pp. 115-116.
16. Etxeberria, J. et al, (1996), Toward turning monitoring by the processing of indirect measures using neural networks, XI Congress of Machine Tools and Manufacturing Technologies, San Sebastian.
17. Fei, J., et al, (1993), A fuzzy classification technique for predictive assessment of chip breakability for use in intelligent machining systems, 2nd IEEE Conf. March, San Francisco, pp. 1275-1280.
18. Frank, V., et al, (1996), Automated construction methods for press brake tools, 4th Int. Conf. On Sheet Metal, Enschede, pp. 427-434.
19. Geiger, M. et al, (1995a) Neural networks in metal forming, 2nd Int. Workshop on Learning in Intelligent Manufacturing Systems, Budapest, pp. 527-548.

20. Goldberg, D.E., (1989), Genetic algorithms in search, optimization and machine learning, Addison-Wesley, Reading, MA.
21. Grabec, I., et al, (1994), Characterization of manufacturing process based upon acoustic emission analysis by neural networks, *Annals of CIRP*, 43, 1, pp. 77-80.
22. Gu, Z., et al, (1995), Genetic form feature recognition and operation selection using connectionist modeling, *J. of Intelligent Manufacturing*, 6, pp. 263-273.
23. Hatamura, Y., et al, (1993), Development of an Intelligent machining center incorporating active compensation for thermal distortion, *Annals of CIRP*, 42, 1, pp. 549-552.
24. Holland, J.H., (1975), Adaptation in natural and artificial systems, MIT Press, Cambridge MA.
25. Jang, J. S. R., (1997), Neuro-fuzzy and soft computing, Prentice-Hall.
26. Jurgen, S., (1999), Knowledge-based scheduling Techniques in industry, Knowledge-based intelligent techniques in industry (Ed. Jain L. C.) The CRC Press, pp. 55-83.
27. Kowal, Z., et al, (1996), Artificial Intelligence in the shaping of thermal properties of machine tools, VII workshop on Thermal behavior, Intelligent diagnostic and supervising of machining systems, Wroclaw Technical University.
28. Kruth, J. P., et al, (1995), New trends in automatic control of electro-discharge machining, 4th Int. Conf. On Monitoring and Supervision in Manufacturing, Aug., Warsaw.
29. Kumara, S. R. T., et al, (1992), Application of adaptive resonance theory to conceptual design, *Annals of CIRP*, 41, pp. 213-216.
30. Li, Y., et al (1994), Grinding wheel selection using a neural network, 10th Nat. Manufacturing Research Conf., Sept. Loughborough, pp. 594-601.
31. Merchant, M.E. (1984), Analysis of existing technological forecasts pertinent to utilization of Artificial Intelligence and Pattern Recognition Techniques in Manufacturing Engineering, 16th CIRP Int. Sem. on Manufacturing Systems, 14:1, pp11-16.
32. Moldskred, L. R., (1999), A promising genetic algorithm approach to job shop scheduling problem (JSSP), Master thesis, NTNU, Trondheim.
33. Monostori, L., (1993), A step towards intelligent manufacturing: modeling and monitoring of manufacturing processes through artificial neural networks, *Annals of CIRP*, 42, 1, pp. 485-488.
34. Monostori, L., et al, (1994), Modeling and monitoring of milling through neuro-fuzzy techniques, 2nd IFAC/IFIP/IFORS Workshop on Intelligent Manufacturing Systems, June, Vienna, pp. 381-386.
35. Monostori, L., et al, (1996a), Hybrid AI solutions and their application in manufacturing, 9th Int. Conf. On Industrial and Engineering Applications of Artificial Intelligence and Expert systems, June, Fukuoka, pp. 469-478.
36. Osman, K., (1998), Improving the efficiency of vehicle water-pump design using Genetic Algorithms, Proceedings of ANNIN '98, Smart Engineering System Design, St. Louis, ASME Press, pp. 291-296.
37. Pfeifer, T., et al, (1993b), Pattern classification with multi-layer perceptrons in quality assurance, Proc. of the EUFIT '93, Aachen.
38. Revilla, J. and Arana, R. (1996), Thermal deformations compasation by neural networks, VII workshop on supervising and diagnostics of machining systems, 10-15 March, Karpacz, pp. 1-14.

39. Rossgoderer, U., et al, (1996), Optimization of production systems using genetic algorithms, 2nd World Automation Congress, Montpellier.
40. Rowe, W. B., et al, (1994), Applications of Artificial Intelligent in Grinding, *Annals of CIRP*, 43, 2, pp.521-531.
41. Sakakura, M, et al, (1992), A neural network approach to the decision making process for grinding operations, *Annals of CIRP*, 41, 1, pp. 353-256.
42. Simon Hykin, (1994), Neural Networks, a comprehensive foundation, Macmillan College Publishing Company, NY.
43. Tansel I. N., et al. (1998b), Identification of cutting conditions by using an analytical model and genetic algorithms for micro-end-milling operations, Proceedings of ANNIE '98, Smart Engineering System Design, St. Louis, Missouri, ASME Press, pp. 779-784.
44. Tansel I. N., et al., (1998a), Wear estimation in micro-end-milling with wavelet transformations and probabilistic neural networks, Proceedings of ANNIE '98, Smart Engineering System Design, St. Louis, Missouri, ASME Press, pp.755-760.
45. Teshima, T., et al, (1993) Estimation of cutting tool life by processing image data with neural network, *Annals of CIRP*, 42, 1, pp. 59-62.
46. Teti, R., (1995a), Fuzzy logic approach to sensor monitoring in machining, WILF '95, Sept. Naples, pp. 189-199.
47. Teti, R., et al, (1995b), Tool ware monitoring in turning using signal frequency analysis, II AITEM Conf. Sept. Padua, pp. 353-364.
48. Teti, R., et al, (1997), Intelligent NDE of composite material components, 12th Int. Conf. AIENG '97, July, Capri.
49. Thou, G. et al, (1997), Evolutionary computation on multicriteria production process planning problem, Proc. of 1997 IEEE conf. On Evol. Comp., pp. 419-424.
50. Van Laarhoven, P. and Aarts, E., (1987), Simulated annealing, Theory and application, D. Reidel, Dordrecht.
51. Wang, K, et al. (1998), Genetic algorithms for constructing feed-forward neural network in a centrifugal pump condition monitoring, Proceedings of ANNIE '98, St. Louis, Missouri. ASME Press, pp. 303-310.
52. Wang, K., et al., (1997), Application of Artificial Neural Networks (ANN) to process planning for machining process, International Conference on Industrial Engineering Actual Activities, Tallin, Estonia.
53. Wang, M., et al, (1985), Fuzzy pattern recognition of the metal cutting states, *Annals of CIRP*, 34, 1, 133-136.
54. Wang, Y. et al, (1998), An application of Genetic algorithms to selection of sequences of machining operation, Intelligent Design, Manufacturing, and Management, Nordic-Baltic Summer School '98, Riga, Latvia.
55. Wang, Z., et al, (1992), In-process tool wear monitoring using neural networks, Japan/USA Symp. On Flex. Auto. ASME, 1, pp. 263-270.
56. Warnecke, G., et al, (1994), Application of artificial neural networks for the identification of cutting processes, Production Engineering, II, 1, pp. 65-68.
57. Weatkamper, E., et al, (1996), Modeling in grinding process with regression models and artificial neural networks, Annals of German Society for Production Eng., 3, 1.
58. Westkamper, E, et al, (1997), Development of an adaptive simulation system, *Annals of the German Society for Production Eng.*, 4, 1.
59. Xu, G., (1998), A method for assemble line balancing using Genetic Algorithms, Proceedings of ANNIN '98, Smart Engineering System Design, St. Louis, ASME Press, pp. 329-334.

60. Zhang, F. et al. (1997), Using Genetic algorithms in process planning for job shop machining, IEEE trans. On Evil. Comp. Vol.1, No. 4, pp. 278-289.
61. Zhu, J. Y., et al, (1982), Control of machine tools using fuzzy control techniques, *Annals of CIRP*, 31, 19, pp. 347-352.

Computer Applications in Agile Manufacturing

M. A. Pego Guerra and W. J. Zhang*

Department of Mechanical Engineering [+]
University of Saskatchewan, Saskatoon, SK S7N 5A9, Canada.

1. INFORMATION TECHNOLOGY AND INFORMATION SYSTEMS

The term *data* is used to represent unstructured facts, *information*, in contrast, is the proper selection and presentation of data that satisfies a given purpose established by the information recipient. Information has meaning, has to be analyzed and it is addressed to a target, data on the other hand, is meaningless in its isolation [1]. Alternatively, information can be associated to the existence of uncertainty [2]. As the level of information about certain aspects of the environment increases, the degree of uncertainty decreases. The term *Information Technology* has been commonly used to describe a very wide conglomerate of microprocessor-based technologies, including both hardware and software applications. These technologies are essentially aimed at providing more efficient ways to manipulate, process, collect and distribute information. Hardware applications comprise computers, fax machines, "smart" printers, networks, communication devices and so on. Software applications involve the development of computer programs and systems ranging from stand-alone single purpose to networked, distributed and multipurpose applications. These applications also include Artificial Intelligence, database technologies and a great variety of techniques for supporting decision making.

The term *Information System* has been widely used for many years. It is in a fact a combination of two different terms – each one with its own meaning – information and system. Systems Theory (ST) and especially, the term system has been extensively used in a variety of fields. O'Sullivan (1994) defined the term *system* as "an identifiable, complex dynamic entity composed of discernibly different parts or subsystems that are interrelated to and interdependent on each other and the whole entity with an overall capacity to maintain stability and to adapt behaviour in response to external influences". In addition, it has been Avison and Fitzgerald (1988) pointed out that a system always has a well-defined purpose or objective to satisfy. One important characteristics of a system is its emergent properties. Systems are able to attain goals that none of the components can achieve independently.

---

*To whom all correspondence should be addressed.
[+] This research was funded by the Natural Science and Engineering Research Council of Canada (NSERC), the Atomic Energy of Canada Limited (AECL) and the Department of Mechanical Engineering of the University of Saskatchewan.

During the past years, several definitions of Information System have emerged [1, 4, 5]. In this work we adopt the definition provided by Laudon and Laudon (1998). They defined *Information Systems* as a set of interrelated and integrated components or subsystems that work together with the aim of providing and processing information for an organization. The information processing capabilities deal with the ability of the system to collect, process, store and distribute information. Furthermore, the information provided is used for planning, control, coordination, analysis and decision making tasks. Although there is a large variety of information systems, this work focuses exclusively on Computer-Based Information Systems (CBISs). It is well known that computers are very efficiency in performing repetitive data processing tasks. They are therefore usually able to provide information promptly, accurately and with the desired level of detail. It should be noted that CBISs are one of the representatives of software applications of Information Technology. Therefore, advances in Information Technology have directly or indirectly boosted the development and advancement of Information Systems.

Advances in Information Technology (IT) are one of the most important changes that have taken place in human history [6]. They have significantly influenced the way we live, work, study and communicate. Three major advances have markedly shaped this trend; increasing microprocessor computing power, the development of networking and communication technologies and the appearance and growth of the world wide web (WWW) and Internet [6, 7]. A larger microprocessor computing power practically means that the amount of information and the speed at which it can be processed have significantly and steadily increased. Networking and communication technologies make possible an easy and more reliable access and distribution of information. Indeed, they allow for a more productive concurrent and cooperative work in physically distributed facilities. Finally, the doubling of the number of web-servers in use every 18 months has increased the sources and amount of information to unmanageable numbers [7]. Furthermore, its Hypertext Transfer Protocols (HTTP) and other latest development such as Java and XML (eXtensible Markup Language) have made possible a seamless exchange of information among diverse computer systems. In this context, the usefulness of Information Systems as entities related to all the aspects of information processing becomes clear. Although, Information System are still considered in their infancy [1], these advantages in Information Technology have made possible the development of more powerful and versatile applications and this trend is expected to continue in the future. Computer Based Information Systems, among other information technology applications, have experienced an evolution from simple data processing in the middle of this century to knowledge processing systems [1, 7]. These knowledge processing systems are used for the competitive advantage of host organizations, influencing short and long term market and business goals [1, 6, 7]. There are also disadvantages associated with Information Technology, one of the most visible and important one is the information explosion [7, 8]. Indeed, it has been noted that this phenomenon has a detrimental impact on decision-making capabilities of an enterprise [7]. Information explosion is a term used to characterize the fact that a huge amount of information is provided as a result of an information query. Thus, Information Systems can be used as filters of "relevant information". This capability reinforces the importance and utility of Information Systems as a mechanism for furnishing the intended target with "valuable" information.

Manufacturing systems are large, complex and costly. They involve, among others, a considerable amount of financial and human resources. Their distributed and agile nature has fueled the need for models capable of providing first time capabilities[1] in their design and implementation [9]. Nowadays products are so complex that one independent company can hardly produce them in isolation [9, 10]. Companies are expected to quickly offer customized responses to emerging market opportunities.. The advantages derived from the use of Information Systems in an agile manufacturing environment are immeasurable since design and manufacturing are information intensive tasks. Information Systems contribute to more efficient information access, sharing and distribution. They increase the company's competitive advantage in the market place through the achievement of better product performance, shorter product life cycles, a higher degree of responsiveness and an improved level of customer satisfaction.

2. GLOBAL MANUFACTURING

Traditionally, manufacturing firms have been separated islands that do not exchange design and manufacturing information concurrently. They have seen themselves in isolation in a legal and operational context [11]. Physically distributed manufacturing systems are characterized by the geographical distribution of the enterprises' facilities. These manufacturing facilities are usually under the same management umbrella, that is, there is a headquarters-subsidiary relationship. Companies usually do the majority of the things by themselves and the design and manufacturing processes are sequential or not completely concurrent. There is not concurrent access to and distribution of information and more importantly, Information Systems are designed to capture specific process or manufacturing situation and usually do not share information among them.

Kadar *et al.* (1998) stated that physically distributed manufacturing environments have been characterized by a high degree of centralization. Hierarchical control structures have been proven to be inefficient in manufacturing systems characterized by production in small batches in a dynamic environment. Several agent-based distributed manufacturing paradigms have appeared during recent years. They help companies to face the demands of a global manufacturing environment. Some of the widely applied paradigms include, Bionic Manufacturing, Random Manufacturing, Holonic Manufacturing [12, 13] and Fractal Manufacturing [14]. In general, agent-based distributed systems are aimed to reduce centralization and rigidity while increasing flexibility. In addition, they can support concurrent design and manufacturing processes. Furthermore, they provide a more robust response to changes and easy reconfiguration [12]. These distributed manufacturing approaches seem to be directed to solve internal needs of enterprises rather than to address the interactions of an enterprise with partners and subcontractors.

A global manufacturing environment represents the highest and most developed level of a physically distributed manufacturing system. Physically distributed manufacturing environments may or may not be global but, all the global manufacturing systems are physically distributed. The evolution from physically distributed to global manufacturing systems has been influenced by a variety of circumstances. The global nature of markets and business competition, deregulation of international laws, rapid technological

---

[1] First time capabilities refer to the possibility to produce a product or put a system to work in a first attempt.

advances and the ubiquitous availability and distribution of information [10, 15] constitute examples of such factors. Better product performance, larger product variety, lower production cost, flexibility and shorter product life cycles are some of the new challenges encountered by manufacturing companies [2, 8, 10, 15]. To confront these challenges, companies are reassessing their business strategies and adopting new product, services, manufacturing, engineering and management approaches. Agility, the ability to rapidly react to market changes, has been identified as one of paradigms capable of helping companies face these challenges.

Global and agile manufacturing systems have coexisted together with other forms of structural organizations for years [8, 10]. Major benefits from fully distributed or global systems include the reduction in complexity, a high level of modularity and therefore a high degree of flexibility [16]. Other benefits include a better performance-cost ratio, extensibility, availability, scalability, reliability, speed, conceptual clarity and simplicity of design. Finally and more importantly, global manufacturing systems reduce the information content needed to manufacture a given product or provide a service from an individual company perspective. Therefore, companies are able to attain business, production and market goals faster. This approach enhances the organizational structure since superfluous and non-value-adding activities are eliminated and the focus is shifted to customers rather than to the management hierarchy.

Globalization has considerably increased competition among enterprises. There is a substitution of the competition among individual companies by a competition among extended enterprises [11]. Furthermore, the competition at a national or domestic level is being replaced by a global one [17]. This migration from a closed to an open market society [6] is forcing enterprises to operate according to global standards. In addition, companies are increasingly focusing on their core competencies and are adopting new strategies that allow them to compete at a global level. Indeed, since there is no longer a need for centralized manufacturing facilities [6], companies are adopting new organizational structures that allow a more efficient information and knowledge management.

Preiss (1997) noted as another consequence of globalization the fact that companies have started to consider themselves as parts of valued-adding chains with the capability to adapt to a variety of market situations. Thus, the dependency of manufacturing enterprises on other companies such as subcontractors and suppliers has also increased [9, 18]. This trend has substantially influenced the emphasis enterprises place on timely delivery [19]. Companies need to effectively coordinate and have access to physically and globally distributed manufacturing facilities and resources. New techniques such as distributed project coordination, distributed decision making and manufacturing control have emerged as means to cope with this new reality. There is also a renewed and augmented focus on customers. Customers are the driving force in a global and agile manufacturing environment [6, 8, 15]. The ubiquitous access to information allows customer to get involved in the design and manufacturing process from early stages and enables their almost complete satisfaction.

In a global manufacturing environment two well-defined types of interactions can be identified; *inter* and *intra enterprise interactions* [15]. The former includes the interactions between the companies and customers, companies and the marketplace and with other companies [6]. The latter refers to the interactions among developing teams.

Teams need not belong to the same organization and can be geographically apart. The interactions between customers and enterprises allow the identification of the functional requirements that a given product or solution has to satisfy. The interactions of the enterprises with their environments relate to determining and interpreting of customer needs and the analysis of the market trends.

Although an agile environment does not necessarily have to be global, it is becoming increasingly apparent that agile environments need to become global or at least be physically distributed. This understanding is supported by a comparison between the factors that lead to the establishment of subcontracting and suppliers ties in traditional and agile environments. Traditionally, the relationships among manufacturing companies have been rooted in factors such as location, cost of services or sub-products and habits [20]. In an agile manufacturing environment, these relationships are established on the basis of complementary core competencies, regardless of where these competencies are located [15, 20]. As geographical constraints become less influential, the possibility of companies engaging in collaborative relationships with qualified but distant partners increases. This in turn influences the distributed and global nature of the working environment.

Nowadays, information and knowledge, rather than labor, are the constraining factors in complex global enterprises [16]. There exists a need of a concurrent development of products and their supporting manufacturing system infrastructure [8]. Such a degree of concurrency cannot be achieved without the implementation of an efficient infrastructure for information access, sharing and distribution among all the interested parts. These capabilities enable the integration and cooperation among enterprises. Therefore a greater degree of coordination and communication can be achieved. Indeed, they permit to carry out the work from globally distributed facilities and a distributed problem solving approach.

Information Technology plays an enabling role in achieving a fully capable global manufacturing environment [6]. The appearance and deployment of global communication networks and other technological advances have boosted a higher level of interaction and cooperative problem solving capabilities. Enterprises have no other choice but make Information Technology an integral part of their environments in order to compete and achieve a higher degree of cooperation, integration and communication. The major issues related to the use of Information Technology in an agile manufacturing environment are analyzed next.

3. JUSTIFICATION OF INFORMATION TECHNOLOGY IN AGILE MANUFACTURING

As stated above, nowadays products are so complex that no independent company can produce them in isolation. The existence of distributed facilities, concentration on core competencies and outsourcing of a variety of business functions unveil the need for information sharing, cooperation and coordination capabilities for globally distributed working teams and their working environments [22, 23]. Complexity and uncertainty greatly influence the structure of information processing systems of an organization [12]. The current industrial situation is characterized by an extensive, but not always efficient, use of information technologies during most of the stages of the design and manufacturing processes [24]. There are also a variety of information needs associated with each and everyone of these stages. A novel feature of an agile environment is the supportive role that Information Technology plays in the establishment of collaborative and cooperative link among enterprises [23]. Efficient information processing capabilities is the supporting infrastructure that allows companies to undertake changes in their organizational structures [25].

In addition to excellent technical capabilities an agile enterprise needs from *integration* and *cooperation*. Sharing company's resources – human, materials, equipment and so on – is a must according to the agile manufacturing paradigm [15]. Cooperation, on the other hand, is the foundation on which integration rests. Integration is the enabling factor that allows the achievement of the global manufacturing benefits. This integration has to be considered from structural, behavioural and informational perspectives [26]. The need for integration and cooperation has become increasingly essential since individual companies are realizing that they lack all the capabilities needed to take advantage of market opportunities that can arise.

Goldman *et al.* (1995) have stated the need for a redefinition of the design process in the agile enterprise. Design has to become a concurrent and holistic production process, which includes the participation of the complete supply chain, from suppliers to customers. An electronic access to information from all the interested parts – customer, partners, suppliers, different units of the enterprise and so on – is becoming an imperative to achieve and sustain a competitive advantage in a manufacturing environment [25]. The complete integration of the value-adding chain in an agile manufacturing environment demands a transformation from a linear to a concurrent information exchange mechanism [17]. Organizations must have the capability to adapt to a new market situation characterized by frequent changes and a great degree of uncertainty [15, 17]. Gardiner (1996) suggested the consideration of three basic elements in the decision making process associated to the realization of activities. These elements are: value, learning and improvement. An activity should always add value for the customers. Furthermore, it should provide learning or knowledge enhancing opportunities. Alternatively, an activity can be performed for the purpose of serving as a foundation of future improvements. Fortunately, Information Technology provides these three elements to an agile manufacturing environment.

Some of the major advantages of cooperative work reside in speed of responses and efficient use of resources [17]. Cooperation makes possible shorter times to market once opportunities have been identified. In addition, it allows for the use of risk and cost sharing strategy in market environments characterized by hostility and uncertainty [8]. Indeed, cooperation enables knowledge sharing and makes possible the development of learning organizations [3, 20].

Information Technology is the enabling mechanism for achieving a high degree of integration of human, technical and information resources as well as operation in a global manufacturing environment [27]. As the migration from selling product to selling solutions takes place [15], a second migration from isolated to integrated enterprises is indeed occurring. Furthermore, there is a new shift of enterprise management systems towards considering companies as part of a global communication network [11]. Sharing human, technological and informational resources allows companies to build flexible extended enterprises capable of adapting to new market situations dynamically. Informational integration demands the establishment of communication networks capable of promptly distributing information and knowledge. This distribution of information in the enterprise enables a distributed decision-making process along the enterprise [26].

There is an increasing need for mechanisms with the ability to link accurate information sources, computers and user efficiently [24]. Design and manufacturing related technologies such as CAD/CAM, CIM and CAPP needs from an efficient information and data exchange capabilities.

As examined above, it is information rather than labor or capital what is becoming the major factor limiting the productivity of global enterprises. Indeed, it is Information Technology that markedly influences the way in which the information processing is performed. Enterprises need of reliable information to operate and adapt to their environment. The advantages of the use of Information Technology in an agile manufacturing environment are of many folds. Information Technology benefits the companies' competitive advantage, communication mechanism, cooperative work and more importantly, its strategic position in the marketplace. In short, Information Technology supports the majority if not all the operations performed by agile and globally distributed enterprises [13].

Information is one of the fundamental elements of a process. It makes possible the implementation of feedback and control strategies and the process modeling [8]. In addition, it allows for a better project coordination, control in distributed manufacturing environments [18] and enhances the strategic planning of the enterprise [19]. A successful logistics strategy certainly relies on an efficient use of information. Industrial and market examples show how in many cases, a logistics strategy reinforces the enterprises' competitive advantages and customer satisfaction [8].

Information Technology also allows companies to store, organize, process and successfully retrieve information and knowledge. Certainly, information and knowledge are two of the resources that company can sell in an agile manufacturing environment [11, 15]. Furthermore, through an effective and timely information and expertise sharing, individual companies can take advantages of other partners experience and knowledge [28]. Information Technology helps to achieve more robust distributed decision making capabilities. These capabilities are instrumental in the success of tasks related to design and manufacturing of products. These tasks demand superior information processing capabilities since a considerable amount of information is involved in their fulfillment.

Value adding capabilities is a skill that has to be developed and enhanced rather than a static characteristic [19]. These capabilities can then be added to the company core capabilities repertoire thus improving the company's competitive advantage and market differentiation. A ubiquitous and timely access to information influences a company abilities to effectively design and manage value-adding chains capable of reaching higher levels of performance.

Studies of the industrial practice have identified that some of the primary difficulties associated with physically distributed design and manufacturing working environments relate to the information inconsistency, availability, storage, maintenance and retrieval and to the achievement of first time capabilities of the products that are manufactured [29]. Information transferred along the supply chain is lost, omitted or misinterpreted and it is not always sufficiently clear. Furthermore, the scarcity of reliable information impedes an appropriate decision making process of the most critical business parameters. In addition, it is for the lack of proper information distribution mechanisms which cause poor propagation of local changes to other members of the supply chain. This statement supports Fox and Gruninger's (1998) understanding of the need for efficient tools and models capable of determining the impact of changes in all the components of the enterprise.

Other critical issues are connected with the creation and maintenance of information repositories with the ability to cope with long product life cycles, increased complexity and variability and instability of the enterprises' human resources. Regarding the first time capabilities, Whitney *et al.* (1995) analyzed how the genuine causes of unsatisfactory performances are not usually promptly identified due to insufficient or incomplete information resources. Information is widely disperse over the distributed environment and it becomes extremely difficult to gather all the essential information to take the appropriate corrective decisions. It is not redundant to state that first time capabilities can become one of the company's competitive advantages in a global manufacturing environment.

Ham and Kumara (1997) estimated that future global manufacturing environments will be strongly dependent on Information Technology in general and computers in particular. Indeed, one of the most important elements of the factory of the future is the establishment of fully capable cooperation and communication mechanisms that allow each entity of an enterprise to successfully contribute to achievement of the enterprise desired goals [27]. Everyday more and more companies are implementing Intranets as a vehicle to address their intra and inter enterprises need of information access, sharing and distribution [7, 25]. Intranet is a cost-effective solution for enterprises interested in attaining a higher level of flexibility and responsiveness. An Intranet allows a ubiquitous access to information and enables a concurrent, cooperative and efficient work among the enterprise units. It can make possible the seamless integration of data and information from different departments of manufacturing enterprises. Lau (1998) reported that as a result of this approach, notable reductions in lead-time have been accomplished. An electronic access to information resources reduces the overall production cost internally as well as externally. The implementation of a company's Intranet is the primary step in the integration to the company to the Internet and the achievement of fully global manufacturing capabilities.

One of the major scientific challenges ahead lies in how to design and operate agile manufacturing environments [26]. This has also been recognized by Campbell (1998) who stated: "the critical challenges for many [European] business organizations will be in transforming their existing structures, processes and services to meet the demands of agility". Undoubtedly, Information Technology has an instrumental role to play in this transformation.

4. A FRAMEWORK FOR THE INFORMATION SYSTEMS REQUIREMENTS FOR AGILE MANUFACTURING

The term framework has been widely used in both scientific and no-scientific literature. Unfortunately, its definition has been at times loose and inconsistent. In this work, we adhere to the definition originally adopted by the CAD Framework Initiative (CFI) [31]. This definition is modified to take into consideration the characteristics of Information Systems (ISs). An Information System *framework* therefore can be defined as the infrastructure which provides a common operating environment that allows access, distribution and sharing of information by all the parts involved. These capabilities have to be provided as efficiently as possible and with the minimum use of technical infrastructure. Integration and support are at the center of an Information Systems' (ISs) framework. The framework should allow the coherent and seamless integration of information and ISs from different sources. Furthermore, it should provide an efficient support of the processes, activities and users. Frameworks provide the foundation on which further developments of ISs can be based.

Reithofer and Naeger (1997) proposed the partner and interface problems as the two fundamental issues that need to be addressed in an agile manufacturing environment. The use of a scenario approach reveals that these two problems are located at opposite extremes. The first extreme arises from having a well defined interface between two given points of a process chain. The problem then can be reduced to finding the best partner able to fill the existing gap, *i.e.* interface fixed, select partner company. The second extreme deals with the case of two or more well-defined partners, *i.e.* two or more parallel process chains that have to interact at some point of the process. This second issue addresses how the partners have to communicate in order to successfully attain the desired goals, *i.e.* partner fixed, select the interface. According to this analysis, Information Systems fall in the second extreme due to the communication and information access and sharing requirements they are expected to fulfill. Since we are mainly concerned at this point with the interface problem, the partner selection problem will not be considered here. Furthermore, the partner problem has been a focus of other research efforts such as [10, 32]. The analysis of the most relevant characteristics of ISs being used in industry will be performed next. It provides insights about what IS applications have achieved and their limitations.

Traditionally, Information Systems for production management have been targeted at supporting the manufacturing of anonymous products [33]. Anonymous products are those manufactured in large lot sizes and their parts are generally interchangeable. In this kind of environments many of the production management oriented techniques assume the existence of the repetitions of activities, implicitly or explicitly.

Hierarchical Information Systems just like hierarchical organization structures, are not flexible enough. They are unable to achieve the degree of responsiveness demanded by an agile manufacturing environment [16]. Indeed, this situation is more the norm than the exception in the majority of the Information Systems currently in use in industry and business. Many Information Systems instances are inefficient and do not support proper distribution mechanisms in agile environments [28]. Difficulties involved in achieving seamless electronics communication networks that allow for a lean exchange of information have also been reported [34]. The assumption that perfect and complete information is available is one of the major drawbacks of Information Systems used for production planning [33].

It should also be understood that at the time many of the ISs currently being used in industry were deployed the demands for information sharing at inter-enterprise level did not exist. That means, many of these systems were targeted at fulfilling particular needs of the host enterprises. Due to their narrow focus, they could even use contradictory views and semantics to represent similar concepts, data and information. This situation has been openly recognized in some of the previous research related to inter-enterprise information sharing and cooperative work [23, 26]. It has also been ignored by some of the recent information system and knowledge design research projects such as [21]. New designs of Information System frameworks therefore need to deal with these legacy systems and allow the exchange of information among them. Some of the challenges here reside in enabling the communication among a set of heterogeneous systems and preserve the companies' autonomy while achieving the integration of their information resources.

The migration to global manufacturing environments, the adoption of agile strategies and the increasing focus on distinctive treatment and satisfaction of customer have proven true the forecast made by Wortmann (1992) about the appearance of a new type of production paradigm. This new paradigm is characterized by the manufacturing of products in small batch sizes. In contrast with the production of anonymous product, this new type of products, known as one-of-a-kind products (OKP), are targeted to satisfy the needs of specific customers. Indeed, they are not required to have interchangeable characteristics. Similar to the manufacturing of the OKP, an agile manufacturing environment acknowledges no completeness of information. Furthermore, the dynamic, uncertain and risky nature of today's market place can make the most complete information obsolete in a short period of time. Indeed, it is highly probable that, as in OKP, information becomes complete once the manufacturing of the product is finished and the market opportunity has passed. This reality puts an enormous pressure on the capabilities that have to be provided for Information Systems of any type.

Camarinha-Matos et al. (1998) identified the lack of a common reference model or architecture able to drive the design and deployment of alliances of enterprises in agile manufacturing environments. Based on the supportive role that ISs play to the organizational structure [35], it can be concluded that the same situation applies to the design and development of ISs for these environments. The development of a framework becomes therefore instrumental. In that regard, it is useful to identify of the functional requirements that such a framework needs to fulfill.

The analysis of the requirements of the framework for Information Systems in an agile manufacturing environment can be performed by considering three different perspectives. This analysis needs to take into account the environment, the technical infrastructure and the information management requirements. The requirements related to information and knowledge sharing are the most important one since they enable other functional requirements such as the control, coordination of processes and the management of the supply chain [21].

Knowledge and information management systems for an agile manufacturing environment are directly related to the characteristics of agile processes, their environments and technologies associated with information management and sharing [21]. There is no other viable alternative but to consider that environment exist and cannot be changed [8]. It has to be taken into consideration during the whole design process. These systems have to consider an integrated approach in handling information and data sharing requirements. They also need to address issues associated to data management and distribution channels.

Although the complete analysis of the functional requirements of agile manufacturing environments is beyond the scope of this work, it is important at this point the analysis of the most relevant functional requirements of this kind of environments. This is based on the fact that Information Systems always play a supportive role in the environment they are developed for. An agile manufacturing environment should achieve rapid responses to customer demands in an uncertain market environment. This requirement considers issues related to the kind of information that has to be shared and with whom and how. Chen et al. (1998), suggested that the aforementioned requirement can be satisfied through a set of enabling capabilities that allows the proper management of products' life-cycle variability. The proposed life-cycle variability includes the product requirements, the production volume variability, the utilization and integration of supply chain and its processes and a redesign of organizational structures. Shorter product life cycles are supported by the use of concurrent and integrated design and manufacturing development processes, the creation of efficient supply chains and the development of outsourcing and partnership capabilities [10, 15, 28].

Another functional requirement of agile manufacturing environments deal with information capabilities. Information functional requirements need to address issues related to coping with dynamic and concurrent information demands and how to improve information access, completeness and distribution speed [6]. Furthermore, there should be an information repository for holding companies' knowledge and expertise [9, 33]. This information repository should carefully maintain information about previous projects to avoid duplication of effort and easy retrieval.

The structure of the agile environment in which an Information System will be implemented should also be taken into consideration. Star-like, democratic alliances and federations have been identified as the three major agile environment structures [23]. The star-like structure is characterized by a dominant role of one of the members, who imposes its own protocol of information and communication exchange. Star-like structures are the most common implementations because of the impetus and funding provided by the strongest member of the agile alliance [34]. Democratic alliances, on the other hand, depict a collaborative and egalitarian environment where each member keeps its autonomy. Members get together as a result of complementary core competencies that all of them bring to the alliance. Federated alliances are an expansion of collaborative alliances based on the need for a common management of resources and skills. Such an expansion relies on previous mutual alliance successes and the benefits obtained from this kind of management coordination structure. This kind of alliances seldomly occur in industry. Agile environments can also be composed of floating or fixed memberships [15, 23]. The former shows a variable and dynamic nature that allows for an opportunistic membership (joining or leaving the alliance) based on the market factors or the phase of the business process. The latter refers to alliances that remain stable in relation to the members, suppliers and clients. Also considered here is whether or not the members take part in one or more alliances concurrently. The environment in which the alliance operates, *i.e.* in a monopolistic or open market situation, is examined at this point too.

Camarinha-Matos *et al.* (1998) identified as the basic infrastructure requirement for an agile manufacturing environment the need to enable real-time information exchange. Real-time information exchange brings the inter-enterprises cooperation to a higher level. Indeed, it allows a conglomerate of autonomous enterprises to behave as single monolithic units [11]. Ottaway and Burns (1997) suggested that one of the most important features of any system that plays a supportive role to processes, as Information Systems do, is to be able to adapt when processes undergo a reinvention, improvement or are reengineered. Unfortunately, many of the Information System in industry today are unable to keep the pace of changes of the processes they support.

To deal with the heterogeneous character of an enterprise Information Systems, an infrastructure has been proposed to address these issues at both internal and external levels [23]. The internal level allows enterprises to continue using their particular Information Systems independently of the other member enterprises. The external level makes possible the cooperation and information sharing among companies. This is achieved through the addition of a new layer that acts as a coordinator for the information exchange. This layer "translates" the information and data from a particular company to a universal format. The information then can be "translated", accessed or distributed to other member companies that may need it. Uptown and McAfee (1996) have pointed out that in order to attain a lean information exchange infrastructure, there is an unavoidable need for making the Information Systems more flexible, easily extensible (open) and user friendly. These systems should be able to accommodate members with a diverse degree of sophistication of information technology and computer systems. Furthermore, the system should be capable of coping with the dynamic nature of an agile environment, characterized by a constant variation of its membership and provide a selective access to information source while guaranteeing security. The design and implementation of Information Systems should therefore allow for its quick reconfiguration when the need arises. Supporting plug and play principles from early design stages can enable these capabilities. This approach has been successfully applied to the management of agile systems [16, 36].

The incorporation of a coordinator or administrator for handling a variety of organizational issues has been widely suggested. The coordinator is in charge of registering and assisting new and old members of the conglomerate of enterprises. Other of its functions deal with the maintenance of the informational links and information distribution and broadcast to all the member enterprises. This type of infrastructure is similar to the information broker [34, 37] or an integrator acting as an intermediary [6]. In addition to the aforementioned infrastructure requirements other requirements that make technically possible the achievement of the infrastructure functional requirements can be mentioned. They are mainly related to hardware that link and enables communication, the physical exchange, access, distribution and sharing of information among computer systems. These requirements were partially analyzed in Section 24.1, when the advances in Information Technology were examined.

Information and knowledge management in an agile environment is a challenging issue [21, 23]. Success in the implementation of an agile environment knowledge management system relies on the ability to properly integrate and manage processes, activities and the resources of each member enterprise [21]. The management of information should primarily allow the distribution of information among all the member enterprises and its control. In addition, ISs are expected to enable an efficient cooperation and collaboration among globally distributed working teams. They should indeed support a high level of interaction among member companies. The integration of a variety of manufacturing and design related information is another relevant requirement [23, 24] since design and manufacturing are information intensive activities. These requirements are substantially influenced by the dynamic nature of the information in an agile environment.

Information Systems should be oriented to support the smooth development of engineering and manufacturing tasks. The unavailability of reliable data and information on the shop floor is considered one important source of uncertainty [33]. It is the authors' opinion that Information Systems should have the ability to deal with incomplete, inconsistent and ambiguous information. Primary sources of inconsistency and ambiguity of information reside in the need of taking reasonable assumption that can be proven incorrect during the design and manufacturing process. Therefore, a framework for ISs in agile manufacturing environments is expected to provide intelligent capabilities that allow companies to deal with the aforementioned information shortcomings.

Information Systems should contribute to the improvement of the processes they support. In this regard, the analysis of the different ways in which Information Systems can enhance the agility of an organization become of paramount importance for their industrial success. The analysis of the most relevant forms in which Information Systems can widen the agile behaviour of organizations is performed next.

5. ROLE OF STRATEGIC INFORMATION SYSTEMS IN ENHANCING THE AGILITY OF AN ORGANIZATION

An enterprise's competitive advantage is an issue of a singular importance. A company's ability to distinctively differentiate itself from its competitors in the marketplace significantly influences its chances of survival. The industrial organization approach recognizes the universal influence of the environment on the enterprises strategic goals [19]. It considers the strategic decision-making process in an enterprise a result of the variations in the marketplace and the need for maintaining the companies' competitive advantages.

In comparison to their operational counterpart, strategic goals are characterized by the need for external data. In such cases, the completeness of the information is not a major concern [5]. Strategic goals can target both internal and external operations [4, 5]. Some examples of external strategic goals are the launching of new products or services, the establishment of new relationships with customer, suppliers or value-adding chain. Internal strategic goals can focus for instance, on searching for more effective and efficient internal management, quality improvement or better customer service strategies. In principle, any initiative that contributes to the company's differentiation and success in the market place can be considered strategic.

Information Systems are very important in the collection, analysis and distribution of information about the changes in the market place. They are particularly useful in identifying opportunities and hazards to the company's competitive well being. It has been widely recognized that an objective and adequate strategic planning leads to an understanding of the enterprises competitive and market environments. Such an understanding heavily depends on the availability of up to date and accurate information [19].

Strategic Information Systems are targeted to help companies establish long-term or strategic goals. Additionally, they contribute to the decision making process associated with the achievement of these goals. Fawcett et al. (1997) examined the instrumental role of information in enterprises' strategic planning and decision making processes. Enterprises use information as input in the identification, analysis and fulfillment of their long-term market goals. Furthermore, the value of information as an enabling mechanism for strategic control in globally distributed of enterprises has become increasingly apparent. A systematic strategic planning performed on informed basis substantially influences the profitability and growth of an enterprise. It also involves substantial efforts and investment. Through a proper strategic planning, enterprises are able to analyze and evaluate their internal and external environments, state the companies strategic objectives and determine and evaluate the core capabilities that enable the achievement of these objectives [5, 19, 28]. In addition new value adding capabilities for the enterprise can be identified. In fact, the lack of strategic planning is one of the major factors that prevent companies from achieving agile responses [28].

Laudon and Laudon (1998) distinguished four basic strategies companies have traditionally used to gain a competitive advantage in the marketplace. In an agile environment these four strategies can be described as follows; decrease cost, concentrate on core capabilities, differentiate from competitors and nurture customer trust and loyalty. On the other hand, Goldman et al. (1995), have identified four universal strategic dimensions of agility. These dimensions apply regardless of the company size, field of operation or other differentiating factors. An agile enterprise should enrich its customer, enable internal and external cooperation as a mechanism for increasing its competitiveness, be able to successfully cope with uncertainty and changes, and achieve a market differentiation through a creative and distinctive use of human and informational resources. The creative and distinctive use of a company's human and informational resources is in the authors' opinion, the most important of the four strategies. An efficient utilization of these two resources significantly contributes to the fulfillment of the other strategic dimensions. The level of success that a company can achieve in customer enrichment, competitiveness, cooperation and capability to adapt to uncertain conditions depend on the appropriate use of both informational and human resources. Information Systems that allow gains in any of these dimensions are certainly improving the agility of a company. Also, it should be noted that these four dimensions are intertwined. Improvements in customer enrichment can definitely provide a company competitive edge in the marketplace. Moreover, these improvements can be a result of a better utilization of the company human and informational resources. The high level of interconnection among these dimensions makes possible the achievement of direct and indirect benefits from the implementation of long-term goals.

The possibilities for a company to enrich its customers are almost unlimited, cost reduction, better customer service, better supply and variety are some of the most visible examples. Cost reduction can be achieved by focusing on providing high quality products or services at lower prices than their competitors do. Information Systems used to improve the companies internal operations allow savings in production and overhead cost. Selling solutions instead of selling isolated products was previously identified as one of the major paths to customer enrichment [15]. Cost saving can also be a result of reduction in administrative cost and shorter product development cycles [4, 15]. The reduction of the elapsed time from and idea to its materialization allows companies to deliver more cost-effective solutions and beat their competitors [15].

An efficient customer service is one of the forms to improve customer satisfaction [4]. Information Systems that keep customer service data such as repairs, defects, returns and so on, help companies improve quality and to provide a distinctive customer treatment and satisfaction. Therefore, improving the company's strategic position in the market place. Customer loyalty and trust is another aspect that benefits from the use of strategic Information Systems. Trust is built on the company's ability to deliver high quality solutions on time. Loyalty, on the other hand, rests on the customer's trust that they benefit from the solutions a company provides to them.

An efficient management of the enterprise's supply or value-adding chain can be achieved through the use of Information Systems [4, 5]. The benefits associated with efficient management supply chains can be instantly recognized when their influence in the overall cost is analyzed. Fifty to ninety five percent of the overall shipping cost of manufacturing companies comes from the suppliers [11]. In these cases even small improvements can substantially enhance the company's strategic performance. Information Systems enable the use of reliable and timely information. This makes possible the coordination of globally distributed activities and the integration of globally distributed value-added chains [19]. The company delivery capabilities can be improved by the use of efficient Information Systems. Lower inventory cost, better responsiveness and more efficient supply and provision of products are some of the most significant advantages in this case. Creative and more efficient information exchange and communication mechanisms make possible the achievement of such benefits [5, 28].

Significant strategic advantages can be obtained through quality improvements [4]. Information Systems can make possible the simplification of the development process of a given product. As a result of a simplified design and manufacturing process, it is possible to achieve shorter product life cycles and an early identification of design and manufacturing problems. It is possible to improve manufacturing precision and specification of a product too. Simplified design and manufacturing processes reduce the possibility of human errors and make the design and the product operation more robust. All of the aforementioned gains contribute to the company differentiation in the market place and the satisfaction of its customers.

Information Systems are very useful for information processing and decision-making [5]. They can be particularly useful in performing unstructured decisions. Unstructured decision are those in which the decision making process is not well understood. Furthermore, there is not a procedure in place to reach the solution. Structured decisions, on the other hand, are repetitive decisions in which the procedures to reach a solution are known. Information Systems have different objectives depending on the kind of decision making process they are expected to support. In structured decisions, their major role is the improvement of information processing capabilities. The objective of these systems in supporting unstructured decision making process is the improvement of the organization and presentation of the information to the decision-maker.

One special type of strategic Information Systems is the one used by company executives, the so-called Executive's Information Systems. Traditionally, this kind of system has concentrated on accounting data. King (1996) identified the migration of these systems towards a more extensive use of internal and external information sources. He also suggested, as an instrumental characteristic of this new trend, its the capability of Information Systems to support the analysis enterprise's competitive capabilities. There is an increasing need for providing enterprise executives with business-intelligence information [7]. Information Systems that furnish company executives with timely and accurate information in their decision making process therefore, improve companies adaptation and performance in the marketplace.

The next section describes an industrial implementation of an Information System that provided a substantial improvement in the company strategic goals, responsiveness and customer satisfaction.

6. CASE STUDY

The manufacturing of made-to-order (MTO) products is characterized by productions of small size batches driven by customer's orders. In these cases, product architectures have a higher degree of modularization and parts or components are standardized to a larger extent. As in other instances of modular designs, the design can be reduced to the determination of the appropriate product configuration. Obviously, mass individualization – from a production perspective – cannot be afforded by managing each customer in complete isolation or by redoing all the design from scratch. It is well-known the fact that companies custom-build products by reusing up to 80% of their basic modules. Goldman et al. (1995) suggested that product individualization can be achieved through reconfiguration and reuse of modules in distinctive and unique ways. Despite the reusability of modules, made-to-order products can generate thousands of product families and variants [38].

The need for enterprises to concentrate on their core capabilities in an agile manufacturing environment and its benefits has been previously explained. The production of made-to-order products in this kind of environments certainly entitles the use of some outsourcing or subcontracting strategy. In this case, the complexity of the management of design and manufacturing related information substantially increases. Heterogeneity and autonomy of each member of the value-adding chains are the fundamental reason for this. In addition, the demands for maintaining the information consistency, an efficient data retrieval and the generality of the data definition grow. In this IS an additional requirement was considered. It allows for the selection of the subcontractors or partners to whom to engage in a collaborative relationship and it takes into consideration data from partner companies.

This application focuses on a star-like agile environment. The master company can be defined as the company that is in charge of the overall product development process. It is also responsible for subcontracting design and manufacturing steps partially or fully. Partner companies, on the other hand, are the companies that have been granted the subcontract or outsourcing of some of the design and manufacturing steps.

The management of information needs four data types [38]. They are: data about the product, data of the manufacturing on each part or component, data on the partner company in charge of producing an specific component and the data that allows the monitoring of the project. The data about the manufacturing of the components is not further analyzed. This information is kept by partner companies individually and does not play a very significant role from a master company perspective.

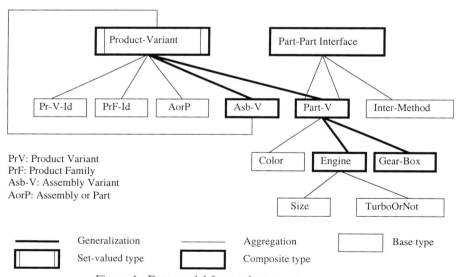

Figure 1. Data model for product structure.

Figure 1 shows a data model proposed by [38] that addresses the aforementioned requirements. The "Product Variant" was defined as a generic type. Classes "Asb-V" and "Part-V" are specialized types of the "Product Variant". The bold lines that connect the "Product Variant" to both "Asb-V" and "Part-V" are of has-a link type. They represent the hierarchy of the product under consideration. In seeking of simplicity and efficiency in data retrieval, the attribute "AorP" is used. This attribute makes it possible to know whether or not a specific product variant instance is a product or an assembly. The search is then directed towards "Asb-V" for assemblies or "Part-V" for parts.

This model represents both the product family structures and product variants in one single definition. The "PrF-id" and "PrV-id" attributes serve this purpose. They respectively identify the product family or the product variant. The "Part-Part Interface" type class explicitly represents the situation in which a product differs from another product in the connections among parts. Furthermore, the model uses an implicit representation for cases in which the differences among products are at the family or type level. This situation arises when a product is obtained as a result of increasing or decreasing the capabilities of another product. This representational decision is based on the fact that, the structure of the data model does not represent the product family structure.

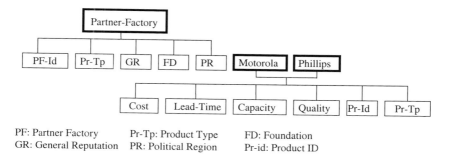

PF: Partner Factory   Pr-Tp: Product Type   FD: Foundation
GR: General Reputation PR: Political Region   Pr-id: Product ID

Figure 2. Data model for partner companies behaviour and performance.

Figure 2 illustrates the data model used for representing partner behaviour and performance [38]. Master companies pay particular attention to these criteria when carrying out the selection process. The "Partner Factory" object type represents a general information about the prospective partner company. The "Pr-Tp" attribute allows for the representation of cases in which a partner company provides more than one part, product or service to the master company. There exists a many-to-many relationship between partner companies and part variants. Partner companies may provide different variants of the same part or different companies can supply the same product variant. The performance and behaviour of a partner company for each part or product variant do not necessarily have to be the same. The representation of individual companies as classes captures this many-to-many relationship.

The models described above were successfully implemented for CLL a Hong Kong based company [38]. CLL is in the field of small personal health appliances. The

company manufactures a large variety of products. Each product also has a considerable number of variants. In addition, CLL has a large number of potential partners with globally distributed manufacturing facilities around the world. The system served the company's competitive advantages in several ways. The company was able to enhance its internal operations, since the information access and retrieval process were considerable improved. Furthermore, the system makes possible a more consistent maintenance of information. Difficulties associated with information inconsistency can range from using obsolete designs to losing the design rationality in products and process developments. In that regard, the system contributes to the achievement of shorter the product life-cycle, avoid common human errors and increase customer satisfaction. All of these improvements enhance the company's competitive advantages in the marketplace. The incorporation of the partner selection capabilities into the system made possible the consideration of different design alternatives after the original design of a product needed to be modified. The system contributed to the attainment of a higher level of information integration, since partner companies need not to change their own
product and part classification schema in order to share the information with the master company.

7. SUMMARY

The chapter started with the analysis of the relationship between Information Technology and Information Systems. It was shown that Information Systems are one of components of the software applications of Information Technology. Next, a brief analysis of the physically distributed manufacturing systems was performed. The major drawbacks of this kind of systems include the existence of centralized facilities and the lack of support for concurrent and collaborative work. This situation does not allow companies to achieve a higher degree of flexibility and responsiveness. Global Manufacturing Systems were also examined. They are an enhancement of physically distributed systems. In this case, globally distributed enterprises achieve a higher degree of communication, integration and collaboration. Enterprises share information, human and technical resources. The justification of investments in Information Technology by agile manufacturing companies was analyzed next. It was shown that Information Technology helps companies to achieve a high degree of concurrency and collaboration in their operation. Information Technology improves the collection, storage, maintenance, exchange and distribution of information among companies in a global manufacturing environment. The identification of the requirements of an IS framework followed. This section examined the current situation of Information System used in industry, their advantages and disadvantages. An Information Systems framework should allow for concurrent and a cooperative work, a ubiquitous access and exchange of information. The strategic role of Information Systems in enhancing the agility of an organization was also analyzed. It was shown that Information Systems contribute to the company's competitive advantage by improving customer services and quality, reducing product life-cycles and distinguishing the company from its competitors. Finally the analysis of an Information System implemented in an industrial setting was performed. The system supports the fabrication of made-to-order product in an agile environment. It also made possible the selection of partner companies to engage in a subcontracting or outsourcing relationship. The major advantages derived from the use of this system were explained.

REFERENCES

1. D. E. Avison and G. Fitzgerald, Information Systems Development. Methodologies, Techniques and Tools, Blackwell Scientific Publications, Oxford, 1988.
2. G. Rzevski, A Framework for Designing Intelligent Manufacturing Systems, Computers in Industry, Vol. 34, 1997, pp. 211-219.
3. D. O'Sullivan, Manufacturing Systems Redesign: Creating the Integrated Manufacturing Environment, Prentice Hall, New Jersey, 1994.
4. K. C. Laudon, and J. P. Laudon, Information Systems and Internet. A Problem Solving Approach, The Dryden Press, Forth Worth, Texas, 1998.
5. H. C. Lucas, Jr., The Analysis, Design and Implementation of Information Systems, 4th Edition, McGraw-Hill, Watsonville, California, 1992.
6. I. Ham, and S. R. T. Kumara, Global Collaboration for Customer Oriented Manufacturing, CIRP International Symposium, August 21-22, 1997, Hong Kong, pp. K1-K12.
7. D. King, Intelligent Executive Information Systems, IEEE Expert, Dec. 1996, pp.31-35.
8. K. M. Gardiner, An Integrated Design Strategy for Future Manufacturing Systems, J. of Manufacturing Systems, Vol. 15, No. 1, 1996, pp. 52-61.
9. C. H. Fine and D. E. Whitney, Is The Make-Buy Decision Process a Core Competency. 1996, http://web.mit.edu/ctpid/www/Whitney/papers.html.
10. L. Wildeman, Alliances and networks: the next generation, Int. J. Technology Management, Vol. 15, Nos. 1/2, 1998, pp. 96-108.
11. K. Preiss, The Emergence of the Interprise, Keynote Lecture, IFIP WG 5.7 Working Conference, Ascona, Switzerland, Sept. 15-18, 1997.
12. B. Kadar, L. Monostori, E. Szelke, An Object-Oriented Framework for Developing Distributed Manufacturing Architectures, J. of Intelligent Manufact. (1998) 9, pp. 173-179.
13. W. Reithofer and G. Naeger, Bottom-up Planning Approaches in Enterprise Modeling-The Need and the State of the Art, Computer in Industry 33 (1997) pp. 223-235.
14. H. J. Warnecke, The Fractal Factory, A Revolution in Corporate Culture, Springer-Verlag, Berlin, 1993.
15. S. Goldman, R. N. Nagel, K. Preiss, Agile Competitors and Virtual Organizations: Strategies for Enriching the Customer, Van Nostrand Reinhold, New York, 1995.
16. T. Ottaway and J. R. Burns, Adaptive, Agile Approaches to Organizational Architecture Utilizing Agent Technology, Decision Sciences, Vol. 28, No. 3, Summer 1997.
17. P. J. Sackett, J. M. Gilles, P. A. Lowenthal, K. K. Sehdev, Accurate Response Manufacturing: A Model for Co-operative Competition, CIRP International Symposium August 21-22, 1997, Hong Kong, pp. 455-460.
18. X. Liu, W. J. Zhang, C. A. van Luttervelt, A New Methodology and Computer Aid for Manufacturing System Design with Special Reference to Partner Factories Selection in Virtual Enterprises, Proc. of the Int. Conference of World Manufacturing Congress, 1997, New. Zealand, pp. 61-66.
19. S. E. Fawcett, R. Calantone and S. R. Smith, Delivery capability and firm performance in international operations, Int. J. Production Economics, 51 (1997), pp. 191-204.
20. Y-M. Chen, C-C. Liao, B. Prasad, A Systematic Approach of Virtual Enterprising Through Knowledge Management Techniques, Concurrent Engineering: Research and Applications, Vol.6 No.3, Sept. 1998, pp. 225-244.

21. R. Bentley, W. Appelt, Buscach U., E Hinrichs, D. Keer, K. Sikkel, J. Trevor and G. Woetzel, Basic Support for Cooperative Work on the World Wide Web, Int. J. Human-Computer Studies, Vol. 46, 1997, pp. 827-846.
22. L. M. Camarinha-Matos, H. Afsarmanesh, C. Garita, C. Lima, Towards an Architecture for Virtual Enterprises, J. of Intelligent Manufacturing Vol. 9, 1998, pp. 189-199.
23. T. Kjellberg, Umbrella Project: Information Systems for Virtual Manufacturing, http:// www.psm.kth.se/woxen/projects/virtual.html
24. H. Lau, The New Role of Intranet/Internet Technology for Manufacturing, Engineering with Computers, Vol. 14, 1998, pp. 150-155.
25. M. S. Fox, and M. Gruninger, Enterprise Modeling. AI-Magazine, Vol. 19. No. 3, 1998, pp. 109-121.
26. M. E. Merchant, Some Thoughts on What Lies Ahead in Manufacturing, CIRP International Symposium August 21-22, 1997, Hong Kong, pp. K31-K35.
27. G. C. Kunkle, Moving Small Firms Toward Agility: Agile Business Practices in Agile Web Firms, 1999, www.agileweb.com/busprac.html.
28. D. Whitney et. al., Agile Pathfinder in the Aircraft and Automobile Industries – A Progress Report, web.mit.edu/ctpid/www/agile/atlanta.html.
29. A. Campbell, The agile enterprise: assessing the technology management issues. Int. J. Technology Management, Vol. 15, Nos. 1/2, 1998, pp. 82-95.
30. P. van der Wolf, Architecture of an Open and Efficient CAD Framework, Ph. D. Thesis, Delft University of Technology, June 1993.
31. W. Bailey, Masson, R., R. Raeside, Choosing successful technology development partners: a best-practice model. Int. J. Techn. Manag., Vol. 15, Nos. 1/2, 1998, pp. 124-138.
32. J. C. Wortmann, Production Management Systems for One-of-a-kind Products, Computers in Industry, Vol. 19, 1992, pp. 79-88.
33. D. M. Upton, and A. McAfee, The Real Virtual Factory, Harvard Business Review Vol. 74, No. 4, 1996.
34. K. M. van Hee, Information Systems Engineering. A Formal Approach, Cambridge University Press, 1994.
35. S. Gosain, Applying Plug-and-Play Design Phylosohpy to Virtual Organizing. Newsletter, Vol. 2, No.4, December, 1998. (www.virtual-organization.net)
36. Q. Zhou and C. B. Besant, Information management in production planning for a Virtual Enterprises, Int. J. Prod. Res., Vol. 37, No. 1, 1999, pp. 207-218.
37. W. J. Zhang and Q. Li, Information Modeling for Made-To-Order Virtual Enterprise Manufacturing Systems, Computer Aided Design Vol. 31, 1999, pp. 611-619.

Secure Communication in Distributed Manufacturing Systems

István Mezgár[a] and Zoltán Kincses[b]

[a] Computer and Automation Research Institute, Hungarian Academy of Sciences,
Budapest 1111 Kende u. 13-17, Hungary
E-mail: mezgar@sztaki.hu

[b] Ph.D. student, University of Eötvös Loránd, Dept. of Informatics
Budapest 1117 Pázmány Péter st. 1/D. E.424
E-mail: kincses@elte.hu

In the industry the communication security has got a bigger role due to the increasing number of distributed production system (DPS) realizations. As the basic characteristics of DPSs is the communication on the Internet, and during these transactions huge amount of extremely valuable technical data, and information (product, process data besides the business information) are moving between the members of the networks, the security has became a vital problem of this field. Typical representative of DPSs are the virtual enterprises (VE) as all functions of the production (planning, manufacturing, etc.) are present in them in a distributed form.

The chapter introduces shortly the main characteristics of VEs and some VE reference architectures and also the communication architectures and protocols VEs can use. In the second part the hardware and software security methods, approaches and tools are described that can be applied during the transfer of technical related data, information on the network, and on the computer while storing or handling the information. The approaches introduced are general enough to be used in other cases, environments as well. As the topic is too wide to give a full overview in such a very limited extent, the main goal of the chapter is to give a short overview on the possibilities and future on the security in (computer) network communication and to call the attention of the industrial community to the importance of this field with some examples.

1. INTRODUCTION

The developments in the fields of information technology, telecommunication and consumer electronics are extremely fast. The digital technology now makes possible to forward information and services in different forms (voice, data, or pictures) using both traditional and new communication services through different types of networks (computer network, cable TV, telephone).

The ability of different network platforms to carry essentially similar kinds of services and the coming together of consumer devices such as the telephone, television and personal computer is called "technology convergence". But this convergence is not just about technology. It is also about services and about new ways of doing business and of interacting within the society. The impact of the new services resulting from convergence can be felt in the economy and in the society as a whole as well as in the relevant sectors themselves.

Today the global nature of communications platforms (in particular, the Internet) is providing a key that is opening the door to the further integration of the world economy. At the same time, the low cost of establishing a presence on the World Wide Web, is making it possible both for businesses of all sizes to develop a regional and global reach, and for consumers to benefit from the wider choice of goods and services on offer. Globalization is therefore the key theme in developments.

Computer network technologies themselves as one of the main drivers of convergence and globalization are integrated into all fields of the economy, in different applications of industry, banking, health care, etc. Network connections are not limited only for one enterprise (Intranet), or for a country, or for a certain sector of economy, but for many functions and for the whole world. This globalization trend can be identified in most sectors of the economy. The functional integration and the globalization have resulted the integration of material-, information-, and money flows, which are the three basic components of complex production and service processes.

Secure identification, authentication of the users and communication security are main problems in network communication. Encouraging best business practices in areas related to data protection and security of digital signatures may be supported by standardization as well. In the industry the communication security has got an increasing role since the growing number of different types of distributed production systems.

Agility is the main characteristic of distributed systems. Agility means the ability to thrive in an environment of constant and unpredictable change. The expression of "agile manufacturing" includes among others the competitiveness through cooperation, integrated flexibility, flexible management structure, and virtual corporations.

There are different expressions to define such types of networked, flexible production architectures, like extended enterprise, virtual enterprise and virtual corporation. In this chapter the authors have selected the virtual enterprise (VE) to represent this type of production system.

The spread of VEs and VE-like manufacturing architectures is on the doorstep of the industry. There are many applications, but because of the lack of proper cooperative, effective and secure software tools in the filed the massive spread of real VEs has been pushed into the close future. As the basic characteristics of VEs are the communication on the Internet, and during these transactions huge amount of extremely valuable technical data and information (development, product, process data beside business information) is moving through the networks, the security has become a vital question.

The chapter aims to introduce the existing networked manufacturing architectures, their communication protocols and standards, the connections among the different fields of the networking technology. The description of the connected applications is focusing on the VE, on the importance of security in "manufacturing communication", and outlines the possible secure forms of communication for VEs.

The chapter doesn't intend to give a full overview, or a detailed description on networking, or on security, rather wants to flash the dangers of sending valuable information through networks and how to avoid these traps, and push the users of manufacturing community into the direction of secure communication.

As the chapter covers a very broad area it is not possible to introduce all aspects in detail. References for each important part are given (also WEB sites). Some general networking protocols are also introduced in order to give a full overview, to give the possibility for the reader to compare directly, to find the place of the manufacturing and security protocols, and security techniques in the general networking terms.

2. VE CHARACTERISTICS AND VE REFERENCE ARCHITECTURES

2.1. Network-based distribution systems

Distribution systems provide the transportation of goods, or services to the consumers according to their orders and payments. It is natural that commercial networks which operation are based on computer network (e-commerce) belongs to distribution systems (DS), but VEs also can be classified as DS as the connections among the nodes of a VE can be handled as a consumer- supplier relation with some restrictions.

The e-commerce represents today in the best way the efforts of convergence and globalization described in the introduction, and from information point of view the future of VEs moves in the same direction. E-commerce is the "fore-runner" of VE, as they have many similarities in their functions in connecting the "three flows". E-commerce is the ability to perform transactions involving the exchange of goods or services between two parties or more using electronic tools and techniques. E-business is a combination of the broad reach of the Internet with the vast resources of traditional information technology systems. Its scope is broad–spanning private intranets, shared extranets, and the public Internet. E-business is the general strategy of that e-commerce is an important part. According to past statistics and the predictions, the future of e-commerce (and the e-business) is breathtaking.

The VE can be defined as a coordinated network of autonomous production units (factories, firms) with the goal of producing a product while changing all the needed information via computer network. This temporary production network has been formed also by using the Internet or even the wireless communication.

The main reason of not developing even faster the e-commerce is the lack of secure communication (nobody likes to send his/her credit card number through an unreliable connection). As a result of a big joint effort, the SET standard has been developed but the use of this secure protocol is not general yet. [1] Security is the key issue of e-commerce as it makes users trust and use e-commerce.

In case of VEs the situation is very similar, but the question of security is even more important as high-value development/research results, product, technology, financial and management information and data are exchanged via network regularly. In case of e-commerce the user risks the value of the selected goods plus the sum on his bank card, in case of a VE the firm risks the thief of the result of a long development work, a product, or a technology probably worth millions of dollars.

2.2. Main characteristics of VE

Most VE theorists refer to the holonic doctrine, first introduced by Arthur Koestler in his book "The Ghost in the Machine" [2]. This theory is based on the *holon* concept, where holons are defined as independent entities sharing some basic features; as *openness* (they are able to cooperate with each other to reach a common goal), *flexibility* (each of them can easily re-configure itself in response to an external stimulus), and *similarity* (they share the same basic principles, values and purposes).

A holon is said to be "a whole into itself and a part of other wholes", so holons are defined as independent entities that are capable of coordinated behavior. A system including entities with such characteristics, along with the necessary links to support their mutual interaction, is called a *holonic system*. In its most common representation, a holonic system is seen as a network graph, where nodes represent holons and arcs indicate interaction links between the nodes. In the last years the holonic doctrine has been applied to the manufacturing domain, leading to the concept of VE. This new organizational paradigm is founded on the assumption that the manufacturing environment is going to transform itself into a holonic system.

To be part of such a system, individual enterprises have to change into holons, that is, to become flexible and open enough to fit the above definition. At the same time, the environment must support the integration of these enterprises within an evolving system, taking the form of a multi-layer network. This is obtained through efficient communication and transportation means, as well as through the spread of principles, values and know-how across the network. There are different forms of the distributed enterprise, the VEs are one of the most up-to-date form of production. Based on the different definitions of VE it can be stated that the intensive use of computer networks and the high-level organization flexibility are main parameters of VEs.

Enterprises forming a holonic manufacturing system are potentially enabled to cooperate with each other to achieve a common goal. This happens in reaction to an external stimulus, taking the form of a new business opportunity which can be better exploited by more joined enterprises than by an individual firm.

In these circumstances a *virtual enterprise* is created. This structure can be seen as a holarchy, in that it is a temporary, goal-oriented aggregation of several individual enterprises. Each VE is created to pursue a specific business objective, and remains in life for as long as this objective can be pursued. After that, the individual nodes resume their independence from each other. Node resources that were previously allocated to the expired business are re-directed toward the node individual goals, or toward other VEs it may have joined.

To allow this kind of dynamic re-configuration of the whole system in response to market changes, significant requirements must be met. On one side, individual enterprises must improve their flexibility and extend their connections with the other members of the system. On the other side, the manufacturing infrastructure must support fast interaction as well as information sharing between the nodes.

The reasons of creating VEs are such as: rapidly evolving markets, reduction of design and manufacturing times (because of shorter product life cycle), increased efficiency of communication and transportation means. In the practice there are two main ways to form a VE; decompose a large company into smaller units, or aggregate little firms (e.g. Small and Medium size Enterprises - SMEs) into the form of a VE [3]

The VEs formed by the two approaches have different requirements as both the inherited characteristic and the goals of the original manufacturing units are very different. The common requirements of environmental factors that make possible the VE realization are the fast transport and communication means, and the spread of principles, know-how, and business practice to all enterprises in the VE.

Importance of safe communication in VE.
The basic characteristic of VE is the flexibility both in information- and in material flow. All main events of its life cycle are connected to communication on the network The communication requirements for a VE can be summarized in the followings:
1. Integration of different communication forms and resources
    Communication through connected telephone-, computer- and cable networks, and application possibilities of different protocols, connecting wired- and wireless equipment.

2. Reliable and high quality communication services
Reliability covers the high on-service time (technical reliability), the high availability (well designed/balanced network – resource reliability), the HW and SW security, both for equipment and communication lines (access reliability), well controlled/organized networks (organization reliability), all with reasonable cost.
3. Global time coordination
It is essential the exact coordination of the different actions in time during the life cycle of the VE, so a "general time" has to be declared for communication.
4. Traceable communication
Traceability means to document and audit the communication in a way that fulfills the requirements of bookkeeping (e.g. delivery report and receipt notification) and legal aspects (e.g. digital signature).

As the goal of the present chapter is the description of the VEs from security aspects, in the following the HW and SW security of equipment and communication lines (access reliability) and the control/organization of networks (organization reliability) will be discussed. The security requirements for a VE can be listed as follows:

1. Protection of all types of enterprise data (for all company forming the VE)
Privacy and integrity of all types of documents during all phases of storage and communication (Data and communication security – Certification, Encryption),
2. To enable companies confidential access control,
3. Authorization and authentication of services (digital signature).

These services need to be flexible and customized to meet a wide array of security needs, including specific high level requirements. In order to fulfill the communication and security demands some basic aspects have to be taken in consideration while selecting security and communication technologies:

1. Platform independent SW tools have to be applied,
2. Standards have to be applied (accepted and "de facto" standards as well),
3. Appropriate architectures with ability to integrate different resources.

Fulfilling all types of the introduced requirements for individual enterprises would be very hard if not possible, so different general network- and organizational structures have been developed, that have been carefully designed and tested. These structures can be defined as reference architectures, and they are available both for the organization and for the information infrastructure of VEs.

2.3 Models and Reference Architectures for virtual enterprises

There are different projects to aim some type of general (reference) VE model, or a more complete reference architecture. A general model offers wide possibilities but it can have some special characteristic at the same time.

The network architecture can be defined as the collection of design principles, physical configuration, functional organization, operational procedures, and data formats used as the bases for the design, construction, modification, and operation of a communications network. A reference architecture is a combination and arrangement of functional groups and reference points that reflect all logically possible network architectures in a structured, flexible, and modular way. A reference architecture gives the general representation of the architecture from which all individual architectures can be derived.

The reasons, or goals to develop reference architectures are to give a complete, and well defined frame (applying the standards, the ready-to-use of new technologies) for the practical applications. By using the reference architecture the actual individual architectures can be configured easily, and through applied technologies and communication protocols they are platform independent.

2.3.1 The PLENT VE model

In the frame of the PLENT project a general model of VE for SMEs has been developed. This model is a typical example for the "aggregation-based" development of VEs [4].

In a distributed manufacturing environment the production planning has a key role. The coordination of the orders, the optimal assignment of the different resources in a cooperative production of several SMEs is a very difficult task. In SME networks there is a strong need of a network model that is based on the planning strategy Another important goal during the software development was the product independence; i.e. the network model has to be general enough, to be able to apply it in a wide range of different products, in different production environments (a reference network model). The main goal of the project was to support organizations that manufacture mechanical parts and products in SME-like distributed production environments.

In order to achieve the required network timeliness and coordination neutrality, the PLENT network organization must be supported by expressly designed planning tools. In particular, three software modules are required: a coordination module, a local planning module and a performance evaluation module. With this module structure it is possible to realize the flexible and general model of VE.

2.3.2 The NIIIP Reference Architecture

The National Industrial Information Infrastructure Protocols (NIIIP) project initialized by the National Institute for Standards and Technology (NIST) is one of the most complete realizations of VE architectures. It intends to bring together the product realization process integration efforts, by developing general global protocols for the technical standards of product data definition, communication, and object technology and workflow management. The NIIIP doesn't intend to develop a new system, rather applying existing standards to consolidate, harmonize, and integrate the many sets of existing protocols. The main goals of the NIIIP reference architecture is to help establishing and operating of VEs in the industry, by applying standardized solutions for [5]:

- VE connectivity
- Industrial information modeling and exchange
- Management of VE projects and tasks.

The IETF, the Object Management Group (OMG), STEP, and WFMC (Workflow Management Coalition) are defining solutions in these areas. NIIIP's aim is to adopt these technologies; develop, reuse, and integrate implementations; and help to accelerate standards development and adoption. The NIIIP Consortium has developed a reference architecture and reference implementation for the technology needed for industrial VEs. It consolidates, harmonizes, integrates, and extends existing protocols. By means of scenarios for industrial VEs, pilot projects, and demonstrations, NIIIP will prove the efficacy of the recommended approach. More detailed description can be found at [5].

Technology Categories and Interaction

NIIIP's goal is to define and demonstrate, via the OMG and the Internet, an integrated infrastructure based on object technology that:

- Connects industrial VEs with extended security and availability, via Internet

- Provides transparent access to other application environments
- Supports industrial information sharing, via STEP
- Provides VE work and knowledge management., via OMG

The NIIIP reference architecture is based on the application of three methodologies and technologies:
Communication Technology – Internet
Object Technology- Application inter-operability
Information Technology - information exchange based on STEP

2.3.3 PRODNET-II - an open platform with adequate IT protocols for VEs

The main goal of the PRODNET project was the development of infrastructures to support industrial VEs through the design and development of an open platform and the adequate IT protocols and mechanisms [6]. PRODNET was focused mainly on SMEs in order to support them with tools to inter-operate with other networks. The architecture employs the new emerging standards and advanced technologies in communication, cooperative information management, and distributed decision making. PRODNET deal with a number of VE environment requirements and necessary steps, listed below:

(1) study and structure of the business data and the information that needs to be communicated between partners in a Virtual Industrial Enterprise,
(2) design and develop a software infrastructure to provide an environment for these data and information to be exchanged, shared and managed in the virtual industrial network,
(3) to promote the utilization of international standards such STEP, EDIFACT, Internet WWW / Java, ISO 900X series to assist SMEs to maintain and improve their competitiveness,
(4) support the implementation of open, standards-based software components, that are easy to use, low in cost and provide a high value-added benefit.

Description of the infrastructure

The PRODNET Cooperation Layer (PCL) contains the basic functions to connect the company network. It represents the communication role and works as the interlocutor of the company within the net.

The PRODNET Communication Infrastructure module is responsible for handling all communications with the other nodes in the network. It includes functionalities such as: selection of communication protocols and channels; basic communications management; privacy mechanisms (cryptography); authentication and safety: This module is responsible for the implementation of safety and authentication mechanisms, at the VE level. It has to check access rights, handle key management.

3. NETWORK REFERENCE MODELS AND PROTOCOLS IN MANUFACTURING

3.1 Reference models of network architectures

Open Systems Architectures (OSA) became an important approach to develop flexible, adaptable sets of methodologies, standards and protocols for structured communication systems. OSA is a layered hierarchical structure, configuration, or model of a communications or distributed data processing system that enables system description, design, development, installation, operation, improvement, and maintenance to be performed at a given layer or layers in the hierarchical structure, allows each layer to provide a set of accessible functions that can be controlled and used by the functions in the layer above it, enables each layer to be implemented without affecting the implementation of other layers, and allows the alteration of system performance by the modification of one or more layers without altering the existing equipment, procedures, and protocols at the remaining layers .

An OSA may be implemented using the Open Systems Interconnection-Reference Model (OSI-RM) as a guide while designing the system to meet performance requirements. A good and detailed description of computer networks is in [7].

Open Systems Interconnection - Reference Model (OSI - RM)

The Open Systems Interconnection (OSI) - Architecture is a communication system architecture that adheres to the set of ISO standards relating to open systems architecture. The Open Systems Interconnection-Reference Model (OSI-RM) is an abstract description of the digital communications between application processes running in differing systems. The model employs a hierarchical structure of seven layers. Each layer performs value-added service at the request of the neighboring higher layer and, in turn, requests more basic services from the next lower layer. The layers are as follows:

Physical Layer (Layer 1), Data Link Layer (Layer 2), Network Layer (Layer 3), Transport Layer (Layer 4), Presentation Layer (Layer 6), Application Layer (Layer 7).

Transmission Control Protocol/Internet Protocol - TCP/IP

The TCP/IP is two interrelated protocols that are part of the Internet protocol suite. TCP operates on the OSI Transport Layer and breaks data into packets, controls host-to-host transmissions over packet-switched communication networks.

Internet protocol (IP) was designed for use in interconnected systems of packet-switched computer communication networks. IP operates on the OSI Network Layer and routes packets. The Internet protocol provides for transmitting blocks of data called datagrams from sources to destinations, where sources and destinations are hosts identified by fixed-length addresses. The Internet protocol also provides for fragmentation and reassembly of long datagrams, if necessary, for transmission through small-packet networks.

3.2 Communication protocols for manufacturing networks

Information technology (IT) is increasingly determining growth in automation technology. It has changed hierarchies, structures and flows in the entire office world and now covers all sectors – from the process and manufacturing industries to logistics and building automation. The communications capability of devices and continuous, transparent information routes are indispensable components of future-oriented automation concepts. Communication is becoming increasingly direct, horizontally at field level as well as vertically through all hierarchy levels.

In this chapter a short overview is given of manufacturing communication, while concentrating on the protocol architectures and on the network reference models and on protocols that are applied in manufacturing industry. The goal of this description is to give the possibility for a direct comparison with the security risks in each layer and their avoidance described in subchapter 4.3. MAP, TOP, CNMA, OSACA, PROFIBUS and Fieldbus have been selected for introduction. These architectures, or protocols use, or are based on the ISO/OSI reference model, and partly also use its protocols. Recently the techniques and tools supported by OMG are also involved in manufacturing automation.

Manufacturing Automation Protocol (MAP)
The MAP networking protocol has been specifically developed for computer communications in a factory environment and was expected for the high data rates, while improving noise immunity and to provide a common standard for all equipment to simplify integration.

MAP was a great promise, but because of some technical and organizational problems (e.g. difficulties have arisen getting countries and vendors to agree on specific standards; standards are so broad that they have become very complex and hard to develop hardware and software for, thus driving up the costs) it has not fulfilled the expectations. Inspite of this it is applied in different industrial companies.

Communications Network for Manufacturing Applications - CNMA
CNMA has been developed in the frame of a series of ESPRIT projects in Europe with participation of CIM users and developers. CNMA aims to standardize the factory automation in cooperation with MAP and MMS based on MAP 3.0 specification.

Technical and Office Protocols - TOP
TOP is designed for office environment. The preferred LAN is IEEE 802.3 Ethernet, but allows for Token Bus (IEEE 802.4) and Token Ring (IEEE 802.5). Fibred optical media will be added when they are defined. TOP uses same ISO standards for network, transport, session and presentation layers that MAP does.

Open System Architecture for Controls within Automation Systems - OSACA
The OSACA is the European initiative to define a vendor-neutral, open controller architecture in order to improve the competitiveness and flexibility of suppliers and users of control systems: machine tool builders, control vendors and end users. To cope with these aims the requirements for a new generation of control systems were analyzed. Out of this preparatory work, an application programming interface (API) for control applications and an appropriate infrastructure (the so-called system platform) was specified and implemented. The benefits of the OSACA architecture can be interoperability, portability, scalability and reusability of applications.

PROFIBUS Technology
PROFIBUS is a vendor-independent, open field bus standard for a wide range of applications in manufacturing and process automation. Vendor-independence and openness are ensured by the international standards EN 50170 and EN 50254. PROFIBUS allows communication between devices of different manufacturers without any special interface adjustment. PROFIBUS can be used for both high-speed time critical applications and complex communication tasks. The structure of the PROFIBUS Protocol Architecture is based on recognized international standards. The protocol architecture is oriented to the OSI (Open System Interconnection) reference model. More details on PROFIBUS can be found at [8].

Fieldbus technology
Fieldbus is an all-digital, serial, two-way communications system that interconnects measurement and control equipment such as sensors, actuators and controllers. At the base level in the hierarchy of plant networks, it serves as a Local Area Network (LAN) for instruments used in process control and manufacturing automation applications and has a

built-in capability to distribute the control application across the network. FOUNDATION Fieldbus is used in both process and manufacturing automation applications. FOUNDATION Fieldbus [9] technology is controlled by the non-profit organization Fieldbus Foundation.

4 POSSIBILITIES OF SECURE COMMUNICATION IN A VE ENVIRONMENT

4.1 Computer system and network security

Security can be defined as the state of certainty that computerized data and program files cannot be accessed, obtained, or modified by unauthorized personnel or the computer or its programs. Security is implemented by restricting the physical area around the computer system to authorized personnel, using special software and the security built into the operating procedure of the computer. When applied to computer systems and networks denote the authorized, correct, timely performance of computing tasks. It encompasses the areas of confidentiality, integrity, and availability.

In the "Annual Computer Crime and Security Survey" of Computer Security Institute [10] there are significant data about security and losses of attacks against computer systems. While approx. 60% of the organizations participated in the survey answered 'yes' to the question "Has your organization experienced unauthorized use of computer systems in the last year?", 20-20% answered 'no' or 'do not know' to this question. Theft of proprietary information and financial fraud have a share of more than 30-30% each, while insider abuse of net access go up to 6% of the total loss (100% equals $123,779,000 loss!). The survey states, that "It is clear that computer crime and other information security breaches pose a growing threat to U.S. economic competitiveness and the rule of law in cyberspace. It is also clear that the financial cost is tangible and alarming", and the firms "that want to survive in the 'Information Age' simply have to dedicate more resources to staffing and training of information security professionals". These data clearly show the importance of taking care of security from physical level to the information level. Security has its own cost, but it is possible to calculate, while losses can not be predicted!

Security is a conscious risk-taking, so in every phase of a computer system's life cycle must be applied that security level which costs less than the expense of a successful attack. With other words security must be so strong, that it would not be worth to attack the system, because the investment of an attack would be higher than the expected benefits. At different levels different security solutions have to be applied, and these separate parts have to cover the entire system consistently.

In Table 1 the main practical fields of security are summarized in order to better understand the content of the following chapters. The abbreviations of SW and HW have a broader purport in the table not only referring to the computer science. In the field of security standards and quasi standards have an important role. In the followings some of the most relevant ones are introduced shortly, only to show the directions and status of these significant works.

In order to classify the reliability and security level of computer systems an evaluation system has been developed and the criteria have been summarized in the so-called "Orange book" [11]. Its purpose is to provide technical hardware/firmware/software security criteria and associated technical evaluation methodologies in support of the overall ADP system security policy, evaluation and approval/accreditation responsibilities promulgated by DoD Directive 5200.28.

Table 1.
Main fields of computer security

	Organization security	Personal security	Network (channel) security	Computer (end point) security
SW security	Definition of security policy (e.g. access rights)	Employment of trained and reliable staff	Using tested network SW tools, and continuously checked communication channels and well configured network elements	Using tested application SW tools, and continuously checked operation system, and properly configured HW systems
HW security	Placing the computers in secure location of the building and offices	Physical identification technologies (fingerprints, etc.)	Prevent direct, or close access to network cables, or application of special technologies	Prevent direct physical access to computers by unauthorized persons, or a close access in electromagnetic way

The ISO/IEC 10181- [12] multi-part (1-8) "International Standard on Security Frameworks for Open Systems" addresses the application of security services in an "Open Systems" environment, where the term "Open System" is taken to include areas such as database, distributed applications, open distributed processing and OSI. The Security Frameworks are concerned with defining the means of providing protection for systems and objects within systems, and with the interactions between systems. The Security Frameworks address both data elements and sequences of operations (but not protocol elements), which may be used to obtain specific security services. These security services may apply to the communicating entities of systems as well as to data exchanged between systems, and to data managed by systems.

The ISO/IEC 15408 standard [13] consists of three parts, under the general title "Evaluation Criteria for Information Technology Security" (Part 1: Introduction and general model, Part 2: Security functional requirements, Part 3: Security assurance requirements). This multipart standard defines criteria, to be used as the basis for evaluation of security properties of IT products and systems. This standard originates from the well-known work called "Common Criteria" (CC). By establishing such a common criteria base, the results of an IT security evaluation will be meaningful to a wider audience. By the time there are available "Protections Profiles" created for computer systems and for smart cards also based on CC guidelines. The standard is useful as a guide for the development of products or systems with IT security functions and for the procurement of commercial products and systems with such functions.

4.2 Security problems and solutions in a VE

4.2.1 Security in Distributed Environments

Distributed systems and collaborative environments, such as widely distributed supercomputers and large-scale storage systems, data sharing in restricted collaborations, network-based multimedia collaboration channels and distributed production systems give rise to a range of requirement for distributed access control and the overall security of the systems.

In all of these scenarios, the resource (data, instrument, computational and storage capacity, communication channel) has multiple owners, and each owner will impose use-conditions on the resource. All of the use-conditions must be met simultaneously in order to satisfy the requirements for access. Furthermore, today it is the norm that the members (nodes) of such distributed networks tend to be diffuse, being geographically distributed, and multi-organizational. Therefore the security/access control mechanism must accommodate these special circumstances.

The goal for security in such distributed environments is to reflect, in a computing and communication based working environment, the general principles that have been established in society for policy-based resource access control. Each involved entity/node should be able to make their assertions without reference to a mediator and especially without reference to a centralized mediator (e.g. a system administrator) who must act on their behalf. Only in this way will computer-based security systems achieve the decentralization needed for scalability in large distributed environments.

The resource access control mechanisms should be able to collect all of the relevant allegations and make an unambiguous access decision without requiring entity-specific or resource-specific local, static configuration information that must be centrally administered.

In order to be the security a successful part of the distributed environment -- providing both protection and policy enforcement -- each principal entity should have no more nor less involvement than they do in the currently established procedure that operates in the absence of computer security. Only the form has to be changed, e.g. digital signature instead of signing a paper. In case of such system this sort of a security infrastructure should provide the basis of automated management of resources that proceed the construction of dynamically, and just-in-time configured systems to support different user defined application-oriented requirements.

The expected advantage of computer-based systems is in maintaining access control policy, but with greatly increased independence from temporal and spatial factors (e.g. time zone differences and geographic separation), together with automation of redundant tasks such as credential checking and auditing.

The security architectures represent a structured set of security functions (and the needed hardware and software methods, technologies, tools, etc.) that can serve the security goals of the distributed system. In addition to the security and distributed enterprise functionality, the issue of security is as much (or more) a deployment and user-ergonomics issue as technology issue. That is, the problem is as much trying to find out how to integrate good security into the industrial environment so that it will be used, trusted to provide the protection that it offers, easily administered, and really useful.

4.2.2 Security solutions in reference architectures for virtual enterprises

The critical security points for virtual enterprises are the access points, the improperly configured systems, the software bugs, the insider threats and the physical security. The following two examples shows how part of these problems are solved in the VE reference architectures.

Security in the NIIIP
In NIIIP, secure communication can be implemented at three levels:
- IP level — discussions of protocol-level security are underway within industry organizations. The adoption and deployment of this extension to IP is, however, years away.
- OMG level — a Request For Proposal (RFP) for the Security Object Service has been issued. NIIIP participates in these activities, defining solutions, and influencing their direction.
- NIIIP level — solutions for VE security. NIIIP is proposing an object interface with an extra security tag to activate the enforcement of secure communications over networks. These security requirements can be implemented using well-known data encryption methods available on the Internet. These include Public Key Cryptography from RSA Data Security Inc. and Data Encryption Standard (DES).

Internet provides services that facilitate the locating and accessing of data distributed throughout the network. These facilities are essential for NIIIP. NIIIP data directory and information facilities will use and extend Internet services, such as Domain Name Server (DNS), and provide gateways for NIIIP users for the utilization of emerging services supporting electronic commerce.

Security in PRODNET
The goal of the PRODNET Communication Infrastructure (PCI) is to fulfill the security and the legal requirements besides the functional ones. By using the PCI it is guaranteed that no one, other than the owner, can access to a document (privacy), the content of a document can't be changed without detection (integrity), each received document is unambiguously connected to an identifiable sender (authentication) and the logging information is maintained for auditing and communication management. [14]

4.3 Security in the network - methods, technologies and tools

At the beginning of networking there was a need mainly for the reliable operation, but the secure and authentic communication has became a key factor for today. According to Internet users, security and privacy are the most important functions to be ensured and by increasing the security the number of Internet users could be double or triple according to different surveys. The main reason of the increased demand is the spread of electronic commerce through the Internet, where money transactions are made in a size of millions of dollars a day. It is not just the question of our letters content or our user account, it is the question of money.

Every part of a network could be attacked, just the level of expected success and effort determines the targets (Table 2.). From server break-ins (change a Web-page, create false user account, steal information) through DNS spoofing till password or other information stealing every element of network is dangerous without attention to security.

There are several solutions to secure the network, just security is in inverse proportion to usability and the most of the security tools are patches, extra solutions and rather stand-alone techniques [15]. Telnet program is on every system, SSH on more and more, but is still in minority. When using telnet or SSH it can't see big differences, just telnet was the first, and if there is not a visible difference, why to use the other – asks users. For an attacker there is a visible difference, because the attacker will use a sniffer program, and will get the password too! After that the attacker could enter to the account, and will see the e-mails, files, projects and maybe credit card numbers too. With SSH, where the password and other information are traveling in an encrypted form, the attacker will see just an unusable data flow.

The same situation is with FTP, where with ftpdump is possible to dump whatever the attacker want to dump, even the password. The most successful service of the Internet the World Wide Web allows to users to use the mail and file transfer services beside surfing. When surfing on the Web, the user may find something he/she wants to buy. He/she just type in his/her card data, and the account is in danger even if the server is a trusted one, and the owner is a real merchant or bank. If the submitted data travels in plain text format without encryption then between the user site and the target server the data could be stolen, and used later by the attacker.

Experts are arguing with each other and with outsiders too about the strength of each crypto algorithm, but in the first step every algorithm is better than nothing. Instead of FTP there is SFTP (secure FTP) or SCP (secure copy), instead of HTTP there is SHTTP which is HTTP over SSL (Secure Socket Layer). Instead of simply e-mail there is the PGP (Pretty Good Privacy) signed e-mail. With these techniques it can be guaranteed that the information in e-mail, file or on Web page will be reached only by authorized parties. These solutions are SHOES (security help-tools over existing solutions) which helps not to walk barefoot on the information superhighway.

In cryptology based algorithms a key-role has the key-management from issuing state through storage and use till revocation of keys. At issuance it is possible to sign a key by the issuer, and then the issuer must be a well-known party for others who want to check the signature on this key. If a key is not signed, it is up to us the trustee decision. On each way there is the problem of revoke, because these keys are not forever. Sometimes there are time-stamps on the keys or signatures, sometimes they are used for just one time (for example the One Time Password system), and exists algorithms with merging these two ideas. In this case the One Time Password is a ticket, which has an expiration time, like in Kerberos system [16]).

The portability and the secure storage of these keys is solved by smart tools, called smart cards which has a computer logic on a 2 mm^2 area of silicon. These ISO/IEC 7816 standard based tools have the ability of storage and process different cryptography algorithms providing secure and mobile access not only from a computer, but from different computers and other tools, like stand-alone door-access systems or even mobile phones with dual slot system. Basically a smart card is a small computer with the same HW structure, just in a smaller capacity and pocket size.

4.4 Security on the computer - methods, technologies and tools

The first line of defense is the physical access of a computer (See Table 3.). There is a phrase in security expert community that every computer is possible to break in, if its console is accessible. Electromagnetic emission is another key factor in securing physically a computer from being monitored from distance with specific antennas. Here the tempest room is applied based on the well-known Faraday-cage idea.

The password protection is the first step in computer access management. As the next step the private files have to be protected from other users and even from the supervisor, who can access private files on a server even if the owner have not logged in. Therefore in high security systems and today for PC users too, there are different CFS solutions (Crypto File System) available where the content of the hard disk or home directory is encrypted. After the logon procedure the user has to enter another password which will deactivate the lock algorithm of HD or home directory.

The passwords gives the answer to the "what you know?" question, but the sure identification is ensured with biometry, giving the answer to the "who you are?" question.

Today beside the fingerprint recognition systems the iris or retina based authentication systems are available also for PC users. The patterns are recorded on a smart card not accessible for the outside word. The pattern is coded with a one way function and only the card's operating system is able to match the given sample with the stored one. In case of matching the user is the owner, and it is allowed to start the required functionality or application.

In case of logging in to a computer system it is recommended to use an interactive memory-resident virus scanner (under some operating systems, like Windows) to protect our resources from viruses coming from Internet or intranet or from a colleague's floppy.

And the computer still is not in safe, because always the virus is the first, and after that comes the virus killer. The relevance of a new security hole is the first always and the patch for that security hole will came after a few days. Therefore a well constructed and written security policy have to be developed and maintained which describes not only the prevention, but also the data management of sensitive data (hot and cold background, archived database, etc.) and the disaster plan if the unwanted trouble became a reality.

During the work it is worth to be the events logged. After the crash the log files are useful for the future in order to find the security holes on the system, and in detailing some weak parts of the security policy. Logging recognized events save us from familiar problems, but there must be added heuristic logging too. The administrator can listen to the differences or to divergent events, or both of the events can be logged just when they happen simultaneously. The reaction is pre-programmed and the result of the reaction could be an alert or suspension of the respective account. The events must be well defined to avoid the high number of false alarms.

The art of reaction is the background alert, when the attacker can go on with working, just all of his operations are sniffed and tracked to get more information about the knowledge and aim of the attacker. This requires a professional knowledge, but in most cases this is the only way to know more about the attack than just the fact that it happened.

The best is when the prevention techniques are applied on different levels. Routers help us in selecting the information by its source of IP address. The firewall filters different services and protocols. The server filters the break-in trials, and user operations from outside and inside too. Local programs are scanning for viruses. There are solutions to filter viruses on the firewall too, but these elements of security services are occasional solutions in specific systems. The required security technologies have to be applied on each level to result the whole system consistent evenly (also called robust).

The training of users has a significant role in that domain. Users will never apply a security technology if there is any less complicated procedure (which is insecure!) than the secure solution. It must be explained for the users, what they can lost without security.

4.5 Future solutions

Databases play a key role in the world of information. The databases used daily are constructed to store their clients/members under a unique key. Social insurance numbers, passport or other official personal identification numbers can be brought on a common base, this is the digital signature. The digital signature can be used from PC access codes through signing e-mails, till the bank transaction processing with wireless phone or other devices like Personal ATM. This is the very close future. To be portable, the data for digital signature will be put on a smart card, which is appropriate to be inserted and used in a mobile phone, in a

Table 2.
Place and role of security protocols, methods, technologies and tools in the network

Layer Number	Layers of the OSI reference model	TCP/IP Protocols	SECURITY PROTOCOLS	Security method, technology, tool, etc.	What type of security activities are done on the level
7.	Application	FTP, SMTP, TELNET, SNMP, NFS, Xwindows, NNTP, IRC, HTTP, WAP	S-HTTP, SET / S/MIME, PEM, PGP, MOSS / SMTP	-Firewall (typical) - application level to check digital signatures - authentication protocols, - encryption protocols, - Virus scanner (memory resident)	- identification of the user, - authenticate messages - encryption of messages - virus scanning in active mode.
6.	Presentation	ASCII, EBCDIC, ASN1, XDR	SSL, SSH	Firewall - max. filter of images, like Netscape "show images" checkbox filtered by the HTTP server!	filter, or hide of information (e.g. at password typing)
5.	Session	RPC		Firewall - filtering the query/request	filter of disallowed requests/services
4.	Transport	TCP, UDP	TLS (Transport Layer Security Protocol), WAP/WTLS	Firewall - coded/encrypted transportation Screening router (filtering)	digitally coded/encrypted transport after authentication of the next transmission party
3.	Network	IP	IPv6	Screening router (filtering) - Firewall - NW level, mainly in router to filter false/untrusted/not authentic IP addresses	encryption and DNS filter
2.	Data link	X.25, SLIP, PPP, Frame Relay	Electromagnetic Emission standard (89/336/EEC - European Economical Community guideline)	Screening router (filtering)	Link encryption
1.	Physical	LAN, ARPANET		Screening router (filtering) e.g., without valid Ethernet card address declined access, or by an address in a specified domain: limited access	physical security methods and tools, mainly not information techniques!

laptop or PC. The operation of smart card is based on common standards that are referring both for physical access and for the mode of communication.

There is no 100% secure solution, just the striving process to reach the conscious risk management. The continuous control of the systems (both bigger and PC-based ones) is the best to maintain the required security level. The virus developers, the crackers and hackers and all other members of the "dark" side of the information society are working continuously but the "good guys" are active as well. There are mailing lists and WEB sites from where even the "simple" users can collect information, tips how to avoid e.g. being infected by viruses, or hacked. Because of the fast changing methods, techniques of the computer field this is, and will be also in the future one way to distribute important information on security.

To check the system against known problems can be useful surfing in the archives of "Bugtraq" security mailing list [17]. There are different security lists, tools and archives for respective operating systems for detailed information, but the Computer Emergency Response Team [18] and the System Administration, Networking and Security [19] sites are good starting points. For some interesting undocumented but discovered features of operating systems worth to check the Eastern Eggs archive [20].

As it was declared in the CSI/FBI survey, more than 90% of companies use access control, physical security, firewalls and anti-virus software, while only 61% of them are using encrypted files and reusable passwords. Less than a half of them are using encrypted login, one third uses digital ID's, and only 9% uses biometrics identification. It is important to mention, that on the market there are various solutions in these fields, their application is just a question which one to be used.

A consistent security starts with physical security and access control based on complementary use of tokens, digital ID's (what you have?) and passwords or PIN's (what you know?) supported by biometrics in a high security system. Maybe in the near future biometrics became a basic tool in any access system from Cyberspace through workplace till private life. The second step is the consistency of information flow, which could be resolved by different cryptology systems. In this way the information flow can't be read or modified by others between the end points. The third step will be when these systems will be applicable as wide as possible without space, time and system limitation.

With the broad spread of wireless communication combined with the technological convergence there is a strong need for a similar flexible and portable security means. The solution could be the portable and easy to use tokens, called smart cards. These small computers give the possibility of flexible and secure programming, mobile crypto-key management, and a good base for biometric recognition systems. When talking about biometry as part of security systems, the aspects of the disabled (elderly) people as potential users have to be taken into consideration. Their requirements have to be fulfilled as well, so this is an important factor in the design phase of future complex security systems.

During the 25 years of smart card life there were always platform dependency problems. Due to Java Card [21] these cards could be programmed in Java, and due to the eXtended Markup Language standard for smart cards [22] a very high compatibility can be reached between old and new cards and readers too. Taking into consideration the Wireless Application Protocol standard (WAP) [23], and its Smart Card Expert Group's work, it is not so far when mobile phone can became a very intelligent and widely used smart card reader/writer where phoning possibility will play secondary role. In the field of enterprise communication smart cards can be applied too in the very close future [24].

Table 3.
Place and role of security methods, technologies and tools on the computer and its environment

Levels	Function of the Level	Example	Security method, technology, tool, etc.	What type of security activities are done on the level
User interfaces	To help the user to use the computer HW and SW possibilities (USEABILITY)	Xwindow, pop-up menus, sensitive surfaces (e.g.. HTML, Windows help)	password protected screen saver	secure access to the information displayed on the screen filtered access to sensitive data (Excel cell hiding)
Applications	To help the user in solving the given tasks through different program packages (FUNCTIONALITY)	Word processors, image editors, Excel, MatLab, etc.	Cryptography SW, password protected appearance of programs or information	Secure use of applications and applications related files
Basic SW and communication	To manage data, applications and communication tasks.	Networking SW, WWW browsers, file managers, archivation programs	Password protected archives, and file systems,	Secure use of SW and the SW related files
Operation system	To solve OS dependent tasks by a specific HW based, more specific SW.	DOS, Windows versions, UNIX versions, VMS, Mainframe, Macintosh	user authorization file (SYSSUAF.DAT on VMS. /etc/passwd on UNIX - /etc/shadow on secure UNIX...) and ACL files (Access Control List) and different rights for different groups/entities.	Secure use of OS and OS related programs, and files.
Hardware	To help in extending computer's capabilities: printing, scanning, presenting on monitor or by a miller machine in different materials, store data, etc.	printer, monitor, mouse, scanner, plotter	Physical security, tokens, smart cards, HW locks	To guarantee the secure physical access to the computer itself.
Environment	To extend the computer's capacity in connection with the outside world: phone-modem, ATM-line, ISDN-line, Internet, telescope or other tool's control, etc.	Ethernet card, modem, camera, fax, microphone, head-set	Security policy, environment security, security and disaster plan, education...	To guarantee the secure physical access to the computer environment

In future communication and computer systems become more integrated, more distributed and more personalized. As the dynamism is a main factor of all economic systems also in geographic meaning, the personal mobility and through this the mobile systems will take a big sector. The complex, personalized security will have an increasingly important role in all application fields, including the special needs of distributed production systems.

5 CONCLUSIONS

The chapter has shortly introduced the main characteristics of VEs, their architectures and communication problems and requirements. The general basic networking communication architectures (in the field of manufacturing as well) also have been presented to help navigating the readers in the very broad space of the security techniques.

The security techniques and technologies have become high priority as networking of different kinds are approaching to each other, sometimes are integrated. This gradual integration is called convergence, and the technology convergence is based on the common application of digital technologies to systems and networks associated with the delivery of services. Technological convergence is under way, and continuing advances in technology will further consolidate the process along the different elements of the value chain.

Today the global nature of communications platforms, particularly, the Internet, are providing a key which opens the door to the further integration of the World economy. The distributed production systems with different sizes will play a definite role, but originating from their openness and flexibility their information systems will be a security risk. They will need complex, flexible security systems that are user friendly and platform independent at the same time. The developments of hardware and software elements of such systems are going on and the potential users have to get acquainted with them. The main goal of this chapter was to help this process.

ACKNOWLEDGEMENTS

Part of the work included in this chapter has been done with the support of the OTKA (Hungarian Scientific Research Found) project with the title "The Theoretical Elaboration and Prototype Implementation of a General Reference Architecture for Smart Cards (GRASC)" (Grant No.: T 030 277).

REFERENCES

[1] Secure Electronic Transactions standard - http://www.setco.org
[2] A. Koestler, *The Ghost in The Machine*, Arkana Books, London, 1989.
[3] Mezgár, I.. Communication Infrastructures for Virtual Enterprises, position paper at the panel session on "Virtual Enterprising - the way to Global Manufacturing," in the Proc. of the the IFIP World Congress, Telecooperation, 31 Aug.- 4 Sept. 1998, Vienna/Austria and Budapest/Hungary, Eds. R. Traunmuller and E. Csuhaj-Varju, pp 432-434.
[4] Mezgár, I., Kovács, G. L., PLENT: A European Project On SME Co-operation, In Human Systems Management, Volume 18., No. 3-4, 1999, IOS Press, Amsterdam, pp193-201.
[5] The NIIIP reference architecture, final document, http://www.niiip.org/public-forum/index-ref-arch.html

[6] L.M. Camarinha-Matos, H. Afsarmanesh, C. Garita, C. Lima, Towards an architecture for virtual enterprises, Keynote paper, Proc. 2nd World, Congress on Intelligent Manufacturing Processes and Systems, Springer, Budapest, Hungary, June 1997, pp. 531-541. - The PRODNET II project, http://www.uninova.pt/~prodnet/
[7] Tanenbaum, A.S., Computer Networks, Third Edition, Prentice-Hall, 1996.
[8] PROFIBUS Technical Overview, http://www.profibus.com
[9] The Fieldbus technology, http://www.fieldbus.org/ftwhatpg.htm
[10] "Issues and Trends: 1999 CSI/FBI Computer Crime and Security Survey," Final report of Computer Security Institute, San Francisco, http://www.gocsi.com/ prelea990301.htm,
[11] Trusted computer system evaluation criteria, Orange book, DoD 5200.28-STD, Department of Defense, December 26, l985, Revision: 1.1 Date: 95/07/14.
[12] ISO/IEC 10181-1:1996 Information technology -- Open Systems Interconnection -- Security frameworks for open systems: Overview.
[13] ISO/IEC 15408, 1999, Evaluation Criteria for Information Technology Security.
[14] Osório, A. L., PRODNET: Safe communications for VE, PRO-VE'99, Prodnet Working Conference On Infrastructures For Industrial Virtual Enterprises, Porto, Portugal 27-28 October 1999,
[15] Hare, C., Siyan, K., Internet Firewalls and Network Security, Second Edition, New Riders Publishing, Indianapolis, 1996.
[16] Timestamp-ticket based secure access algorithm on the Internet and Web - http://nii.isi.edu/info/kerberos and ftp://ftp.pdc.kth.se/pub/krb for the source code
[17] The security mailing list - http://www.securityfocus.com/bugtraq/archive
[18] Computer Emergency Response Team - http://www.cert.org
[19] System Administration, Networking, and Security - http://www.sans.org
[20] The Easter Egg Archive can be reached at: http://www.eeggs.com,
[21] Java Card Forum - http://www.javacardforum.org,
[22] smartX white paper 1999, http://www.smartxml.com
[23] Wireless Application Protocol - WAP white paper, 1999, http://www.wapforum.org,
[24] Vaeth, S., The next Smart Card Frontier – Enterprise security, Card Forum International, May-June, 1999, pp 22-24.

Part IV

SUPPLY CHAIN MANAGEMENT IN AGILE MANUFACTURING

Agile supply chain management

J. Sarkis[a] and S. Talluri[b]

[a]Graduate School of Management, Clark University, 950 Main St., Worcester, Massachusetts 01610, USA

[b]Department of Information Systems and Sciences, H323D, S. J. Silberman College of Business, Fairleigh Dickinson University, 1000 River Road, Teaneck, New Jersey 07666, USA

The growth of inter-organizational theory and relationships has caused a revolution in the practices and management of supply chains within organizations. In this chapter we provide a review of the practices and issues facing organizations that wish to compete within an agile supply chain context. A number of issues are presented with a specific focus on selection and general practices in agile supply chains.

1. INTRODUCTION

Organizational relationships within the agile environment are expected to become more complex. This complexity is due to the greater need for rapid integration among members of agile relationships. Yet, this complexity is not meant to preclude rapid, timely, and opportunistic response to unexpected change. The complexity arises from the variety of relationships and partners that will need to be managed. No longer is the worry on just managing a one-to-one relationship among a variety of organizations, but how to manage a web of partners integrated as a single organization, with the ultimate goal of a globally optimal relationship meant to address the ultimate customer's needs.

Given this general environmental context, this chapter investigates the many evolving issues related to developing and maintaining an effective and efficient, as well as agile, supply chain. The chapter will begin with a discussion of various characteristics of the supply chain. Relationships and implications of these characteristics with respect to agility are then discussed. An important aspect of strategic supply chain management, that of partner selection, with a number of issues and models, is then presented. Future directions and evolving inter-organizational relationships conclude the chapter.

2. SUPPLY CHAIN MANAGEMENT

Supply chain management is the management of activities and processes associated with the flow and transformation of goods from raw materials through the end user and to disposal or back into the system. Materials also include related information flows.

The literature has provided a number of classifications used to show the evolving relationship and elements of the procurement/supply management and logistics functions [1-5]. Within these classifications the procurement function acquires a more strategic role within the business organization. Logistics, production and operations management, and marketing field evolution have profoundly influenced the procurement function and supply chain management. What have traditionally been operational functions (purchasing and distribution) that had a short term and structured vision, have grown to an integral strategic function, requiring long term planning with broad implications across the organization.

The growth of the procurement and logistics functions, especially in complex product and component environments, not only influences the relationships within the enterprise, but also relationships external to the enterprise. As well, the traditional concept of just managing the direct relationship between a supplier and customer, has grown to include the management of relationships that includes the chain of suppliers (i.e. the suppliers' suppliers) as well as the chain of customers. This growing set of relationships is part of the integrated linkage that is needed for most complex product organizations to maintain a competitive advantage. The complex relationships has led to the concept of supply chains and supply chain management where an interdisciplinary focus incorporating a number of functions across the organization is critical to the supply chain management.

Managerial and organizational changes that can be traced to TQM and JIT philosophies have also influenced the purchasing and procurement function. The development and adoption of technology such as electronic data interchange (EDI), enterprise resources planning (ERP), distribution requirements planning (DRP), computer integrated manufacturing (CIM) systems, and electronic commerce, to name just a few, have also provided a foundation for evolution to supply chain management. Thus, technological, organizational, and operational factors have all come together to influence and form the area of supply chain management.

Customer-supplier (buyer-seller) alliances have also begun to evolve. Where, Teece [6] defines the customer-supplier alliance as a "constellation of agreements" typified by a commitment between two or more partner firms to reach a common goal that involves pooling of resources and activities. This might include: exclusive purchase agreements, exclusionary marketing/manufacturing rights, technology swaps and possible joint R&D co-development agreements. This evolution is further discussed in our final section.

3. IMPLICATIONS OF SUPPLY CHAIN MANAGEMENT IN AGILE MANUFACTURING

The Agile Enterprise, is one of the latest organizational paradigms characterized by organizations whose ability to respond to frequent and unpredictable change, is being promoted as the successful company of the future. As competitive forces in the manufacturing environment continue to intensify, agility is touted as a condition for success and even survival. The ability to respond rapidly to changing market opportunities by utilizing agile business practices is a key attribute of an agile enterprise.

Supply chain management evolution has provided a number of practices that directly relate to improving agility within and between organizations. Each of the major performance factors of agility, robustness, time, cost, flexibility, dependability, are all affected by the management of the supply chain. An agile supply chain is a necessary prerequisite for the formation of virtual enterprises. Virtual enterprises are at the end of the spectrum of the customer-supplier relationship that range from commodity, to partnerships, to alliances and ultimately to virtual enterprises [7].

We shall look at four areas (supplier relationships, customer relationships, internal organizational processes, and system) of the supply chain to present these emerging and "best" practices. Even though there is some commonality among the practices in each of the areas, their uniqueness arises from their application along different links within the supply chain. Practices for four areas, supplier relationships, customer relationships, organizational processes, and the overall system, are discussed.

3.1. Supplier Relationships

The literature acknowledges the need for a good customer-supplier relationship for distribution and purchasing functions to be cost-effective in an organization. Firms have started discovering that close partnership relationships with important suppliers can produce agile benefits. A good supplier relationship may be based on the following factors: low price, excellent quality, on-time delivery, reactive and helpful, technically innovative, and trust. Much of these successes can be seen on an international scale, especially when observing lean practices of partnerships [8,9,10].

The following have been defined as exemplary practices for strategic supplier relationships:

1. Development of long-term, performance-oriented supplier partnerships
2. Continuous quality improvement and joint learning by both the customer and its supplier base.
3. Focus on total cost of ownership, not just on price.
4. Companies are taking a boundaryless view of their participation in the value chain.
5. Long-term contracts
6. Multi-level relationship across the organization including bi-company project teams;
7. Critical buying decision based on value;
8. Early involvement in marketing, design and product development cycle;
9. Exchange of information including not only information on work-in-progress but information on basic costs and insight into long-term strategy;
10. Integrated quality control;
11. Mutual support and joint problem solving;
12. Joint teams sharing information and expertise and sharing in the benefits;
13. A genuine insight into the buy decision and market forces both up and down the value chain; and
14. Two or three suppliers at most, with single sourcing agreements common, thus enabling the purchasing, engineering, production, and quality personnel to work more closely with the surviving few.

3.2. Customer Relationships

Customer relationship issues focus on linking the customer to the supply chain. Much of the traditional functional responsibility for customer strategies has been within the domain of the marketing function. In agile organizations this relationship needs to be extended to include engineering, logistics, manufacturing and purchasing, among other functions.

Linking the customer to the supply chain will require data and communication integration. A customer service strategy can be established with a simple methodology that includes the following major steps: 1) external audit; 2) internal audit; 3) evaluation of customer perceptions; and 4) identification of opportunities [11].

Similar to the focus on a smaller, closer set of suppliers and building stronger relationships in that manner, agile suppliers are also reducing their customer base to get rid of "bad" customers, although this may not always be feasible. The key to attracting good suppliers is to be a good customer. Being a good customer reduces the frustration level and improves product quality while reducing purchasing costs. A good customer-supplier relationship: 1) recognizes that a partnership is a two-way street; 2) rewards the best suppliers, and 3) encourages supplier involvement in product development [12]. A concept such as JIT II, developed by BOSE Corporation, where the organizational buyer is a customer employee, is one way of integrating organizations more fully [13].

3.3. Organizational Process Issues

Some internal management issues that should be considered are the roles of purchasing as a strategic function and as a support mechanism for other strategic areas. For this to occur purchasing has to go through an evolution to include the development of new skills (education), having top management commitment, and reductions of organizational barriers.

Supplier development programs are an integral aspect of supply chain management practices. A supplier development program can be defined as any systematic organizational effort to create and maintain a network of suppliers. Before beginning a formal supplier development program, a purchasing organization must review those aspects of its own operations that can adversely affect supplier performance, such as purchase specifications, communications, training, and organizational roles.

Supply chain management personnel will necessarily have to be cross-functional if the strategic aspects (which cross functional boundaries) occur. A multi-level relationship across the organization including bi-company project teams needs to exist, not just at one level, which is usually operationally oriented.

Supplier selection is an activity requiring a significant amount of managerial resources. Supplier selection is discussed later when we review partnership formation and supplier selection. Supplier or vendor selection identifies and selects the supplier of materials and products. Deming [14] lists fourteen requirements for a business to remain competitive. Requirement number four states that there should be an end the practice of awarding business on the basis of price tag. Instead, minimize total cost and move toward a single supplier for any one item, on a long-term relationship of loyalty and trust. The two primary motivating factors behind the fourth principle, quality improvement and cost reduction, must be pursued simultaneously. A team approach is recommended in selecting suppliers and establishing long-term relationships. Senior managers should lead the organization in improving quality, productivity, and competitive position.

Related to supplier selection is the organizational process of supplier certification. Supplier certification programs have typically been limited to one level of supplier, evaluating suppliers' suppliers have not been common in the literature or practice. This may be difficult to do since this may infringe on the autonomy that most enterprises enjoy in selecting their suppliers. One method of attaining a consistency in supplier certification is through the ISO 9000 (or similar) type standards. Where quality associated with suppliers are measured by third party auditors for ISO 9000 certification. A standardized certification program has greatly enabled development of agile supply chains.

3.4. Supply Chain System

Central to management of the supply chain system is continuous improvement, a major tenets of TQM. Continuous improvement is essentially the feedback loop of the supply chain management system, where feedback among and between organizations means strengthening the supply chain links. The continuous improvement link allows for full system analysis. A difficulty with a complete supply chain improvement is that optimizing relationships link by link does not guarantee an optimal improvement among the full supply chain. The needs of organizations will vary, as are the needs among various relationships. In some cases an almost complete supply chain improvement may be possible such as in governmental contract and subcontract relationships or when one of the major manufacturers is very powerful as in large automobile manufacturers, but this is not the case for many supply chain relationships.

In summary, continuous improvement:
1) Necessitates a multidimensional, multiorganizational effort simultaneously focused on quality, productivity, innovation, inventory, etc.. In other words, continuous improvement is an integrated system that requires a number of performance measurement improvements be combined into a single strategy.
2) Advances in stages where dedication, is followed by sustenance, and continuity.
3) Requires a model to guide implementation.
4). Should be custom-fit to the circumstances within and between organizations.
5) Requires change management including group efforts, facilitating leadership, and a foundation on appropriate performance measurement [5].

Electronic Commerce and EDI hold promise of increased productivity, less waste for companies and as part of the linkage for rapid feedback for an interorganizational continuous improvement framework. Through these mechanisms, companies work with customers and suppliers to reap the efficiencies that they try to attain internally across departments or divisions. EDI is a new way of doing business as existing customer and supplier relationships move into cooperatively developed trading partner roles. This understanding must be spread broadly throughout a company implementing these systems.

3.5. Supply Chain Agile Enablers

The review of various emerging and best practices in supply chain management targets a number of issues. The Agility Forum has also provided a listing and discussion of four major categories of agility enablers for customer-supplier relationships. These categories are shown in Table 1 and include internal company, industrial system, company technology, and natural resources. Clearly, this is a different set of dimensions than those presented here, but a significant amount of overlap exists with some complementary categories covered. Overall, it shows that the issues related to managing an agile supply chain can be quite complex and require a broad set of practices, policies and tools.

4. STRATEGIC ALLIANCES AND PARTNER SELECTION

Strategic partnerships and alliances exist in a number of industries and are on the rise. A "partnership" type relationship can be described as a voluntary agreement that commits both the supplier and customer to mutual openness, productivity, and quality in the service of the customer's customer. It is an agreement that involves sharing proprietary information, risks, and rewards.

Although choosing the right supplier for a given job is the most fundamental and important decision a buyer makes, it may also be one of the most difficult ones, it becomes more complex as the selection evolves to the level of partnerships. Supplier evaluation and motivation plans typically involve the utilization of one or more of the following approaches: 1) Formal quantitative rating systems; 2) In-depth performance reviews, and 3) On-going communications and development of business partnerships. The selection process and its maintenance needs to incorporate these three important dimensions.

To be able to complete a comprehensive evaluation of suppliers a number of criteria can be used. For example, the supplier could be screened technically on a number of variables, some of these have included: 1) Emphasis on quality at the source; 2) Design competency; 3) Process capability; 4) Declining nonconformities; 5) Declining WIP, Lead-time, space, flow distance; 6) Operators cross-trained, doing preventive maintenance; 7) Operators able to present SPC and quick setup; 8) Operators able to chart problems and process issues; 9)

Hours of operator training in TQC/JIT; 10) Concurrent design; 11) Equipment / labor flexibility; 12) Dedicated capacity; 13) Production and process innovation.

Table 1:
Categories and Agility Enablers for Agile Customer-Supplier Relationships (Adapted from [7]).

Category	Agility Enabling Subsystems
Internal Company	Continuous Learning
	Customer Interactive Systems
	Empowered Individual in Teams
	Organizational Practices
	Performance and Accounting Metrics
	Benchmarking
Industrial System	Enterprise Integration
	Evolving Standards
	Factory America Network
	Global Multi-Venturing
	National Infrastructure Support
	Pre-Qualified Partnering
	Rapid Cooperation Mechanisms
	Streamlined Legal Role
	Technology Adaption and Transfer
	Wide Area Broadband Network
Company Technology	Distributed Databases
	Groupware
	Human-Technology Interface
	Intelligent Control
	Intelligent Sensors
	Integration Methodology
	Knowledge-Based Systems
	Modular Reconfigurable Process Hardware
	Representation Methods
	Simulation and Modeling
	Software Prototyping and Productivity
Natural Resources	Energy Conservation
	Waste Management and Elimination
	Zero-Accident Methods

Buyers must analyze and document the significance of several of the above mentioned factors, converting instinctive qualitative indicators to concise empirical measures. Models and some elements to address these issues are presented next.

4.1. Strategic Partnership/Alliance Formation Models

There is a myriad of models by which strategic partnering type relationships between buyers and their key suppliers begin, are developed, and are maintained. Ellram [15], suggests a five stage model for the development and implementation of "purchasing partnerships". The five phases are: 1) Establish strategic need, form team, confirm top management support; 2) Identify potential partners; 3) Screen and select; 4) Establish relationship:- provide high attention level, give prompt feedback; and 5) Evaluate relationship: continue, expand or reduce.

As part of an overall strategy for agility and building virtual relationships, firms are gaining experience through formation of strategic alliances between customers and suppliers, within R&D consortia, or as equal partners. These relationships are an effective way for firms to develop new technologies and products, procure critical resources, investigate new markets and complement core competencies and incompetences. Relationship theory has been considered as a way to filter the numerous possibilities of potential partners down to a manageable number of potential partners to assess. Partnering, strategic alliances, and virtual enterprises form a spectrum of relationships that may exist among enterprises, with the relationships of the inter-enterprise business processes becoming more unified along the spectrum.

Currently, the literature and case studies have focused on partnerships among suppliers and customers, with the literature focusing on strategic alliances lagging far behind. The qualitative literature has surpassed the quantitative decision modeling for the partnership selection problem. One of the more popular research areas (quantitatively) has been the vendor selection problem. A number of techniques have been used to solve this problem including matrix or weight approaches, mathematical programming, fuzzy set models, and the analytical hierarchy process (AHP) (see [16] for a review). Yet a strategic alliance formation involves more then consideration of operational characteristics. A holistic systemic evaluation model that considers strategic and operational factors, is needed for more strategically oriented relationships. We look at these factors in an AHP-based model.

The literature on forming strategic alliance has been qualitative with few decision tools and a focus on methodological approaches. For example, Lorange, et al. [17] suggest a two-phase formation process. The initial analytical phase deals with assessing the match, whereas the more intensive phase addresses questioning partners on market potential, worst-case scenarios, and competitive advantages of the alliance. Another methodology is a four phase approach proposed by [18]: 1) Strategic Decision - includes situation analysis, identification of strategic cooperation potential, and evaluation of shareholder value potential, 2) Configuration - decisions regarding the field of cooperation and intensity of cooperation 3) Partner Selection - fundamental, strategic, and cultural fit, and 4) Management - contract negotiations, coordination interface, learning, adaptation, and review.

Speckman [19] has determined four sequential steps in the alliance planning process: *Strategy Development, Partner Assessment, Contract Negotiations and Control/Implementation.* Within his research of the strategic alliance process, a number of phases were identified. These phases included anticipation, engagement, valuation, coordination, investment, stabilization, and decision.

The initial development of a strategy has certain complex issues, which must be addressed. These issues include the identification of major strategic challenges, evaluation of business risks, and consideration of resource strategies in terms of production, technology and people. Furthermore, in assessing a partner numerous issues are involved, which range from understanding the partner's management style and organization to creative strategies in how to merge two different corporate cultures. A business case (justification) analysis is a useful tool in helping to evaluate the particular strategy a company should embark upon as well as determining a strategic partnership. Part of business justification is to evaluate the partners on multiple criteria, financial, quantitative (tangible), and qualitative (intangible) [20].

4.2. Core Competencies Based Supplier Selection

An organization's core competencies can be focused on process, function, or even technology. There is an implication of vendor certification and supplier selection procedures to help in the determination and selection of a set of preferred suppliers and vendors. The models recommend a wider scope of responsibility of the purchasing function of the organization. This points towards a larger strategic influence and role of supplier management focus. The use of group or team management has been recommended for good practices among vendors and suppliers. There is an emphasis on both cost and quality of the deliverable goods. Instead of looking for low cost items, the attempt is to look for high "value" goods. To help in the core competencies based selection and partnership formation process, an overview of factors that can be used to evaluate the various organizations is presented in Table 2 (see [15,21-23] for sources and explanations of these various factors). This listing can be used to form a hierarchical decision framework using AHP.

4.3. A Simplified AHP Example

As we have mentioned, a tool to help in evaluating qualitative and quantitative factors for strategic supplier selection is AHP. We provide a brief example using the criteria shown in Table 2. We have decided to use two levels of criteria and an alternative set of three suppliers. Including the object of supplier selection, the decision hierarchy has four levels (see Figure 1). The steps in the process are to 1) develop a decision hierarchy; 2) evaluate the various factors using pairwise comparisons; 3) aggregate the results; and 4) analyze the results.

The first step of the process is already complete. We have selected a subset of factors and decided to evaluate them at one additional level. The level 1 factors include Financial, Technology, Quality, and Flexibility. The level 2 factors are derived from Table 2 and are sub-factors of the level 1 factors.

The second step in the process is to calculate the relative importance weights. This step is first completed by compiling a pairwise comparison matrix. In this pairwise comparison matrix the decision maker is asked to various factors and their relative importance to their controlling factor. For example, the controlling factor for the major criteria is the "select supplier" goal. The first level criteria in this case are compared to each other to determine the relative priorities. Table 3 provides the pairwise comparison matrix for this relationship. To complete this table a series of questions is asked for each relationship. The question is in the form "How much more important are Financial Factors than Technology Factors". The response can be anywhere from extremely less important (typically scored as 1/9) to extremely more important (typically scored as 9). There are many stages in between and the scaling is based on this 1-9 range in either direction (a score of 1 is an indifference in relative importance). In the example in Table 3 we see that Financial factors are slightly more important than Technological factors with a score of 3 recorded. Once the matrix is complete the relative importance weights are determined (these weights appear in the last column, which is a vector of weights). There are a number of algorithms and heuristics to solve for the relative importance weights (see [16]). The execution of this problem was completed on the Internet on Web HIPRE3+ [24] an interactive decision analysis tool that has a number of scoring approaches and is located at (http://www.hipre.hut.fi/). The results in Table 3 show that Quality factors are by far the most important (with a score of .542) when selecting suppliers.

Table 2
Hierarchical list of factors for supplier/partner evaluation

1. Financial and Economic Issues
   1.1 Profitability of Supplier
   1.2 Financial Records Disclosure
   1.3 Performance Awards
   1.4 Financial Stability
   1.5 Cost Analysis
      1.5.1 Low initial price
      1.5.2 Compliance with Cost Analysis system
      1.5.3 Cost Reduction Activities
      1.5.4. Compliance with Sectoral Price Behavior
2. Organizational Culture and Strategy Issues
   2.1 Feeling of Trust
   2.2 Management Attitude/Outlook for the Future
   2.3 Strategic Fit
   2.4 Top Management Compatibility
   2.5 Compatibility among levels and functions
   2.6 Suppliers organizational structure and personnel
3. Technology Issues
   3.1 Technological compatibility
   3.2 Assessment of future manufacturing capabilities
   3.3 Suppliers speed in development
   3.4 Suppliers design capability
   3.5 Technical capability
   3.6 Current manufacturing facilities/capabilities
      3.6.1 Production Planning System
      3.6.2 Maintenance Activities
      3.6.3 Plant Layout and Material Handling
      3.6.4 Transportation, Storage and Packaging

4. Consistency and Quality
   4.1 Conformance Quality
      4.1.1 Rejection Rate
         4.1.1.1 Rejection in Incoming Material
         4.1.1.2 Rejection in Production Line
         4.1.1.3 Rejection from Customer
      4.1.2 Lot Certification
      4.1.3 Sorting Effort
      4.1.4 Defective Acceptance
   4.2 Consistent Delivery
      4.2.1 Compliance with Quantity
      4.2.2 Compliance with Due Date
      4.2.3 Compliance with Packaging Standards
   4.3 Quality Philosophy
      4.3.1 Management Commitment
      4.3.2 Quality Planning
      4.3.3 Continuous improvement
      4.3.4 Quality Assurance in Supply
      4.3.5 Statistical Applications
      4.3.6 Inspection and Experimentation
   4.4 Prompt Response
5. Relationship
   5.1 Long-term Relationship
   5.2 Relationship Closeness
   5.3 Communication Openness
   5.4 Reputation for Integrity
6. Flexibility
   6.1 Product volume changes
   6.2 Short Set-up Time
   6.3 Short Delivery Time
   6.4 Conflict Resolution
7. Service
   7.1 After-sales Support
   7.2 Sales rep's competence
8. Other Factors
   8.1 Safety Record of the supplier
   8.2 Business references
   8.3 Suppliers customer base.
   8.4 Product Liability

The next step is to aggregate the weights to determine a composite set of relative importance rankings for the suppliers. This result is determined by taking the weighted sum of the factors and how well each supplier performs on them. The aggregation equation is shown in (1):

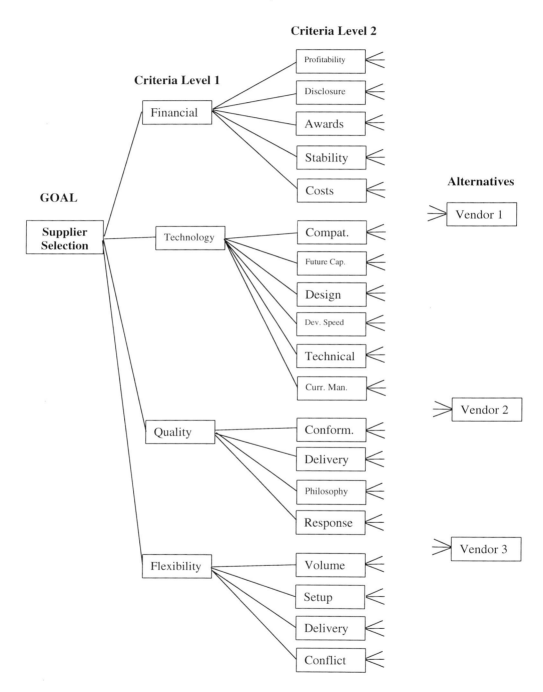

Figure 1. AHP decision hierarchy for supplier selection.

Table 3. Pairwise comparison matrix and relative importance weight results for criteria 1 level factors.

Supp. Sel.	Finan.	Tech.	Qual.	Flexibility	w
Finan.	1	3	1/2	3	0.271
Tech.	1/3	1	1/6	1/2	0.027
Qual.	2	6	1	6	0.542
Flexibility	1/3	2	1/6	1	0.109

$$S_i = \sum_k w_{kdec} \left(\sum_{j_k} w_{ij_k} w_{j_k k} \right) \qquad (1)$$

where: S_i is the relative composite score for supplier i
w_{kdec} is the relative importance weight for the first level criteria k on the decision
$w_{j_k k}$ is the relative importance weight of second level criteria j of first level criteria k on criteria k.
w_{ij_k} is the relative importance weight of supplier i on second level criteria j of first level criteria k.

Web Hipre3+ was used to provide us with the final results, shown in Figure 2. These results show that Supplier 2 with a score of .361 is most preferable, based on decision maker preferences. The comparison can also be completed on each of the three first level criteria. Which show that Supplier 1 is perceived to be best on Flexibility and Financial factors and Supplier 3 is best on Quality and Technology. Surprisingly, Supplier 2 did not do best in any among any of the first level criteria factors, but was still the overall most preferable supplier.

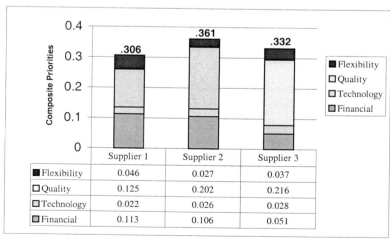

Figure 2: Composite relative ranking scores for Suppliers.

5. NETWORK ORGANIZATIONS

In the late 1950s and early 1960s, management theorists proposed decentralization of large companies in order to better manage the workplace. Subsequently, in the 1970s, more and more firms disaggregated their operations for reasons of economics and efficiency. This finally led to the development of the new organization forms and supply chain networks (SCN) of today, such as network organizations, virtual corporations and value-added partnerships. A brief background and future directions of development of these networks is now presented.

Initially, Miles and Snow [25] introduced the concept of external groups, which they termed as dynamic networks. According to them, a dynamic network is a combination of independent business processes with each contributing what it does best to the network. This concept of network organizations gained importance in latter part of 1980s. Since then, conceptual and empirical research in the area of network organizations has received increased attention. Miles [26] studied the effect of networks on labor relations and human resources development. Lawless and Moore [27] studied the application of dynamic networks in private and public industries. Although these types of networks are more common in private industries, these studies provided evidence of such network forms in public services such as fire fighting and health. Kensinger and Martin [28] suggested financing policies, procedures, and strategies in network organizations. Managerial processes for designing, operating, and care-taking a network were suggested by Snow et al. [29] and Snow and Thomas [30].

Snow et al. [29] illustrated three types of network organizations: internal, stable, and dynamic. In an internal network, firms own most of their assets in the business, and they do not become involved in outsourcing. In this type of network, various divisions are formed within the organization, which specialize in a particular component of business. For example, in the 1980s, General Motors significantly reduced its component divisions with each division retaining expertise in a particular automotive system or subassembly.

A stable network engages in a moderate level of outsourcing. Usually, in this type of network, a set of vendors supports the lead firm. These vendors either provide inputs to the firm or distribute its outputs. For example, BMW can be considered to be operating with a stable network. It is estimated that 55 % to 75 % of total production costs at BMW are incurred from outsourcing. Certain divisions at BMW called the research groups are constantly working towards identifying best vendors to link with.

Dynamic networks are formed by a group of independent companies. The lead firm, in this network, identifies potential partners who own a large or sometimes the entire portion of the assets in the SCN. Each partner contributes its 'core competency' to the supply chain. In some cases, the lead firm contributes the core skill, such as manufacturing in the case of Motorola, research and development in the case of Reebok, and design and assembly in the case of Dell Computers. It is also possible for the lead firm to undertake a pure broker role in designing and forming a network of highly efficient business processes. For example, at Lewis Galoob Toys, a handful of key executives select potential partners to design, manufacture, and sell children's toys. Thus, Galoob acts as a pure broker in forming an effective supply chain of highly competent partners.

Virtual corporations are very similar to network organizations. They are an alliance of independent business processes or enterprises with each contributing 'core competencies' in areas such as, design, manufacturing, and distribution to the network [31-36]. Virtual corporations are formed in the event of a market opportunity and would be dissolved when the opportunity passes. Similar to broker-led dynamic network organizations, they do not own any of the individual business processes that design, produce, and market the product. However, virtual corporations always indulge in *temporary* relationships to take advantage of a specific market opportunity. For example, Apple Computer and Sony Corp. engaged in a similar temporary alliance to manufacture PowerBook notebooks. TelePad Corp. collaborated with more than two dozen partners in bringing its pen-based computer to market. IBM, Apple Computer, and Motorola have become involved in an inter-firm alliance to develop an operating system and microprocessor for a new generation of computers. This idea of temporary alliances has gained prominence in service industries also. For example, InterSolve Group Inc., a Dallas-based management-consulting firm, assembles 'just-in-time' talent to solve problems or implement strategies for a large number of clients.

According to Johnston and Lawrence [37] the value-added partnerships are "a set of independent companies that work closely together to manage the flow of goods and services along the entire value-added chain". Each company focuses on performing a single step of the value-added chain. For example, Japanese auto companies are a typical form of value-added partnerships, producing only about 20% of the value of their automobiles. Similarly, Chrysler's resurgence may be attributed to the creation of such partnerships with its suppliers.

Although several conceptual models for network supply chain models have been proposed and discussed in the literature, research efforts in this area are lagging behind in the development of formal decision models for SCN design. However, one key issue that is emphasized by researchers in designing SCNs is the selection of highly efficient and compatible partners.

5.1. Network Supply Chain Management

Research in the development of frameworks for supply chain management and decision making have mainly focussed on the tactical and operational levels rather than the strategic level. Management of the supply chain primarily involves three planning levels: strategic level planning, tactical level planning, operational level planning. Strategic level planning involves SCN design, which determines the location, size, and optimal numbers of suppliers, plants, and distributors to be used in the network. It also to some extent includes sourcing and deployment plans and actions for all plants, distributors, and customers. This phase can be summarized as determining the *nodes* and *arcs* of the SCN and their relationships. Strategic level planning is long-range planning and is typically performed every few years, when firms need to expand their capabilities.

Tactical level planning involves supply planning, which includes the optimization of flow of goods and services through a given SCN. Tactical level planning is medium-range planning, which is typically performed on a monthly basis. Finally, operational level planning is short-range planning, which involves production scheduling at all plants on an hour-to-hour basis. It is at this level of analysis that much of the modeling and research effort has been targeting.

Geoffrion and Graves [38] performed some of the initial work in this area. They proposed a multicommodity logistics network design model for optimizing annualized finished product flows through the entire supply chain, which involved all the nodes from vendors to factories to warehouses to customers. Their method utilized Benders decomposition in such a way that a separate classical transportation problem can be solved for each individual product. Their work is based more on the operational aspects rather than strategic aspects of SCN design.

Cohen and Lee [39] proposed a pair of models for the network design problem, which is based on Geoffrion and Graves' [38] work. Their multicommodity manufacturing network design model optimizes the product flows from raw material vendors to end customers. The other nonlinear model emphasizes scale economies of production. In a subsequent paper,

Cohen and Lee [40] proposed a set of approximate stochastic sub-models and heuristic methods for developing stationary long-term operational policy for supply chains. In their next paper, Cohen and Lee [41] propose a deterministic model for a global manufacturing and distribution network. Their model included value markups and costs, exchange effects, and before and after tax profitability estimation. These models concentrate more on the operational aspects of SCN design as well.

Davis [42] describes a global supply chain analysis from raw materials to finished products by highlighting uncertainty at all levels. His work addresses a framework developed at Hewlett-Packard for dealing with uncertainty that adversely affects performance of supplier, manufacturing, and transportation processes. However, he provides very few details about his stochastic models, although it seems that his procedures are applied at a more tactical level rather than at the strategic level. Lee and Billington [43] proposed an operational model for material management and inventory control in decentralized supply chains, and demonstrate its use in a new product introduction project at Hewlett-Packard.

Arntzen et al. [44] present a global supply chain model, which has been utilized at Digital Equipment Corporation for designing a production, distribution, and vendor network. Their mixed-integer linear program minimizes cost and/or weighted cumulative production and distribution times subject to a variety of demand and capacity constraints. However, their model does not consider efficiencies of SCN processes, and primarily addresses issues related to reengineering an existing supply chain.

Talluri et al. [45] proposed a framework based on Data Envelopment Analysis (DEA) and multi-criteria decision models for value chain network design. However, their approach utilizes the basic CCR (Charnes, Cooper and Rhodes [46]) model for efficiency evaluation, which has certain limitations in terms of unrestricted weight flexibility. Also, their models do not consider issues relating to capacity and location constraints, and deployment plans. Finally, their methods identify a single best supplier-manufacturer combination, and do not specifically address issues relating to designing the entire SCN with several nodes at each value-added stage.

5.2. Supply Chain Organizations and Their Growth

The development of SCN is relatively new, but a number of organizations and groups of organizations have started to develop strategies in the formation of SCN. One such example is a former contract manufacturer in the electronics industry, Solectron Corporation. Solectron has grown considerably in recent years from a multi-million dollar organization to a billion dollar revenue generator. The goal of Solectron is to position itself as the world's premier supply-chain facilitator for customized electronics technology, manufacturing and service solutions.

One of their strategies is to serve as the virtual manufacturing arm of its customers and not only as a preferred subcontractor. Two examples of such developments include recent agreements with Zhone Corporation and Alcatel Inc. In the Zhone relationship, Zhone has agreed to focus its resources on the design and development of new products by having Solectron provide product life cycle and supply-chain management. To develop this arrangement, Solectron purchased Premisys Communications, Inc., a wholly owned subsidiary of Zhone, from Zhone. While Zhone agreed to retain the engineering assets of Premisys. Under the agreement, Solectron will work closely with Zhone as it builds its technology base and product portfolio. In a similar move Alcatel (a developer of next-generation networks and data-voice technology) sold its manufacturing assets to Solectron. Solectron then provides a full range of manufacturing services to Alcatel including the prototyping and new product Introduction management.

To help in further completing it's supply chain facilitator package, Solectron has also positioned itself as an "electronic" commerce conduit. It has expanded its value-added network capabilities to incorporate both Internet and EDI linkages. Offering linkages between customers and suppliers using these technologies. Thus, Solectron not only serves as manufacturing arm of a SCN, but also the possibility of serving as a broker and systems integrator across the supply chain.

An example of the more complete network organization is the "Agile Web" located in Pennsylvania [47]. In this situation, a broker is the major contact point for customers. The broker than chooses from among a list of 19 member companies. The decision is based on capabilities ranging from design and technological expertise, to whether or not enough capacity exists at the time of a customer request. As well, the work the customer may seek may range from a complete design to delivery of a product, to a manufacture of a single simple part. The Web is flexible enough to offer this wide range of capabilities. They state that this flexibility is one of their big advantages, not offered by traditional organizations.

As we have seen with these two examples, the formation of supply chains and networks can be completed by diverse organizations joining together as a broker or as part of an individual organization that offers some of the services to help facilitate the supply chain network formation. The examples of such supply chain networks are increasing including those from medical health care products to publishers forming virtual publishing houses.

6. FUTURE ISSUES AND CONCLUSION

The development of supply chains and supply chain management has taken the focus of operations from the shop-floor to extend to higher levels of management and across functional and organizational boundaries. We have seen that there exist a number of possible avenues that organizations can pursue in improving their supply chain operations. We have briefly described the various practices from the perspective of the internal operations of an organization, its relationships with customers and suppliers, and the improvement of the system overall. Some requirements were described. There is ample additional detail to these practices, but specifying each would require a significantly longer exposition. One of the most important supply chain practices, selecting partners, was then discussed. AHP, a potential tool to evaluate possible partners from a strategic or operational level of analysis was presented. This simple exposition of the model was supported with a comprehensive listing of criteria by which organizations can evaluate potential partners. Clearly, there are a wide variety of partner selection ranging from simple checklists to advanced mathematical programming and decision theoretic approaches. Network organizations, an area of profound interest to emerging supply chain practices and organizational structure was also described. This evolving area also has a number of issues including technology integration, operations optimization and partner selection.

Given that supply chain management and the study of interorganizational networks is a relatively new area, the need to identify and develop good practices and enablers is greater then ever. There is a fertile bed for developing and applying new knowledge in this area, but with opportunity come risks. Decision models and research on what works and what doesn't is evidently needed. To help in defining some needs in the area of network organizations and virtual enterprises a group from the Agility Forum put together a listing of key need areas for agile virtual enterprises [48]. A summary of these areas is shown Table 4. The major categories include cultural, business and technical dimensions. Even though this work is from the mid-1990's, few of these enablers have been widely developed and adopted.

Table 4: Summary of Key and Specific Need Areas for Integrating the Agile Virtual Enterprise (Adapted from Assava and Engwall [48]).

Cultural	Business	Technical
Empowered People Shared Vision Trust Flexible Work Hours and Telecommuting Fully Supported Contingent Workforce Career Growth in a Transient Employee Environment **Total Quality Management** Quality Assurance versus Control Quality Metrics Quality Perspectives Problem-Solving Tools **Value-Based Compensation** Consensus-Driven Value Perceptions Management and Organizational Practice Performance Measurements and Benchmarks Streamlined Legal Practices **Concurrency** Team Communications Information Surety Product Data Exchange Project Planning Processes Education and Training	**Design for Customer Delight** Information Highway Customer Education Customer Involvement Common Protocols Performance Measurements and Benchmarking **Synchronized Processes** Business Process Reengineering Resource Inventory of Skills and Competencies Standards for Product/Process Integration Manufacturing Product/Process Integration Management Philosophy **Dynamic Multiventuring** Certification Process for Organizations Code for Forming and Dissolving Enterprises Legal Requirements Distributed Cross-Functional Teams Synchronized Processes elements **Refined Accountability** Strategic Planning Process-Based Performance Measures Self-Empowered Work Force Continuous Education and Training **Info. System Infrastructure** Vision-Based Leadership Dynamic Multiventuring Adaptable Standards Natural Language-Based Information Models	**Rapid Response** Accommodate for the Changing Environment Adaptive Infrastructure Standardized Interfaces Concurrency Risk Management **Information Management** Recognition Information is a Critical Asset Mathematical Underpinning End-User Centered Deployment of IT Education and Awareness Programs for Users Standards for Product/Process Integration Vision-Based Leadership **Enabling Tools and Techniques** Useful Tools with Effective Human Interfaces Effective Global Communications Incorporation of New Software and Hardware Virtual Reality Systems Simulation and Modeling **Adaptive Infrastructure** Modularity Extensibility Distributed Information Monitoring Support **Connectivity** Access Affordability Openness

Initial steps to improve supply chain practices are beginning. Integrating these practices with agile organizations is even less established. Where the future will guide these developments will be determined by future generations of visionaries in practice and in academia.

REFERENCES

1. Bhote, K. (1989). *Strategic Supply Management*. New York, N.Y., AMACOM.
2. Burt, D. N. and M. F. Doyle (1993). *The American Keiretsu: A Strategic Weapon for Global Competitiveness*. Homewood, Illinois, Business One Irwin.
3. Ellram, L. M. (1991). "Managerial Guideline for the Development and Implementation of Purchasing Partnerships." *International Journal of Purchasing and Materials Management* **27**(3): 2.
4. Landeros, R. and R. M. Monczka (1989). "Cooperative Buyer/Supplier Relationships and a Firm's Competitive Posture." *Journal of Purchasing and Materials Management* **25**(3): 9.
5. Poirer, C. C. and W. F. Houser (1993). *Business Partnering for Continuous Improvement*. San Francisco, CA, Berrett-Koehler Publishers.

6. Teece, D. J. (1992). "Competition, Cooperation, and Innovation, Organizational Arrangements for Regimes of Rapid Technological Progress." *Journal of Economic Behavior and Organization* **18**(1): 1.
7. Agility Forum (1994), *Agile Customer-Supplier Relations*, The Agile Manufacturing Enterprise Forum, Monograph RS94-01, Bethlehem, PA.
8. Dyer, J. H. and W. G. Ouchi (1993). "Japanese Style Partnerships." *Sloan Management Review* **35**(1): 51.
9. Newman, R. G. (1988). "The Buyer-Supplier Relationship Under Just-in-Time." *Production and Inventory Management Journal* **29**(3): 45.
10. Newman, R. G. and K. A. Rhee (1990). "A Case Study of NUMMI and Its Suppliers." *Journal of Purchasing and Materials Management* **26**(4): 15.
11. Sterling, J. U. and D. M. Lambert (1987). "Establishing Customer Service Strategies within the Marketing Mix." *Journal of Business Logistics* **8**(1): 1.
12. Sheridan, J. H. (1991). "Are You a Bad Customer." *Industry Week* **240**(16): 24.
13. Isaacson, B., and Shapiro, R., (1994), Bose Corporation: The JITII Program, Harvard Business School Case 9-694-001, Havard Business School Publishing, Cambridge, MA.
14. Deming, W.E., (1986), *Out of the Crisis*, MIT Press, Cambridge, MA.
15. Ellram, L. M. (1990). "The Supplier Selection Decision in Strategic Partnerships." *Journal of Purchasing and Materials Management* **26**(4): 8.
16. Saaty, T. L., (1988) *Decision Making: The Analytic Hierarchy Process*, Pittsburgh, PA, (1988).
17. Lorange, P and Roos, J., (1992), *Strategic Alliances: Formation, Implementation, and Evolution*, Blackwell Publishers, Cambridge, MA.
18. Bronder, C., and Pritzl, R., (1992),"Developing Strategic Alliances: A Conceptual Framework for Successful Co-operation", *European Management Journal*, **10**(4): 412.
19. Speckman, R.E., (1994), "Alliance Planning Process", Quality Customer/Quality Supplier Program, CAM-I, Arlington, TX.
20. Meade, L., Sarkis, J., and Liles, D., (1997), "Justifying Strategic Alliances and Partnering: A Prerequisite for Virtual Enterprising." *OMEGA*, **25**(1): 29.
21. Barbarosoglu, G., and Yazgac, T., (1997), "An Application of the Analytic Hierarchy Process to the Supplier Selection Problem." *Production and Inventory Management Journal*, **38**(1): 14.
22. Choi, T.Y., and Hartley, J.L. (1996), "An Exploration of Supplier Selection Practices Across the Supply Chain." *Journal of Operations Management*, **14**(4): 333.
23. Weber, C.A., Current, J.R., and Benton, W.C. (1991), "Vendor Selection Criteria and Methods." *European Journal of Operational Research*, **50**: 2.
24. Mustajoki, J., and Hämäläinen, R.P., (1999), "Web-HIPRE - A Java Applet for AHP and Value Tree Analysis," *5th International Symposium on the Analytic Hierarchy Process* (ISAHP'99), August 12-14, 1999, Kobe, Japan.
25. Miles, R. E and Snow, C. C., (1984), "Fit, Failure, and Hall of Fame," *California Management Review*, **27**: 3.
26. Miles, R. E., (1989), "Adapting to Technology and Competition: A New Industrial Revolution System for the 21st Century." *California Management Review*, **32**(2): 9.
27. Lawless, W. L. and Moore, R. A. (1989), "Interorganizational Systems in Public Service Delivery: A New Application of the Dynamic Network Framework. " *Human Relations*, **42**(12): 1167.
28. Kensinger, J. and Martin, J., (1991), Financing Network organizations. *Journal of Applied Corporate Finance*.

29. Snow, C. C, Miles, R. E. and Coleman, H. J., (1992), "Managing 21st Century Network Organizations." *Organizational Dynamics*, **20**(3): 5.
30. Snow, C. C and Thomas, J., (1992), "Building Networks: Broker roles and Behaviors," in Peter Lorange, et al., Editors, *Strategic Processes: Designing for the 1990's*, Basil Blackwell, London.
31. Byrne, A. J., (1993), "The Virtual Corporation," *Business Week*, February: 98.
32. Goldman, L. S., (1994), "Co-operating to Compete." CMA Magazine, **68**(2): 13.
33. Iacocca Institute, (1991), *21st Century Manufacturing Strategy*, Lehigh University, Bethlehem, PA.
34. Porter, A. L., (1993), "Virtual Companies Reconsidered." *Technology Analysis & Strategic Management*, **5**(4): 413.
35. Presley, A., Barnett, B. and Liles, D.H., (1995), "A Virtual Enterprise Architecture," Fourth Annual Agility Forum Conference, 2, 3-12.
36. Sheridan, J. H., (1993), "Agile Manufacturing: A New Paradigm." *International Productivity Journal*, **11**: 39-48.
37. Johnston, R. and Lawrence, L. R.,(1988), "Beyond Vertical Integration - The Rise of the Value-Adding Partnership." *Harvard Business Review*, **66**(4): 94.
38. Geoffrion, A.M. and Graves, G. W., 1974, "Multicommodity Distribution System Design by Benders Decomposition." *Management Science*, **20**(5), 822.
39. Cohen, M. A. and Lee, H. L., (1985), "Manufacturing Strategy: Concepts and Methods." in *The Management of Productivity and Technology in Manufacturing*, edited by P. Kleindorfer, Plenum Publishing Co., Chapter 5, 153.
40. Cohen, M. A. and Lee, H. L., (1988), "Strategic Analysis of Integrated Production-Distribution Systems: Models and Methods." *Operations Research*, **36**(2), 216.
41. Cohen, M. A. and Lee, H. L., (1989), "Resource Deployment Analysis of Global Manufacturing and Distribution Networks." *Journal of Manufacturing and Operations Management*, **2**(2), 81.
42. Davis, T., (1993), "Effective Supply Chain Management." *Sloan Management Review*, **34**(4): 35.
43. Lee, H. L. and Billington, C., (1993), "Material Management in Decentralized Supply Chains." *Operations Research*, **24**(5), 835.
44. Arntzen, B. C., Brown, G. G., Harrison, T. P. and Trafton, L. L., (1995), "Global Supply Chain Management at Digital Equipment Corporation." *Interfaces*, **25**(1): 69.
45. Talluri, S., Baker, R. C., and Sarkis, J., (1998), "A Framework for Designing Efficient Value Chain Networks." *International Journal of Production Economics,* **62**: 133.
46. Charnes, A., Cooper, W.W. and Rhodes, E., 1978, "Measuring the Efficiency of Decision Making Units." *European Journal Of Operational Research*, **2**(6): 429.
47. Sheridan, J.H., (1996), "The Agile Web: A Model for the Future." *Industry Week*. **245**(5): 31.
48. Asava, R.G., and Engwall, R.L., (1994), *Key Need Areas for Integrating the Agile Virtual Enterprise*, Agility Forum, Monograph, AR94-04, Bethlehem, PA.

Engineering the Agile Supply Chain

Denis R.Towill

Cardiff University, Logistics Systems Dynamics Group
PO Box 907, Cardiff, CF10 3YP, United Kingdom

1. WHAT IS A SUPPLY CHAIN?

Our preferred working definition of a supply chain is that described by Stevens (1) as:
> "*a system whose constituent parts include material suppliers, production facilities, distribution services and customers linked together via the feedforward flow of materials and the feedback flow of information*"

A supply chain consists of a number of echelons (business units) operating sequentially. A simple supply chain can consist of just four echelons (raw materials supplier: manufacturer: distributor: and retailer) and thus on to the end customer. *Typically there will be many more echelons or levels* (i.e. a number of discrete businesses each adding value during their manufacturing operations), and at each echelon there are usually a variety of businesses operating in parallel with each other. For example there may well be multiple-sourcing as vendors supply goods to the Original Equipment Manufacturer (OEM) with the latter responsible for production of such products as automobiles, computers, and vacuum pumps. Rationalising and integrating such a vendor base is a key factor in agile supply.

In a supply chain material flows downstream and information traditionally flows upstream from echelon to echelon. Our definition of supply chain applies right across the business spectrum ranging from global enterprises right down to a number of related sequential activities undertaken under one roof but covering a number of independent cost centres. So proposals for improving supply chain performance based on material flow control are equally applicable to internal re-engineering as much as to global networking. Streamlined flow enables seamless operation of the supply chain. *Here "seamless" is defined as a supply chain in which all "players" think and act as one (Towill, 2), and which is a necessary condition for agile supply.*

Each echelon within a supply chain embraces the following constituents (Towill, 3):
(a) *perceived demand for products,* which may be firm orders or just sales department forecasts;
(b) *at least one 'production' or added value process;*
(c) *information on current performance,* which may be 'stale' or alternatively 'distorted' or both;
(d) *'disturbances'* due to machine/equipment breakdowns etc;
(e) *'interference'* between products in different supply chains competing for the same resources;
(f) *decision points* where information is brought together and acted upon;
(g) *transmission lags,* which occur for both value added and other activities;

(h) *decision rules* (based on company procedures) for changing stock levels, placing new orders, etc., in the light of available internal and external information and forecasting mechanisms.

The potentially complex operation of supply chains implied above is made worse because there is uncertainty associated with perceived demand, with the quality of information and with the time associated with the many transmission lags (both for material flow and information flow). As we shall see later there are enormous benefits to be obtained by improving information and material flow throughout the supply chain. Both flows are much enhanced via time compression of value added activities and the elimination of non-value added activities. Such focussed re-engineering is particularly helpful in reducing uncertainty, which is itself an important goal in the agile supply chain. Time compression dramatically reduces uncertainty as Table 28.1 demonstrates (Watson, 4). Hence the time compression principle features strongly throughout this Chapter. However, to achieve agility, the improved forecasts must additionally be made transparent and transmitted without distortion or delay throughout the supply chain, (Mason-Jones and Towill, 5). Such visibility is an outstanding feature of our Dell Computing and Benetton Case Studies.

It is important to note that an individual business can simultaneously be a part of many supply chains at the same time. As well as deliberate customer policy to multiple source important components a business may also supply particular products to multiple outlets. Handling such "interference" between individual "value streams" (specific product routes within supply chains) is a major challenge; streamlined material flow is a proven technique for minimising such unwanted effects (Burbidge and Halsall, 6). Note that at each echelon in the chain, the business upstream is its immediate supplier, and the business downstream is its immediate customer. As the Case Studies later in this Chapter will ably demonstrate, the agile supply chain requires that a holistic view be taken, so that every business in the chain is focussed on delivery to the end-customer i.e. the marketplace.

Time Horizon	Accuracy of Forecast
1 month	$\pm 5\%$
2 months	$\pm 20\%$
3 months	$\pm 50\%$
4 months	"Toss a Coin"

Table 1
The Need to Compress Cycle Time: Typical Effect of Time Horizon on Accuracy of Forecast Marketplace Demand (Watson, 4)

2. A FRAMEWORK FOR ENABLING AGILE RESPONSE

Agility in supply chains can be best achieved by the integration of organisations, people, and technology into a meaningful unit by deploying advanced IT and flexible organisational structures to support highly skilled, knowledgeable and highly motivated people (Gunasekaran, 7). To achieve this goal it is helpful to define a framework which if properly followed will give a maximum guarantee of success in moving from the "traditional" adversarial supply chain to the agile supply chain. A suitable proven framework already exists, and relates to the convergence noted between management consultancy organisations practising Business Process Engineering. Werr et al (8) thereby established the following four important shared principles:

1. *An Holistic View of Organisations* leading to a systems model of the enterprise. (Our target here is the Seamless Supply Chain as discussed in Section 3)
2. *Time as the Explicit Improvement Target* reflected in the performance metrics used during the diagnosis and subsequent re-engineering of the organisation undergoing change. (In Section 4 we shall demonstrate the crucial role of Total Cycle Time Compression in enabling agile response.)
3. *A Focus on Learning* especially during the engineering change process but also followed by competence transfer within the total supply chain. (In Section 6 we describe a Supply Chain Change Model which shows the route from adversarial to seamless behaviour and against which learning may be judged.)
4. *Highly Structured Methods* which seek to identify and support the many steps in the change process. (We believe these methods should be centred on process mapping and process re-design based on the proven set of material flow simplification rules as also described in Section 6. These enable seamless operation and hence agile response.)

Conventional Thinking (as found in traditional supply chains)	Systems Thinking (as found in agile supply chains)
Seeing individual items	Seeing material, information, capacity and cash flows
Seeing individual businesses	Seeing the whole supply chain
Seeing only linear cause-and-effect	Seeing inter-relationships
Seeing only static snapshots of supply chain behaviour	Seeing trends and patterns of process change
Seeing people as powerless reactors	Enlisting people as active participants
Seeing boundaries as brick walls	Realising interfaces are to be streamlined and managed
Managing by reacting to the present	Managing by creating the future
Managing by fire-fighting	**Managing via innovation**

Table 2
The Paradigm Shift; Systems Thinking Applied to the Agile Supply Chain
(Source: Author, adapted from Senge, 11)

A real-world supply chain has many interfaces. Information, instructions, materials, and capacities flow across these boundaries and these activities need careful planning, co-ordination, and execution. Hence the justification for using systems concepts as outlined by Kidd (9) when designing agile supply chains linking customer need right through to that need being satisfied. If the word "network" is substituted for "system", then our definition becomes identical to that adopted by Christopher (10) and since widely used in the logistics community. This interchange of words is regarded as important, since the approach to agile supply chain engineering to be outlined later in this Chapter is essentially based on business systems engineering (Towill, 11). In practice this change to agile mode requires the paradigm shift i.e. mindset change indicated in Table 2 which is based on Peter Senge's "Fifth Discipline" (12).

Our route to agility is via the seamless supply chain, which may be regarded as a practical example of systems thinking, and particularly of systems engineering. The comparison between "conventional thinking" and "systems thinking" made in Table 2 summarises the ultural differences between conventional (adversarial) supply chain operation and that observed in agile enterprises.

These differences really do describe a paradigm shift and hence definitely requires a change in mindset. Unfortunately, as Morris and Brandon (13) point out, it is very difficult for old and new paradigms to exist alongside each other. It is therefore unlikely that "agile" and "traditional" organisations will co-exist within the same supply chain. Indeed the four dimensions of the agile company specified by Goldman, Nagel, and Preiss (14) are seen as diametrically opposed to "traditional" operations. These differentiating definitions are as follows:

(a) *An agile enterprise is perceived as enriching customers*
(b) *An agile enterprise has an organisational strategy based on internal and external co-operation*
(c) *An agile enterprise is organised to master change and uncertainty*
(d) *An agile enterprise maximises the impact of people and information on its business operations*

Thus it is important that the agile organisation sees change and volatility as opportunities, not threats, and has a culture to take full advantage of the situation (Berry et al, 15).

3. THE "SEAMLESS" SUPPLY CHAIN – AN ESSENTIAL CONDITION FOR AGILE RESPONSE

The Seamless Supply Chain (SSC) is an important concept of the ideal situation in which all "players" think and act as one (Towill, 2). It requires that all operations are carried out in minimum time, every operation is performed "right first time", all operations are synchronised, Decision Support Systems (DSS) are optimum for the particular product range, all information is accurate and undistorted and transparent to all players. Finally, there is a common goal for satisfying the end customer which is shared and actively supported by all players in the chain. Thus the SSC can be seen as a minimum entropy i.e. minimum confusion system which in this sense is diametrically opposite in philosophy to the traditional i.e. adversarial supply chain.

SSC is not a new concept since it reasonably describes advanced delivery processes which are thought to be well developed amongst "Supply Chain Predators" (Schmidt, 16). For example, Farmer and Van Astel (17) describe "Pipeline Management" as necessitating the continuing controlled flow of goods on demand and make use of the analogy with water flowing though a hydraulic system. They correctly attribute the origin of the concept to Jay Forrester (18) following his dynamic simulation of production-distribution systems, and believe that the pipeline to be controlled covers inboard logistics, production and distribution. It may be further argued that free flow of the right information in the right place at the right time is a necessary characteristic of "seamless supply". A practical example is the agile reengineering programme of the Carpenter Technology Inc. speciality steel products supply chain which identified JIT information availability and transparent information flow throughout the company as key enablers for effective change. Consequential business improvement included quality up by 75%; on-time shipments doubled; cycle times down by 25-75%; and WIP down by 50% (Goldman, Nagel, and Preiss, 14).

The benefits of moving towards the SSC are not restricted to large companies, as Table 3 readily demonstrates. In contrast to Carpenter Technology, Shalibane are a small UK automotive components supplier. The company has improved its competitive position enormously via streamlining the internal supply chain by concentrating on material flow. Note that every performance metric of interest has been bettered following the re-engineering programme. Shalibane is now much more suited to being an integral part of an agile supply chain – provided the OEM has a transparent information visibility policy! (Stalk and Hout, 19).

A complimentary vision to the SSC concerns the "Integrated Supply Chain" (Stevens, 1). Here the goal is that the entire supply chain including material flow and the associated flow of the related planning information should be managed as a single business process which should be simplified, streamlined, and optimised to reduce lead times, inventories, and non-value added activity throughout the system. Provided that the various companies and organisations which comprise the supply chain are regarded as members of the same team, then the supply chain can perform at a high level of effectiveness (Schmidt, 16). All "players" within the chain are then potentially in a win-win situation, as the next section illustrates.

PERFORMANCE METRIC	PERFORMANCE BEFORE RE-ENGINEERING	PEFORMANCE AFTER RE-ENGINEERING	RELATIVE CHANGE
Throughput Time	4 weeks	4 days	Down 86%
Set-up Time	50 mins	15 mins	Down 70%
Production runs p.a.	12	52	Down 70%
Rejects/Millions Parts	200	80	Down 60%
Overdue Orders/week	120	30	Down 75%
Sales p.a.	£m6.0	£m8.0	Up 33%
Return on Investment	15.6%	19.4%	Up 24%

Table 3: Improved Business Results Obtained Via Streamlined Material Flow Controls as Implemented In UK Automotive Products Supplier (Burbidge and Halsall, 6)

4. BENEFITS OF MOVING TOWARDS THE SEAMLESS SUPPLY CHAIN

A major factor in enabling agile supply is the compression of total cycle time (TCT) i.e. the time taken between customer need identified and that customer need being satisfied. This is convincingly demonstrated via our Case Studies. By concentrating on TCT reduction via properly analysed and implemented re-engineering programmes, the business bottom-line is leveraged by improving all the important management metrics. This is illustrated in Fig. 1, where typical results are shown on the impact of TCT reduction on productivity indices, WIP, stockturns etc. (Thomas, 20). These results confirm the power of utilising time metrics as explicit goals in re-engineering programmes. We regard the Time Compression Paradigm as a necessary, but not sufficient condition for agility. Yusuf Sarhadi, and Gunasekaran (21) highlight this fact by emphasising that agility is just not doing things fast, but requires massive structural and infrastructure changes as clearly visible in all three of our Case Studies. Some specific publicly quoted cost benefits potentially achievable by better pipeline management and hence moving towards the SSC include:

- United Health Company estimate that £K50 p.a. is saved for every day by which their "total order fulfilment cycle" is reduced (Evans et al, 22).
- Hewlett-Packard suggest that between 25% and 50% of total system stocks may be eliminated (even in well run chains) by strategic re-distribution of stocks (Davis, 23)
- The U.K. Institute of Purchasing and Supply argue that some companies spend up to one third more than necessary by adopting the wrong purchasing policy such as price rather than total worth (Cassell, 24)
- Phillips (USA) experience is that companies can save up to 10-15% of their annual revenues presently tied up in supply chains (Schmidt, 16)
- Elders (New Zealand) estimate they can save $K150 transport costs by better design and management of outbound logistics (Kosta, 25)
- McKinsey and Co. argue that between 10 and 40% of total supply chain costs are due to complexity much of which can be eliminated by streamlining products, processes, flows and inventories (Child, Diederichs, Sanders and Wisniowski, 24)
- UK retailers could treble profit margins (which are well known to be under constant threat) by more efficient management of their supply chains (Gilchrist, 27)

These actual and potential savings are impressive and importantly cover a wide range of market sectors. They provide a helpful way of influencing businesses to plan and implement change and can be seen as providing a useful adjunct to management consultants sales pressure by providing independent evidence that collaborative movement towards the SSC can provide a powerful competitive edge. However, as Womack and Jones (28) have stressed, the agreement between partners has to be "shared pain and shared gain". This is particularly true in agile supply as the trust necessary to manage the entire chain effectively can be destroyed if strong "players" selectively grab the profits but attempt to pass on any losses to their more vulnerable brethren.

Engineering the Agile Supply Chain

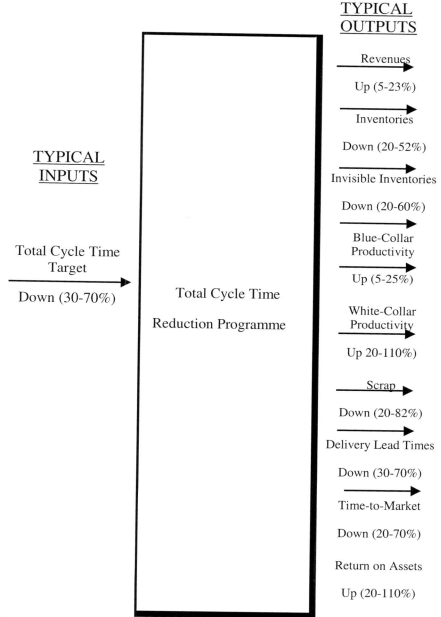

Fig. 1: The Power of Total Cycle Time Compression: Input-Output Diagram Showing Range of Business Performance Improvements (Source: Thomas, 20)

5. THE NEED FOR EFFECTIVE SUPPLY CHAIN MANAGEMENT

The benefits quoted in Section 4 do not occur just by chance or by a process of osmosis: they have to be earned, designed, engineered, and managed according to the vision of the SSC. This requires dedicated Supply Chain Management to pro-actively enable "seamless" operations. Does such dedication currently exist in the present industrial climate? To answer this question the author and his team recently conducted an in-depth site based survey of European Automotive Suppliers. Only 10% of that particular sample even closely approach the SSC, although a further 20% exhibit some degree of good practice (Towill, Disney and Childerhouse, 29). Whereas many of the best of these suppliers can readily move into the rapid response mode any idea of converting some of the laggards to "agile supply" can be discounted. Many of the reasons for poor performance encountered during this survey are to be found in Table 28.4 and will be immediately recognised by industrialists as the operating characteristics of adversarial supply chains. Generally the latter have not been properly thought through; designed; and implemented. Instead they have more likely grown like Topsy. The way forward is to eliminate the offending causes at source: a suitable structured methodology already exists which enables this action to be speedily undertaken (Towill, 31). But there is a change process to be undergone to enable agility which can only reasonably be expected from a supply chain which has already learnt to operate in a "seamless" manner (Victor and Boynton, 32). Fortunately there is a proven comprehensive conceptual Model for Supply Chain Integration which will be described in detail in the next section and which plots the route from adversarial to seamless operation.

Finally, it must be emphasised that even in the Year 2000 inventory management is still proving surprisingly difficult despite recent advances in electronic communications. Although inventory systems are continuously improving, the need for expediting late shipments never seems to disappear entirely. There are always delays in shipments for a variety of reasons, and as many companies have discovered, having inventory buffers is doubly costly if we additionally stock the wrong products (Belk and Steels 33). In the agile supply chain this situation can be potentially disastrous as managers seek to achieve

ESTABLISHED CAUSES	TYPICAL EXAMPLES
1. Waiting for processing	Waste time spent in queues
2. NVA activities	Redundant quality inspection of even top vendor supplies
3. Resistance to parallel operations	Poor work practices - "we've always done it this way"
4. Repetition of activities	Rework generating "interference" to value stream
5. Batching	Order release patterns generate boom-and-bust capacity needs
6. Excessive controls	Unnecessary clearance signatures delay system flows
7. Poor synchronisation	Material shortages alternating with excesses in typical boom-and-bust behaviour

8. Ambiguous objectives	No vision of company role within supply chain
9. Poor procedures	Value added process delays due to awaiting clearance
10. Outdated technology	MRP system not up to required standard
11. Lack of information	Important system states not monitored by anyone
12. Poor communication	System states available but not channelled to those who need them
13. Limited co-operation	System states available but deliberately withheld from those in need
14. Limited co-ordination	Organisation not operating holistically within the supply chain
15. Inadequate training	Employees not trained in SCM practices

Table 4: Typical Causes of Poor Real-World Supply Chain Performance (Source: Author, following Handfield and Nichols, 30)

minimum reasonable inventory levels in the presence of yet higher targets for availability of the right product at the right time of the right quality and in the right marketplace. As we have already seen reducing total cycle time over the complete supply chain is the best way forward since this will reduce uncertainty throughout the system.

6. A CONCEPTUAL MODEL FOR ENGINEERING THE SEAMLESS SUPPLY CHAIN

A very powerful conceptual and practical model for engineering supply chain integration has been proposed by Stevens (1). He suggests that four distinct stages may be identified as follows:

Model Stage One – Base Line – is typified by the company which vests responsibility for different activities within the supply chain in separate, almost independent compartments. Hence there are stand-alone and often incompatible control systems and procedures covering sales, manufacturing, material flow and purchasing. Managers still operate within functional silos.

Stage Two - Internal Integration – develops factory functional integration by focusing principally on the inward flow of goods and hence on supplier performance. The emphasis is on total cost reduction rather than total business system performance improvement. Customer service still tends to be reactive, and inventory will be suffering as a consequence of the continuing interfaces between discrete business functions.

Stage Three – Local Integration – recognises that there is very little point in just focusing on the flow of goods within the organisation. The management of flow to the customer is therefore also improved. This requires *local integration* under the control of the company. Consequently, there is a goods outward management integrating supply and demand functions along the company's own chain. Frequently a JIT manufacturing philosophy will be applied to the execution of the material plan derived from an MRPII system.

Stage Four – Seamless Operation – full supply chain integration is achieved by extending the scope of management outside the company to embrace the suppliers and customers. It embodies a change of focus away from being product-oriented to being customer-oriented.. Thus there is penetration deep into the customer organisation to understand the products, culture, market and organisation. Integration to include all suppliers is also undertaken, and the whole supply chain is capable of responsive (demand driven) operations.

This Conceptual Model is frequently cited in the literature as the way forward and can therefore be recommended with considerable confidence. It links in well with the concept of movement towards the agile supply chain. For example, by using Likert scales questionnaires may be developed which allow "scores" to be codified thus assessing our movement towards the SSC. In this respect the four Stevens stages outlined above broadly correspond with the four performance categories identified within the European automotive suppliers sector, Towill, Disney, and Childerhouse (29). It is also possible to relate these model stages to definite phases in real world Business Process Re-Engineering Programmes in the electronic products industry (Berry, Naim and Towill, 34), and in the mechanical precision products industry (Towill and McCullen, 35).

This Conceptual Model requires underpinning with a suitable structured approach to supply chain analysis and design. Table 5 is a suitable proven checklist of rules based firmly on process mapping followed by streamlined material flow control. The Checklist fully acknowledges the relevance of Ashby's Law of Requisite Variety to supply chain design. It is based on a combination of simulation modelling results and practical guidelines tested in a wide range of market sectors (Towill, 31), including the rating of present practice in the European automotive sector survey of Section 5. However, these material flow rules must be interpreted in the light of the particular agile environment required. Shewchuk (36) has defined four different types of manufacturing systems needed to enable agility as related to

RULE 1	Only make products which you can quickly despatch and invoice to customers.
RULE 2	Only make in one period those components needed for assembly in the next period.
RULE 3	Minimise the material throughput time, i.e. compress all lead times.
RULE 4	Use the shortest possible planning period, and the smallest run quantity.
RULE 5	Only take deliveries from suppliers in small batches as and when needed.
RULE 6	Synchronise "Time Buckets" throughout the Chain.
RULE 7	Form natural clusters of products and design processes appropriate to each.
RULE 8	Eliminate all process uncertainties.
RULE 9	Understand, document, simplify and only then optimise (UDSO) the supply chain.

RULE 10	Streamline and make highly visible all information flows.
RULE 11	Use only proven simple but robust Decision Support Systems.
RULE 12	The Business Process target is enabling the Seamless Supply Chain.

Table 5: The Structured Approach - Twelve Proven Rules For Simplifying Material Flow Necessary in Agile Supply Chains (Source: Towill, 31)

specific business requirements. The type required determines, for example, where strategic stocks are to be located.

7. A PRACTICAL CHANGE PROGRAMME TO ACHIEVE AN AGILE RESPONSE

In working with both customers and suppliers to enable an integrated supply chain, "product champions" should take the following actions (Stalk and Hout, 18);
- *Firstly,* they work to provide each company in the chain with better and more timely information about orders, new products and special needs
- *Secondly,* they help members of the chain, including themselves, to shorten work cycles by removing the obstacles to time compression that one company often unwittingly imposes on another
- *Thirdly,* they synchronise lead times and capacities among the levels or among tiers of the supply chain so that more work can flow in a co-ordinated fashion up and down the chain.

These actions require a substantial and lengthy investment of leadership and management effort and the careful selection of all partners forming the chain. This must be backed up with supporting engineering and training programmes extending throughout the chain.

Follow-up to all these actions is vital, as is the establishment of simple visual but key measures of performance to monitor the effectiveness of change.

A typical example of real-world transition from mass production via lean production to agile response is summarised in Table 6. The BPR phases shown relate to the actual Change Programmes initiated by the Original Equipment Manufacturer (OEM) in moving from their Baseline (traditional adversarial supply chain) right through to Integrated Sales and Research enabling an agile response to market demand (Mason-Jones, Naylor and Towill, 34). The associated cultural change meets the "interprise" criteria laid down by Goldman, Nagel, and Preiss (14) as it exhibits the required integration with customer, business process, more co-operation with suppliers, and an entrepreneurial internal requirement. Note that the predicted improvement in supply chain performance (requiring vision and dedicated change management over a period of a decade) is so huge as to be beyond the imagination of an observer armed only with the traditions and knowledge of the Baseline state. These supply chain enhancements include lead time reductions by 10:1; greatly improved response to real marketplace demands; much higher stock turnover; and much more effective use of suppliers resources.

BPR Phase	Company Description	Operating Characteristics	Major Manufacturing Philosophy		
			mass	lean	agile
0	Baseline	1. Many adversial echelons 2. Upstream demand out-of-phase with marketplace. 2. Severe demand amplification. 4. Business as independent "fiefdoms"	↓		
1	Continuous Flow Manufacturing (CFM) including JIT.	5. Cellular concepts applied 6. Kanban controls introduced 7. Multi-skilling/Team working. 8. Quality issues "owned"	↓	↓	
2	Global Material Planning System (MPS).	9. MRP Logic integrated throughout chain 10. Total visibility of orders 11. Orders sent via EDI 12. Total stock and WIP control			
3	Vendor Based Integration Into Global MPS	13. Vendor rating programmes introduced 14. Vendor base slimmed down 15. Suppliers linked via EDI 16. JIT supply strategy introduced		↓	↓
4	Integration of Sales and Research and strategic use of the decoupling point.	17. Focus on reducing "Total Cycle Time" 18. Total visibility of marketplace 19. Distribution lead times reduced			↓

Table 6: Transition from Mass to Agile Production Observed During the Various Stages of an Electronics Products Supply Chain BPR Programme (Source : Mason-Jones, Naylor, and Towill, 37)

8. CASE STUDIES OF AGILE SUPPLY CHAINS

The three Case Studies included here have been specially selected from different market sectors. Of the three, Benetton (Case Study A) are the oldest established in agile mode, and have set the performance benchmark for ensuring that the right fashion goods in the right colour and size are on the retailers shelves as judged against current customer demand. Our Case Study B, CARPETNET has found a way to provide made-to-order domestic carpets installed in the customer premises within a one week time window. This agile response was made possible via a major technological breakthrough which reduced the actual production process from 4 weeks to 3 days. In other words this was an innovation way beyond the capability of continuous improvement programmes associated with the lean production paradigm and required "breaking the china" to make the necessary step change in production technology, which did not come cheaply. In the third Agile Case Study, C, Dell Corporation have taken a further step to embrace customised agility. Hence Dell go way beyond satisfying current marketplace demand by individualising the product to meet specific customer requirements. By shrinking and integrating an already agile supply chain, coupled with exploiting IT at the marketing end of the chain, the customised computing system is delivered direct to the customer within 7 days of placing the order.

Case Study A – United Colours of Benetton

A. *Background*

Benetton is a world leader in the design, manufacture, and marketing of distinctive casual apparel for men, women, and children. The product range is characterised by a wide range of colours, fashionable Italian design and youthful image. Products are offered on a world-wide basis to accommodate the needs of many markets. They also supply complementary products such as underwear, shoes, and fragrances thus providing "one-stop" shopping. Benetton is seen as an agile enterprise which has been operating in this mode for nearly two decades. It is a prime example of achieving agility via time compression.

B. *Actions*

The Benetton agile enterprise model is based on external production; indirect sales organisation utilising agents; widespread indirect retailing network, and centralised management and operations. In 1999 the agile enterprise produced 85 million garments composed of 7,000 different items with an average of five colours and four sizes. 90% were manufactured in Europe (80% in Italy). More than 400 sub-contractors are integrated into the production process. The origin of this agile enterprise can be traced ack to the mid-1980's when Benetton collapsed the replenishment cycle time (orders received in Europe to goods on shelves in USA) by a factor of at least 4 to 1. So shops could keep Benetton products on shelves in the currently selling colours and styles without carrying piles of inventory in the pipeline. The major engineering breakthrough was achieved by a combination of rapid information feedback to producers, organising cut-and-sew and dye houses to work in fast-turnaround, small-lot cycles, thus leveraging the huge reduction in replenishment cycle time.

Seamless flow of materials is a key factor in Benetton enabling agility throughout the supply chain. Flow is engineered and organised by the six Production Divisions handling wool products, jeans/trousers, coats/jackets, cotton products, shirts, and shoes. Each Division is responsible for production planning, material requirements definition and procurement, and prepares production batches for sub-contractors. All materials purchased for production are stocked in Division warehouses. The average raw materials stock is 20 days. At the end of the production process, finished goods are stored in customer addressed packages.

United Colours of Benetton (continued)

Manufacture is characterised by made-to-order production. A single job contains only one style with many colours and sizes and can be the aggregation of orders from many customers. Finished order product-picking is eliminated as the job matches the orders. In exceptional circumstances "an internal order" (libero) is permitted, but finished products remain as WIP.

Benetton is an advanced IT user which is seen as a competitive advantage in five major areas of business. These are *Collection* (assures integration between design and production); *Commercial* (linking between manufacture and the marketplace); *Production* (providing programming and control of all the external manufacturing units); *Distribution* (manages the centralised distribution of goods throughout the world); *Finance* (supports the centralised management of the cash flows throughout the agile enterprise). The present IT system connects the Benetton business with 15 subsidiaries, 400 travelling salesmen, 120 agents, and numerous stores in 28 countries.

The current Benetton trend is to encourage more decentralisation. The consequence is that a particular sub-contractor may be expected to carry out more operations and take responsibility for a bigger share of the action. As a consequence these sub-contractors will become larger with further rationalisation of the supplier base. Foreign plants act as first level sub-contractors with instructions and raw materials supplied centrally, but local sub-contractors scheduled by the plant. Finished goods are always forwarded to the Central Automated Warehouse (and not delivered directly to the stores). The mainframe computer connected to the warehouse manages goods-in, inventory, delivery schedules, and invoicing. When the decision to deliver is communicated to the warehouse, invoices, delivery data, and packing lists are automatically despatched. Hence there is seamless operation from the marketplace via order generation (by the Division), then production, then distribution, and finally right through to the cash management system.

C. Observations

The present way of achieving customer service level requires store managers to commit 80% of their orders seven months in advance of the spring/summer and autumn/winter seasons. These are then produced and shipped on a 20-day total order cycle time. The remaining 20% of orders, which can result from forecast errors, surprise hot sellers, or small fashionable collections, can be satisfied by agile response in 7-8 days total order cycle time. There is a monthly stock rotation, making it possible to have entirely new collections in the shops every few months. The daily influx of orders and sales data updates in line with market conditions enables divisions to immediately adjust manufacturing in line with actual demand. Note that the 20 day "normal" order cycle time usually means land/sea transport, whilst the 7-8 day "agile" order cycle time normally requires air freighting.

Source; Author based on:
(1) B. Zuccaro, Keynote Address, EUROMA Conference on Managing Operations Networks, Venice, June 9, 1999.
(2) M. Christopher, "Marketing Logistics", Butterworth Heinemann, Oxford, 1997.

Case Study B - CARPETNET Virtual Supply Chain

A. Background

Carpet manufacture and supply using traditional processes has a cumulative total lead-time of sixteen weeks. This involves five basic processes (make fibre: spin yarn: tuft carpet: dye carpet: back and shear) prior to warehouse stock piling. A total of five truck movements are also required in transit from the make fibre process to the retail store. Although much NVAT could be eliminated from the chain, traditional production processes still required up to four weeks to execute. But as far back as 1990 the marketplace was demanding immediate delivery of cut-piece orders. Once an end-customer has selected a style and colour they expect delivery and installation "now". In practice this may be interpreted as laying the exact required carpet in the customer residence within seven days of placing the order. Thus the problem facing the supply chain is providing an agile response within the one week window of opportunity. In this Case Study the "product champion" was the upstream fibre manufacturer who determined that the only way to provide responsiveness to customers was to take control of a very poorly organised value chain.

B. Actions

The fibre manufacturer decided the only option was to re-engineer the total value stream to create a Demand Driven Logistics System (DDLS). This required manufacture to be linked so tightly in production that the discrete businesses became virtually one entity. Within the new DDLS, when a customer orders a carpet from the retailer, the latter places an order directly upstream at the fibre manufacturer and simultaneously reserves capacity at the mill. The customer order then pulls the requirement through the value chain
resulting in the specific carpet size and colour being made and installed in one week. To achieve this 1-week Total Cycle Time goal, the cumulative actual production plus transport times cannot exceed five days. The key enablers here are minimal lot sizes and minimal cycle times plus fully integrated communications up and down the supply chain. To meet the target production cycle time a technical breakthrough on the manufacturing process was required. This necessitated a focus on what might be possible, rather than on how things are presently done. In management consultant jargon this has become known as "breaking the china" with the clear intention of seeing if it can be put back together in a better way. In this case the technological breakthrough (needed and achieved) required the fibre to be dyed uniformly before it was woven as a rug rather than dyeing the rug itself. The latter process, although not complex is very labour intensive and slow with the drying process alone taking 24 hours. The process scientists responded with the required new modus operandi they termed Solution Dyed Nylon.

In order to set up the Virtual Supply Chain (VSC) the "product champion" selected a few carpet mills as potential strategic partners against the following criteria;

        (a) trust   (d) investment record
        (b) attitude and philosophy   (e) improvement potential
        (c) competitive position

Each metric was estimated following a detailed analysis of mill performance. Joint Task Forces were then set up between the fibre manufacturer and selected mills. In business

CARPETNET Virtual Supply Chain (continued)

terms the VSC goals were partitioned according to fibre manufacturer/ mill market share and geographical penetration. Where both are high, the aim is to increase both mix and margin via JIT-type deliveries. In areas where both had a low share joint merchandising programmes are being introduced. Finally in the remaining market quadrants the stronger partner would promote the weaker.

The Lead-Time Reduction Team determined that some product rationalisation was essential. It recommended that the one-week Total Cycle Time service value stream should be concentrated on just the 10% of the product lines that yielded 52% of the mill's volume. Downtime was to be reduced by 50%, and average changeover times reduced from 4 hours to 15 minutes, quality costs reduced, and streamlined material flow engineered throughout the chain. The VSC developed an integrated computer networking/scheduling system spanning the entire supply chain. This is known as CARPETNET, from which the VSC takes its name. The introduction of CARPETNET was supported by training courses at all levels in the VSC thus enabling a uniformly high standard of IT competence.

C. Observations

The CARPETNET VSC is an agile system which for the value stream covering top selling items delivers the selected carpet fitted in the customer's residence within the total allowable time window of one week. In addition to engineering lean delivery processes, synchronising material and information flows, and providing rapid and transparent communications, the whole project has been enabled via a huge technological breakthrough. Developing the Solution Dyed Nylon production process has greatly reduced total cycle time by enabling dyeing to precede carpet manufacture, reducing the latter to an operation now measured in hours rather than many days. The outstanding achievement is turning a 16 week lead time "traditional" supply chain with large inventories and much waste into an agile VSC capable of meeting the customer one week delivery time window. Projected benefits from engineering the CARPETNET VSC include a 48% reduction in inventory with an associated one-off cost reduction of £M12. Customer service level is predicted to rise from 65% to 99%. Quality costs will be reduced by about 30%; also obsolescence is eliminated by making to order with pre-dyed yarn. Finally the new carpet mill layout substantially reduces material handling and associated costs so that the whole chain remains competitive in the new environment. Without the benefit of the VSC programme the likelihood is that these "players" may well have joined the many carpet supply chain companies who have already gone out of business.

Source; Author based on:
Johansson, H J, McHugh P, Pendlebury, A J, and Wheeler, William A III, "Business Process Re-Engineering - Breakthrough Strategies for Market Dominance", John Wiley and Sons, NY, 1993.

Case Study C - Dell Computer Corporation

A. Background

The recent phenomenal success of Dell Computer Corporation rests squarely on its innovative and finely tuned distribution channel for direct sales to customers This competitiveness is built around the concept of customised agile response. Customers have direct access to the computer manufacturer through various toll-free telephone numbers, over the Internet *(www.dell.com)*, and face-to-face in the case of large corporate and institutional accounts. Customers can receive product information, place orders, or speak directly to a Dell representative who can Fax back a quotation within minutes. Far less expensive than traditional channels, which use resellers and middlemen, the direct approach eliminates reseller mark-up and speeds up the flow of inventory. Under this system, the company can turn around a customer's order in five days, typically at a cost 10 percent to 20 percent lower than the customer would pay at a retail outlet. It also allows the company to move new technology such as advanced Pentium chips into its product lines faster.

A. Actions

Dell has become the master of the vertical distribution channel by being the sole distributor of its products and services. A customer initiates the sales process by contacting the company via telephone or the Internet. There are three ways of selling: face-to-face; ear-to-ear; and keyboard-to-server. A customer can order from Dell on-line 24 hours a day or by phone from early morning until late in the evening. A Dell representative is available to make suggestions and help customers determine what systems will best meet their needs. Through the Web site, customers can access product information and receive price estimates instantaneously (during the first quarter of 1997, the company sold £1 million worth of products *daily* on its Internet site). Dell then confirms the order and verifies the financial credit charge. Usually the representative promises that the computer will arrive within five business days although the customer often receives the product quicker than this. The Dell factory receives a printout of the order and begins manufacturing within hours. Each computer is customer-built and put through several hardware and software tests in less than one day. After a final inspection the computer is boxed by Dell and sent to a distribution centre that ships it by carrier in time to arrive with a monitor that is built ahead of time by a separate supplier.

Thus customer orders are satisfied by an agile execution-based, direct model driven business operation without any finished inventory. One way Dell Computer beats its competitors on prices is by also keeping component inventories to a minimum. Vertical integration has helped the company further reduce costs. Monitors, for example, are relatively standardised and built by an agile supplier. So today Dell may need as many as 8,742 monitors from the supplier; tomorrow they might need as few as 962. One of their carriers simply picks the monitors up at night, matches them up with the PCs by purchase order, and the next day delivers them to the customers. The rule on working with suppliers is to keep it simple, with fewer than 40 vendors providing 90% of material needs. This justifies close working relationships and reduces cost and further speeds upnew products to market. A financial analysis showed that "supplier proximity pays". Hence as Dell became a global manufacturer their preferred suppliers were expected to follow suit. Fast feedback from customers enables Dell suppliers to rapidly change product mix and maintain their

> **(Dell Computer Corporation continued)**
>
> inventory velocity. To make the required breakthrough on supply strategy, Dell share their goals and objectives with suppliers. By shipping as required (hourly or daily depending on the product), Dell bought more components and assemblies from the suppliers faster and paid them quicker, so everyone benefited.
>
> **B. *Observations***
>
> Examination of the present Dell "direct model" supply chain shows that it embraces our simplicity-by-design rules in abundance. We particularly highlight synchronisation of material flows: time compression: information transparency: simple and robust DSS; and echelon elimination. Thus vertical integration and direct marketing are the keys to Dell Computer's success. The company has totally integrated the distribution channel by clearly identifying its markets and by designing products and services to fit the needs of its customers. Hence the direct contact with customers gives Dell minute-by-minute input from the largest customer down to the individual purchaser in terms of what products they want and what new services they would like to see Dell develop. This information on present demand and future requirements is shared with Dell suppliers in a real-world example of effective supply chain partnering.
>
> Source; Author based on :
> (1) M. Dell and C. Fedman, "Direct from Dell : Strategies that Revolutionised and Industry", Harper Collins, London, 1999,
> (2) R. Hiebeler, T.B. Kelly and C. Ketteman, "Best Practices : Building Your Business with Customer-focussed Solutions", Simon and Schuster, NY, 1998.

9. SUMMARY AND CONCLUSIONS

Agile manufacture is hardly profitable when practised within an adversarial supply chain. However, it is a fundamental part of the agile (nimble and quick) supply chain. The latter is targeted at satisfying actual current marketplace demand using "pull" procedures. This contrasts within the traditional supply chain, where products are "pushed" towards the marketplace on the basis of long term forecasts. Thus the philosophy in the agile supply chain is to make what is selling, not to sell what has been made, often many months previously. Manifestly the Benetton, CARPETNET and Dell supply chains are agile exemplars. Indeed Dell are already seen as a benchmark company for electronic shopping in the 21st century (Kare-Silver, 38). Particularly noteworthy are their standards of marketing, logistics, and systems within the customised agile environment.

All three of our exemplars are "seamless" in that the players in the chain think and act as one. However, the route to agility was different, even if the vision is the same. Benetton has reached agility via a long established programme of relatively small step changes. In contrast CARPETNET reached agility only after a massive production technology breakthrough. Finally, Dell moved to the agile mode following a business appraisal leading to the implementation of a direct sales only policy and fundamentally re-engineering the supply chain to take full advantage of this new marketing opportunity.

The Change Programme Model needed to support the move to agile response is well established. It requires a holistic (systems) view of organisations targeted at the seamless supply chain. Then all the business processes throughout the enterprise must be designed for agility. This includes Total Cycle Time Compression so as to meaningfully reduce the time taken between customer need being identified and that specific need bring satisfied. There has to be a focus throughout the supply chain on the Learning Organisation involving everyone, and a suitable Change Model exists for monitoring progress from adversarial to agile operation. Finally structured methods of system analysis and synthesis are required. Process mapping and the twelve material flow simplification rules are helpful in establishing where we are and where we must go next.

In all three Case Studies (and indeed in our experience of every successful move towards agility), a product champion has to develop and market the vision, change the culture from adversarial to collaborative, and lead the physical and organisational re-design needed. Like all long lead time projects, engineering such a change requires an understanding of time scale. The planning has to be thorough and include meaningful but transparent measures of performance to ensure that rapid feedback corrects any movement off course. The real paradigm shift required is to create the environment where change is the norm and is a challenge to be overcome, not a challenge to be resisted.

REFERENCES

1. J. Stevens, *"Integrating the Supply Chain"*, Int. J. Phy. Dist. Mat. Man. 19, (1989), 8, pp 3-8.
2. D.R. Towill, *"The Seamless Supply Chain"*, Int. J. Tech. Man., 13, (1997), 1, pp 37-56
3. D.R. Towill, *"Supply Chain Dynamics – The Change Engineering Challenge of the Mid-90's"*, Proc. Inst. Mech. E. on Eng. Man., (1992), 206, pp 233-245.
4. G. Watson, *"Business Systems Engineering"*, John Wiley. Inc. New York, 1994.
5. R. Mason-Jones and D.R. Towill, *"Total Cycle Time Compression and the Agile Supply Chain"*, Int. J. Prod. Econ. 62, (1999) pp 61-73.
6. J.L. Burbidge and J. Halsall, *"Group Technology and Growth at Shalibane"*, Int. J. Prod. Plan. Cont., 5, (1994), 2, pp 213-218.
7. A. Gunasekaran, "Agile Manufacturing : A Framework for Research and Development", Int. J. Prod. Econ. 62 (1999), pp 87-105.
8. A. Werr, T. Stjernberg and P. Docherty, *"The Functions of Methods of Change in Management Consultancy"*, J. Org. Man.Change, 10, (1997), 4, pp 288-307.
9. P.T. Kidd, *"Agile Manufacturing: Forging New Frontiers"*, Addison-Wesley Pub. Co. New York, 1994.
10. M. Christopher, *"Logistics and Supply Chain Management"*, London, FT Pitman Publishing 1992.
11. D.R. Towill, *"Successful Business Systems Engineering"*, IEE Eng. Man. J. 7, (1997) Pt. I, 1, pp 55-64, Pt. II, 2, pp 89-96.
12. P.M. Senge, *"The Fifth Discipline"*, Century Business Books, London, 1993.
13. D.C. Morris and J.S. Brandon, *"Re-engineering Your Business"*, McGraw-Hill, New York, 1994.
14. S.L. Goldman, R.N. Nagel and K. Preiss, "Agile Competitors and Virtual Organisations", Van Nostrand Reinhold, New York, 1995.

15. J.B. Naylor, M.M. Naim and D. Berry, (1997), *"Leagility: Integrating the Lean and Agile Manufacturing Paradigm in the Total supply Chain"* J. Prod. Econ. 62, (1999) pp 107-118,
16. J.D. Schmidt, *"Achieving the Elusive Integrated Supply Chain"*, Proc. 2nd Ind. Eng. Res. Conf. (1993) pp 138-141.
17. D.Farmer and R.P. Van Amstel, *"Effective Pipeline Management: How to Manage Integrated Logistics"*, Gower Press, 1991.
18. J.W. Forrester, *"Industrial Dynamics"*, Cambridge MA, MIT Press, 1961.
19. G.H. Stalk Jr., and T.M. Hout, *" Competing Against Time: How Time Based Competition is Reshaping Global Markets"* Free Press, New York, 1990.
20. P.R. Thomas, *"Competitiveness Through Total Cycle Time"*, McGraw-Hill, New York, 1990.
21. Y.Y. Yusuf, M. Sarhadi and A. Gunasekaran, *"Agile Manufacturing: The Drivers, Concepts, and Attributes"*, Int. J. Prod. Econ. 62, (1999), pp 33-43.
22. G.N. Evans, D.R. Towill and M.M. Naim, *"Business Process Re-engineering the Supply Chain"*, Int. J. Prod. Plan. Cont. 6, (1995), 3, pp 227-237.
23. T. Davis, *"Effective Supply Chain Management"*, Sloane Management Review, Summer (1993) pp 35-46.
24. M. Cassell, *"Lament of the Big Spenders"*, Financial Times, 10th January 1994.
25. A. Kosta, *"Elders Pastoral Distribution"*, Int. J. Phy. Dist. Mat. Man., 21, (1991), 4, pp 15-20.
26. P. Child, R. Diederichs, F-H Sanders and S. Wisniowski, *"The Management of Complexity"*, Sloan Management Review, Fall (1991), pp 73-81.
27. S. Gilchrist, *"Stores Aim to Reclaim Buried Pots of Gold"*, The Times, 7 September, 1994.
28. J.P. Womack and D.T.Jones, *"From Lean Production to the Lean Enterprise"*. Harv. Bus Rev. (1994), March-April, pp 93-103.
29. D.R. Towill, S. Disney and P. Childerhouse, *"The Diffusion Dynamics of Supply Chain Management"*, Proc. EUROMA Int. Conf. Man. Oper. Net. Venice (1999), pp 321-328.
30. R.B. Handfield, and E.L. Nichols, Jr., *"Introduction to Supply Chain Management"*, Prentice Hall, New Jersey, 1999.
31. D.R. Towill, *"Simplicity Wins : Twelve Rules for Designing Effective Supply Chains"*, IOM Control, 25, (1999), 2, pp 9-13.
32. B. Victor and A.C. Boynton, *"Invented Here : a Practical Guide to Transforming Work"*, Harvard Business School Press, Cambridge, Mass. 1998.
33. K. Belk and W. Steels, *"Case Study : APS BERK from Arbitration to Agility"*, Log. Inf. Man. Vol. 11 (1998), 2, pp 128-133,
34. D. Berry, M.M. Naim, D.R. Towill, "Business Process Re-engineering of an Electronics Products Supply Chain", IEE Proc. Sci. Meas. Tech. 142, (1995), 5, pp 395-403.
35. D.R. Towill and P. McCullen, *"The Impact of an Agile Manufacturing Programme on Supply Chain Dynamics"* Int. J. Log. Man. 10, (1999), 1, pp 83-96.
36. P. Shewchuk, "Agile Manufacturing: One Size Does Not Fit All", Proc. Int. Conf. Manuf. Value Chains, Troon, (1998), pp 143-150.
37. R. Mason-Jones, J.B. Naylor, and D.R. Towill, *"Engineering the Leagile Supply Chain"*, to be published in Int. J. Agile Man. Syst, Spring 2000.
38. M. de Kare-Silver, *"E-Shock 2000; the Electronic Shopping Revolution – Strategies for Retailers and Manufacturers"*, MacMillan Books London, 2000.

Information Technologies for Virtual Enterprise and Agile Management

Nijaz Bajgoric
Bogazici University, Department of Industrial Engineering
80815 - Bebek, Istanbul, Turkey
E-mail: nijaz@boun.edu.tr

This paper aims at presenting conceptual models of virtual enterprise (VE) and identifying the role of information technology (IT) in establishing a virtual business. After a short introduction, section 2 proposes a set of models of virtual enterprise. Section 3 introduces the concept of Virtual Enterprise Information System - VEIS and defines the requirements against IT. In that context, IT-related contingency factors that determine the process of establishing a VEIS are explained. These factors, in other words VEIS technological platforms (hardware/operating system, communications-networking, applications and user interface), represent major IT technologies that support different aspects of virtual business. Additionally, through the concept of Agile Management Support System - AMSS some aspects of agile management are discussed.

1. INTRODUCTION

Contemporary corporations today, more than ever, are faced with tremendous competition in a rapidly changing environment. Markets are becoming more dynamic and customer oriented. To cope with increasing market demands, many corporations are turning to information technology in order to boost establishing of virtual enterprises. Such enterprises can be formed in different ways, depending mainly on the primary goal of entering virtual business. The main requirement is to connect several geographically separated entities involved in VE business (people, companies) in a way that they can communicate among each other and run the business. That is where information technology comes in, providing a set of resources for an efficient and effective virtual business.

Virtual reality in contemporary business has also changed the way managers do their job. The term "agile management" has been introduced with the concept of virtual enterprise with an emphasis on information agility or informational efficiency. In virtual enterprises, information agility represents the major prerequisite for agile management and means eliminating inefficiencies in accessing, exchanging and disseminating of all kinds of information. Information technology provides managers with a set of tools that eases access to corporate information.

2. VIRTUAL ENTERPRISE AND ITS INFORMATION ARCHITECTURE

Recent interest in the topic of virtual enterprises has been enormous, and continues to grow. Almost since the Internet era began, reserchers and IT-professionals have considered its application in business to be one of Internet's most valuable contributions. Not surprisingly, since the term "virtual enterprise" and its several aspects has become one of the hottest topics in the second half of the '90s. This section briefly summarizes some definitions of virtual enterprise. Several authors approach virtual enterprise by describing "virtuality" from different aspects.

The concept of "virtual enterprise" has been introduced as a result of an explosive growth of computer communications in the '90s, particularly with the emergence of Internet and Web technologies. In fact, it was the concept of Electronic Data Interchange (EDI) which " ... made a first step towards automating online business-to-business commerce, but fell short of providing the comprehensive communication environment needed." [1]

Camarinha-Matos et al. [2] describe the steps towards an architecture for virtual enterprise. They provide several definitions of the term "virtual enterprise" and other similar terms that are used in this context such as: extended enterprise, supply chain management, electronic commerce, cross border enterprise, network of enterprises, virtual corporation. All these terms represent related concepts. Byrne [3] defines a virtual corporation as a "temporary network of independent companies - suppliers, customers, even rivals - linked by information technology to share skills, costs and access to one another's market". In a similar definition, Jacoliene van Wijk et al. [4] say that "within the network, all partners provide their own core competencies and the co-operation is based on semi-stable relations. The products and services, which a Virtual Organisation provides are dependent on innovation and strongly customer based." A broader definition given by Jansen, Steenbakkers & Jägers [5] explains a virtual organization as a "Combination of various parties (persons and/or organizations) located over a wide geographical area which are committed to achieving a collective goal by pooling their core competencies and resources." This definition introduces a possibility of applying the concept of virtuality in all organizations, not only enterprises, including such organizations consisting of several persons. In our work we will mainly rely on this definition of VE.

Goldman, Nagel, and Preiss [6] defines the six reasons why a company would form a virtual enterprise:

- Sharing infrastructure, risks, and costs.
- The linking of complementary core competencies.
- A reduction in the concept to cash time.
- An increase in facilities.
- Gaining access to other markets.
- The transition from selling products to selling solutions.

Song and Nagi [7] propose a framework for an agile manufacturing information system (AMIS) and introduce the concept of Virtual Information System for Agile Manufacturing (VISAM).

Virtual enterprises can be formed in different ways. The most common approaches are:

- **Groupwork-based VE.** Group of people working together separated geographically can form a virtual enterprise (Figure 1). Doing such a kind of business is possible thanks to contemporary communications-networking and Internet technology (phone, fax, GSM, e-mail, Web, talk, videoconferencing).
- **Business specialization-based VE.** Group of small or large companies that specialize in some specific business activities can establish a new company on temporary or permanent basis (Figure 2).
- **Virtual manufacturing-based VE.** Business based on outsourcing some manufacturing operations can be done within a virtual framework (Figure 3). Actually, this is a new approach within agile manufacturing philosophy which is

called virtual manufacturing. Virtual enterprise framework enables such a business.
- **Extranet-based VE.** Virtual enterprise can also be the case in which a company decides to form special relationships with its customers and/or suppliers through an extranet infrastructure (Figure 4).
- **Distributed computing-based VE.** We can also consider a case in which distributed objects are used between two or more companies usually connected through Internet, as some sort of virtual business. These are applications created using distributed computing standards such as CORBA and DCOM (Figure 5).

In the section that follows, these models of virtual enterprises are briefly described. In either case, the main prerequisite for virtual enterprise is an efficient and effective IT infrastructure which is based on fast and stable communication backbone. Virtual Enterprise Information System (VEIS) is then established to support virtual business.

Groupwork-based virtual enterprise is very simple type of VE and its information architecture is based on two or more computers connected through Internet. Users from geographically separated sites share data and applications by using communication technology.

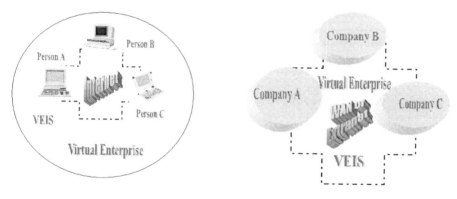

Figure 1.   Figure 2.

In the second case (Figure 2), a new company - virtual company is established from the partnering companies A, B, and C which enter a partnering business. Virtual enterprise consists of the segments of the existing businesses. Communication-network infrastructure can be based on either WAN or Extranet.

The solution from Figure 3, the virtual manufacturing-based VE model can be applied when companies enter a partnering business which requires much closer relationships among companies. Virtual enterprise is set as superset of the existing businesses. Model makes use of the concept of virtual manufacturing and in short means that a manufacturing company may outsource some of its work to subcontractors. Communications are provided either by Wide Area Network, Extranet or even standard Internet connection.

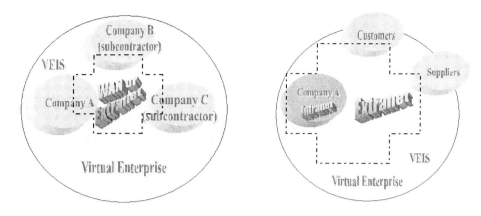

Figure 3.   Figure 4.

In Figure 4. the Extranet-based model of virtual enterprise is presented. In this model, Company A forms a virtual enterprise with its customers and suppliers, mainly through a well established Extranet. This model is sometimes called "extended enterprise" in which a dominant enterprise "extends" its boundaries to all or some of its suppliers [8].

Distributed computing (Figure 5) is a special case of virtuality in running computer applications. It is a new, object-oriented and Internet based framework in which different modules of an application may be running on separate computers on Internet. Applications can also share some objects so that they ask for some services from them. As can be seen from Figure 5, several distributed objects from different locations communicate among each other participating within the same application or coming from different applications.

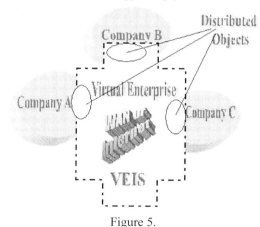

Figure 5.

3. VIRTUAL ENTERPRISE INFORMATION SYSTEM

The appropriate design of a Virtual Enterprise Information System, referred to here as VEIS, is of central importance in applying IT in virtual enterprises. In any type of VE model, its VEIS must provide a set of IT facilities that support VE business efficiently and effectively. The organizational requirements depend on the type and complexity of VE, therefore, a contingency approach should be respected. VE-related contingency factors can be extracted from the models defined above (group of people or group of companies, temporary or permanent business, virtual manufacturing, relationships with customers and suppliers, etc.). On the other side, the following requirements are expected to be fulfilled by information technology:

- Multiple hardware platform support (different hardware platforms can be used in companies involved in VE).
- Data and application integration support (support for different data formats and application platforms).
- Support for high-speed LAN and WAN technologies.
- Support for multiple hardware and software communication protocols.
- Efficient remote access for telecommuters.
- Secure data transfer.
- Efficient access to all kinds of information.

These requirements can be grouped into the following IT dimensions or IT-related contingency factors for establishing an efficient and effective VEIS architecture:

- Hardware/operating system platform
- Communications-networking platform
- Application platform
- User Interface platform

3.1. Hardware/OS Platform

Today's computer platforms are classified mainly with respect to the type of computer system configuration and operating system which is running on it. Apart from desktop computing, when mission-critical applications are considered, they are installed either on the old, "big-iron" mainframe computers, proprietary minicomputers, or contemporary enterprise server systems. The following platforms are most commonly used today:
- Mainframes running mainly IBM MVS/OS390.
- UNIX servers (HP HP-UX, Sun Solaris, IBM AIX, Compaq Tru64UNIX, SGI IRIX, etc.).
- Proprietary systems running single-vendor operating systems (VAX OpenVMS, Alpha OpenVMS, AS/400, HP MPE, etc.).
- Intel servers running Windows NT Server, Novell NetWare, OS/2 Warp Server, or Linux.

In the last decade there has been a lot of discussions on which platform is more suitable for running business-critical applications. Although there are many requirements that should be taken into account when speaking about that suitability (hardware platform support, application support, applications development tools support, network support, systems management), major issue is almost always related to so-called RAS model (Reliability, Availability and Scalability) of a specific platform. From that viewpoint, it is well known that mainframes have been delivering exceptional systems availability and reliability for long time. This means that end users can use data and applications and system continues running even when systems or application software is being upgraded or backed-up. In addition to this, mainframes have the possibility to support parallel databases - a platform which is very important for fast and efficient access to large amounts of data (data warehousing applications, OLAP systems, integrated decision support applications, ERP systems).

With new approaches in multiprocessing technologies such as SMP (Symmetric Multiprocessing), Clustering, and NUMA (Nonuniform Memory Access), UNIX vendors and vendors of some proprietary systems can also provide mainframe-like uptime environment.

Just to mention some examples of such platforms: a well-known and 20-years present OpenVMS clustering system, Silicon Graphics' servers with NUMA ("cluster in a box") technology, new Sun's High Performance Computing ClusterTools which allows connecting up to 16 Sun UE10000 servers, each of them can work with up to 64 UltraSparc II RISC processors, HP's Enterprise Parallel Server with support of several dozens 64-bit V-class servers, each of them supporting up to 32 processors. Also, with new facility called Dynamic Loadable Kernel, UNIX operating system can now be upgraded without the need of shutting it down. In short, UNIX advantages include: performance, clustering, robust systems management tools, widespread applications support, the ability to handle very large databases, and scalability.

On the other side, Windows NT Server has many advantages at the workgroup and small business size level. NT's key advantages over UNIX are: ease of use and administering, price, support and integration with API Windows application development environments. Most major application vendors have written NT versions of their packages, and some 45% of the implementations of SAP's R/3 enterprise resource planning package these days are on NT, according to SAP. PlugIn Datamation/Cowen findings indicate that NT leads in the Web server market: 50% of survey respondents are using NT as a Web server today, and that number will climb to 62% in the next one to two years. NT also has made a big impact in file and print: some 43% of those surveyed are using NT as a file-and-print server today and 56% will use it as a file-and-print server in the next year or two [9].

In the context of hardware/OS platform, it is noteworthy to see some results from a survey made by Gallup Organization. According to them, the availability of mainframes can be as high as 99,999%, which corresponds to a downtime of less than five minutes per year. The average downtime for a PC server is measured at 1.6 hours per week, while UNIX systems with RAID devices and clustering technologies can have close to 99,99% availability, which, in terms of the system downtime can be as much as an hour or more per year [10].

From VE perspective, the hardware/OS platform is important from the following aspects:

- Application servers, data servers, e-mail servers, and Web servers should run on reliable machines with high availability ratio.
- Servers must be scalable enough, because as time goes on, new companies with new users may wish to enter VE business.
- Servers supporting VE business must support open hardware-software communication protocols in order to be able to exchange data.
- Servers must support remote access capabilities.

3.2. Communications-Networking Platform

With the enormous growth in computer networking, business data communications, and particularly Internet, the demand for fast, reliable and cost-effective data communications backbone has also been growing. For purposes of a virtual business, several high-speed communication technologies are available, such as Fast Ethernet, FDDI, leased lines, ISDN, ATM, xDSL, cable modems, wireless connections, etc. Most of these technologies are costly to implement and maintain, therefore the selection of the appropriate one is a critical point in creating a VEIS architecture. Many factors can be involved, some of them are: type of the VE, geographical distances among the companies involved in VE business, hardware platforms of partners involved in VE business, application platforms of

Information Technologies for Virtual Enterprise and Agile Management

partners involved in VE business, number and characteristics of remote users, availability of specific LAN/WAN technologies that are provided by national Telecom company.

From the viewpoint of VE, the technologies that improve WAN performances and remote access are of most importance. In the section that follows, these high-speed networking technologies are shortly explained.

3.2.1. Technologies for WAN infrastructure

Leased lines have been traditionally used as a WAN backbone for establishing inter-company network infrastructure. In VE business, they are also suitable for connecting companies that form Virtual manufacturing-based or Extranet-based models of VE. From technological perspective, they are telephone lines that are leased for private use forming a dedicated phone line between two points. Leased lines are capable of carrying data at several rates, ranging from 56 Kbps, up to 1, 2, or more Mbps. T1 lines are widely used as major data transfer backbone in USA and have a capacity of 1.544 Mbps. T3 lines transmit data at thirty times that rate. For small businesses with few users who rely on standard utilization of e-mail messaging system and Internet a 56Kbps leased line would be enough. Businesses that rely on heavy e-mail messaging traffic and heavy use of Web technologies should select a T1 or even T3.

FDDI (Fiber Distributed Data Interface) is both LAN and WAN technology. It is mainly used as a network backbone connecting two or more LAN segments. A simple backbone might connect two servers through a high-speed link consisting of network adapter cards and cable. Fiber channel refers to a relatively new technology with the most common usage being in connecting clustered servers in distributed computing environment. FDDI and Fiber channel support data transmission speeds of 100 Mbps. This technology may be a choice in connecting information systems of partnering companies in Virtual manufacturing-based model of VE.

X.25 and Frame Relay packet-switched network protocols. X.25 is a simple, commonly used and inexpensive WAN technology. Although it is widely available, X.25 is slow compared to newer technologies. Frame relay works at the data-link layer of the OSI model and provides data transfer rates from 56 Kbps to 1.544 Mbps. Frame relay services are typically provided by telecommunications carriers. This technology is less expensive than other WAN technologies because it provides bandwidth on demand, rather than dedicating lines whether data is being transmitted or not. A version of Frame Relay called International Frame Relay is suitable as a WAN backbone for those VE companies with partners abroad.

Cell Relay or ATM (Asynchronous Transfer Mode). ATM is also both LAN and WAN technology which is usually implemented as a backbone technology. ATM is very scalable networking platform, with data transfer rates ranging from 25 Mbps to 2.4 Gbps.

Synchronous Optical Network. Synchronous Optical Network (SONET) is a WAN technology that works at the physical layer of the OSI model. It provides data transfer rates from 51.8 Mbps to 2.48 Gbps.

VPN. Virtual Private Network (VPN) is a way or organizing a WAN infrastructure by using public switched lines with secure messaging protocols. Actually, public Internet infrastructure is used for business data communications. It should be noted here that the attribute "virtual" does not mean necessarily that it is a dedicated platform for virtual business. VPN infrastructure can be used in any type of business, not only virtual business. With the VPN, users from remote locations (branch-offices) not only access a company messaging system (e-mail and faxing), its intranet, they also can use applications running on servers. WAN-VPN platforms are usually established, maintained and managed by telecom companies or ISPs. They are then outsourced to companies willing to use this type of WAN. If a company wants to keep control over its WAN-VPN infrastructure it may chose to build its own VPN, instead of outsourcing it. This approach is cost effective for companies with a number of remote offices, not only because of making an efficient network connection but also because of the possibility of a centralized network management.

The VPN-based WAN usually includes:
- Gateway that encrypts data packets and authenticates users. VPN gateways sit behind firewalls that at most sites are incorporated into the routers.
- VPN management software that lets network managers configure and manage VPNs from a single computer. This software is usually sold in the form of integrated suite which integrates hardware, software, and services, in order to simplify deployment of VPNs. VPN system requires Firewall and Tunneling software with LZO compression utility which improves dial-up connection.
- Client software for users to connect remotely. It allows telecommuters, mobile workers, and other remote users to take advantage of dialed Internet connections for convenient, low-cost, secure remote access.

VPN concept has though some disadvantages. VPN-based WANs are slow because of data compression, less robust and vulnerable to hackers.

3.2.2. Technologies for remote access

56 Kbps Dial-Up Modem Connection. New technology called x2 introduced by U.S. Robotic, together with V.90, a data transmission recommendation developed by ITU (International Telecommunications Union), provides a specification for achieving data transfer speeds of up to 56 Kbps over the standard public telephone lines. With standard V.42 compression, 56K modem technology can download at speeds up to 115 Kbps.

ISDN. ISDN is a set of protocols that integrate data, voice, and video signals into digital telephone lines. ISDN offers data transfer rates between 56 Kbps and either 1.544 Mbps or 2.048 Mbps, depending on the country telecom infrastructure. It requires special equipment at the users' site; user can talk on the phone and make file transfer at the same time. In addition to remote access, ISDN can also be used as WAN backbone.

Cable Modem. Cable modem technology makes use of cable-TV infrastructure by hooking up computer to a local cable-TV line. It is expected to replace standard dial-up and ISDN connections very soon since data transfer speed can reach 1.5 Mbps.. The cable system is a shared medium, which is a fact that should be taken into consideration when thinking of such a type of connection. The cable modem usually has two ports: Ethernet port for attaching to a standard Ethernet card in the computer, while the other port is Coaxial port which is used for plugging in-coming cable-TV wire.

ADSL. In the recent years several versions of DSL technologies (Digital Subscriber Link) have emerged. Because of several possible models, this technology is often referred to as xDSL technology. The xDSL is a digital packet technology like ISDN, but technology that usually uses a dedicated rather than a switched connection. With the appropriate devices, it can deliver signals at the speeds in the range from 1.5 to 6 Mbps over the current telephone wiring system. Therefore, Asymmetric DSL or ADSL, is often considered as an alternative to dial-up and even ISDN.

Wireless LAN and Wireless Internet. With a wireless LAN technology, mobile users can connect to a local area network through a radio connection. A wireless LAN is a data communication system that uses electromagnetic waves for transmitting data over the air. It can be implemented either as an extension to the existing standard LAN, or as an alternative for it. It has gained a public interest with the emergence of remote access computing devices such as notebooks, hand-held computers, and personal digital assistants (PDAs). These devices can be used for more efficient and effective communication among users, as well as for data exchange with host systems. Wireless Internet access is also supported. For example, hand-held PCs running Windows CE operating system include the Pocket Internet Explorer browser for remotely accessing Web or company's Intranet.

3.2.3. LAN technology - Ethernet and Fast Ethernet

Ethernet protocol is a typical LAN technology. Standard Ethernet-based local area networks transmit data at speed up to 10 Mbps. New Ethernet cards known as Fast Ethernet represent high-speed LAN technology as it can provide data transfer rates as high as 100 Mbps. Two new Ethernet standards that are currently being developed are Gigabit Ethernet (up to 1000 Mbps) and 10 Gigabit Ethernet (with data transfer rate of 10,000 Mbps).

As Ethernet cards are used for connecting computers to LANs, they are in the same time an entry point in establishing connection to a WAN and Internet, hence their importance in VE business. Additionally, Ehernet cards are today used in a combination with other communication devices for remote access, e.g. cable-modem technology.

3.2.4. Communication protocols and applications for virtual enterprises

Communication protocols and applications are in fact what drives a virtual business. Protocols are, in short, sets of hardware/software rules that communication end-points must follow in order to exchange some sort of information. Starting from e-mail, other Internet services (telnet and FTP), Web-based technologies, videoconferencing technology, transactions - oriented applications such as EDI and electronic commerce, together with both hardware and software communication protocols, these applications enable companies to organize virtual business. Communication applications usually come in pairs with software protocols that enable them. For virtual enterprises, the following combinations are the most important: TCP/IP-Internet, e-mail/SMTP, Web/HTTP-HTML, WAP (Wireless Application Protocol). Web technology, for example, provides a platform for establishing intranet and extranet applications through which companies create virtual enterprises.

As a supplement to standard e-mail messaging technology, videoconferencing technology enables remote users not only to communicate among each other, exchange standard data, but organize virtual meetings, exchange video and audio data, and share data and applications as well. In order to be able to use this technology, in addition to a standard PC, additional set of hardware-software facilities is needed. It includes a camera that is usually installed on top of a PC, speakers, a microphone and a videoconferencing software. Using videoconferencing technology in contemporary business is on the rise, as prices for equipment and communications fall. This technology is especially suitable for Groupwork-based model of virtual enterprise, but can be used in other models as well.

Making a Business-to-business electronic commerce may take many forms, depending on technology that is used. Over the last decade, Electronic Data Interchange - EDI has been used as a form for exchanging business documents over private networking infrastructures, using a pre-defined data-document format. With the explosion of Internet and Web technology, EDI has been replaced partially by Internet-based electronic commerce applications which include several forms like: online catalogs, virtual malls, online buying and selling, etc. The main advantage of e-commerce over EDI is that no additional equipment is needed, transactions can be made over public network infrastructure, whereas the primary concern with e-commerce is still security.

3.3. Application platform

Basically, there are four types of business applications: (1) transaction processing applications - applications that capture business data in the course of doing all business operations, (2) business intelligence applications that aim in improving decision making performances (3) messaging and collaboration applications and (4) document management applications. Though these systems are separate, they are inter-related and some sort of integration is a prerequisite. Therefore, enterprise application integration tools play a critical role in every information system.

Consequently, in a virtual enterprise information system, the application platform consists of several applications such as (Figure 6):

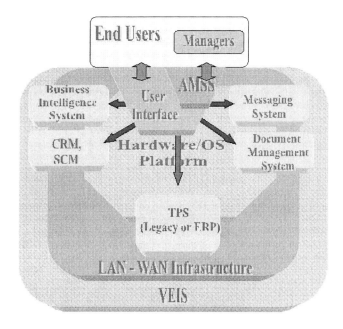

Figure 6.

- Transaction Processing System
- Messaging and Collaboration System
- Document Management System
- Business Intelligence System
- CRM/SCM systems, as special cases of TPS for managing relations with customers and suppliers
- Enterprise Application Integration Tools.

In the area of application platform the six technologies that emerged in the last couple of years are of most importance for the concept of virtual enterprises:
- Web-based access to legacy systems
- Object-oriented application development paradigm and Distributed Computing
- Web-enabled applications
- ERP integrated suites
- Middleware (Enterprise Application Integration Tools)
- ASP model of running ERP applications.

Web-based access to legacy systems. According to Gartner group (www.gartner.com) 74 percent of all corporate data still resides on legacy mainframes. Meta Group also estimates that more than 70% of corporate data in the world is still on mainframe systems (www.metagroup.com).

Legacy Systems (Legacy Data or Legacy Applications) refer to older or mature applications which were developed from the late '50s to the early '90s. Such systems are mainly mainframe systems, or distributed systems where the mainframe plays the major processing role and the terminals and/or PCs are used for application running and data uploading-downloading. Most companies are still relying on these platforms because mainframe systems are more secure, available, reliable, and scalable than UNIX and especially NT systems.

In information technology (IT) history, the invention of Graphical User Interface (GUI) was a revolutionary step in improving both efficiency and effectiveness of IT end-users. GUI interface has become dominant both in operating systems (e.g. MacOS, Windows 9x, OS/2 Warp) and in application software. After introducing Web technology in 1994, it turned out that Web browser is the most convenient way of using computers for end-users since it is completely based on a "mouse-click" operation. This became possible thanks to HTTP protocol, HTML language, and other Internet/Web related facilities

The job of IT people, both IT vendors and IS staff in organizations is to make access to information seamless and easy, especially for managers. In contemporary conditions, it is not reasonable to expect the decision makers to spend their time for specialized training in order to be able to use any software. From the perspective of easiness of use it is the Web technology that can help in that sense. Web-to-host access tools are software products that ease the process of connecting to several types of data: legacy data, c/s applications, e-mails, faxes, documents, etc.

Web-to-host access tools as a specific subset of Web technology are used to improve and ease access to several types of information: legacy data, messaging system, electronic documents, business intelligence, and so on. Access to legacy data through user friendly applications (standard client/server applications and Web-based applications for Intranets and Internet) requires a processing layer between the applications and the data. Web-to-host technology makes it possible for users to access the data stored on legacy hosts just by clicking a web link. What's more, it cuts the costs of software ownership through centralized management.

Distributed computing paradigm is a new framework within the object-oriented software engineering paradigm in which different parts of an application may be running on separate computers on LAN, WAN, or even Internet. There exist two standards that are used for providing such application environments:

CORBA (Common Object Request Broker Architecture) is an architecture and specification for creating, distributing, and managing distributed program objects in a network. It allows programs at different locations and developed by different vendors to communicate in a network through an "interface broker." CORBA was developed by a consortium of vendors through the Object Management Group (www.omg.org).

DCOM (Distributed Component Object Model) is a set of Microsoft concepts and program interfaces in which client program objects can request services from server program objects on other computers in a network. DCOM is based on the Component Object Model (COM), which provides a set of interfaces allowing clients and servers to communicate within the same computer. Creators of CORBA and Microsoft have agreed on a gateway approach so that a client object developed with DCOM will be able to communicate with a CORBA server, and vice versa.

By introducing new approaches in application development, object-oriented paradigm has contributed to the overall agility within information systems. Applications have become more flexible, application components more reusable and scalable, as Dove [11] pointed out "… Adaptability (Agility) actually became a reasoned focus with the advent of object-oriented software interests in the early '80s."

Web-enabled applications. With the increased deployment of client/server applications, system administration and version control have become a significant problem for IT professionals. One of the main advantages of legacy mainframe systems was that the application resided on a single system. There was no need of any software on the client side. This made the process of software upgrades much easier to manage. Web applications bring in a very similar advantage. In such a platform, the only software that is needed on the user's PC is the Web browser. Web-based application development paradigm can be considered in some sense as an environment very similar to a mainframe paradigm. Web server plays the role of legacy system whereas the Web browser replaces a text-based terminal.

ERP integrated application suites. Enterprise Resource Planning software is today considered as a system that aims to serve as an information backbone for the whole organization. The crucial point is in an efficient integration of all business processes with an emphasis put on reporting and business intelligence capabilities important for management. Enterprise Resource Planning (ERP) provides comprehensive information management for organizations. Even though ERP projects are costly and need several months to be implemented, in the last couple of years many companies have replaced their legacy systems with these new integrated suites. ERP systems employ client/server technology within, mainly, three-tier c/s platforms. This means that a user runs an application on a client computer (first tier) that accesses information from an ERP application (finance, HR, logistics, etc.) which is on application server (second tier). Applications use data from the third tier - data server which is running one or more databases under DBMS. ERP software vendors are rearchitecting their applications to be used over the Internet, through Web-browser interface.

The major advantages of an ERP system include: integration of data from all business processes, easier information access for managers, Y2K compliant and Euro enabled, stable and reliable data structure, customizable and adaptable application platform, module based infrastructure, GUI and Web-based user interface. Disadvantages include: expensive and lengthy to implement, many hidden implementation costs, maintenance is costly and time consuming, commitment to a single application vendor for mission-critical applications.

ERP is an enterprise-wide solution. The successful deployment of the ERP suite results in an enterprise that has an integrated information flow between different business functions, and even different business locations. ERP applications include support for many aspects of an internationally-based business (local currencies, taxes, etc.), therefore, they are very suitable for virtual enterprises. Norrish [12] refers to ERP system as an "agile software for agile manufacturing"

Middleware. An efficient access to legacy data is important from application developer's perspective as well. The development of new c/s applications which will exchange data with existing legacy systems requires a sort of middleware that overcomes the differences in data formats. Different data access middleware products exist and they are specific to a single platform, e.g. RMS files on OpenVMS machines, IBM mainframes, or different UNIX machines.

ISG Navigator (www.isg.co.il) is an example of data access middleware tool that provides an efficient data exchange between Windows platform and several host platforms such as OpenVMS for Digital Alpha and VAX, Digital UNIX, HP-UX, Sun Solaris and IBM AIX. ISG Navigator enables access to non-relational data in almost the same way that relational data is accessed. What is more important, application developers can build new Internet-based applications which will use data from legacy systems by using data integration standards such as: OLE DB, ADO, COM, DCOM, CORBA. Other examples include: *ACUCOBOL* from *Acucorp* (www.acucorp.com), ClientSoft's *ClientBuilder Enterprise* (www.clientsoft.com).

While middleware products serve as a data gateway between legacy systems and Windows-based c/s and desktop applications, Web-based application development products support building Web-enabled c/s applications (e.g. *Microsoft's Visual Studio, Inprise-Borland's Delphi, C++Builder and JBuilder*).

Many Web-to-host products offer APIs to host systems that developers can use to build custom intranet applications, especially for reporting. This model usually includes taking data from host systems, converting them into HTML format and placing onto Windows NT IIS which acts as Intranet server.

Another important area of business applications in which integration is needed is implementation of ERP systems. This activity usually means a lot of customization efforts. Many companies decide to continue with some legacy code and seek for a way of connecting ERP with legacy systems. New generation of enterprise integration application (EIA) or enterprise application integration (EAI) products provide that kind of integration (e.g. *Prospero* - www.oberon.com, *CrossWorlds* - www.crossworlds.com, *Level8* EAI products - www.level8.com, etc.). ERP vendors like SAP, Oracle, Baan and others are developing their own front-office solutions in order to achieve a higher level of integration. Also there is a need of an efficient and effective integration of ERP systems with messaging system, document management system, business intelligence system, etc. Therefore, the vendors of these applications provide ERP-gateways to integrate their programs with ERP systems.

After the successful implementation of the ERP software, companies usually add business-intelligence tools to their ERP systems to enhance access to data and improve organizational decision making. ERP vendors provide such business-intelligence products the core of which is always a data warehouse (*SAP Business Information Warehouse, Oracle Business Warehouse, PeopleSoft Enterprise Warehouse*).

ASP model of ERP implementation. "Rent, don't buy" is a new approach in deploying corporate-wide business-critical applications. Application Service Providers (ASPs) are the companies that rent applications-running platforms, mostly those applications that are very complex and hard to implement (ERP, data warehousing, electronic commerce, customer relationship management). Actually, they emerged recently as a result of an effort to make ERP suite an application platform for small and mid size companies.

Traditional approach in implementing ERP packages was based on a single license for this software that could cost thousands of dollars per seat (mainly between $2000 and $4000), but the real expense was in implementing these programs (consulting, process rework, customization, integration, testing). ERP implementation costs should fall in the range of $3 to $10 per dollar spent on the software itself. Unlike ISP (Internet Service Providers) - the companies that provide Internet access and standard Web hosting, ASPs help companies in such a way that they install, implement and manage complex applications on their sites and bill these services usually on monthly basis. They are providing application hosting services mostly by partnering with software vendors and networking companies. Rental fees include software customization, integration with other back-end systems and ongoing maintenance of the apps at fault-tolerant data centers.

ASP model of renting ERP and other business-critical applications represents an example of virtual business, in fact, this is the version of Business specialization-based model of VE. Companies prefering to focus on their business only may decide to outsource running business applications to another company which is specialized in it - application service provider - ASP. Application and data servers are usually located in ASP company, while applications and data are accessed on remote basis. The main prerequisite for this model of VE is a high-speed and reliable communications backbone.

3.4. User Interface In A Virtual Enterprise and Agile Management Support System

Access to corporate information today is determined firstly by a device that is used by end users. From that viewpoint, the types of information access are:
- Terminal-based access
- PC-based access
- Portable devices-based access.

The vast majority of end users access corporate information from desktop computers, mainly by using the following programs:
- PC-Terminal emulation programs
- PC-X Windows programs
- Standard client programs within c/s application platforms
- Web-to-Host access tools
- Web-enabled client programs within c/s application platforms
- Enterprise Information Portal applications.

As mentioned above, there are many types of user interface in use today, ranging from character-based, menu-based, standard GUI, and on up to the Web browser-based interface. Also, there are several devices that are used to connect to hosts. In the context of an VEIS, we can view the entire user interface as a system that provides users with an efficient and effective access to all kinds of information. In that sense, an Agile Management Support System (AMSS) is defined as a subsystem of user interface that provides managers with an efficient and effective access to the information they need (Figure 7).

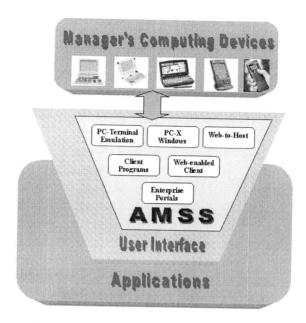

Figure 7. An Agile Management Support System

The primary design goal of an AMSS is that the information it contains be easily accessible and retrievable by managers at the time they need it. Actually, AMSS does not have to contain all that information, rather it should provide a way of accessing it, no matter where that information is stored, and which device user connects from. Early efforts to develop such systems were limited to the implementation of terminal emulation access tools and PC/X Windows emulation programs. An extended scope of AMSS began with the advent of Internet and Web technologies. The structure of AMSS is very dynamic since it is based on the available data access technology. The lowest level of AMSS is based on using PC-terminal emulation tools, whereas the most sophisticated solutions include enterprise portal solutions. The technologies that determine the level of AMSS are explained bellow.

3.4.1. Traditional data access: Terminal, PC-Terminal Emulation, PC-X emulation, and Client programs

Access to corporate data has always been determined by type of information architecture which an information system is built on (Figure 8). In the mainframe environment, the processing is done by a mainframe computer, while the users work with "dumb" terminals. The terminals are used to enter or change data and access information from the mainframe. This was the dominant architecture until the late 1980s. A version of this architecture is an architecture where PCs are used to connect to host machines through so-called PC-terminal emulation programs. Traditionally, the access to legacy data has been confined to dumb terminals and PC-based terminal emulation software. However, today as more users wish to standardize on Web browser-based client access, software vendors are being pressured to provide so-called Web-enabled versions of their applications.

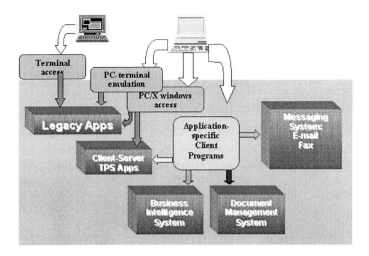

Figure 8. Traditional information access

PC/X Windows software serves as an emulation program that emulates UNIX GUI based on X-Windows standard on desktop computers. With this software, GUI-based UNIX applications can be used directly from PC desktops.

A client/server-based information architecture divides processing into two major categories: clients and servers. A client is a computer such as a PC or a workstation attached to a computer network consisting of several dozens (hundreds or thousands) clients and one or more servers. A server is a machine that provides clients with services. Examples of servers are the database server and SMTP server that provides e-mail services. Each client/server application has its own client program that needs to be installed on all client machines.

3.4.2. Web-enabled Access

PC-terminal emulation tools and PC/X-based access can help end users, particularly managers, in improving informatioal agility by streamlining information access from their desktops. Another set of activities important for extending virtuality in modern business is based on implementation of so-called Web-to-Host access tools. In contemporary conditions, there is a requirement to standardize a client software as much as possible.

Web-to-Legacy tools. Access to legacy data through user friendly applications (standard client/server applications and Web-based applications for Intranets and Internet) requires a processing layer between the applications and the data. Web-to-host technology makes it possible for users to access the data stored on legacy hosts just by clicking a web link. With the emergence of Web technology and Web browser as a unique GUI interface, independent software vendors (ISV) started working on Web-based gateway or middleware products which should provide browser-based access to corporate legacy data. Another reason for using Web browser interface for host access are cost savings that can be achieved in the total-cost-of-ownership model for client systems. According to a recent Gartner Group study, businesses realize 15% savings in software costs such as distribution and maintenance and 15% in technical support from replacing IBM 3270 terminal emulation software with a Web browser (www.gartner.com).

All these Web-to-host tools are created for different host/OS platforms: IBM OS/390, IBM OS/400, Digital/Compaq OpenVMS, or for specific application platforms (e.g. COBOL apps, RMS-based apps, etc). Some of them provide only access to host data, whereas some programs figure as middleware or gateway in a way that they enable adding GUI capabilities, integrating with c/s apps, converting non-DBMS data into DBMS format, e.g. *ISG Navigator*.

Programs are mainly based on the host-emulation server - software that runs on any Web-server platform. The emulation server is used to download Java or ActiveX applets to browser. The applets permit the browser, to establish the connection to the host using appropriate terminal emulation protocol: TN3270 for IBM mainframes, TN5250 for IBM AS/400 systems, and VT100-400 for Digital VAX/Alpha systems.

A recent report by International Data Corporation (www.idc.com), found that the worldwide market for Web-to-host browser license shipments is exploding: from 67,000 desktop licenses in 1996 to an estimated 17 million in 2002. IDC also predicts that shipments of Web-to-host gateways will surge from 2,200 units in 1996 to more than 330,000 in the year 2001, representing sales in excess of $1 billion.

Web-to-Reporting programs go a step further and provide reporting facilities applied on host data. For example, *Report.Web* (www.nsainc.com) is a Web-to-Reporting program, actually an intranet report distribution tool from Network Software Associates Inc. At the heart of Report.Web is the Enterprise Server, a powerful and robust engine which automates the entire process of delivering host-generated reports to the Web - from almost any host, including IBM mainframes, AS/400s, DEC VAXes, and PC LAN servers, to the corporate intranet/extranet. Report.Web also supports distributing ERP-generated reports across the corporate Intranet, without deploying ERP clients at every desktop.

Many other applications are "webified" too. Just to mention some examples:

- **Web-to-Mail or Mail-to-Web** program is a service which lets users to use their POP3 email accounts through an easy Web interface.
- **Web-to-Fax** program, which is very similar to Web-to-Mail, gives an opportunity of sending and receiving fax documents from Web browsers with no additional software.
- **Web-to-GSM** software allows users sending GSM messages through a Web-browser interface.
- **Web-to-Document Management and Work Flow Systems.** They provide a Web-based access to user documents supporting at the same time an efficient integration with company's messaging system.
- **Web-to-Business Intelligence Systems.** The browser-based access to Business Intelligence Systems is very important for decision makers because of its easiness of use.
- **Web-enabled Desktop DSS Tools.** Decision modeling by using desktop DSS tools is also available via Web browser. Decision maker can use a model which is already created and stored on a server from his/her computer through Web browser.
- **Web-enabled EIS**. Executive information systems or integrated reporting applications provide a user-friendly access to corporate data.
- **Web-to-ERP Systems.** ERP vendors are also working on releasing Web-enabled ERP application suites. New versions enable ERP users to use applications via Web browser. For example, SAP has introduced *System mySAP.com,* an add-on to its mySAP.com Web portal product introduced in May 1999.

3.4.3. Portal-based Access

Enterprise portal is a new approach in Intranet-based applications, therefore is often referred to as next-generation intranet. It goes a step further in the "webification" of applications and integration of corporate data. There have already been several "portal-based" products, particularly from business intelligence area. Business intelligence portal is a next trend in enterprise-wide decision support. Examples include:
- Information Advantage's *MyEureka* business intelligence suite was the industry's first business intelligence portal, (now Sterling Software - www.sterling.com).
- *WebIntelligence* from Business Objects (www.businessobjects.com) includes a business-intelligence portal that gives users a single, Web entry point for both WebIntelligence and BusinessObjects, the company's client-server reporting and OLAP system.
- *Brio.Portal* from Brio Technology (www.brio.com) is another example of an integrated business intelligence portal software capable of retrieving, analyzing and reporting information over the Internet.

The concept is later extended to "enterprise information portal" which describes a system that combines company's internal data with external information. White [13] defines several types of information portals from an evolutionary perspective, starting from the most basic form - the intranet portal, collaborative portal, and decision processing portal.

An integrated portal solution on enterprise level provides an efficient Web-based interface to all kinds of data coming from all relevant business applications (TPS, messaging system, document management system, and business intelligence system). Also, it adds an access to external information such as news services and customers/suppliers Web sites. Brick and Henry [14] propose such an architecture which emphasizes access from several computing devices, not only PCs, to all business applications. Gartner Group lists eight components it identifies as critical for a complete EIP solution: security, caching, taxonomy, multy-repository support, search, personalization, application integration, and a metada dictionary [15]. *The Hummingbird Enterprise Information Portal (EIP)* is an example of an integrated enterprise-wide portal solution. It provides companies with Web-based interface to structured and unstructured data sources and applications. The Hummingbird EIP promotes Enterprise Agility, enabling the entire organization to be flexible and to react quickly to changing market conditions [16].

3.5. Portable devices-based access

The ultimate goal in using portable computing devices that are designed as companion products to personal computers is again in improving information access of mobile users or teleworkers, firstly just for accessing and downloading data, but later on for uploading data as well. Currently, these devices are used mainly by managers and service workers for managing their schedules, contacts and other business information. They have the utility for synchronizing information with a personal computer. In addition to standard office scheduling needs it is a Customer Interaction Software - CIS (Customer Relationships Management - CRM) that drives the PDA market. These are applications like: sales force automation, customer support, service support, maintenance, etc. In the same time, both ERP and CIS/CRM vendors are already working on introducing non-PC links to their sites (PDAs, Windows CE-based hend-held PCs, GSM).

There are three different forms of portable devices:
Standard hand-held devices or hand-held PCs (H/PCs). Provide the user with a screen and a small but useable keyboard. Data entry and access are provided via keyboard, function buttons, and even a mouse. These devices run mainly Windows CE operating system. Windows CE incorporates many elements of the well-known Windows 95/98 OS platform. Basic Windows CE programs for hand-held PCs include pocket versions of Microsoft Office suite. By using Microsoft ActiveSync™ technology, Windows CE Services component automatically synchronizes information between a hand-held PC and the desktop.
Palm-held devices or Personal Digital Assistants (PDAs). These are the keyboardless devices that rely on function buttons to activate applications and access or enter information. They run either Windows CE or 3Com's PalmOS.
Cellular telephone-based devices. Even these standard phone communication devices are being enhanced from visual information access perspective enabling users with keyboards and small screens. Some GSM vendors: Nokia, Ericsson, Motorola and Psion announced forming a joint venture called Symbian which will standardize creating wireless information devices, such as smartphones and communicators. They will be running EPOC operating system as an operating system for mobile wireless information devices and applications designed by Starfish Software (www.starfish.com).

4. CONCLUSIONS

The concept of virtual enterprise represents a new paradigm in contemporary business. Companies are entering virtual business in order to compete in a rapidly changing business environment. For that purpose, participating companies need several technological improvements in their information infrastructures. At first, they need an efficient communication backbone in order to be able to communicate efficiently, exchange all kinds of data, share data, applications and resources, improve relationships with suppliers and customers. Additionally, they need a reliable, continuously available, and scalable computer systems that can support doing business in such a distributed environment, integrated application suites covering operations on dispersed sites and an efficient user interface that not only eases the access to information, but provides users with the right content of a certain, users' requirements-oriented context as well. That is where information technology comes in.

Virtual Enterprise Information System - VEIS is such information system which supports virtual business. A VEIS should be structured with respect to requirements based on a specific model of virtual enterprise. In the paper, these models are described and accordingly, major IT-related aspects needed for an effective VEIS design are explained.

In case of hardware/OS platform, the current status of most commonly used platforms is briefly explained and some directions related to VE are proposed. The second set of IT facilities consists of several high-speed networking technologies that can be used in improving communications in virtual business. In that sense, the three groups of technologies are identified as crucial: WAN technologies, technologies for remote access, and communication-networking applications. Within the third aspect of the IT framework, application platform is described with six technological trends important from the VE viewpoint. User interface as the fourth component in our VEIS model is considered mainly from manager's perspective. This resulted in defining a model of AMSS (Agile Management Support System) as an information system that supports informational agility of managers.

REFERENCES:

[1] Dunn, J.R., Varano, M.W., Leveraging Web-based Information Systems, Infromation Systems Management, Fall 1999

[2] Camarinha-Matos, L. M., Afsarmanesh, H., Garita, C., and Lima, C. Towards an architecture for virtual enterprises. Journal of Intelligent Manufacturing, Volume 9, Number 2, April, 1998.

[3] Byrne, J. A., The Virtual Corporation, in: Business Week, February 1993.

[4] Jacoliene van Wijk, drs Daisy Geurts, ing René Bultje, 7 steps to virtuality: Understanding the Virtual Organisation processes before designing ICT-support, http://www.cs.tcd.ie/Virtues/ocve98/proceedings/005.html

[5] Jansen, W., Steenbakkers, G.C.A. & Jägers, H., Coordination and use if ICT in Virtual Organisations, PrimaVera working paper Series, Universiteit van Amsterdam, May 1998. Presented at the EGOS Conference, Maastricht, 9-11 July 1998.

[6] Goldman, S. L., Nagel, R. N., & Preiss, K., (1995). Agile Competitors and Virtual Organizations: Strategies for Enriching the Customer, New York, NY: Van Nostrand Reinhold.

[7] Song, L., Nagi, R., Design and Implementation of a Virtual Information System for Agile Manufacturing, IIE Transactions on Design and Manufacturing, special issue on Agile Manufacturing, 1997, Vol. 29(10), pp. 839-857

[8] Camarinha-Matos, L. M., Afsarmanesh, H., Garita, C., and Lima, C. Towards an architecture for virtual enterprises. Journal of Intelligent Manufacturing, Volume 9, Number 2, April, 1998.

[9] Datamation online
http://www.datamation.com/PlugIn/issues/1998/may/05nt.html

[10] Datamation online
http://www.datamation.com/PlugIn/issues/1999/january/01main1.html

[11] Dove, R., Design Principles for Agile Production,
http://www.parshift.com/Essays/essay014.htm

[12] Norrish, D., Agile Software for Agile Manufacturing, APICS - Agile Manufacturing, Volume 6, Number 12, December 1996

[13] White, C., Decision Threshold, Intelligent Enterprise, November 16, 1999

[14] Brick, B., Henry, J., Enterprise Portals: Not Just Information Delivery, Intelligent Enterprise, November 16, 1999

[15] Hummingbird
http://www.hummingbird.com

[16] Hummingbird
http://www.humingbird.com

Early Supplier Involvement: A Design-Based Sourcing

Shad Dowlatshahi
The University of Missouri-Kansas City

1. INTRODUCTION

Since the advent of the JIT philosophy, there has been a growing interest in the role and function of suppliers in manufacturing operations. The role of suppliers and purchasing professionals has gained prominence. The ability of a firm to secure the appropriate raw materials, supplies, and parts on a timely basis has been viewed as a strategic impotence to many organizations. The ability of an organization to remain competitive is largely dependent upon on the amount, quality, cost, and timing of its materials and supplies and the effectiveness of its supply chain. Strategic alliances with national and international suppliers have been considered as the key to success by wise and future-oriented managers. The scarcity of worldwide materials in so many categories have made the role of suppliers even more important. The simple fact remains that the most technologically oriented countries and companies can not survive without a flow of quality materials and supplies.

2. CONCURRENT ENGINEERING ENVIRONMENT

The concept of concurrent engineering has recently gained prominence in US manufacturing circles. Concurrent engineering considers and includes various product design attributes such as aesthetics, durability, ergonomics, interchangeability, maintainability, marketability, manufacturability, safety, procurability, reliability, remanufacturability, schedulability, serviceability, simplicity, transportability, and testability in the early stages of product and system design. The greatest impact and benefits of concurrent engineering are realized at the design stage of product development. Design decisions made in the early phases of product design and development will have a significant impact upon future manufacturing and logistical activities.

The concept and applications of concurrent engineering will gain even more importance when the design, manufacturing, marketing, financing, purchasing, supplier, and distribution activities of a manufacturing/commodities firm become international and meet with the usual difficulties associated with geographic location, culture, language, operational practices, etc. There are, therefore, tremendous opportunities to utilize methods, procedures, and rules to plan, analyze, select, and optimize the entire manufacturing system and its related issues under the umbrella of concurrent engineering. These activities are interdisciplinary in nature.

Concurrent engineering has been defined in various ways. These definitions are mostly the reflection of an individual's background and experience. Vasilash (1987) suggests terms such as simultaneous engineering, life cycle engineering, process driven design, team approach, and design for manufacture. Other terminologies such as concurrent design, Unified Life Cycle Engineering (ULCE), total quality management, and parallel engineering have also been used.

The natural focus of concurrent engineering is on product design. A decision concerning product design tends to have a number of significant manufacturing and non-manufacturing impacts upon the life cycle of a product. Figure 1 depicts the role of purchasing and supplier's function in a concurrent engineering environment.

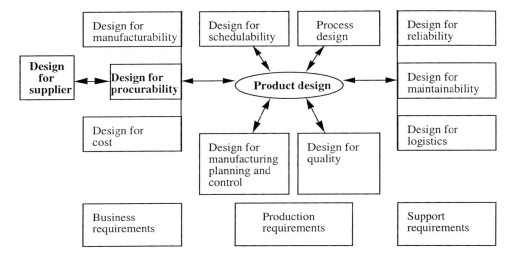

Figure 1. The role of the supplier in a concurrent engineering environment

2.1. Advantages of Concurrent Engineering

The proper consideration and inclusion of design attributes early in the design process yields these advantages:
- Reduction in product development cycle time
- Avoidance of costly future redesigns
- Reduction in duplication of effort
- Better communication and dialogue
- More efficient operations and higher productivity
- Overall cost savings resulting from:

  - Reduction in the number of parts manufactured
  - Better machine utilization time
  - Easier manufacturable parts
  - Fewer reworks and scraps
  - Greater use of standard features, thereby reducing tooling costs
  - Fewer changes in process planning

- Avoidance of product recalls
- Lower maintenance costs
- More reliable products
- Better customer satisfaction
- Improved bottom-line

3. REVIEW OF LITERATURE

Although there have been a number of articles published on supplier development, supplier evaluation and certification, supplier-purchasing relationships, etc., there has been very little work that focuses on the relationship among product designer, supplier, and purchasing professionals. The role of supplier in a concurrent engineering environment, especially at the early stages of product design, has received cursory attention in practitioner journals. For interested readers a summary of related research is presented:

1. Dickson (1966) conducted a survey in which he identified the factors considered in supplier selection. The author attempted to present the most important factors in supplier selection.
2. White (1978) conducted a survey in which he attempted to underscore the relationship between product categories and buying situations.
3. Gregory (1986) presented a prescriptive matrix model to weigh supplier selection at Texas Instruments. This was accomplished through predetermined benchmarks.
4. Timmerman (1986) developed a model to rate suppliers on a number of equally weighted factors. He used a cost ratio method by which all internal costs associated with conducting business with a supplier as a percentage of the supplier's costs were quantified. Then a linear averaging approach was used to rate suppliers on a number of factors, which are ranked based on the degree of their importance.
5. Soukup (1987) developed a payoff matrix for purchasing personnel to evaluate potential supplier's performance under different scenarios.
6. Hahn et al (1990) presented a conceptual model for the supplier development program. The program is designed to create and maintain a network of competent suppliers and to improve various supplier capabilities that are necessary for purchasing professionals to meet their competitive challenges.
7. Thompson (1990) used the Monte Carlo Simulation Technique to evaluate supplier performance under different scenarios. A modified weighted point approach was used. This approach rates suppliers on a number of factors.
8. Dowlatshahi (1992) presented the role of purchasing in a concurrent engineering environment. Purchasing's contributions and concerns in the early product design phases were discussed and the areas of collaboration between product design and purchasing were explored. The paper did not, however, directly include suppliers in the process.
9. Brill (1993) stated that unless early collaboration between designer and purchasing professional is planned and implemented, problems such as increase costs, lost time, and inferior product quality or production efficiency may arise. The author further suggests that the use of vendors' catalog numbers is often satisfactory unless the vendor has changed the materials. The author cautions against the potential problems caused by purchased parts from suppliers with whom sound business practices are not developed.

4. RATIONALE AND JUSTIFICATIONS FOR EARLY SUPPLIER INVOLVEMENT

This paper proposes an Early Supplier Involvement (ESI) in the early stages of the product design and development process. The objective of this paper is to explore the relationship and impact of supplier involvement on the product design under the umbrella of concurrent engineering. This paper intends to present a number of specific rules and suggestions by which a successful interface between the designer and the supplier can take place.

The following examples, reasons, and justifications signify the importance of such an approach:
1. A study shows that 70 percent of the life cycle cost of a product is determined at the design stage. The life cycle cost here refers to cost of materials, manufacture, use, repair, and disposal of a product (Nevins and Whitney 1989)
2. On the average, more than half of every dollar received from sales revenue of manufactured goods is spent on the purchase of materials, suppliers, and equipment required to produce the goods. For example, Ford Motor Company spent 60% of every sales dollar on materials, supplies, and equipment. Air Reduction Company's share was almost 60% (Heinritz, et al 1991). The sheer volume of materials purchased requires development and early involvement of the suppliers.
3. More and more OEM firms outsource their parts and materials from outside suppliers than in-house. The trend is clearly toward outsourcing. ESI is mandatory if an organization attempts to have well integrated products.
4. ESI is a good business practice. Purchasing and design teams should not wait for a major problem or emergency to occur before developing and involving suppliers in the design process.
5. The design stage is where opportunities for improvements are most beneficial and tangible. Suppliers must be involved early at the conceptual phase of product design. If suppliers wait for design specifications and a bill of material, it will be too late to have a significant impact on the design process once the design is frozen.
6. Design-stage sourcing is the most important sourcing activity. Although the process is a complex one, the benefits will outweigh potential costs.
7. Design-stage sourcing is a form of brainstorming. It requires involvement of as many experts as possible.
8. The technical difficulty of designing and manufacturing most products is rapidly rising. Advance technologies and a rise in product complexity necessitates ESI and multidisciplinary approaches to product design. It is virtually impossible for any one firm to possess all the technical expertise needed to develop a complex product.
9. Because of the expansion in international trade, there is a growing rate of products and processes. This phenomenon requires the exchange of technical information and the involvement of various expertises.
10. Assembly operations make up a large percentage of the total manufacturing cost of product development. Design-stage sourcing, if accomplished properly, will be most effective at the assembly stage of operations.
11. Purchasing may have to develop and involve a new supplier because there is no established source of supply for a particular part or material. Although this must be done due to necessity, this is not the all encompassing reason for supplier development and involvement.
12. Design-stage sourcing provides unique challenges and adventures into esoteric sources of materials and new technologies.

5. DESIGNER-BUYER-SUPPLIER INTERFACE

The relationship and collaboration areas among designer, buyer, and supplier are presented in three planning horizons (strategic, intermediate, and tactical). The three planning horizons encompass two areas of interface with supplier relations, and cost and time to market. The underpinnings of such an interface are based on design-stage sourcing and early supplier development. Table 1 presents the specific areas of interface and collaboration among designer, purchasing, and supplier. Table 1 also contains the level of product design and organizational involvement at each area of the interface.

Table 1. Design-purchasing-supplier interface in concurrent engineering

Planning Horizon	Product design level	Organizational level/ personnel	A. Supplier relations	B. Cost and time to market
1. Strategic (long-term)	Entire Product lines	Top management/ VP of manufacturing or operations, manufacturing executives, purchasing executive	1.A.1 Long-term strategic alliance 1.A.2 Supplier R&D investment and financial strength 1.A.3 Confidential relationships 1.A.4 Reduction in the number of suppliers	1.B.1 Strategic cost of new products to market 1.B.2. Strategic time of new products to market
2. Intermediate (medium-range)	Individual Product	Division operations/ production manager, plant manager, manufacturing engineer, factory superintendent, program manager, materials specialist, purchasing professionals	2.A.1 Information sharing 2.A.2 Supplier plant visitation 2.A.3 Supplier selection, evaluation, and certification 2.A.4 Supplier training/ meetings 2.A.5 Sourcing 2.A.6 Supplier relations	2.B.1 Total throughput cost 2.B.2 Total throughput time
3. Tactical (short-term)	Part	Factory-shop floor level/ foreman, team leader, crew chief, department supervisors, purchasing agent, buyer, expediter, shipping specialist	3.A.1 Inspection and receiving policy	3.B.1 Cost of purchased materials 3.B.2 Delivery time of purchased materials

6. SPECIFIC DESIGN RULES AND SUGGESTIONS FOR THE INTERFACE

Now that the importance of the collaboration among designer, buyer, and supplier has been underscored and the framework for collaboration has been developed, one may develop the specific rules and suggestions governing such collaboration. These rules and suggestions pertain to every area of the interface in Table 1.

6.1. Long-term Strategic Alliance

1. For fear of becoming a meaningless slogan, strategic alliances with suppliers are still the most crucial aspect of sourcing.
2. Strategic alliances call for technological links between suppliers and purchasing-design teams.
3. A strategic supplier relationship is largely based on a supplier's contributions to design simplicity, effectiveness, and compatibility with existing parts and products.
4. A captured buyer-supplier relationship--such as in the Japanese approach--is the ultimate strategic alliance between the buyer-supplier.
5. Strategic alliances may lead to the solicitation of information and/or contact regarding foreign businesses from American embassies and consulates, the U.S. Department of Commerce, Trade Directories of the World, etc.
6. Early supplier involvement (ESI) must be at the heart of strategic alliances with suppliers.
7. The supplier's production and distribution facilities, in reality, become an extension of the buyer's manufacturing facilities. Purchasing-design teams should use their suppliers' parts as well as their expertise.
8. The purchasing-design teams' relationship with suppliers should be long-term and stable.
9. The buyer and supplier have a joint opportunity/responsibility to collaborate on mutually agreed upon product design goals.
10. ESI is essential in the application of new technology and the shortening of a product's time to market.
11. The supplier-buyer relationship is from the cradle to the grave with the supplier's voice being heard in engineering and production activities.
12. Strategic alliances require strategic sourcing.
13. The sole source supplier is much more accommodating when it comes to satisfying a buyer's long-term requirements.
14. In design-stage sourcing, the focus should be on product design, a sound technological base, a prospective supplier's capacity, and a long-term commitment.
15. ESI efforts must be formal.
16. In return for genuine and effective assistance in design facilitated by ESI, purchasing should promise suppliers long-term contracts.
17. ESI tends to favor "full-service" suppliers that are willing to cooperate with purchasing-design teams on a number of fronts. ESI usually leads to the selection of a single source of supply.
18. Most large Japanese manufacturers require ESI as a matter of practice. These manufacturers have generally enjoyed parts and supplies that are of high quality, of reasonable cost, and are delivered in time.
19. A supplier's attitude is very important. This requires genuine and close communication and commitment to the purchasing-design teams by the supplier.
20. Purchasing-design teams should evaluate the supplier's current level of commitment to judge its ability to commit to them on a long-term basis.
21. Developing long-term suppliers are analogous to developing customers in a

marketing effort. This is called reverse marketing.
22. Organizational downsizing typically requires more cooperation and outside expertise from suppliers. Organizational downsizing has recently proven to be practical by a large number of corporations.
23. Design-purchasing teams must be as willing, prepared, and as professional as possible in their dealings and relationships with suppliers. This is a two-way commitment.
24. Major capital investments in machines, processes, and new technologies are not cost effective or economical for suppliers if they are not utilized and dedicated to a small number of buyers on a long-term and continuous basis.
25. Nowadays engineers' acceptance of worldwide and design-stage sourcing is better than in the past.

6.2. Supplier R&D Investment and Financial Strength

1. The existence of a long-term partnership encourages a supplier to make a large R&D investment for the betterment of the buyer's product design.
2. A supplier's specific suggestions with regard to new product design should be viewed as equivalent to the R & D invested by the buyer's company.
3. The expansion of international trade necessitates joint R & D investment by both purchasing and the supplier. This not only creates synergy for new product designs, but also saves scarce R & D dollars for both parties.
4. R & D are an important part of purchasing's responsibility.
5. The financial stability of the supplier firm should be a major concern to the buying firm.
6. The financial strength of a supplier is largely dependent upon the economical and effective investment in capital budgeting. This kind of investment by the supplier usually requires a long-term contract with the buyer.
7. A formal evaluation of a supplier's financial strengths and weaknesses should be initiated by purchasing. This process should be all encompassing and future-oriented.
8. Long-term contracts should not be based on a fixed price. They should be based on shared financial responsibility and risks between the buyer and the supplier.

6.3. Confidential Relationships

1. Confidence that suppliers will not leak out any confidential information is the main building block of the buyer-supplier partnership. The reverse must be true as well.
2. Suppliers should swear to keep the information related to a buyer's design and operations as confidential as possible.
3. Confidential partnership can not be realistically developed and maintained if the relationship is short-term, limited, or a one-time event.
4. Shared benefits of a partnership require coordination, cooperation, trust, and loyalty between parties.
5. A great deal of communication and goodwill is required for a trusting and long-term partnership between parties.

6.4. Reduction in the Number of Suppliers

1. There are three main reasons to reduce the number of suppliers: (a) supplier development is costly--so they must be limited to a manageable number, (b) a

close and workable relationship is only achievable with a limited number of suppliers, and (c) only a limited number of suppliers can be rewarded with substantial business.
2. The use of a JIT philosophy and ESI normally leads to a major reduction in the number of suppliers.
3. Suppliers should be involved in the developmental efforts of purchasing-design teams and ESI programs only when the number of suppliers is drastically reduced.
4. Design-purchasing teams must make certain that a reduction in the number of suppliers does not lead to the inhibition of creativity for the remaining suppliers.
5. Purchasing should attempt to eliminate bureaucracy as it relates to supplier relations and operations.
6. Purchasing's effectiveness should be partially measured by top management on supplier-base reduction.
7. Engineering should know that purchasing does not intend to change suppliers every 12 months. Long-term partnerships must be advocated throughout the company.
8. The trend is toward fewer suppliers who handle the sale of a fewer number of materials and parts.
9. A supplier, who has parts that can be used, substituted, or who is willing to develop parts for the buyer's products is most likely to become the buyer's sole source.
10. A single-source supplier is more likely to offer generous managerial and technical expertise to the buyer than a multiple-source supplier.
11. Pontiac Fiero single sourced more than half of its body parts without bids (Dowst 1985).
12. Purchasing-design teams should identify the potential problems and weaknesses of sole sourcing. Sole sourcing requires nurturing and continuous care.
13. Design-purchasing teams should make certain that the existence of a sole supplier does not make them complacent regarding improving their processes and operations. Eliminating waste and improving quality must be an ongoing commitment for design-purchasing teams. As a matter of fact, the buyer should actively and continuously solicit the supplier's inputs as how to improve processes.
14. The sole supplier is not a yes-man supplier. The sole supplier should challenge design specifications, manufacturing processes, tolerances, etc. The sole supplier should be as much a part of the project's success as is the buyer.
15. The design-purchasing team must notify a sole supplier well in advance of major changes in its product line, models, or manufacturing processes. Doing otherwise may be detrimental to the sole supplier's operations and its future cooperation with the design-purchasing team. It may also tarnish the image and reputation of the buyer in its industry for some time to come. A long-term partnership between the supplier and the purchasing-design team must be genuine and based on mutual benefits.

6.5. Information Sharing
1. Effective information sharing must take place between and within different functional areas that contribute to product design and development.
2. Effective communication and dialogue with the supplier is essential. This includes professional sharing of timely, relevant, and accurate information.

3. Designer-supplier collaboration and information sharing is most effective if it takes place in the early (conceptual) stages of product development.
4. Information requirements of management, in terms of depth and breadth, are growing rapidly. The supplier should contribute to the fulfillment of these information requirements, especially the ones with strategic implications.
5. Electronic Data Interchange (EDI) should be used whenever possible to facilitate the exchange of routine information.
6. An effective dialogue should take place between marketing and the buyer/supplier group. This is essential for the buyer/supplier group to be aware of forecasting for future demand and volume of operations, inventory levels, product lead times, and the like.
7. Suppliers should be contacted by the buyer's marketing group as to how the company's products are marketed. The information regarding the performance of parts and products supplied, is of vital importance to the supplier and its future contributions.
8. Nowadays suppliers are more expert in parts and materials than purchasing agents. Purchasing's role as a catalyst between designer and supplier is essential.
9. Information sharing must be targeted to the improved flow of ideas/operations.
10. Buyers and suppliers should share their collaborative successes with design and manufacturing engineers to gain their cooperation.
11. The supplier may need to adopt the same policies and practices as the buyer. These policies and practices need to be clearly communicated to the supplier.
12. The buyers should share the results of their research, innovations, and ideas with their suppliers and vice versa.
13. There must be formal and effective channels of communication between and among suppliers, designers, and the buyers who collaborate on the same product/project.
14. ESI requires advanced and developed programs and channels of communication in order to be effective.
15. Purchasing-design teams should provide an interdepartmental liaison for information sharing with suppliers.
16. Suppliers should be required to make formal presentations on the advantages and limitations of their products and processes.
17. International trade and global competition requires a different and more effective exchange of information.

6.6. Supplier Plant Visitation

1. Supplier plant visitation is a serious matter and should be taken seriously by all parties. Supplier plant visitation should have the following characteristics: (a) determine the objectives, write them down, share them with the visitation team; (b) prepare an agenda and do not deviate from it; (c) become familiar with the operations of the supplier, review company materials, and conduct a thorough literature review. Concentrate on these activities before the visit so that the team will not be distracted from its primary tasks; (d) visiting team members should have different skills and backgrounds capable of observing, asking questions, and evaluating suppliers in a variety of areas; (e) attempt to observe such intangible criteria such as: organizational culture, managerial attitude and technical capability, employee attitude toward the company and one another, employee commitment toward quality and continuous improvement, etc.; (f) take notes of

the entire visit, compare the notes with other team members for their accuracy and feedback, share the results of non-confidential information with the supplier for its verification and response, suggest constructive criticism, and offer advice to the supplier to improve its operations.
2. Purchasing should take the product designer to the supplier's plants. This facilitates an understanding of the supplier's products and processes. It will further encourage additional discussion and information sharing.
3. The supplier plant visitation should reveal how much of the supplier's raw materials, equipment, and storage facilities is exclusively devoted to the buyer's business. If the arrangements are not satisfactory, further discussion and action must be taken.
4. Purchasing should make the visitation to a supplier's plant a contractual requirement for JIT suppliers. Purchasing-design teams should review and gather relevant information about the strengths and weaknesses of the supplier's plant.
5. The design-purchasing team should register their first impressions of the supplier's plant in terms of its appearance and housekeeping.

6.7 Supplier Selection, Evaluation, and Certification
1. A buyer should avoid evaluating suppliers based on quantitative criteria. Instead, a long-term relationship and qualitative evaluation of the suppliers should be emphasized.
2. Buyers should not rely on quotes and bids if they want to involve suppliers in the early design stage of product development.
3. Purchasing should evaluate the managerial practices of the suppliers in terms of management style, product quality, continuous improvement, processes, technical skills, financial strength, and the use of SPC.
4. Purchasing should be allowed more latitude in the use of qualitative criteria in evaluating and qualifying suppliers for new technologies, materials, and products.
5. The buyer should use the supplier's measurement and certification within a larger plan to leverage the development of joint resources with suppliers.
6. Suppliers should be evaluated on the number as well as the extent of their relationships with other companies' buyers. If these relationships are extensive, they can inhibit their ability to become a sole supplier or to enter into a long-term partnership.
7. Suppliers should be evaluated on their ability to follow instructions, to respond accurately and effectively to purchase orders, and to interpret judiciously the future directions of the buyer's company.
8. The criteria for selection of the supplier should be clearly delineated. Suppliers should have a reasonable and clear picture as to what is required to become a preferred supplier and how to maintain this status.
9. Supplier selection criteria should be fair and reasonable. They should emphasize on long-term contributions rather than on short-term objectives. They should also be flexible and allow room for interpretation.
10. The supplier's cooperative attitude is very important. Intangible factors such as creativity, innovativeness, and the like must also be considered.
11. Information as to whom is certified should be publicly disseminated internally and externally. Certified suppliers should have reasonable and ordinary access to a buyer's facilities and resources to conduct their operations.
12. Maintaining competitive pricing on a long-term basis by suppliers is highly valued

by the buyer. These criteria generally encompass life-cycle costing.
13. Supplier certification should typically be initiated by the buyer, not the supplier. Sophisticated supplies may, however, target a buyer as a potential customer. A buyer should be able to pass reverse certification criteria if it finds this relationship to be beneficial.
14. As a part of supplier evaluation, the buyer should require suppliers to produce parts with no or little variability from one shipment to the next. The supplier must be required to use relevant quality control techniques.
15. Purchasing-design teams should anticipate potential design problems, product requirements, and design changes. They should select suppliers accordingly.
16. Purchasing-design teams should develop a system to evaluate a suppliers' design suggestions and improvements.
17. Purchasing-design teams should evaluate suppliers on quality initiatives, delivery times, lead times, schedule development, long-term costing, training programs, employee treatment and development, maintenance schedules, and response time to new product design.
18. Prior approval of certain factors such as managerial practices, technology, delivery, competence (cost, technical) should be initiated by the buyer. This preliminary approval is essential for further development of the partnership.
19. In evaluating a supplier, the purchasing-design team should consult other companies' buyers who have conducted business with the particular supplier regarding its overall performance. They should check its credit references, and evaluate its facilities. The purchasing-design team should use these as a predictor validity of the supplier's performance in the future.
20. The buyer should not exercise hastily the right to refuse a long-term supplier solely based on quality and cost problems. Unless there is an overwhelming reason to the contrary, the buyer should work closely with the supplier to remedy the problems.
21. The buyer should avoid merely issuing report cards. Instead, programs should be developed to eliminate waste and/or to improve supplier performance.
22. The purchasing function should hire more buyers with technical and engineering-based training. Purchasing should then rotate their agents in design, engineering, quality, and manufacturing areas. This process makes the buyers more proactive. The buying function has become more technical and complicated nowadays. This facilitates good supplier relations.
23. The buyer may require ten consecutive, high quality, defect-free shipments as a certification requirement.
24. Maintaining adequate databases and records on a supplier's performance, history, and its future potential to contribute to product design can be of tremendous value to the buyer's company in evaluating a particular supplier for continued cooperation or lack thereof.

6.8. Supplier Training/Meetings

1. Supplier training programs should be formal. The programs should be evaluated for their results and impact. Changes need to be made to keep them current and effective as the designer-buyer-supplier relationships evolve.
2. The role of education and training is important. Purchasing-design teams should be trained as how to provide effective training for suppliers. Top management should provide trainers with the necessary tools to accomplish the stated

objectives.
   3. The level, frequency, and personal involvement of the designer and buyer largely determines the success of supplier meetings and training programs.
   4. A great deal of communication and goodwill is required between the supplier and the buyer in their interactions and meetings.
   5. Purchasing professionals should organize and run initial meetings and orientation programs, making certain that the suppliers are aware of purchasing's training programs and receive the necessary information and knowledge to effectively conduct their business with design-purchasing teams. These meetings should be targeted toward improving product design.
   6. Purchasing-design teams should conduct training programs at a supplier's site or wherever the results seem to bear more fruit.
   7. Suppliers should initiate their own training and development programs designed to improve the quality and productivity of the buyer's operations.
   8. Purchasing professionals should offer a course or a training session for designers regarding the operations of the purchasing department or offer regular meetings to prevent potential design-purchasing problems.
   9. Purchasing should create a supplier advisory board.
   10. Purchasing-design teams should create a joint study group with the supplier.
   11. Caterpillar established a Quality Institute to advance the cause of supplier training and development (Cayer 1990).

6.9. Sourcing
   1. Sourcing is truly long-term and worldwide.
   2. Long-term sourcing does not rely on quotes and bids in selection of the suppliers. It attempts to create fits in products, processes, and technologies with potential suppliers.
   3. Sole-sourcing does not rule out the existence of qualified-supplier lists. Purchasing should maintain a list of qualified suppliers who are capable and willing to cooperate with the designer, if the company attempts to enter a sole-supplier relationship for any of its parts.
   4. Sourcing activities may entail development of a minority supplier. The time, effort, and resources invested in such ventures may well pay for themselves in the long run with a better image, reputation, and higher sales for the buyer's company.
   5. Design-stage sourcing is a form of brainstorming. It requires an innovative and creative process for designers, buyers, and suppliers.
   6. For design-stage sourcing, design-purchasing teams should focus on technology, a prospective supplier's production capacity, and long-term commitment.
   7. The role of purchasing in sourcing is to select economic alternatives, to evaluate a supplier's financial strengths and weaknesses, and to identify sole-source problems.
   8. Purchasing-design teams should secure a long-term supplier commitment for parts and materials that are difficult to obtain or are critical to product design.
   9. Purchasing typically initiates the identification of the best sources for particular products. Purchasing should monitor costing of the items as well as changes in parts, models, colors, and materials needed for a given product.
   10. Purchasing-design teams may consider prototype sourcing.
   11. Suppliers should provide inputs to purchasing-design teams concerning the availability of parts and sourcing decisions.

12. A good supplier proposes other procurement alternatives other than itself. This mature relationship only exists when there is a professional, mutually beneficial, and long-term partnership between the buyer and supplier.
13. Design-purchasing teams should consider branching out of their usual sourcing domain for new distributors and suppliers.
14. Half of the parts used in the Pontiac Fiero were sourced without bids (Dowst 1985).

6.10. Supplier Relations

1. Purchasing, while enjoying top management support, should be able to establish its independence as a fully functional unit of the organization. Buyers' relationships with suppliers are largely affected by the perception and practice of such an independence.
2. Suppliers view the personal integrity, honesty, and fair dealings of the purchasing and design professionals as one of the most important factors in their relations. This relationship must be indeed mutual and long-term.
3. The buyer should share information on projected annual quantities along with the products' projected life cycle with the suppliers in order to develop the best manufacturing methods and processes.
4. Purchasing-design teams should make a genuine effort to sell suppliers on their own company by: (a) offering economic as well as non-economic incentives. In the long-term, non-economic factors may be the pivotal factors; (b) identifying mutual problems and arriving at reasonable and usable solutions. Suppliers find a partner in the buyer in such an environment; (c) sharing plant visitation information with the supplier and offering tangible assistance to improve the processes and operations of the supplier. The more improved the operations of the supplier, the more long-term benefits to the buyer's company.
5. Purchasing-design teams should be involved in all aspects of their firm's operations, especially those areas that they represent. They should show competence and a willingness to buy the supplier's materials. They should have the authority to represent the buyer's company. Purchasing-design teams should always handle themselves in a professional and courteous manner in all dealings with suppliers.
6. Suppliers should be assured that purchasing has a good working relationship with the other functional areas of the company. This in-house relationship is essential for continued support and business by the suppliers. Purchasing frequently acts as the agent of change and idea generator within the organization in order to favorably affect external relations with the suppliers.
7. Investing time and effort in an effective booklet to attract quality suppliers is an important task undertaken by purchasing-design teams. An effective booklet should contain: (a) policies and procedures for awarding contracts on a long and short-term basis; (b) purchasing authority and power in evaluating and awarding contracts; (c) areas of interest and concerns that suppliers can contribute to such as product design, value analysis, etc.; (d) ethical standards and expectations; (e) the booklet should reflect the expertise and expectations of all stakeholders who are affected by the supply chain management; (f) areas of potential needs and collaboration other than the ones mentioned in (c) above should be delineated in fairly reasonable detail; (g) assurances of confidentiality for both current and future suppliers; and (h) the booklet should be attractive, well developed and

written. Although the size of the booklet does not matter, it should serve as a reference booklet for potential suppliers.

6.11. Inspection and Delivery Policy

1. Delivery and receiving activities must be orderly and disciplined. Schedules of deliveries must be planned well in advance and they must be compatible and consistent with the level of future production activities at the buyer's plant.
2. The size, weight, and dimensions of packages and boxes should be compatible with the buyer's physical facilities and equipment. This includes temporary storage areas, warehouses, lanes, elevators, cranes, forklifts, bins, pallets, etc.
3. The scheduling, routing, transportation, and delivery activities, in terms of cost and timing, should be preferably based on acceptable and effective operations research techniques.
4. Once supplier certification is successfully completed for a long-term supplier, there may be no need for inspection of the supplier's shipments.
5. If the shipments are to be inspected, the most effective sampling and statistical quality control methods should be used.
6. There must be acceptable procedures between the buyer-supplier regarding late delivery, defective parts, returns, follow-up on shipments, and the like.
7. There must be acceptable procedures and standards to coordinate the activities of the receiving, shipping, and inspection stations.
8. Design-purchasing teams may attempt to impose pre-determined penalties on suppliers for non-conformance regarding delivery and performance issues.
9. There must be an in-house procedure to promptly unpackage and deliver the items to their respective workstations without loss of time and damage to the items.
10. The receiving department should not deal with suppliers' shipments that are without an order number, a packing list, etc.

6.12. Strategic Cost of New Products to Market

1. An effective measure of strategic cost is the buyer's total product costs. These costs should be competitive on a worldwide basis.
2. For a supplier to remain globally competitive, long-time cost control is more important than the short-term pricing of parts and materials.
3. Overconcern about individual cost items may lead to shortsighted designs. Strategic benefits derived from a good design are more important than the individual cost of several parts.
4. Design costs should not dominate design goals. Overall design will determine a product's performance, cost, and its success in the long run.
5. Initial material cost must be weighted against the service life of the part/product. Inexpensive materials may have an inverse relationship with longer life. On the other hand, higher cost materials do not guarantee longer life.
6. A long-term, cooperative sole supplier may contribute to the buyer's JIT environment and design goals by stocking parts/products for the buyer. This reduces total costs for the buyer.
7. Total strategic costs include, but are not limited to: price, cost of quality, capacity, delivery schedules, parts reliability, and post-sales services.
8. The strategic cost of introducing a new product to market includes its life cycle cost. Short-term cost savings may result in higher rework, replacement, and recall costs.

9. The buyer-supplier relationships, as it pertains to cost, must go beyond mere cost control and focus on value added to products.
10. Preventive measures rather than corrective measures reduce long-term costs.
11. Inventory reduction and inventory turnover are important measures of an effective strategic cost system.
12. Purchasing should be advised that importing or purchasing low-cost items may not be beneficial to the buyer's company in the long run.
13. The structure and objectives of the strategic cost system must be discussed and agreed upon in advance by both buyers and suppliers.

6.13. Strategic Time of New Products to Market

1. There must a concurrent consideration of various functional areas in the early phases of product design in order to shorten the strategic time of products to markets.
2. Purchasing should define the areas of collaboration between the designer and supplier with respect to parts that are currently being purchased or manufactured in order to accelerate the introduction time of products to market.
3. The strategic time of new products to market includes a strategic logistics system by which parts and materials are secured and finished goods are delivered to their final destination in an effective and expeditious way.
4. Product life cycles are getting shorter. Time is a major competitive weapon. There must be an overlap between introducing a new product and phasing out an existing one. This requires close cooperation/coordination between designer and supplier.
5. The essence of strategic time to market is on "time-based competition".
6. Shorter lead times are the result as well as a major objective of ESI.
7. Inter-and intradepartmental communication and dialogue in both the supplier and buyer's organizations is essential regarding timetables of operations/products and in expediting products to market.
8. The supplier and buyer's operations should be organized so that they are more flexible and able to respond promptly to market forces.
9. Purchasing should consider teaming arrangements by pulling the resources of two or three suppliers together for difficult-to-design products. This has the potential of creating synergy to shorten product time to market.
10. The supplier must be willing to apply new technologies and expedite the introduction of new products. This is a gauge of supplier effectiveness.

6.14. Total Throughput Cost

1. In conducting cost comparisons, the design-purchasing team should consider production quantity. The larger the initial investment in tooling special machinery, the lower the unit cost. This type of investment may require different suppliers, a longer-term supplier relationship, or at least a different mix of materials and supplies.
2. It is usually too expensive to change a process after a product has been released to manufacturing.
3. ESI is capable of lowering costs. In fact, lowering the throughput cost is a major objective of ESI.
4. Compression of a new product development cycle, a result of ESI, should ideally lead to the reduction of a total throughput cost.

5. The cost orientation throughout design-production cycle should remain on reducing the total throughput cost and not necessarily individual cost items.
6. The initial cost of materials should be placed in a larger context and should include: cost of manufacturing, installation, operations, and maintenance.
7. Direct costs decrease as a result of judicious automation and an elimination of no-added-value activities. This is generally initiated and accomplished by the design.
8. Suppliers, in collaboration with purchasing, should develop a costing mechanism and schedule for materials and parts to determine cost estimates and ranges.
9. Minimizing the cost of assembly significantly minimizes the total throughput cost. Minimizing the cost of assembly is clearly a design function.
10. The total throughput cost should include non-conformance quality costs. These include: scraps, reworks, recalls, etc.
11. Suppliers should be asked to contribute to reducing the total throughput costs by lowering direct material costs.
12. The structure and objectives of total throughput costs must be discussed and agreed upon in advance by buyers and suppliers.

6.15. Total Throughput Time

1. Design and manufacturing activities must be overlapped to the extent possible in order to reduce total throughput time.
2. There must be firm and achievable timetables for the introduction of new products from design to production to marketing. Designers, buyers, and suppliers must be aware of these timetables and collaborate closely to achieve them.
3. The formation of multidisciplinary teams is required to reduce the lead times in the supply chain as well as in production processes. Lead times must be relentlessly pursued, investigated, and reduced whenever possible.
4. It is usually too time-consuming to change a design after it has been released to manufacturing. ESI should result in faster production cycles and shorter lead times.
5. Suppliers should be evaluated on their ability and effectiveness to reduce the lead times and decrease total throughput time.
6. An effective and realistic delivery schedule should be developed to meet the needs of both the supplier and the buyer. A commitment to enforce the timetables is essential.
7. Suppliers should be able to propose procurement lead times and JIT coordinating policies.
8. The Pontiac Fiero single sourced all its chassis' parts and completed the product two years earlier than the scheduled time (Dowst 1985).

6.16. Cost of Purchased Materials

1. Suppliers should be able to offer savings in components and materials for their best buyers whenever possible.
2. Purchasing may establish benchmark costs. In this case, suppliers should meet or be effectively competitive with them.
3. ESI is inherently a cost savings strategy. Generally, ESI can lead to a 10% to 15% cost savings (Remich, Jr. 1989).
4. The higher and tighter the surface roughness and tolerances of the purchased materials, the higher the cost of the items.
5. In addition to the initial cost of purchased materials, design and manufacturing

engineers should include the cost of unit stiffness, the cost of unit strength, and the like.

6.17. Delivery Time of Purchased Materials

1. Each shipment and delivery schedule should be considered as a part of a larger strategic logistic system jointly developed by the supplier and buyer.
2. Transportation policies and delivery schedules of suppliers should be known and agreed upon by the buyer.
3. If purchasing does not operate on a JIT mode, suppliers should make deliveries on a timely basis and as expeditiously as possible so as not to disrupt the operations at the buyer's plant.
4. Different modes of transportation and delivery systems must be considered and their effectiveness, timing, and costs ascertained.
5. Supplier response time to emergency shipment requests by the buyer should be reasonable and made within an acceptable time frame.

7. FACILITATING THE DESIGNER-SUPPLIER INTERFACE

The role of purchasing in a designer-supplier interface is crucial. The purchasing function should be given sufficient respect, visibility, and the authority to volunteer a supplier's services instead of being intimidated. Purchasing essentially functions as a catalyst between the supplier and designer. Figure 2 represents this relationship.

Figure 2. Supplier-buyer-designer communication link

The full arrows represent direct communication between parties. The dashed arrows represent indirect and supervised communication and collaboration between the supplier and designer. Effective supplier involvement requires that purchasing-design teams be disciplined in their relations with suppliers. This process necessitates plans and procedures to systematically include and solicit advice/information from the supplier. Suppliers are generally eager to participate. They must, however, be invited to attend meetings and exchange information as full partners. It is important to note that purchasing has the ultimate responsibility for sourcing, supplier selection, and supplier certification. Developing a network of suppliers should be purchasing's responsibility. There must be a clear understanding of purchasing's authority for all internal and external constituents. Purchasing should conduct the investigative aspects of supplier development and certification. Purchasing should act as a liaison, facilitator, agent, or coordinator to facilitate the bringing of designer-supplier together at the early design stage. Suppliers' technological capabilities are often unknown to the designer. It is, therefore, purchasing's responsibility to make certain that the designers are aware of these capabilities and opportunities. Purchasing must typically make preliminary contact with the supplier and then facilitate further contacts of a technical nature (designed-based) with the design and manufacturing engineers. Purchasing must establish the ground rules, outline objectives and timetables, and involve suppliers early on, especially in long-term projects. The role of purchasing should, therefore, be one of coordination rather than inhibition. This process lends itself to the advanced resolution of any potential problems that might occur in the downstream process. Organizations that value this function of purchasing should attempt to locate purchasing centers where design and manufacturing groups are located. This arrangement allows buyers, designer, suppliers and manufacturing engineers to be cognizant of one another's constraints and opportunities. In fact, some companies have attempted to place purchasing agents with engineering design backgrounds in the design and manufacturing teams so that the best result of such an interface can be achieved (Dowst 1984).

Design and manufacturing engineers appreciate it when purchasing assumes the detailed and time-consuming responsibility of developing new sources and obtaining samples. This relieves the designers of a great deal of hardwork and paperwork. If this role is not properly accomplished by purchasing, engineers must often contact suppliers directly. Samples are sent directly to engineers and they must choose parts and materials, perhaps without any prior knowledge of the supplier. This process places all parties (engineering, supplier, and purchasing) in a disadvantaged and precarious position. Purchasing must make certain that the proper supplier is in contact with the engineer. Design engineers should be reminded that a supplier is certified for Part A but not necessarily for Part B. Objectivity and professional judgment of purchasing is essential for design engineers' dealing with a potential supplier. Purchasing has the final coordinating role among different groups of engineers who might be working with different suppliers on different projects. Purchasing has to view these contacts and collaborations on a broader perspective since purchasing is ultimately responsible for the economic and business aspects of such relationship.

Purchasing and design engineers, on the other hand, must look attractive for suppliers as well. This relationship must be based on long-term mutual benefits, trust, and respect. This relationship must be a win-win situation for all; otherwise it will not last for a long period of time.

8. CONCLUSIONS AND ASSESSMENT

This paper has presented a novel approach to the interface and collaboration among designers, buyers, and suppliers in a concurrent engineering environment. The focus of the paper has remained on design-based sourcing as well as early supplier involvement in the product design process. It was argued that the most significant impact and benefits of such an interface are obtained when the contributions and constraints of the supplier and purchasing are known and presented in the conceptual phase of the product design. To underline this significance, a large number of design rules and suggestions were presented. The interface and collaboration among designers, buyers, and suppliers would be best served by the timely implementation of the relevant design rules and suggestions.

A concurrent engineering environment essentially advocates the early involvement of all relevant functional areas in the design process. Care must be taken so that the effects of suppliers and purchasing professionals are considered and coordinated with respect to other functional areas. This process inherently requires a trade-off among various functional requirements. Regardless of the mechanism and the nature of such trade-offs, the focus should remain on the early stages of the product design.

To accomplish the intended purposes of this paper, the following requirements must be met:
    1. There must be genuine and consistent top management support for the dialogue and collaboration among designers, buyers, and suppliers. Top management has to understand the process and nurture it economically and otherwise. There must also exist a corporate and organizational culture conducive to a concurrent

engineering environment. The amount and nature of interactions and cooperation between designer-buyer, designer-supplier, and buyer-supplier are largely dependent upon the attitude of top management.
2. Purchasing professionals should be given an essential role as the key player in the process. This process is certain to fail, if legitimate authority and power is not delegated to the purchasing function. Top management should genuinely encourage purchasing involvement. Purchasing in turn needs to prove that it deserves the new status and is willing to exercise the authority and power judiciously and competently.
3. All collaboration and interface must be done on a long-term basis unless circumstances dictate otherwise.
4. Supplier development and involvement should be perceived as a good business practice and, as such, should be vigorously pursued by small and large firms. This must not be done as a last resort due to an emergency or dire necessity.

The prevalent feeling and current practice is one of pessimism and lack of necessity for supplier development and involvement. Traditionally, there is an air of reluctance and suspicion regarding supplier development and involvement. Persuading suppliers may seem ironic to some purchasing agents in this era of supplier overcapacity. The attitude and short-sightedness of some purchasing practitioners in viewing suppliers as rivals and competitors does not contribute to a healthy climate of partnership and cooperation either. Pitting one supplier against another is still considered a part of a healthy competitive environment in which to do business. In short, the obstacles are numerous, but for those who are willing to venture into the process the benefits outweigh the costs.

REFERENCES

1. Brill, Harold A. (1993) "A common sense approach to handling: an informed buyer has the expertise to properly deal with potential suppliers," *Electronics Design*, 2 September 1993, vol. 41, no. 18, pp. 69-72.
2. Cayer, Shirely. (1990). "Welcome to Caterpillar's Quality Institute, " *Purchasing*, 16 August 1990, pp. 80-84.
3. Dickson, Gary W. (1966). "An analysis of supplier selection systems and decisions," *Journal of Purchasing*, vol. 2, pp. 5-17.
4. Dowlatshahi, S. (1992). "Purchasing's role in a concurrent engineering environment," *International Journal of Purchasing and Materials Management*, Winter 1992, vol. 28, no. 1, pp. 21-25.
5. Dowst, Somerby (managing editor). (1984). "Better-forged links bring in better designs," *Purchasing*, 6 September 1984, pp. 67-75.
6. Dowst, Somberby, (managing editor). (1985). "Purchasing at GM moves closer to the front lines," *Purchasing*, 13 June 1985, pp. 82-85.
7. Gregory, Robert E. (1986). "Source selection: a matrix approach," *Journal of Purchasing and Materials Management*, vol. 22, no. 2, Summer, pp. 24-29.
8. Hahn, Chan K., Charles A. Watts, and Kee Young Kim (1990). "The Supplier Development Program: A Conceptual Model," *Journal of Purchasing and Materials Management*, Spring 1990, vol. 26, no. 2, pp. 2-7.
9. Heinritz, Stuart, Paul V. Farrell, Larry Giunipero, Michael Kolchin. (1991). *Purchasing: Principles and Applications*. 8th edition, Prentice-Hall: Englewood Cliffs, NJ, pp. 4-5.
10. Nevins, J.L. and Whitney, D.E. (1989) *Concurrent Design of Products and Processes*. McGraw-Hill. pp. 2-3.

11. Remich, Jr., Norman C. (editor). (1993) "A purchasing pro brings supplier expertise to DFM," *Appliance Manufacturer*, October 1989, pp. 55-56.
12. Soukup, William. (1987). "Supplier selection strategies," *Journal of Purchasing and Materials Management*, Summer, pp. 7-12.
13. Thompson, Kenneth N. (1990). "Supplier profile analysis," *Journal of Purchasing and Materials Management*, Winter, vol. 26, pp. 11-18.
14. Timmerman, Ed. (1986). "An approach to supplier performance evaluation," *Journal of Purchasing and Materials Management, Winter*, pp. 2-8.
15. Vasilash, G. S. (1987). "Simultaneous Engineering: Management's New Competitiveness Tool," *Production.* Vol. 99, No. 7, pp. 36-41.
16. White, Phillip D. (1978) *Decision making in the purchasing process: a report.* New York, AMACOM.

Information Technologies for Supply Chain Management

Alexander V. Smirnov[a,b] and Charu Chandra[a]

[a]Industrial & Manufacturing Systems Engineering Department
University of Michigan - Dearborn
4901 Evergreen Road, Dearborn, Michigan, 48128, U.S.A.

[b]St.Petersburg Institute for Informatics and Automation of the Russian Academy of Sciences
39, 14th Line, St. Petersburg, 199178, Russia.

The interest in Supply Chain (SC) decision-making problem is fast growing. There is an increasing use of information and knowledge based engineering and management technologies in supply chain problem solving. Furthermore, the trend is towards product data and information & knowledge globalisation in the SC network. The purpose of a SC is to transform incomplete information about the market and available production resources into *co-ordinated* plans for production and replenishment of goods and services in the network formed by *co-operating* units. The value of this information is critical to managing the network, built on principles of a co-operative supply chain (CSC). The goal of SC information & knowledge management is to facilitate knowledge transfer and sharing in the context of SC structures, thereby integrating the customer and the supplier. Ontology-oriented knowledge management tools for SC configuration utilise reusable components and configure knowledge as needed, in order to interactively assist users (agents) in decision-making. This chapter discusses a generic development methodology for information support and decision-making for supply chain management.

1. INTRODUCTION

Knowledge is a critical resource for any business activity. Firms utilise industrial knowledge to manage change in its environment due to product life cycle-time & cost reductions, and variations in product & process specifications. Every firm has collective knowledge in the form of its technological competencies. In order to capitalise on this knowledge base, firms have to organise and manage it in creative and useful ways. As a result, new information technologies, such as Knowledge Management, and Agent Technology are increasingly being used for the purpose.

As firms compete in a global economy, new models of transacting business, such as business-to-business (goods & services), and business-to-consumers have emerged. Novel and creative business partnerships, such as supply chain (SC), and virtual enterprise networks are being formed. These are highly flexible and co-operative business arrangements among partners in order to stay competitive in a dynamic environment. SC configuration is an approach to "network enterprise" creation and reuse that considers enterprises as assemblies of reusable components (units) defined on a common domain knowledge model, such as a "product – process - resources" model described later in this chapter.

The objective of this chapter is to describe elements of a generic information technologies framework for scalable co-operative supply chain management and its implementation through various tools and techniques.

The rest of the chapter is organised as follows. First, a scalable co-operative supply chain is defined as a special class of a supply chain network that offers flexibility, reusability, and extensibility across problem domains. An information technologies framework is described next. A systems approach applied to managing it describes a supply chain management system. For efficient management of a supply chain, its design must be capable of being configured flexibly. A scheme for supply chain configuration is described to accomplish this. Configuring a supply chain involves creating an information & knowledge base, and utilising it in the decision-making process, which are described next. An example of information technologies implementation for a CSC is described following this. Finally, conclusions on this research and plans to extend it are presented.

2. SCALABLE CO-OPERATIVE SUPPLY CHAIN

The concept of supply chain is about managing co-ordinated information and material flows, plant operations, and logistics [1]. It provides flexibility and agility in responding to consumer demand shifts with minimum cost overlays in resource utilisation. The fundamental premise of this philosophy is synchronisation among multiple autonomous business units represented in it. That is, improved co-ordination *within* and *between* various supply chain members. Co-ordination is achieved within the framework of commitments made by Members to each other. Members negotiate and compromise in a spirit of co-operation, in order to meet these commitments. Hence, the label co-operative supply-chain (CSC). Increased co-ordination can lead to reduction in lead times and costs, alignment of interdependent decision-making processes, and improvement in the overall performance of each Member, as well as the supply chain network (Group) [2-4].

The purpose of this type of *organization* is to transform incomplete information about the market and available resources into coordinated plans for production and replenishment of goods and services in the *network* formed by cooperating units. One of the major characteristics of this organization type is scalability. *Scalability* can be categorized in two groups.

- *Operation Scalability* deals with operationalising the enterprise for efficiency and effectiveness. This means adopting standards, procedures and policies that enhance the performance of the enterprise. Issues encountered pertain to developing common -- standards, procedures, methods, goals, policies and objectives. For example, issues of operational scalability for the approximately 26,000 members of the United States textile industry may pertain to adopting common manufacturing standards across sectors represented by fibre manufacturer, textile manufacturer and garment manufacturer, where manufacturing practices are as diverse as flow, cellular, and job shop manufacturing, respectively. Even within a sector, manufacturing practices can be diverse, depending on the size and volume of an operation.
- *Implementation Scalability* deals with implementing concepts, such as flexibility, modularity, usability, and extensibility in the enterprise. Flexibility pertains to designing and modelling enterprise components to accommodate diverse structures that it may adopt

in a changing market environment, such as virtual networks, alliances, value chains, supply chains, etc. Flexibility also pertains to the decision-environment under which the network has to operate, for example batch vs. on-line, deterministic vs. stochastic, optimal vs. Pareto-optimal. Modularity implies that an enterprise can be modularised according to specialisation of knowledge or technology, and then these modules can be replicated throughout the enterprise, thus providing extensibility of knowledge. Extensibility also implies that information in the enterprise can be shared via Internet, Intranet, and Extranet for business-to-consumer, business-to-business service, and business-to-business goods transactions. Issues related to this category are – co-ordination, co-operation, negotiation & compromise, and synchronisation throughout the enterprise echelon. For example, in the case of the large and diverse United States textiles industry, co-ordination for sharing information related to demand forecasts, capacity allocation, or synchronisation of actions among business partners is a problem.

The above categories of scalability are highly interdependent. For example, development of an operational standard, such as common order formats, requires common implementation standards for electronic-data-interchange or Internet-based communication. Yet, these categories are independent by themselves because of the nature and propensity of issues that the system developer has to deal with in designing scalability in supply chains.

3. INFORMATION TECHNOLOGIES FRAMEWORK FOR SCALABLE SUPPLY CHAIN MANAGEMENT

Enterprise integration issues have been addressed with a multi-disciplinary research focus. Techniques and approaches are being incorporated from economics, management science, industrial engineering & operations research, systems sciences, information sciences, and computer science & artificial intelligence fields. For the purposes of this framework, research areas have been roughly classified as those belonging to *systemic*, *reductionist*, and *analytic* approaches. Systemic approaches pertain to study of nature of systems. Reductionist approaches pertain to application of these systems in unique ways, capturing behaviours observed through systemic approaches. Analytic approaches describe how these systems can be utilised efficiently and effectively. This framework offers combining these approaches to advance system rooted constructs to develop scalable enterprises. For purpose of illustration, figure 1 depicts a template that is used to describe below interconnectedness between above approaches and major technology areas in order to create the fusion, necessary for scalability of enterprises.

The coupling of relational data models, and Internet-based tools provide shared representation models / schemes for development of a *kernel* for scalable enterprise systems. To accomplish this, however, the modeller has to overcome difficulties in integration of the enterprises, due to incompatibility of conceptual models for Enterprise Resource Planning with the above kernel. To overcome this hurdle, the methodology offered in this paper incorporates above three approaches. Research thrusts in these approaches are described below with notation of level number marked from (1) to (6). For this illustration, a bottom – up approach has been used.

Systemic Approach: This approach incorporates the abstract level. This level consists of relational data model (1), object-oriented "product-process-resource" model (2), and dynamic knowledge problem domain model (3).

Reductionist Approach: This approach incorporates the activity level. This level consists of dynamic knowledge problem domain model (3), internal state, and goals and objectives of the enterprise *units*, e.g., producer, plant, department, supplier, vendor, etc. (4), and strategic management models (5).

Analytic Approach: This approach incorporates the implementation level. This level consists of internal state, and goals and objectives of the enterprise units (4), and strategic management models (5), and shared goals and objectives of the enterprise (6).

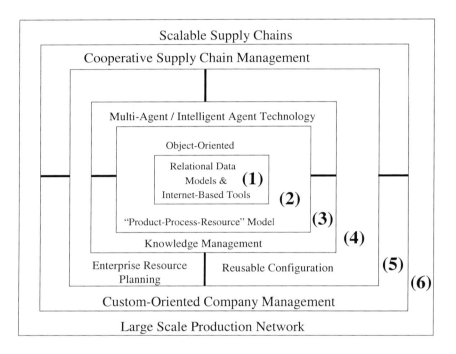

Figure 1. Interconnectedness of problem-solving approaches and technologies for scalability of enterprises.

Supply chain management is the art and science of managing this complex network of interrelated systems and their components. It encompasses identifying goals, and objectives of the supply chain and outlining policies, strategies, and controls for its effective and efficient implementation. A formal mechanism to organise these cohesively requires a systems approach. A supply chain management system (SCMS) is proposed for this purpose. It is a system (S) that describes Member (M) units in a supply chain and their relationship (R) to one another. Notationally, $S = (M, R)$, where $M = \{m_1, m_2, ..., m_n\}$, and $R = \{[(m_1,m_2) (r_1,r_2,...,r_n)], ..., [(m_1,m_n) (r_1,r_2,...,r_n)], ..., [(m_{n-1},m_n) (r_1,r_2,...,r_n)]\}$. Likewise, such a relationship can also be expressed for components of a Member (m_i); at a function (or business) m_i (b_i); process $\{(m_i,b_i) (p_i)\}$; and activity $\{(m_i,b_i,p_i) (a_i)\}$; $i = 1,2,...,n$. Relationships in a SCMS describe, (a) actions between its members (and their components) involving exchange of information, controls, and resources, and (b) meta (or logical) systems (that is, Members and their components), decentralisation, and specialisation of autonomous Members [5-7].

Figure 2 depicts a decentralised view of a textile supply chain. This CSC is comprised of a Group and more than one Member.

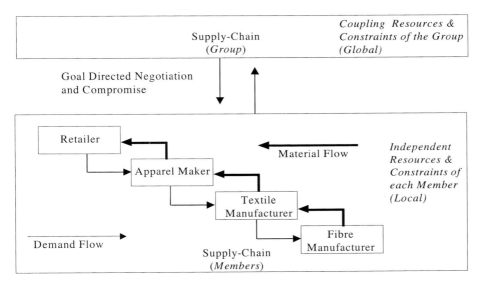

Figure 2. A co-operative supply chain decentralised enterprise view.

The network is arranged in the order of flow of materials, processes, and information between its Members. In this example, consumer demand is relayed by retailer to apparel maker, textile manufacturer, fibre manufacturer, and to cotton grower. Similarly, flow of material occurs in transforming cotton to yarn by fibre manufacturer, fibre to fabric by textile manufacturer, textile to apparel by apparel maker, and a name brand garment by retailer. The interaction between Members occurs as a *consumer* and a *provider*. Thus, an apparel maker assumes the role of a provider (of apparels) in its dealings with retailer (a consumer of apparels). However, the apparel maker acts as a consumer of fabric while dealing with a textile manufacturer (a provider of fabric).

A decentralised CSC is a physically and logically distributed system of interacting components and elements of autonomous business units (Members) [8-9]. In this distributed problem-solving environment, the task of solving a problem is divided among a number of modules or nodes (autonomous business units and their systems). Members co-operatively decompose and share knowledge on the problem and its evolving solutions. Interactions between Members in the form of co-operation and co-ordination are incorporated as problem-solving strategies for the system. Unit Group is responsible for co-ordination throughout the supply chain. Unit Member brings specialised expert knowledge and product and process technology(ies) to the supply chain. The decision-making process is centralised for the Group. The Group enforces common goals and policies for the supply chain on Members. However, decision-making at Member is decentralised. Each Member pursues its own goals, objectives, and policies conceptually, independently of the Group, but pragmatically in congruence with Group goals. A common knowledge base supports the CSC structure [2]. Knowledge is assimilated for an *activity* (the lowest level of information) in a specific domain and aggregated for various decision-making levels in the enterprise. Main concepts in *activity modelling* in a CSC decentralised enterprise are described below.

- *Activity* represents transformation of input (in the form of a technology) to an output (or product) through use of resources of an enterprise, such as a supply chain.

- *Process* is a collection of activities representing various forms of technologies mobilised by the enterprise in generating an output.

- *Supply Chain Management* comprises of activities or processes. These units when associated to a user assume unique ontological forms.

- *Ontology* is a unique form of representing knowledge applied in various domains. It is useful in creating unique models of a CSC by creating specialised knowledge bases specific to various supply chain problem domains.

- *Representation*. An *activity* represents the lowest level of interaction in the supply chain model. It is synonymous with a "Member" for modelling business process, and an "agent" for knowledge management environment. It is classified into various *activity types* depending on unique *service*(s) they provide. Activity (ies) is used in relation to an aggregation. An activity possesses *attribute*(s), which describe its characteristics or features. An attribute assumes *parametric* values in relation to an aggregation model. Activities communicate with each other by exchanging *message*(s). Communication occurs based on a *protocol* whose boundaries are set by a *control* matrix prescribing level of *resource*(s) to be utilised by an activity, policies to be pursued, and objectives to be met in providing the service(s).

- *Aggregation*. Aggregation represents a *system* form, depicted in figure 3 for a textile operation. It has seven components -- input, process sequence, output, mechanism, agent, environment, and function, described in [10], which are defined by four matrices, namely, resource, performance, technology, and input/output. Each aggregation (system) has its own control matrix to define relationships between its components. Aggregation can take on many forms manifested by the orientation it is based upon, that is aggregation "within" system(s), or aggregation "between" systems. For example, a material-life-cycle-flow and order-life-cycle represent horizontal aggregation between systems. Building decision models across the enterprise represents a vertical aggregation between systems. Similarly, aggregating all activities within a Member function represents "within" systems integration.

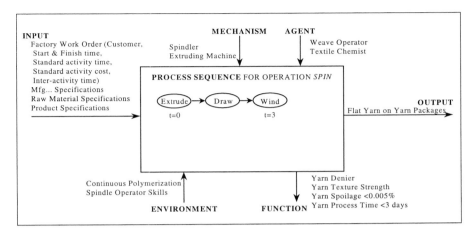

Figure 3. A generic system representation in a co-operative supply chain.

- *Protocol.* Protocols for each aggregation (system) describe conventions governing communication between activities, services rendered by activities to one another, and controls for that system.
- *Communication* between activities occurs in the form of *message*(s) exchanged to request a service.
- *Services* are of resource and information types.

As the business environment is changing rapidly, supply chains are being designed innovatively to offer competitive advantage. Some of the potential uses of flexible supply chain are (i) dealing with changing production demand, and (ii) trading product models on business networks.

The efficient management of this SC, therefore, requires that its design be properly configured. An important facet of configuration is to understand the relationship between supply chain system components that define its structure, that is products, and associated processes & resources.

The objective of designing SC utilising configuration principles is to generate customised solutions based on standard components, such as templates, baselines, and models. There are two aspects to configuration, (i) configuring / reconfiguring, and (ii) configuration maintenance.

Configuring deals with creating configuration solutions and selecting components and ways to configure these. Configuration maintenance deals with maintaining a consistent configuration under changing environment. This requires consistency among selected components and decisions. When a decision for selected component changes, configuration maintenance must trace all related decisions and revise them, if necessary. Thus, consistency among components and decisions is maintained. A corporate knowledge management (KM) spanning the supply chain is desired to implement SC configuration.

4. SUPPLY CHAIN KNOWLEDGE MANAGEMENT

An important requirement for collaborative system is the ability to capture knowledge from multiple domains and store it in a form that facilitates reuse and sharing [11-12]. KM could be identified by four factors behind successful KM systems [13]:

1. An understanding by employees as to why knowledge sharing is important,
2. Recognition by employees,
3. Legacy of existing practices, and
4. Support mechanism or safety net that allows employees to experiment.

KM is 90 per cent people and 10 per cent technology. General functions of KM are -- externalisation, internalisation, intermediation, and Cognition [14]. These describe the relation "user – knowledge / databases".

The approach suggested in this paper is limited to designing SC configurations for manufacturing systems and focused on using ontological descriptions. It is based on GERAM, the Generalised Enterprise Reference Architecture and Methodology [15]. This approach is one of several key approaches evaluated and recommended by the IFAC/IFIP Task Force on Enterprise Integration that facilitates developing an overall definition of a generalised architecture. It could be implemented using a variety of Enterprise modelling languages, such as ARIS, CIMOSA, GRAI/GIM, IEM, and the IDEF family of languages.

Applying the GERAM approach enables forming the conceptual model of the SC system. This is accomplished by modelling its product, process, and resource components to satisfy manufacturing constraints in its environment.

The SC configuration stage is represented by the following relationship:

"*configuring product* (product structure, materials bill) → *configuring business process* (process structure, operation types) → *configuring resource* (structure of system, equipment and staff types)".

The implementation of SC approach is based on the shareable information environment that supports the "product - process - resource" model (PPR-model) used for integration and co-ordination of user's activity. This model is studied from different viewpoints or user groups as depicted in figure 4.

Enterprise models may be defined in various ways. In increasing order of formality, generic enterprise models may be defined as [15]:

- Natural language explanation of the meaning of modelling concepts (*glossaries of terms*),
- Some forms of meta models for example, *entity relationship, meta schemas, conceptual models of terminology, and component modelling languages* describing the relationship among modelling concepts available in enterprise modelling languages, and
- Ontological theories defining the meaning (*semantics*) of enterprise modelling languages to improve the analytic capability of engineering tools and through them the usefulness of enterprise models.

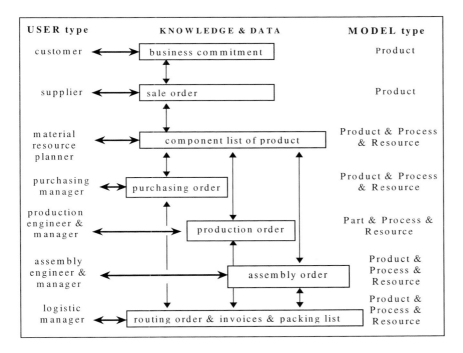

Figure 4. Links of users to knowledge and data in supply chain.

Ontologies are created similar to how knowledge domain models are built. Formal steps in creating ontologies are:
- Defining entities,
- Assigning attributes to entities with regard to its domain,
- Identifying relationships between entities and domain,
- Identifying relationships between entities,
- Describing levels of constraints, and
- Labelling ontologies by classes of entities.

Ontologies are managed by translation and mapping between different types of attributes and entities.

Ontological theories are formal models of concepts that are used in SC model representations. They capture rules and constraints of the domain of interest, allowing useful inferences to be drawn, analyse, execute, cross check, and validate models.

Ontological translation of an enterprise, such as a supply chain is necessary because networks are multi-ontology classes of entities. Various ontologies for an entity describe its unique characteristics in context with the relationship acquired for a specific purpose or problem. For example, an entity "textiles" may have a multi-ontology representation for a user with a marketing perspective, and for another user with a design perspective, respectively. For the user interested in the marketing perspective of textiles, its attributes of size, denier, and style are important. However, the same textiles' characteristics may be represented by size, quality, and finish for the user interested in its design specifications.

5. OBJECT-ORIENTED DYNAMIC CONSTRAINT NETWORK AS TOP-LEVEL ONTOLOGY PARADIGM FOR SUPPLY CHAIN CONFIGURATION

It is widely accepted that engineering & management activities can be regarded as search involving techniques to satisfy constraint [16-18]. Constraint satisfaction is a fundamental SC problem. Conventional constraint satisfaction procedures are designed for the problem with one constant set of constraints. For example, in engineering for manufacturing systems, such as design for productivity, layout, and scheduling, it is often necessary to solve a dynamic constraint satisfaction problem where applicable constraints depend on the design aspect [19].

The domain knowledge model of SC contains entities (objects), which can be of different types (classes). Multi-level and multi-aspect/viewpoint are used for PPR-model description. Obviously, each user works with its own ontology-oriented constraint network. KM tools support the conversion of PPR-model, from one ontology to another.

An abstract PPR-model is based on the concept of ontology-oriented constraint networks. Networks are represented by multi-ontology classes of entities, the logic of attributes and the constraint satisfaction problem model. This abstract model unifies main concepts of languages, such as standard object-oriented languages with classes, and constraint programming languages. It supports the declarative representation, efficiency of dynamic constraint solving, as well as problem modelling capability, maintainability, reusability, and extensibility of the object-oriented technology.

Ontology-oriented constraints network model is denoted $A = (St, C)$, where St — an ontology structure, C — a set of ontology constraints. To deal with the concept schema of configuring process defined in terms of constraints, a dynamic constraint network (DCN) model is applied. A static constraint network (SCN) $A_i = (V_i, D_i, C_i)$, involves a set of variables $V_i = (v_{i1}, v_{i2}, \ldots, v_{iN_i})$, each taking value in its respective domain $D_i = D_{i1} \times D_{i2} \times \ldots \times D_{ij} \times \ldots \times D_{iN_i} = \underset{j=1}{\overset{N_i}{\times}} D_{ij}$, and a set of constraints $C_i = \{c_{i1}, c_{i2}, \ldots, c_{ik}\}$. A DCN N is a sequence of SCNs, each resulting from a change in the proceeding one imposed by "the outside world". For description of top-level ontology the following abstract model based on integration of DCN model and object-oriented model [20] could be used:

Model: $M_c = = [< A_1, F_{11}, \ldots, F_{1n} >, \ldots, < A_n, F_{n1}, \ldots, F_{nn} >]$,

where, A_i — an ontology, F_{ik} a conversion from an ontology A_i to A_k, and F_{ii} — an identity.

Ontology: $A_i = (V_i, D_i, C_i)$ is a SCN.

Conversion: $F_{ik} = \underset{j=1}{\overset{N_i}{\times}} v_{ij} = E_{D_{ij}} \to \underset{l=1}{\overset{N_i}{\times}} v_{kl} = E_{D_{KL}}$, where, $v_{ij} \in V_i$, $j = 1, \ldots, N_i$, $D_{ij} \subseteq D_i$, and $\underset{j=1}{\overset{N_i}{\times}} v_{ij} = E_{D_{ij}}$ is a set $\{(v_{i1} = e_{i1}, \ldots, v_{iN_i} = e_{iN_i}) | e_{ij} \in E_{D_{ij}}\}$ describing the labelled Cartesian product, built on the structure (V_i, D_i). The term $E_{D_{ij}}$ designates the values set of D_{ij}.

Constraints: $C_i: \underset{j=1}{\overset{N_i}{\times}} v_{ij} = E_{D_{ij}} \to Boolean = \{true, false\}$.

Instance: $<d_1,\ldots,d_n>$, where $d_i = <A_i, (v_{i1}=e_{i1},\ldots,v_{iN_i}=e_{iN_i})>$ is a description in ontology A_i, $<v_{i1}=e_{i1},\ldots,v_{iN_i}=e_{iN_i}>$ — a tuple of equalities in $\underset{j=1}{\overset{N_i}{\times}} v_{ij} = E_{D_{ij}}$.

The above Ontology Management approach is based on two mechanisms:

1. Class inheritance mechanism, supported by inheritance of class ontologies (attributes inheritance) and by inheritance of constraints on class attribute values, and
2. Constraint inheritance mechanism for inter-ontology conversion, supported by constraint inheritance for general model (constraints strengthening for "top-down" or "begin-end" processes).

6. INTEGRATION OF ONTOLOGY AND AGENT

An *agent* is a surrogate unit that encapsulates and inherits properties of the real (physical or logical) unit. Through its surrogate status, the agent adapts to behaviour in its environment, thereby offering flexibility, modularity, and adaptability in designing agile systems.

A recent development in information technologies that adapts co-ordination and interaction in distributed groups, such as SC, is multi-agent systems (MAS) technology. MAS utilises intelligent agents for its implementation approach. Intelligent agent is an autonomous software unit that can navigate heterogeneous computing environment and can, either alone or working with other agents, achieve some goals [21]. It possesses following properties:

- *Autonomy*: agents act without any human or other global external control, and their states and actions are operated by a local control,
- *Social behaviour*: agents interact with each other either by message passing or by direct communications and synchronisation of their actions,
- *Reactivity*: agents can perceive external information and react on it. External information is received from their environment, which may be the physical world, a user via a graphical user interface, a collection of other agents, or perhaps all of these combined, and
- *Initiativity*: agents may act not only in response to demands of it surrounding, but according to their own plans and aims.

MAS technology allocates following agents [21]:

1. Regulation agents which react to environment conditions and always know what to do, and
2. Planning agents that possess properties of regulation agents and additionally, they can plan their actions depending on external conditions.

MAS technology can be considered as the basis for group decision support system. Intelligent agents are widely used in Co-operative E&M. Mobile agents are widely used in Internet search systems [22].

Supply Chain Management Systems as described in section 3, is a special class of a complex autonomous adaptive systems (CAAS), whereby each component (Member or its class) is autonomous and adaptive. Components Member and Group can be classified as an agent.

An adaptive system is multi-structural. This system changes its internal structure, in order to keep up to changes in its environment. A CAAS models complexity by discerning patterns of behaviour of its components (or agents). It decentralises functionality and encourages diversity and specialisation. Thus, the supply chain philosophy with inherited properties of a CAAS, offers a flexible approach to building complex systems that meet criteria of appropriate structures, controls, and optimisation in a multi-criteria decision-making environment. A distributed problem-solving hypothesis that recognises decentralised and autonomous nature of hierarchy in the supply chain can be applied to implement the SC configuration framework. Aggregation of system components preserves internal structure of *parts* within a *whole*. *Co-operation* in the supply chain is achieved through synchronisation and co-ordination of interactions between demand and supply in the supply chain. Integration in the model is realised through feedforward and feedback mechanisms incorporated in the decision processes.

The implementation of the basic principle for the SC, i.e., its collaborative nature is based on distribution of procedures between different users (or different agents). For this purpose, it is obvious to represent the SC configuration KM as a set of interactive autonomous agents. Different configuration problems are treated as separate agent-oriented tasks with embedded constraint satisfaction and consistency support facilities [23-24]. Agent-based distributed constraint satisfaction problem is a problem in which variables and constraints are distributed among agents. Agents communicate by sending messages. The chart of co-operation cycle for agents has the following structure:

1. Application domain knowledge is shared between agents; knowledge about constraints (user requirements and artefacts) are transmitted to particular agents,
2. Agents solve local configuration tasks on the basis of shared knowledge, and
3. Results are collected and transmitted to all interested agents.

These stages are continuously repeated until the solution is globally confirmed or consensus can not be reached, and constraint relaxation methods are to be applied.

Co-ordination of concurrent agent activities is supported by Communication Network depicted in figure 5. Each consistency arc is associated with a message to the shared KB. An internal arc associates states of the agent (or ontology)-oriented decision making process. General rules of above network are as follows:

Rule 1. Two arcs l_1 and l_2 are simultaneous arcs if $\neg(l_1 \rightarrow l_2) \wedge \neg(l_2 \rightarrow l_1)$.

Rule 2. Two simultaneous arcs l_1 and l_2 are data-race free, <u>iff</u> all the following conditions are true:

1) $\text{WRITE\_SET}(l_1) \bigcap \text{WRITE\_SET}(l_2) = \varnothing$
2) $\text{WRITE\_SET}(l_1) \bigcap \text{READ\_SET}(l_2) = \varnothing$
3) $\text{READ\_SET}(l_1) \bigcap \text{WRITE\_SET}(l_2) = \varnothing$, where $\text{WRITE\_SET}(l_1)$ and $\text{READ\_SET}(l_1)$ are sets of shared variables written and read in arc l_1, accordingly.

n_{km}^i - node (agent) associated externally defined predicate from DCN N_{km}^i;

l_r^i - variable from $V^i \subseteq V$, which is consistent with shared DCN N;

i, j – ontology numbers, k - task number, m – a state number

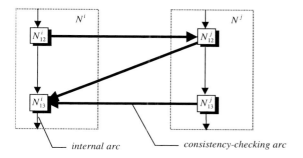

Figure 5. Communication network of agent activities.

7. COALITION GAMES AS A SUPPLY CHAIN NEGOTIATION MODEL

The following formal description of the SC model in term of resources dependency network could be used. SC resources dependency network depicted in figure 6 shows a result of conversion from SC Technology Model to SC Resource Model. Two types of nodes represent goods and SC participants (units). A network is a directed acyclic $graph(B,L)$, where $B = Un \bigcup G$, Un – a set of SC units (agents), Un - a set of producers, suppliers, customers, logistic companies, etc., G – a set of goods (parts, subassembly, products, orders, goals, etc.), L - a set of edges L, connecting units (agents) with goods they can use or produce.

As a result, an agent-based SC network that describes connections of SC local tasks (partial technologic/business processes or unit's resources) as the task allocation problem is developed. This problem could be solved by an agent-based technology for decentralised task allocation.

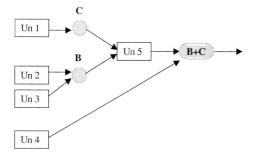

Figure 6. Example of supply chain resources dependency network.

Un 1, Un2, Un 3 – produce units (Un 2, Un 3 – could produce same goods);
Un 4 – could trade unit for subassembly (B+C); Un 5 – sub-assembly unit

In current approaches, objects in manufacturing systems (customer or supplier) are assigned decision-making intelligence and autonomy, which enable co-ordination and control. This is generally accomplished by mean of negotiation [22, 25, 26], market-based mechanisms [27-30], auction-based mechanisms [31-32], coalition formation [33], and game theory [34].

The economic or market-oriented co-operation of SC agents / units is motivated by the endeavour of its participants to increase their individual profit via co-ordinated activities. The coalition game theory offers results showing the general form of possible co-operation, and conditions under which achievement of these targets is feasible. In many cases, there exists relatively large number of achievable co-operation patterns and profit distribution schemes. For practical reasons, it is desirable to suggest exactly one of them as the best, or most unbiased. This problem could also be investigated by the coalition game theory, and its approach represented by the concept of the values of coalition games, depicted in figure 7.

The real bargaining usually takes place before the application of co-ordinated strategies and when agents (partners) have only a vague idea of the expected coalition profits. This means that results of the bargaining, including distribution of the profit also can be only vague. Using game theory, it is possible to study this vagueness and follow the path from the vague input idea about the expected profits via the bargaining process to possibilities of vague output distributions of individual pay-off.

The deterministic coalition game with side payments is defined as pair (I,V), where I is a non-empty and finite set of players, sub-set of I are called coalitions, and V, which is called a characteristisctic function of the game, is a mapping of set 2^I to a set R^I, prescribing to every coalition $\bullet \subset 2^I$ a real number $V(K) \subset R^I$, which satisfies following conditions:

$V(K)$ – a finite set;
if $x=(x_i)_{i \in I} \in V(K)$, $y=(y_i)_{i \in I} \in R^I$, $y_i \leq x_i$ for $\forall i \in K$, then $y \in V(K)$;
$V(K) \neq \emptyset$;
$V(K) = R^I \Leftrightarrow K = \emptyset$.

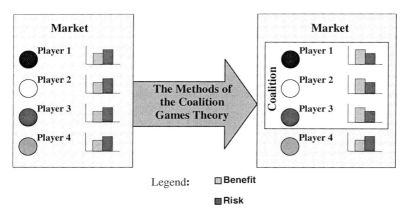

Figure 7. A coalition game approach.

In CSC case every solution about coalition reguires an added constraint – a condition of integrity for technologic/business process:

$$\sum_{i \in K} a_{ij} \geq Q_j \quad \text{for all j, where}$$

a_{ij} – a volume of partial technologic/business process (or a technology operation) j realized by player (unit) i;

Q_j – a volume of partial technologic/business process j, needed for order realization.

$i \in K$ – all players from the coalition (or CSC).

This model describe SC management problems as a declarative problem and could be solved by using agent-based distributed constraint satisfaction technologies [19, 23].

8. AGENT-BASED GROUP DECISION SUPPORT SYSTEM

A SC configuration management GDSS is based on dynamic constraint networks model for knowledge representation, object-oriented programming technique, and co-operative decision making. It provides users with the ability for requirements specification indexes sampling and weighing, and multi-level configuration synthesis. Different configuring problems are treated as separate tasks with embedded constraint satisfaction and consistency support facilities.

The implementation of the basic principle for the SC co-operation is based on distribution of procedures between different users (or different agents) concurrently in the common knowledge space. It is natural to represent the configuration management knowledge as a set of interacting autonomous agents.

The structure of multi-agent environment supported by a SC GDSS consists of three types of agents [35], namely (i) SC manager agent, (ii) SC unit agent, (iii) SC co-ordinator agent. Depending on the information exchanged, different agents perform different functions. Using agent classification Unit and Co-ordinator agents may be classified as regular agents, while the Manager Agent is a planning agent.

Figure 8 depicts the step by step procedure in the decision-making algorithm applied to the GDSS. It shows following sequence of steps on which a unit's private opinions have been generated, and group decisions have been formed:

1. The task becomes accessible for processing when the manager prepares a list of attributes for evaluation.
2. The co-ordinator agent receives incoming message (M_{inp}) about task readiness for processing. It defines and prepares necessary information for the unit's local database. Using internal set of function (F), it prepares messages for unit agent.
3. The unit manager's agent checks existing tasks (analyses incoming messages M_{ij}, where j is a unique number of the unit manager and i is a number of the process step) and develops a set of actions F_j.
4. Unit agent increments readiness degree of the task on the local level when the unit manager finishes (transition S from readiness degree N into readiness degree $N+1$). It sends a message (M_{oj}) to co-ordinator agent.
5. The co-ordinator agent analyses whether similar messages have been received on readiness from all other members of the group. If someone is not ready, there is a wait on

the entire group readiness. The readiness of the task is incremented when everybody is ready (transition S from readiness degree N into readiness degree $N+1$). As a rule, after a stage of opinions making by each unit manager, a stage of the conformation follows.
6. The co-ordinator agent sends the message (M_{out}) to the manager agent that a stage of the conformation follows.
7. Depending on the co-ordination type, that is (a) automatic control by the co-ordinator agent, or (b) personal control by the SC manager, there are two possible outcomes:
   - an automatic check of conformation is received, or
   - the task is stopped and the SC manager using symbolical and graphical tools of data representation, analyses whether it is necessary to advance the task readiness or to return back a step.
8. Appropriate decision is sent to the co-ordinator agent which either advances the task per unit further, or returns the unit agents on to the previous stage.

The unit managers use local data which are representation of data prepared by the SC manager, while each agent at local and group levels works with their own internal representations.

9. DESO PROJECT: OBJECT-ORIENTED TOOL FOR CORPORATE ONTOLOGY CREATION AND MAINTENANCE

The goal of DESO (DEsign of Structured Objects) project is to develop a methodology and tools for automated re-use of industrial experience from large collection of knowledge & data in the engineering and business domains. Figure 9 depicts the DESO architecture. It aims at establishing a knowledge platform that enables manufacturing enterprises to achieve reduced lead-time and cost based on customer satisfaction by means of improved availability, communication and quality of product information. DESO follows a decentralised method for intelligent knowledge and solutions access. The configuration process incorporates following features: order-free selection, limits of resources, optimisation (minimisation or maximisation), default values, and freedom to make changes to the SC Model.

An ontology-oriented SC model depicted in figure 10 is an object library comprised of the following:

- A set of object classes structured as a taxonomy, i.e., each object is linked to one or more other objects by a sub-class / super-class relationship.
- For each object class, a set of relations is defined linking it to other object classes, as well as a constraint set for each relation, and
- For each object class, a set of attributes and a definition of the intended meaning of each attribute are provided.

The "product-process-resource" model serves as a knowledge repository for manufacturing system. The structure of the database formed by it is presented in figure 11.

Information Technologies in Supply Chain Management 453

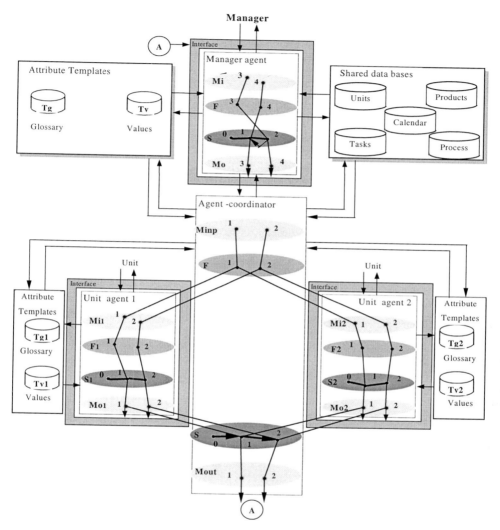

Figure 8. A general structure of interacting agents.

Basic units of the DESO tool database are objects, such as "part", "machine", "operation", "plant" etc. The highest level of object abstraction is object Class, which captures an object's nature. Classes are defined and entered into database by the developer at the installation stage. The next level of object abstraction is object Type. Each type belongs to one of the classes of the system and contains objects with similar properties. At this level, object attributes (one type of object possesses identical set of attributes) are defined. The user of the tool may create, edit and delete types. At the level of objects, values for each attribute may be entered and edited. For example:

Class = Machine, Type = Extruding Machine.

Attributes, objects and values are shown in Table 1.

Table 1
DESO: Object Properties

Attributes: Objects	Max Diameter mm	Precision	Price, USD
M1415	1500	4	55,800
M1335	3500	3	63,400

The database structure allows setting relations between classes and types of *objects*, e.g., machines of "Extruding_type_1" and "Extruding_type_2" can perform the extrusion operation. At the level of *types*, complex constraints for values of attributes can be set, e.g., the length and the diameter of a part should not exceed maximum values allowed for turning machines. An example depicted in figure 12 is described as follows:

- Operation "123T – A" can process only part "O – 123T", and
- Operation "123T – B" can process both part "O – 123T" and part "I – 123T".

Constraint for parts «Yarn» and operations «Extrude, Draw, Wind»:

- Part.L1 • Operation.L1, and
- Part.L Ref • Operation.LREF.

The hierarchy editor is a tool for creating, editing and managing hierarchical relations between objects, e.g., "part > technological process > operation > machine". These relations may show structures of objects, sequence of operations for a part production, and possible alternatives of accommodation etc. The hierarchy editor supports inheriting subordinate objects, that allows creating complex hierarchical systems of objects by some stages and using templates automating the user's work.

The user could evaluate systems costs for every unit, such as time, efficiency, fixed and variable production costs. Every cost centre could include several machine tools.

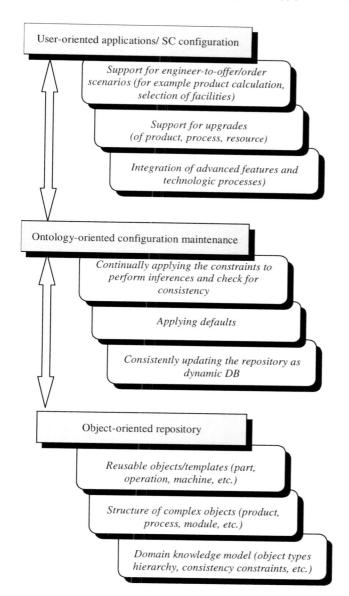

Figure 9. DESO architecture.

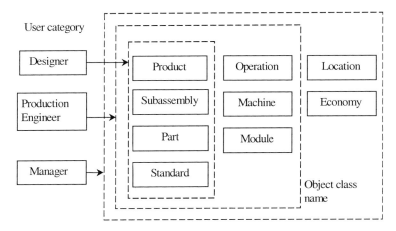

Figure 10. Sharing ontology example for the "product-process-resource" model.

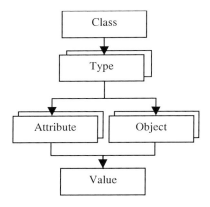

Figure 11. DESO: database structure.

Properties for parts are given in Tables 2, and 3.

Table 2
Properties for parts «Yarn»

Attribute	O – 123T	I – 123T
L1	1300	1500
L Ref	1200	1450

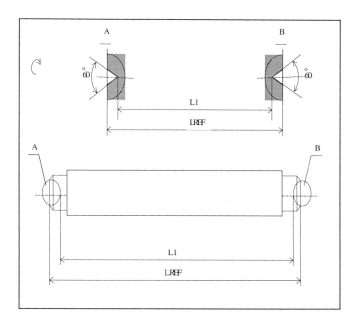

Figure 12. DESO: The use of constraints
(upper figure is operation, lower one is part).

Table 3
Properties for operations «Extrude, Draw, Wind»

Attribute	123T – A	123T – B
L1	1300	1500
L REF	1400	1600

10. CONCLUSIONS

Implementation of information technologies for scalable co-operative supply chain management enables achieve flexibility, modularity, usability, and extensibility in the networking organization. Scalability deals with extensibility of supply chain knowledge. This implies that information in the enterprise can be shared via Internet, Intranet, and Extranet for business-to-consumer, business-to-business service, and business-to-business goods transactions. Issues related to this category are – co-ordination, co-operation, negotiation & compromise, and synchronization throughout the networking organization's echelon.

The universality of the knowledge representation by object-oriented dynamic constraint networks for all kinds of SC models makes it feasible to provide powerful interactive tool for DB & KB maintenance. Knowledge management environment comprises means for SC project consistency control and allows multiple aspect / ontology-oriented collaborative engineering. Ontology-based DESO architecture supports co-operation among agents, thereby reducing time and increasing the quality of SC configuration process. Ontology-based knowledge management technology is an innovative technology in the domain of SC. Using this technology enables fewer people to make faster and better quality decisions on SC configurations from SC unit templates, under constraints networks with reduced variance. Implementation of this technology will enable realise increased quality, reduced cost, reduced errors, and better SC configuration solutions.

Implementation of the supply chain management methodology for decision support of scalable networking organizations would fundamentally change global business environment by enabling constructive collaboration among its network members to achieve shared targets. This approach satisfies the need for increasingly complex business relationships as well as a considerably complex underlying technology infrastructure that companies are implementing as part of their global strategies.

Future plans for research on this topic are to develop and experiment generic SC network configurations based on, (a) divergent problem solving strategies, (b) functional or operational policies of Members and / or Group, and (c) levels of co-operation among members of the SC.

ACKNOWLEDGEMENTS

Some components of the above approach have been developed for the "Demand Activated Manufacturing Architecture project for the United States Integrated Textile Complex", supported by a grant to the second author from the United States Department of Energy; and the project "Affordable Cost Structure", supported by a grant from Ford Motor Company to the first author.

REFERENCES

1. Lee, H. L., and Billington, C. 1993. Material management in decentralised supply chains. *Operations Research* 41(5): 835-847.
2. Chandra, C. 1997. Enterprise architectural framework for supply-chain integration. In Proceedings of the Sixth Annual Industrial Engineering Research Conference, 873-878. Norcross, Georgia: Institute of Industrial Engineers.
3. Poirier, C. C. 1999. *Advanced Supply Chain Management: How to build a sustained competitive advantage*. San Francisco, California: Barrett-Koehler Publishers, Inc.
4. Tzafestas, S., and Kapsiotis, G. 1994. Co-ordinated control of manufacturing/supply chains using multi-level techniques. *Computer Integrated Manufacturing Systems* 7(3):206-212.
5. Bond, A. H., and Gasser, L., eds. 1998. *Readings in Distributed Artificial Intelligence*. San Mateo, California: Morgan Kaufmann.
6. Gasser, L. 1991. Social conceptions of knowledge and action: DAI foundations and open systems semantics. *Artificial Intelligence* 47:107-138.

7. Moulin, B. and Chaib-Draa, B. 1996. *An overview of distributed artificial intelligence.* O'Hare, G. M. P., and Jennings, N. R. eds. New York, New York: John Wiley & Sons. 3-55.
8. Hillier, F. S., and Lieberman, G. J. 1990. *Introduction to Operations Research.* New York, New York: McGraw-Hill Publishing Company.
9. Taha, H. A. *Operations Research: An Introduction.* 1987. New York, New York: Macmillan Publishing Company.
10. Nadler, G. 1970. *Work Design: A Systems Concept.* Homewood, Illinois: Richard D. Irwin.
11. Neches, R.; Fikes, R. E.; Finin, T.; Gruber, T.; Patil, R.; Senator, T.; and Swartout, W. R. 1991. Enabling Technology for Knowledge Sharing. *AI Magazine* 12(3):16-36.
12. Patil, R. S.; Fikes, R. E.; Patel-Schneider, P. F.; MacKay, D.; Finin, T.; Gruber, T.; and Neches, R. 1992. The DARPA Knowledge Sharing Effort: Progress Report, In Proceedings of the Annual International Conference on Knowledge Representation. Cambridge, MA.
13. Donkin, R. 1998. Disciplines complete for a newcomer. *Financial Times* October 28:14.
14. Delphigroup. 1998 www.delphigroup.com.
15. ISO TC 184/SC 5/WG 1. 1997. Requirements for enterprise reference architectures and methodologies, http://www.mel.nist.gov/sc5wg1/gera-std/ger-anxs.html
16. Giachetti, R. E., Young, R. E., Roggatz, A., Eversheim, W., and Perrone, G. 1997. A methodology for the reduction of imprecision in the engineering process. *European Journal of Operational Research* 100:277-292.
17. Hirsch, B. 1995. Information System Concept for the Management of Distributed Production. *Computers in Industry* 26:229-241.
18. Tsang, J. P. 1991. Constraint Propagation Issues in Automated Design, In: *Expert Systems in Engineering: Principles and Applications,* Gettlob, G., and Nejdl, W., eds. 462:135-151. Berlin: Springer-Verlag.
19. Smirnov, A. V. 1994. Conceptual Design for Manufacture in Concurrent Engineering. In Proceedings of the Concurrent Engineering: Research and Applications Conference, 461-466. Pittsburgh, Pennsylvania.
20. Royer, J. 1992. A new set interpretation of the inheritance relation and its checking. *OOPS Messager* 3(3):22-40.
21. Franklin, S. 1996. 1996. Is it an agent, or just a Program?: A Taxonomy for Autonomous Agents. In Proceedings of the Third International workshop on Agent Theories, Architectures, and Languages, Springer-Verlag.
22. Van Aeken, F., Demazeau, Y., and Kozlak, J. 1997. Migration, Mobile and the Palindrome Problem. In Proceedings of the International Workshop "Distributed Artificial Intelligence and Multi-Agent Systems" (DAIMAS'97), St. Petersburg, Russia. 74 - 80.
23. Sheremetov, L. B., and Smirnov, A. V. 1997. A Model of Distributed Constraint Satisfaction Problem and an Algorithm for Configuration Design. *Computación y Sistema* 2(1):91-100.
24. Smirnov, A. V., and Sheremetov, L. B.. 1998. Agent & Object-based Manufacturing Systems Re-engineering: a Case Study. In Proceedings of the Conference on Integration in Manufacturing, 369-378. Göteborg, Sweden: IOS Press.

25. Jennings, N. R. 1994. Cooperation in Industrial Multi-agent Systems, 43:World Scientific Series in Computer Science. World Scientific Publishing Co. Inc.
26. Wooldridge, M. J., and Jennings, N. R. 1995. Intelligent Agents: Theory and Practice (Knowledge Engineering Review). 10:2,115-152.
27. Jamali, N., Thati, P., and Agha, G. An Actor-Based architecture for Customizing and Controlling Agent Ensembles. 1999 (March/April). *IEEE Intelligent System.* 38 – 44.
28. Park, S., Durfee, E., and Birmingham, W. 1999. An adaptive agent bidding strategy based on stochastic modeling. In Proceedings of the third international conference on Autonomous Agents. Seattle, WA, USA, May 1-5. 147-153. ACM Press.
29. Preist, C. 1999. Commodity trading using an agent-based iterated double auction. In Proceedings of the third international conference on Autonomous Agents. Seattle, WA, USA, May 1-5. 131-138. ACM Press.
30. Walsh, W., and Wellman, M. 1998. A market protocol for decentralized task allocation. In Proceedings of International Conference on Multi Agent Systems. Paris, July 3-7. 325-332.
31. Davis, R., and Smith, R. 1983. Negotiation as a metaphor for distributed problem solving. *Artificial Intelligence* 20:63-109.
32. Fisher, K., Müller, J. P., Heimig, I. and Scheer, A. 1997. Intelligent Agents in Virtual Enterprises, Institute für Wirtshaftsinformatik.
33. Larson, K., and Sandholm, T. 1999. Anytime coalition structure generation: an average case study. In Proceedings of the third international conference on Autonomous Agents. Seattle, WA, USA, May 1-5. 40-47. ACM Press.
34. Wu, S., and Soo, V. 1999. Game theoretic reasoning in multi-agent coordination by negotiation with a trusted third party. In Proceedings of the third international conference on Autonomous Agents. Seattle, WA, USA, May 1-5. 56-61. ACM Press.
35. Pashkin, M., Smirnov, A., and Rakhmanova, I. 1996. Multiexpert: an Internet bases support environment for solution evaluation. In Proceedings of the Special Day on pan-European Co-operation and Technology Transfer, Zakopane, Poland.

Enterprise Integration and Management in Agile Organizations

F.B. Vernadat

LGIPM and MACSI-INRIA
ENIM/University of Metz, Ile du Saulcy, F-57045 Metz cedex 1, France

This Chapter discusses Enterprise Integration techniques in the context of agile management of stand-alone or networked organizations, mainly for the manufacturing sector but also for the service industries. The goal is to provide support to cope with required system reactivity and needs for managing change. One of the key points is to think the business in terms of a large set of business processes (or projects) and a large set of business entities (or organization units), all interacting with one another and which have to be timely coordinated to achieve the desired business mission.

1. INTRODUCTION

Agility is the ability to timely react to changing conditions. For an enterprise, it means to be able to respond quickly and successfully to change (be it market, regulation, environment, management or technological change) [1-3]. Indeed, management of change is the challenge.

At the turn of the 21st century, the manufacturing world is facing major changes due to globalization of markets. Fierce competition and new customer attitudes lead to increased product variety and reduced margins. At the same time, technology is evolving very quickly, systems become highly complex and never before has access to information, technology and capitals been made so easily available world-wide. Furthermore, new regulations concerning environment impact and official labor times are imposed to industrial companies. As a consequence, many enterprises must become more flexible, responsive and efficient to sustain their growth and adapt to their environment, be innovative and capture new markets.

Nowadays, new trends in industrial development can be summarized by:
- Globalization
- Customer orientation
- Process/project orientation
- Continuous improvement and productivity enhancement
- Life cycle perspective
- Teamwork and interdisciplinary teams

Globalization concerns both markets and manufacturing systems. While trade barriers are torn down and products can be sold in various areas of the world, spatial location of manufacturing plants, enterprise departments, retailers and distribution centers has to be optimized. However, there is no steady-state situation and this configuration must periodically be reconsidered as business evolves.

Customer orientation means that the business must be focussed on customer requirements, which are characterized by the well-known Quality-Cost-Delay (QCD) paradigm.

Process/project orientation means that the whole business must be organized in terms of business processes or projects, i.e. sequential or concurrent sets of activities handling the various flows (of materials and/or information/decisions) circulating throughout the enterprise. This is in opposition to the traditional hierarchical approach for organization and management of enterprises which has been the rule allover the 20th century.

Continuous improvement and productivity enhancement should be the essence of management of change. The goal is to ever find out how to do things better and in a more cost-effective way. This concerns all sectors, all processes and all individuals of the company. This requires the use of pertinent performance indicators at all decision levels.

It is also important for a company to develop a sound life cycle culture to achieve better management of change. This means that the product life cycles as well as the manufacturing system life cycles must be well understood and formalized within the company in order to monitor their performances, identify opportunities for potential improvements and modify them whenever required. Models of these life cycles need to be maintained.

In this context, teamwork and interdisciplinary teams, organized on the basis of competencies of individuals (or groups of individuals), are essential factors to get the required flexibility and reactivity to design, evaluate, reengineer, realize and maintain processes as well as products and services of the enterprise.

Consequently, manufacturing agility equally concerns product development, manufacturing systems engineering and management of overall production operations.

In this Chapter, we show how Enterprise Integration and its management can provide support to cope with required system reactivity and needs for managing change in manufacturing or service companies. The approach resolutely adopts a model-based engineering and management viewpoint. The enterprise model highlights two fundamental types of enterprise entities (business processes and functional entities) and their coordination.

2. ENTERPRISE INTEGRATION

2.1 Definitions

Enterprise Integration (EI) deals with increasing interoperability among people, machines and applications to enhance synergy within an enterprise in order to better realize the business mission [4]. In other words, EI is mostly concerned with breaking down organizational barriers and making the enterprise, or a network of enterprises in the case of the extended enterprise, interoperable to improve enterprise efficiency and reactivity. EI is essentially concerned with facilitating decision-making in distributed environments, simplifying business processes across the enterprise and providing the right information at the right place at the right time [4-6].

It is now well-understood that EI has at the same time an organizational, a social and a technological dimension.

Organizational imperatives: EI is most of all an organizational challenge. If information and decisions are not timely and properly transmitted, understood, managed or processed, then the enterprise will not operate correctly. A relevant organizational structure has first to be devised before any technological change can be made.

Social imperatives: The role of human resources remains central and essential to successfully plan, implement and then operate any EI solutions. Core competencies of the enterprise rely on knowledge, know-how and behavior of people. Improvements and changes to be made to the enterprise will be decided and shaped by people using 'the gold they have in their mind', an essential facet of enterprise agility.

Technological imperatives: From a technological point of view, to make integration possible there is the need to facilitate information exchange and sharing as well as interoperation throughout the company or all along the supply-chains. Such *interoperability* assumes the availability of an integrating infrastructure, i.e. a middle-ware facility made of a set of computer services, which will support the flow of messages, information or data objects and will send requests to remote systems and/or receive responses from them. In addition, there exists a logistics system to cater for the materials flows.

Two major types of enterprise integration problems are usually distinguished: intra-enterprise integration and inter-enterprise integration.

Intra-enterprise integration, or 'inside the four walls integration', typically refers to the classical CIM concept but extended to the whole enterprise [7]. Technologically, it could be achieved if company standards are imposed on the enterprise in terms of a limited number of compatible and interoperable pieces of equipment, standard interfaces and normalized procedures as well as data structures to be used. Although feasible, this would be a too rigid approach to meet the need for reactivity imposed nowadays by modern business and technological conditions as well as by business made in partnership. The CIM concept had to evolve towards EI to take into account emerging requirements in terms of agility, human factors as well as distribution of knowledge and decision-making.

Inter-enterprise integration concerns have emerged with the concepts of extended enterprise or networked organization. The challenge is to face such realities as global markets, the rise of mass customization and the need to develop environmentally benign products and processes [8]. The solution is to build networks of enterprise entities to go beyond 'the four walls of the manufacturing plants', i.e. to control the entire supply chains to better serve the customers world-wide and to have access to resources (finance, knowledge or technology) outside the enterprise. Inter-enterprise networking is the key. Again, this is first of all an organizational and social problem (for instance regarding the choice of partners, the topology of the business network, legal issues or responsibilities) before being a technical problem (i.e. to deal with system connectivity, data exchange and process coordination).

2.2 A general EI framework

Traditionally, three levels of integration are differentiated: physical or system integration, application integration and business integration [7]. Table 1 provides an indication of integration aspects covered by each of these levels while Figure 1 provides an indication of technologies involved. Physical system integration is mostly concerned with system interconnection and message passing using computer communications networks. Integration services, distributed computing environments and object request brokers can then be implemented on top of these. Application integration builds on physical systems integration and is more concerned with application interoperability and distributed co-operative applications and requires data and information exchange using integration platforms (such as CORBA, NIIIP, AIT-IP or OPAL). Finally, business integration goes beyond application integration in terms of business process co-ordination and enterprise networking. It must take into account organizational issues and human factors and requires enterprise-wide knowledge sharing. This assumes a precise modeling of intimate enterprise details concerning enterprise procedures, organization, knowledge and know-how in the form of a shared enterprise model. Part of this model can be enacted using workflow engines and computer support for collaborative work (CSCW).

Table 1 Integration levels and related aspects

Integration levels	Integration aspects	
Business	Business process coordination Enterprise-wide knowledge sharing Interoperation; enterprise networking	Organization and human issues
Application	Distributed cooperative applications Data/information exchange Application interoperability	Technological issues
Physical	Basic computer communication Message passing Interconnection	

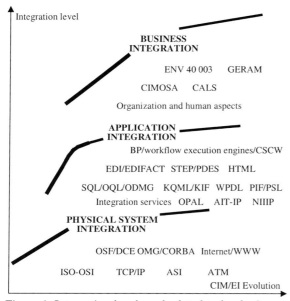

Figure 1. Integration levels and related technologies

Figure 1 shows the diversity and complexity of technologies involved in EI. These are briefly surveyed in this section. Details can be found in [4]. Historically, they have been first tackled from the bottom level, i.e. from a technical angle due to evolution of the technology. However, from a methodological viewpoint and taking advantage of recent technology advances (concerning high speed networks, distributed computing environments, object-oriented technology, hypermedia technology, workflow management and CSCW), EI should be first addressed at the business level (concentrating on business process coordination and organizational issues) and then at the technological levels (application and system levels).

Physical system integration: The system integration layer has been the focus of tremendous research efforts over last two decades both from the traditional EDP standpoint and the CIM standpoint. OSI (MAP/TOP) and non-OSI communications protocols (e.g. Ethernet, SNA, TCP/IP) are now widely available both for local and wide area networks. The current trend is on one hand to improve the performance of computer networks (in terms of band-with and speed) and on the other hand to have simple, fast and low-cost communication at the equipment level. The state of the art in manufacturing environments is to use field-buses (based on FIP, Profibus or ASI protocols) at the equipment level, local area networks based on TCP/IP at the station, cell or plant levels and ATM technology at the cell, plant and enterprise levels. Furthermore, Internet and the web technology are rapidly penetrating the manufacturing area from design offices to shop floors. This is made possible by the emergence of the object-oriented technology and distributed computing environments such as OSF/DCE and OMG/CORBA.

Application integration: For many companies it is important that their applications can inter-operate, i.e. work together and not in isolation whatever the operating systems, database systems or communication network systems utilized are. Therefore, application integration concerns data/information exchange and sharing as well as application interoperability. Three complementary technologies are currently under development in this area [4]:

Data exchange formats and data and knowledge manipulation languages: Manufacturing and engineering data exist under various forms in an enterprise, including text, drawings, images and signals. Central to these are product and process data. Application integration thus relies on standards such as EDI for administrative data exchange, ISO STEP for product and process data exchange and HTML or XML for technical and hypertext document exchange. In addition, access to data can be made via SQL to relational databases, OQL for object-oriented databases and QKML/KIF can be used for knowledge exchange. PIF and PSL are languages developed as process interchange formats [9] while WPDL is the workflow process description language of the Workflow Management Coalition (WfMC) [10].

Integration services and integration platforms: The client/server architecture provides a basic model for the communication between two or more (remotely located) interacting agents: the client and the server. The communication infrastructure linking agents in a client/server architecture is called an integration platform. An integration platform is a set of computer services together with a middle-ware layer built on top of communications systems to facilitate message and request exchange among human or technical agents (i.e. functional entities) of a system. These services can deal with data transmission, data access security, transaction management, concurrency control, clock synchronization, binding a name to an object, trading a request to a client, data presentation to an application, computer network access, to name a few. Examples are provided in section 6.

Workflow management and enactment: Application packages are often used according to predefined business procedures called business processes and made of a chain of activities to be executed in a given order to perform a business task. To control the execution of these processes, they can be modeled in the form of a workflow which can then be executed under the control of special systems, called workflow execution engines.

Business integration: This level is concerned with interoperation of people and systems. It therefore mostly deals with organization and social aspects. Especially, the many business processes of the enterprise have to be coordinated and monitored. No mature technology exists at the moment to support this integration level. Only general frameworks are so far available such as CALS [11] supported by the US DoD (Department of Defense), EU funded CIMOSA architecture [7], CEN Framework ENV 40 003 [12], PERA (Purdue Reference Architecture) [13] or GERAM (Generalized Enterprise Reference Architecture and Methodology) of the IFAC-IFIP Task Force on Architectures for Enterprise Integration [14].

3. ENTERPRISE MANAGEMENT AND AGILITY

Management of the integrated manufacturing enterprise must recognize two major types of life cycles existing in any manufacturing enterprise: *manufacturing system life cycles* and *product life cycles*. They are to some extent 'orthogonal' to one another in the sense that each product life cycle can only exist within the operation phase of a given manufacturing system life cycle (Figure 2).

Furthermore, with respect to the system life cycle, enterprise management must recognize the co-existence of two tightly coupled and complementary environments: the *Enterprise Engineering Environment* and the *Enterprise Operation Environment*.

In the Enterprise Engineering Environment, business processes and their supporting organization elements (i.e. required technical and human resources, information systems, communication systems, logistic systems, decision responsibilities and authorities, etc.) are defined, engineered, analyzed, tested and validated before being released for implementation. This engineering process, referred to as *Enterprise Engineering*, may be based on a structured approach as proposed by CIMOSA, PERA as well as GRAI-GIM, another methodology proposed for CIM and EI [15]. Such an approach usually comprises a phase for requirements definition, design specification and implementation description. To assist business users in this process, *Reference Architectures*, such as CIMOSA and PERA, are also proposed. They contain guidelines, methodological rules, modeling support and even partial models to help users to develop their own particular enterprise models (Figure 2).

In the Enterprise Operation Environment, business processes and supporting functional entities are put into operation to perform the day-to-day operations of the enterprise. At this level, each manufacturing entity can be seen as being made of a physical system (the people and the facilities), a management or decision system (set of decision centers organized in a hierarchical or in a holonic fashion) and an information system. The functional entities are integrated by means of an integrating infrastructure in both environments.

The link between the two environments is materialized by the so-called release process which consists in installing system components (i.e. new pieces of equipment, new procedures, new database systems, new job functions, etc.) and testing their operation.

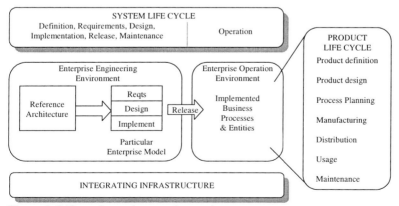

Figure 2. Life cycles and enterprise environments

Figure 3 presents the details of the GERAM life cycle for manufacturing systems engineering. This process encompasses tasks such as business entity identification and conceptualization (i.e. definition of business mission, values and strategies), requirements definition, system design, implementation, operation and decommissioning. Of course, this engineering process allows for many iteration loops among phases and for reengineering of any part of an existing system all over the life history of the manufacturing system [14].

The generalized structure of Figure 3, which is an appendix to the ISO TC 184/SC5/WG1 draft international standard on Requirements for Enterprise Reference Architectures and Methodologies (ISO DIS 15704), has been obtained from a generalization of the top three specific architectures for CIM and EI, namely CIMOSA, GIM and PERA.

The structure has three dimensions, which explains its cubic structure.

The *life cycle axis* defines the various phases to go through to completely engineer or reengineer and then operate an integrated enterprise (Identification, Concept, Requirements, Design, Implementation, Operation and Decommission). This is by no means a strict waterfall approach and loop-backs are allowed among the phases of this axis.

The *instantiation axis* defines three layers, namely a *generic* layer (in which generic modeling constructs and engineering rules are provided), a *partial* layer (in which partial models of data, processes, organization or resource structures are provided to be freely reused), and a *particular* layer representing the models of a particular enterprise. This particular architecture is split into an equipment architecture (physical part of the system), an information and control architecture (control part) and a human architecture (human aspects).

The *view axis* suggests that at least four fundamental aspects of the enterprise must be taken into account in the models: function, information, resource and organization aspects.

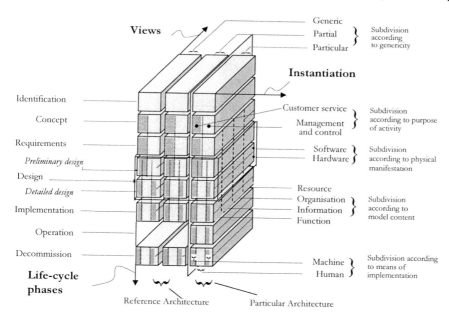

Figure 3. The Reference Architecture of the GERAM framework

4. ENHANCING AGILE MANAGEMENT BY ENTERPRISE INTEGRATION

Agile enterprise has been defined as 'the capability of surviving and prospering in a competitive environment of continuous and unpredictable change by reacting quickly and effectively to changing markets, driven by customer-designed products and services' [1]. Thus, more than ever before, the strategies and processes of world class manufacturing companies must be flexible and adjustable. Enterprise Integration can help in this respect.

To successfully integrate an enterprise, the rule is first to conceptually 'disintegrate' the company, i.e. to organize it in terms of a large set of autonomous units, and then to physically integrate these units, i.e. providing them with efficient ways to communicate, to cooperate and to be mutually coordinated. Enterprise Modeling (EM) plays a cornerstone role in this process [4-6].

Autonomous units represent business entities. Their size, functions or responsibilities are decided by the business user. Thus, they can represent a decision center (one person or a group of persons), a service, a department, a temporary aggregation of people (a team), or one or more technical agents (machines or computer applications). They must communicate to one another to cooperate and their actions need to be coordinated [16].

Their role is to plan, control and execute business processes of the enterprise. Business processes represent the flow of actions to be executed in the enterprise. They need to be synchronized. They manage the various flows of the enterprise (physical material flows, information flows and decision flows). They are described in the form of a control flow.

For a better management of the agile enterprise, we advocate to first model and define the autonomous units, called functional entities, and the business processes separately. Then, coordination mechanisms need to be defined to synchronize required actions (i.e. functional operations) of the processes with actions really performed by functional entities to achieve the desired business goals. As illustrated by Figure 4, functional entities (i.e. the doers) can be allocated to process steps (i.e. 'what has to be done') on the basis of their capabilities and/or competencies (depending whether these are technical or human agents).

This separation principle is justified by the fact that behavior of processes and functional entities have a different nature. The behavior of a process is mostly modeled in terms of a sequence of steps (process models), units of which are activities. The behavior of functional entities is mostly modeled in terms of change of states (state transition models). Processes need to be synchronized over time while functional entities need to be coordinated [17].

Modeling of these elements and related concepts is the topic of the next section.

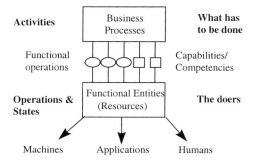

Figure 4. Separation principle between business processes and functional entities

5. CIMOSA: AN ENTERPRISE MODELLING METHOD FOR AGILITY

To model and engineer business processes and functional entities of the enterprise, we propose to use the CIMOSA approach developed by the ESPRIT Consortium AMICE [7,18]. CIMOSA provides a comprehensive framework for enterprise integration, and as such, has largely influenced the work of standardization bodies dealing with this field (notably, CEN and ISO). CIMOSA provides an Enterprise Modeling framework, a structured approach for Enterprise Engineering and an Integrating Infrastructure. In this section, principles of the modeling framework are presented with respect to agile manufacturing management.

5.1 Business process modeling

A *business process* can be defined as a partially ordered set of process steps executed to achieve one of the business objectives, i.e. to provide a defined end-result.
Process steps are either sub-processes or elementary steps called enterprise activities.

An *enterprise activity* is the locus of action, i.e. it transforms inputs into outputs using time as well as capabilities and/or competencies provided by functional entities (be they technical or human agents). Activities can be further detailed into *functional operations*, which are the atomic actions that a functional entity can perform. In other words, functional operations are primitive services or commands that a functional entity can execute.

For instance, a manufacturing process is made of a number of machining, part handling or assembly activities. Machining activities are made of machining operations such as lathe, turning, drilling, tapping, reaming, etc.

Inputs and outputs of each activity are enterprise objects (such as products, orders, resources, forms, data files, etc.) in a given state before and after being employed by the activity. Resources are enterprise objects used in support to the execution of an activity. The set of all resources of the enterprise comprises all the functional entities of the enterprise.

Enterprise activities are described in terms of their inputs, outputs, transformation function, required capabilities and ending statuses. CIMOSA defines six types of inputs and outputs:
- function input (i.e. objects to be processed or transformed by the activity)
- function output (i.e. objects processed, transformed or produced by the activity)
- control input (i.e. control objects used but not modified by the activity)
- control output (i.e. ending status or events produced by the activity)
- resource input (i.e. objects used as resources)
- resource output (information on resource status at the end of the activity).

An activity A_i can be formally defined by a 14-tuple as follows:

$$A_i = \langle Aid_i, Aobj_i, Adesc_i, FI_i, CI_i, RI_i, FO_i, CO_i, RO_i, Atf_i, Asrc_i, Aes_i, Pre_i, Post_i \rangle$$

where Aid_i is the unique identification of activity A_i, $Aobj_i$ states its objective(s), $Adesc_i$ is a textual description, FI_i is the set of function inputs, CI_i is the set of control inputs, RI_i is the set of resource inputs, FO_i is the set of function outputs, CO_i is the set of control outputs, RO_i is the set of resource outputs, Atf_i is the transformation or decision function employing the functional operations such that $(FO_i, CO_i, RO_i) = Atf_i (FI_i, CI_i, RI_i)$, $Asrc_i$ is the set of required capabilities and competencies and Aes_i is the set of all possible ending statuses for activity A_i. Pre_i and $Post_i$ are pre-conditions and post-conditions defined as logical statements on input and output objects, respectively.

An ending status is a 0-argument predicate (or information item) indicating the condition upon which the activity has been completed (e.g., 'OK', 'failed', 'tool broken', etc.).

A process is a chain of activities put in sequence or in parallel. It describes the flow of control (also called workflow in computer sciences) for the execution of some activities to fulfil part of the mission of the enterprise. A process is activated by some triggering conditions. These conditions are based on occurrences of events.

An *event* depicts a change in the state of the enterprise domain considered. It can therefore be modeled by a predicate which evaluates to true if its condition is met, false otherwise.

Examples of events are:
- arrival of a customer order (note that the customer order is an enterprise object, the fact that the customer order has arrived is the event)
- 5 pm o'clock (this is called a timer or planned event)
- machine breakdown (expected but unplanned event for which exception handling procedures may exist)
- fire alarm (unexpected event)

An event E_i can be formally defined by a 4-tuple as follows:

$$E_i = \langle Eid_i, Epred_i, Esource_i, Eov_i \rangle$$

where Eid_i is the unique event identification, $Epred_i$ is the event predicate, $Esource_i$ is the source of the event (outside environment, activity or functional entity) and Eov_i (optional) is an object view on the enterprise object associated to the event (for instance, the customer order in the first example above).

A process P_i can now be formally defined by a 5-tuple as follows:

$$P_i = \langle Pid_i, Pobj_i, Pdesc_i, Pev_i, Pflw_i, Pes_i \rangle$$

where Pid_i is the unique process identification, $Pobj_i$ states the process objective(s), $Pdesc_i$ is a textual description of the process, Pev_i defines the triggering condition of the process in the form of a list of events involved in the triggering of P_i, $Pflw_i$ is a set of behavioral rules defining the process workflow and Pes_i is the set of ending statuses of P_i.

CIMOSA behavioral rules are all defined by the very simple, extendible and formally verifiable syntax: 'WHEN (condition) DO action'.

The condition part of each rule is a logical expression made of two types of predicates: process triggering conditions (prefixed by the reserved word 'START' and combining one or more event occurrences) and ending status checks noted 'ES(S) = status', where S is a process step, ES(S) is a function returning the ending status of S when S is completed, and 'status' is an ending status value to be matched. These predicates can be combined using the AND and NOT logical operators.

The action part of the rules defines the action(s) to be started next when the condition part of the rule evaluates to true. When several downstream process steps have to be initiated in parallel, the operator '&' is used for asynchronous parallel execution; if synchronous parallel execution is required, the action clause is prefixed by the reserved word 'SYNC'.

The various types of behavioral rules available in CIMOSA include:
- Process start (for instance, a process triggered by two events, EV1 and EV2):
  WHEN (START WITH EV1 AND EV2) DO NextStep
- Process start (triggered by a calling process at a higher level):

  WHEN (START) DO NextStep

Enterprise Integration and Management in Agile Organizations 471

- Regular sequential flow:

  WHEN (ES(PreviousStep) = OK) DO NextStep
- Forced sequential flow (requires the use of the reserved word 'any' in the condition clause to replace any ending status value of the finishing upstream process step)

  WHEN (ES(PreviousStep) = any) DO NextStep

  This reads as 'whatever the ending status of PreviousStep is, next move to NextStep.
- Conditional flow (or case statement):

  WHEN (ES(PreviousStep) = case1) DO NextStep1
  WHEN (ES(PreviousStep) = case2) DO NextStep2
  ...

  This assumes that PreviousStep has 'case1', 'case2', ... as possible ending statuses.
- Asynchronous parallel flow:

  WHEN (ES(PreviousStep) = Done) DO NextStep1 & NextStep2 & ...
- Synchronous parallel flow:

  When (ES(PreviousStep) = Done) DO SYNC (NextStep1 & NextStep2 & ...)
- Rendez-vous or synchronization flow:

  WHEN (ES(PreviousStep1) = value1 AND ES(PreviousStep2) = value2 AND ...)
  DO NextStep
- Process end (the action clause only contains the reserved word 'FINISH'):

  WHEN (ES(PreviousStep) = Done) DO FINISH

There is no specific rule for loop-back situation. This can easily be described using a conditional flow.

Figure 5 presents a graphical representation of the concepts presented as well as a fictive example of a business process with loops. The process is triggered by event e_1 and generates events e_2 and e_3 in the course of its execution (s_{xy} represent ending statuses of the various process steps). Step B is a so-called non-structured sub-process because it is made of three activities but their order of execution is unknown. The two-way arrow linking step BP_1 to step BP_2 indicates that these two steps exchange messages during their execution.

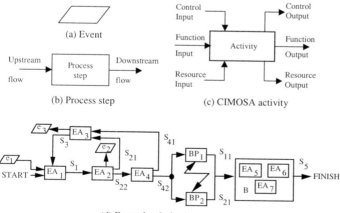

Figure 5. Graphical notation used in the function view constructs of CIMOSA

The translation of this process into CIMOSA behavioral rules is given by:

WHEN (START WITH e_1) DO EA_1
WHEN (ES(EA_1) = S_1) DO EA_2
WHEN (ES(EA_2) = S_{21}) DO EA_3
WHEN (ES(EA_3) = S_3) DO EA_1
WHEN (ES(EA_2) = S_{22} DO EA_4
WHEN (ES(EA_4) = S_{41} DO EA_3
WHEN (ES(EA_4) = S_{42}) DO SYNC (BP_1 & BP_2)
WHEN (ES(BP_1) = S_{11} AND ES(BP_2) = S_{21}) DO B= {EA_5, EA_6, EA_7}
WHEN (ES(B) = S_5) DO FINISH

Figure 6 shows a the control flow of a business process for manufacturing plan execution in a mechanical shop producing mechanical parts using a flexible manufacturing cell. The process starts with an occurrence of event EV-5 (Execution-Request) which initiates sub-process BP-51, the goal of which is to schedule manufacturing operations. When this is done, activity EA-51 displays the schedule on a computer screen and the foreman decides to accept or modify the schedule. It may happen that the schedule is not realizable with the current capacity of the shop. In this case an event EV-6 (Shop-Floor-Overload) is issued and this process stops. A loop goes on between activities EA-52 and EA-51 until a schedule is finally accepted or a request to reschedule is issued (EV-6). Once a schedule has been found, control is passed to BP-52 which corresponds to producing the parts by the cell.

It is interesting to note that events generated in this domain process (DP-5) can be used to trigger processes in other domains of the enterprise. For instance, occurrences of event EV-7 (Request-for-Maintenance) will trigger processes in the maintenance domain.

With respect to agility, the process model of CIMOSA enforces the *principle of separation of functionality and behavior*. Indeed, enterprise behavior is primarily modeled by the flow of control of business processes while enterprise functionality is modeled by the activities. This justifies the use of two separate constructs in the modeling language. The main advantage in terms of agility is that enterprise functionality can be changed without impacting the logic of business processes. Conversely, the flow of control of business processes can be reorganized without changing the implemented functionality.

CIMOSA constructs are represented by templates or 'fill-in-the-blank' structures. They can easily be implemented using object-oriented tools. For instance, the template of the event EV-5 and the template of the activity EA-52 are given hereafter.

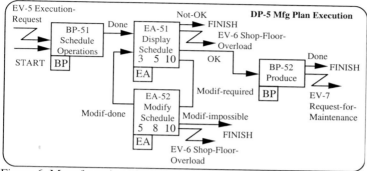

Figure 6. Manufacturing plan execution process

EVENT
    Type: Shop event
    Identifier: EV-5
    Name: Execution-Request
    Description: Request to execute a manufacturing plan
    Triggers: DP-5 / Mfg Plan Execution
    Source: DM-2 / Production Planning and Control
    Object View: OV-30 / Mfg-Plan
ENTERPRISE ACTIVITY
    Type: Scheduling Activity
    Identifier: EA-52
    Name: Modify Schedule
    Description: This activity modifies a schedule on request by the operator
    Objectives: To 'optimize' a proposed schedule by hand according to shop floor foreman preference
    Constraints: To minimize throughput; To respect manufacturing order due dates
    Declarative Rules:
    Function Input: OV-35 / Schedule
    Control Input: OV-30 / Mfg-Plan, OV-70 Shop-Capacity
    Resource Input: FE-35 / Foreman
    Function Output: OV-36 / Modified-Schedule
    Control Output: EV-6 / Shop-Floor-Overload
    Resource Output:
    Ending Statuses: Modif-done: Modification made, Modif-impossible: Schedule incorrectly modified
    Required Capabilities: CS-5 / Schedule-Modif-Capabilities
    Where-Used: DP-5 / Mfg Plan Execution

Other constructs exist in the function view of CIMOSA. An interesting one regarding enterprise agility is the domain construct (See DP-5 in Figure 6). It is used to group a set of business processes as well as related objects into one large autonomous organizational unit which is only defined for the outside world by its relationships (i.e. event or object exchanges) with other domains. This is very useful to model large supply-chains.

5.2 Information and enterprise object modeling

Function inputs, control inputs and function outputs of enterprise activities are enterprise objects in a given state. These are called object views because they represent the objects at a certain stage in their life cycle. Therefore, CIMOSA provides two essential constructs in the information view: *enterprise objects* and *object views* [4].

Enterprise objects represent entities of the enterprise such as products, orders, people, machines, data stores, documents, etc. They are described by their attributes, i.e. descriptive properties, and are subject to generalization and aggregation mechanisms.

Object views are manifestations or states of enterprise objects. They can be defined over one enterprise object or a group of enterprise objects. For instance, a part drawing produced with AutoCad (enterprise object) can have several object views (e.g. the drawing on a computer screen, the CAD file on a diskette or the drawing plotted on paper). Object views are characterized by their embodiment: they can exist as information entities (information view) or as concrete objects (physical view). This distinction is fundamental in CIMOSA because depending on the nature of the object views (information or physical nature), it is possible to clearly differentiate in the model the information flows from the material flows.

An example of an object view definition (OV-30 / Mfg-Plan) concerning enterprise object Manufacturing-Plan (EO-30) and associated to event EV-5 (Execution-Request) and used as control input in activity EA-52 described above is given next.

OBJECT VIEW
   Type: Order
   Identifier: OV-30
   Name: Mfg-Plan
   Nature: Information
   Leading Object: EO-30 / Manufacturing-Plan
   Related Objects:
   Properties:
      Issue-date: Date
      Shop#: String [8]
      Production Period: String[3]
      List of Orders: Listof [1: 30] Production-Orders

5.3 Resource and organization modeling

A special class of enterprise objects concerns resources. Resources are objects used in support to the execution of enterprise activities. CIMOSA defines two fundamental classes of resources: functional entities and components.

Functional entities are active resources in the sense that they can perform actions (or services) on request (i.e. by receiving a message). They can receive, send, process and even store information. In computer science, they are called agents. CIMOSA defines three generic classes of functional entities, each one requiring a different form of management:

- *Humans*, all human beings playing a role in the business system
- *Machines*, all pieces of equipment or devices which can be accessed via its controller
- *Applications*, all computer systems and software applications including databases

Functional entities are interconnected by means of an integrating infrastructure.

Components are passive resources, i.e. resources which cannot perform any action unless they are used by a functional entity.

Functional entities can be combined or aggregated with other functional entities together with components to form higher level functional entities. For instance, a manufacturing shop is made of machines, tools, carts, robots and material handling systems. The granularity of functional entities is therefore decided by the business user.

Functional entities are mostly described in terms of their capabilities (especially in the case of technical agents) and their competencies (especially in the case of human agents) as well as in terms of the services they can provide, i.e. the list of functional operations they can perform to execute activities of business processes.

The behavior of functional entities also needs to be modeled to indicate how the resource reacts to external stimuli (requests), what are all its possible states and how the use of its various functional operations as well as internal actions provoke state changes. We model this behavior using state machines such Harel's state-charts [17].

CIMOSA provides only two constructs for the resource view: *resource* and *capability set*. Resource can both describe functional entities as well as components. Capability set is used to both describe capabilities and competencies provided by resources as well as those required by activities [4].

Once each process, activity, input/output object and resource have been modeled for each domain, CIMOSA provides two other constructs to describe the organization and its control structure at different decision levels: *organization unit* and *organization cell*. An organization unit is a decision center (a workplace) while an organization cell is an aggregation of organization units (e.g. department or division). Both have responsibilities and authorities.

6. TOOLS USED IN ENTERPRISE INTEGRATION AND MANAGEMENT

6.1 Integrating infrastructures

Integrating Infrastructures (IIS) or integration platforms are made of a standardized set of software services (collectively named middle-ware) to be installed on top of basic Information Technology (IT) services, i.e. operating systems of computers connected by computer and communication networks, to facilitate application interoperability as well as communications among functional entities of a business system [4].

Figure 7 provides an illustration of the mediating role of such platforms between business application systems (accessed via their API or Application Program Interface) and other functional entities of the system.

The role of an integrating infrastructure is to support communications (i.e. sending messages and passing responses) among functional entities of the enterprise system, wherever these entities are geographically located.

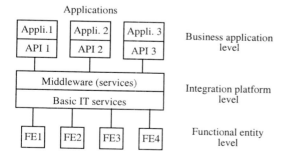

Figure 7. Integration platform principle

The CIMOSA architecture has first proposed a generic definition of an IIS for the manufacturing industry [7]. The CIMOSA IIS is made of five service entities as illustrated by Figure 8. These are:

- Business services: they deal with model enactment and workflow control in the execution of business processes. Thus, they are made of business process control services, resource management services and activity control services.
- Information services: they support activity execution and ensure system-wide data exchange, and especially uniform data management, to access enterprise data in the various databases across the distributed business environment.
- Presentation (or front-end) services: these are services making the interface between the neutral languages and data formats used within the IIS and the proprietary or dedicated languages of all types of functional entities used in the particular enterprise architecture. Three types of dialogue services can be found in this entity: machine front-ends, human front-ends and application front-ends.
- Common services: these are IT services dealing with IT related issues such as communication network access, distributed naming service, clock synchronization for the various computers involved in the IIS, etc.

To illustrate this, three integrating infrastructures recently developed by large EU and US industrial consortia for design and manufacturing environments are briefly presented.

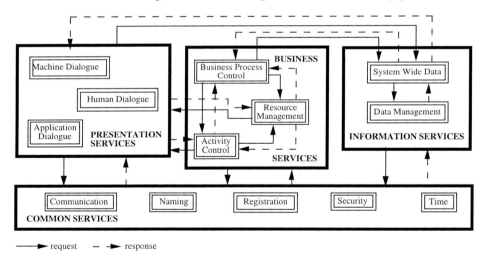

Figure 8. CIMOSA Integrating Infrastructure

6.2 AIT-IP

AIT is a European initiative launched in 1993 for Advanced Information Technology in design and manufacturing activities [19]. It is jointly sponsored by all major European manufacturers of the automotive and aerospace industries and by the ESPRIT Program of the European Commission (UE EC). Its aim is to identify industry requirements for future IT solutions and to influence IT product developments for design and manufacturing activities on the basis of a consensus built among IT users.

Among major requirements identified by AIT partners there is the need for an integration platform making independent the business support applications or BSAs (e. g. CAD/CAE, CAPP, MRP, CAM, SFC) from the IT basic infrastructure proposed by IT vendors, thus allowing their independent evolution (principle of application isolation). The integration platform must enable integration of business support applications using a 'plug and play' strategy to guarantee IT vendor independence and protection of previous investments (principles of modularity and interoperability).

To specify the requirements for such an integration platform, a reference model, called AIT ITRM for AIT IT Reference Model, has been defined using the CIMOSA IIS philosophy [20]. The AIT ITRM is an open, standard-based, non proprietary reference model defining middle-ware requirements for integration in large distributed manufacturing environments. The ITRM envisions integration as a set of interoperable integration platforms being installed at different user sites and made of a number of integration services. The ITRM is organized around a software bus (preferably a CORBA-like object request broker [21]) as depicted by Figure 9. Integration services include generic execution services (GESs), application front-ends (AFEs), machine front-ends (MFEs), human front-ends (MFEs), data server front-ends (DFEs) and communication front-ends (CFEs). In addition, system-wide actor directory management services to record all functional entities (humans, machines or applications) used in the enterprise architecture and system-wide information management services to access information models containing the enterprise model are mandatory.

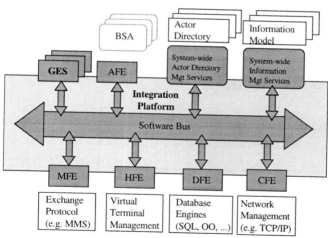

Figure 9. The AIT integration platform

These integration platforms with the necessary services can be installed on various sites of a company to integrate specified application systems. They can then be interconnected to one another to form a system-wide network making up an integrating infrastructure for some part of a distributed enterprise or for a group of enterprises (networked enterprise).

6.3 OPAL

OPAL is a software platform for integrated information and process management dedicated to design and manufacturing environments [22]. OPAL has been developed as an ESPRIT project as part of the AIT initiative. As opposed to AIT-IP, which is developed as a generic approach, OPAL takes advantage of modern existing technologies for information management and sharing including object request brokers (ORBs), object-oriented database technology, neutral product and process data exchange formats (such as ISO STEP and WfMC WPDL), workflow engines, hypermedia technology, HTML and the world-wide web (WWW) and web browsers.

The global system architecture of OPAL is presented by Figure 10. It consists of a data repository, an execution environment based on a workflow engine as well as a middle-ware layer (object request broker) and central services for system-wide message and object exchange and interoperability.

The enterprise repository contains business process, organization and data models to be used to control business support applications. Process models and data models are stored using a common object-oriented data structure. Each application package (e.g. CAD, PDM or MRP system) must be encapsulated using its application program interface (API) to become a functional entity of the integrated system (thanks to the encapsulation services). It can then be accessed by other modules via the middle-ware layer (in this case ORBIX, a CORBA compliant platform) and the central services (providing naming services, time services, communication services and information services). Since this integrating infrastructure is dedicated to engineering and manufacturing data management, it also provides so-called virtual folder system services and uses WWW-browser facilities to deal with complex product data and compound hypertext documents.

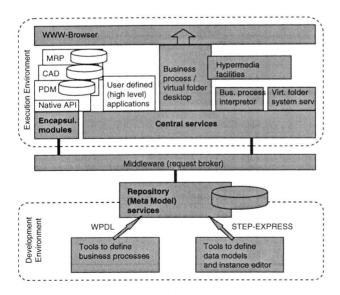

Figure 10. OPAL general architecture

Virtual folders are used to collect and transmit technical documents (e.g. CAD drawings, bills-of-materials, product documentation, simulation data, text files, etc.) put together as one document and to be exchanged by users or application systems across the system. Hypermedia facilities and web browsers (such as Netscape or Internet Explorer) provide end-users with the ability to navigate through the whole set of product and process data to which they have access, wherever they are located in the enterprise system.

The specification phase of OPAL took place in 1996, services were developed in 1997 and a prototype and pilot cases made available for the end of 1998. Two industrial test cases have been demonstrated, one at PSA in France and one at Robert Bosch in Germany.

6.4 NIIIP

NIIIP (National Industrial Information Infrastructure Protocols) is a US project, the aim of which is to establish new standards for integrating manufacturing applications, focusing on application interoperability and manufacturing execution systems [23]. In fact, NIIIP has similar objectives and Reference Architecture than AIT-IP and adopts the same philosophy as OPAL. More specifically, NIIIP building blocks to enable the virtual enterprise are:
- Internet for communication connectivity
- OMG-CORBA for application interoperability
- ISO STEP for information sharing
- Workflow technology for process management

In addition, NIIIP has taken the lead in defining the requisite protocols among these technologies to achieve interoperability. NIIIP also includes a Manufacturing Execution System (MES). MES is a layered mechanism for releasing work orders to the factory floor in which MES activities and tasks are organized in three layers: Plant Management, Process Management and Asset Management. This technology has been deployed in several industries and market segments.

7. PERFORMANCE MANAGEMENT

Once integrated, the operational performance of the enterprise (be it a stand-alone or an extended enterprise) needs to be monitored and controlled.

Performance management requires a system of performance indicators to be put in place. This system may reflect the hierarchical structure of the decision system because indicators used at one decision level may not be relevant for another decision level. This poses two problems:
- definition of relevant and reliable performance indicators
- aggregation of performance indicators, wherever suitable, to match the decision hierarchy

A *performance indicator* is either a scalar value (e.g. 2% daily scrap out of total production) or a fuzzy value (e.g. product quality is rather good) which proves to be useful to make decisions.

Performance indicators are built or computed on the basis of performance variables. These must be measured and processed to be converted into useful information. Their variation depends on performance drivers, which have to be identified.

Performance drivers are elements impacting performance variables (for instance, the degree of wear of a cutting tool impacts the cutting finish).

The activity-based costing (ABC) method as proposed by Cooper and Kaplan and described by Brimson [24] is an interesting approach which nicely complements the model-based engineering approach presented before because it is also based on the activity concept. However, rather than focusing on object flows, the ABC method focuses on cost drivers.

8. CONCLUSION

The aim of this Chapter was to explain how Enterprise Integration techniques, especially Enterprise Modeling, Enterprise Engineering and Integrating Infrastructures, can be profitably used to increase and favor agility in manufacturing and service industries.

Indeed, although examples used have been extracted from manufacturing case studies, all concepts and methods presented can be applied to other economic sectors such as banking, health care, insurance companies or even government and non government agencies.

The major lesson to be remembered from this Chapter is first to think the business entity both in terms of its processes and functional entities, then to model these as separate entities, next to figure out relevant coordination mechanisms adapted to objective criteria to be fulfilled, and finally to put into operation the necessary integration platform to make them inter-operate and favor their coordination.

REFERENCES

1. *Int. J. Agile Management Systems*, 1:1 (1999) whole issue.
2. R. Booth, Agile manufacturing, *Engineering Management Journal*, 6:2 (1996) 105-112.
3. A. Gunasekaran, Agile manufacturing : enablers and implementation framework, *Int. J. Prod. Res.*, 36:5 (1998) 1223-1247.
4. F.B. Vernadat, *Enterprise Modeling and Integration: Principles and Application*, Chapman & Hall, London, 1996.
5. C. Petrie (ed.) *Enterprise Integration Modeling*, The MIT Press, Cambridge, MA, 1992.
6. K. Kosanke and J.G. Nell (eds.), *Enterprise Engineering and Integration: Building International Consensus*, Springer-Verlag, Berlin, 1997.
7. ESPRIT Consortium AMICE, *CIMOSA: Open System Architecture for CIM, 2nd revised and extended version*, Springer-Verlag, Berlin, 1993. http://www.cimosa.de.
8. J. Browne, P. Sackett and H. Wortmann, Industry requirements and associated research issues in the extended enterprise, in *Integrated Manufacturing Systems Engineering* (P. Ladet and F. Vernadat, eds.), Chapman & Hall, London, 1995, 2-28.
9. NIST, PSL: Process Specification Language, 1999. http://www.mel.nist.gov/psl.
10. WfMC, Workflow Management Coalition Interface 1: Process Definition Interchange, WfMC TC-1016, 1997.
11. J. Lyu, CALS: an enabling strategy for agile management systems, *Int. J. Agile Management Systems*, 1:1 (1999) 41-44.
12. CEN, ENV 40 003: Computer-Integrated Manufacturing - Systems Architecture. - Framework for Enterprise Modelling. CEN/CENELEC, Brussels, January 1990.
13. T.J. Williams, *The Purdue Enterprise Reference Architecture*, Instrument Society of America, Research Triangle Park, NC, 1992.
14. IFAC-IFIP Task Force, GERAM: Generalized Enterprise Reference Architecture and Methodology Version 1.6.3. Also ISO TC 184 SC5 WG1 N398 (1999).
15. D. Chen, B. Vallespir and G. Doumeingts, GIM: GRAI Integrated Methodology, a methodology to design and specify integrated manufacturing systems, Proc. Of ASI'96, Toulouse, France, 2-6 June 1996, 265-272.

16. T.W. Malone and K. Crowston, The interdisciplinary study of coordination, *ACM Computing Surveys*, 26:1 (1994) 87-119.
17. G. Berio and F. Vernadat, Formal foundations for a process/resource approach in manufacturing systems behaviour modelling. Proc. 13th IFAC World Congress'99, Beijing, P.R.C., July 5-9, 1999. Vol. J, 181-186.
18. Special issue on CIMOSA: Open Systems Architecture Evolution and Applications in Enterprise Engineering and Integration (K. Kosanke, F. Vernadat, M. Zelm, eds.), *Computers in Industry*, 40:2-3 (1999) whole issue.
19. E.J. Waite, AIT – Advanced Information Technology for Design and Manufacture, *Enterprise Engineering and Integration: Building International Consensus* (K. Kosanke and J.G. Nell, eds.), Springer-Verlag, Berlin, 1997, 256-264.
20. AIT Consortium, AIT Integration Platform, ESPRIT Project EP 22148, Deliverable, Dec. 1997. http://www.ait.org.
21. OMG, CORBA The Common Object Request Broker: Architecture and Specification, Version 2.0, OMG, 1995. http://www.omg.org.
22. R. Bueno, Integrated information and process management in manufacturing engineering, Proc. IFAC 9th Symposium on Information Control in Manufacturing (INCOM'98), Nancy-Metz, June 24-26, 1998, Volume I, pp. 109-112. http://opal.ceit.es.
23. R. Bolton, A. Dewey, A. Goldschmidt, P. Horstmann. NIIIP – The National Industrial Information Infrastructure Protocols for industrial Enterprise Integration: Enabling the Virtual Enterprise, *Enterprise Engineering and Integration: Building International Consensus* (K. Kosanke and J.G. Nell, eds.), Springer-Verlag, Berlin, 1997, 293-306.
24. J.A. Brimson, *Activity Costing – An Activity Based Costing Approach*, John Wiley & Sons, New York, 1991.

Agility, Adaptability and Leanness: A Comparison of Concepts and a Study of Practice

Hiroshi Katayama
Department of Industrial and Management Systems Engineering,
Waseda University, Japan

David Bennett
Technology and Innovation Research Centre,
Aston Business School, Aston University, UK

This paper deals with three concepts of concern to manufacturing management; agile manufacturing, adaptable production and lean production. These concepts are described and compared within the context of the modern competitive situation in Japan. A survey of Japanese firms is described where the concepts are explored through a number of questions concerned with strategy, action programmes and performance measures. Many companies have responded to the change in economic conditions through a modification of their production operations and by changing their cost structure. The results suggest that companies are trying to realise their cost adaptability through agility enhancement activities.

1. INTRODUCTION

Many writers have identified the importance of adopting a strategic approach towards manufacturing [1, 2]. During the 1980s and early 1990s the strategic trend in Japanese manufacturing was to expand market share. This trend led towards a cycle that not only comprised the traditional measures of price competition and cost reduction but also included the proliferation of new products [3]. This in turn resulted in higher fixed costs, increased break-even points and lower profits, thereby causing the cycle to be perpetuated (Figure 1).

Coupled with this trend was the development of the 'lean production' concept [4] largely based on the 'Toyota' production system [5]. This approach involved fewer resource inputs together with increasing pressure for higher output performance (Figure 2).
In adopting lean production, the aim of firms was to support their strategy of reducing costs and gaining a higher market share. To achieve this aim a wide choice of products was offered to short delivery times and at increasingly competitive prices. In this way the cycle described in Figure 1 could be perpetuated, provided the economic and social conditions were favourable.

In Japan, the cycle based on market share expansion could be sustained during the period of the 'bubble economy' when economic growth was fast, interest rates and the value of the Yen were low, exports were high and younger workers were available in abundant quantities. However, most manufacturing companies today, including those in Japan, are experiencing dramatic changes to their competitive environment as a result of greater globalisation, market instability, information intensive technological development etc. On top of this Japan has recently witnessed an increase in the value of the Yen, a depressed economy and a reduction in the number of younger people wishing to work in manufacturing jobs [6]. There has also

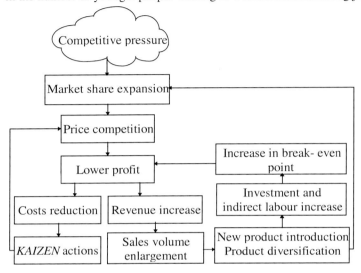

Figure 1. The past cyclic trend in Japanese manufacturing based on a market share expansion strategy

been criticism about the environmental cost associated with having frequent deliveries of materials to factories. As a result of these changes a new situation has emerged where pressure for survival has replaced the competitive pressure based on market share expansion. Consequently this has demanded new approaches to production which are better suited to the new conditions that prevail.

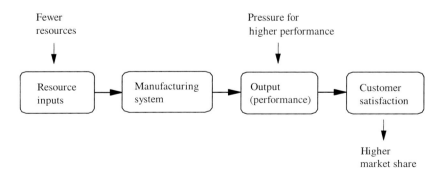

Figure 2. The essential elements of lean production

This paper considers two new approaches; 'agile manufacturing' and 'adaptable production' and explores the way in which, together with some of the underlying principles of lean production, they can be used to answer the demands on modern manufacturing enterprises.

The term 'agile manufacturing' was coined by a U.S. government sponsored research programme at Lehigh University and, latterly, MIT [7]. It seeks to cope with demand volatility by allowing changes to be made in an economically viable and timely manner [8]. Although the word 'manufacturing' is used with this concept, the principles of agility can equally apply to other functions of a business and to service industries.

'Adaptable production' is based on increasing cost sensitivity by shifting some of the company's fixed cost base towards variable costs. This is in contrast to the more typical approach under lean production, whereby the variable cost element is lowered by reducing unit resource consumption. In revising their cost structure and reducing fixed costs companies are thereby able to make an appropriate strategic response to the pressure for survival by improving their profitability (See Figure 3) . This is in contrast to the response of expanding market share which characterised the earlier paradigm [9].

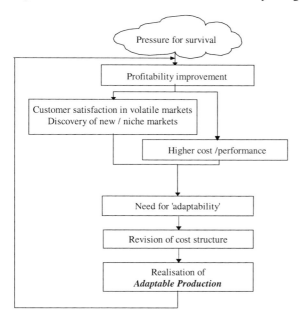

Figure 3. The cycle in manufacturing based on a strategy of survival

Later in this paper these new production concepts are compared and contrasted through an analysis of survey data performed in 1996 in Japan, where many large scale companies have responded to the change in economic conditions by restructuring aspects of their production operations. The main features of both will be examined to assess their sustainability in an era of globalised and uncertain markets. Before the survey is described the purpose and characteristics of all three production concepts; agility, adaptability and leanness are discussed.

2. THE RELATIONSHIP BETWEEN AGILITY, ADAPTABILITY AND LEANNESS

In this section the agility, adaptability and leanness concepts are examined in terms of their overall purpose and characteristics. The relationship between them is also considered.
Agility relates to the interface between the company and the market. Essentially it is a set of abilities for meeting widely varied customer requirements in terms of price, specification, quality, quantity and delivery. Agility has been expressed as having four underlying principles [10]. These are:

- delivering value to the customer,
- being ready for change,
- valuing human knowledge and skills, and
- forming virtual partnerships.

Of course, there also need to be some concrete technological changes to realise agility [11].

Adaptability is a feature of the company's production system. It is the inherent ability to adjust or modify its cost performance according to demand. In the current competitive environment surrounding manufacturing industries, a defensive approach in response to the pressure for survival is through changing the company's or factory's cost structure from being flat with a big fixed cost function to a small fixed cost function even though the variable cost element becomes steeper. Figure 4 shows the essential difference in cost structure between systems with high and low fixed costs.

When demand is reduced adaptable production becomes more cost effective because it is characterised by a cost function with a lower break-even point (BEP_A) compared with lean production which has a higher break-even point (BEP_L). Moreover, adaptable production is more cost sensitive than lean production. For example, for a particular distribution of demand (f_D) the cost distribution for lean production is narrow (f_{C_L}), while that for adaptable production is much wider (f_{C_A}). Adaptable production therefore has much greater flexibility in terms of cost than does lean production.

Adaptability may be realised through a number of different organisational and technological solutions [12]. In earlier studies we have identified several features of adaptable production which can enable the cost structure to be modified [9]. These are shown in Table 1.

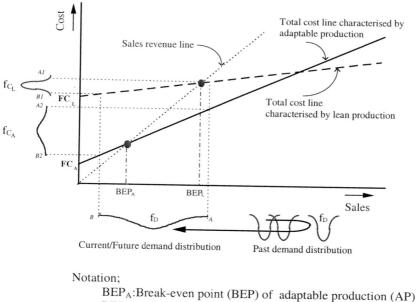

Figure 4. The cost structure and sensitivity for systems with high and low fixed costs

Notation;
BEP_A: Break-even point (BEP) of adaptable production (AP)
BEP_L: BEP of lean production (LP)
f_D : Demand distribution
f_{C_A} : Cost distribution of AP
f_{C_L} : Cost distribution of LP
FC_A : Fixed cost of AP
FC_L : Fixed cost of LP

'Lean production' is described as being different from mass production in that it uses less of everything ".... half the human effort in the factory, half the manufacturing space, half the investment in tools, half the engineering effort to develop a new product in half the time. Also it requires keeping far than less inventory on site, results in many fewer defects, and produces a greater and ever growing variety of products" [4]. Implementing lean production therefore means eliminating waste in the production system, be it in the form of materials, labour or plant capacity. However, we would argue that lean production is not an alternative to mass production but is a means of enhancing it. Mass production means simply the manufacture of items in large numbers, thereby exploiting 'economy of scale' principles. In fact the waste elimination concepts of lean production can equally be employed in other production systems including projects, batch production (in job shops) and continuous processing. Thus, in theory, leanness is an over-arching concept that is compatible with any production system and complements the other approaches described in this paper. However, in practice, a difficulty arises from the use of 'labour productivity' as a measure of leanness as is often advocated by the lean production proponents [13]. This naturally presumes a high level of automation which in turn increases the company's fixed costs, making it less sensitive to changes in demand. Critics of this emphasis on labour productivity argue that more labour intensive production systems can also be lean while at the same time offering other benefits [14, 15].

Table 1. The characteristics of adaptable production

- Production costs are more sensitive to changes in demand
- Systems enable production rate to be adjusted to accommodate changes in demand
- System software can support changes in production rate and product mix
- Lower fixed costs on new product development activities and the acquisition of new production facilities
- Use of human operators as a flexible resource
- Prevalence of mechanisms to support manual work
- Production systems support job enlargement and job rotation
- Use of technological solutions to increase the variety of upstream products and flexibility of upstream processes
- Grouping of parts and products into families to reduce work-in-process variety and shorten set- up times
- Modularisation of product designs to enable efficient production of greater product mixes
- Planned mixing of different product complexities to smooth production load
- Extensive use of kaizen activities and methodologies such as TQM, TPM etc.

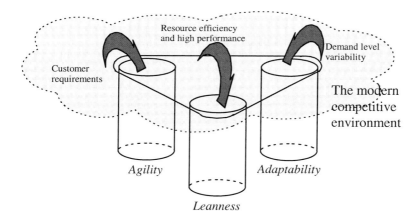

Figure 5. The three pillars of agility, adaptability and leanness

To summarise this comparison, agility, adaptability and leanness are not alternatives, but are mutually supporting concepts. As is shown in Figure 5 they form the three pillars to improve competitiveness and the prospects of survival in an increasingly volatile and global business environment. Together they ensure:
- responsiveness to customer requirements,
- cost sensitivity at different levels of demand, and
- resource efficiency and high performance.

3. THE SURVEY AND ITS CONTEXT

A survey carried out during 1996 was used to investigate the concepts addressed in this paper. The survey was administered among Japanese domestic companies with the objective of identifying the respondents' critical problems as well as their strategic direction and action programmes. Relatively large companies or business units were selected because of their capability to devote resources to their strategy making, the execution of action programmes and the scale of influence on their industries. The total number of survey respondents was 182 with the majority (157) coming from the manufacturing sector (electronics and electrical equipment, machinery, automotive and transport equipment etc.). The remainder came from manufacturing related service industries (software supply, manufacturing consultancy, logistics etc.).

The cycle of behaviour of companies is presumed to derive from their competitive priorities based on their business objectives, overall operations strategy, action programmes based on the competitive priorities and their results performance based on the action programmes.

A company's competitive priorities can be described as the emphasis assigned to each element of its each business strategy, i.e. cost/price competitiveness, quality competitiveness, delivery competitiveness, flexibility competitiveness and service competitiveness. Action programmes are companies' activities aimed at accomplishing the objectives defined by their strategy. The survey included nineteen such issues including: 'giving workers a broad range of tasks', 'reorganising plant networks', 'cross-functional teams', 'outsourcing manufacturing'. Results performance was measured by reported sales revenues, resource utilisation rates etc.

4. RESULTS AND DISCUSSION

4.1. The role of agility in a company's strategy

Analysis of strategic priorities

Operations strategy is considered to be the set of strategic priorities that guide companies' towards attaining their objectives [16]. In order to assess their strategic priorities respondents were asked to indicate on a scale of 1 to 7 the importance of each priority for their organisation when competing in the marketplace over the next two years and to say which priority best described their current strength relative to their primary competitor.

The procedure of 'gap analysis' was used to determine the difference between future and past/current scores and reveal the strength of direction towards the future. Figure 6 shows the future directions and current strengths of Japanese domestic companies' competitive priorities.

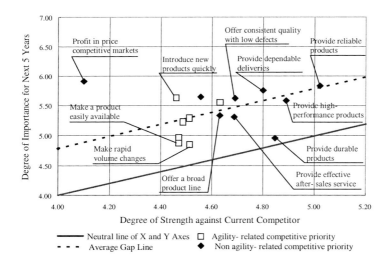

Figure 6. Gap analysis between current strength and future importance of competitive priorities

Table 2. List of Agility-related competitive priorities

- Making rapid design changes
- Introducing new products quickly
- Making rapid volume changes
- Making rapid product mix changes
- Providing fast delivery
- Making a product easily available
- Customising products

This figure is obtained by taking the average score of the companies' current strength as the horizontal axis and future importance as the vertical axis. The solid line is the 'neutral line' where scores on both axes are the same (i.e. no change in past and future priorities). The dotted line is parallel to the neutral line and passes through the centre of the distribution of data points. Therefore, this line indicates the average strength of direction towards the future.

It can be noted from this figure that all the priority gaps are positive toward the future. This is most evident if the priorities are weaker currently, then much stronger in the future. Similarly it is also evident where the priorities are already strong and are maintained in the future. Both these situations are common in Japanese companies because they usually try to make efforts to eliminate their weaknesses or retain their advantages against competitors. Typically they tend to be comprehensive rather than niche firms. Also a significant gap is evident in price (cost) competitiveness, which has again become an emerging issue after the collapse of the 'bubble economy' in Japan.

In the analysis, competitive priorities are categorised into two types, i.e. agility-related and non agility-related issues. It is noticeable that the average scores of current strength for the agility-related priorities, which are plotted on the horizontal axis, are relatively low and concentrated between 4.45 and 4.65, whereas those for the non agility-related direction are significantly higher, namely between 4.55 and 5.05 with the exception of the priority of "profit making in price competitive markets". On the other hand, scores of future importance for the priorities are widely varied between 5.0 and 6.0 without any significant distinction between the two categories. This means that companies now tend to be more clearly aware of the importance of agility than in the past and conflicts of priority with conventional issues seem to be occurring. The distinctive agility-related priorities are "introduction of new products quickly" and "offering a broad product range", which could be summarised as a 'product-focused direction'.

Analysis of action programmes

Action programmes are the set of methodologies in which companies invest to realise their competitive priorities. Again, for the two types of questions respondents were asked to indicate on the same scale as for the competitive priorities the extent of any payoff resulting from these programmes or activities in the last two years, together with the relative degree of emphasis their organisation will place on each action programme over the next two years.

Figure 7 shows the results of the gap analysis for the action programmes from which it can be noted that scores of both agility-related and non agility-related action programmes are widely dispersed with no significant difference between the two categories. This implies that companies' action policies are still item-by-item investments which focus on separate improvement islands, e.g. customer partnerships, CAD/CAE, integration of information systems etc. This way of working is not very effective for agility realisation despite companies having had many successes in implementing various action programmes already. This is because the critical issue for agility is a company-wide ability to respond quickly and efficiently to customer requirements.

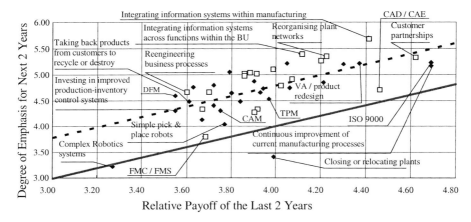

Figure 7. Gap analysis between last two years' scores and next two years' scores for action programmes

Table 3. List of Agility-related action programmes

- Giving workers a broad range of tasks
- Reorganising plant networks
- Worker training
- Management training
- Supervisor training
- CAD/CAE
- CIM
- Cross-functional teams
- Functional teamwork
- Integrating information systems within manufacturing
- Integrating information systems across functions within the business unit
- Integrating information systems with suppliers and distributors
- JIT
- FMC/FMS
- Concurrent engineering
- Reengineering business processes
- Outsourcing manufacturing
- Supplier partnerships
- Customer partnerships

Analysis of operational performance

The next issue to report is operational performance and the time series behaviour of a number of performance measures categorised into 'operational processes', 'supply processes', 'order fulfilment processes' and 'product development processes'. These are described in Table 4 which is constructed from the respondents' own assessment of their performance in each category. The base year for this table is 1988 for which a value of 100 is assigned for each measure. The assessed performance for every other subsequent year is shown in relation to the base year. For every measure an increasing number means that performance is improving.

Table 4. Trends in agility-related performance measures (normalised by base year 1988)

Related process	Performance measure	Year				
		1988	1990	1992	1994	1996
Operational processes	Set up time	100	113.00	127.69	137.50	151.51
	Operational cycle time	100	108.00	119.23	132.62	150.83
	Variety of products that can be offered	100	115.00	129.03	143.29	172.91
Supply processes	Procurement lead time	100	106.00	115.96	125.83	137.33
Order fulfilment processes	On-time delivery to customers	100	109.00	116.41	124.50	132.78
	Delivery lead time	100	109.00	121.75	129.65	142.17
Product development processes	Speed of new product development	100	109.00	113.47	120.63	136.03

From this table, it is noticeable that improvements in every agility-related performance measure were sustained despite the collapse of the 'bubble economy' in the early 1990s and the subsequent continuing recession. This is especially the case for product variety which shows an impressive growth in the 'product focused direction'.

4.2. Relationship between agility policies and cost adaptability

To investigate the relationship between agility and cost adaptability from the survey data, companies were classified into two groups based on the distribution of scores indicating the company's average agility-related competitive priorities in the past two years. Figure 8 shows the distribution of agility-focused and non agility-focused companies based on each company's average score. The two groups are distinguished by being above or below the overall average.

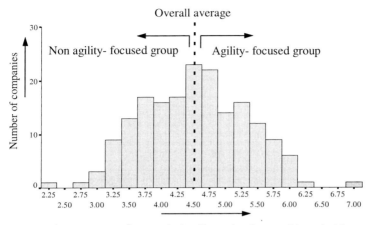

Average score of companies' agility-related competitive priorities

Figure 8. Distribution of scores indicating agility-related competitive priorities in the past two years

For each group the past and future cost management policies have been separated and compared, i.e. treatment of fixed cost, variable cost, break-even point, and strength of focus on each cost issue. The results of this analysis are summarised in Tables 5 and 6. The first three columns of numbers in the tables refer to the average cost management policy score for each group of companies (agility and non agility-focused). Table 5 shows the average value for fixed cost, variable cost and BEP in the last two years (using the score against the scale of 1 to 7), while Table 6 shows the average value expected for the next two years (against the same scale). The last two columns of each table also shows the relative emphasis between fixed and variable cost for each group expressed as percentages. In both tables the values can be compared for the agility-focused and non agility-focused groups.

Table 5. Score for cost management policy (*CMP*) in the last two years classified by current agility priority (*AP*)

AP	*CMP* Average score (Scale 1-7)			Relative cost emphasis
	Fixed cost	Variable cost	BEP	Fixed : Variable
Agility focused	4.521	4.342	4.545	73.61% : 26.39%
Non agility-focused	4.840	4.320	4.775	70.77% : 29.23%

Table 6. Score for cost management policy (*CMP*) in the next two years classified by current agility priority (*AP*)

CMP AP	Average score (Scale 1- 7)			Relative cost emphasis
	Fixed cost	Variable cost	BEP	Fixed : Variable
Agility focused	4.025	4.062	3.932	64.29% : 35.71%
Non agility-focused	3.515	3.631	3.563	81.82% : 18.18%

It can be seen from Table 5 that agility-focused companies have been directing their attention to fixed cost rather than variable cost and trying to reduce it along with the BEP in accordance with the argument described earlier and shown in Figure 5. On the other hand, looking at the companies' future cost management policies in Table 6, it can be seen that companies that are not agility-focused are still aware of their slow progresses concerning cost management activities because their intention in the next two years is to operate with much lower costs and BEP (Figure 4).

5. CONCLUDING REMARKS

In this paper the nature of, and relationship between, agile manufacturing, adaptable production and lean production has been described. Two of these concepts, agility and adaptability have been investigated by analysing survey data on strategy and performance collected from a sample of major Japanese companies. The principal results from this investigation are summarised as follows:

An agility-related manufacturing strategy, as determined by the companies' competitive priority scores, is relatively less common but emerging very rapidly. A product-focused direction is the current priority of most companies. On the contrary, action programmes related to agility are very varied and there is currently no significant difference in priority between these and the non agility-related activities. This means that companies are starting to become aware of the importance of agility but have not yet linked the concept to concrete actions.

So far the strategy of agility-focused companies seems to have been to move their organisation toward cost adaptability, which aims to reduce fixed cost and lower the break-even point rather than to concentrate on variable cost. On the other hand, non agility-focused companies are only just becoming aware of the importance of this strategic direction. Therefore, it should be possible to establish the hypothesis that companies are trying to realise their cost adaptability through agility enhancement activities.

Acknowledgements

This research has been accomplished with the support of Waseda University Grants for Special Research Projects and collaboration between the authors has been made possible through visiting fellowships funded by the Japan Society for Promotion of Science and the UK Royal Academy of Engineering. The authors sincerely appreciate this support and also acknowledge the contribution of Mr. Yoshikazu Azami, a student at the Department of

Industrial & Management Systems Engineering, Waseda University, for his valuable assistance in preparing this paper.

REFERENCES

[1] Hayes, R. H., Wheelwright, S. C. and Clark, K. B., 1988, Dynamic Manufacturing: Creating the Learning Organization, Free Press, New York.
[2] Ettlie, J. E., 1988, Taking Charge of Manufacturing, Jossey-Bass, San Francisco.
[3] Deschamps, J-P. and Nayak, P. ,1995, Product Juggernauts, Harvard Business School Press, Boston.
[4] Womack, J. P., Jones, D. T. and Roos, D. ,1990, The Machine that Changed the World, Rawson Associates, New York.
[5] Monden, Y., 1993, Toyota Production System (2nd Edition), Industrial Engineering and Management Press, Norcross, GA.
[6] Miyai, J., 1995, The Redesign of Japanese Management Systems and Practices, APO Productivity Journal, Summer.
[7] Iacocca, 1991, 21st Century Manufacturing Enterprise Strategy: An Industry Led View, Iacocca Institute, Lehigh University, Bethlehem, PA.
[8] Kidd, P., 1994, Agile Manufacturing: Forging New Frontiers, Addison-Wesley, Wokingham, UK.
[9] Katayama, H. and Bennett, D. J., 1996a, Lean Production in a Changing Competitive World: a Japanese Perspective, International Journal of Operations and Production Management, 16 (2), pp 8 - 23.
[10] Goldman, S., Nagel, R. and Preiss, K., 1995, Agile Competitors and Virtual Organizations, Van Nostrand Reinhold, New York.
[11] CEST, 1997, OSTEMS Agility Mission to the U.S.- Findings and Recommendations, Centre for the Exploitation of Science and Technology, London.
[12] Katayama, H. and Bennett, D. J., 1996b, Adaptable Production and Its Realisation: Some Cases of Japanese Manufacturers, in Voss, C. (Ed), Manufacturing Strategy: Operations Strategy in a Global Context, proceedings of the European Operations Management Association 3rd International Conference, London Business School, London, pp 357 - 362.
[13] Oliver, N., Jones, D. T., Delbridge, R., Lowe. J., Roberts, P. and Thayer, B., 1994, Worldwide Manufacturing Competitiveness Study: The Second Lean Enterprise Report, Anderson Consulting, London.
[14] Skorstad, E., 1994, Lean Production, Conditions of Work and Worker Commitment, Economic and Industrial Democracy, 15, pp 429 - 455.
[15] Berggren, C. , 1993, Lean Production - The End of History, Work, Employment and Society , 13 (7), pp 22 - 56.
[16] Skinner, W., 1978, Manufacturing in the Corporate Strategy, John Wiley & Sons, New York.

Part V

OPERATIONS OF AGILE MANUFACTURING SYSTEMS

Computer control for agile manufacturing systems

Robert W. Brennan

Department of Mechanical and Manufacturing Engineering, University of Calgary, 2500 University Dr. N.W., Calgary, Alberta, Canada, T2N 1N4[*]

1. INTRODUCTION

Global competitive pressures in manufacturing have resulted in fundamental changes in the manufacturing environment. This can be seen in the current trend towards highly automated systems that are intended to adapt quickly to change while providing extensibility through a modular, distributed design. In order to realise the flexibility and productivity that these advanced systems promise, the computerised control system has become a central element in the design of agile manufacturing systems.

The basic responsibilities of a manufacturing control system are the sequencing and scheduling of orders, monitoring and execution of detailed plans, and monitoring of system status. The main challenge in manufacturing systems control research has been to develop a system that can not only deal with these requirements, but is also responsive to disruptions in the manufacturing environment (e.g., machine failures, operator absences, material shortage, changes in demand, and absent or inaccurate status information). Ideally, the control system will have this responsiveness embodied in its design in such a way that when changes do occur on the shop floor, changes do not have to be made to the control software (or, the changes that have to be made are minimal).

Experience has shown that traditional centralised forms of control where information is stored and calculations take place in one location or control computer cannot meet these requirements, particularly in an environment that is characteristically concurrent, asynchronous, and decentralised. As a result, recent work in manufacturing systems control has focused on moving away from centralised forms of control, towards a decentralised approach, involving the use of a number of interacting decision-makers. This has led to the question of how one can best organise these systems and has led industrial and academic researchers to the development of a spectrum of decentralised control architectures ranging from hierarchical to non-hierarchical or heterarchical control architectures [11].

A number of questions follow from the research that has been conducted in the area of manufacturing control architectures. Of primary importance is the fundamental question concerning the choice of control architecture: i.e., is it possible to determine whether a specific control architecture is appropriate for solving a given manufacturing system control problem?

---

[*] The author wishes to thank the Natural Sciences and Engineering Research Council of Canada for their general support of this research, under the grant OGP-019-7339.

In order to answer this question, it is first important to understand the characteristics of alternative control architectures and to be able to compare their relative performance while controlling given manufacturing systems [4].

Recent work in this area has pointed in the direction of control architectures that fall between the two extremes of the control architecture spectrum and contain elements of both hierarchies and heterarchies [5,40,41]. Although, these "partial hierarchies" provide some insight into the possible structure of control architectures for agile manufacturing, they are static in nature. In order to develop control architectures that are capable of adapting in an environment that is continuously in a state of change, an adaptable and continuously evolving control system is required. As a result, new approaches to manufacturing systems control are required.

This Chapter will focus on recent developments in production planning and control systems to achieve the goal of rapid, adaptive response to change in the manufacturing environment. In Section 2, the area of production planning and control in advanced manufacturing systems is investigated more closely. The background on existing capacity planning, scheduling, and control techniques discussed in this section sets the context for the current research on production planning and control systems for agile manufacturing discussed in the remainder of the chapter.

Unresolved issues in manufacturing system control are the focus of Section 3. In this section, the central theme is that future manufacturing control systems are required to be responsive to the continuously and unpredictably changing environment characteristic of agile manufacturing systems. Development of this type of system will not only require researchers to investigate new approaches, but will also require them to think about these systems in new ways.

Emerging research in production planning and control is the topic of the next section. In the search for the most appropriate control architecture for manufacturing systems, two closely related paradigms have generated a considerable amount of attention: multi-agent systems (MAS) and holonic manufacturing systems (HMS). Section 4 will focus on current research in these promising new areas, which focus on achieving "stability in the face of disturbances, adaptability and flexibility in the face of change, and efficient use of available resources" [19].

A brief case study is provided in Section 5 that focuses on the author's research into objective comparisons of alternative control architectures. This section ties together much of the material presented in previous sections by illustrating how the traditional manufacturing systems discussed in Sections 1 and 2 perform for a typical manufacturing scenario and how systems of the type discussed in Section 4 appear to show the most promise.

The chapter concludes with a brief summary and discussion of the future direction of research into production planning and control systems for agile manufacturing systems in Section 6.

2. BACKGROUND

When considering the design of a manufacturing control system, one of the main goals is to achieve a design that is flexible enough to adapt to changes in the manufacturing environment. Ideally, the control system will have this flexibility embodied in its design in such a way that when changes do occur on the shop floor, changes do not have to be made to the control software (or, the changes that have to be made are minimal). In this section, a brief review of some of the approaches that have been taken towards achieving this goal will be discussed as well as some of the unresolved issues in this area.

2.1. Manufacturing planning and control

Typically, manufacturing planning and control systems are viewed in terms of a hierarchy of decision levels that range from long-term, strategic decisions at the highest level to short-term, detailed execution and control at the lowest level. The National Bureau of Standards (NBS) [21] and the International Standards Organisation (ISO) [20] models are two widely accepted examples of this view of manufacturing.

The NBS architecture is a five-level hierarchical planning and control model where high level goals or tasks are decomposed into subtasks that are then passed to lower levels. At the lowest levels, primitive tasks are executed and status information is fed back to higher levels. The five levels of the NBS architecture can be summarised as follows [1]:

1) Facility control (months to years): manufacturing engineering, information management, and production management,
2) Shop (weeks to months): responsible for real-time management of jobs and resources on the shop floor (scheduling and monitoring),
3) Cell (hours to weeks): sequencing of batches and material handling facilities,
4) Workstation (minutes to hours): co-ordination of shop floor equipment is done at this level,
5) Equipment (milliseconds to minutes): "off-the-shelf" control systems are linked directly to individual pieces of equipment in a workstation.

Similar to the NBS architecture, the ISO architecture is a temporal decomposition of the production planning and control problem that covers decision spanning months to years down to decisions that are made in the range of milliseconds. The six-level ISO hierarchy can be summarised as follows [1]:

1) Enterprise (months to years): concerned with the achievement of the mission of the enterprise,
2) Facility/plant (weeks to months): manufacturing engineering, information management, production management, scheduling, production engineering,
3) Section/area (days to weeks): provision and allocation of resources and co-ordination of production,
4) Cell (hours to days): sequencing of jobs through stations, resource analysis and assignment, job routing and monitoring,
5) Station (milliseconds to hours): co-ordination of "relatively small, integrated workstations",
6) Equipment (milliseconds to minutes): realises the physical execution of tasks on machines.

Since the focus of this chapter is on manufacturing systems control, we will be mainly concerned with the lower four levels of both the NBS and the ISO hierarchies. As will be discussed in the next section, much of the work in manufacturing systems control has concentrated on these levels where sequencing and scheduling of orders and monitoring and execution of detailed plans are of primary importance.

2.2. Manufacturing control architectures

Recent work in manufacturing systems control has focused on moving away from centralised forms of control where information is stored and calculations take place in one location or control computer. This centralised strategy typically involves the classical control theoretic techniques of analysis and design for small-scale systems, which, as Sandell et al.

[37] note, "rest on the common presupposition of centrality". When large-scale systems such as manufacturing systems are considered, the problem becomes difficult, if not impossible, to solve using classical control theoretic techniques. The solution to this inadequacy of centralised control of large-scale systems is the decentralised approach, which involves the use of a number of interacting decision-makers in place of a single centralised one.

The question that is raised at this point is how does one organise these independent decision-makers? This leads to the question of appropriate control architectures for these systems and has led industrial and academic researchers to the development of a spectrum of decentralised control architectures ranging from:
a) a hierarchical decomposition as shown in Figure 1(a), to
b) an oligarchical approach where communication paths are less rigid as shown in Figure 1(b), to
c) a completely decentralised approach where individual controllers are assigned to subsystems and may work independently or may share information as illustrated by Figure 1(c).

The hierarchical approach to large-scale systems control involves decomposing the overall system into small subsystems that have weak interactions with each other. In order to control a system in this manner, the two most common approaches of hierarchical control are spatial separation (i.e., the multilevel technique described by Mesarovic et al. [31] and Singh [38]), and temporal separation (i.e., the multilayer, or frequency decomposition approach described by Gershwin [16] and by Kimemia and Gershwin [23]).

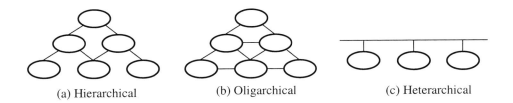

(a) Hierarchical   (b) Oligarchical   (c) Heterarchical

Figure 1: The Control Architecture Spectrum

In the manufacturing domain, Jones and Saleh [22] have combined both the spatial and temporal approaches at the Automated Manufacturing Research Facility (AMRF) in their multilevel/multilayer hierarchical controller design for manufacturing cells.

More recently, an alternative approach has been followed to deal with the complexity of automated manufacturing systems that is based on the concept of distributed data processing systems (DDPS) described by Enslow [15]. The basic definition that is given for a DDPS is that the system must support a "high degree of distribution in all dimensions, as well as a high degree of co-operative autonomy"[15]. The majority of the work in this area is based on the negotiation metaphor and has stemmed from the work of Davis and Smith [9] on distributed sensor systems. Their distributed problem solving approach requires the "co-operative solution of problems by a decentralised, loosely coupled collection of problem solvers"[9] and differs from distributed computing in that it is concerned with a "single task envisioned for the system"; the goal is to create an "environment for co-operative behaviour".

Parunak [33] extended the concept of the contract net from the information domain to the manufacturing domain with the development of the YAMS (Yet Another Manufacturing System) software. This system uses Davis' and Smith's negotiation metaphor to achieve coordination of a network of distributed computing nodes. The YAMS system uses an "open system model" that allows for the addition and removal of agents and separates knowledge from the control structure through information hiding.

Duffie and Prabhu [14] have investigated this distributed approach with a "heterarchical", or non-hierarchical opportunistic scheduling approach for machine cell control. One characteristic of this approach that sets it apart from traditional hierarchical control structures is that part flow is achieved by part-oriented requests as opposed to machine-oriented requests. This requires the parts to be "intelligent enough to know when they are done at one machine and need another"[13]. Similar work in the area of bidding-based control of manufacturing cells has also been investigated by other researchers [39,42].

An alternative approach to distributed control that uses the negotiation metaphor is the flexible routing simulation system developed by Lin and Solberg [26,27]. To achieve flexible routing, all routes are considered in real-time in this system. System modelling is based on a market-like model where resource agents act as vendors, setting their processing charge according to their status.

3. UNRESOLVED ISSUES

A number of questions follow from the research that has been conducted in the area of manufacturing control architectures. Of primary importance is the fundamental question concerning the choice of control architecture: i.e., is it possible to determine whether a specific control architecture is appropriate for solving a given manufacturing system control problem? In other words, can methods be generated to determine the "best" control architecture for a given problem? In order to answer this question, it is first important to understand the characteristics of alternative control architectures and to be able to compare their relative performance while controlling given manufacturing systems. The review of the literature in this area has shown that existing industrial and academic research on manufacturing system control has focused on qualitative comparisons of alternative structures and, although this approach has proven the concept of heterarchical control, it is agreed among researchers that quantitative comparisons of the various control architectures are still required [11,12,39].

Various key research questions are involved in answering the primary question noted above. Firstly, the trade-off between planning ahead and reacting on-the-fly is important in control architecture design. In other words, what is the most appropriate level of planning? Planning too far ahead in a highly uncertain environment is pointless since plans can quickly become obsolete due to disruptive events. However, not planning at all fails to take advantage of knowledge about the structure and constraints of a particular manufacturing environment, which could result in unnecessarily poor performance. A second key issue concerns decision-maker co-ordination and autonomy levels within a control architecture. It is possible to increase the level of autonomy of decision-makers within a control structure (i.e., broadening the scope of the decisions each is able to make without consulting, or being constrained by, others), but what is the appropriate level of autonomy that ensures "optimal" co-operation of decision-makers? Finally, when deciding between hierarchical and heterarchical approaches, the issue of local v. global information must be addressed. The trend has been towards distributing information resulting in decision-makers with limited views of the entire system. The important question that follows is how can globally optimal coherence be obtained with only local information?

Import as is the question of appropriate choice of control architectures for given manufacturing control problems, it is also important to recognise that system control requirements are never static in the manufacturing domain. Globally competitive pressures in manufacturing have forced firms to recognise the need for agile manufacturing systems: i.e., manufacturing systems that are capable of adapting in an environment continuously in a state of unanticipated change. To meet these requirements, the manufacturing control system plays a central role in integrating the components of these advanced systems.

As Dilts et al. [11] point out, manufacturing control systems are required that are reliable, fault tolerant, have a software design that is both modifiable and extensible, are capable of quickly allowing system reconfiguration, and can adapt to unanticipated changes on the shop floor. Clearly, control architectures for advanced manufacturing systems need to meet a long list of requirements. In order to take advantage of current technological advances in manufacturing, a new approach to manufacturing systems control is required. More specifically, a continuously changing environment needs an adaptable and continuously evolving control system. In the next section, we will look at emerging research in production planning and control that has been motivated by this need for manufacturing systems that are capable of rapid, adaptive response to change.

4. EMERGING RESEARCH IN PRODUCTION PLANNING AND CONTROL

Recently, two promising manufacturing control paradigms have generated a considerable amount of attention: multi-agent systems (MAS) and holonic manufacturing systems (HMS). As Bussmann [6] notes, "both paradigms have different, but compatible views on manufacturing control and ... a combination is beneficial to both paradigms".

Holonic Manufacturing Systems is one of the Intelligent Manufacturing Systems (IMS) program's six major projects resulting from a feasibility study conducted in the beginning of the 1990's [19]. The objective of the work of the HMS consortium is to "attain in manufacturing the benefits that holonic organisation provides to living organisms and societies, e.g., stability in the face of disturbances, adaptability and flexibility in the face of change, and efficient use of available resources" [19].

The term, holon, was coined by Arthur Koestler [24]. He observed a dichotomy of wholeness and partness in living organisms and social organisations and stated that "wholes and parts in the absolute sense do not exist anywhere". To understand these systems, Koestler used the "Janus Effect" as a metaphor for this dichotomy of wholeness and partness observed in many such systems: i.e., "like the Roman god Janus, members of a hierarchy have two faces looking in opposite directions". To explain this, he suggested a new term to describe the members of these systems: "holon" from the Greek (*holos* meaning "whole" and the suffix "on" implying "part" as in "proton" or "neutron"). In order to attain the benefits of Koestler's holonic organisations, an HMS "consists of autonomous, self-reliant manufacturing units, called holons" [6] that co-operate to achieve the overall manufacturing system objectives.

Like HMS, multi-agent systems also consist of co-operative and autonomous manufacturing units, but unlike HMS, MAS can be thought of as a "general software technology that was motivated by fundamental research questions" [6]. Research in MAS has lead to numerous advances in distributed systems (e.g., multi-agent negotiation [9]) that has contributed to much of the work described previously in this section as well as much of the work on HMS [40,41].

An example of earlier work on the application of multi-agent system concepts to the manufacturing systems control domain is the MetaMorph I architecture developed by Maturana and Norrie [28,29,30]. This architecture takes its name from its ability to change its structure and activities to adapt dynamically to tasks that emerge in the manufacturing system. A key element that enables this behaviour is the use of mediator agents to facilitate the coordination of the heterogeneous resource agents that are associated with the physical manufacturing system.

The mediator agents facilitate co-ordination by "promoting co-operation among intelligent agents and learning from the agents' behaviour" [28]. As well, mediators do not interfere with low-level decision-making in this architecture unless a critical situation occurs which requires their attention. The result of intelligent agent co-operation is the creation of "dynamic hierarchical decision trees, which occur naturally, without predefined organisational structures" [28].

Recently, a considerable amount of attention has been placed on the extension of general MAS concepts to the development of holonic manufacturing systems. In the area of manufacturing systems control, research has been conducted on the general requirements of these systems [7,10,25,35], decentralised planning and scheduling [17,18] and on the evaluation of the performance of holonic control systems [2,5,8]. Although this work is still in the preliminary stages, it appears that this manufacturing-specific, multi-agent paradigm shows considerable promise for the future of agile manufacturing systems.

5. THE PERFORMANCE OF HYBRID CONTROL ARCHITECTURES

As discussed previously, to meet the requirements of agile manufacturing, control systems are required that can automatically adapt to changes in the controlled system. Since these systems will be expected to operate automatically with little, or no, human intervention, these control systems need to be adaptable in an emergent sense: i.e., they should be intelligent enough to be able to automatically adapt and reconfigure based on the ever-changing requirements of the controlled system. Rather than a fixed architecture, the resulting structure of the new control system will be metamorphic in nature: i.e., to meet the dynamic requirements of agile manufacturing, the control system's architecture will be capable of constant metamorphosis.

Before such a metamorphic control system can be developed though, we must return to the issue of evaluating the relative performance of alternative control architectures for manufacturing systems, as noted previously. This limitation of the current manufacturing control research has acted as motivation for recent research [3] into providing an objective and quantitative comparison of alternative control architectures for manufacturing systems. In this section, a brief introduction to this research will be given. First, the experimental test bed developed to evaluate the relative performance of alternative control architectures will be described, then the results of experiments with this system are reported.

5.1. An Experimental Test Bed for Evaluating Alternative Control Architectures

As was noted previously, the work in this area was initially motivated by the need to answer questions such as: can methods be generated to determine the "best" control architecture for a given control problem? In particular, two more specific questions were asked that relate to this broader question: (i) what factors characterise a control architecture, and (ii) how do various control architectures perform for various constraints? The first question is concerned with determining appropriate metrics for evaluating the characteristics and relative performance of alternative control architectures. Since the literature on manufacturing systems control provides few quantitative comparisons of alternative control architectures, very few measures are reported that can be used to compare their relative performance. This gap in the literature led to the evaluation and development of a number of control system parameters and measures that could be used to characterise and gauge the performance of alternative control architectures [3,4].

The development of control system metrics allowed the second question to be tackled. In order to provide an objective comparison of the relative performance of alternative control architectures, a subset of the full spectrum of control as shown in Figure 2:
a) constrained hierarchical (CH): this test control architecture is intended to represent those control architectures firmly at the hierarchical end of the control architecture spectrum,
b) unconstrained hierarchical (UH): this test control architecture is intended to represent those control architectures that fall between the two extremes and have characteristics of both hierarchical and heterarchical control, and
c) non-hierarchical (NH): as the name implies, this test control architecture lacks any formal hierarchies between decision-makers.

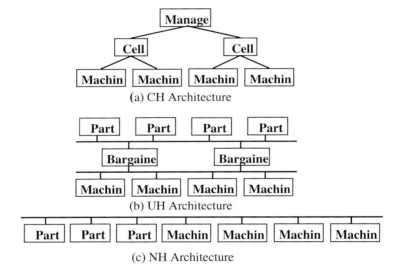

Figure 2: The test control architectures

Although the choices of test control architectures are each static structures, they are intended to be representative of the full spectrum of control architectures shown in Figure 1 and provide a basis for comparison between the different types of control architectures that have been investigated in the literature.

A modular simulation-based experimental test bed was developed [36], as shown in Figure 3, to evaluate the relative performance of these three test control architectures. The basic goal of the experimental test bed is to support the investigation of any control architecture when applied to any manufacturing system. Furthermore, it should permit the collection of performance data on both the controlled system (i.e., the manufacturing system), and on the control system (i.e., the modules of decision-making comprising the control architecture) itself (since we wish to gain insight into the nature of the different architectures too). To best meet this goal, the approach taken is to de-couple the manufacturing system and the control system as fully as possible, resulting in the basic structure of the modular experimental test bed as shown in Figure 3.

This figure shows the two main modules of the test bed, which can be identified as:
a) an emulation module, composed of a discrete-event simulation model, together with a communication model, which is intended to emulate the behaviour of the manufacturing system being controlled, and
b) a state/control module, which is used to implement alternative decision-making schemes.

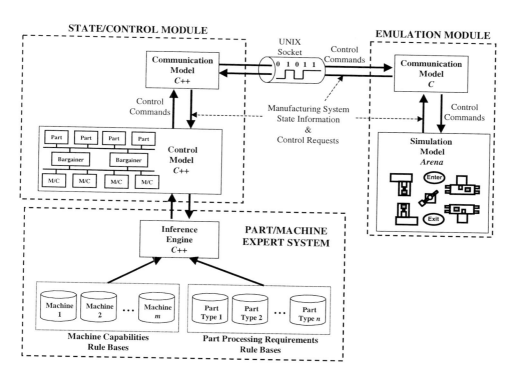

Figure 3: The Experimental Test Bed

The simulation model is written in the SIMAN simulation programming language [34], which is the language underlying the Arena simulation package, and can be modified relatively easily to represent alternative manufacturing system configurations. Arena's in-built statistics collection facilities make it easy to collect and analyse a wide range of performance information on the emulated system while its animation capability is valuable for debugging and visualisation of system behaviour. Additionally, the capabilities of the Arena "Professional Edition", which allows developers to design their own model building blocks, were used to define an Arena "module" to simplify the emulation-control system communication. A custom, but generically reusable, modelling building block was designed that allows status update messages from the emulation, and control messages from the control architecture, to be transferred seamlessly between the two processes.

The Arena emulation model is further augmented with additional ANSI C routines, which carry out the low-level communication functions via input/output streams (implemented as UNIX or INET sockets). These allow any important state changes in the emulated system to be reported to the control module and reactive control decisions to be communicated back to the system.

The state/control module, which is used to implement the test control architectures, is implemented in the C++ programming language. The modular nature of this experimental test bed has resulted in a system that has the ability to deal with any type of manufacturing system (i.e., by changing the Arena model) and any type of control architecture (i.e., by changing the collection of decision-making objects present in the control model).

The majority of the control architecture work described in the literature has been concerned with part scheduling in flexible machining systems (FMS) [16,23] and cellular manufacturing systems [14,22]. As a result, a similar type of system is investigated for this research dealing with the scheduling of parts through a simple manufacturing cell. The simulated manufacturing cell is similar in structure to the manufacturing cell that is investigated by Duffie et al. [12].

The system that is investigated here consists of a number of automatic machines connected by a material-handling robot as illustrated in Figure 4. Each of the machines in the system is capable of performing various operations depending upon its tooling set-up, which is limited by the number of tools that can be held at the machine's local tool storage area. For some experimental scenarios, each machine in the investigated system is prone to failure. The parts that are introduced to the manufacturing cell require that all of their processing should be performed within the manufacturing cell. Since each machine is capable of performing a number of operations, there is a degree of routing flexibility that is available to the part. Additionally, since machines may fail, and their local tooling set-up may change, this routing flexibility is dependent upon the machine status and set-up of the manufacturing cell.

In Figure 4, the two machines shown on the right represent automatic milling stations, while the two machines on the left represent automatic boring stations. A material-handling robot, which loads and unloads all machines, is shown at the centre of the figure.

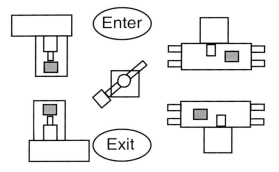

Figure 4: The Simulated Manufacturing System

In the next two sections we will look at preliminary investigations with this experimental test bed. The experiments that are reported here focus on the issue of planning v. reacting noted previously by comparing the manufacturing system performance of the UH and NH control architectures.

5.2. The Experimental Test Scenarios

As was discussed previously, the experimental test bed is designed to allow various changes to be made to the emulated manufacturing system and the control software in order to allow each of the test control architectures to be evaluated under various operating conditions. In this section the different operating scenarios that are used for the simulation experiments will be presented. Each of the control architectures is tested under the same set of operating conditions in order to provide an objective comparison of the resulting manufacturing system performance as well as the relative control architecture performance. Table 1 provides a summary of the basic operating conditions that were used to evaluate each of the test control architectures.

The manufacturing system under these circumstances operates in a deterministic fashion that, although unrealistic, should provide a basis for comparison with the other test scenarios. This base test scenario is intended to be used to provide a basis of comparison between a deterministic scenario and two test scenarios that introduce successively greater uncertainty into the manufacturing environment. Since one of the primary design goals for manufacturing control systems is achieving a system that is capable of coping with changes that occur on the shop-floor, introducing uncertainty to this environment should allow the relative effectiveness of each control architecture at achieving this goal to be determined.

As noted in the previous section, the part service time is based on machine agent bids. A single part type, requiring 8 operations to be completed by the machines is manufactured by the cell. When all of the required tools are available at each of the machines, the forward reasoning method used by each of the control architectures calculates a minimum processing and set-up time for all 8 operations of 9 minutes. All of the operations can be completed by visiting two machines.

The base scenario described above is compared with two stochastic scenarios. For these scenarios, part inter-arrival times and machine service times, mean-time-between-failures (MTBF) and mean-time-to-repairs (MTTR) are modelled by the gamma distribution. Table 2 provides the details of the two stochastic scenarios (C.V. in the figure stands for "Coefficient of Variation"). The main difference between stochastic scenario #1 and stochastic scenario #2 is that only stochastic scenario #2 has processing time variability.

Table 1
Base Scenario Parameter Values

Parameter	Description
Inter-arrival time	4.0 minutes
Machine Breakdowns	none
Service time	deterministic
Robot load time	0.05 minutes
Robot move time	0.5 minutes

Table 2
Stochastic Scenario Parameter Values

Parameter	Scenario #1	Scenario #2
Mean Inter-arrival time	4.0 minutes	4.0 minutes
Inter-arrival time C.V.	1.0	1.0
Machine MTBF	120 minutes	120 minutes
Machine MTBF C.V.	1.0	1.0
Machine MTTR	5 minutes	5 minutes
Machine MTTR C.V.	1.0	1.0
Service time C.V.	0.0	0.5
Robot load time	0.05 minutes	0.05 minutes
Robot move time	0.5 minutes	0.5 minutes

In order to evaluate the relative performance of the UH and NH control architectures, each control architecture was evaluated for each of the three test scenarios described above. A non-terminating simulation analysis was used with a simulation run length of 55,000 minutes and a 5,000 minute warm-up period (i.e., collected statistics were cleared at 5,000 minutes into each simulation run). Common random number streams were used for the arrival process, the machine breakdown process, and the machine service process, in order to better discriminate between the performance of each architecture.

5.3. Experimental Results

The relative performance of the UH and NH test control architectures for the first three test scenarios is shown in Figure 5. This figure shows the test control architectures' tardiness and flow-time performance respectively. Based on point estimates from the simulation runs, the average work-in-progress (WIP) for the base scenario is 2.8 for the UH control architecture, and 3.8 for the NH control architecture.

For the three test scenarios, tardiness is evaluated by assigning a constant arbitrarily chosen flow time allowance to the parts. A time of 20 minutes has been assigned for these experiments in order to create a situation where all of the control architectures would have some tardy jobs. By using mean tardiness as a measure we are assuming that early jobs bring no rewards and that there are penalties for late jobs.

5.4. Discussion

An interesting result follows from the experimental analysis of the UH and NH test control architectures described in the previous section. It appears that control architectures that contain properties of both hierarchical and non-hierarchical control architectures (i.e., the UH control architecture) show better flow time and tardiness performance than architectures that sit on either extreme of the control architecture spectrum.

One likely explanation for this difference in the control architecture performance is related to issues of global v. local information and planning v. reacting discussed previously. The UH control architecture's bargainer agents provide both global information and planning capabilities, both through bargainer agent schedules.

Although both architectures can plan for all part operations, only the UH architecture actually does. The UH architecture will generate a schedule for all part operations on multiple machines that can be either followed explicitly or disregarded. This provides the UH architecture bargainer agents the opportunity to either plan or react respectively. The NH architecture tries to schedule as many operations as possible on a single machine. Since it does not, as implemented, have a mechanism for scheduling on multiple machines, it is forced into a reaction mode. The results shown in Figure 5 reflect this unfair advantage of the UH control architecture.

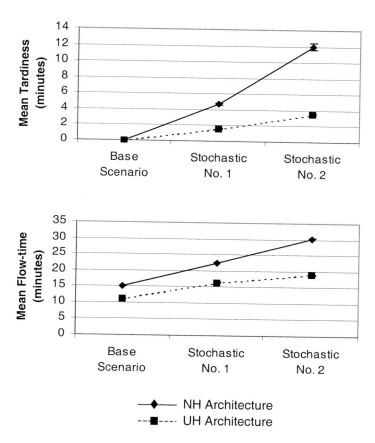

Figure 5: Test Control Architecture Tardiness & Flow Time Performance

In order to gain further insights into the relative performance of alternative control architectures such as the UH and NH architectures, a deeper understanding of the underlying structure of alternative control architectures is required. For example, questions such as (i) what factors characterise a control architecture, and (ii) how do various control architectures perform for various constraints, will need to be addressed. Work is currently being conducted by the author that is focused on identifying key parameters of the manufacturing system control problem that can be used to characterise alternative control architectures [4].

Larger control systems than the test systems studied here will most likely be hybrids of these smaller architectures, composed of clusters of hierarchies and heterarchies that are dynamically re-configurable (i.e., both the control architecture's organisational structure and co-ordination modes should be re-configurable). If these "partial hierarchies" are to be realised, it is important that the underlying properties of alternative control architectures are first identified in order to allow decisions to be made concerning the most appropriate type of control architecture for a given manufacturing control problem.

6. SUMMARY

In this chapter, we have looked at recent developments in the area of manufacturing systems control that are focused on the goal of achieving rapid, adaptive response to change in the manufacturing environment. A primary concern of recent research in this area has been the choice of architecture, or in other words, how the decision-making agents in a control system should be organised to effectively achieve overall manufacturing system objectives.

To achieve the reconfigurability and adaptability required of agile manufacturing systems, it appears that the control system cannot be limited to a single static structure. As noted in the previous section, experimental results have indicated that control architectures that are hybrids of heterarchical and hierarchical control architectures offer reconfigurability as well as better manufacturing system performance than either extreme of the control architecture spectrum. Similarly, recent work in the area of multi-agent systems and holonic manufacturing systems also point in the direction of hybrid architectures that provide the benefits of centralised elements and functionality while maintaining the robustness and agility of heterarchical architectures [40,41].

If these partial hierarchies are to be adaptable and reconfigurable, the control system will also have to be capable of responding to changes in the manufacturing environment. In other words, partial "dynamic" hierarchies will result that should not only be thought of as hybrids of heterarchical and hierarchical control architectures, but should also be thought of as metamorphic control systems. That is, these systems will adapt their architecture to respond to changes in the manufacturing environment.

REFERENCES

1. Bauer, A., Bowden, R., Browne, J., Duggan, J., Lyons, G. *Shop Floor Control Systems: from Design to Implementation.* New York: Chapman and Hall, 1991.
2. Bongaerts, L., Indrayadi, Y., Van Brussel, H., Valckenaers, P. "Predictability of hierarchical, heterarchical and holonic control," the 2nd International Workshop on Intelligent Manufacturing Systems, 1999, pp. 167-176.
3. Brennan, R.W. "Appropriate control architectures for automated manufacturing systems," *Ph.D. Thesis*, The University of Calgary, 1996.
4. Brennan, R.W. "Metrics for evaluating alternative manufacturing control architectures," the Proceedings of the Third International Conference on Industrial Automation, 1999. pp. 19.5-19.8.
5. Brennan, R.W., Norrie, D.H. "The performance of partial dynamic hierarchies for manufacturing," the 2nd International Workshop on Intelligent Manufacturing Systems, 1999. pp. 5-13.
6. Bussmann, S. "An agent-oriented architecture for holonic manufacturing control," First Open Workshop on Intelligent Manufacturing Systems Europe, 1998, pp. 1-12.

7. Bussmann, S., McFarlane, D.C. "Rationales for holonic manufacturing control," the 2nd International Workshop on Intelligent Manufacturing Systems, 1999, pp. 177-184.
8. Cavalieri, S., Bongaerts, L., Macchi, M., Taisch, M., Wyns, J. "A benchmark framework for manufacturing control," the 2nd International Workshop on Intelligent Manufacturing Systems, 1999, pp. 225-236.
9. Davis, R., Smith, R.G. "Negotiation as a metaphor for distributed problem solving," *Artificial Intelligence*, vol. 20, pp. 63-109, 1983.
10. Deen, S.M. "Cooperation issues in holonic manufacturing systems, in *Information Infrastructure Systems for Manufacturing*. Elsevier Science, 1993, pp. 401-412.
11. Dilts, D., Boyd, N., Whorms, H. "The evolution of control architectures for automated manufacturing systems," *Journal of Manufacturing Systems*, vol. 10, no. 1, pp. 79-83, 1991.
12. Duffie, N.A., Chitturi, R., Mou, J. "Fault-tolerant heterarchical control of heterogenous manufacturing system entities," *Journal of Manufacturing Systems*, vol. 7, no. 4, pp. 315-328, 1988.
13. Duffie, N.A., Piper, R.S. "Nonhierarchical control of manufacturing systems," *Journal of Manufacturing Systems*, vol. 5, no. 2, pp. 137-139, 1986.
14. Duffie, N.A., Prabhu, V.V. "Real-time distributed scheduling of heterarchical manufacturing systems," *Journal of Manufacturing Systems*, vol. 13, no. 2, pp. 94-107, 1994.
15. Enslow, Jr, P.H. "What is a 'distributed' data processing system?" *Computer*, pp. 13-21, 1978.
16. Gershwin, S.B. "Hierarchical flow control: a framework for scheduling and planning discrete events in manufacturing systems," *Proceedings of the IEEE*, vol. 77, no. 1, pp. 195-209, 1989.
17. Gou, L., Luh, P.B., Kyoya, Y. "Holonic manufacturing scheduling: architecture, cooperation mechanism, and implementation, *Computers in Industry*, vol. 37, p. 213-231, 1998.
18. Hino, R., Moriwaki, T. "Decentralized scheduling in holonic manufacturing systems," the 2nd International Workshop on Intelligent Manufacturing Systems, 1999. p. 41-47.
19. Holonic Manufacturing Systems Consortium. *Holonic manufacturing systems overview*. http://hms.ifw.uni-hannover.de/public/overview.html, 1999.
20. ISO *The Ottawa Report on Reference Models for Automated Manufacturing Systems*. ISO, Geneva, Switzerland, 1986.
21. Jones, A., McLean, C. "A proposed hierarchical control model for automated manufacturing systems," *Journal of Manufacturing Systems*, vol. 5, no. 1, 1986.
22. Jones, A., Saleh, A. "A multi-level/multi-layer architecture for intelligent shopfloor control," *International Journal of Computer Integrated Manufacturing*, vol. 3, no. 1, pp. 60-70, 1990.
23. Kimemia, J., Gershwin, S.B. "An algorithm for the computer control of a flexible manufacturing system," *IIE Transactions*, vol. 15, no. 4, pp. 353-362, 1983.
24. Koestler, A. *The Ghost in the Machine*, Arkana, 1967.
25. Langer, G. "HoMuCS: A methodology and architecture for holonic multi-cell control systems," Ph.D. Thesis, Department of Manufacturing Engineering, Technical University of Denmark, 1999.
26. Lin, G.Y., Solberg, J.J. "Flexible routing control and scheduling," *Proceedings of the Third ORSA/TIMS Conference on FMS: Operations Research Models and Applications*, Elsevier Science Publishers, 1989, pp. 155-160.

27. Lin, G.Y., Solberg, J.J. "Integrated shop floor control using autonomous agents," *IIE Transactions*, vol. 24, pp. 57-71, 1992.
28. Maturana, F.P. "MetaMorph: An adaptive multi-agent architecture for advanced manufacturing systems," *Ph.D. Thesis*, The University of Calgary, 1997.
29. Maturana, F.P., Balasubramanian, S., Norrie, D.H. "Intelligent multi-agent coordination system for advanced manufacturing," Proceedings of SPIE: Architectures, Networks, and Intelligent Systems for Manufacturing Integration, 1997, pp. 202-212.
30. Maturana, F.P., Norrie, D.H. "Multi-agent mediator architecture for distributed manufacturing," *Journal of Intelligent Manufacturing*, vol. 7, pp. 257-270, 1996.
31. Mesarovic, M.D., Macko, D., Takahara, Y. *Theory of Hierarchical, Multilevel, Systems*, Academic Press, 1970.
32. National Research Council - Committee on Analysis of Research Directions and Needs in U.S. Manufacturing, *The Competitive Edge: Research Priorities for U.S. Manufacturing*, National Research Academy Press, 1991.
33. Parunak, H.V.D. "Manufacturing experience with the contract net," *Distributed Artificial Intelligence*, vol. 1., London: Pitman, 1987.
34. Pegden, C.D., Shannon, R.E., Sadowski, R.P. *Introduction to Simulation Using SIMAN*. McGraw Hill, 1995.
35. Rahimifard, S., Newman, S.T., Bell, R. "Distributed autonomous real-time planning and control of small to medium enterprises," *Proceedings of the Institution of Mechanical Engineers*, vol. 213, p. 475-489, 1999.
36. Rogers, P., Brennan, R.W. "A simulation testbed for comparing the performance of alternative control architectures," the Winter Simulation Conference Proceedings, 1997, pp. 880-887.
37. Sandell, N.R., Varaiya, P., Athans, M., Safonov, M.G. "Survey of decentralized control methods for large scale systems," *IEEE Transactions on Automatic Control*, vol. AC-23, no. 2, pp. 108-128, 1978.
38. Singh, M.G. *Dynamic Hierarchical Control*. North-Holland, 1980.
39. Upton, D.M. "An organic structure for a computer-controlled manufacturing system," *Graduate School of Business Administration Report*, Harvard University, Boston, MA, 1991.
40. Valckenaers, P, Van Brussel, H, Bongaerts, L, Wyns, J. "Holonic manufacturing systems," *Integrated Computer Aided Engineering*, vol. 4, pp. 191-201, 1997.
41. Van Brussel, H, Wyns, J, Valckenaers, P, Bongaerts, L, Peeters, P. "Reference architecture for holonic manufacturing systems: PROSA," *Computers in Industry*, vol. 37, pp. 255-247, 1998.
42. Veeramani, D. "Task and resource allocation via auctioning," the Winter Simulation Conference Proceedings, 1992, pp. 945-954.

Computer Aided Process Planning for Agile Manufacturing Environment

Neelesh K. Jain[a] and Vijay K. Jain[b]

Mechanical Engineering Department, Indian Institute of Technology, Kanpur
Pin Code: 208 016 (U.P.), India

[a] E-mail: nk_jain@mailcity.com; nkjain@iitk.ac.in
[b] E-mail: vkjain@iitk.ac.in; Tel: 91-512-597 916 (Off.); Fax: 91-512-590 007, 91-512-590 260

Agile manufacturing environment demands quick and effective response to unexpected and sudden changes in the products and services, and their demands. Being a crucial intermediate and integrating activity between CAD and CAM, CAPP plays an important role in the implementation of agile manufacturing environment by translating design requirements of an improved and/or redesigned products into the manufacturing instructions. Role of CAPP, its strategies/techniques, implementation framework, and other related issues in the context of agile manufacturing environment, have been discussed in this chapter along with the preliminary discussion of agile manufacturing environment and CAPP itself.

1. INTRODUCTION

Globalization and diversification of market, increasing international competition, shortening of product life, increasing product variety, decreasing production volume, and strong need for rapid and dynamic product innovation cycle, have changed the nature of market from manufacturer oriented to customer oriented. This has resulted in greater and greater attention being paid to customer satisfaction. To survive in such a diversified, globalized, customer oriented, and vibrant market, the need was felt for the development of methods and means that are reconfigurable and are capable of developing (rendering) products (services) rapidly and cost-effectively. This has led to the emergence of concept of agile manufacturing. The concept of agile manufacturing was floated in 1991 at Lehigh University at the end of a government sponsored research work [1].

With the changed market concept from manufacturer oriented to customer oriented, the ultimate success of a manufacturer, nowadays, depends on its ability not only to meet but also to excel the customer's demands and expectations. Therefore, a manufacturer should be well equipped to identify the needs and expectations of the customer and must possess strategies, initiatives and flexibility to react efficiently, rapidly, and cost effectively to the unexpected, unpredictable and sudden demands of the customer. Today's manufacturing organizations, therefore, need to be customer oriented.

2. AGILE MANUFACTURING ENVIRONMENT

2.1. Definition

Conventionally, *agile* means fast moving. Agility is an extension of flexibility. Flexibility is the ability to respond rapidly and adapt to changes. It includes dimensions of volume, product, process, mix, delivery, and operations. But, agility goes beyond flexibility and merges the components of flexibility, quality, cost, and reliability. Agility implies being flexible with high quality, low cost, superior service, and greater reliability. As depicted in Table 1, agility represents a drastic divergence from traditional mass production-based system [2]. Agility in action represents a paradox as firms compete and cooperate simultaneously. In business world, to be *agile* means to master changes and uncertainty, and to integrate employees and information tools in all aspects of production. It is capability to survive and prosper by reacting quickly and effectively to a continuously and unpredictably changing, customer-driven, and competitive environment. Agility is not only a performance issue, but a key competitive strategy also. Agility is a comprehensive and strategic response to the fundamental and irreversible changes that are undermining economic foundations of mass production-based competition [1].

Agile or quick response manufacturing means production of highly customized products and quick responses to customer demands without associated higher costs, through efficient and effective use of flexible and programmable machinery, and reconfigurable production facilities. Agile corporations are able to rapidly reorganize and even reconfigure themselves so as to capitalize on immediate and temporary market opportunities. Agile manufacturing is not simply concerned with being flexible and responsive to current demands but also requires an adaptive capability to be able to respond unpredictable and sudden future changes. This requires development of internal capabilities within the manufacturing system, and ability to reconfigure company's physical and intellectual assets. Agile and lean are not synonymous. One of the biggest differences between the two is in terms of supplier relationship. Lean manufacturers believe in finding the best supplier by searching the open competition market (i.e. crossing the border), which may not be true with agile manufacturer.

Agile manufacturing is a concept to standardize common manufacturing data, research data, CAD/CAPP/CAM structure, and integrate them into a network. A *common manufacturing database* and a standardized research database are very crucial for agility and can significantly reduce the product design period, planning period and even research period. Agile manufacturing and agile equipments sharply reduce the cost and time span from initial design to consumer-ready products and have become stronger and cost-effective tools to meet unexpected, unpredictable and sudden customer demands [3].

Table 1
Comparison of traditional and current focus on the manufacturing [1].

Traditional practice	**Current practice**
Standardized or uniform products,	Customized or highly variable products,
Longer market life,	Shorter market life,
Produced to forecast,	Produced to order,
Low information content,	High information content,
Priced by manufacturing unit cost +margin,	Priced by customer perceived value,
Specific market niche,	Multiple market niche,
Self contained,	Open for upgrades/ information/ services.

2.2. Principles/strategies and objectives

According to Agile Manufacturing Enterprise Forum, agile manufacturing has major characteristics like rapid introduction of new and modified products, product customization, upgradable products, dynamic reconfiguration of production processes, etc [5]. Agility has following *four* underlying ***principles/strategies,*** or alternatively agile manufacturing enterprise can be defined along these four dimensions [1, 2, 4]:

- Value based pricing strategy that enriches the customer by delivering value to it,
- Cooperation to enhance the competitiveness by forming Virtual Enterprise (VE),
- Organizational mastery of handling changes and uncertainty, and
- Valuing human knowledge and skills by making investments that reflect their impact.

Objective of agile manufacturing is to create an open and scalable manufacturing infrastructure, and to demonstrate its effectiveness in pilot production. Agility fulfills different objectives from different viewpoints. For the customer, it translates into customer enrichment. While, for the businessman, agility translates into cooperation that enhances competition. An agile manufacturer has to present a solution to its customer's needs on a continual basis and not just a product that is sold once. For this, the producer must understand both stated and implied needs of a customer, i.e. producer must learn what a customer needs now and what will need in future [2]. Agile enterprises cross company borders to work together by integrating and coordinating core competencies of their organizations to reduce time-to-market. Thus, agile manufacturers can respond quickly and effectively to the situations of rapidly changing markets, global competitive pressures, needs of decreasing time-to-market of new products, increasing inter-enterprise cooperation, interactive value-chain relationship, global sourcing/marketing/distribution, and increasing value of information/service [1].

2.3. Components/ingredients

Synthesis of innovations in the fields of manufacturing, information technology (IT) and communication technologies along with radical organizational redesign and new marketing strategies, have made the agility possible [1]. Ingredients of the agile manufacturing system include small batch size, minimal buffer stock, improved work processes, redesign of workflow, total quality control, elimination of waste, setup reduction, preventive maintenance, and use of Kanban system.

2.4. Enabling philosophies/techniques/tools

Broadly speaking, both Computer Integrated Manufacturing (CIM) and Concurrent Engineering (CE) are enabling philosophies for agile manufacturing environment. These philosophies should be considered more than collections of tools and techniques for manufacturing management. A company committed to both of these philosophies is well positioned to qualify as an *agile manufacturer*. But, vice-versa is not true, i.e. an agile manufacturer may use neither CIM nor CE. Also, it is possible for a manufacturer to be a "CIM organization" without employing CE or "CE organization" without CIM [4].

CIM can be defined as interface of CAD, CAM and Direct (or Distributed) Numerical Control (DNC) with logistic information system. Its definition also includes a group of intelligent machine cells or Flexible Manufacturing Systems (FMS) constituting a small local network. Concept of CIM is based on integrating computer technology and Artificial Intelligence (AI) into a machine tool, while agile manufacturing is more focused on the networking. Therefore, it can be regarded as *macro CIM* system [3].

CE is a concept that refers to the participation of all functional areas of the firm, including customers and suppliers, in the product design activity so as to enhance the design with inputs from all the key stakeholders. This process ensures that final design of the product meets all the needs of the stakeholders and ensures that the product can be brought quickly to the market while maximizing quality and minimizing associated costs. Table 2 presents enabling philosophies, tools, or technologies of agile manufacturing, along with their functions or objectives and the means of achieving them.

2.5 . Implementation framework and implications of agility

Agile manufacturing environment should be implemented in a consistent and systematic manner. Agile companies must be innovative, highly responsive, constantly experimenting to improve the existing products and processes, and striving for less variability and greater capability. Manufacturing practice for managing agility includes: enterprise integration, shared database, multimedia information network, product and process modeling, intelligent process control, virtual factory, design automation, super-computing, product data standards, paperless transactions via Electronic Data Interchange (EDI), high speed information highway, etc. Suggested order of introduction of agility on shop floor should be adopting cellular layout followed by reduction in number of setups, paying attention to integrated quality, preventive maintenance, production control, inventory control, and finally improving relations with the suppliers. Different areas of an enterprise, which are affected by the implementation of agile manufacturing environment include design and production, marketing, distribution, waste disposal, management, organization, and its people. A conceptual framework for design and implementation of agile manufacturing system is shown in Figure 1.

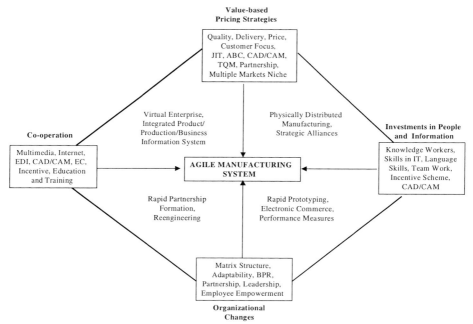

Figure 1: A framework for the development of agile manufacturing system [1].

Table 2
Enablers of agile manufacturing, their functions, and means.

Enabler	Functions or objectives	Means of achieving
Virtual Enterprise (VE) formation	* To facilitate reconfiguration of the organization, as a single organization is not able to develop sufficient internal capabilities to respond quickly and effectively to changing production needs.	* A system embracing virtual design, virtual manufacturing, and virtual assembly by extending capabilities of existing CAD/CAM system [1], * Internet assisted manufacturing system consisting of CAD, CAPP, CAM, and (CAA) integrated via Central Network Server (CNS) [3].
Physically distributed teams and manufacturing	* To support agility with the objective to reduce time-to-market.	* Electronic mail (e-mail), networks, * Graphical User Interface (GUI), * Video conferencing.
Rapid partnership formation (Partnership formation precedes VE formation and it is a sub-function of VE formation)	* To provide the firm with new technologies, products, markets, critical resources, and core competencies, * To position a company in the competitive global manufacturing spectrum by combining its technical and marketing skills with those of the leader in manufacturing.	* Analysis of strategic and operational opportunities of potential partnering firms, *Alignment of business, manufacturing, and operational strategies, and pooling of core competencies, * Tools: Quality Function Deployment (QFD), Benchmarking, Internet, Multimedia, Microsoft Project, Electronic Data Interchange (EDI), Case Tools, etc [1].
Concurrent Engineering (CE)	* To include customers, suppliers, all functional areas of the firm in design process of the product so as to eliminate non-value adding activities in engineering, production, distribution, accounting, and customer service, * To combine connectivity of CAE, CAD, and CIM with DFM, and to facilitate agility in all areas of VE.	* Functional analysis, * Solid modeling, * Finite Element Analysis (FEA), * Design for Manufacturing (DFM), * Design for Cost (DFC), * Design for Assembly (DFA), * Design for Reliability (DFR), * Design for Ergonomics (DFE), * Failure Mode and Effect Analysis (FMEA), * Optimization, * Value and robust engineering, * CAM and NC verification.
Integrated product/ production/business information system	* To economically achieve configurability of agile manufacturing system.	* Multimedia, * Internet, * EDI.
Rapid Prototyping (RP)	* To reduce product development time and non-value adding activities.	* CAD and solid modeling, * CAE, CE * Various RP techniques.
Electronic Commerce (E-commerce)	* To reduce cycle time, delivery time, response time, and time-to-market.	* Internet, * WWW, * EDI.

3. COMPUTER AIDED PROCESS PLANNING (CAPP)

3.1. Process planning
Process planning (PP) is very crucial intermediate and integrating activity between design and manufacturing of a part. Process planning can be defined as the systematic determination of methods and means to manufacture a component economically and competitively. It translates various qualitative and/or quantitative requirements (i.e. functional, physical, ·mechanical, aesthetic, etc.) of a product into manufacturing instructions. It also identifies various resources of manufacturing a product.

Process planning is the first step in building up a manufacturing plan for a product. But, it does not take into account availability of and load on the industrial facilities. In the global structure of an industrial enterprise, process planning can be considered as a part of *production planning*. Production planning uses technical information of process planning as its input. New trends in the development of process planning are to integrate more and more functions of production and process planning to achieve better productivity [11].

3.2. Tasks and levels of process planning
Tasks involved in the process planning include **identification** of appropriate manufacturing processes, their sequence, appropriate machine tools, cutting tools, fixtures, and product routes. Followed by **calculation/selection** of process parameters for individual manufacturing process (i.e. cutting speed, feed, depth of cut, material removal rate (MRR), etc. in case of machining processes), and total manufacturing time and cost. Finally, **documentation** of the output in the form of route card, process planning sheet, instruction sheets, and/or NC/CNC part program depending on level of process planning and type of machine to be used for production, i.e. conventional or NC/CNC. Task of process planning can be carried out at following *three* different levels:

- **Level 1. Operation Planning:** At this level attention is paid to individual surfaces. Typically, a *route card* is generated at the end of this stage. In general, tasks to be carried out at this level involve identification of necessary manufacturing processes and their sequencing to meet the objectives like minimizing production time (or maximizing production rate) and cost, and/or maximizing profit earned. Particularly, for ***conventional machining processes*** it involves following tasks:

  [1] Determination of direction of tool access/approach for each surface,

  [2] Formation of work holding setups,

  [3] Identification of total volume to be removed in each setup, and its decomposition into *elementary volumes*. Grouping of elementary volumes to form *machinable volumes* with the objective of minimizing volume removal cost, tool and fixture utilization cost, and other associated costs. Some algorithms to accomplish this task are already available in the literature [14],

  [4] Identification of required machining operations to remove machinable volumes,

  [5] Sequencing of identified machining operations.

- **Level 2. Preliminary process planning:** Selection of manufacturing resources like: raw stock, machine tools, cutting tools, fixtures, etc. Usually output of this stage is presented in the form of *operation sheets*, and

- **Level 3. Detailed process planning:** Calculation of different process parameters of the manufacturing processes to generate output in the form of *instruction sheets or NC/CNC part program* depending on the type of machine tool being selected.

3.3. Need of CAPP

Traditional approach to process planning requires a lot of preparatory work to be done before making a final decision about the manufacturing plan of a component. In traditional process planning, process planner first studies the part drawing and manufacturing specifications, then identifies whether a *similar* part has been produced in the past or not, if a previously developed process plan exists then it is retrieved otherwise a new process plan is developed. To develop a new process plan or to edit an existing one, process planner has to refer many standards, databases, and books on tolerances, surface roughness, material codes, machine tool capabilities, cutting tool specifications, and cutting conditions. Analysis of activities of process planners by F. A. Logan (as reported in [11]) has revealed following approximate distribution of time:

- 15 % in technical decision making,
- 40 % in referring databases and tables of standards and in making calculations, and
- 45 % in documentation and text preparation.

However, above time distribution would be completely different if Decision Support System (DSS), computerized database, and computer-assisted calculations are used. For the same requirements of a part, there can be different manufacturing alternatives, therefore process planning is more or less an iterative process rather than being a straightforward process. Task of process planning is labor-intensive, time consuming, and tedious. In most of the developing countries, it is carried out manually. *Manual process planning* suffers from the following disadvantages associated with it [8]:

- **Commitment to personal experience, skills, intuitions, preferences, and prejudices:** If different process planners are asked to develop a process plan for the same part, they would probably come up with different process plans and there is no guarantee that any one of them will constitute an optimum method to manufacture the part. Process planners make their decisions based upon global understanding of the problem without figuring out individual parameters. They know the problem and its solution, but they cannot pinpoint the controlling parameter(s) that causes problems. Therefore, they can apply same solution to many similar problems, even if the controlling parameter(s) is (are) not same.

- **Non-uniformity and inconsistency:** A process plan developed for a part during a current manufacturing program may be quite different from the old plan and it may never be used again. This causes wastage of effort, and many inconsistencies in routing, tooling, labor requirements, costing, and purchasing requirements.

- **Static nature:** Current process planning practices should reflect the changes in batch sizes, availability of new technologies, equipments, processes and tools. Unfortunately, manually prepared process plans do not reflect a consistent view towards changes in manufacturing technologies and processes, and do not take into account recent developments in manufacturing technologies, equipments, tools, etc.

These drawbacks lead to conclude that manual/traditional process planning still remains as an art rather than being a technology. They also compel to automate process planning by exploiting potentials of computers. Idea behind CAPP is to utilize power and capabilities of computers to assist process planners in clerical work, and dividing the

work between process planners and computers such that each performs the task it knows best. CAPP serves the following objectives [10]:
- Overcome the drawbacks of manual PP as stated above,
- Reduce pre-production lead times and can increase process planner's productivity,
- Capture experience, logic, and skills of process planners in algorithms and heuristics,
- Relieve task of process planning from personal experience, skills, prejudices,
- Generate uniform, consistent and dynamic process plans,
- Provide intelligent assistance to the user rather than fully automated reasoning,
- Integrate CAD and CAM, and play a pivotal role in development of FMS and CIMS.

3.4. CAPP approaches and implementation strategies

Detailed descriptions about different CAPP approaches can be found in the references [11-14]. These approaches can be classified in the following categories:
- **Variant approach:** This approach is similar to traditional or manual process planning in which process plan for a new part is created by retrieving and then modifying an existing process plan for a similar part from the computerized databank of process plans. This approach is derived from Group Technology (GT) concept in which parts are classified and coded into part families. A *part family* is a collection of parts having similar design and/or manufacturing attributes. Each part family is represented by a *master component,* which is a hypothetical component combining all possible design and manufacturing attributes of that part family. There exists a standard process plan for the *master component* of each part family. Example part is coded using a *coding and classification system*, and process plan of the corresponding master component is retrieved and suitably modified. Most commonly used tools for part family formation are *production flow analysis, and coding and classification system.* This approach mostly uses *backward or bottom-up planning,* tracing the workpiece from finished to raw stage. Quality of process plan still depends on the knowledge of process planner. Computer is merely used to assist in manual process planning activity. Decision tree, decision table, and expert system can be used to enhance this approach [11].
- **Generative approach:** This approach synthesizes process information to automatically create process plan for a new component using technology algorithms, decision logic, formulae, and geometry database. It does not refer to the predefined or stored process plans and uniquely make many processing decisions required to convert a part from raw material to finished stage [15]. Generative approach is usually fully automatic and mostly uses *forward or top-down planning* tracing the workpiece from raw to finished stage. It does not rely on the expertise of a specific process planner. Purely generative CAPP systems are complex and difficult to develop, as they require in depth knowledge of manufacturing processes and mechanical engineering [11].
- **Hybrid or semi-generative approach:** It is combination of variant and generative CAPP. Almost all practically developed CAPP systems belong to this category. System's working steps are same as that of a generative system, but final process plan and intermediate stages of generative process plan can be edited or modified [11].
- **Expert system:** An expert system is a specialized computer program, which emulates human capacity of reasoning in a particular field of expertise and is rule-driven rather than algorithm or data-driven. Expert system exhibits same level of problem solving skills as an expert for a narrow problem domain. Expert system combines knowledge and reasoning capabilities to draw quality conclusions. Expert systems are developed for those problems for which precise series of solution steps may not be known, and it may be necessary to search through a space containing many alternative paths, some of

which lead to solutions and some of which do not [11]. They mostly use *Artificial Intelligent (AI) techniques* in which thinking process of an expert is coded in computer programs using logical programming languages like PROLOG, LISP, etc. which allow pattern-driven invocation of programs as required by expert systems.

4. CAPP IN AGILE MANUFACTURING ENVIRONMENT

4.1. Role of CAPP in agile manufacturing system (AMS)

Process plan of a part has relatively more influence on its manufacturing strategy, quality, cost, and production rate as compared to any other factory activity. Activities of product designing and process planning together largely determine the final cost and quality of a product. Process planning also affects production planning, production efficiency, company competitiveness, and other factory activities. It is a major factor in determining manufacturing lead-time, work-in-process (WIP), and associated costs. With the changed market perspectives, primary goal of any industrial activity related to producing a product or rendering a service, has focussed on reducing *lead-times and associated costs*. To cope up with challenges of quick and effective response to unexpected and unpredictable changes in products and their demands, development of an efficient, effective, flexible, and reliable interface between design and manufacturing is required. CAPP is considered as a logical means to achieve factory integration by bridging the gap between design and manufacturing in a production environment.

Role of CAPP becomes even more significant in an agile environment, which is characterized by *high product variety and low volume production of customized products/services having shorter market life.* In such a volatile and vibrant environment, products may be subjected to frequent modifications and/or redesigning process. Once a product design is modified or redesigned, its manufacturing strategy outlined in its process plan may also change. CAPP system can play an important role in responding quickly and effectively to meet the challenges of such changes by not only regenerating process plan for the modified and/or redesigned product, but also by providing other possible process plans. A properly designed and developed CAPP system plays a crucial role in attributing flexibility and reliability to AMS.

4.2. How to incorporate agility in CAPP: strategies for CAPP in AMS

A CAPP system for agile manufacturing environment should be developed so as to incorporate elements of agility, i.e. flexibility, quickness, quality, cost, and reliability. Traditional process planning generates rigid process plans, and is not flexible enough to cope up with the dynamic, stochastic, and unforeseen bottlenecks like breakdown and overloading of machine, non-availability of machine, cutting tool and fixture, etc. It is authors perception that AMS can be considered as extension of Flexible Manufacturing System (FMS). In AMS, flexibility should be attributed to the CAPP system while maintaining cost-effectiveness, quality, and reliability at the same time. Therefore, carefully and properly designed and developed CAPP system, which can meet the requirements of FMS, can also cope up with AMS. Following strategies or guidelines are suggested to develop a CAPP system suitable for agile manufacturing environment:

1. Selection of Appropriate Part Modeling Technique for CAPP: Manner of part description as input to automated process planning system is one of the measures of integration and automation that can be achieved through CAPP system. A *product model* is a structured representation of the product data and is capable of meeting the product data requirements of various engineering tasks encountered throughout life cycle of the product. A competent product-modeling scheme should provide complete and unambiguous geometrical and topological information and/or data as well as should capture the design and manufacturing content of the part in totality. Input part description suitable for CAPP system should contain the information about material of the part, geometric model and nominal dimensions of the part, dimensional, geometric, and positional tolerances, surface roughness specifications, description of joining and heat treatment techniques, information about batch size, delivery date, form and dimensions of raw stock, and other extra information useful for planning and control activities [11]. Various part modeling/representation schemes can be divided into following groups [15]:

- **Conventional Methods:** like *Natural Language Description, Freehand Sketches, Physical Models, and Engineering Drawings*. Of these, engineering drawings are most commonly used. But they have the limitations like: difficulty in automatically identifying machinable surfaces, incompleteness of drawings, dependence on personal experience for their interpretation, etc.,
- **Non-CAD models:** like *GT code, Descriptive Languages, Menu Driven Input*,
- **CAD Models:** like *Wireframe models (2D, 3D), Surface Models, and Solid Models [Constructive Solid Geometry (CSG), Boundary Representation (B-rep), Sweep]*, and
- **Feature-Based Design:** It has been described in brief in the next item.

Of the various part modeling schemes described above, a *proper* CAD model of the part has potential of eliminating human effort of translating design into a code or other descriptive form necessary for process planning system as its internal data representation can provide another computer-compatible format. Thus, use of a CAD model paves the way of integration of CAD system with CAPP system. There are many commercially available CAD systems (like CADDS, CATIA, AutoCAD, IDEAS, Pro/Engineer, etc.) belonging to any category of the CAD systems as described above. Systematic determination of requirements, careful study of the capabilities, limitations and applications of various commercially available CAD systems, and proper selection of a flexible, efficient and easily integrating CAD system, are required to select a suitable CAD system for agile manufacturing environment.

2. Capabilities and Requirements for CAD System: The most helpful feature of a CAD system from CAPP system's point of view is its ability to graphically display geometric shapes, allow their manipulation, comparison, and regeneration. Some CAD systems (i.e. IDEAS) also have capability to perform kinematic, strength, stress, flow, FEM, manufacturability and other types of analysis; all of these are very helpful in product designing, planning, and manufacturing functions. Graphical display is more needed for the designers and draftsman than the process planners, and production management. An ideal CAD system, that can be helpful for a CAPP system and suitable in the agile manufacturing environment, should [11]:

- Assist in translating concepts into engineering design,
- Retrieve existing drawings and designs by attributes,
- Allow automatic design of sub-systems and tooling,
- Allow automatic change of design,
- Check and recommend design for manufacturability and assembly,
- Check and recommend surface roughness and tolerances,
- Generate such a data format that is easily transferable to a CAPP system,
- Enable control of FMS,
- Allow to use product data as input to vision (robotic) system to enable path planning,
- Check and enforce company standards.

Majority of these features can be accomplished provided they are specified as the objectives to a CAD system designer. Unfortunately, old and well established drawing standards are not being used as basic concepts for the development of CAD systems. This may be due to the fact that computer experts, mathematician, and management experts are trying to do the job of an engineer. Also, most of the CAD systems are not designed to serve for CAPP systems. These systems contain geometrical and topological information and/or data about the part being modeled. But, they do not store and possess the supplementary information and/or data necessary for CAPP. Another limitation of CAD models is that the data and/or information stored are in the form of geometric features (vertices, lines, arcs, circles, and surfaces). This low-level information cannot be directly used for process planning and manufacturing related applications unless they are pre-processed further to provide interpretation in the form of manufacturing information [15]. Existing popular CAD techniques, that can be useful to develop or select a CAD system suitable for agile manufacturing environment, have been described below:

- **Parametric Modeling:** CAD systems based on parametric modeling can be very helpful in agile manufacturing environment as they have capability of *flexible product modeling*. Parametric modeling allows user to create product models with variational dimensions. Dimensions are linked via conditional expressions. Bi-directional connectivity between model and dimensioning scheme allows automatic regeneration of a product model and automatic updating of related dimensions, once the product dimensions are changed. Most of the CAD systems offer 2D-parametric modeling while only a few CAD systems with 3D-parametric modeling have been introduced. Both CAD systems are based on boundary representation (B-rep) and enable feature-based modeling with a limited set of primitive features, which can be combined to make more complex features [11].
- **Feature-Based Modeling:** Features can be defined from different perspectives like: *design, engineering, geometry, or manufacturing.* An engineering feature is defined as *a physical element of a part or component having some specific engineering significance (i.e. function, its manufacturing method, etc).* In addition to this, an engineering feature should be mappable to a generic shape, and should have predictable properties [11]. Part design can be considered as *structured compilation of features*, where each feature has specific geometry and some feature attributes [16].

Feature-Based Modeling (FBM) is based on the idea of designing with *higher-level building blocks or primitives,* in which part geometry is defined directly in terms of features instead of low-level analytical shapes like box, cylinder, sphere, cones, etc. This allows a designer to specify more about design than just geometry and topology, and to express design content and constraints at different levels of abstraction.

Limitations associated with this approach have been discussed in item 8 of this subsection. An important dilemma of this approach is that there can be infinite number of possible features and system should know how to handle them. While set of features offered by present generation of feature-based modelers is too limited to use them for industrial applications. Need for geometric modeling of features on a higher level of abstraction has initiated development of *non-manifold topology based solid modelers,* which enable geometric processing of *incomplete* geometry [11].

3. Capabilities and Approaches for CAPP System: Due to dynamic, agile, and vibrant nature of the modern manufacturing and marketing environment, a competent manufacturing planning system needs to have variety of capabilities, knowledge, and intelligence to accomplish the task of overall planning in general and that of process planning in particular. Agile manufacturing demands an ***exhaustive, flexible, responsive, and integrated*** CAPP system. Most of the existing CAPP systems fail to meet these requirements. There can be no single approach that can fulfil all these requirements. Different approaches, strategies, techniques, and methodologies from the various fields of science and engineering are to be incorporated to develop such a CAPP system suitable for agile manufacturing environment. Following are some approaches and guidelines, which can be adopted to develop such a CAPP system:

- *Use Flexible Process Planning:* It ensures ability to generate all possible alternative process plans using different possible routes a product can follow without compromising its quality and economy. Current process planning approaches do not take into account the availability and loading of resources at the shop floor due to lack of integration. In real-life, production suffers from unforeseen, stochastic, dynamic disruptions and bottlenecks. To overcome such disruptions and bottlenecks, re-planning is to be carried out which is time and money consuming. **Linear process planning** generates only one process plan at a time for a component/part by finding optimal solution through static optimization criteria like minimizing number of setups, tool changeovers, etc., thus lacking flexibility and is not suitable for FMS and AMS. If all possible process plans are simultaneously generated considering various feasible routes and manufacturing operations to produce a product then it can optimize the time and money investments. It is reported that F. Glusti, and A. Kusiak have developed CAPP systems for FMS, while J. P. Kruth has developed a CAPP system for non-linear process planning [15]. Two commonly used approaches to consider alternative routings are matrix method, and non-linear process planning. **Matrix method** involves formulation of alternative routings as a function of available machinery in a matrix format. It establishes network of possible routings while deferring the decision of choosing the path to a later stage. Choice of path may be changed after each operation. Thus, scheduling problem becomes to employ available routing alternatives so as to meet production specifications without any disruptions or bottlenecks and minimizing operation costs. In **non-linear process planning,** process planning is given a tree structure defining different routings, and scheduling is decided only after certain stage of advancement of process planning [11]. An effective interface between job shop control and planning is essential to realize the benefits of non-linear process planning. For this purpose, use of *fuzzy logic* has been suggested. *Petrinets* can be used for representation of alternative process plans. Petrinets are *directed graphs* and are powerful tools for modeling logical and temporal relationships between manufacturing operations. In process planning context, workpiece states are preconditions and manufacturing operations are events. Actually, a hierarchy of Petrinets is to be created to represent large number of objects relevant to shop floor [15].

- *Avoid Open-Loop Process Planning:* Most of the CAPP systems use open-loop process planning approach in which once process plans are prepared, no feedback is taken from the execution phase of the developed process plans. As a result of this, they do not have the capability of making on-line intelligent decisions and lack self-adaptive capability. It is impossible to generate optimal process plan in the first iteration itself due to complexities and interdependence of manufacturing activities [15]. To develop an adaptive or flexible, responsive, and on-line intelligent decision-making CAPP system, open-loop process planning approach should be avoided.

- *Avoid Micro-Level Process Planning:* Micro-level process planning approaches perform narrow range of planning function in isolation of other planning activities and hence have limited practical applicability. Most of the CAPP efforts are focussed on developing specific kind of planning system for specific range and type of parts. Integrated planning should be developed in such a manner so as cover all three types of planning, i.e. operation, process, and production planning, as well as other functions of manufacturing and industrial engineering in a logical and integrated fashion. More detailed description in this regard has been presented in the item 6.

4. Intelligent CAPP Assistant instead of Fully Automated CAPP: With the emergence of agile manufacturing and marketing environment, need is being felt of having a *good assistant* as an alternative to a system that performs in *fully automated* mode. If difficulties in the existing CAPP systems are viewed and summarized, they are largely related to the inability of the system to solve a global problem in a robust manner since most of the CAPP systems have decision logic, database, and other information relevant to the specific class of components. Therefore, efforts should be directed towards developing an **intelligent CAPP assistant** with the following characteristics:

♣ It should be user oriented rather than computer oriented,

♣ It should help the user to save time by organizing the choices in a systematic manner,

♣ The user should always be able to override any decision made by the system.

Also, many manufacturing parameters such as jobholding, tooling, fixturing, machine tool characteristics, etc. are sometimes beyond the scope of full automation. Consequently, most CAPP systems need to be more *assistants* and less *fully automatic* systems. User needs to be shown the processing choices available and be allowed to edit the generated process plan. A better CAPP system should adopt following approach [9]:

♣ Use accessibility information in identifying sequencing of machining operations.

♣ Build a sequencing editor to allow the engineer to edit the process plan efficiently.

♣ Communicate the subset of faces being machined in the same process as a feature.

5. Integration of CAD System with CAPP System: Effective communication is the cornerstone of human progress. Same technology, which makes our life easier, can lead to reverse effects due to ineffective communication. CAD, CAPP, and CAM are classified as *islands of automation*. True benefits of these technologies can be realized when they are integrated together. Integration of CAPP with CAD and CAM enhances design and/or production flexibility. Lack of generic integration of CAPP with CAD is a major 'bottleneck' in the implementation of CIM, one of the major enabling philosophy of agile manufacturing concept. Ideally, component description should be extracted from CAD database and its output should be used as input for process planning system. Generic integration of CAPP and CAD systems becomes more mandatory in AMS, in which product design is more frequently subjected to modifications and/or redesigning process. Integration should be generic in a sense that effective data communication is achievable irrespective of format of data being used. Realization of benefits of flexible, quick, and efficient CAD and CAPP systems is hampered by lack of integration between them.

CAPP systems should be integrated with the CAD system at the front end through neutral file format. A *neutral file format* is independent of specific system standards, and acts as an agent to connect dissimilar computer systems that normally cannot communicate with each other due to format incompatibility. It is an acceptable, commonly agreed, and consistent format within an organization or a group of organizations. Various companies are using neutral file formats to integrate different manufacturing systems. Different reported neutral file formats include Computer Graphics Interface (CGI), Computer Graphics Metafile (CGM), Cutter Location DATA (CLDATA), Industrial Robot DATA (IRDATA), Graphics Kernel System (GKS), Programmers HIerarchical Graphics System (PHIGS), Standard Exchange et de Transport (SET), Data Exchange Format (DXF), Standard for Transfer and Exchange of Product model data (STEP), Initial Graphics Exchange Specification (IGES), and Product Data Exchange Standard (now using STEP) (PDES) [16].

Of these, last four formats are supported by major CAD systems. **DXF** is suitable for exchanging *geometric information* only. While, **IGES** is an *engineering data or product definition data* exchange format, which is supported by major CAD systems. It is not an exhaustive description of data exchange standard and does not include the data related to CAPP systems, but it shows broad support of data technology [11]. More details about IGES are available in [17].

STEP is an ISO standard (ISO-10303) for exchange of *product data*. Product data include all the data relevant to the entire life cycle of a product from its concept to design, analysis, manufacturing, quality assurance, testing/verification, and support [16, 17]. STEP includes a series of International Standards with the aid of defining data across the full engineering and manufacturing cycle. It has ability to share data across various applications. Within STEP, there are many Application Protocols (AP) designed for different application purposes. STEP AP203 interface (also called Configuration Controlled Design) is widely supported in manufacturing sector. Major CAD packages like CADDS, CATIA, AutoCAD, IDEAS, Pro/Engineer, etc. have conversion program to convert CAD data into STEP AP 203 format. STEP AP203 is designed for mechanical assembly components and is appropriate for mechanical design and manufacturing industry. It is believed that IGES will eventually be replaced by STEP as a neutral file format for exchange of data between dissimilar CAD/CAPP/CAM systems since IGES covers design data only while STEP stores complete product life cycle information [16].

PDES [17] is another international standard for the exchange *product data*. PDES is an extension available to CAD systems, where additional and retrieved data are stored in a special physical PDES file. It does not interfere with the CAD system itself [11]. It is intended that PDES and STEP will be identical. PDES represents US interests in STEP efforts. To emphasize this intention, acronym PDES now stands for Product Data Exchange using STEP [17].

6. Integration of CAPP with other Manufacturing and Planning Activities: Process planning is a series of decisions that uniquely specify the processes. Once a decision is made in process planning, it becomes a constraint for the subsequent decisions. Though these constraints are artificial, and exist only because of series of decisions being made. Different sequences of decisions will result in different sets of constraints [11]. In majority of the systems, planning is viewed as hierarchically structured activity. Process plan is skeletal basis for the design evaluation, resources planning, scheduling, routing, and manufacturing planning and control activities in a CIM environment. Therefore, CAPP should be considered in an integrated mode instead of an isolation mode [15].

Though goals and scope of *three* types of manufacturing planning activities (i.e. ***operation planning, process planning, and production planning***), are distinct, but they are highly interrelated, and are built upon each other. Production planning uses technical information of process planning via ***routing***, which prescribes flow of a product in the plant and lists the sequence in which workstations are required to produce it. Conversely, process planning should use information from production planning regarding availability and loading of various production facilities of the shop floor in order to generate feasible and economic process plans. Without ensuring feasibility of process plans, it is meaningless to prepare production management and controls plans. Conversely, it is uneconomical to design a process plan which needs machines, tools, equipments not available in the company, or which are more vital for other manufacturing operations. ***Scheduling***, which mentions time schedules for processing of a product, is another basic function of production management that is also to be well matched with process planning.

It can be concluded that decisions of production planning should be supported by those made at process planning level, which in turn, should be supported by detailed specifications made at operation planning level based on the physical models of the considered processes. Production planning has also to be tied along with long-term strategic factors to arrive at economically, technically, and socially optimal decisions. Therefore, three main factors of manufacturing operation, i.e. *physics of the processes for operation planning, geometry of the product for process planning, and factory resources for production planning,* along with different production management and control functions, should be dealt in an integrated fashion. In a truly adaptive, effective, flexible, and integrated planning system, boundaries between these activities should vanish [15].

7. Development of CAPP Considering Global Spectrum of Manufacturing: Literature reveals that most of the research and development efforts in the area of CAPP have centered around the *conventional material removal or machining processes* considering either axisymmetric or prismatic parts. While some research and development of CAPP systems have been carried out for *conventional additive and forming processes*, but areas of *unconventional manufacturing processes,* whether material removal or machining processes, material additive processes, or formative processes, are still a neglected lot.

Ever-growing demand for better product performance has resulted in the development of numerous **difficult-to-machine materials** like carbides, ceramics, composites, variety of diamond and glass, refractory materials, semiconductors, quartz, etc. The very properties of these materials like very high strength and stiffness at elevated temperatures, extreme hardness and brittleness, high strength to weight ratio, very good oxidation and corrosion resistance due to chemical inertness, poor machinability, etc., make them commercially attractive. But these properties are also responsible for imposing great challenges in shaping and/or machining these materials. Conventional manufacturing and shaping processes result in high costs and moreover degradation of some useful properties. These factors along with the requirements like precision and miniaturization requirements (i.e. micro or nano-machining), complexity of shape and size, machining of inaccessible areas, surface integrity, etc. have led to the development of various *Advanced (or Non-traditional or Unconventional) Material Removal or Machining Processes*. Development of most of these processes took place during the period of 1945-90 and still continuing with the development of hybrid processes by combining two or more processes for specific type of requirements. These processes use new cutting tools, innovative energy sources, and unfamiliar material removal concepts.

Major obstacle to mass manufacturing is large manufacturing-lead-time. This varies according to type of industry. It is reported to be approximately 6-months for *garment* industry, 9-months for *shoe* industry, 3-4 years for *automobile* industry, 4-10 years for *aerospace* industry, 10-15 years for *submarines*, and may be up to 25 years for some defense applications. Sometimes it turns out to be so long that by the time product is ready to market it becomes outdated or even obsolete. In view of this, strong need for the development of automated manufacturing processes was felt to produce the prototype of a product very quickly and effectively so that different types of tests (i.e. form, fit, function, etc.) can be performed on it. Also, there has been a demand for manufacturing different types of products of *advanced materials or difficult-to-machine materials* without using special tooling. These demands and the demands from toys, jewelry, aerospace, medical (specially orthopedic) and defense applications, and other industries for manufacturing of intricate shapes, free-form surfaces, and assembled product itself, have brought about the second revolution in the physical and chemical manufacturing processes. As a result of this, many **unconventional material addition or accretion or deposition processes** or so called ***Rapid prototyping (RP) processes*** have been developed during last decade of 20th Century. As mentioned earlier, with their immense potential, various RP techniques constitute one of the most important enablers of agile manufacturing environment.

Some **unconventional forming methods** have also been developed that employ laser energy to deform materials inducing thermal stresses. Accounting all these developments, the overall spectrum of different manufacturing processes is being presented in Table 3.

With increasing developments and subsequent use of advanced materials, and emergence of customized products, use of advanced manufacturing processes is also becoming economical and unavoidable. Therefore, while developing a CAPP system, its knowledge base, decision logic, algorithms, and other related systems for agile manufacturing environment, the entire spectrum of manufacturing processes should be considered. Detailed methodology for developing a CAPP system for unconventional machining processes has been presented in the next Section.

8. Minimization of Reliance on the Features in Process Planning: Feature-driven CAPP is the dominant model of CAPP. Task of transformation of geometrical features of design into corresponding manufacturing features is called *feature recognition*. It involves two major tasks: identification of individual surfaces, and geometric reasoning to merge two or more surfaces to identify a feature [15]. There are some inherent limitations associated with feature-based process planning approach. These limitations are [9]:

[1] Despite more than a decade of efforts, there appears to be no consensus on universal feature nomenclature, and consequently feature identification approaches are generally limited to specific classes of components, and CAPP approaches in full generality are still missing.

[2] Attempts of incorporating features at the design stage itself have also not been successful since *design feature* and *manufacturing feature* may not be same. Designer's features are more concerned with functionality and mating surface requirements, while manufacturing features are focussed on the manufacturing operations. Also, use of feature-based design as input to a CAPP system requires part modeling in terms of features which may restrict integration of CAPP system with CAD and other systems. It may eventually require intervention of user having fairly good expertise in feature identification.

[3] A part cannot be designed with feature primitives only, it also needs intermediate surfaces to connect feature primitives [15]. Some simple geometrical shapes can be well captured by features, while more complex shapes and surfaces such as lofted or ruled surfaces, bi-cubics or bi-splines, free-form surfaces, etc. are not often possible to capture in terms of simple general geometric primitives.

[4] Feature-based CAPP works well in the manufacturing shop, but it is relatively difficult to integrate it with CAD, production planning, production management, and other functions of manufacturing and industrial engineering.

[5] Grouping of part surfaces into "features" before manufacturing process identification and sequencing restricts number of choices available for process sequencing. Hence, feature-driven CAPP systems may generate limited choice of processing sequences based on a set of pre-determined heuristics [9].

But, at the same time, features are still familiar shorthand to communicate with shop floor, and features remain an important issue in the area of the CAPP interface. Also, selection process of machine tool, cutting tool, and fixture is based on part features.

As an *alternative* to feature-based process planning, possible tool access/approach directions can be used to determine visibility/accessibility status of individual surface.

Table 3
Classification of manufacturing processes.

Type of Process	Traditional	Advanced/Non-traditional /Unconventional
Material Removal (Subtractive)	**Axisymmetric Parts** Turning, Taper turning, Facing, External threading. **Prismatic Parts** Shaping, Planning, Milling. **General** Drilling, Boring, Reaming, Tapping, Grinding, Broaching, Hobbing, etc.	**Mechanical Type** Ultra-Sonic Machining (USM), Abrasive Jet Machining (AJM), Water Jet Machining (WJM), Abrasive Water Jet Machining (AWJM), Abrasive Flow Machining (AFM), Magnetic Abrasive Finishing (MAF) **Chemical and Electro-Chemical Type** Chemical Machining (ChM), Electro-Chemical Machining (ECM) **Thermo-Electric Type** Electro-Discharge Machining (EDM), Electron Beam Machining (EBM), Ion Beam Machining (IBM), Plasma Arc Machining (PAM), Laser Beam Machining (LBM)
	Before **1945**	**(1945-1990)**
Material Addition or Accretion or Deposition (Additive)	Welding, Brazing, Soldering, etc.	**Liquid Based** Stereo Photo (Thermal) Lithography (SLA(T)), Solid Ground Curing (SGC), Solid Object Ultraviolet-laser Plotter (SOUP), Solid Creation System (SCS), etc. **Solid Based** Fused Deposition Modeling (FDM), Laminated Object Modeling (LOM), Multi Jet Modeling (MJM), Selective Adhesive and Hot Press (SAHP), etc. **Powder Based** Selective Laser Sintering (SLS), Ballistic Particle Manufacturing (BPM), Three Dimensional Printing (TDP), Multi Jet Solidification (MJS), Direct Shell Production Casting (DSPC), etc.
		After **1990**
Forming (Formative)	Drawing, Extrusion, Rolling, Forging, Bending, Deep-drawing, Sheet-metal forming, Casting and Molding	Laser Bending, 3D Laser Forming
		After **1990**

Then, different accessible surfaces from a particular tool access direction are combined into features according to their geometrical and topological definitions so as to identify associated manufacturing process, since there is a loose correspondence between geometric features and manufacturing processes. This approach is more flexible and takes into account other constraints that may not be addressed by the heuristics defined for feature-based approach. Although, features are used in such a system, but it can be called *featureless* in a sense that features are used only for detailing the manufacturing process and partially for process sequencing. Complete choice of process operations is based on accessibility of part surfaces by the cutting tool, as opposed to any pre-defined set of features. Other aspect of this approach is that features that emerge also incorporate visibility status of the surfaces in addition to their local geometry. Featureless approach, by providing a complete set of low-level choices, is an important issue in designing *intelligent CAPP assistants*. An operation planning system for axisymmetric components based on such approach has been successfully implemented [8].

5. CAPP FOR ADVANCED MACHINING PROCESSES

5.1. Objectives

Objective of developing a comprehensive, integrated, intelligent, interactive, and user-friendly CAPP system for non-traditional machining environment can be achieved by fulfilling the following specific sub-objectives:
- Make a decision about the necessity of using Advanced Machining Processes (AMPs) and to identify those surfaces and manufacturing features, which require use of AMPs. This identification process should use CAD model of the product as input.
- Rank suitable AMPs according to the different work material properties, shape and operational requirements of the application, and process economy.
- Optimize the process parameters of the highest ranked AMP for each manufacturing feature.
- Calculate the process performance parameters like MRR, Tool Wear Rate (TWR), machining time and machining cost.
- Perform process dependent auxiliary tasks like tool design, machine tool selection, etc.
- Generate Cutter Location Data (CLD) file and/or NC/CNC-code if required.

5.2. Solution methodology

A CAPP system, developed to fulfill the objectives mentioned above, should use suitable neutral file format of CAD model of the example part as input and generate CLD file or NC/CNC code as output (if applicable), thus integrating itself with CAD and CAM systems. Following methodology can be adopted for developing such a CAPP system:
- CAD model of the part, created in a suitable CAD software like IDEAS or Pro/Engineer, should be used as input to process selection module.
- Information required to identify those surfaces and features, which require use of advanced machining processes, can be extracted from CAD model of the part.
- Information related to material characteristics, operational requirements, and shape requirements can be given *interactively* through pull down menus, if it is difficult or not possible to extract it from the CAD model data.
- A comprehensive *selection* of advanced machining process can be performed at the following **three levels** using elimination strategy:

- ❖ **Level 1:** Elimination of some processes based on ***material characteristics*** like metallic or non-metallic, ductile or brittle, electrically conducting or not, , optically reflective or not, etc.
- ❖ **Level 2:** Further elimination can be carried out based on ***operational requirements*** and ***shape requirements*** of the part. Operational requirements are *quantitative* in nature and are specified in approximate ranges. They should be considered along with their relative weightage. Shape capabilities of AMPs are *qualitative* in nature and need to be quantified and computerized. ***Fuzzy logic*** can be an appropriate tool for this level of process selection because (1) Key elements in human thinking process are NOT numbers but linguistic terms, (2) Human beings are comfortable making imprecise verbal statements and these can be evaluated using fuzzy theory, (3) Fuzzy logic controllers do not use chain-inference mechanism, i.e. consequent of a decision rule is not applied to the antecedent of another rule, (4) Process capabilities of AMPs are mostly expressed in *ranges* (quantitative) and *linguistic terms* (qualitative), (5) In most of applications there is *flexibility* in requirements.
- ❖ **Level 3:** Short-listed processes can be finally ranked according their economy. It requires economic analysis and data collection regarding initial investment cost, maintenance cost, tooling cost, operating cost, cost of auxiliaries, etc.
- **Non-linear multi-objective optimization** of process parameters of the highest ranked AMP should be performed without any major approximations of the objective function(s) and/or constrain(s). ***Genetic Algorithms*** (GA) can be a suitable tool because (1) They work with population of points instead of a single point therefore a number of optimal or sub-optimal solutions can be obtained simultaneously, (2) Use coded strings instead of variables directly thus discretizing search space and a **discontinuous or discrete function** can be handled, (3) Do NOT require objective function to be **unimodal, continuous and/or linear,** (4) GA operators are *probabilistic* in nature while traditional optimization methods use deterministic algorithms, (5) Traditional methods **cannot** be *parallelized* as they use serial algorithm, (6) Constraints of any nature can be handled by GA, and (6) ***GA*** give a set of pareto-optimal solutions for multi-objective optimization.
- Development of an auxiliary module to perform the tasks of selection of machine tool, cutting tool, electrolyte in ECM, dielectric in EDM, etchant in CHM, or any such medium, horn design in USM, nozzle design in AJM, WJM, AWJM, tooling design in AFM, etc. This will involve collection and preparation of related databases,
- Development of modules to generate CLD file and/or NC/CNC-code,
- Experimental verification of optimized process variables and CNC/NC program.

6. CONCLUSIONS

Unfortunately, nowadays computer and AI experts are dominating field of CAPP. They are developing the framework of CAPP systems. This has resulted in proper attention not being paid to manufacturing and engineering content of process planning. Current CAPP approaches, and manufacturing planning and control methods are *information-intensive* and are unable to meet the challenges of agile, dynamic and vibrant environment. *Intelligent and integrated planning* is *knowledge-intensive*. It will be paradigm for CIM in the future and will be main driving force in the evolution of industry from present information-intensive stage to knowledge-intensive stage in future. Therefore, intelligent and integrated planning approach seems to be a feasible solution to meet the challenges of agility, and for effective planning and control of the factories of future. AI-based techniques are capable of capturing, representing, organizing, and utilizing knowledge on computers, and hence they will be the key technology for intelligent and integrated planning systems of the future. At last, technology alone is not sufficient to make an enterprise agile. Every company has to find right combination of work culture, business practices, and technology that are necessary to make it agile.

REFERENCES

[1] A. Gunasekaran, "Agile Manufacturing- Enablers and an Implementation Framework", *Int. J. of Prod. Research*, 36(5) (1998) 1223-1247.

[2] S. G. Deshmukh, "Agile Manufacturing: An Overview", *Proceedings of 6th SERC School on Advanced Manufacturing Technology*, held at I. I. T. Kanpur during March 15-26, 1999, (Editor: Prof. V. K. Jain) 1-7.

[3] Z. Y. Wang, K. P. Rajurkar and A. Kapoor, "Architecture for Agile Manufacturing and its Interface with Computer Integrated Manufacturing", *J. of Materials Processing Technology*, 61 (1996) 99-103.

[4] J. D. Hall and M. Usher, "Computer Integrated Manufacturing and Concurrent Engineering: Complementary Philosophies Enabling Agile Manufacturing" in *Simultaneous Engineering Methodologies*, Gordon and Breach Sc. Publishers, 1998.

[5] G. H. Lee, "Design of Components and Manufacturing Systems for Agile Manufacturing", *Int. J. of Prod. Research*, 36(4) (1998) 1023-1044.

[6] A. Kusiak and D. W. He, "Design for Agile Assembly: an Operational Perspective" *Int. J. of Prod. Research*, 35(1) (1997) 157-178.

[7] V. A. Cruz-Machado and J. J. Pamies-Teixeira, "Modeling for Agile Manufacturing", *Japan-U. S. Symposium on Flexible Automation* held in Japan, July 1994, 483-486.

[8] Neelesh K. Jain, "Visibility Check Based Operation Planning of Axisymmetric Components", *M. Tech. Thesis*, May 1995, Deptt. of Mech. Engg, I. I. T. Kanpur.

[9] Amitabha Mukerjee and Neelesh Kumar Jain, "Featureless Process Planning", *Proceedings of 17th AIMTDR Conference*, Jan. 1997, REC Warangal, India (Editors: R. L. Murthy, G. Srihari, and C. S. P. Rao) 125-130.

[10] Neelesh K. Jain and Vijay K. Jain, "Computer Aided Process Planning Approach for Advanced Machining Processes", *Industrial Engineering Applications and Practice: Users' Encyclopedia (on CD-ROM)*, (Editor-in-Chief: A. Mittal), 1999.

[11] G. Halevi and R. D. Weill, "*Principles of Process Planning: A Logical Approach*" Chapman and Hall Publication, London, 1995.

[12] H. P. Wang and J. K. Li, "*Computer-Aided Process Planning*", Elsevier Science Publishers B. V., Amsterdam, 1991.

[13] T. C. Chang and R. A. Wysk, "*An Introduction to Automated Process Planning Systems*", Prentice-Hall Inc. Englewood Cliffs, New Jersey, 1985.

[14] A. Kusiak, "*Intelligent Manufacturing Systems*", Prentice-Hall, Englewood Cliffs, New Jersey, 1985.

[15] P. K. Jain, "Computer Aided Process Planning" *Proceedings of 6th SERC School on Advanced Manufacturing Technology*, held at I. I. T. Kanpur, March 1999, (Editor: Prof. V. K. Jain) 175-191.

[16] H. Lau, and B. Jiang, "A Generic Integrated System from CAD to CAPP: a Neutral File-Cum-GT Approach", *Computer Integrated Manufacturing Systems*, Vol. 11, No. 1-2, 67-75.

[17] I. Zeid, "*CAD/CAM Theory and Practice*", McGraw-Hill, Inc. New York, 1991.

Aggregate capacity planning and production line design/redesign in agile manufacturing

Z.-S. Hua[a] and P. Banerjee[b]

[a]Department of Management Science, University of Science and Technology of China, Hefei, Anhui 230026, P. R. China

[b]Department of Mechanical Engineering, University of Illinois at Chicago, Chicago, IL 60607-7022, USA

In this chapter, roles of aggregate capacity planning in agile manufacturing are previewed. By reviewing the relationship among aggregate capacity planning, production line design/redesign and manufacturing agility, it is suggested that aggregate capacity planning, production line design/redesign, and disaggregate decision problems be integrated in flexible manufacturing environment to realize better customer responsiveness and business goals. Models of the suggested integration for electronic industry and automobile manufacturing industry are presented. Cases from electronic industry are described to show the integration process, in which a model of scenario-based line capacity expansion problem for printed wiring board (PWB) assembly systems at the aggregate level is developed. The model brings together the bill of components of product families and machine flexibility, thus manufacturing flexibility and customer response can be explicitly considered in the model.

1. INTRODUCTION

Because capacity decisions have a major influence on all issues of manufacturing competitiveness such as manufacturing cost, diversity of products, manufacturing lead time, introduction of new products, and market share, etc., choice about how and what type of capacity to install have a strong direct impact on agile manufacturing strategy (AM).

There are many approaches to accomplish agile manufacturing, which can be methodologically classified as enabling technologies and organization flexibility. Enabling technologies include concurrent engineering (CE), virtual manufacturing, information and communication infrastructure, etc. Organization flexibility is the ability to rapidly reconfigure a manufacturing system for efficient production of new products as they are introduced by the way of integrating manufacturing resources from inside and outside of the organization. Manufacturing resources refer to people, plants, parts, processes, and planning and control system [1]. This chapter tries to present an integration of parts, processes, and planning and control system in favor of agile manufacturing.

In the next section, hierarchy framework of production planning and control is previewed and discussed with the view of manufacturing competitiveness. Section3 discusses the relationship between capacity planning and customer responsiveness. In section 4, we summarize and develop models of integrating capacity planning and production line design with production planning for electronic industry and automobile manufacturing industry. A case from electronics industry is presented in section 5.

2. INTEGRATING CAPACITY PLANNING AND LINE DESIGN WITH PRODUCTION PLANNING

Capacity planning is usually motivated by reducing operation costs, meeting market demand fluctuation and/or pursuing better customer service. The major decisions in current capacity planning literature are machine replacement decisions and capacity expansion decisions. In machine replacement literature [2], "demand for capacity is generally expressed in equipment units" [3]. An example of this class is a firm such as United Parcel Service (UPS), which delivers packages of various sizes across the world, planning to replace and possibly expand part of its large vehicle fleet with demand measured in number of vehicles. In

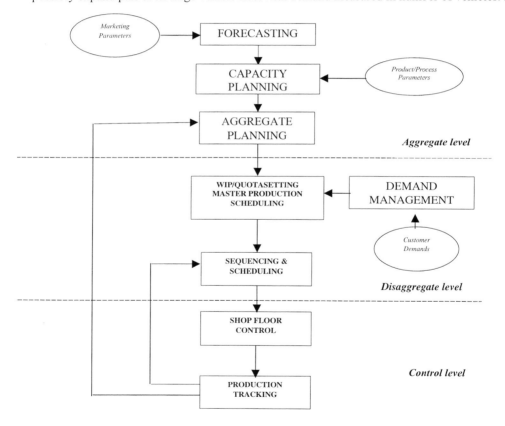

Figure 1. A production planning and control hierarchy

the capacity expansion literature, demand for capacity is usually expressed in output units [3-8]. An example of this scenario is a fertilizer plant that could be built in one of several sizes, with demand measured in thousands of pounds of fertilizer. In these literatures, capacity is treated as discrete or continuous variable, but structure of products (bill of component) is not addressed.

From the point of view of a production planning and control (PPC) hierarchy [9], capacity/facility planning based on demand forecasted works as the main input of all other production planning issues, e.g., aggregate planning, demand management, sequencing or scheduling and shop floor control as they are shown in Figure1. In Figure 1, each rectangular box represents a separate decision problem. The ovals represent inputs to decision problems that are generated outside the planning hierarchy.

The PPC hierarchy of Figure 1 is divided into three basic levels, corresponding to long-term (aggregate level), intermediate-term (disaggregate level), and short-term (control level) planning. The basic function of the aggregate planning is to establish a production environment capable of meeting a firm's overall goals. This begins with a forecasting of future demand based on marketing information. Capacity planning uses these demand forecasts, along with descriptions of process requirements for making the various products, to generate capacity plan. Aggregate planning makes rough predictions about future production mix and volume according to demand forecast and capacity plan.

Taking plans from the aggregate level, disaggregate planning generates a general plan of action that will help the firm prepare for upcoming production. The WIP/quota setting problem decides the quantity of work in process or periodic production quotas according to aggregate plan and capacity's characteristics (e.g., setup time, changeover time among batches, production process type etc.). Based on orders on hand and demand forecasts, the planner prepares a budget master production schedule. The shop floor control problem controls the real-time flow of material through the plant in accordance with the work schedule, while the production tracking measures actual progress against the schedule.

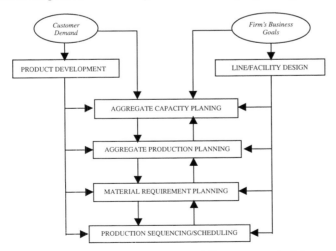

Figure 2 Illustration of integration aggregate and disaggregate decision problems

As shown in Figure 1, production and process parameters are usually treated as inputs outside of the planning hierarchy. In the face of shorter product lifecycles, higher product variety, increasingly unpredictable demand, and shorter delivery times, manufacturing facilities dedicated to a single product line cannot be effective any longer. Investment efficiency now requires that manufacturing facilities be able to shift quickly from one product line to another without major retooling, resource reconfiguration, or replacement of equipment. Investment efficiency also requires that manufacturing facilities be able to simultaneously make several products mixes and volumes can be more easily accommodated.

All these requirements can be interpreted two kinds of suggestions: increase manufacturing flexibility, integrate capacity planning and line design with production planning. Many researches have suggested that a firm must link its technology choice to its total manufacturing strategy and business unit's goals to gain competitive advantage [10], and that flexible manufacturing system is desirable to response to a wide variety of future demand [11-12]. Manufacturing flexibility does not only depend on machine flexibility, but is also constrained by decisions made at system design stage and planning decisions made during preproduction setup [13]. Thus to accomplish manufacturing agility, capacity planning (decide capacity in equipment units or output units) and production line design (decide production line structure relative to products' structure and processes' requirements) should be integrated with production planning. Specifications of products and process are desirable to be considered in capacity planning. In shorter, the decision of product and process parameters should included in the PPC hierarchy as a decision module as shown in Figure 2.

Figure 2 is different from Figure 1 mainly on two aspects: 1) product and process parameters are not exogenous; 2) the decision method that first decide aggregate planning and then take it as the input of disaggregate planning is not necessary or sufficient for agile manufacturing. Rather, aggregate and disaggregate planning decisions are tightly correlated and should be integrated to attain better customer responsiveness and business goals.

Integrating aggregate capacity planning with disaggregate production planning is usually a difficult problem. Yet for flexible manufacturing systems (FMS), there still are some researches on integrating these two levels of planning problem, for example, [14-16]. While their results are interesting and insightful, machine flexibility and the configuration of production capacity are not incorporated in their models. Benjaafar and Gupta [17] developed a model to explicitly express the number of facilities, the number of products assigned to each facility and their corresponding capacities. Eppen and Martin and Shrage [4] proposed a scenario based approach for capacity planning, which integrated the overall level of capacity, the type of facility (e.g., the level of flexibility) and production assignment. Although product parameters are not explicitly involved in their models, their researches can be viewed as initial attempts of integration of aggregate and disaggregate planning.

3. AGGREGATE CAPACITY PLANNING AND MANUFACTURING FLEXIBILITY

Several taxonomies for classifying manufacturing flexibility are available in the current literature. In this chapter, we use the framework proposed by Stecke and Raman [13] with some modifications especially with regard to aggregate flexibilities. Similar to [13], we identify three levels of manufacturing flexibility: (i) *basic flexibilities* which include machine flexibility, material handling flexibility and operation flexibility; (ii) *system flexibilities* which include volume flexibility, routing flexibility and process flexibility; (iii) *aggregate flexibilities* which include production flexibility and market flexibility. These flexibilities and their bases and constraints are briefly listed in Table 1. From Table 1, it can be observed that aggregate flexibilities are the most visible measures of overall system flexibility because of their impacts on parameters that are immediately measurable, such as machine utilization, range of products manufactured, customer order turnaround time, and new product introduction frequency. Yet it can also be observed that aggregate flexibilities come mainly from system flexibilities and basic flexibilities or their integration. As Stecke and Raman [13] pointed out, "Decisions made at the design stage provide an upper bound of the levels of

Table 1
Flexibilities and their bases and constraints

Dimension of flexibility	Measure of flexibility	Bases and constraints of the flexibility
1. Machine flexibility	A machine can process various operations	Equipment selection decision
2. Material handling flexibility	Different part types can be transported and positioned at the various machine tools in a system	The way by which machines to be connected
3. Operation flexibility	Alternative operation sequences can be easily used for processing a product type	Product development and production system's layout decisions
4. Volume flexibility	System can be operated profitable at different volumes of the existing product types	Quotas setting decisions and machine flexibility
5. Expansion flexibility	System can be built and expanded incrementally with a reduction in the marginal capital investment	Modular capacity (i.e., long-term economies of scale)
6. Routing flexibility	A product can effectively follow the alternative paths through the system for a given process plan	Sequencing or scheduling, machine flexibility and capacity utilization.
7. Process flexibility	A system can produce many product types without incurring any setup	Machine flexibility, production system's layout decisions and routing flexibility
8. Production flexibility	A system can produce many product types without major investment in capital equipment	Layout of the machine lines, expansion flexibility.
9. Market flexibility	A system can efficiently adapt to changing market condition (changes in customer orders and needs)	Operation flexibility, material handling flexibility and production flexibility

flexibilities achievable, the realized levels depend also substantially on planning decisions". Unlike traditional capacity expansion problems concern only expansion sizes, times and locations (and/or capacity types) [18], capacity planning for FMS should concern much more.

From the viewpoint of a firm, machine flexibility is determined when specific machines are purchased from machine vendors and selected from available machine types. Equipment selection decision is usually made after capacity expansion decision (see, for example[19]). From Table 1, we can see that machine flexibility is an indispensable base for aggregate flexibilities, thus it is necessary to be considered in capacity planning. Material handling flexibility depends on the way by which machines to be connected. There are two typical machine connection methods in FMS, i.e. automatic guided vehicle (AGV) and other automated material handling devices such as pallets or automatic conveyor [20]. The operation flexibility is necessary when new product is introduced. This flexibility depends not only on material handling flexibility and machine flexibility, but also on capacity utilization and production systems layout. Volume flexibility is determined by machine's setup time, which comes from machine flexibility, quotas setting and product mix decision which are parts content of disaggregate production planning. The requirement that capacity can be expanded modularly is a kind of long-term economies of scale [9]. It is a profitable effect, which should be pursued in capacity planning. Routing flexibility is usually determined in sequencing or scheduling, and is determined by machine flexibility and capacity utilization, which are also tightly related to capacity planning. Process flexibility depends on machine flexibility and routing flexibility. At the plant level, it is also affected by the structure of production lines. As Jordan etc. [21] pointed out, "limited flexibility (i.e., each system builds only a few products) configured in the right way (i.e., configured to chain products and production systems or plants together to the greatest extent possible), yields most of total flexibility (i.e., each system builds all products)". Production flexibility is determined by layout of the machine lines, expansion flexibility. Market flexibility is the main objective of agile manufacturing, which is an integration of operation flexibility, material handling flexibility and production flexibility.

In the era of information technology and exquisite competition, integration of competitive advantages outside and inside a firm is necessary and possible. The integration is a wide-ranging and complex research topic, which may include virtual manufacturing, relative organization changes and many others. This integration works as part a firm's manufacturing strategy, and should be reflected into capacity decision.

To sum up: 1) Manufacturing flexibility is important base of manufacturing agility, which should be considered in capacity planning. 2) For FMS capacity planning problems, three objectives are necessary of pursuing, i.e., modular capacity, products and production lines chaining, and subcontracting or outsourcing. 3) We suggested that bill of components of products and parameters of production process should be explicitly considered in FMS capacity planning.

4. AGGREGATE CAPACITY PLANNING AND CUSTOMER RESPONSIVENESS

Customer responsiveness refers to shorter lead time at all stages of manufacturing, increasing diversity of products, lowering manufacturing price, increasing market share, rapid introduction of new products. We attribute diversity of products, manufacturing price and market share mainly to manufacturing flexibilities. This section discusses the relationship between aggregate capacity planning and lead-time (i.e., capacity expansion lead time and production cycle time), and the relationship between capacity planning and the introduction of new products.

4.1. Capacity planning and capacity expansion lead time

Capacity expansion lead-time is the time elapsed from when capacity is ordered until it can be utilized. Influences of this lead-time are three folds: It may have influence on capacity timing which will lead to capacity lead demand, meet demand or lag demand; It may have influence on cycle time; It may have influence on capacity expansion decision itself.

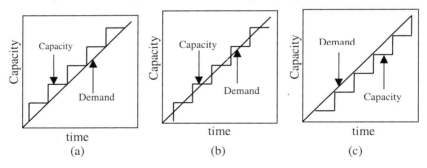

Figure 3. Illustration of capacity timing

Because of the uncertainty of capacity expansion lead-time, a firm's practical capacity based on capacity planning may lead demand, meet demand or lag demand as shown in Figure 3. As Hopp and Spearman [9] summarized, when capacity lead demand, it may create "capacity cushion" (i.e., planned capacity underutilization), accommodate surges in demand

and attract new business (i.e., take market share from competitors) by paying extra cost of capacity expansion and maintenance. When capacity meet demand, there is a likelihood of falling short roughly equal to likelihood of having too much capacity and it could require overtime, extra shifts, scrambling, etc. to make up difference. When capacity lag demand, it

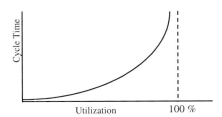

Figure 4. Illustration of the relation of capacity utilization and cycle time

may lead to negative capacity cushion, assure high utilization and higher return on investment, it will lead the firm to be conservative with regard to forecast, *but*, it can also lead to erosion of market share.

Capacity expansion decisions related to flow time are (a) find best configuration given fixed budget and (b) minimize cost to attain performance specs such as decrease mean time to repair (MTTR), reduce setups, smooth demand, reduce operator-related variability, reduce move batches and process batches (setups), improve synchronization. From the theory of *Factory Physics* [9], capacity decisions impact on cycle time by the way of capacity utilization as it is shown in Figure 4. Figure 4 indicates that congestion (cycle time) increases with utilization. So cycle time and thus customer responsiveness are tied to capacity decisions.

Capacity expansion lead-time may have influence on capacity expansion decisions itself [22]. When there is zero capacity expansion lead time and demand must always be met, the timing of the expansion is no longer an independent decision variable. So it is often possible to formulate an equivalent deterministic problem that recursively sizes each expansion, as a function of the installed capacity at the time that it is first fully utilized. When the timing of an expansion is an independent decision variable and there is a strictly positive lead-time, it is usually unable to identify an equivalent deterministic problem [22].

Capacity expansion lead-time should be considered in the decision model when demand is random and correlated over planning periods, while it can be set to zero if demand is deterministic or random but independent over planning period.

4.2. Cycle time and capacity planning

Cycle time reduction is an important issue to improve customer responsiveness. To reduce cycle time, it is essential to identify the components of cycle time. As Hopp and Spearman [9] did, the average cycle time for a station can be expressed as follows:

Cycle time = queue time + process time + wait for batch time
+ move time + (wait to match time)   (1)

Queue time is the time jobs spend waiting for processing or move resources, it is aggravated by variability and high utilization as it is seen in the previous subsection. Process time is the time jobs spend actually being worked on. Waiting for batch time is the time jobs spend waiting to form a process batch or a move batch, it is proportional to batch size. Move time is the time jobs spend being moved between workstations. Wait to match time occurs at assembly stations where components wait for their mates in order to allow the assembly operation to occur. In equation (1), wait to match time is placed in parentheses because this term is relevant only to assembly stations.

Cycle time of a line (or a routing) is made up of the sum of cycle times at the stations. It is common that the sum of queue time, wait for batch time and wait to match time comprise 90 percent or more of total manufacturing cycle time [23]. When all products and production process parameters can be explicitly considered in capacity planning problem, it will obviously improve *wait to match time*. Because unlike traditional capacity planning treats production and process parameters as exogenous, this capacity planning will take into considerations one machines capacity synergy with other machines and with bill of components of products.

4.3 Capacity planning and frequency of new product introduction

The frequency of new product introduction is certainly related to a firm's ability of product development. *What we try to expound here is that the frequency of new product introduction is also related to capacity planning.* Capacity planning can help new product introduction on three aspects. Firstly, for flexible manufacturing system, if capacity is expressed in the part types it can process rather than the products it can make (product is consisted of certain part types), then capacity decisions may accommodate more products, e.g., three part types can composite numerous product types theoretically by varying quantities of each part type. Secondly, new product production can be facilitated by integrating manufacturing resources inside and outside of the firm. Thirdly, when technology breakthrough or advance can be taken into account in the capacity planning model, for example, [10][24], then facility for producing new product is considered impliedly.

5. AGGREGATE CAPACITY PLANNING IN DIFFERENT INDUSTRIES

Based on our understanding on capacity planning for agile manufacturing, models of aggregate capacity planning for electronic industry and automobile industry are presented in this section.

5.1 Aggregate capacity planning for electronics industry

Electronic systems play an increasingly important role in today's consumer and industrial products. The most common production process for electronic systems is assembling components on a printed wiring board (PWB). Interest in PWB assembly is mainly driven by the desire to optimize the utilization rate of expensive component placement machines [25]. The surface mount technology (SMT) placement machine is essential equipment in a surface mount PWB assembly line. In general, an SMT placement machine is composed of a body base, a board handling system, component feeders, placement heads, placement tools, and a vision system. It places components (parts) reliably and accurately to meet quality requirements. The PWB assembly system consists of inserting or mounting electrical components, such as resistors, transistors and capacitors, into prespecified positions on a PWB. Each machine can insert certain part types on a PWB, some machines form a machine line which can produce some product families, where families are defined as sets of PWB types (items) which share similar operational characteristics and can be easily determined by exploiting special clustering techniques [26]. A firm usually has certain machine lines. The demand forecast at the family level is suitable and easy to operate for aggregate planning. A machine line makes insertions on PWBs with the only restriction that all the insertions for a product family be done on the same machine line. The placement operations for a given product family are distributed across the available machines in the line. Figure 5 shows an

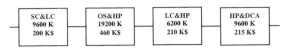

Figure 5. Production line illustration

illustration of such a machine line, with four assembly processes. In this example, each machine is capable of assembling two different types of parts from the list of five categories: small chips (SC), large chips (LC), odd-shaped (OS) parts such as connectors and pins, high-precision (HP) placements such as fine- pitch integrated circuits, and direct chip attach (DCA) parts that require an additional flux dispensing operation.

In Figure 5, each text box represents an SMT placement machine. The first machine can insert part types SC and LC. Its designed capacity is 96,000 k placements per quarter, and its purchasing cost is 200 k dollars.

5.1.1. A scenario based model

For capacity expansion problem, there are two decision stages. The first is to determine which machine types to be installed at which assembly lines, the second is to decide how to meet market demand. Note that in the above process, determining a good (or optimal) line capacity configuration is complicated by the fact that the line capacity decision affects future production planning decisions. The production planning decisions are further complicated by the fact that the demands of product families are random variables whose value are revealed after the line capacity configuration decision. A scenario-based model is formulated below; the notation used here is given in Table 2.

(P0)

$$\min \sum_{t,m,l} C_a(t) Y(t,m,l) + \sum_{t,n,j} p(t,n) * (\sum_i C_{sc}(t,i) S(i,j)) * X_{sc}(t,n,j)) + \sum_{t,n,i,j,m,l} p(t,n) C(t,n,m,l) X(t,n,i,j,m,l) \quad (2)$$

$$s.t. \sum_{i,j} X(t,n,i,j,m,l) \leq (N(t,m,l) + Y(t,m,l)) * Q(m) \quad n=1,...,N, m=1,...,M, l=1,...,L, t=1,...,T \quad (3)$$

$$\sum_j X(t,n,i,j,m,l) \leq (N(t,m,l) + Y(t,m,l)) Q(m) F(i,m) \quad n=1,...,N, j=1,...,J, m=1,...,M, l=1,...,L, t=1,...,T \quad (4)$$

$$\sum_m X(t,n,i,j,m,l) * S(k,j) = \sum_m X(t,n,k,j,m,l) * S(i,j)$$

$$(S(i,j) \neq 0)), \quad n=1,...,N, i,k=1,...,I (i \neq k), j=1,...,J, l=1,...,L, t=1,...,T \quad (5)$$

$$\sum_{i,m,l} X(t,n,i,j,m,l) + X_{sc}(t,n,j) = d(t,n,j) \quad t=1,...,T, n=1,...,N, j=1,...,J \quad (6)$$

$$\sum_{t,m,l} C_p(t,m) * Y(t,m,l) \leq B \quad (7)$$

$$X(t,n,i,j,m,l) \geq 0, X_{sc}(t,n,j) \geq 0, \text{int } Y(t,m,l) \quad (8)$$

In the objective function (2), the first term is cost of installing machines, the second term is expected subcontracting cost, and the third is expected production cost. Formulae (3) and (4) are capacity constraints which ensure that a production assignment does not exceed a machine's total capacity and the capacity of inserting part type i respectively. Formula (5) is machine line constraint that requires all insertions for a product family be completed on one machine line. Formula (6) is demand constraint that requires all demand be met by production or subcontracting. Constraint (7) is the capital investment budget limitation, and formula (8) is nonnegative and integer constraint.

Table 2
Notation

Indices	Capacity Parameters
t time periods, $t=1,2,\ldots,T$	$F(i,m)$ capability of machine m inserting part type i, $F(i,m)=1$ if machine type m can insert part type i; 0, otherwise.
n demand scenarios, $n=1,2,\ldots,N$	
i part types, $i=1,2,\ldots,I$	
j product families, $j=1,2,\ldots,J$	$Q(m)$ number of insertions machine type m can make in one period
m machine types, $m=1,2,\ldots,M$	
l machine lines, $l=1,2,\ldots,L$	$N(t,m,l)$ number of machine type m in machine line l in initial capacity layout in period t
Demand Parameters	
$d(t,n,j)$ demand for product family j under scenario n in period t	
$S(i,j)$ number of part type i in each of product family j	B capital limitation on new equipment investment
	Cost Parameters
$p(t,n)$ the probability that scenario n occurs in period t	$C_a(t)$ cost of installing a machine to a machine line in period t
Decision Variables	$C_p(t,m)$ the sum of costs of purchasing and setting up a machine type m in period t
$X(t,n,i,j,m,l)$ number of part type i for product family j produced on machine m in machine line l under scenario n in period t	
	$C_{sc}(t,i)$ cost of subcontracting part type i in period t
$X_{sc}(t,n,j)$ amount of subcontracting product family j under scenario n in period t	$C(t,n,m,l)$ average unit insertion cost of machine m under scenario n in period t, line l
$Y(t,m,l)$ number of machine type m added to machine line l in period t	

Some explanation about the motivation and rationale of model (P0) is as follows. The total production costs include two parts: variable production cost (depends on production quantity) and fixed production costs (independent of production quantity, mainly depend on machine depreciation and setup costs). Thus for a specific machine m in a machine line l, its average unit production cost is a nonlinear function of its production assignment $\sum_{i,j} X(t,n,i,j,m,l)$:

$$C(t,n,m,l) = \begin{cases} C_{\text{var}}(t,n,m,l) + \dfrac{C_p(t,m)*(N(t,m,l)+Y(t,m,l))}{\rho(m)*\sum_{i,j} X(t,n,i,j,m,l)} & \text{if } \sum_{i,j} X(t,n,i,j,m,l) > 0 \\ 0 & \text{if } \sum_{i,j} X(t,n,i,j,m,l) = 0 \end{cases} \quad (9)$$

In formula (9), setup cost of a machine is incorporated into $C_p(m)$, and $C_{\text{var}}(m)$ is unit variable production cost, $\rho(m)$ is duration of service of machine type m expressed in number of time periods ($\rho(m) > T$).

Because demand is exogenous, thus from formula (6),

$$X_{sc}(t,n,j) = \sum_{i} S(i,j) \times d(t,n,j) - \sum_{i,m,l} X(t,n,i,j,m,l) \quad n=1,\ldots,N \quad (10)$$

Objective function of formula (2) can be rewritten as:

$$Z = \sum_{t,m,l} C_a(t)Y(t,m,l) + \sum_{t,n,i,j,m,l} p(t,n)*(C(t,n,m,l) - C_{sc}(t,i))*X(t,n,i,j,m,l)) \quad (11)$$

The requirement that all insertions of one product family be completed on one machine line (constraint (5)) can be roughly explained as: The parts are to be assembled on a board (product family) in the right proportion with respect to each other. For any $j(j \in \{1,2,...,J\})$, when parameters $S(i,j)$ $(i=1,...,I)$ $(S(i,j) \neq 0)$ do not have a common divisor that is greater than one, formula (5) is the exact expression of the requirement. Otherwise, there is some difference between the requirement and formula (5). In this case, the maximum difference caused by formula (5) in each assembly line is $(\alpha-1)\sum_i S(i,j)$ placements (α is the maximum common divisor of parameters $S(i,j)$ $(i=1,...,I)$ $(S(i,j) \neq 0)$), which is minor compared with the number of placements assignment to the line. Also, the production decision variable $X(t,n,i,j,m,l)$ is practically an integer variable, but since the capacity of a placement machine represented as placements it can make in one time period is usually very large, it is reasonable to treat $X(t,n,i,j,m,l)$ as continuous variable.

5.1.2. Solution approach

Problem (P0) is a large-scale nonlinear mixed integer-programming problem. The basic idea of solving it is to limit its searching space. An approximate solution procedure includes three main steps:

Step 1. Identify the rough set of machines that needs to be added for every machine line under demand scenarios;
Step 2. Get the rough sets of adding machines for problem (P0);
Step 3. Solve the capacity expansion problem, (P1), which is the same as problem (P0) but all capacity expansion decision variables, $Y(t,m,l)$, except those determined in Step 2 are set to be zeroed.

Let $\{A(t,m,l) | \text{machine type } m \text{ is considered to be added to machine line } l \text{ in period } t, \forall t,m,l\}$ be the rough addition set for problem (P0). Then the corresponding problem (P1) is as follows:

(P1)
$$\min Z = \sum_{t,m,l} C_a(t)Y(t,m,l) + \sum_{t,n,i,j,m,l} p(t,n)*(C(t,n,m,l) - C_{sc}(t,i))*X(t,n,i,j,m,l))$$

s.t. (3), (4), (5), (6),(7) and
$$X(t,n,i,j,m,l) \geq 0, X_{sc}(t,n,j) \geq 0, \text{int } Y(t,m,l) \in \{A(t,m,l)\} \quad (12)$$

5.1.2.1 Identifying rough addition set and rough removal set under a demand scenario

Let $\{A'(t,n,m,l)|$ number of machines of type m added to machine line l under demand scenario n in period t, $\forall t,n,m,l\}$ be the rough addition set under demand scenario n in period t. Under a certain demand scenario, problem (P0) is deterministic. Hua and Banerjee [27] described an approach of determining rough addition set when demand is deterministic, which is summarized in Figure 6.

Problem SPACU formed in Step 2 of Figure 6 is a recursive programming problem [28], which has following properties.

Proposition 1: Solution to problem SPACU formed in Step 2 of Figure 6 is convergent.
Proof. See Appendix A.

Proposition 2: $|\{A'(t,n,m,l)\}| \leq \left\lceil \dfrac{I}{\min_{m}\sum_{i}F(m,i)} - 1 \right\rceil$ for any t, n and l. $|\bullet|$ is the size of a set, $\lceil \bullet \rceil$ is a ceiling integer function.

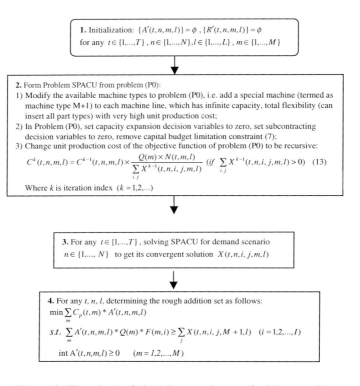

Figure 6. Flowchart of obtaining rough sets of adding machine

Proof. The motivations of adding machines for SPACU is to make fuller utilization of current equipment (current equipment may be not fully utilized because of insertion constraint (5)).

Since demand is assumed deterministic and no budget limitation is involved in model SPACU, the worst case is that one machine type in the current capacity layout needs to be matched to improve its utilization by adding other machine types. Because one machine can produce at least $\min_{m}\sum_{i}F(m,i)$ part types and all product families are made of I pare types; thus at most $I - \min_{m}\sum_{i}F(m,i)$ part types need to be produced on added machine types. So at most $(I - \min_{m}\sum_{i}F(m,i))/\min_{m}\sum_{i}F(m,i)$ machine types are possibly to be added to the line, i.e.

$$|\{A'(t,n,m,l)\}| = \left\lceil \frac{I - \min_m \sum_i F(m,i)}{\min_m \sum_i F(m,i)} \right\rceil = \left\lceil \frac{I}{\min_m \sum_i F(m,i)} - 1 \right\rceil \quad (14)$$

When I=5, $\min_m \sum_i F(m,i) = 2$, Proposition 2 indicates that at most two machine types are possibly be added to a machine line.

5.1.2.2. The rough addition set for problem (P0)

The above subsection provides a way to determine the rough addition set for each market demand scenario in each period. To explain the approach of determining rough addition set for random demand (multiple demand scenarios) problem, new definitions and some relative properties are elaborated next.

Definition 1: For any time period, a rough addition set is said to be *complete* for problem (P0) if and only if the optimal objective value of (P0) equals the optimal objective value of corresponding problem (P1).

Because rough addition set for deterministic problem by solving SPACU is not complete [27], hence we introduce:

Definition 2: For any time period, a rough addition set is said to be *ε-complete* for problem (P0) if and only if the relative deflection of the optimal objective value of the corresponding problem (P1) compared with that of problem (P0) is within ε (0≤ε≤1).

An intuition way of determining $\{A(t,m,l)\}$ is as follows:

$$\{A(t,m,l)\} = \{A'(t,n,m,l)| \text{ if } A'(t,n,m,l) > 0 \text{ for any } n\} \quad (15)$$

Formula (15) is actually a way of getting the maximum of rough addition sets under all market demand scenarios (This is termed "Approach 1" of determining rough addition set for P0, and it will be abbreviated as "Approach 1" afterward). This approach is obviously better than the approach which selects a rough addition set under a demand scenario as the rough addition set for problem P0 by making comparison among the expected costs of all possible selections. The problem of this approach is, for any time period t, if $\{A'(t,n,m,l)\}$ is complete under demand scenario n (n=1,…,N), whether $\{A(t,m,l)\}$ is complete for problem (P0) or not? To answer this question, we should go back to looking into continuous demand function.

Lemma 1: Given cumulative probability distribution function $D(x, j)$ (The probability that demand of product family j is less than or equal to x, $1 \le j \le J$, $x \ge 0$) of market demand, and $d(x, j)$ is its probability density function. Function $d(x, j)$ ($1 \le j \le J$) is sampled with small-enough step size and each sampling can be treated as a demand scenario. If a complete rough addition set for each demand scenario can be obtained, then the rough addition set determined by getting the maximum of rough addition sets under all demand scenarios is complete for problem (P0).

Lemma 1 is true because capacity expansion decision of (P0) depends on the random characteristics of demand density function. Since the objective function of problem (P0) is continuous, thus if demand scenarios can provide sufficient information about the density function of market demand, i.e. $d(t, x, j)$ ($1 \le j \le J$) is sampled with small-enough step size, the maximum of rough addition sets under all demand scenarios will reflects all potential capacity expansion requirement.

Given demand scenarios $d(t,n,j)$ $(n=1,...,N, j=1,...,J)$, the algorithm of determining rough addition set for problem (P0) can then be conceived based on lemma 1 as follows (This is termed "Approach 2" of determining rough addition set for P0, and it will be abbreviated as "Approach 2" afterward):

Algorithm 1
Step 1. Order $d(t,nn,j)$ from the most pessimism case to the most optimism case, and let
$\{A(t,m,l)\} = \phi$, $NN = N$;
Step 2. Solve SPACU under all NN demand scenarios and update $\{A(t,m,l)\}$:
$$\{\tilde{A}(t,m,l)\} = \{A'(t,nn,m,l)| \text{ if } A'(t,nn,m,l) > 0 \text{ for any } nn\} \quad (16)$$

$$\{A(t,m,l)\} = \{A(t,m,l)\} \cup \{\tilde{A}(t,m,l)\} \quad (17)$$
Step 3. Generate $NN-1$ new artificial demand scenarios as follows:
$$d'(t,nn,j) = \frac{d(t,nn,j) + d(t,nn+1,j)}{2}, nn = 1,2,...,NN-1 \quad (18)$$
Step 4. Solve SPACU under all newly generated $NN-1$ artificial demand scenarios:
$$\{\tilde{A}(t,m,l)\} = \{A'(t,nn,m,l)| \text{ if } A'(t,nn,m,l) > 0 \text{ for all } nn = 1,2,...,NN\text{-}1\} \quad (19)$$
Step 5. If $\{\tilde{A}(t,m,l)\} \subset \{A(t,m,l)\}$, STOP; otherwise, update $\{A(t,m,l)\}$:
$$\{A(t,m,l)\} = \{A(t,m,l)\} \cup \{\tilde{A}(t,m,l)\} \quad (20)$$
Step 6. Insert newly generated artificial demand scenario $d'(t,nn,j)$ $(nn = 1,2,...,NN-1)$ between demand scenario $d(t,nn,j)$ and $d(t,nn+1,j)$ to obtain a new ordered demand scenario list which consists of $(2NN-1)$ demand scenarios;
Step 7. Let $NN = 2NN-1$, go to Step 3.

Proposition 3. Algorithm 1 is convergent.
Proof. See Appendix B.

Proposition 4. For each time period, if model SPACU can provide complete rough addition set for each demand scenario, then algorithm 1 will provide complete rough addition set for the corresponding problem (P0).

Since algorithm 1 is a constructive algorithm based on lemma 1, proposition 3 is obviously true under definition 1. As mentioned above, in each time period, the rough addition set for each demand scenario that determined by model SPACU is *ε-complete* (ε is 0.05 on the worst case based on numerical experiments [27]), so algorithm 1 can provide *0.05-complete* rough addition set for problem (P0) in each time period.

5.1.2.3. Capacity expansion decisions

For single period problem, problem (P1) for practical PWB manufacturing firm has about 10 integer capacity expansion variables and thus can be directly solved by using standard mixed integer programming code on PC computer or workstation. For multiple-period problem, approximate searching strategy is still necessary to be developed. But according to the result of previous subsections, searching space can be significantly reduced.

5.2. Aggregate capacity planning for automobile industry

Automobile manufacturing is a typical modern industry that can apply the suggested integration of this chapter. Eppen and Martin and Schrage [4] described a model for General Motors to aid in making decisions for four of their auto lines. Although their paper was written mainly to making tradeoffs between risk and profit in the capacity planning decision, its basic idea is consistent with this chapter.

Some assumptions are made for their model, which can be summarized as follows:

A1: Demand is realized before the production decision;
A2: No inventory is carried from period to period;
A3: There are some production sites configured in one of some possible configurations which are capable of producing some products;
A3: The scenario probabilities are independent from period to period;
A4: Specified percentages of unsatisfied demand were diverted:
  1) to other products considered in the model, 2) to other products not in the model,
  3) to other products not produced by the firm.

A scenario-based model is formulated below; the notation used here is given in Table 3.

$$\max \sum r_{ikhmt} p_{mt} x_{ikhmt} + \sum r_s \tau_{is} p_{mt} z_{imt} - \sum F_{kht} w_{kht} - \sum C_{kht} y_{kht} \quad (21)$$

$$\sum_{k=1}^{K} \sum_{h=1}^{H_k} x_{ikhmt} + z_{imt} = D_{imt} + \sum_{j=1}^{N} \tau_{ji} z_{jmt} \quad (22)$$

$$\sum_{i=1}^{N} a_{ikh} x_{ikhmt} \leq G_{kht} y_{kht} - L_{kht} w_{kht} \quad (23)$$

$$\sum_{h=0}^{H_k} y_{kht} = 1 \quad (24)$$

$$y_{kht} - y_{kh,t-1} \leq w_{kht} \quad (25)$$

$$y_{kh1} \leq w_{kh1} \quad (26)$$

$$\sum_{h=0}^{H_k} \sum_{t=1}^{T} w_{kht} \leq 1 \quad (27)$$

$$w_{k0t} \leq y_{k0t} \quad (28)$$

$$x_{ikhmt}, z_{imt} \geq 0; y_{kht}, w_{kht} \in \{0,1\} \quad (29)$$

The first terms of (21) are revenue terms. Two terms are required because the model does not explicitly consider all products produced by the firm. It only considers a set of products that are produced in the given set of facilities. The first term accounts for the revenue from the products 1 to N. The second term accounts for revenue from demand for these products that is not satisfied by production in the given facilities but leads to demand for, and hence, revenue from other products produced by the firm. The last two terms in formula (21) are costs terms that account for the retooling costs and the fixed cost of operating a facility in a particular configuration. Constraint (22) is demand constraint. Constraint (23) is capacity constraint. Constraint (24) is the constraint that each site must always be in one configuration. Formulae (25) and (26) are retooling forcing constraints for periods 2 to T and period 1 respectively. Equation (27) is the constraint that each site may retool at most once in planning horizon. Formula (28) is shutdown forcing constraint if the variable indicates such a move. Constraint (29) is nonnegtive and integer constraint.

This model was created to look at capacity. However, in evaluating various alternatives the model consistently recommended that a specific subset of the products not be in the product mix [4]. The described model is a good model which can produce insights not directly related to the original question by integrating capacity planning and product parameters.

Table 3
Notation

Indexes	Demand and Probability Parameters
t planning time period, $t=1,...,T$ m demand scenarios, $m=1,...,M$ i product type, $i=1,...,N$ k production sites, $k=1,...,K$ h_k possible configuration of the capacity at site k, $h_k = 0,1,...,H_k$ ($h_k=0$ imply the plant is shutdown)	D_{imt} demand for product i under scenario m in period t • $_{ij}$ proportion of the unsatisfied demand for product i transferred to product j. For $i=j$, •$_{ij}=0$. p_{mt} the probability that scenario m occurs in period t
Capacity Parameters	Cost Parameters
G_{kht} production capacity of site k in configuration h, period t a_{ikh} amount of capacity required to produce one unit of product i at site k under configuration h L_{kht} capacity that is lost during period t if site k is retooled into configuration h during period t	F_{kht} fixed cost of changing the configuration at site k from 1 to h period t. $h=1$ is the initial configuration. C_{kht} fixed cost per period of having site k in configuration h during period t r_{ikmht} the variable contribution margin from production one unit of product i at site k with configuration h under scenario m in period t r_s a variable contribution margin. This contribution accrues when demand for a product that can be produced in this model is unsatisfied and is transferred to product s.
Decision Variables	
x_{ikhmt} amount of product i produced at site k that is in configuration h under scenario m in period t y_{kht} 1 if site k is in configuration h in period t, 0 otherwise. z_{imt} amount of unsatisfied demand for product i, scenario m and period t w_{kht} 1 if site k is retooled into configuration h in period t, 0 otherwise.	

6. CASE STUDIES—a case from electronics industry

A computational study was done to evaluate the performance of the aggregate capacity planning model described in the previous section for PWB assembly systems. No lose of generality, two market demand scenarios and a single time period (a quarter) are assumed for the simplicity of developing further searching strategy on rough addition sets (*So we can remove the time dimension from all notations of this section for clarity*). Suppose a firm has 12 assembly lines. The initial line capacity at the beginning of the budget period is shown in

Table 4. Three product families are produced from five part types (shown in Table 5). Market demand for product families is random, and predicted and expressed in two demand scenarios, i.e. S1 and S2 as shown in Table 6 (demand scenarios S3 to S9 are artificial demand scenarios). There are 20 machine types available, and their characteristics are shown in Table 7. The addition sets based on S1 to S9 are depicted in Table 8.

From Table 8, it can be observed that algorithm 1 is convergent after artificial demand scenario S3, and the rough addition set for (P0) is: $\{A(m,l)\} = \{A(5,4), A(6,2), A(1,7), A(5,2), A(6,8), A(8,7), A(1,4), A(5,7), A(8,4)\}$.

When demand scenarios are shown in Table 6 and rough addition set for problem (P0) is obtained from Table 8, the corresponding problem (P1) can be solved directly on PC machine. Suppose the subcontracting cost for product families depends on part types they are consisted of, and subcontracting costs for SC, LC, OS, HP, DCA are 6.0, 7.0, 9.0, 11.0 and 12.0 (10^{-3} \$/placement) respectively. Cost of installing a machine is \$20,000. The probability of market demand scenarios S1 and S2 are 0.35, 0.65 respectively. Capacity expansion decisions based on different rough addition set are shown in Table 9.

From Table 9, it can be observed that rough addition set based on algorithm 1 provides larger but better searching space for problem (P0). The recommended rough addition sets lead to less ratio of subcontracting, less expected total cost and higher expected capacity utilization under the assumption that all market demand be met.

Table 4
Initial line capacity layout

m/c#	l_1	l_2	l_3	l_4	l_5	l_6	l_7	l_8	l_9	l_{10}	l_{11}	l_{12}
m_1	1	0	0	0	0	1	0	0	1	1	1	0
m_2	1	2	0	0	0	0	1	1	0	1	0	1
m_3	1	0	1	1	1	0	0	0	1	0	0	0
m_4	0	1	1	0	1	1	0	2	0	1	1	1
m_5	1	0	0	0	1	0	1	0	0	0	0	1
m_6	0	0	0	2	0	0	0	1	0	0	0	0
m_7	1	1	1	0	0	0	0	1	1	0	0	0
m_8	1	1	0	0	0	2	0	0	0	1	1	0
m_9	0	0	0	1	1	0	1	0	0	0	0	0
m_{10}	0	1	1	0	0	0	0	1	0	0	0	0
m_{11}	0	0	0	1	0	0	0	0	1	1	1	1
m_{12}	0	0	0	0	2	1	0	0	0	0	0	0
m_{13}	0	0	1	0	0	0	1	1	0	0	0	1
m_{14}	0	1	0	0	0	0	0	0	1	1	1	0
m_{15}	0	0	0	1	0	0	0	0	0	0	0	0
m_{16}	0	0	1	0	0	1	0	0	1	0	0	1
m_{17}	1	0	0	0	1	0	1	0	0	0	1	0
m_{18}	0	0	1	0	0	0	0	1	1	0	0	0
m_{19}	0	0	0	0	0	1	0	0	1	0	0	0
m_{20}	0	0	0	1	0	0	1	0	0	0	1	0

Table 5
Structure of the product family, in terms of the number of placements

	SC	LC	OS	HP	DCA
Product Family 1 (F1)	300	80	30	150	5
Product Family 2 (F2)	300	50	20	100	0
Product Family 3 (F3)	200	80	0	150	7

Table 6
Market (artificial) demand scenarios for the product families (in thousands)

	Demand Scenarios								
	S1	S2	S3	S4	S5	S6	S7	S8	S9
F1	307	437	372	340	405	325	356	389	420
F2	650	420	535	593	477	624	564	506	450
F3	400	700	550	475	625	438	513	588	630

Detailed analyses on machine lines (l_2, l_4 and l_7) to which additional machines to be added show some further rationality of algorithm 1. From table 7, it can be observed that expected ratios of capacity utilization of those lines are lower than the overall expected ratio of capacity utilization no matter by which approach the rough addition set for P0 is determined. This indicates that Approach 2 made right choice among all machine lines to be expanded. It can be also observed that capacity expansion decisions for line l_2 are the same, but different for lines l_4 and l_7 based on Approaches 1 and 2. By making comparison of the ratios of capacity utilization of the two Approaches, it can be concluded that machines added to lines l_4 and l_7 improve not only the ratio of capacity utilization of line l_4 and l_7 but the

Table 7
Available machine types

m/c	Capacity (k placements per quarter)	Flexibility					Purchasing Cost (k $)	Unit variable production cost ($10^4$$/placement)
		SC	LC	OS	HP	DCA		
m_1	12000	1	1	0	0	0	556	29
m_2	6000	1	1	0	0	0	288	30
m_3	12000	1	0	1	0	0	317	17
m_4	6000	1	0	1	0	0	317	33
m_5	12000	1	0	0	1	0	614	32
m_6	6000	1	0	0	1	0	345	36
m_7	6000	1	0	0	0	1	384	40
m_8	6000	1	1	1	0	0	336	35
m_9	12000	0	1	1	0	0	576	30
m_{10}	6000	0	1	1	0	0	307	32
m_{11}	12000	0	1	0	1	0	614	32
m_{12}	6000	0	1	0	1	0	326	34
m_{13}	12000	0	0	1	1	0	634	33
m_{14}	6000	0	0	1	1	0	326	34
m_{15}	6000	0	0	1	1	1	432	45
m_{16}	6000	0	0	0	1	1	413	68
m_{17}	9000	0	0	0	1	1	590	41
m_{18}	6000	0	1	0	0	1	393	41
m_{19}	6000	0	1	1	1	0	403	42
m_{20}	9000	0	1	1	1	0	547	38

Table 8
Rough addition sets based on Approach 2

Demand Scenarios	Rough addition sets
S1	$A'(1,5,4)$, $A'(1,6,2)$
S2	$A'(2,1,7)$, $A'(2,5,2)$, $A'(2,5,4)$, $A'(2,6,8)$, $A'(2,8,7)$
S3	$A'(3,1,4)$, $A'(3,5,7)$, $A'(3,6,2)$, $A'(3,8,4)$
S4	$A'(4,1,4)$, $A'(4,6,2)$, $A'(4,8,4)$, $A'(4,8,7)$
S5	$A'(5,5,4)$, $A'(5,5,7)$, $A'(5,6,2)$
S6	$A'(6,1,4)$, $A'(6,6,2)$, $A'(6,8,4)$, $A'(6,8,7)$
S7	$A'(7,5,4)$, $A'(7,6,2)$, $A'(7,8,7)$
S8	$A'(8,1,4)$, $A'(8,5,7)$, $A'(8,6,2)$, $A'(8,8,4)$
S9	$A'(9,1,4)$, $A'(9,5,7)$, $A'(9,6,2)$, $A'(9,8,4)$

Table 9
Capacity expansion decisions based on different Approaches

		Approach 1	Approach 2
Rough addition sets $\{A(m,l)\}$		$\{A(5,4), A(6,2), A(1,7), A(5,2), A(6,8), A(8,7)\}$	$\{A(5,4), A(6,2), A(1,7), A(5,2), A(6,8), A(8,7), A(1,4), A(5,7), A(8,4)\}$
Capacity expansion decision		$Y(6,2)=1$	$Y(1,4)=2$, $Y(5,7)=2$, $Y(6,2)=1$
Ratio of Capacity utilization*	Demand realization S1	88.5793%	82.6097%
	Demand realization S2	90.7100%	98.3214%
Expected ratio of capacity utilization		89.9999%	92.8223%
Ratio of subcontracting*	Demand realization S1	7.7292%	7.8822%
	Demand realization S2	17.6578%	4.4577%
Expected total cost**($)		-21995278	-22671219

* Capacity and subcontracting are expressed in number of parts.
** Please c.f. the definition of expected total cost which is expressed in formula (11).

Table 10
Comparison of capacity utilization

	Approach 1			Approach 2		
	l_2	l_4	l_7	l_2	l_4	l_7
Capacity after expansion	48000	63000	60000	48000	87000	84000
Production assignment under S1	48000	52440	39330	48000	52440	39330
Ratio of capacity utilization under S1	1.0000	0.8324	0.6555	1.0000	0.6028	0.4682
Production assignment under S2	28200	52440	39330	48000	87000	84000
Ratio of capacity utilization under S2	0.5875	0.8324	0.6555	1.0000	1.0000	1.0000
Expected ratio of capacity utilization	0.7318	0.8324	0.6555	1.0000	0.8610	0.8139

ratio of capacity utilization of line l_2 also. This should be attributed to chaining effect of flexible machine lines [21].

7. SUMMARY AND CONCLUSIONS

This chapter has focused primarily on impacts of aggregate capacity planning on all issues of manufacturing competitiveness such as manufacturing cost, diversity of products, manufacturing lead time, introduction of new products, and market share, etc. Our main points can be summarized as follows:
1. *Manufacturing flexibility is an important base of manufacturing agility*, which should be considered in capacity planning.
2. For FMS capacity planning problems, *three objectives are necessary of pursuing*, i.e., modular capacity, products and production lines chaining, and subcontracting or outsourcing.
3. To pursue agile manufacturing, *parameters of products and process are necessary to be explicitly considered in capacity planning*.
4. We suggest that *aggregate capacity planning, production line design/redesign, and disaggregate decision problems be integrated* in flexible manufacturing environment to realize better customer responsiveness and business goals.
5. *Models of the suggested integration for electronic industry and automobile manufacturing industry are presented*.
6. *Case from electronic industry is described to show the integration process*, in which a model of scenario-based line capacity expansion problem for printed wiring board (PWB) assembly systems at the aggregate level is developed. The model brings together the bill of components of product families and machine flexibility, thus manufacturing flexibility and customer response can be explicitly considered in the model.

Integration of aggregate capacity planning, production line design/redesign, and disaggregate production decisions is a difficult research topic. This chapter discusses some measures of agile manufacturing and their relation with aggregate capacity planning, while other measures such as setup time are not involved in this chapter. Also, further cases from non-flexible industries and even service industry may need to be studied.

REFERENCES

1. R.B. Chase and N.J. Aquilano and F.R. Jacobs, Production and Operations Management: Manufacturing and Services, Beijing: McGraw-Hill,1998.
2. P.C. Jones and J.L. Zydiak and W.J. Hopp, Parallel Machine Replacement, Naval Res. Logist., l(1991):351-365.
3. S. Rajagopalan, Capacity Expansion and Equipment Replacement: A Unified Approach, Operations Research,6(1998), 846-857.
4. G.D. Eppen and R.K. Martin and L. Schrage, A Scenario Approach to Capacity Planning, Operations Research, 37(1989): 517-527.
5. C. Gaimon, Subcontracting versus Capacity Expansion and the Impact on Pricing of Services, Naval Research Logistics. 41(1994): 875-892.
6. C. Gaimon and J.C. Ho, Uncertainty and the Acquisition of Capacity - a Competitive Analysis, Computers & Operations Research. 21(1994): 1073-1088.
7. S. Rajagopalan, Flexible Versus Dedicated Technology: A Capacity Expansion Model, International Journal of Flexible Manufacturing System, 5(1993): 129-142.
8. S. Rajagopalan and A.C. Soteriou, Capacity Acquisition and Disposal with Discrete Facility Sizes, Management Science, 40(1994): 903-917.

9. W.J. Hopp and M.L. Spearman, Factory Physics—Foundations of Manufacturing Management, Irwin/McGraw-Hill, 1996.
10. A.V. Roth and C. Gaimon and L. Krajewski, Optimal Acquisition of FMS Technology Subject to Technological Progress, Decision Sciences Journal, 22(1991): 308-334.
11. C.H. Fine and R.M. Freund, Optimal Investment in Product-Flexible Manufacturing Capacity, Management Science, 36(1990): 449-466.
12. P.T. Kidd, Agile manufacturing: forging new frontiers, Wokingham: Addison Wesley. 1994.
13. K.E. Stecke and N. Raman, FMS Planning Decisions, Operation Flexibilities, and System Performance, IEEE Transactions on Engineering Management, 26(1995), 82-90.
14. U.S. Karmarkar and S. Kekre, Manufacturing Configuration, Capacity and Mix Decisions Considering Operational Costs, J. Manufacturing Systems, 4(1988): 315-324.
15. D.J. Vander Veen and W.C. Jordan, Analyzing Trade-offs between Machine Investment and Utilization, Management Science, 10(1989): 1215-1226.
16. S.-L. Li and J. Qiu, Models for Capacity Acquisition Decisions Considering Operational Costs, The international Journal of Flexible Manufacturing Systems, 8(1996):211-231.
17. S. Benjaafar and D. Gupta, Scope versus Focus - Issues of Flexibility, Capacity, and Number of Production Facilities, IIE Transactions. 30(1998): 413-425.
18. H. Luss, Operations Research and Capacity Expansion Problems: A Survey, Operations Research, 10(1982): 907-947.
19. C. Liu and C.R. Emerson and K. Srihari, An Expert System Approach to Surface Mount Pick-and Place Machine Selection. Handbook of Expert Systems Applications in Manufacturing Structure and Rules, Chapman & Hall, 303-320(1996).
20. J.C. Ammons etc., Component Allocation to Balance Workload in Printed Circuit Card Assembly Systems, IIE Trans., 29(1997): 265-275.
21. Jordan, W. C., Graves, S. T., "Principles on the Benefits of Manufacturing Process Flexibility, Management Science, 41(1995): 577-594.
22. S.A. Andreou, A Capital Budgeting Model for Product-Mix Flexibility, Journal of Manufacturing and Operations Management, 3(1990): 5-23.
23. L.J. Bradt, The Automated Factory: Myth or Reality, Engineering Cornell Quarterly,3(1983).
24. S. Rajagopalan and M.R. Singh and T.E. Morton, Capacity Expansion and Replacement in Growing Markets with Uncertain Technological Breakthroughs, Management Science, 44(1998):12-30.
25. B.A. Peters and G.S. Subramanian, Analysis of Partial Setup Strategies for Solving the Operational Planning Problem in Parallel Machine Electronic Assembly Systems, International Journal of Production Research, 34(1996): 999-1021.
26. P.J.M. Van Laarhoven and W.H.M. Zijm, Production Preparation and Numerical Control in PCB Assembly, International Journal of Flexible Manufacturing System, 5(1993): 187-207.
27. Z.S. Hua and P. Banerjee, Impact of Feedback on Equipment Changeover and Production Assignment for Flexible PWB Assembly System, Technical Report, University of Science and Technology of China (submitted to the International Journal of Production Economics) 1999.
28. R.H. Day and A. Cigno, Modeling Economic Change—The Recursive Programming Approach, North-Holland, 1978.
29. R.H. Day and A. Cigno, Recursive Decision Systems: An Existence Analysis, Econometrica, 38(1970): 666-681.
30. W.L. Winston, Operations Research—Applications and Algorithm, Duxbury Press, 1994.

APPENDIX A. PROOF OF PROPOSITION 1

Proposition 1: Solution to problem SPACU formed in Step 2 of Figure 6 is convergent.

Proof. With the indexes I, J, M, L defined in model ECSP and SPACU, we define:
p=I*J*M*L is the number of variables of SPACU in each period
q=I*J+M*L+I* (I-1) *J*L is the number of constraints of SPACU in each period
Without considering feedback and iteration, model SPACU of each period can be rewritten as

$$\min\{cr | Br \geq b, r \geq 0\} \qquad (A1)$$

Its dual can be written as

$$\max\{sb | B^T s \leq c, s \geq o\} \qquad (A2)$$

In model (A1) and (A2), c and r are p-dimensional vectors, s and b are q-dimensional vectors, B is a p×q matrix.
Let us define:

v=(r,s) $\qquad (A3)$

w=(c,B,b) $\qquad (A4)$

Let v be a (p+q)-dimensional vector which corresponds to a feasible solution to model (A1) and its dual model (A2). Let V={v} be called the decision space; w be a (p+1)(q+1)-dimensional vector and W={w} be called the state space.
We define correspondences:

$$\Gamma : W \rightarrow V \qquad (A5)$$

and

$$\Gamma w = \{(r,s) | Br \geq b, B^T s \leq c, (r,s) \geq 0\} \qquad (A6)$$

$$\Psi_M : W \rightarrow V \qquad (A7)$$

and

$$\psi_M = \{(r^*, s^*) | (r^*, s^*) \in \Gamma w, cr^* - s^* b = 0\} \qquad (A8)$$

Formula (A6) states that for a given state $w \in W$ the image Γw is the set of possible decisions corresponding to a given state w; correspondence Ψ_M defines an optimizer in V that is feasible and that minimizes the objective function for a given w. Here, the optimizers are LP models defined in formula (A1) and (A2).
If we specify the convergence condition for model SPACU as:

$$|Z^k - Z^{k-1}| \leq \varepsilon \qquad (A9)$$

In formula (A9), Z^n is the value of objective function of model SPACU at the n th iteration, ε is a small positive number.

From the definitions of formula (A3) to (A9) and the basic properties of linear programming problem, it can be easily verified that model SPACU has the following properties:
1) V is nonempty and convex. After adding the special machine (infinite capacity and total flexibility) to each machine line, the static LP problems of SPACU are always feasible.
2) Γ and Ψ are nonempty. Because each static LP problem of SPACU has optimal solution, Γ and Ψ, as defined in (A6) and (A8) respectively, are nonempty.
3) Γ is a continuous correspondence, as shown in (A6).
4) Γ is convex and objective function cr is concave (more specifically, cr is linear).
5) Objective function is a continuous numerical function on V×W.
6) The feedback function defined in formula (7) and (8) is single valued and continuous,

Since model SPACU is a first-order recursive programming problem and has an optimal solution at each iteration, according to the theorem introduced by Day (1970, p.676), we can see that the solution to SPACU is convergent if V is closed and bounded in addition to properties 1) to 6).

For each iteration of the SPACU model, the corresponding LP problem always has an optimal solution. When the iteration process converges, it is still an optimal solution. Thus, V is closed. Now we need to check that V is bounded, i.e. r is bounded.

In model SPACU, we know that for $t=1,...,T, n=1,...,N, i=1,...,I, j=1,...,J, l=1,...,L$:

$$0 \leq \sum_{i,j} X^k(t,n,i,j,m,l) \leq N_l(t,m,l) * Q(m) \quad \text{for } m = 1,2,...,M \quad (A10)$$

$$0 \leq \sum_{l} X^k(t,n,i,j,M+1,l) \leq d(t,n,j) * S(i,j) \quad (A11)$$

As it is defined in (A1) that $r=(X(t,n,i,j,m,l))_{q\times 1}$ is the variable vector of model (A1), from (A10) and (A11), we know that r is bounded. According to the properties of dual programming, s is bounded. Thus, for arbitrary $v=(r,s)\in V$ is bounded, and V is bounded.

Therefore, the solution to the SPACU is convergent.

APPENDIX B. PROOF OF PROPOSITION 3

Proposition 3. Algorithm 1 is convergent.

Proof. According to proposition 1, solution to the recursive linear programming problem SPACU formed according to step 2 of the flowchart illustrated in Figure 6 is convergent. So we can assume that:

Given a deterministic market demand vector $\underline{d} = (d_1, d_2,...,d_J)^T$, solving model SPACU will get a convergent solution $\underline{x} = (X(t,i,j,m,l))$ ($t=1,...,T, i=1,...,I, j=1,...,J, m=1,...,M, l=1,...,L$). Under deterministic market demand vector $\underline{d}' = \underline{d} + \Delta\underline{d}$, solving model SPACU will also generate a convergent solution $\underline{x}' = (X'(t,i,j,m,l))$.

Proposition 3 is equivalent to the following formula:

$$\lim_{\Delta\underline{d}\to 0} \underline{x}' = \underline{x} \quad (B1)$$

To verify formula (B1), we denote $\underline{x}_k$ ($k=1,2,...$) as solution to the LP problem (termed as model M_k) corresponding to kth iteration under demand $\underline{d}$, $\underline{x}'_k$ ($k=1,2,...$) as solution to the

LP problem (termed as model M'_k) corresponding to k th iteration under demand $\underline{d}^{iä}$, we need to prove that for any k ($k = 1, 2, ...$):

$$\lim_{\substack{\text{ædjüo}}} x_k^{iä} = x_k \tag{B2}$$

When $k = 1$, there is change on resource side of model $M_1^{iä}$ compared with model M_1. Thus formula (B2) is obviously true according to general sensitive analysis theory of LP problem [30].

When $k = 2$, there are changes both on resource side and on coefficient of objective function of model $M_2^{iä}$ compared with model M_2. But because the difference between $x_1^{iä}$ and x_1 is approaching to zero, difference between coefficients of objective function of model $M_2^{iä}$ compared with model M_2 is also approaching to zero because the feedback defined in formula (12) is a continuous function. So formula (B2) is also true according to general sensitive analysis theory of LP problem. For $k > 2$, the proving process is similar to the case $k = 2$.

The control problems of Agile Manufacturing

B. Ilyasov, L.Ismagilova and R.Valeeva

Technical Cybernetics Department, Ufa State Aviation Technical University,
12 K.Marx Street, Ufa, 450000, Russia

In the given chapter some control problems of agile manufacturing system (AMS) as complex dynamic object are under discussion. There are considered existing approaches to AMS dynamics research, it is emphasized appropriateness of system approach realization to development of dynamic models of the given class of systems. After analyzing of modern conceptions of control systems construction of agile manufacturing (AM) it is emphasized orientation necessity in intelligent control to usage of enterprise ontology including all totality of knowledge about laws, principles, and functioning peculiarities of the given system. There are classified all the types of uncertainties arising in control of agile manufacturing system. Intelligent control of agile manufacturing behavior in competitive market conditions, during product realization with the purpose of getting of prescribed profit is under consideration. There is presented example of effectiveness estimation of intelligent control during manufacturing system functioning in conditions of non-stable economics. Development perspectives of control systems of agile manufacturing are discussed.

1. AGILE MANUFACTURING SYSTEMS AS COMPLEX DYNAMIC CONTROLLED OBJECT

Recently agile manufacturing systems gain wide acceptance among modern classes of manufacturing systems (MS). This class presents further development of flexible manufacturing systems. It is characterized by not only more high level of integration, but also more high level of intellectualization and self-organization, and also capacity to adapt quickly and effectively to unpredictable changes of an environment. In these systems organizational knowledge and experience, human factor and modern advanced technologies are combined optimally.

These properties of agile manufacturing systems are reached due to their capacity:
- to adaptation (structural and parametric, by means of restructuring production among them);
- to development, by means of forming of optimal strategic schedules;
- to fast change of the current productive schedules according to changes of both market situation, and interior state of manufacturing system;
- to fast reorientation to change of product mix;
- to fast change of behavior tactics in competitive market conditions;
- to making of intelligent control decisions in uncertainty conditions;

- to creation of temporary organizational structures which ensure faster effective reaching of the finite purpose.

Thus, agile manufacturing systems are characterized by high level of dynamicability, and non-equilibrium functioning regimes are dominating for this class of systems. Simultaneously AMS fall in the class of organizational, hierarchical, multifunction, multiple objective, multiply connected nonlinear dynamic complex systems which consist of subsystems set of different physical nature (structural subdivisions), agents, having own local purposes and containing the man as active element. However, study of dynamic properties of such complex open system as agile manufacturing systems is connected with solution difficulty of problems series such, as:

- development of adequate models of a system both in equilibrium, and in non-equilibrium functioning regimes;
- authentic and complete enough change description of an interior system state and an environment state in various situations which are generated by action of random and uncertain factors;
- forming of such control actions on AMS in uncertainty conditions which would ensure either functioning effectiveness of a system, or achievement of the local and global purposes, or preservation of stability at presence of failures, loss of information and other factors which lead to origin of unnominal (critical, emergency, crisis) situations;
- development of efficient algorithms for rebuilding of organizational structure of agile manufacturing system according to change of production and market situations;
- co-ordination of the local purposes of subsystems and global purpose;
- taking account human factor which introduces subjective and uncertain character in the functioning process of agile manufacturing system.

To construct control system of AMS as of complex object it is necessary to solve the following tasks:

- choice of the global and local purposes and their concordance functioning effectiveness criterions of a system;
- timely deriving of the authentic information about a current state of object and environment;
- exposure of state variables, controlled output coordinates and control factors (actions) which ensure purposeful motion of a system;
- synthesis of decision making algorithms for forming control actions: their form, magnitude, sign, and also time and point of application;
- security of timely and qualitative fulfillment of made decisions.

Summary. Agile manufacturing system represents complex dynamic controlled object, badly-formalizable, with unpredictable behavior, especially in uncertainty conditions. Therefore, to control such object it is necessary to apply intelligent algorithms realized by means of modern information technologies.

2. APPROACHES TO DYNAMICS INVESTIGATION OF AGILE MANUFACTURING SYSTEMS

Today there are a few approaches to description of dynamics (motion) of manufacturing systems. In our opinion, one can distinguish the following approaches.

The first approach is based on description of complex system as a whole by means of state variables. Dynamics is described as a system of nonlinear differential equations in a normal form of Cauchy. Description can be both in continuous, and in discrete form. The main problem at such description is exposure of state variables and establishment of relation between them for a wide range of system state changes.

The approach to description and dynamics research of manufacturing systems with the help of Petri nets is widely enough spread. This approach is realized in some modern information technologies, for example, in IDEF systems on the basis of CASE methodology. Petri nets are productive enough at discrete description of manufacturing systems. To use Petri nets it is necessary to have large a priori information about manufacturing system functioning, which describes a set of possible states, transitions and resource expenses (time among them) for implementation of these transitions. However, many these parameters frequently have mean statistic nature, and the appearance possibility of new states, as the violation result of a usual production rhythm requires revising a net structure. And such process as, for example, interaction of manufacturing system with an environment (with the competitors in market conditions) is difficult for description with the help of Petri nets.

The third approach to dynamics study of manufacturing system is based on its description with the help of lags, which represent some time delays necessary for fulfillment of those or other technological and organizational operations. The realization difficulty of the given approach is in the fact that the magnitudes of lags have statistic nature and therefore they are defined approximately, and dynamic stability of a system in a large measure depends on concrete values of delays. However, some processes in manufacturing systems have mostly inertial (forcing, integrating) nature than nature of dead time. Therefore, dynamics representation of manufacturing system only with the help of lags not always is adequate to actual processes. It should be noted, that also lags don`t allow to describe interaction dynamics of a system and the competitors in market conditions because of appearing uncertainties.

The fourth approach to construction of dynamic models of manufacturing systems is in the fact that there is calculated a set of dynamic ally equilibrium system states with the help of available a priori information about potential possibilities of a system and expected states of an environment. If these states correspond to extremal values of chosen functionals for given restrictions a set of these states is optimal in the sense of chosen criterion.

The given conception of MS dynamics study is widely used in investigation practice of manufacturing systems and allows to obtain positive results in many cases. However, the given approach doesn`t take into account, in the first place, influence of random and uncertain factors which act on a system and violate a scheduled dynamic equilibrium. Secondly, it doesn`t take into account influence of organizational control system on violation of dynamic system equilibrium because of action of such factors as errors in decision making, errors connected with inaccuracy of information processing about a current state of manufacturing system and environment, and also errors connected with untimely and poor-quality fulfillment of made control decisions.

Next dynamics description conception of complex systems as of open systems is based on functional-structural form of system approach, when complex system is represented in the form of a set of separate subsystems, agents, strongly interconnected and interacting with each other, having own local purposes and functions. Subsystems function independently, a part from them can be integrated in more large independent organizations (to form coalitions) and to subordinate the local purposes to achievement of the global organization purpose. Subsystems can also enter among themselves into conflict and competitive relations. In addition each from them should have own local organizational-control structure which ensures for subsystem the best orientation in fully formed situation. Study of dynamic ally properties of a system for the given system approach is reduced to study of its dynamic ally equilibrium and dynamic non-equilibrium states. The given conception is the basis of construction and investigation not only agile manufacturing systems, but also such classes of manufacturing systems as holonic, multi-agent systems and also virtual enterprises. All these classes of manufacturing systems are open systems which are capable to self-analysis and self-organization due to man participation and application of intelligent information technologies that forms approach generality for research of their behavior.

Dynamically equilibrium, or steady-state regime describes system functioning under effect of planned control and calculated disturbing actions. Calculation of motion trajectory of manufacturing system in steady-state regime is relative to the task of functioning process planning of manufacturing system. Requirements of fast reaction on changes of demand, prices, competitor`s behavior are presented for to planning. These requirements can be achieved by means of creation of planning systems which are able to change values of volumes and (or) product output nomenclature according to market needs. Dynamically non-equilibrium regime corresponds to motion of manufacturing system on non-calculated trajectory generated as the action result of disturbances unaccounted in planning process, in production and in market.

The authors distinguish two types of dynamic functioning regimes of controlled system: stabilization regime and control regime. Problem is in the elaboration of such dynamic control algorithms, which, in the first case, would ensure with the least dynamic errors stabilization of manufacturing system motion relatively established calculated trajectory under action of disturbing factors, and in the second case - transition from one established trajectory to another under action of control factors for a given time and also with small dynamic distances.

Summary. To elaborate dynamic models of agile manufacturing systems it is appropriate to represent it as a set of interconnected and interacting subsystems (agents) having own local purposes and functions, which are realized by own organizational information-control systems. Such models offer possibility to describe most adequately complex situations and clashes appearing in real systems at their functioning.

3. FORMATION CONCEPTIONS OF CONTROL SYSTEMS OF AGILE MANUFACTURING

To construct organizational control system of complex dynamic objects, and agile manufacturing concerns to them, it is necessary to observe known system principles and regularities, their violation or ignoration inevitably lead to construction of low-effective information-control systems.

System principles and regularities can envelop all the types of creative man`s activity which are connected with creation of control systems and concern to designing, research, modeling, planning, control, testing, production etc. both system as a whole and its separate subsystems. We shall consider below some principles, which are used in the elaboration of structures and algorithms of control systems.

Principle of coherence is one from such principles, according to it complexity level of control part of a system should be adequate (corresponding) to complexity level of controlled object (CO). For example, if controlled object is hierarchical, control system too should be also hierarchical with adequate number of levels. Special case of coherence principle is *coordination principle* of subsystems functioning. According to this principle subsystems must fulfill projected functions in time and concordantly with each other, i.e. must function with coordinated fulfillment rates of planned activities.

According to *principles of controllability and observability* control system should be constructed so that it will enable to act purposefully on any state variable or will allow to estimate a current value of any state variable by means of measurement of physical output coordinates.

Principle of disturbance-based control allows to construct within a system additional information-control channel, in which additional control action is formed, it is directed on compensation of possible consequences caused by action of disturbing factors. Such approach is especially effective if possible disturbing factors are well predicated and there is a possibility of action consequences estimate of these factors.

Principle of feedback is widely used in the construction in information-control systems and is based on estimate of a current state of controlled object under measured information and comparison of this state with desired state in the given instant with the purpose of error exposure between these states. The received error is used later on in the formation of control decisions with the help of some algorithms, their implementation should lead a current state of a system to a desired one. For example, control system of agile manufacturing should observe, on one hand, implementation accordance of output plans of projected product mix. On the other hand, it should observe continuously all changes of market conjuncture and to correct current plans according to market needs and results of product sale in market.

Principle of adaptive model-based control generates two-level structure of a system and is based on comparison of a current state of controlled object with a state, which is defined by functioning model of a system. Model reflects desired course of production process or desired behavior of a system. According to this deviation control decisions are being formed, they are directed either on correction of production program, or change of parameters or structure of control system, including production restructuring. Model-based control imparts adaptation property to a system, ensuring more flexible operative response on market needs. Sometimes principle of adaptive control of a system by means of enterprise restructuring is named *principle of self-organizing*. In the broad sense principle of self-organizing makes possible changing of organizational structure both all system as a whole and its separate subsystems, and also structure both control, and production components of a system. At that there is change possibility of relations (connections) between subsystems (agents) including business partners. Principle of self-organizing is widely used in multi-agent and holonic manufacturing systems.

Implementation of *self-training principle* requires to construct three-level system, at its upper level there is formed a new model of functioning (behavior) of a system, which is adequate to originated situation and is the analysis result of available knowledge, experience, current state of a system and environment. Later on, at the second control level (adaptation level) there is solved a task of system adaptation in a new situation with the help of new model of its functioning.

Principle of situational control provides such organization of a system, which allows to analyze thoroughly originated situation connected with interior system state and environment state, and to form, as the result of this analysis, such control actions on a system that their implementation ensures its effective functioning in a given situation. Situations may be standard, known in advance, and then standard control decisions for them are stored in knowledge base of information-control system. Efficiency of control systems of agile manufacturing constructed according to a given principle is developed in control in conditions of uncertainty, resources deficit and possible appearance of nonstandard, critical situations by means of use of intelligent control algorithms. Formation of intelligent control is realized on the base of computer-aided integration of knowledge due to systematization and structuring of knowledge in different subject areas. This knowledge is used later for decision making both separate subsystems (agents, holons), and decision makers (DM).

Principle prediction-based control requires information to estimate possible (with some probability) future states of a system. Estimate of future states is necessary to form strategic development plans of a system. The higher is accuracy of prediction, the greater efficient is strategic plan of development. *Principle of control based on finite result* is close to above-mentioned principle, there is estimated efficiency of finite result instead of efficiency of undertaken actions and plans for achievement of this result.

Principle of self-development focuses on construction of such system, which is able not only to elaborate strategic plans of development, but also to accumulate purposefully potential (resources) and to distribute and expend optimally this potential for the purpose to ensure desired development process. To realize this principle in information-control system computer-aided integration of knowledge about past, present and future system states is necessary. Knowledge about development laws of dynamic systems are also necessary, they reflect regularities of qualitative changes within systems in realization process of their life cycle.

The above-listed principles don`t enclose all the variety of existing system (synergetic) principles. But, in our opinion, they have considerable utility in the construction of organizational information-control systems for agile manufacturing.

Let us next consider description of functioning complexity of production-control system for agile manufacturing. Functioning complexity of a given class of control systems is defined by necessity of quick and rational (optimal) solution of tasks sequence of decision making in conditions of uncertainty, resources deficit and appearance of critical situations. Some decision-making tasks are listed below, one should solve them in the construction and control of agile manufacturing system.

- What control factors and in what combination must take part in control of AMS at present instant of time?
- To what subsystems, when and in what sequence one must apply control actions?
- What must be value, shape and sign of control action?
- What parameters and combined indexes are most informative in present situation with point of view of fast, sharp and authentic estimate of current system state?
- What control factors are most efficient in present situation with point of view of controllability for a system?
- What information about system state and environment, in what amount and in what form are required in present situation for decision making?
- What factors facilitate and what factors hinder to implementation of effective system control in present situation?
- What contradictions or conflicts may arise during functioning process of a system and when they may arise?
- What way one must realize production restructuring or reorganization of administrative-control component of a system with the purpose of raising functioning efficiency of a system?
- What way one may solve adaptation problem of agile manufacturing system to changed environment conditions?
- What way one may use in concrete situation accumulated knowledge and experience, which are relating to other analogous situations had arisen in the past?
- What decision-making algorithms are in nonstandard situations?

- What way one must realize informational, organizational, resource interaction between the higher and the lower levels to ensure stability in the vertical for hierarchical system?
- What way one must ensure development process stability of a system?

These questions don't cover certainly all those problems and tasks, which one must solve in control of such complex object as agile manufacturing system.

General conception of forming control systems of complex dynamic objects is constructed as a rule with the help of three-level scheme. System principles and regularities are arranged at a upper level, they reflect systematic construction character of material universe, man`s activity and his thinking. At the second level there are arranged models adequately reflecting studied properties of investigation object. Methods and techniques of solution and investigation of tasks are at the greatest lower level. This three-level system is destined for receipt of new knowledge during solution process of task set.

Totality of knowledge about laws of construction, functioning and development of complex dynamic systems, and basic concepts connected with them represents *ontology* of agile manufacturing systems. Knowledge of ontology is necessary for elaboration of conception of AMS intelligent control and computer-aided representation and proceeding knowledge [1-5]. Thus, knowledge of above-mentioned system principles and regularities is constituent part of enterprise ontology.

Problem of creation of control systems for complex objects is polyhedral, all the manifold of approaches to systems investigation is reflected in contents of those models, which are available for investigator.

One can distinguish, with this point of view, the following approaches to AMS investigation as of complex dynamic system:

- dynamic, when to elaborate models one uses approaches which allow to describe, with required degree of adequacy, regularities of equilibrium and non-equilibrium functioning regimes of a system;
- functional, when a system is divided into subsystems according to functional sign, and there are elaborated functional models reflecting local and global purposes of a system;
- structural, where there is studied structure (construction) of a system and its subsystems with the purpose of choosing of efficient structures;
- cybernetic, when investigator operates by dynamic models for the purpose of control algorithms synthesis or efficiency analysis of control processes;
- informational, when there are studied receipt, processing, transition, transformation, storage and search processes of information;
- resource-purposeful, when one estimates with the help of models resource security of purpose achievement process with the point of view of efficient system functioning.

Summary. To construct control system of agile manufacturing it is necessary to orient on use of general system conception taking into account all the manifold of having solved problems and tasks, which origin in the designing of a given class of systems.

Observance of general system (synergetic) principles and regularities allows to form such control of agile manufacturing, which will be adequate to having changed situation.

Orientation on ontology of enterprise allows to transfer knowledge and experience, accumulated in control of complex systems of some class (technical, technological), to complex systems of other physical nature (social-economical, organizational etc.). Ontology of agile manufacturing system as complex organization allows to form methodology (conception) of its research, to develop models adequate to real systems, to synthesize efficient control algorithms both in standard, and nonstandard situations, and also to construct correctly strategy of its development.

4. UNCERTAINTIES ARISING DURING CONTROL OF AGILE MANUFACTURING SYSTEM

Effective control of agile manufacturing system as a complex dynamic object is difficult to realize in practice because of necessity of decision making in conditions of uncertainty, which reason is associated often with deficit of information about real state of a system or about real change of environment. In the process of accumulation of knowledge about agile manufacturing system functioning in different regimes and at various conditions level of uncertainty will be reduced. Knowledge of types of uncertainties, their nature and origin reasons is very important at working out of dynamic models of agile manufacturing system.

According to origin nature uncertainties ought to be divided into three classes: cognitive, internal and external. Cognitive uncertainties are caused by our lack of knowledge about laws (or principles) of controlling, functioning and development of agile manufacturing system, especially in critical (crisis, emergency) situations, and also by incorrect using of these laws, by lack of knowledge about market price formation mechanisms and mechanisms of demand formation in one or another social-economic environment, by lack of knowledge and skill to manage person or group of persons, included in AMS as organizational system. To reduce influence of this uncertainty class cognition of development laws of nature, thinking and society, i.e. development laws of biosphere, noosphere, psychosphere as united system is necessary, because only intelligent activity of a human becomes determinative factor of development of created by him artificial systems, in particular, AMS. Internal uncertainties are caused by lack of information about real state of agile manufacturing system, its elements and subsystems that reflects in constructing of their models. These uncertainties, in their turn, can be divided to uncertainties related with manufacturing as controlled object, uncertainties related with planning system, with decision making system, with executive system, with information system, with communication system. It should be noted among the factors, creating uncertainty in production modeling as controlled object are the following:

- random deviations in technologies, leading to their violation,
- failures of technological equipment,
- lag of technical and technological maintenance level,
- lag of automatization and robotization level,
- failure of resource supply schedule inside agile manufacturing system.

Influence of these factors leads to failure of production rate, to decreasing of quality of output products with uncertain rate.

To uncertainties, concerning elements of organizational-controlling part of AMS the next one's, can be referred:

- fuzziness and subjectivity in formation goals of system functioning and development;
- contradictoriness and changeability of goals of separate subsystems and system at all;
- fuzziness of prediction of future states or consequences because of made control decisions, that leads to fuzziness in strategic and operative planning, to formation of erroneous ineffective algorithms of control;
- fuzziness in control decision formation because of complexity of simultaneous deciding choice tasks of number of regulated and regulating factors, tasks of choice of points of acting

to system, tasks of choice of moment, level, sign and form of control actions among the set of alternative decisions;
- fuzziness in formation of control decisions because of influence of social-psychological factors to person, that leads to errors in identification and analysis of appearing situations, in prediction of their development, in choice of direction and path to goal; in estimation of existing situation;
- fuzziness of current situation estimation because of inaccuracy, inauthenticity and inobjectivity of information in view of measurements and data about state of a system and its elements, about state of environment, about available resources, and also because of delay, losing and distortion of information during its processing and transmission;
- fuzziness in implementation of made decisions because of human factor, which is expressed as in untimely and inexact implementation of decisions as in some voluntarism and low professionalism of actions of executor, and also in incorrect perception of instructions of higher organs;
- fuzziness in functioning of all system because of violations in communication system and connections between subsystems.

Complexity of compensation of foregoing factors is in the fact, that they are interconnected, difficult to be divide, high dynamical, and they provides general indirect influence, which is significantly changed during time both quantitatively and qualitatively, to state variables.

Internal uncertainties are caused by lack of information about state and behavior of environment. Here can be pointed the next types of uncertainties which are caused by:
- unpredictable behavior of competitors, their purposeful counteraction;
- appearance of new technologies, constructions, material and so on, and on their base -new high qualitative products;
- inaccuracy and inobjectivity of state estimation of both products and resource market; including changes of demand on goods and resources;
- violation of rules of behavior in market and violation of sale organization conditions;
- violation of transport and communication system functioning;
- incorrect actions of partners ;
- incorrect actions regarding to partners;
- sudden and unpredictable change of prices of resources including energy;
- inflation of monetary and as a consequence sudden reducing of purchasing power;
- change of state fiscal policy.

Complexity of accounting of external undefined factors is in difficulty of prediction of moment and form of their revealing, and also their consequences of their influence. Influence of external factors can not only decrease effectiveness of agile manufacturing system functioning, but also lead it to unstable state or even destroy it. Therefore general task of control of agile manufacturing system in uncertainty conditions is maintenance of its stable functioning and development by means of change motion goal, system restructuring, change of parameters and algorithms of control, and also form and way (trajectory) of motion to new goal.

Classification of uncertainties, acting to AMS, can be made also according to other characteristics. For example, according to their acting nature these factors can have the next character: physical, social, organizational, psychological, financial and so on. According to type of influence they can be signal, parametrical, structural. According to method of mathematical description the next factors can be noted: deterministic and stochastic, discrete and continuous, linear and nonlinear, and also logical, stationary and unsteady and so on. According to degree of influence consequences the next factors can be noted: destroying (leading to catastrophes, for example, erroneous control decisions); life-dangerous (leading to crisis, critical situations); very unpleasant (leading to reduction of system functioning effectiveness); not dangerous for life (leading to liquidation or limitations of implementation of some functions); weak (leading system to insignificant deviations from equilibrium state). According to effect of influence to system there can be noted the next factors: stimulating (positive), inhibitory (negative) and neutral. It is extremely important to reveal timely accidental positive factors (fortunate opportunity) and to use them to increase efficiency of control of AMS. According to time of influence to system the next factors can be noted: rapidly proceeding • slowly proceeding undefined factors. Among the first rapid change of situations can be named, and among the second - ageing of output products or used technology or ageing of equipment. It must be noted that one type of uncertainty can be referred to different classes simultaneously. It depends on, what features of agile manufacturing system are studied and in what situation it functions.

Summary. In formation of control actions to agile manufacturing system it is necessary to take into account influence of numerous undefined factors. As prediction of influence of these uncertainty will be more exact, as the presentation of agile manufacturing system will be more effective. Analysis of consequences from these factors and effectiveness of made control decisions can be estimated by simulation method not excepting statistic efficiency estimates methods of made decisions.

5. INTELLIGENT CONTROL OF AGILE MANUFACTURING

As noted above, solution of control problem of agile manufacturing system as of complex dynamic object (CDO) in uncertainty conditions is connected with a series of difficulties, their overcoming is not possible with the help of classical methods of control theory. Therefore, control system of agile manufacturing should be designed in the class of intelligent systems.

Here we shall consider as intelligent control system (ICS) such system which can control CDO in conditions of uncertainty, resource deficit and origin of unnominal situations not less effectively, than it could be made by very competent (high professional) practiced manager.

It should be noted that the control problem of AMS should be solved simultaneously with modeling problem of processes of its functioning and development. The solution complexity of this problem is in the elaboration of models of agile manufacturing systems and models of its elements which adequately reflect their structure, organization laws, functioning and development dynamics. In so doing models are rational forms of knowledge representation about agile manufacturing. They form together with concepts, patterns, principles, and laws etc. ontology of agile manufacturing. In the whole ontology of enterprise is destined not only for obtaining, storage, manipulation and representation of knowledge about enterprise in computer systems, but also for realization of communications between separate functional subsystems (agents, persons, computing systems etc.), and also for logical choice justification of tools of enterprise. The conception of computer integration of knowledge necessary for construction of intelligent control system is founded on ontology.

General functional diagram of agile manufacturing intelligent control system is represented in figure 1. Intelligent system has three channels of control decision formation. In one channel

decision is formed on the base of analysis of a current situation with use of mathematical (probabilistic) methods of efficiency estimations of functioning and development of a system as a whole. In the second channel decision is formed with the help of expert systems which use experience and knowledge of the experts and are based on logical inference rules. And in the third channel it is formed on the base of simulation results of possible system behavior in given situation or close to it situation. In so doing in modeling the following intelligent control algorithms can be used:
- algorithms based on fuzzy logic (fuzzy set theory);
- algorithms realized in the form of neural networks;
- genetic algorithms;
- combination of these algorithms and probabilistic methods in framework of conception of "soft computations".

Except of these channels using artificial intelligence algorithms, there is also channel of natural intelligence, in which decision is formed by the person responsible for decision making. Further all decision alternatives are to be coordinated among themselves by one of known methods (voting, priority and so on) and as the result unique control decision U^0 is formed, which is recommended to implementation. In decision making it must be taken into account both actual possibilities of its realization, and consequences because of its realization. The peculiarity of the given system is in the fact, that agile manufacturing and environment are considered as united controlled object. At that concept of environment (exterior world) includes not only market infrastructure, but also social, ecological, information one and so on. Therefore disturbances acting on agile manufacturing are not destabilizing factors, but parameters of controlled object, which define its state.

The global function of agile manufacturing is output of high-quality product in definite volume in given time, which satisfies consumer demand and allows to obtain maximum profit in its realization in the originated situation. This global purpose can be divided on a series of subpurposes with separation of functional subsystems providing them. Such decomposition of purposes and functions is corresponded to usage of modular (block) principle in manufacturing system model constructing.

Manufacturing system functioning in steady-state dynamically equilibrium regime can be described according to the next assumptions. Manufacturing subsystem realizes product output according to preset (planned) volume $N^0(t)$ and intensity $\dot{N}^0(t)$. For multiproduct system output volume and output intensity of every product according to available resources $R^0(t)$ are planned. Product output is ensured by normative resource consumption $\dot{R}^0(t)$, and product realization on market at dynamically equilibrium price in condition of demand and supply equality defines planned income and profit at expenses according to planned ones. Described functioning scheme presents undisturbed motion of manufacturing system on calculated (planned) trajectory and is realized idealization of system obtained as a result of planning task decision by optimization methods. However, acting of external factors in form of different types of uncertainties lead to deviations of system motion. These deviations can have "positive" character, when they lead to unexpected getting the profit $P(t)$ more than planned one $P^0(t)$, or "negative" when profit is less.

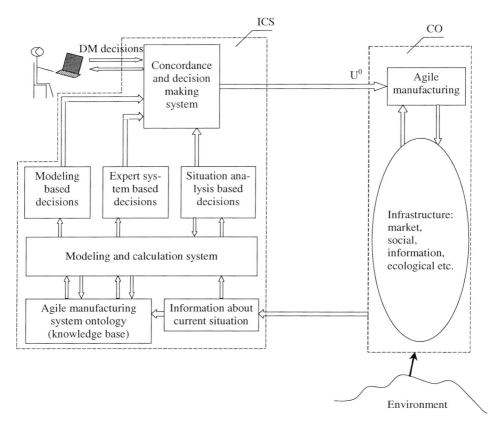

Figure 1. General functional diagram of intelligent control system of AMS

In constructing of mathematical model of manufacturing system being designed on the base of functional diagram in the accepting modeling methodology it must be taken into account two principal peculiarities.

The first peculiarity of dynamics description is in the fact that all the processes (manufacturing, product sale, forming of expenses and profit), characterizing functioning of manufacturing system, are modeled in rates or intensities. It corresponds to mean of system motion description accepted in control theory, and it don't contradict to economical-mathematical methods of dynamics modeling of manufacturing systems. Such approach allows to separate disturbed motion from dynamically equilibrium one, to estimate character and value of deviations, and to define required in particular situation control action. Using of product output intensity $\dot{N}(t)$, intensity of realization (sale) $\dot{N}_{sal}(t)$, intensity of cost $\dot{C}_{ps}(t) = f(\dot{N}(t), \dot{N}_{sal}(t))$, connected with product manufacturing and its realization, allows to define gain intensity $\dot{G}(t) = \dot{N}_{sal}(t) * C(t)$ and profit intensity $\dot{P}(t) = \dot{G}(t) - \dot{C}_{ps}(t)$. Here $C(t)$ is individual product price. Modeling of manufacturing system motion using intensities does not exclude usage of volume indexes in situation analysis and decision making.

The second peculiarity is in the fact that every model variable consists of static and dynamic components. Static component reflects equilibrium motion of a system under action of specified external forces, i.e. motion on planned, calculated trajectory. Dynamic component characterizes motion in non-equilibrium state under action on a system of unforeseen disturbances and deviations, having uncertain character.

These peculiarities reflect dynamic approach in the developed general methodology of research, modeling and control of manufacturing systems and correspond to the conception of agile manufacturing.

The functional diagram of control system of manufacturing is presented in figure 2, it reflects interconnection of production processes and product sale processes and also interconnections of controlled and control variables.

Here the following designations are used: $N^0, \dot{N}^0$ - planned product output and its intensity; $N, \dot{N}$ - factual product output and its intensity; $\dot{R}^0, \Delta \dot{R}$ - planned resource consumption and correcting value of resource consumption; $\dot{N}_{com}$ - competitor's supply intensity; $\dot{N}_{dem}$ - demand intensity; $C, \Delta C_i$ – individual product price and price addition; N_{sal}, $\dot{N}_{sal}$ - product sale volume and intensity of sale; $\dot{N}_{sup}$ - supply intensity; C_{adv} - advertising cost; P^0, P - planned profit and factual profit; $\dot{C}_{ps}$ - intensity of summary cost; $\dot{T}$ - taxes; Δu_1, Δu_2 - control actions; $\Delta N^0, \Delta \dot{N}^0$ - correcting volumes of planned indexes.

Functional diagram of MS control system should contain the following components named models (figure 3): manufacturing control; product sale in market; cost formation; profit intensity definition.

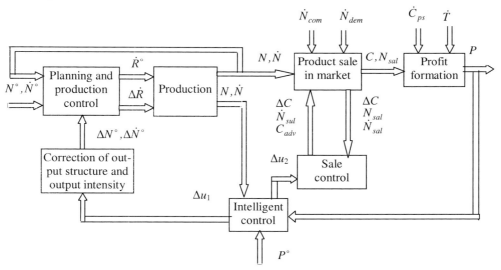

Figure 2. Functional diagram of manufacturing control system

Manufacturing control model describes production process as transformation process of planned resource volume $R^0(t)$ neccesary for product manufacturing into planned output $N^0(t)$ and output intensity $\dot{N}^0(t)$, and also process of production control. Distance-based control in the form of feedback is in the correcting of factual output intensity $\dot{N}(t)$ according to distance value $\varepsilon_N(t) = \dot{N}^0(t) - \dot{N}(t)$ by means of variation of resource consumption $\Delta\dot{R}(t)$. Resource consumption is restricted in principle both on volume, and on intensity by possibilities of manufacturing, therefore task of distance-based control is setted up as the task with restrictions.

Essence of control of material production is in resources consumption. Types of resources, their relation, usage means, degree of influence on production and control processes are defined by technological peculiarities, type of product and information support.

In the paper it is introduced concept of generalized resource, which includes all the types of resources, necessary for product output of a given type and represented in monetary form. Planned generalized resource $R^0_{gen}(t)$ is destined for realization of prescribed schedule of product output. Generalized resource defines economical controllability of a system, and usage intensity (consumption) $\dot{R}_{gen}(t)$ of generalized resource defines productivity of a system. At that the questions connected with organizing of allocation process of generalized resource between separate structures concern to tasks of lower level of hierarchical control system of MS and are not considered in the paper.

Model of product sale in market consists of block of demand and supply definition, product price formation block according to disyance ε between demand and supply, sale rate formation block according to distance ε and difference ε_c between producers's prices, gain rate definition block. Input coordinates of model are factual product output intensity, cost $\dot{C}_{adv}$ connected with advertisement and product extension in market, they are factor influencing on product realization in significant degree. Output coordinates are factual intensity of sale $\dot{N}_{sal}(t)$ and gain intensity $\dot{G}(t)$.

In cost formation model there are calculated manufacturing costs $\dot{C}_p$, which depend on factual consumption of generalized resource. Sale costs $\dot{C}_s$ include transportation expenses and expenses connected with product extension in market. Input coordinates of model are factual intensity of sale $\dot{N}_{sal}(t)$, product inventories consumption $\dot{N}_{str}(t)$, output coordinates are factual intensity of summary cost $\dot{C}_{ps}(t)$, and also ratio $\dot{C}_{adv}/\dot{C}_{ps}$.

In *profit rate definition model* there are calculated values of profit volume $P(t)$ and profit intensity $\dot{P}(t)$. At that for purposes of control it is not essential, what profit indexes one uses, and what payments from income one must make. These payments are considered as constant in dynamics modeling of manufacturing system, because period of their change is essentially larger than model time.

In control of manufacturing systems planning, prediction and decision making tasks are decided at various control levels and in different functional subsystems. Revealed peculiarities of manufacturing systems, functioning in market environment and also peculiarities of their control in conditions of uncertainty allow to choose methods of artificial intelligence for decision of these tasks.

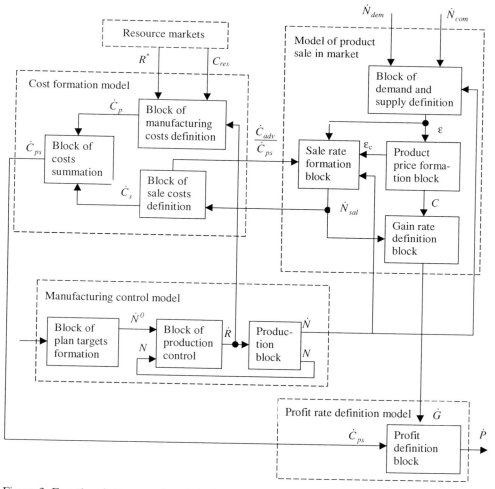

Figure 3. Functional diagram of model of control system

In the paper planning task decision according to demands is realized with use of neuronet modeling[6-10]. At that there are decided:

• planning task of demand in resources, which is in formation of resource volumes necessary for output of products of planned nomenclature and volume. The task is decided in conditions of uncertainty of resource market state and limitations of resource availability in market, and also limitations of enterprise financial possibilities;

- product output volume planning task, which is in definition of planned output that will be capable to ensure effectiveness (profitability) of enterprise. The task is decided in conditions of uncertainty of product market state, changeability of product demand and limitations of enterprise technological possibilities.

The given approach allows on the base of inexact data to create automated planning systems, which are rapidly responsible to changing market and manufacturing situation and are differ from known economical-mathematical methods in plan calculation quickness.

Analysis of existing methods of prediction tasks decision of manufacturing system allows to assert that for increase of prediction effectiveness it is necessary to use methods, which allow to reveal inexactitude of data received in conditions of information incompleteness and uncertainty of controlled object behavior. One of such methods can be method based on use of neural networks. At that additional procedure of neuroanalysis, unlike planning tasks, is working up of learning sample with the help of experts of subject area. In control of manufacturing system there are decided tasks of production and product sale processes indexes (price and volume of sales) prediction and task of environment characteristics (demand and supply of competitors) prediction.

Use of neural networks allows to make predictions on the base of incomplete and inexact data and to take into account influence of various unfavorable factors, to reveal among them most influencing on to sale and manufacturing processes, and to take into account their action in further predictions.

The developed operative control algorithms based on combined use of neural networks and fuzzy logic allow to decide in automated regime tasks of situation identification and classification, tasks of making and choosing of decisions by methods excepting subjectivity and uncertainty in situation analysis and decision making.

The proposed methodology of modeling of dynamics and manufacturing systems control is in the fact that complex organizational systems can be studied with the help of different types of hybrid models which are applicable for description of base functional subsystems. In modeling of manufacturing process the dynamic characteristics of process of material resource transformation into finite product and process of information signals transformation into control decisions become most informative. Reactivity, inertia, time delay of response to input signal are peculiar to all these processes.

Dynamic manufacturing control model is represented as aggregate of dynamical units, their mathematical models are described by nonlinear differential equations reflecting base properties of considered dynamic processes.

Production process is modeled by dynamical units which simulate change of base production indexes during time: output intensity and volume (controlled variables) in changing of generalized resource consumption (controlling actions). At that production process inertia, time delay, necessary for production preparation, intensity of production are accounted. Choice of product output intensity as one of base production indexes is based on the fact that it reflects production process dynamics and it is basic for account of other production-economical indexes (labor productivity, productive capacities and production spaces, prime cost and so on). Correction of $N^0(t)$ and $\dot{N}_0(t)$ values is made by decision making system on the base of information about market state of given production type and it represents control decision of control system external contour. Actions of environment on production process, which are typical for production-economical systems and are conditioned by man participation, are modeled as random disturbances $F(t)$. Question about adequacy of model to certain production is decided on the base of statistic data about manufacturing system functioning during long term.

Further we shall consider one of approaches to agile manufacturing and market control intelligent system constructing as a whole object on base of developed dynamical models and decision making models based on fuzzy logic.

Application of fuzzy logic gives possibility to decide wide range of tasks where data, goals and restrictions are complex, badly defined and are difficult to be exactly mathematically described. By virtue of its orientation to decision making process modeling in conditions of uncertainty such approach provides designing of control algorithms with artificial intelligence methods. Foundations of choice of this approach are also obviousness, non-traditional statement of control task, correctness and availability of used body of mathematics, understandable interpretation and enough simplicity of worked out algorithms. Perspectiveness of control method of complex objects, organizational systems among them, with use of fuzzy logic algorithms is confirmed by different investigations in given domain [11].

Manufacturing system control task in automated regime with use of situational approach and analysis of system dynamics is described in details in authors`s papers [12-14]. Manufacturing and market state identification is made by vector of variables $\overline{B} = \{P, \dot{P}, \Delta P, \varepsilon, \dot{\varepsilon}\}$ which reflects system dynamics by means of analysis of profit rate $\dot{P}$ and volume P, deviation of factual profit from planned ΔP, deviation of demand from supply ε and also rate $\dot{\varepsilon}$ of its changing. As base for decision making the next hypothesis is used: change of every state variable in admissible boundaries defines possible situation classes which require different control algorithms. It is possible, for example, to choose boundary "base" situations defining at qualitative level (when every variable from chosen ones takes value: more, less, or zero). However in case of such approach intermediate values of state variables are not considered.

System transfer from one state to another is made by action on one or more control variables. Making of control decision is presented by badly formalizable procedure of value change of control influence directed to variable values change. At that influence degree are defined by decision-maker, which takes into account recommended admissible range of variable values. In case when correction value is known to decision-maker, calculation of new value is not difficult. In situation when correction value is not known beforehand, fuzzy algorithms are advisable. In this case decision making system model can be represented in view of aggregate of rules describing system dynamics, regularities of interconnection of model parameters and control variables.

For realization of such approach, values of linguistic variables were defined, membership functions and logical rules were designed.

Every linguistic variable, chosen for system state description in vector $\overline{B}$, can take values from range:

profit P = {"losses", "satisfied", "planned", "high", "extraprofit"};

profit rate $\dot{P}$ = {"very low", "low", "normal", "high", "very high"};

deviation of profit from planned one ΔP = {"large negative", "small negative", "normal", "small positive", "large positive"};

ε = {"large negative", "small negative", "equilibrium", "small positive", "large positive"};

$\dot{\varepsilon}$ = {"very low", "low", "normal", "high", "very high"}.

As controlling variables the next ones can be chosen: volume N, product output rate $\dot{N}$, price C, advertising expenses $\dot{C}_{adv}$, sale organization expenses $\dot{C}_{sal}$, product transportation expenses $\dot{C}_{tr}$, storing rate $\dot{N}_{str}$ and so on. At that every variable can be presented as linguistic one which takes values from its range:

volume N = {"decrease", "few decrease", "leave at previous level", "few increase", "increase"};

output rate $\dot{N}$ = {"decrease", "few decrease", "leave at previous level", "few increase", "increase"};

price C = {"increase price to maximum price C_{max}", "increase price to prices of competitors C_{com}", "increase price to equilibrium price C_{equ}", "leave price at previous level", "decrease price to equilibrium price C_{equ}", "decrease price to prices of competitors C_{com}", "decrease price to minimal price C_{min}"}. (Minimal price of product is set as equal to its prime cost, maximum price is defined by purchasing power of consumers, equilibrium price is set in conditions of system equilibrium).

For other variables values are determined similarly.

Algorithm of transformation of one situation S_i to another S_{i+1} (unfavorable one to favorable) at that approach can be designed in the form of logical rule Pr_n: - {IF S_i THEN D}. For that purpose unfavorable situation S_i is identified by the next variable values:

S_i: {profit P = "satisfied", profit rate $\dot{P}$ = "very low", deviation of profit from planned one ΔP = "large negative", ε = "large positive", $\dot{\varepsilon}$ = "high"}.

Rules Pr_n of making decisions D is presented as the next particular decisions, which are logically tied up by conjunction operations:

D:- {volume N = "increase"} & {output rate $\dot{N}$ = "increase"} & {price C = "increase price to maximum price C_{max}"} & {advertising expenses $\dot{C}_{adv}$ = "decrease"} & {expenses to sale organization $\dot{C}_{sal}$ = "increase"} & {transportation expenses $\dot{C}_{tr}$ = "increase"} & {storing rate $\dot{N}_{str}$ = "decrease"}.

Decision of intelligent control task with application of fuzzy logic algorithms is possible use of simulation software. Analysis of existing simulation software based on fuzzy logic, such as CubiCalc, RuleMaker, FuziCalc showed that these software products at first are narrowly directed to concrete task. At second they are not suitable for embedding them into general MS simulation system because of their hardware or software peculiarities. At third they are not easy available because of their high price.

Therefore there was developed system of simulation of dynamics and manufacturing control decision making and product sale processes in uncertain market conditions, which uses fuzzy logic algorithms. System allows to get controlling variable values on base of data about state of manufacturing and market, which is represented in view of linguistic variables.

The peculiarity of proposed simulation system is that in the system different types of membership functions can be set. It can function in client-server mode, process data from other subsystems and from environment, i.e. allows to serve in interactive mode several external programs, adjusted to current task; to give results both in digital and graphical views and activity levels for every rules. In the course of program work, depending on changing market and manufacturing situation it is possible to realize trimming of functions, to change linguistic variables or their values.

In proposed simulation system for every variable membership functions are constructed with defining of range of admissible values of linguistic variables on base of expert knowledge about data domain.

Here knowledge base has been created which consists of rules of Pr_n type and in which logical relations between variables (difference between demand and supply, between output and profits and similar) are tied up. Digital values of variables, reflecting dynamics and defining concrete situation in manufacturing and market, are transmitted from dynamic models which is described in details in [15]. In result of simulation both rules leading to decision making activity level and variable digital value obtained in result of defuzzification are calculated. Thus, set of possible control decision in concrete situation is formed on base of set of variable values.

In course of active dialog with simulation system user gets estimation of market and manufacturing state at given time moment and best digital value of output variable. At the same time he has possibility to choose his own decision from set of recommended decisions. Variable digital value, which is recommended to be accepted with purpose of transformation of described unfavorable situation into more favorable one, is defined taking into account calculated activity levels of rules on base of known, realized in simulation system, max-min and max-product methods and on base of different defuzzification methods. Having set new value of this variable in condition that it is admissible, in course of simulation experiment we can contend that transition into new situation allows increase manufacturing system profitability.

Using of given approach gives possibility to act directly to system for its transition on new motion trajectory.

Proposed control algorithms, as experiments showed, allows not only to increase system effectiveness, but also decrease losses of profit in influencing of unfavorable disturbances from the side of environment due to making more effective decisions.

Curves of factual rate $\dot{N}(t)$ and product output volume $N(t)$, and also factual rate of sale $\dot{N}_{sal}(t)$ and sale volume $N_{sal}(t)$, constructed according to data for one from products produced by machine-building enterprise in the framework of conversion program, are represented in figure 4 and figure 5. Significant changes of these curves are stipulated in particular by an essential non-stability of Russian resources and goods market. There are represented curves of output rate $\dot{N}_{rec}(t)$ and output volume $N_{rec}(t)$ recommended by intelligent control system as the result of decision making task in control of output rate and volume.

Figure 6 illustrates effectiveness of made control decisions which is expressed in this case in decrease of profit losses of enterprise by means of correction of output rates and volumes according to market needs.

Summary. Developed system allows to implement investigations of manufacturing system dynamics and carry out experiments connected with system behavior simulation in changing conditions of environment.

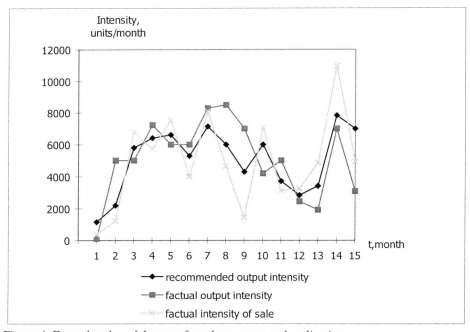

Figure 4. Factual and model rates of product output and realization

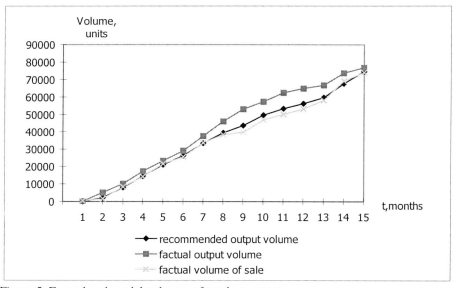

Figure 5. Factual and model volumes of product output

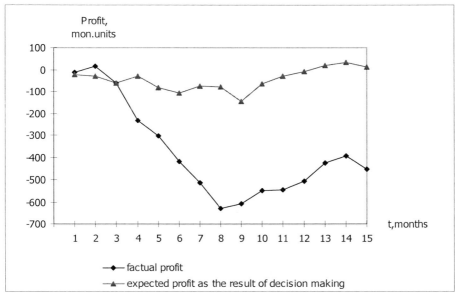

Figure 6. Factual and model profit

Advantage of proposed approach to manufacturing system investigation by method of simulation with use of fuzzy logic consist in the fact that it makes possible to simulate situation accounting its dynamics with enough authenticity, to predict its developing and to estimate effectiveness of one or another decision, at that every experiment can be reproduced with new initial data or decision variants.

6. DEVELOPMENT PERSPECTIVES OF AGILE MANUFACTURING CONTROL SYSTEMS

Agile manufacturing system conception is further development of conceptions directed to manufacturing system organization improvement, and it embraces lot of them, including flexible manufacturing systems, multiagent, holonic, virtual ones.

This conception defines manufacturing development strategy in 21st century and it is directed to creation of self-organizing intelligent manufacturing system based on latest information technologies. Further development of this conception provides also agile manufacturing control system development, which will be directed to improvement of tools, algorithms and methods:

- receiving of new knowledge;
- estimation and control of agile manufacturing internal state and future states prediction;
- constructing of adequate model of exterior world and prediction of its changes;
- effective and purposeful influence to agile manufacturing system, including for purpose of self-organizing;

- formation of optimal development strategy and tactics of manufacturing system behavior in exterior world, at that accounting requirements to careful relation to natural environment and to rational use of natural resources;
- intelligent control (decision making), directed to securing of high effectiveness of agile manufacturing functioning and development in conditions of uncertainty, resource deficit, and critical (crisis) situations appearance;
- self-preservation and self-protection from destructive influences to a system on the base of exact prognosis of extremal (crisis) situations.

It must be noted that agile manufacturing system development also provides for improvement of technical tools, technological equipment, and technologies themselves, communication tools, and professional level of laborers. Intelligent information-control system development above all directed to computer integration of knowledge, functioning in different functional subsystems of agile manufacturing, for the purpose of control improvement of a system as a whole. For a system of computer integration of knowledge it is necessary to develop software and procedural-linguistic tools and also logical tools of their interaction. Here the updating process of knowledge given in result of engineering creative activity is especially important.

Presence of developed knowledge base, in which enterprise ontology is stored, is necessary condition for agile manufacturing system self-improvement. At that agile manufacturing intelligent system, in which knowledge is integrated, gives the possibility to change operatively enterprise organizational structure in conditions of uncertainty and quick-changing situation in order to obtain finite result faster and without unnecessary losses.

Not less important problem is the improvement of account methods of psychological factor influence to engineering creative activity results and to decision making processes in agile manufacturing framework at level of computer systems.

Summary. Development of agile manufacturing control system should transform it at the future into intelligent highly organized system, which would be inalienable part of global life-support system of world information civilization of humankind.

REFERENCES

1. M.Uschold, M.King, S.Moralee, Y.Zorgios. The Enterprise Ontology. - The Knowledge Engineering Review, V.13, No.1, 1998.
2. M.H.Lee. The Knowledge-based Factory // Artificial intelligence in Engineering. 1993. V.8. No.2.
3. V.A.Vittikh. Engineering Theories as a Basis for Integrating Deep Engineering Knowledge // Artificial Intelligence in Engineering. 1997. V.11. No.1.
4. I.A.Budyachevsky, V.A.Vittikh. A Knowledge-based Calculus: a Logical-computational Framework for Engineering Theories// Proc. 11-th Intern. Conf. On the Application of Artificial Intelligence in Engineering, Clearwater, Florida, USA,1996.
5. S.Gallagher, J.Gillespie, M.T.Khorami. Computer integrated manufacturing (CIM) systems are dead. Long live knowledge integrated manufacturing (KIM) systems // Proc. Internat. Conf. "Factory 2000". Great Britain, New York, 1994.
6. Pham X. Liu. Neural Networks for Identification, Prediction and Control.- Springer Verlag, London and Heidelberg, 1995.
7. John Kean. Using Neural Nets For Intermarket Analysis.- Technical Analysis of Stocks & Commodities, November 1992.

8. Elizabeth L. Hartshorn. Mapping the Market's Future with Neural Networks - Research, September, 1995.
9. Anil K. Jain, Jianchang Mao, K.M. Mohiuddin Artificial Neural Networks. -A Tutorial, Computer, V.29, No.3, March 1996.
10. Dean S. Barr, Ganesh Mani. Using Neural Nets to Manage Investments. - AI Expert, February, 1994.
11. A.L.Guiffrida, R.Nagi. Fuzzy set theory application in production management research: a literature survey / Journal of Intelligent Manufacturing V.9, No.1, February 1998, is published by Chapman&Hall, London.
12. B.G.Ilyasov, L.A.Ismagilova, R.G.Valeeva. Intelligent control for dynamic organizational production-market systems. Proceedings of ICI&C'97, St.Petersburg, Russia, V.2,1997.
13. P.P.Groumpos, B.G.Ilyasov, L.A.Ismagilova, R.G.Valeeva. Production control as of a complex dynamic object. Prep. of the 8th IFAC Symposium on Large Scale systems: theory and applications. Rio Patras, Greece, V.1, 1998.
14. P.P.Groumpos, B.G.Ilyasov, L.A.Ismagilova, R.G.Valeeva. Intelligent control algorithms of dynamic manufacturing systems. Pros. of ASI'98 Life cycle approaches to production systems: Management, control and supervision. Bremen, Germany, 1998.
15. B.G.Ilyasov, L.A.Ismagilova, R.G.Valeeva. Modeling of production-market systems. UGATU Press, Ufa (in Russian), 1995.

Contingency-Driving Autonomous Cellular Manufacturing - Best Practice in the 21st Century

S.-J. Song

Department of Information and Intelligent Systems Engineering
Hiroshima Institute of Technology, 2-1-1, Miyake, Saeki-ku, Hiroshima 731-5193, Japan.

This paper is directed toward the development of models, methods and tools suitable for effective, profitable, and autonomous cellular manufacturing as a new vision for 21st Century. The cellular manufacturing is mainly concerned with the efficient integration of four major concerns: (1) planning-oriented predictive cellular layout design; (2) an unexpected order-adapted cellular manufacturing through the integration between manufacturing decision processes and cellular layout, (3) a contingency-driving shop floor adaptation, and (4) worker's high skills oriented cooperative planning between manufacturing and marketing activities for supporting continuous improvement. The possibility of sudden large changes in demand rate and process route is greatly reduced by both the detailed analysis of decision processes and the reconfiguration of existing facilities through worker's high skills. To increase the ability of reaction to shop floor contingencies, real-time information on the conditions of the shop floor will be analyzed to support continuous improvements for sustaining superior performances in unexpected changes. The continuous improvements support the user with five modules: real-time status monitoring, real-data analysis and renewal forecasts, diagnosis through intelligent decision support system, simulation optimization, and results evaluation and performance measurement.

1. INTRODUCTION: What's Changing in Manufacturing?

Today's manufacturing is now facing profound issues of achieving a better productivity and competitiveness in permanently changing market environment [1, 2]. An ultimate manufacturing system for future manufacturing perspectives, which can continuously ensure customer unique requirements in more cost-effective ways, requires satisfying the following vital important characteristics [3, 4]:
1. Increased product diversity in low volume,
2. Shorter lead time through concurrent engineering,
3. Just-In-Time delivery and supply in product variants,
4. Market-adapted flexible manufacturing for low volume, high variety production, and
5. Increased awareness of the social expectations such as workers' skills-based and green production as well as globalization.

It is refreshing to realize that Cellular Manufacturing (CM) is now discernible to satisfy these characteristics with superior performances. In upward trend, concerns of CM have been a fruitful area for improving shop floor during the last two decades. The CM, which is a recent synonym for Group Technology (GT) pioneered by Mitrofanov [5], seeks to exploit many of the efficiencies associated with mass production in less repetitive batch and jobbing production. Despite the insights of "system engineering" approach provided by Arn [6] in Group Technology, Jackson in his book "Cell System of Production" has originally suggested a perspective solution for the implementation of CM in 1978 [7]. These days, for efficiently producing innovative products, Japanese leading companies, Toshiba, SONY, and NEC have now developed an unique cellular manufacturing system which was operated by multi-skilled workers as a basically manual cell, resulting in the dramatic improvements in terms of worker motivation and productivity of shop floor by reconfiguring their automatic transfer line [8-10].

2. ARE REALLY CURRENT WAYS SUFFICIENT FOR CELLULAR MANUFACTURING?

2.1 Where We Have Been

Although hundreds of publications on CM have appeared so far including some excellent reviews and special issues since its inception which address its various aspects [11, 12, 13, 14, 15], many of the issues in the areas of design, planning, and control of CM systems still remain unexplored. It is surprising that most of these early researches have been on the development of techniques to identify part families and / or machine cells as the cell formation procedures. With requirements for short-term control steadily rising, there is very little attention on the design, management, and shop floor control of CM system in an effective way of coping with unpredictable changes in social and customer desires. In addition, most attempts on CM were directed toward "one-shot" problem solving through various objectives reflecting information on similarities with respect to facilities, routing, equipment, and design attributes as a short-term approach, even though layout design would be long-range strategic planning. While some are still solving the block diagonal problem with the binary matrix, flexibly integrated autonomous cellular manufacturing is becoming a new trend beyond traditional manners in CM.

2.2 Changing Trends in the Cellular Manufacturing

A better way to success CM is to change our thinking paradigm beyond "one-shot" problem solving towards "on-going" manner for support continuous improvement of shop floor. Manufacturing contains much more undeterministic attributes than ever before. To ensure more flexible and more profitable in responding to market changes, continually improving facilities, systems and planning which were conducted with optimal or state when first introduced, increasingly surpass their initial capabilities.

An innovative, flexible and more cost-effective cellular manufacturing should dynamically respond to unpredictable, rapid market changes because layout design problem is not only long-range strategic planning, but also more significant issue than the other manufacturing decision to improve productivity 50% or more. It is extremely important to develop a flexible cellular layout in the light of two reasons: firstly, layout configuration influences all of the manufacturing activities such as production and process planning, scheduling, etc.; and secondly, layout change needs to long-term period and is very costly. Therefore, it is impossible to change the layout configuration day-by-day according to the changes of product and market needs. Manufacturing system design and layout configuration is important in the creation of a manufacturing environment that will facilitate sound production management and control. With well-designed layout design, it is possible, in many cases, to move in direction of relatively simple material flow and high factory profitability.

The possibility of sudden large changes in demand rate and process route including seasonal changes is greatly reduced by detailed analysis of decision processes and utilization of existing facilities as well as human resources more efficiently. The primary strategy of the integrated autonomous cellular manufacturing is concerned with the reconfiguration of the existing facilities and the role of human resources to quickly respond unpredictable changes instead of new investment in computer-integrated machines that will transform their factories into highly flexible operations.

3. INTEGRATED AUTONOMOUS CELLULAR MANUFACTURING (IACM)

3.1 Why the Integrated Approach?

The increasing complexity of manufacturing environment can be counteracted by flexible integration that could reconfigure operations, processes, and planning in a cost-effective timely manner. It also implies an attempt to seek the global optimal solution rather than local optimal solution for evaluating the performance of the system. New ways of integration between material flow and information flow related to manufacturing activities are usually given by:
1. Using information technologies such as computer network for sharing and managing product data,
2. Applying industrial organization in terms of cross-functional team and sharing skill and knowledge enhancing technologies.

In analyzing the models of integrated manufacturing system proposed thus far, they can be classified according to whether the information feedback process exist in the solution procedure or not. The monolithic and hybrid models are often solved by the iterative interaction between the upper-level and the lower-level problems. However, hierarchical approach that divides a problem into a hierarchy of subproblems does not have a complete feedback process back to the higher level again. The IACM is to be achieved by the integration of information feedback process between cellular layout design and order reactive planning as well as shop floor control to quickly respond customer unique orders.

3.2 Essentials of Autonomous CM

The integration of various manufacturing decisions through human-based cooperative planning between production activities and marketing behaviors implies the possibility of realizing an autonomous manufacturing system for low volume, high variety production [16]. Integrated autonomous cellular manufacturing (IACM) toward a new trend for 21st Century, is mainly concerned with efficient integration of three major following concerns:

(1) *Long- / Medium-range for flexible layout design.* An effective planning-oriented predictive cellular layout design is essential to stable enterprise-wide operating system over strategic long-range planning period [17-19];

(2) *Short-range planning to unexpected order.* An order-adapted reactive planning is extremely significant for responding quickly to unexpected customer order changes in a short period through an effective integration between manufacturing decision processes and cellular layout [20-23];

(3) *Unpredictable sudden daily changes.* Contingency-driving shop floor adaptation for supporting continuous improvement is vital to success of continuous improvement shop floor control [24, 25].

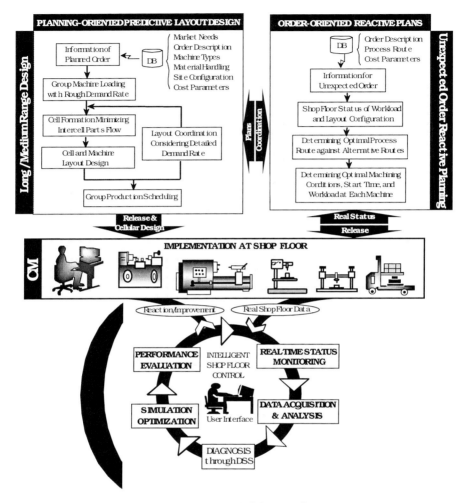

Figure 1 Contingency-Driving Autonomous Cellular Manufacturing (CM): Functions, its Issues, and Structure

3.3 A Successful Way for Developing and Implementing IACM

Modern manufacturing is faced with great challenges, particularly in terms of rapid changes in customer requirements, shorter product design and life cycle, and shorter lead time to customers. In order to compete successfully, companies must now be able to respond much faster to market needs. In an attempt to move beyond the concerns of traditional cellular manufacturing designed with a static nature of demand, capacity, and production attributes, the authors developed a market-adapted flexible cellular manufacturing system in 1993 [21]. The market-adaptive flexible cellular manufacturing system based on autonomous and cooperative manufacturing activities embodies three major changing situations from long-range strategic planning on cellular layout design to real-time shop floor control to respond efficiently to short term variations in dynamic shop floor situations.

Integrated autonomous cellular manufacturing, as depicted in Figure 1, attempts to integrate the planning-oriented predictive design with expected demand pattern and order-oriented reactive planning as well as shop floor control for supporting the continuous improvement of layout configurations for multi-products families with specific due date and production quantity at multi-planning periods. Flexibly integrated and adaptable forms of cooperation among these three phases are an appropriate means of responding quickly to dynamic requirements from markets and customers.

3.3.1 Planning-Oriented Predictive Design

In dynamic market situations associated with the unexpected changes in demand fluctuations, varying product mix, rescheduling for expediting products, defective in-process products, material shortage, and process routing, the market desires are satisfied by collaborative integration between planning-oriented layout design and order-oriented reactive planning as well as contingency-driving shop floor control phase with feedback information flow. The unexpected changes are generally divided into three distinct cases as follows:
1. The planned response to anticipated changes by introducing long-range strategic plans, such as facility investment or foreign direct investment, technologies and process innovation, or man-power planning, which relate to the design of the manufacturing systems to ensure sufficient flexibility and / or agility;
2. Handling of external unanticipated changes efficiently by utilizing existing technologies and resources, for instance, overtime, increasing machining speed, inventory, and changing shifts to increase capacity temporarily, rationalization of material flow etc.;
3. Uncertain changes concerning machine breakdown, defective in-process inventory, material shortage, and all of which influence production scheduling.

To successfully develop the integrated autonomous cellular manufacturing, this paper deals with two major capabilities of external unanticipated and uncertain changing environment with information feedback process among decision-making problems. In planning-oriented predictive design phase, we discuss several problems concerning how flexible cellular layout will be developed efficiently over a long period [17, 18, 19].

3.3.2 Order-Oriented Reactive Planning

For successful implementation of order-oriented flexible cellular manufacturing, it is significant to integrate two major production types: planning-oriented predictive design and order-oriented reactive plans. In order-oriented reactive plans, four decision-making problems which integrate the information flow of both planned products and ordered products are simultaneously analyzed and optimized under given cellular layout configuration [22, 23]. The decision problems relate to production planning, process planning, cellular layout, and production scheduling for multi-product with specific due date and production quantities desired. Also the reactive plan proposes an adaptation that are capable of coping with the high rate of unexpected changes by considering overtime usage, increasing or decreasing machine speeds, and intercell movement for the following two conditions [20, 21]: (1) when the cell does not include a machine required for processing the ordered products; (2) when the whole machine within the cell is not capable of producing total quantities of ordered products.

3.3.3 Effective Improvement of Cellular Manufacturing

Effective improvement activities are vital to success of integrated autonomous cellular manufacturing. In order to improve continually cellular manufacturing systems that were considered to be most favorable (optimal) conditions to dynamic shop floor status, real-time shop floor control introducing user interactive simulation assistance consists of the following *five modules*: real-time status monitoring for data acquisition from shop floor, automated real-data analysis and updating forecasting models, diagnosis supporting the cause and potential solutions to problems or events from the shop floor status data, simulation optimization for seeking the best solutions through combining simulation model and a class of direct optimization techniques called Evolutionary Algorithms [26], and results evaluation and performance measurement for supporting continuous improvements as a way to sustain the optimal or near optimal conditions of factory-wide operating processes in unpredictable changes.

4 MOVING TOWARD CONTINGENCY-DRIVING SHOP FLOOR CONTROL

4.1 Major Requirements and Activities

The goals of contingency-driving shop floor control are to develop and demonstrate a monitoring and decision support concept for improving planning and control of manufacturing resources and materials more efficiently.

Several effective manufacturing systems that have evolved with the evolution of computer technology, such as CIM, CALS, Virtual Manufacturing and philosophical techniques such as Lean or Agile manufacturing are firmly based in the fundamentals defined in the model for a production management system which comprises the basic concepts of planning, controlling, measuring, and evaluating. Effective planning should be future-oriented and flexible. Well-designed and long-range robust planning, in most all cases, is basically necessary for developing an effective production control system when considering the situational contingencies affecting.

The concepts of contingency-driving shop floor control is based on the need to coordinate the plans and control all the way from customer order to shipping products. Successful control process is a matter of establishing a good plan and adhering to it. Thus, good aggregate planning and cellular layout design are a prerequisite to good shop floor control. Key to the operation being successful are accurate forecasting, aggregate planning, inventory control, scheduling and gathering of shop floor status data to allow revision of plans as the status of conditions changes in the shop floor.

Some attempts were made to improve production planning and control or logistic plans by applying computer systems [27, 28, 29, 30, 31]. They provided an overview of current research projects and a commercial system for shop floor control system as well as the state-of-the-art survey. Although both predictive layout design based planned orders and reactive planning for random orders from customers are successfully achieved at strategic planning stage, shop floor control is most significant production activity to provide trend and performance information to management relating to daily quality and cost status, performance to schedule, dispatching effectiveness, and material flow fitness between the various machining cells. A contingency-driving shop floor control involves the actual execution of the monitoring of capacity and work in progress, and reaction to shop floor contingencies as far as possible within the limits of the original plans.

To increase the ability of reacting to shop floor contingencies, real-time information on the conditions of the shop floor will be analyzed to support continuous improvements for sustaining the optimal or near optimal factory-wide operating control in dynamic changing environment. The contingency-driving shop floor control operates through the repetition of the following five sequential functions: monitoring, data acquisition and analysis, diagnosis, simulation optimization, and periodic evaluation and performance measurement. These five functions evolving major shop floor should be integrated closely with long-range planning concepts to control quality, cost, and delivery by quickly spotting the cause of production defects.

In the simulation optimization function combining generic optimizes and simulation model, to be useful in supporting of contingency-driving reactive shop floor control, an alternative plan selected in diagnosis will be examined by all possible simulation parameters through the user interface. Finally periodic evaluation and performance measurement are needed to ensure or predict the consequences of implementation of suggested potential solutions.

4.2 Key Concerns Supporting Effective Shop Floor Control

Unpredictable market changes have required manufacturing companies to find more intelligent ways on the shop floor to be more efficient, more flexible, and more cost-effective. The goals of effective shop floor control are continual and rapid improvement in quality, cost, lead time, and customer service while achieving superior manufacturing performances. Contingency-driving control process, which is accomplishable through the use of intelligent decision support system, will increasingly play a vital role in the future manufacturing.

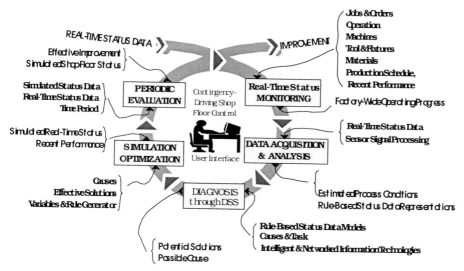

Figure 2 Key Functions Supporting Contingency-Driving Shop Floor Control

Figure 2 illustrates the key functions and a wide variety of real data embedding intelligent shop floor control.

4.2.1 Real-Time Status Monitoring

Monitoring in shop floor control is not a well-defined area. One important element of real-time shop floor control is the activity of monitoring the progress of production and the actual shop floor situation in terms of current status on jobs / orders on machining cell and machines over time, production facilities and its utilization, materials, operators, manufacturing process, recent performance, as well as coordination of specific plans when deviations occur. All of these data items have to be collected in real time. One of primary key to collect real-time status is through the use of sensors, such as noise, optical sensors, process monitoring, vibration, and force signals, to monitor sufficiently quickly and adequately to sudden changes in the process resulting from malfunction [32]. Figure 3 shows a model-based prototype of the machine and job current status, which was developed by Microsoft Visual Basic language and Excel application.

4.2.2 Real-Data Analysis and Renewal Forecasts

Automated data analysis and acquisition from shop floor would be required to ensure that the status of all shop floor entities was updated continuously. Based on shop floor data acquisition, the diagnosis is to support the user in determining the cause and potential solutions to problems or events detected by the data analysis. The accuracy and timeliness of the real-time status data collected from shop floor have a significant impact on the efficiency of manufacturing. A wide variety of real data embedding data acquisition and these collected real data has to be analyzed and compared in order to eliminate undesirable conditions and improve the readability of certain signal features.

4.2.3 Diagnosis through Intelligent Decision Support System (DSS)

The purpose of diagnosis is to support the user in determining the cause and potential solutions to problems or events detected by monitoring factory-wide operating processes (see Figure 4). If faults or breakdowns occur in manufacturing processes despite preventive maintenance periodic, more rapid identification and elimination of the cause of the defect is possible with the aid of a diagnosis through intelligent decision support system. The diagnosis module has a function to accumulate a store of knowledge concerning how best to overcome contingencies in the forms of rerouting work and reallocating production facilities. Usually the number of alternative ways of overcoming a contingency may be large because, in most manufacturing environments, a number of criteria need to be considered simultaneously.

Intelligent DSS, however, is computationally efficient and provide a quick solution through the decision-making process by human interface, which is often required in a constantly changing manufacturing system. Intelligence techniques are often effective for complicated problems containing a substantial degree of uncertainty [33]. The sophisticated DSS, thus, offers an appealing and efficient solution to dynamic process in less time and effort, and offering more intelligent capabilities. A key to success is to integrate various process activities through an intelligent and networked information technology. In addition, a more effective and quicker processing of the contingencies results from interfacing the user and the problem solving mechanism.

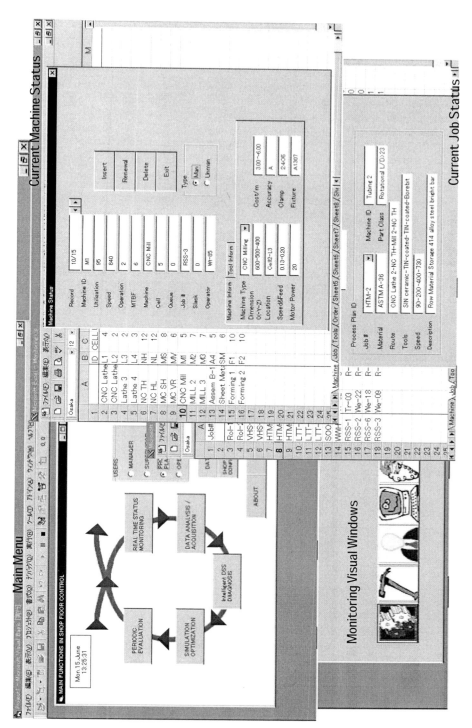

Figure 3 Model-Based Prototype for Developing Intelligent Shop Floor Control

4.2.4 Simulation Optimization Techniques

In the simulation optimization module, as shown in Figure 5, an alternative plan will be examined by combining optima-seeking procedures and all possible simulation parameters (or *decision variables*) through the user interface.

To be useful in supporting contingency-driving reactive shop floor control, simulation optimization techniques [24, 25, 34] would be essential. *Biethahn and Nissen* [35] and *Glover, et al.* [36] reported that evolutionary algorithms are very suitable for simulation optimization. *Akbay* has introduced some successful cases in applying simulation optimization as a decision support tool [37]. Actual performance goals will be dependent on the random variables and dispatching rules for each simulation parameter [38, 39]. The simulation parameters including dispatching rules and random variables can be altered semi-automatically in a good way of changing their values from minimum to maximum limits with desirable small step at each simulation execution. Simulation parameters are divided into five categories. The routing parameters involve the suggestion of alternative routing from a broken machine to an alternative machine or considering alternative process plans for handling intercell part flow more efficiently. Job properties relate to processing and setup times, ratio of processing and material handling times, batch size, due date, process routing, handling coefficient, and order quantity. Material handling includes speed, capacity, and initial position and status of each transporter unit. Material flow category has two major elements: layout configuration and intercell parts flow for calculating volume and travel distances between all pairs of machines. Finally, physical and human resources will be required to consider shift change or on/off sift, man-power capacity, maintenance times, and machine characteristics. On the other hand, performance measures, such as machine utilization, mean flow time, average waiting time, due date deviation and throughput rate are discussed to support continuous improvement of cellular manufacturing.

4.2.5 Results Evaluation and Performance Measurement

Results evaluation and performance measurement is needed to ensure or predict the consequences of implementation of suggested potential solutions. Consequence evaluations create the basis for selecting the best solution, and are to evaluate the actual and simulated performance of the production. Evaluation is made continuously, and any deviations and modification are accumulated in data acquisition module through monitoring activities. Periodic performance measures from the simulation optimization techniques with the various criteria, and is compared to threshold values with specification of upper and lower limits.

4.3 A Model-Based Prototyping Approach

A model-based rapid prototyping approach for contingency-driving shop floor control is suggested by using Microsoft Excel application and AI language, Visual Basic as an appropriate programming environment for the realization of the decision supporting system. As illustrated in Figure 3 and Figure 4, actual shop floor situation and the progress of production are collected in spreadsheets periodically. SimRunner, which is based on Tabu Search [40] and Evolutionary Algorithms, is used in conjunction with ProModel simulation packages for simulation optimization and periodic evaluation of shop floor.

Contingency-Driving Autonomous Cellular Manufacturing   593

Reasoning

	A	B	C	D	E	F	G
1		Date	Cell Location	Type	Ack.		Condition
2	1	Jun 14, 12:10	5 CNC Mill	A			Shotage Raw Material
3	2	Jun 14, 14:20	2 CNC Lathe2	A			High WIP
4	3	Jun 14, 15:02	1 Mill 21	A			Breakdown
5	4	Jun 14, 15:12	3	W			Deviation in Due Date
6	5	Jun 14, 15:32	2 Lathe4	W			too longer in Queues
7	6	Jun 14, 15:56	12 MC HL	S			Bottleneck

Solution

	A	B	C	D	E	F
1		Ack. Date	Cell Location	Cause		Solution
2	1	1998/6/5 10:12	5	transport capacity low		schedule change based Penalty rule
3	2	1998/6/5 11:28	2 CNC Lathe2	utilization low		change dispatching rule
4	3	1998/6/5 12:52	1 Mill 2	worn tool		change tool type TiN-casted bite
5	4	1998/6/5 15:10	3 MH 45	No cause identified		Increase average WIP level
6	5	1998/6/6 10:12	2	lot size too bigger		setup with two operator
7	6	1998/6/6 13:32	12 MC HL	lack standardization		apply classification code rule 41
8	7	1998/6/6 17:18	10 MC VR	No cause identified		change processing route to MC TH
9	8	1998/6/7 9:41	10 Forming 1	tool wareness		tool change, decreasing speed
10	9	1998/6/7 10:42	12	inventory shotage		route change in cell #3
11	10	1998/6/7 11:22	8 CNC Lathe	operator missing		check quality level at location CNC
12	11	1998/6/7 12:12	4 Mill 1	machine trouble		setup on machine using operator YH
13	12	1998/6/7 12:52	6	frequency of setup		increase with one shift on resource
14	13	1998/6/7 14:33	3	too low wip		increase transport capacity of MH4
15	14	1998/6/7 18:11	5 NC TH	bottleneck		estimated due date deviation of ord

Figure 4   Diagnosis Menu for Quick Decision Support

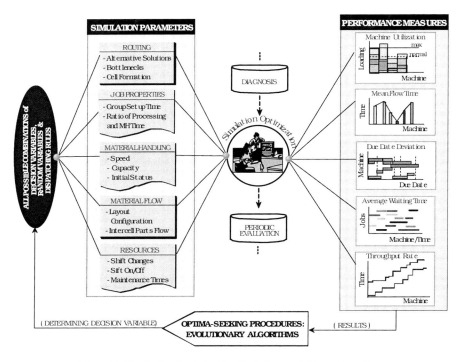

Figure 5 Simulation Optimization Techniques with User Interface

5. WORKER'S SKILLS ORIENTED AUTONOMOUS TEAM MANUFACTURING

5.1 Background to Human-Centered Cellular Manufacturing

Although there are numerous studies that address the problem of optimal or near-optimal machine grouping and parts family classification for cellular manufacturing, little research has been reported that studies a methodology or a concept for designing cellular manufacturing from the viewpoint of human factors as well as flexibility coping with dynamic changes. A large number of studies on human aspects and issues have appeared in the broad field of manufacturing engineering such as organization design, work design, and ergonomics or human factors since the first proposition of occupational biomechanics. *Gillis* introduces background information on human factors collaborative applied research in order to reinforce the benefits of synergy among the three elements of organization, people, and systems [41]. It describes the research project, Manufacturing Organization, People and Systems (MOPS) program, European EUREKA project called Integration of Technology and Organization for Quality Production (INTO - EU860) so as to optimize the full potential of people and modern manufacturing technologies.

In cellular manufacturing, most of researches have dealt with the technical aspects to convert functionally organized manufacturing facility into a cellular layout. Several studies on the human side of cellular manufacturing have attempted to reveal the nature of human resource management through descriptive field research [42, 43]. Although these two empirical studies provide helpful insights into relationships between cellular and functional workers, but there is not enough analysis to figure out a theoretical foundation for cellular manufacturing, particularly in light of human-centered manufacturing issues. This issue concentrates on designing and analyzing an effective incentive scheme that will be performed highly in the pursuit of human-enriching flexible cellular manufacturing. The system will be capable of developing through team-based cooperative planning between production activities and marketing behavior [16].

5.2 Production and Marketing Cooperative Planning with Effective Incentive Mechanism

Every organization has a potential for improving its performance through product and process design change, production cost reduction, demand rate enhancement, and better resource (material, equipment, and manpower) utilization. The crucial issue is whether the people working in the organization is encouraged enough to search for and implement these ideas which can improve the organization's performance. Since Frederick W. Taylor created the first high performance factory system, piecework incentive pay plans were widely popular as devices for increasing production [44]. However, his book Scientific Management notably lack to understanding how companies can dramatically improve their performance by embracing the asymmetric information between individuals and the knowledge dispersion issues, which affects performances, pioneered by K.J. Arrow in 1984.

David M. Upton pointed out in 1995 that achieving low cost and high quality no longer guarantees enough success, and argued that the role of people - both managers and operators is increasingly significant issues for successful development and implementation of advanced manufacturing systems [45]. However, Bassett expressed in 1993 that pay for output, either incentive or piece-work style, can increase cost or decrease product quality if not rigorously administered [46]. Various recent studies on strategic cellular manufacturing have come to recognize that there are two major problems of human-conscious manufacturing and inter-department coordination.

It is desirable to consider creating an incentive design [47] that would encourage workers within each department to make decentralized autonomous decisions. In the traditional cellular manufacturing based on "prior planning," workers are normally assigned to specific operating jobs and are expected to perform those jobs according to predetermined plans. Workers are not authorized to respond to unexpected changes such as machine breakdowns, defective in-process products, and adjusting production and/or logistics planning beyond their job specifications. These changes within a planned period are dealt with by corresponding specialists in each department and buffer inventories. The incentive design focused here is a more flexible and fluid approach to problem-solving based on "posterior information" about local random events affecting individual shop floor operations. Workers are encouraged and stimulated to respond to uncertain changes autonomously within their capabilities.

The self-managed workers in manufacturing department are responsible for increasing the available production capacity, and the self-directed workers in marketing department are responsible for increasing stochastic demand rate for multi-products, as depicted in Figure 6. On the other hand, a coordinator concentrates on establishing production policies based on the schedule of working hours desired. Furthermore, for successful development of human-enriching cellular manufacturing, four major human aspects are also considered: flexible working hours desired in determining available production capacities, higher salaries according to workers effort level exerted, job satisfaction with high salaries and flexible working hours, the elimination of conflict between production and marketing, and empowerment of decision-making.

6. CONCLUSIONS

Contingency-driving autonomous cellular manufacturing has an integration approach of manufacturing decision processes in the major four concerns: planning-oriented predictive cellular layout design, order-oriented reactive plans, contingency-driving shop floor control, and worker's high skills oriented cooperative planning between manufacturing and marketing activities for supporting cell-based effective improvement. For successful contingency-driving shop floor control introducing user interactive simulation assistance, the effective improvement support with following five interactive functions:
(1) monitoring for real data acquisition from shop floor,
(2) automated real-data analysis and renewal forecasting models for the real-time control,
(3) diagnosis through intelligent decision support system for detecting the cause and potential solutions to problems of events from the real time shop floor status,
(4) simulation optimization in combining optima-seeking procedures and all possible simulation parameters, and

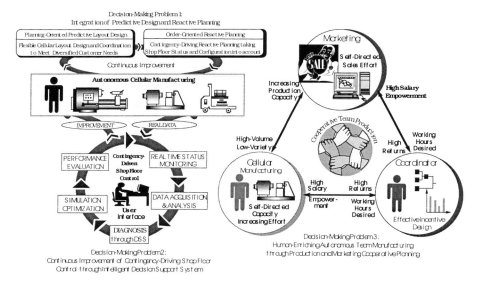

Figure 6 Worker's Skills Oriented Autonomous Cellular Manufacturing: Production and Marketing Cooperative Planning

(5) results evaluation and performance measurement for supporting continuous improvements as a way to sustain the optimal factory-wide operating process in unpredictable sudden changes.

Finally, a prototyping model is developed by incorporating object-oriented language Visual Basic with spreadsheets environment.

REFERENCES

4. Hitomi, K, "*Manufacturing Systems Engineering (2nd ed.)*," (1996), 499-513, Taylor & Francis.
5. Eversheim, W., Klocke, F., Pfeifer, T. and Weck, M. "*Manufacturing Excellence in Global Markets*," (1997), 33-69, Chapman & Hall.
6. Song, S. "Developing and Sustaining Flexible Integrated Manufacturing for Future Manufacturing Perspectives," *Res. Bulletin of the Hiroshima Inst. of Tech.*, Vol. 32, (1998), 107-125.
7. Browne, J., Harhen, J., and Shivnan, J. "*Production Management Systems: an Integrated Perspective (2nd ed.)*," (1996), 3-19, Addison-Wesley.
8. Mitrifanov, S. P., "*Scientific Principles of Group technology* (Russian text published in 1959, translated by Harris E.)," National Lending Library for Science and Technology, U. K, (1966).
9. Arn, E. A., "*Group Technology*," Springer-Verlag, (1975).
10. Jackson, D., "*Cell System of Production: An Effective Organisation Structure*," Business Books, (1978).
11. Nikkei Newspaper, July 10, 1998, p. 2S; June 14, 1998, p.13; Nikkei Business, Sep. 19, 1994, 11-21 (in Japanese).
12. Business & Technology Newspaper, June 8, p.17 and June 24, p.13, in 1998; Factory Management, (1997), Vol. 43, No. 4, 17-63, Business & Technology (in Japanese).
13. Asahi Newspaper, Apr. 16, (1998), p.12; Yomiuri Newspaper, May 21, (1998), p.9 (in Japanese).
14. Wemmerlov, U., "Special Issue on Group Technology and Cellular Manufacturing," *Journal of Operations Management*, Vol. 10 No.1, (1991).
15. Singh, N., "Cellular Manufacturing Systems," *European Journal of Operational Research*, Vol. 69, No. 3, (1993), Sep. 24.
16. Moodie, C. Uzsoy, R. and Yih, Y., "*Manufacturing Cells: A Systems Engineering View*," Taylor & Francis, (1995).
17. Brandon, J. A., "*Cellular Manufacturing: Integrating Technology and Management*," John Wiley & Sons, (1996).
18. Singh, N. and Rajamani, D., "*Cellular Manufacturing Systems - Design, Planning, and Control*," Chapman & Hall, (1996).
19. Song, S. and Choi, J., "Human-Enriching Autonomous Cellular Manufacturing: Production and Marketing Cooperative Plans through Optimal Incentive Design," *Proceedings of the Manufacturing Milestones toward the 21st Century*, Tokyo (Japan), July, (1997), 481-486.
20. Song, S. and Hitomi, K., "Group Machine Loading by Division of the Parts Family," *Trans. of the JSME*, Vol. 56, No. 530, Oct., (1990), 2813-2818.
21. Song, S. and Hitomi, K., "GT Cell Formation for Minimizing the Intercell Parts Flow," Colin Moodie, et al. (ed.), *Manufacturing Cells: A Systems Engineering View*, Taylor & Francis, (1995), 169-184.

22. Song, S. Hitomi, K., "Integrating the Production Planning and Cellular Layout for Flexible Cellular Manufacturing," *Prod. Planning and Control*, 7-6, Nov.-Dec., (1996), 585-593.
23. Song, S. and Choi, J., "Flexible Integrated Manufacturing Systems based on Individual Production and Handling for Ordered Products," *Journal of Advanced Automation Technology*, Vol. 4, No. 6, (1992), 304-313.
24. Song, J. and Choi, J., "Optimization Analysis of Flexible Cellular Manufacturing: Route Selection and Determining the Optimal Production Conditions for Ordered Parts," *European J. of Operational Research*, Vol. 69, No. 3, (1993), 399-412.
25. Song, S. and Choi, J., "Integrating Analysis of Manufacturing Decisions for Ordered Products under the Production for Stock in Cellular Manufacturing," *Trans. of the JSME*, 60-579, (1994), 3997-4005 (in Japanese).
26. Song, S. and Choi, J., "Flexible Integrated Manufacturing (FIM): The Concept, its Design and Optimization Analysis," *Japan Industrial Management Association*, Vol. 45, No. 2, (1994), 417-429 (in Japanese).
27. Song, S. and Choi, J., "A Method of Simulation Optimization for Supporting Flexible Cellular Manufacturing," *Proceedings of 20th ICC&IE '96*, (1996), 1003-1006.
28. Song, S., "Contingency-Driving Shop Floor Control for Supporting Effective Improvement of Flexible Integrated manufacturing," *Proceedings of 1998 Japan-USA Symposium on Flexible Automation*, Vol. 1, (1998), 165-172.
29. Goldberg, D., "*Genetic Algorithms in Search, Optimization, and Machine Learning*," Addison Wesley, 81989).
30. Kerr Roger, "*Knowledge-Based Manufacturing Management*," Addison-Wesley, (1990), 203-267.
31. Bauer, A., Bowden, R., et al., "*Shop Floor Control Systems: From Design to Implementation*," Chapman & Hall, (1991), 35-63.
32. Ullmann, W., "Logistical performance measurement is needed for today's production planning and control," *CIRP 24th Int. Seminar on Manufacturing Systems*, Copenhagen (Denmark), (1992), 303-313.
33. Christensen, J., "*Decision Support for Continuous Development of Factory Lo-gistics*," Institute of Manufacturing Engineering, Technical Univ. of Denmark, Ph.D. Dissertation, (1994).
34. Scherer, E. (ed.), "*Shop Floor Control - A Systems Perspective: From Deterministic Models towards Agile Operations Management*," Springer-Verlag, (1998).
35. Pfeifer, T., et al., "*Manufacturing Excellence: The Competitive Edge*," Chapman & Hall, (1994), 291-313.
36. Chryssolouris, G. et al., "A Decision Making Approach to Machining Control," *J. of Engg. for Industry*, Nov. 110, (1988), 397-398.
37. Bateman R.-E., et al., "System Improvement Using Simulation," ProModel Corporation, (1997), 81-94.
38. Biethahn, J. and Nissen, V., "Combinations of Simulation and Evolutionary Algorithms in Management Science and Economics," *Annals of Operations Research*, Vol.52, (1994), 183-208.
39. Glover, F., Kelly, J. P., and Laguna, M., "New Advances and Applications of Combining Simulation and Optimization," *Proceedings of the 1996 Winter Simulation Conference*, (1996), 144-152.

40. Akbay Kunter, S., "Using Simulation Optimization to Find the Best Solution," *IIE Solutions*, 28-5, May, (1996), 24-29.
41. Harrell Charles, R. and Tumay, K., "*Simulation Made Easy: A Manager's Guide*," Int. Engg. and Management Press, (1995), 135-205.
42. Weck, M., "*Simulation in CIM*," Springer-Verlag, (1991), 51-75.
43. Glover, F., "Tabu Search: A Tutorial," *Interfaces*, Vol. 20, Iss. 4, (1990), 41-52.
44. Gillis, J. D., "Human Factors Collaborative Research within Manufacturing," in *Kidd, P. T. and Karwowski, W. (eds.), Advances in Agile Manufacturing* , (1994), 3-8, IOS Press.
45. Huber, V. L. and Brown, K. A., "Human Resource Issues in Cellular Manufacturing: A Sociotechnical Analysis," *J. of Operations Management*, Vol.10, No.1, (1991), 138-159.
46. Fazakerley, G. M., "A Research Report on the Human Aspects of Group Technology and Cellular Manufacturing," *Int. J. of Prod. Res.*, Vol.14, No.1, (1976), 123-134.
47. Taylor, F. W., "*Principles of Scientific Management*," Harper and Row, (1947).
48. Upton, D. M., "What Really Makes Factories Flexible?" *Harvard Business Review*, July-August, (1995), 74-84.
49. Bassett, G., "*The Evolution and Future of High Performance Management System*," Quorum Books, (1993), 87-108.
50. Campbell, D. E., "*Incentives: Motivation and the Economics of Information*," Cambridge Univ. Press, (1995), 116-136.

Role of IT/IS in Physically Distributed Manufacturing Enterprises

Walter W.C. Chung* and Michael F.S. Chan

*Department of Manufacturing Engineering, The Hong Kong Polytechnic University, Hung Hom, Hong Kong. (E-mail: mfwalter@inet.polyu.edu.hk, Fax: (852) 2362 5267

This paper describes the information system/information technology challenges that small and medium enterprises (SME) must face when they form alliances to compete in the twenty first century. The role of IS/IT should create an environment in which knowledge workers can be aligned together and work collaboratively to create new knowledge within the knowledge-based competition. The key to market success lies in the inspiration of employees to set the pace of change in the creation of wealth and customers. From the perspective of a chief executive officer, IT/IS must be deployed to support the business in the emerging digital world when the job of a SME is to generate value for market and customer creation. The manufacturing strategy must embrace the implementation of information systems to innovate the infrastructure that support a new business model. Decision making authority should be delegated to the lowest levels possible so as to increase organizational responsiveness with respect to their customers.

1. INTRODUCTION

1.1. Manufacturing in the information age: SMEs to compete globally

For our fast-paced economy, in which the industrial era is overtaken by the information age, product life cycles are shortening [1][2][3]. The problem of this is that the value of knowledge embedded in products, services and processes is becoming more shorted-lived. This forces firms to develop continuous streams of innovation. Furthermore, not only does knowledge become obsolete more quickly, knowledge-based innovation embedded in smart products is quickly spotted by competitors; they then imported and improved on that idea, modify their products and launch into the market [4].

---

* The authors wish to express their gratitude to the sponsors of the Manufacturing Information Systems Research Project, The Hong Kong Polytechnic University, and the Industry Department, The Government of The Hong Kong Special Administrative Region. They have provided a technology transfer infrastructure for assisting local small medium enterprises to implement manufacturing information systems and making organizational innovations. The dissemination of best practices and their cross fertilization across industrial companies should speed up the formulation of strategies in building up an extended enterprise for agile manufacturing and gaining competitive advantage.

For this reason, many companies are seeing their unique knowledge as a powerful offering in the market place and place great emphasis in managing knowledge as companies' competencies [5]. To profit from the knowledge innovation, there is a need for firms to create new markets or new customers. However, a single firm alone only have a limited expertise and resources, and is not capable to deliver product to the market responsively. As a result, considerable interests have been paid to inter-organizational collaboration in developing and commercializing new products inter-organizationally [6].

1.2 SMEs competing with large corporations

Large enterprises are becoming more global as competition demands them to have local presence in multiple market niches. SMEs competing in the local market may face tremendous competitive pressure from the large firms as they are able to draw expertise globally to satisfy local needs. And for this reason, collaborations among SMEs are mandatory to successfully address specific market needs.

On the contrary, physically distributed manufacturing enterprises like that of local SMEs will also need to have presence in the global market in order to remain competitive. They will need to continuously search for or develop new markets for the commercialization of innovation into new product. Thus, there is a need for SMEs to move into the direction that provides them with the reach for global customers.

1.3 Leveraging IS/IT strategy for gaining competitive advantage

Information technology is changing the way firm operates. It is affecting the entire process by which companies create their products [7]. SMEs are submerged in today's information age in which the impact of information technology and the rapid pace of technological change have created spawn of new business opportunities [8]. As businesses are becoming more complex, SMEs need to exploit the inextricable tie between information and knowledge for competitive advantage through the deployment of IS/IT.

The challenge of deploying IS/IT for competitive advantage lies not just to SMEs but to many larger corporations in the understanding the integration between technology and people. The challenges are to develop IT/IS infrastructure that continuously stimulate worker to innovate, enhance inter-organizational collaboration to practice knowledge sharing and facilitate responsiveness in delivering new product to customers.

This paper responds to the growing needs to the understanding of issues involved in encouraging people to use information more effectively and to promote collaborative knowledge sharing for developing innovation.

2. LITERATURE REVIEW

2.1 Next Generation Manufacturing give insights to gain competitive advantage

It is important to understand and distinguish a "company" from an "extended enterprise" for the Next Generation Manufacturing (NGM) new business model [9]. "A company is a conventionally defined profit-making unit with management supremacy and clearly established bounds of ownership and liability. The owner controls the company's actions and charges it with responsibility." "An extended enterprise is a group of institutions for developing linkages, share resources and knowledge, and collaborate to produce a product and/or service. This unity maximizes combined capabilities and allows each institution to recognize its strategic goals by providing cost-effective, integrated solutions that fully comply with the customers' needs"[NGM framework].

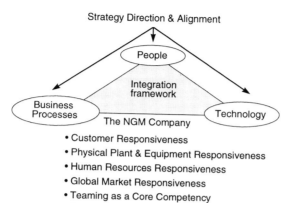

Figure 1. The Next Generation Manufacturing (NGM) Company Model

The key concepts of Extended Enterprise collaboration lie in the way firms compete. Competition will be between value chains, not between firms. Strategic alliances are developed on the basis of teaming and partnering. Firms of NGM attributes will need to be very proficient and measurable in this aspect in order to be suitable for the inclusion of Extended Enterprise.

The proposed NGM Approach to implement the new business model embraces many steps. (a) Understand the new competitive environment and identify the global drivers of the market place. (b) Derive a set of attributes from these global drivers and articulate a vision for the next generation companies and enterprises must possess. (c) Address a series of barriers arising from these attributes and identify the dilemmas for change management focus. (d) Define the imperatives and create the people, business process and technology solutions for resolving the conflicts arising form the NGM company and enterprises and (e) Provide an action plan to implement the NGM company model as shown in Figure 1. However, the transition to NGM Company Model involves the immediate resolution of issues confronted by physically distributed manufacturing enterprises.

2.2 Creating Value-chain for market competition

The value chain is a useful framework for manager to think in terms of how a firm operates. The generic value chain is consisted of nine categories of value activities. See Figure 2 for the diagrammatic representation of value-chain.

For an SME to deliver a service of value to its customers, a business unit may examine the value chain as proposed by Michael Porter [10]. The activities of a business can be grouped under five primary cells and 4 supporting cells as shown in Figure 2. The primary cell activities include Inbound logistics, Operations, Outbound logistics, Marketing and sales, and Service. The supporting cell activities include Firm infrastructure, Human resources management, Technology development and Procurement.

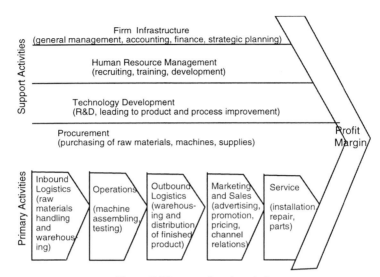

Figure 2. The generic value chain

The activities of an SME are normally confined within one cell (Operations) of a value chain. To develop synergetic benefits in extended enterprise, the linkages and interrelationships between cell activities should be exploited via a networked infrastructure to bring business partners together along the value chain to increase competitive gain. This should facilitate business communication and coordination of activities between partners. A computer based information system (say a web-based information system) may be implemented to provide information for faster decision making to reduce the time to deliver a product to the customer.

2.3 IT/IS in manufacturing to support SME transformation

An SME of the 21st century may embrace the IT/IS strategy in its manufacturing strategy to transform its organization structure. However, the transformation should not be too radical. Hooper and Winters [11] reveals in a survey that as the size of organization decreases, the importance of the success of any implemented change increases. Within the SME, a lack of success can have a dramatic impact on the viability of that organization. Technology alone would not bring about a gain in competitive advantage. It requires employees to share the vision of SME transformation and organizational structure change to leverage the power of technology. It is imperative that an approach be found for SMEs to focus its efforts to engage transformation at all levels and achieve a balance in the use of IS/IT for serving the customers. Chung and Chik [12] identified the key supports SMEs requested for assistance in a computerization project as shown in Figure 3.

It is evident that from the survey showing that nearly half of the SME respondents require substantial supports in the area of Customized Application Training and Reference Site Benchmarking. This is consistent with the suggestion that organization learning is important in SME transformation.

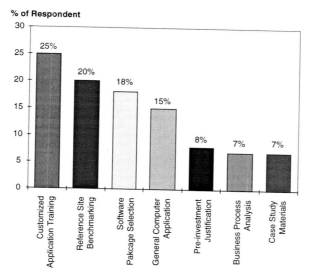

Figure 3. Key supports asked by SME in a computerization project

2.4 Internetworking and web-based technologies

The flow of information is now global. The impact of technology has spawned a wide array of business. While at the same time, the Internet is changing everything. With the growing popularity of the email and the explosive growth in the number of websites, information is flowing a lot more freely. For a SME to use information to compete, its workers must learn new skills related to searching or collecting data, processing them into information and distributing them to the intended users. In short, workers are constantly engaging in activities that can be characterized by one of the facilitating factors that expedite learning.

In the information age, the computer and telecommunication technologies are changing the ways people relate to each other in organizations, at work and at home. The Internet is connecting an individual or a company to others (individuals, companies and businesses) in the public at affordable prices. It is emerging as a powerful "self-service" channel. The Intranet is connecting one department (functional area) to another inside an organization. It is a cost-effective means of delivering new applications across multiple heterogeneous desktop environments, especially for internal communication and collaboration. The extranet is connecting a company to its suppliers, business associates and diverse organizations behind firewalls using web-based technologies. It supports the exchanges of information among partners, suppliers, distributors, contractors and others that operate outside of the physical boundary of the company but are critical to its success.

2.5 Knowledge and knowledge work in SME

Knowledge in general, found in the literature can be distinguished into two types: explicit and implicit. Explicit knowledge can be codified, such as simple software code and market data. This type of knowledge can normally be handled through technology such as computerized information system. Implicit knowledge, however, resides in peoples' heads and is difficult to articulate in writing and is generally acquired through experience. Other scholars have similar definition of knowledge but described in a more holistic manner. They view knowledge as existed on a spectrum [13]. At one end of the spectrum knowledge is totally tacit and held in the minds of people, it is therefore difficult to access. At the other, knowledge is codified, structured and explicit and is easily accessible by others.

The inextricable tie between knowledge and information advocates the need to integrate people with IS/IT. When information is combined with experience, interpretation, and reflection in the context it is applied, it becomes knowledge. This is ready to apply to decision and action. Therefore, knowledge is an important asset for an SME. However, if an SME does not setup a process to manage this knowledge, it will be eventually lost when employees leave.

Knowledge work is a set of activities using individual and external knowledge to produce outputs characterized by information content [14]. The primary activity of knowledge work involves the acquisition, creation, packaging, application and reuse of knowledge. The person who knows how to acquire, create, assemble, apply and reuse knowledge is a knowledge worker. He does not only know how to work (the skill or the "one best way"), but also knows what and why. The acquired knowledge can then be used to develop a new knowledge. Knowledge is diffused when it can be store and shared and becomes the new knowledge with another who works in the same SME. This has become the basis of organizational learning in improving the knowledge competent level of each employee. The right people can acquire the right knowledge to help their own job [15].

3. FRAMEWORK OF SME TRANSFORMATION

It is useful to think of the role of IS/IT as a way to continuously explore new opportunities for organization transformation. Technology alone, ranging from hard technology as computer integrated manufacturing to soft technology as management practices, cannot transform an organization without substantial human behavioral change. In physically distributed manufacturing enterprises, the role of IS/IT should be more proactive in terms of what it can deliver. Managers should think of what IS/IT can do to create value instead of what value should be delivered through IS/IT.

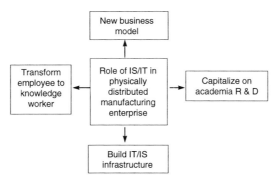

Figure 4. A framework of the role of IS/IT in physically distributed manufacturing enterprise

3.1. A new business model: any time, any place and no matter

In the era of information, SMEs physically distributed in many parts of the world may link up via a web of networks to customize products or services for customers all over the world. For example, the participants may include SMEs (offering small electrical goods, educational toys like electronic dictionaries), banks (offering secured payment services), distribution companies (offering delivery services), a software developer (offering support to customize new software for search engine in data mining), academia (offering support in workshops as well as knowledge work manpower). The players form an extended enterprise and build an IT/IS infrastructure to leverage the network to do work. They are to become a learning organization. They work together by serving the customers and validating a new business model as it evolves - "any time, any where and no matter"[16].

Samson Tam, Chairman of Group Sense International Ltd. [17], shared his insights in the use of information technology and information systems (IT/IS) for competitive advantage. The role of CEO is to create wealth and to create new customers. In this information age, the employers must become knowledge workers capable of making decisions at the appropriate level in the SME. An IT/IS infrastructure must be developed to link up its business partners. The CEO could forge a link in person with a business partner. But he should not make all the decisions by himself. He should only make those vital few decisions (say 10% of his time) that have a lasting impact on the business. The key to success for the SME is to share the new business model of "any time, any place and no matter" with colleagues. They have knowledge and are ready to set the pace of change, to continually re-invent its IT/IS infrastructure, and to leverage the use of best practices in creating a service of value to the customers. However, the employees may be all busy attending to its daily business. The IT/IS is an enabler of business initiatives to innovate new linkages with partners because innovation requires imagination, intuition and creativity.

3.2 Capitalizing on Academic Research

A research framework of the tripartite model is proposed for collaborative learning partnership as shown in Figure 5. Three parties are involved in the tripartite model: SME, Academia and Technology Provider. Each party has its own mission. The SME has a mission to offer goods or services to customers and makes a profit. It continually rolls out new product or services to stay in the competition. It would adopt a best practice or leverages new technology when its implementation could be visualized to reduce lead time, improve quality, reduce cost or increase flexibility with a gain in competitive advantage. The academia has a mission to contribute knowledge to the local community. It delivers ideas, concepts and theories to industry via the students. The students should implement these concepts to make an impact on business performance. The technology provider has a mission to get a return from the technology embedded in a piece of hardware or software. They want to recoup the cost of technology development by marketing its hardware or software to as many users as possible while on the other hand protecting its intellectual asset. Therefore he does not wish to grant the use of a restricted license free of charge.

The tripartite model allows an SME to directly capitalize on academia research and development immediately. Research and development involves the import and assimilation of a best practice to lower the cost of acquiring new technology and compete in the global market. This involves balancing the new technology with sufficient organizational transformation. It leverages the resources of its partners in cultivating an organizational structure that is supported by an IT/IS infrastructure to enable the decision making at the appropriate level for speed.

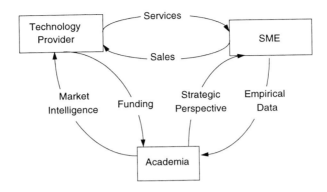

Figure 5. A Tripartite Model of Collaborative Learning Partnership

3.3 Building IS/IT infrastructure

With the rapid changes that are taking place in the information age, the development of a knowledge management system in SMEs for competitive advantage must take care of three aspects in developing a knowledge-based capabilities / workforce.

3.3.1. Technical aspect of knowledge work.

Technological implementation may serve as a driver of a technological driven change. It is easier to visualize the benefit with technology implementation and its adoption avoids having to deal with many cultural factors, even though technological implementation may be costly in both time and other resources. The common technologies (IS/IT tools) implemented for knowledge management have been GroupWare and Intranet (e.g. GroupWise), to provide current awareness services, online databases for desktop access, email for internet access and Office software for word processing and so on. These IT/IS tools can deal with explicit knowledge to some degree. However, when it comes to the knowledge residing in peoples' minds and its enhancement by physical and cultural factors, technology is not very useful [18]. Electronic communication tools are not substitutes for the people relationships, but they are necessary to carry out business. As a result, it is clear that in dealing with tacit knowledge, the human aspects must be involved.

3.3.2 Human aspect of knowledge work.

One key component in determining the success of a knowledge management system is usage. People need training to share knowledge and discharge duties as a knowledge worker. Interpersonal skills may be acquired via software development projects. They may begin with establishing contacts. When employees share knowledge on "who knows what in any department", that information can be configured in a database to be retrieved in several ways. They could create "yellow pages" for retrieval by individual, expertise, department where project was carried out, date of project and so on. Some larger companies (e.g. utility company, finance company) might consider capturing the knowledge of managers, engineers and people with relevant experiences in evaluating a capital expenditure proposal and develop a computer-based expert system to do the retrieval electronically. They might consider selling this knowledge to other companies facing similar situations.

3.3.3. Process aspect of knowledge work.

F. W. Taylor in 1910s advocated planning the work with a view to find "one best way" to schedule the production workers to do the job and increase productivity. In the sense that knowledge workers performed their activities involve lots of skills to capture data, processing them, and distribute the information in reports to the relevant people, they are considered as part of a "skill set" of the coordinator/manager. However, SMEs can do better than making an ad hoc appointment of a person with the right "skill set" to managing knowledge work. A scientific approach could be used to share the know-how in asking "the right question" in order to collect data, process information, extract knowledge and spread the insight in the use of information. The skill of asking the right question is crucial in using information to do knowledge work and should be shared and multiplied by more knowledge workers. The players from different departments trained in using this know-how could be assembled as required to co-design "one best procedure" to structure, sequence, and measure tasks to reach targeted audiences. The specific ordering of tasks across time and place may be depicted as a network diagram. It has a beginning, an end, and clearly identified inputs and output. It is a structure for action. The scientific approach to process knowledge promotes a critical examination of the activities for value creation to satisfy targeted users.

3.4. Transforming employees to knowledge workers

Peter Drucker first coined the term "knowledge worker" in 1960s. His book [19] announces that we are entering "the knowledge society", in which "the basic economic resource" is no longer capital, or natural resources, or labor, but "is and will be knowledge". Included in his definition of knowledge worker is a knowledge executive who knows how to allocate knowledge to productive use, just as the capitalist knew how to allocate capital to productive use. Unlike employees under capitalism they own both the "means of production" and the tools of "production". Making knowledge worker productive would be one of the greatest management challenges of the 21st century.

Employees working in an SME are not aware of their responsibilities in working in the information age. Making them aware of the implications would mean making use of their knowledge in transforming business processes in creating wealth. The challenge to SMEs here is to have them take the ownership of wealth creation whilst maintaining control of the overall business. To achieve this, it requires employees to acquire the ability to take ownership of the problems through action learning. They will be positioned in collaboration projects for teamwork to co-design a computer-based information system making use of IS/IT to support decision making by the group. In the process of implementing such a system, decision authorities are delegated to the lowest level possible.

4. METHOD IN ASSIMILATING BEST PRACTICE

The creation and protection of knowledge assets is a must in the information age. The employees must be capable of managing knowledge and working as a member of a collaborative learning team in the extended enterprise of which SME is a part. The CEO must strive to migrate the SME from a traditional organization structure with decision-making authority centered at the top to the one focusing on the customer with links to partners in many parts of the world. The extended enterprise will proactively set the pace of change and roll out new products and services using the new business model "any time, any place and no matter". This can be facilitated via the formation of a steering team and an implementation team.

4.1 Formation of a steering team

A steering team is formed when academia and SMEs meet and share the vision of manufacturing strategy in the digital age. They understand the reasons behind the need for change in work processes to gain competitive advantage. They recognize where the implementation of a best practice leveraging IS/IT would bring results to a target company. Since the industrial members of the steering team have knowledge of their business, they are familiar of the overall business system in terms of resources needed. The team of senior manager and academics work together to define the initiatives to take and commit the relevant resources to support. Refer to Figure 6.

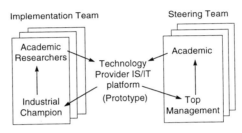

Figure 6. Replication of the Tripartite Model of Strategic Learning Partnership

The steering team typically goes through a process of SME transformation
1. Visit reference sites to benchmark performance
2. Formulate a concept to make use of a best practice to re-invent a business process
3. Define the roles necessary to facilitate the realization of the concept
4. Implement the concept with a view to extract knowledge for replication

One of the problems of transferring best practices from one organization to another is that best practices may be so contextual and specific to an organization that they "don't take" in their new environment [20]. Knowledge has to be adapted to the specific conditions in every case. The academia is seen in a neutral position to advice the steering team members of the collaborating organization to use the relevant criteria to justify a capital expenditure budget on investment in IS/IT infrastructure. The investment should make allowance for training so that a team of technical staff are capable of supporting colleagues to do knowledge work with the IS/IT infrastructure. As a result, an implementation team is required to putting the new practice into the reality.

4.2 Development of Implementation Team

The members of the steering team commit resources to recruit researchers and assign users from both the academic and industrial site for the formation of implementation team. The implementation team acts as a team of change agents and champions. From time to time, they seek advice from the steering team members who caution against engaging on a project in which the scope is too wide for completion within the resources and time limits. The implementation team works out the detailed plan and executes the ideas associated with the vision to streamline work processes to support their daily operations. Together they share a view to using the best practice to do knowledge work.

The implementation team has to carry out several essential functions:
1. Practice the new work method, and develop skills in coaching and learning by doing
2. Acquire techniques in using new technology from technology provider
3. Prepare workshop materials for replication of the best practice in another site
4. Promote business process change through top management support
5. Benchmark new perspective in using new work method with existing ones

Conflicting issues involving many parties during implementation of the best practice will surface. The implementation team will need to attend to them for better understanding of the issues and resolve them for achieving the CEO vision. Once all the important issues are addressed, the communication of CEO vision to subordinates is enhanced.

4.3 Leading change with the Reference Site Methodology

In a learning partnership, the technology provider (say a state-of-the-art software) is persuaded to commit in the development of a prototype to show the power of the technology in use at the "reference site".

The "reference site methodology" [21] is introduced to the implementation team for the development of best practices template. This methodology assumes that people with different academic and industrial backgrounds could communicate with each other and specify the requirements of an information system. A template for best practices is tested with real life data obtained from SMEs. It is revised to make refinements to reflect the real life operating conditions. The implementation team then jointly documents what has been done in a case study report. The academic could extract the experiences gained and lessons learnt in the case study with a view to share the insights in the use of information with other researchers/users. This approach has been used many times in various scenarios in the transfer of a best practice from academia to SMEs for assimilation.

4.4 Transfer research findings into the business system

The problems associated with translating a generic theory for SMEs' own situation makes the implementation of a best practice adoption difficult. The "reference site methodology" puts any research and development effort directly in the environment of the SME business. It overcomes this problem by transferring knowledge of research findings directly into the business system. The detailed workflow design is often left to individual knowledge workers. Therefore, it is important that they share the vision of the new business model and that the strategy in doing knowledge work is to repeat and multiply the use of the new business model for success.

The prototype is developed to articulate a new mode of operation for visualization of what can be possible. It invites more people to replicate the success in other places. It is up to individuals to illustrate to others the initiatives they have taken to make the exiting operation more productive. This multiplier-effect brings out the entrepreneurial spirit for innovation to the SME.

5. CASE STUDIES DEMONSTRATING THE PROCESS OF FORMING AN EXTENDED ENTERPRISE

5.1. Collaborative teamwork to build a value chain

An extended enterprise is shown as a network organization with many partners: SMEs (ASM, Gara Plastic, Group Sense International Ltd, Winkler, etc), Academia (MFG, AMA at PolyU), Microsoft, Grandmass Technology, Bank, CAS, Computer Associates play the role of technology providers. See Figure 7.

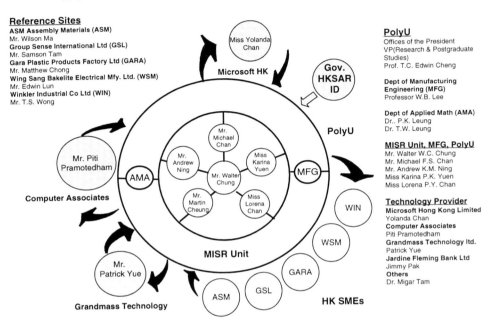

Figure 7. The MISR Unit of MFG at PolyU is the centre of an extended enterprise

Participants of the extended enterprises come together with commitment and initiatives in taking a proactive role to learn from other members. The role of MFG is established as an Extended Enterprise demonstration site to develop a pilot collaborative project for business development, and managerial practice innovations. The demonstration site ensure high-value concept to be transferred into Hong Kong SME for capabilities development and help a company to make successful transition to a new level of competence.

The MISR Unit acts as a host for managing knowledge and deploying them in the form of high-value best practices. These best practices are systematically transferred to Hong Kong SMEs for assimilation depending on their contextual circumstances. With the support of the technology providers, the technological aspect of knowledge management such as the codification, extraction, structuring, distribution of knowledge can be put to practice for empirical testing. New management practices are tested empirically with the support of information technology using SMEs as an experimental reference site.

5.2. Group Sense International – Importing best practices of Activity-based costing (ABC) and Quality Function Deployment (QFD) for assimilation

Group Sense International Ltd. (GSL) is a manufacturer of electronic Chinese-English dictionary and pagers. It was established in the 1980s and since then it has grown into a publicly listed company in Hong Kong.

For more than 2 years, the company has collaborated with the department of Manufacturing Engineering of The Hong Kong Polytechnic University for conducting applied research in the development of IS/IT to gain competitive advantage. The CEO of GSL has his own vision of the company's future and the new business model in which the company is adapting. However, many challenges have to be addressed as the company is confronted with the challenge of organizational alignment and new expertise.

As of Dec 1999, GSL is engaging in applied research in aligning different cluster of culture existed within the organization. Further, it is importing the best practices of activity-based costing, object technology and quality function deployment through an industrially sponsored project. A team of change agents consisted of various grades of research personnel is interacting with different levels of selected employees within GSL. The CEO vision is committed in the form of workshop, personal interviews so on as this amplifies the determination of the board to transform the hierarchical structure of the organization. The employees need to understand the issues involved in order to gain insights for academic research work. They implemented academic theory in the company via change agents in collaborative actions, and adopting best practices into the GSL business system.

The support of top management is crucial for the success of project implementation. Through a publicly held digital factory forum, Samson Tam, the CEO of GSL advocated the importance of the role IS/IT to ensure the continue success of the development of alliance between academia and member organizations. The continuing success of alliance allows him to leverage IS/IT to create wealth to sustain company growth in the future and is the driving force in setting the pace of change to the companies employees.

5.3. Winkler – Learning to replicate a web-based quotation system

The SME was a OEM and makes small electrical appliances like digital thermometer, atomic clock, assembly of walkman radio, cassettes, CD players and so on. The company felt that due to fierce competition, its profit margin was squeezed. It must get its business ready to use the Internet. Therefore, it collaborated with PolyU in 1999 to develop a web-based quotation system. The idea was to put the product in the web. A prospect could use the browser to configure a design of its product. He could, at the "click" of a button, change the color or alter the size of the product. He could get a feel of what the final product would look like before leaving the company's web site. Alternatively, he could submit a request to get a sales quotation of the product from the company.

The company was very happy with the power of the prototype and the potential it could bring to re-design its business process. It was keen to explore the many ways it could link up with a local bank and other business partners to the new customers so that they could form a learning partnership to expand the e-business model. The vision was to let prospects access the product catalogues anywhere in the world (UK, USA or Australia), at any time of the day and enter into business transactions at the "click" of a button.
For references go to website: http://h4u.hongkong.com/technology/internet/winkler

5.4. Gara - Putting an inventory prototype in the business process

The SME is an OEM making toys for Disney World and other buyers. It wanted to leapfrog its competitors by leveraging power of technology and the creativity of its people to gain competitive advantage. It did not possess an accurate stock control system. The company wanted to computerize its inventory operations but it did not have the part number system ready. Being small in size, it did not have enough manpower ready to computerize.

The company collaborated with PolyU in 1999 to leverage its knowledge in implementing manufacturing information systems. The university suggested using three students to undertake Guided Industrial Projects. These were mini-projects for year 4 manufacturing engineering degree students lasting one day a week and for one whole academic year. The students would partner with users of the company to specify the design of a prototype for implementing a best practice. Together the students and the user representatives would choose a best practice for transfer and assimilation into the company. The process involves the implementation team to design, develop, progress-report and finally present to the collaborative learning partners the prototype as "show case" for replication in other parts of the company.

The best practice nominated was Materials Requirement Planning. The prototype to demonstrate feasibility for implementation in the company would link Engineering, Materials Planning, Incoming Quality and Store in the processing of a customer order, monitoring the progress of materials delivery and ensure the materials receipt are in good quality for production in China. The objective of the collaborative learning exercise was to articulate a new procedure to get existing employees involved to redesign its business process using the IT/IS infrastructure to be developed and used. The company would consider employing the students upon their graduations as a part of the longer term management succession exercise.

6. ANALYSIS OF PROCESS IN FORMING AN EXTENDED ENTERPRISE

6.1 Long term development of Tripartite Model

The tripartite Model facilitates the development of extended enterprise by forging a learning partnership between companies. Not only does the academia have a long term perspective in the development of the tripartite model, top management and CEO have allowed themselves to move away from the short-term production-oriented attitude. It has created an environment in which SMEs have put more emphasis on long-term gain for their migration to the information age. See Figure 8

SMEs' commitment in the learning partnership represented a major milestone of breaking their production-oriented attitude in managing the SME destiny. Their commitment in opening up themselves for others to learn develops trust between teaming partners. It helps SMEs to collaborate effectively and responsively in addressing contemporary issues and develop demonstration sites for new initiatives. SMEs' financial commitment in hiring new people to develop new programs represents their willingness to increase provision for future capacity in competing in the world market.

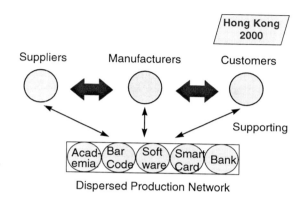

Figure 8. A Value Chain Business Model for the 21st Century

6.2 Sharing knowledge through the networked organization

In the networked organization, the knowledge shared travels in parallel through the host of academia. This delivers a greater value to member organizations competing globally. The sharing of knowledge allows CEO of other companies to realize the rapid pace of change has arrived inevitably. Establishing the urgency for change through demonstration site visits is found effective than just solely to persuade the commitment of resources for a good course. To technology providers, such demonstration is a powerful tool to disseminate knowledge about how their products can deliver value to companies in a convincing manner.

Sharing knowledge through the networked organization becomes the major source of innovation for extended enterprise of the 21st century. When SMEs share knowledge regarding best practices, academia can continue to refine and develop new innovation for the benefit of the extended enterprise. Sharing knowledge can also provide new insight to other partnering members in the other part of the value-chains for increased competitiveness.

6.3 Initiate change through action learning

Organizational transformation of physically distributed manufacturing enterprise requires a decision in setting the pace of change. Member organizations of the extended enterprise will leverage change from external members. Interaction between member organizations of the extended enterprise helps workers see how each other work. Such interactions facilitate the coaching effect developed among the similar communities of practices. Learning is diffused through this channel of community of practices.

Leading change through community of practices for organization transformation can be more effective than the top down process. Such learning allows worker to identify their performance deficiencies and proactively take corrective actions. The corrective actions are articulated by academics as a new idea and embedded into daily working routines. Such activities are called "communicating social change" [22].

6.4 Recruit knowledge workers to sustain change

Through the industrial and academic collaboration, the access to the pool of knowledge workers has been further opened up. SMEs collaborating as members of the extended enterprise selectively hire new graduates from university for new ideas. New graduates with experimental-mind set continue the research work in the organizations concerned and establish new links in the search for new sources of knowledge to help company compete in the knowledge-based global market.

The recruitment of knowledge workers adds new ideas into SMEs for increased levels of creativity and competence. The newly recruited knowledge workers will also exploit the linkage between member organizations within the extended enterprise for mutual benefits through inter-organizational interaction. They also exploit the linkage between value-chain participants for competitive gain by tapping expertise from outside.

7. DISCUSSIONS

7.1 Leading change through strategic alliance in extended enterprise

According to Kotter [23], there are eight common errors found in organizational change efforts. They are
- Allowing too much complacency
- Failing to create a sufficiently powerful guiding coalition
- Underestimating the power of vision
- Undercommunicating the vision by a factor of 10 (or 100 or even 1000)
- Permitting obstacles to block the new vision
- Failing to create short-term wins
- Declaring victory too soon
- Neglecting to anchor changes firmly in the corporate culture

Kotter argued that successful transformation is 70 – 90 percent leadership and only 10 – 30 percent management. Most of these transformation responsibilities are left to the work of the leader. It is difficult for SMEs to lead the change and compete in the turbulent market when CEO because top management themselves have a production-oriented attitude. The strategic alliance derived from the tripartite model have provided SMEs with the necessary benchmarking in community of practices at all level and allows them to learn how to change through managing themselves. When member organizations see each other's company in terms of people who work there, they become more sensitive and responsive to each other's ideas, needs, and constraints.

In developing a value chain, the academia may be a source of best practice for transfer to SMEs. Through the tripartite model of academic industrial collaboration, the relevant technologies (e.g. internet-based electronic quotation, electronic payment, etc) are identified for transfer to SMEs for assimilation. The champions of SMEs are initially selected to form a learning partnership. The team so created work together to identify the business processes in the SMEs for innovation and specify a "show case" prototype to build (e.g. They use the value chain business model for the 21st century to build a web-based EDI prototype). The team and their associates involved in the development of the prototype perform activities that contribute to the creation of an intellectual asset. In the process, they acquire skills in knowledge management. The promotion of the prototype to the public provides a support to the transformation of the business strategy in the SMEs.

7.2 Leading organizational learning through strategic alliance in extended enterprise

Learning is not an easy process. It consists of the development of awareness and insights, which is a change in states of knowledge that expands the range of potential behaviors[24]. In the learning process, the learner is involved in taking on the mindset of a beginner, letting go of the unknown, and being willing to try something new [25]. When the learner tries new behaviors, making mistakes are inevitable. Some authors argued that factors such as managerial resistance or financial constraints may prevent the immediate implementation of new insights or skills, and that learning does not involve a change in behavior but an expansion of the range of potential behaviors [26]. However, organizational learning occurs when new knowledge is transferred across unit boundaries in the organization to others who can benefit from what has been learned [27]. The transition of individual learning to organizational learning requires some form of behavior change, at least to a level that knowledge can be shared with other people within the organization. Therefore, in the context of working in an extended enterprise, organizational learning is enabled when the range behaviors is expanded into each function of the organization.

The flow of information is now global. For a SME to use information to compete in the global market, the role of IS/IT should support workers to learn new skills related to searching and collecting data, processing them into information and distributing them to the intended users as the basis for organizational learning. In short, the role of IS/IT should constantly expose workers with activities that can be characterized by one of the facilitating factors that expedite learning. The ten "facilitating factors" are 1. Scanning imperative; 2. Performance Gap; 3. Concern for Measurement; 4. Experimental Mind-set; 5. Climate of Openness; 6. Continuous Education; 7. Operational Variety; 8. Multiple Advocates; 9, Involved Leadership; 10. Systems Perspective [28].

7.3 Setting the pace of change through organizational learning

The strategic alliance in the form of networked organization and extended enterprise has the potential to develop the ten "facilitating factors". The learning partnership articulated in the tripartite model allows SMEs to develop environmental scanning for new managerial practices. This learning partnership also creates an environment in which the individual enterprises are opened to others. Inter-organizational benchmarking between member organizations can facilitate the awareness of performance gap, encouraging any disconfirming evidence posing as a barrier to learning to be surfaced and resolved.

The hiring of new graduates in the tripartite model facilitates the influx of recruits with similar experimental mind-set into member organizations. Managers need to learn to act like applied research scientists at the same time they deliver goods and services. The involvement of academia also encourages the development of commitment to lifelong education at all levels of the organization. The integration between academia and member organizations would align learning with employees on the job. Industrially sponsored projects will be beneficial to both the company in terms of learning and to academia in terms of applied research. These projects promotes a culture that one is never finished with learning and practicing.

An extended Enterprise provides member organizations with best practices for improving operation performance. New practices stimulate workers with new IS/IT technology to members within the organization. The learning environment of extended enterprise with the compliment of IS/IT technology provides more options in terms of support variation in strategy, policy, process, structure, and personnel with adaptability when unforeseen problems arise.

As mentioned previously, an extended enterprise leads change by inter-organizational benchmarking. Benchmarking between managers among member organizations can help change the mind-set of managers to foster new champions to set the stage for learning. Along, with these new champions, they can implement CEO vision in every level of the organization and act as coach to mentoring others to be champions in various ventures.

Knowledge diffused within the extended enterprise can affect member organizations. The interdependency between member organizations in the extended enterprise in terms of learning allows CEO to initiate change through other channels of communication. Transformation effort can now come from academia or other member organizations in the form of applied research, benchmarking, best practice transfer, etc. Managers could overcome the limits of transformation effort beyond the boundary of their own organization to business and learning systems.

7.4 Role of CEO : Anticipating the future and nurture new style of management

Recent research indicated that the major objective of SMEs in the context of academic-industry collaboration is to address immediate needs and returns. Unlike larger corporations, SMEs do not render themselves as collegial players whose objectives include the enhancement of inter-organizational networks and hiring of recent graduates [29]. From the case studies, their continuing commitment in terms of time and financial support as well as the hiring of recent graduate indicates that the objectives of these SMEs are shifting towards the nature of that of larger corporations. Many come to realize that only through collaboration in building strategic alliance can they reduce uncertainty and exploit opportunity in sustaining competitive advantage. Underpinning the alliance for success is a belief that the partners are willing to share information and knowledge about operational matters.

It is therefore that the CEO has a great job ahead. The challenge facing all SMEs is creating wealth and creating customers in the information age. Competitors are catching up rapidly either by copying the successful products or making innovations on their own. The dynamic market environment is imposing great pressure to create or use more innovative ideas. SMEs are required to anticipate the market and initiate change now. The role of CEO is to facilitate the transition of SME to compete in the information age smoothly.

8.   CONCLUSION

8.1. Implementing a new manufacturing paradigm

In the information era, knowledge is being treated as an asset. The management of knowledge is becoming increasingly important as the market competition moves towards knowledge-based, internet-enabled and innovation driven. SMEs are limited by their size and resources. They face great difficulties in competing against the multinational companies in acquiring knowledge to create new products and new customers. However, there is a window of opportunities for SMEs to collaborate with other organizations and levering the use of IS/IT to form an extended enterprise for competitive advantage. The role of IS/IT enables staffs to explore new approaches and to develop skills for doing knowledge work.

8.2. Tripartite Model to deploy IS/IT

This paper proposes a tripartite model of collaborative learning partnership to enable senior executives and technology developers to leverage an industrial academic collaboration platform and jointly engage into promoting a show case prototype to the public. A case study of an extended enterprise is described to show how such joint projects can be used to facilitate the bench marking of community of practices in managing the transfer of technology for assimilation. This facilitates reference site visits and enables inter-organizational learning so that participants can work together to "learn how to learn faster" in using technology for creating value and serve the customers.

8.3. Further research areas in the role of IS/IT

In the context of developing an extended enterprise for competitive advantage, the role of IT/IS to support organizational innovations opens up many research opportunities. The digitization of a value chain embeds many concepts such as performance measurement, organizational alignment, business process transformation, and IT contribution to knowledge management. In performance measurement, an externally focused performance measurement system is seen as a key for competitive success amongst different member organizations along the value chain. In organizational alignment, a new method of valuation on intellectual asset is required to find out the real cost of a strategic alliance. In business process transformation, the identification of the critical processes for implementing a new business model that uses IT/IS in wealth creation and customer creation puts the spotlight on core competence development. In knowledge management, new business models are required making use the emerging IS/IT technology to support knowledge workers to innovate new products, process and services for customers situated anywhere in the world.

It is a conscious choice in the use of IS/IT for aligning the SMEs in an extended enterprise to exploit the use of information to provide better products and services and competing in the world market. This choice involves organizational learning as to become the major driving force for company transformation and as part of an executive toolbox in managing knowledge within the company.

REFERENCES

1. D.J.Bowersox and D.J.Closs , "Logistic Management: The Integrated Supply Chain Process", McGraw Hill, USA, 1996
2. N.Nahar, "Risks assessment of IT-enabled international technology transfer: case of globlisation of SMEs," Proceedings of the Fifth World Conference on Human Choice and Computers on Computers and Networks in the age of Globalisation, pp.407-418, Geneva, August 1998.
3. N.Nahar, "IT-enabled effective and efficient international technology transfer for SMEs," Proceedings of the Evolution and Challenges in System Development, pp.85-98, Bled. February 1999.
4. Dan Holshouse, "Ten Knowledge Domains: Model of a Knowledge-Driven Company?", Knowledge and Process Management, Vol.6, No.1, pp3-8, 1999
5. Anders Paarup Nielsen, "Outsourcing and the Development of Competencies", Portland International Conference on Management Engineering and Technology, Proceedings of Technology and Innovation Management, 1999, pp.72 – 77.
6. A.Parkhe, "Strategic alliance structuring: A game theoretic and transaction cost examination of interfirm cooperation", Academy of Management Journal, vol3, 4 pp. 794-829, 1993
7. Michael E. Porter, "On Competition", 1998, Harvard Business School Publishing, Boston, Massachusetts.
8. Jeremy Hope, Tony Hope, "Competing in the Third Wave", Harvard Business School Press, 1997, Boston, Massachusetts

9. Patterson, R., Hardt, D. and Neal R. Executive Overview, Next-Generation Manufacturing Report, NGM Project Office, Agility Forum, 1997.
10. Porter, M. E. Competitive Advantage, New York, Free Press, 1985, p.37
11. M.J. Hooper and C.N. Winters, "A Survey of Manufacturing Practice and Organizational Change within the Automotive Rubber Industry", Proceedings of Portland International Conference on Engineering Management, PICMET, 1999, pp. 525 - 532
12. Chung, W.W.C. and Chik, S.K.O. Computerization Strategy for Small Manufacturing Enterprises in Hong Kong, Accepted in 1999 for publication in International Journal of Computer Integrated Manufacturing, (Forthcoming)
13. Christine Fiddis, "Managing Knowledge in the Supply Chain", Financial Times Retail & Consumer, 1998, London.
14. Davenport, T.H. De Long, D.W. and Beers, M.C. Successful Knowledge Management Projects, Sloan Managmeent Review, Winter 1998, pp.43-57
15. Davenport, T.H., Jarvenpaa, S.L. and Beers, M.C. Improving Knowledge Work Processes, Sloan Management Review, Summer 1996, pp.53-65
16. Davis, S. M. Future Perfect, Addison Wesley, 1987.
17. Opening speech by Mr Samson Tan, Chairman, Group Sense International Ltd. To CEOs at the forum "Digital Factory: Manufacturing Blueprint for Hong Kong", jointly organized by Microsoft and The Hong Kong Polytechnic University, Jockey Auditorium, The Hong Kong Polytechnic University, Hong Kong, October 28, 1999.
18. Davenport, T.H., "Working Knowledge" Boston, Massachusetts, 1998
19. Peter F. Drucker, "Post-Capitalist Society", 1993, Butterworth Heinemann, Oxford
20. Thomas H. Davenport, Laurance Prusak, "Working Knowledge", Harvard Business School Press, Boston, Massachusetts, 1998.
21. Chung, W.W.C. "Adaptation of Reference Site Methodology of Organizational Learning", Proceedings of the Sixth International Conference on Human Aspects of Advanced Manufacturing: Agility and Hybrid Automation, Hong Kong, July 5-8, 1998, pp.85-88.
22. Chung, W.W.C., Tam, M.M.C. Saxena, K.B.E., Yung, K.L and David, A.K., "Computer-Integrated Manufacture (Cim) Research in Hong Kong Polytechnic", Prism '94 International Conference, Maui Intercontinental Hotel, Maui, Hawaii, January 3, 1994, pp.153-162
23. John P. Kotter, "Leading Change", Harvard Business School Press, 1996, Boston, Massachusetts
24. Huber, G.P., "Organizational Learning: the contribution processes and the literatures", Organization Science. Vol.2, No.1, 1991, pp. 88 - 115
25. Marty Castleberg and George Roth, "The Learning Initiative at Mighty Motors Inc.", Working paper, MIT Sloan School of Management, 1998, http://learning.mit.edu/res/wp/mmlh.html
26. Gregory E.Osland, Attila Yaprak, "Learning through strategic alliances: Process and factors that enhance marketing effectiveness", European Journal of Marketing, Sep, 1994
27. Hamel, G., "Competition for competence and interpartner learning within international strategic alliances", Strategic Management Journal, Vol.12, 1991, pp. 83 - 103
28. Edwin. C.Nevis, Anothy. J. DiBella, Janet M.Gould., "Understanding Organizations as Learning Systems", 1996, working, http://learning.mit.edu/res/wp/learing_sys.html
29. Michael D. Santoro, Alok K. Chakrabarti, "Why collaborate? Exploring Industry's Strategic Objectives for Establishing Industry-University Technology Relationships", Proceedings of PICMET, pp. 55 – 61, 1999

The Method of Successive Assembly System Design Based on Cases Studies within the Swedish Automotive Industry

T. Engström[a], D. Jonsson[b] and L. Medbo[a]

[a] Department of Transportation and Logistics, Chalmers University of Technology, S-412 96 Göteborg, Sweden.

[b] Department of Sociology, Göteborg University, S-411 22 Göteborg, Sweden.

This chapter presents and illustrates a design procedure applied at several Swedish assembly plants during the last 15 years. It has been used for the design of the Volvo Uddevalla plant in 1984 – 93, the Volvo Truck plant in Tuve in 1988 – 90, the redesign of the Volvo main plant in Torslanda in 1989 – 90 and the design of the Autonova plant located in Uddevalla in 1996 – 1997 as well as for the restructuring of the information system at the Scania diesel engine assembly in the main Södertälje plant in 1998 – 99.

The first four cases concern the design, or redesign, into unorthodox parallel product flow, long cycle time assembly systems, while the fifth Scania case deals with improving the shop floor information in a traditional serial flow assembly system (an assembly line), i.e. improvement of work instructions and product variants codification, as well as handles the organisation of introducing product design change orders.

A more efficient introduction of product design change was an unforeseen consequence of the adoption of parallel product flow, long cycle time assembly systems, which require a restructured information system. This merit was not fully recognised by the management responsible for the first four cases but turned up, more or less, as a side effect of the unorthodox assembly system designs.

The parallel product flow, long cycle time assembly systems, have merits in form of efficient flexible manufacturing compared to traditional assembly systems owing to reduction of losses (due to the fact that fewer operators have to be co-ordinated), increased flexibility (accepting higher product variation, shorter time for introducing product design change orders, etc.) and better space utilisation than for serial product flow assembly systems. Although these merits are mentioned in this chapter and covered by appropriate references the authors have focused on highlighting some specific aspects of general interest by presenting the design procedure used, i.e. the common denominator between the five cases, which has successively been refined during the last years' work. This design procedure is denoted successive assembly system design.

1. INTRODUCTION

The concept of agile manufacturing summarises various trends used for designing and implementing advanced manufacturing systems, thus underlining the integration of technology, organisations and people (Kidd and Krawowski, 1994, pp. 7). Agile manufacturing views manufacturing system design as an interdisciplinary activity, and organisation and people issues need to be addressed in parallel with technical issues, that is, using a concurrent engineering approach rather than a serial engineering one according to Kidd (1995, pp. 46), who states that the design paradigm of agile manufacturing is characterised by: (1) Holistic systems-based approach; (2) Concurrent engineering concepts applied to the design of the manufacturing enterprise; (3) Application of insights from the organisational and psychological sciences to the design of the technology and overall manufacturing enterprise; and (4) Consideration of organisation and people issues at all stages, from the formulation of business strategy right through to the design and implementation of systems. The authors' experiences from assembly system design during the last two decades, support this design paradigm.

Some Swedish assembly plants comprise unorthodox designs in the international perspectiv, since these plants utilise *parallel product flow with patterns* combined *long cycle time assembly work* based on a holistic systems approach. These system designs have merits in form of reduction of losses (due to that fewer operators have to be co-ordinated), increased flexibility (accepting higher product variation, shorter time for introducing product design change orders, etc.), better space utilisation than on the *serial product flow assembly system*, i.e. the assembly line (see e.g. Engström, Jonsson and Medbo, 1999). However, these assembly system designs call for an intriguing design procedure in order to realise their potentials.

Though sometimes viewed as 'social experiments', these plants have successively during the last two decades generated a rational design procedure denoted *successive assembly system design* (Engström, Jonsson and Medbo, 1997). This fact is not always noted in the international discourses concerning merits and malfunctions of various assembly systems.

This chapter will describe and illustrate successive assembly system design which has been refined and used by the authors in five case studies involving long time co-operation between industry and university. The design procedure was initiated and used for the Volvo Uddevalla plant design and was revised in the redesign of the Volvo Torslanda main plant, a redesign which was never implemented (see Engström, Jonsson and Medbo, 1999), and for the Volvo truck plant in Tuve, as well as for the Autonova plant. All these four cases used or use parallel product flow, long cycle time assembly systems. The last case is the reopening of the Volvo Uddevalla plant operating as a joint venture between Volvo and Tom Walkinshaw Racing (TWR), denoted Autonova. This company manufactures exclusive coupés and convertibles for Volvo.

Finally, restructuring of the information system at the Scania diesel engine assembly at the main Södertälje plant utilises the same design procedure, but with the aim to improve the quality of data supplied to the operators on the workstations along the existing assembly line as well as to facilitate the introduction of product design change orders in order to, e.g., ensure product quality (Portolomeos and Schoonderwall, 1998). These five cases, #1 the Volvo Uddevalla plant, #2 the Volvo Torslanda plant, #3 the Volvo Truck plant, #4 the Autonova plant and #5 the Scania assembly line, underline the generality of the design procedure. See Table 1.

Table 1
Summarisation of the five cases on which the successive assembly system design is based.

Case:	Product:	Co-operation [years]	Aim:	Comments:
#1 The Volvo Uddevalla plant:	- Automobiles (Volvo 740-model).	1984 – 93	- Design of assembly systems.	- Parallel product flow assembly system. - Cycle time of 80 – 100 minutes.
#2 The Volvo Torslanda plant:	- Automobiles (Volvo 200-, 700- and 800-models).	1989 – 89	- Design of assembly systems.	- Parallel product flow assembly system - Cycle time of 110 minutes.
#3 The Volvo Truck plant:	- Trucks (F-mod, 6 x 2).	1989 – 90	- Design of assembly systems.	- Parallel product flow assembly system - Cycle time of 240 minutes.
#4 The Autonova plant:	- Automobiles (Volvo 800- and C70-models)	1995 – 97	- Design of assembly systems.	- Parallel product flow assembly system. - Cycle time of 90 – 150 minutes.
#5 The Scania assembly line:	- Diesel engines (Scania D12 engine).	1998 – 99	- Restructuring the information system.	- Serial product flow assembly system. - Cycle time of 4 - 8 minutes.

Since parallel product flow, long cycle time assembly systems have more degrees of freedom than traditional serial product flow assembly systems, the design procedure is more demanding and requires an elaborate theoretical foundation, a foundation not yet fully crystallised and internationally communicated among practitioners as well as researchers.

The *work hypothesis* for design of parallel product flow, long cycle time assembly systems (Engström and Medbo, 1992), demands agreement between (1) operators' perception of the assembly work, (2) the materials display at the work station, and (3) the information at the work station in form of e.g. work instructions and product variants codification.

This will in turn require (a) an assembly-oriented product structure, i.e. a product structure that describes the product and its components from an assembly point of view – in contrast to the existing, traditional design-oriented product structure used within the Swedish automotive industry based on the so-called function group register (codifying a vehicle into the main groups) and (b) a non-traditional materials feeding technique by means of kitting. Since parallel materials addresses are required, resulting in numerous stockkeeping units, there will not be enough space for the materials volumes feed by unit loads.[1]

---

[1] If traditional continuous materials feeding technique, alternatively batching, is utilised in parallel product flow assembly systems, control of the number of components to store, replenishment, and the numerous product design change orders will be complicated to administer and handle. That is, in practice it will prove extremely difficult to secure that the correct components are available at the workstations and fitted to a specific vehicle since it will be impossible to keep track of all small components with continuous materials supply. In kitting the accuracy is increased and the number of stockkeeping units (SKU) are drastically reduced given a parallel flow, in principle just one unit for each part number e.g. Bozer and McGinnis (1992).

A correctly designed parallel product flow, long cycle time assembly system in accordance with these theoretical foundations will result in (c) an increase of the economically viable cycle time, (d) an efficiency potential gain in form of reduction of the production losses, i. e. system, balance and division of labour losses (Wild, 1975), and (e) increased flexibility including rapid market response.

Thus, parallel product flow, long cycle time assembly systems call for a non-traditional materials feeding technique as a precondition for efficiency and flexibility, while the assembly-oriented product structure is needed for assembly system designs that bridge the competence requirements, based on the assumptions that it is impossible to learn the increased work content within reasonable learning times due to the requirement of extensive repetitions of specific work tasks (De Jong, 1956), by turning so-called atomistic learning into holistic learning (Marton, 1970 and 1981). For a more detailed explanation of theoretical and practical frames of references, which is not possible to include in this chapter, see e.g. Medbo (1999).

The elaborate design procedure comprises restructuring of several subsystems as is hinted above, as well as the questioning of traditional knowledge and practice within the automotive industry. This chapter reports on experiences made in assembly system designs by describing successive assembly system design used for work structuring by means of disassembly of products taking advantage of the product information which proved to be already available mainly through the design-oriented product structure but also through the existing product and manufacturing process data.

The assembly system designs considered shall mainly consider work in autonomous work groups, utilising so-called 'collective working' in order to emphasise the fact that operators in an assembly system work together on one or more products, having common responsibility for production output within a so-called *work station system*.

Example 1:
> An example of findings regarding the performance of parallel flow, long cycle time assembly systems is the author's video recordings which showed that the observed assembly time in the defunct Volvo Uddevalla plant was 15 – 20% shorter than the standard assembly time settled by predetermined motion-and-time systems. Actual learning time needed to attain the required assembly competence was substantially shorter than previously reported by some researchers. Also the annual model change cost per automobile for the same automobile model at the Volvo Uddevalla plant proved to be lower than for the serial flow assembly system at the Volvo Torslanda when manufacturing the same product (see Engström, Jonsson and Medbo, 1999).

2 USING PRODUCT DISASSEMBLY AND AVAILABLE PRODUCT INFORMATION IN ASSEMBLY SYSTEM DESIGN

An important work hypothesis involved in the design of parallel product flow, long cycle time assembly systems is that of structural congruence. As stated above, this hypothesis demands agreement between (1) operators' perception of the assembly work (2) the materials display and (3) the information at the work station in form of e.g. work instructions and product variants codification.

In particular, this congruence might be formulated as a need for conformity between (a) a *hierarchical product structuring scheme* used to describe the product as a structured aggregate of components resulting in an assembly-oriented product structure and (b) a *hierarchical assembly work structuring scheme* used to describe the assembly work as a structured aggregate of assembly operations resulting in (c) the *intra-group work pattern*, i.e. the allocation of assembly operations expressed in so-called *work modules*, aggregates of work tasks, to operators within each work group responsible for the assembly[2] as well as (d) materials display at the work station and (e) the layout and product flow pattern within each work station system.

Due to the requirement for structural congruence, the design of one subsystem simultaneously has to take restrictions on other subsystems into account, and the total design procedure is an iterative process which successively crystallises and refines defined subsystems – rather than a linear process proceeding from the design of (a) to the design of (d). For analytical purposes, however, the design procedure may be regarded as starting with the design of a suitable hierarchical product structuring scheme (a).

Although an ideal design procedure of parallel product flow, long cycle time assembly systems ought to be based on total congruence between (a), (b), (c), (d) and (e), this chapter will mainly report experiences on (a) and (b), thus only briefly sketching on (c), (d) and (e) these have been reported elsewhere (see e.g. Engström and Medbo, 1994). The total design procedure developed and used will be discussed below. However, it will mainly be illustrated by the Autonova plant design procedure (case #4) but also by data from the other cases.

Example 2:
> Disassembly of products are shown in Figure 1. These products were used for design of assembly systems and for restructuring of an information system respectively. To the left there is a photograph of a disassembled automobile (Volvo 800-model) from the design procedure of the Autonova plant in 1995 and to the right a photograph of a diesel engine, disassembled at the Scania engine assembly plant in 1998. In principle, these disassemblies constituted an additional development of analyses done for the design procedure of the Volvo Uddevalla plant more than ten years earlier.
>
> The removed components were organised according to so-called work modules in the Autonova case as well as according to the individual operators' work at specific work stations along the existing assembly line in the Scania case. In both cases the components were separated by means of wooden lathes.
>
> Note, for example, in the photograph from the Autonova case the labels fixed to the components representing various product variants as described in example 5 and the small cards guiding the disassembly, positioned on the tables.

---

[2] In the case of the Volvo Uddevalla plant. In this case the modules corresponded to the working position and the position of the automobile body. The modules were designed in accordance with the tilting device which enabled the automobile body to be altered (adjusted in height respectively rotated along the longitudinal axle of the automobile in order to improve ergonomics). Thus, for example one work module was denoted "tilt over" corresponding to assembly work on the upper side of the vehicle as well as on a tilted automobile body.

Disassembly at Autonova (Uddevalla) in 1995 for design of assembly systems (case: #4).

Disassembly at Scania (Södertälje) in 1998 for restructuring of the information system (case: #5).

Figure 1. Disassembly of products.

To support long cycle time assembly work, there is a need for a reformed product perception using product information which has proved to be already available mainly through the design-oriented product structure but also through the existing product and manufacturing process data. This is an essential requirement since it facilitates the design procedure and promotes the introduction of e.g. non-traditional materials feeding techniques (i.e. it is necessary for the function of the new assembly system to communicate with the design-oriented product structure).

To design a hierarchical product structuring and a corresponding hierarchical assembly work, structuring schemes have been developed through disassembly of some automotive products, i.e. the Volvo 200-, 700- and 800-models, as well as the Volvo truck F-model. The methods used, in the case of the Volvo Uddevalla plant design procedure, were in many respects an interactive search process during a period of approximately 8 – 10 months, engaging two of the authors who were involved not only in this activity.[3] This was in almost all respects a tedious manual process, i.e. making notes by hand during the disassembly, using photocopy machines, basing different types of analysis on insufficient and often incomplete data, as well as searching for the appropriate information and personal contacts within Volvo.

Though the process was time-consuming, it certainly resulted in the building-up of knowledge, as well as serving as a method to formalise practitioners' knowledge, e.g. by having Volvo expertise continuously checking the work by e.g. cross-reading constructed alphabetical registers and 'dictionaries' describing product functions, explaining anachronisms, calling for specific documents required for the running in of the plant in the form of assembly instructions and variant specifications, etc.

The development of the hierarchical product structuring scheme and a corresponding assembly work structuring scheme involves a constant change between the components from the disassembled products placed on the floor of an experimental workshop which was at the researchers disposal, and production documents and data printouts placed on large tables. The

---

[3] This procedure has later been speeded up considerably and further refined by the use of e.g. database programmes, personal contacts with expertise within Volvo for the supply of information on diskettes, printouts of specially required labels etc, thus, as was the case in the Autonova plant, making it possible to engage both operators and the plant's engineers in the procedure, thereby bridging the practical gap between practitioners and researchers.

development work was practically performed by moving the physical components around, modifying data printouts, photographs and drawings, using scissors and glue to compose new documents until a logical coherence between physical and logical descriptions was achieved. This was a procedure which also served as an illustration of the work structuring principles and methods successively crystallised during this process.

These work structuring principles and methods have proved to be generally applicable to most vehicles and are based on five characteristics, generic to all vehicles. As illustrated by the photos in Figure 4 there is at least one obviously generic characteristic implying the existence of general structuring principles and methods, i.e. the components are distributed around a symmetrical axle running in the middle of the body, back to front – an organic symmetry where some components are symmetrical in pairs around the mid axle, while others appear only once, almost like a human body. In fact, automobiles and trucks, as well as most other automotive vehicles, show five *generic characteristics, or symmetries*, which form the basis of the work structuring: (1) similarity to the organisms as mentioned above, (2) product functions, (3) plus/minus relationships, (4) generativity and (5) diagonal symmetry (Engström, 1991).

The disassembled products, laid out for long periods of time on the shop floor in the experimental workshop, also explained the unorthodox assembly system design developed, including the materials feeding techniques (e.g. kitting fixtures, kitting process, etc.). The experimental workshop also served as a vital source of information for the management, Volvo expertise and qualified external visitors approved by Volvo during the period 1985 – 1991.

Very briefly described, in general terms, the design procedure used for work structuring contains four phases, denoted A – D, as described below.[4] These *work structuring phases* are:

   A. Collecting information used in later stages.
   B. Preparing for disassembly.
   C. Disassembly and checking through assembly.
   D. Considering the effect of product variants.

Example 3:
   The assembly-oriented product structure separates the components hierarchically into groups that have common characteristics from an assembly perspective, i.e. product function, form, material or modules of the product, thus making it possible to discriminate between the groups of components (see section 4 for a more detailed explanation). Below in Figure 2 the assembly-oriented product structure for the Scania diesel engine is shown.

---

[4] This procedure can of course have different scopes according to e.g. vehicle model and the range of product variants.

 1 Main engine (case: #5).
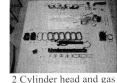 2 Cylinder head and gas exchange (case: #5).
 3 Valves mechanism (case: #5).
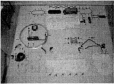 4 Cooling system (case: #5).
 5 Transmission (case: #5).
 6 Lubrication (case: #5).
 7 Fuel system (case: #5).
 8 Components and cables (case: #5).

Figure 2. Disassembled diesel engine at Scania in 1998 expressing an assembly-oriented product structure. The components are positioned on the floor according to their position on the engine. The paper silhouettes illustrate various views of the engine (Portolomeos and Schoonderwall, 1998).

3. COLLECTING INFORMATION USED IN LATER STAGES (WORK STRUCTURING PHASE A)

This work structuring phase mainly includes:

A.1. Collecting the relevant product information including e.g. so-called materials control codes complemented with information about suppliers, materials feeding techniques and quantities, weight, the need for or use of special packaging, etc.

A.2. Collecting the 'correct' translation of product variant codification into true product characteristics.

A.3. Collecting correct component name and descriptions of product systems function including clearing out synonyms and homonyms.

A.4. Obtaining data files stating the assembly times for each detailed assembly task.

A.5. Reconstructing an assembly sequence from various plants or alternatively from some persons familiar with this sequence. The latter might prove to be difficult since overview of the detailed sequence neither was, nor is, within one specific practitioner's knowledge within the Swedish automotive industry.

Collecting the relevant product information in the case of Volvo automobiles requires among other things: (1) to have a data file of the design-oriented product structure containing the component's name/position on the product, variant code, etc. and (2) to get hold of the detailed assembly instruction from the central Volvo product and process department, so-called process and control instruction[5], as well as other types of relevant information such as workshop manuals, service instruction material, interviews, etc.

Collecting the correct 'translation' of product variant codifications and turning these codes into e.g. true, real product characteristics, component names and product properties was (and still is) quite another matter within Volvo since some of the codes (e.g. type of market, type of emission system or type of springing and dampening) are not related to product characteristics relevant on the shop floor. The reason for this is that these characteristics do not influence the product on the shop floor in a logical way. Therefore a deep knowledge of the design-oriented product structure and product and manufacturing process data is necessary to decode this information. The overall knowledge concerning these matters is not available at a single source within the company or promoted as a necessity since it is divided between numerous individuals.

The authors have therefore in all cases of application of this specific phase in the cases mentioned above been required to construct alphabetical registers and 'dictionaries' themselves based on workshop manuals, service instruction material, interviews, etc.

Work in this work structuring phase, in the Volvo Uddevalla case, proved to be time-consuming since the product perception on the shop floor, as well as from the manufacturing engineering's point of view, proved too fragmented to even allow systematic disassembly of an automobile. The general understanding of which components were interconnected or related to other components due to product functions or true product variant characteristics was also lacking. This knowledge is obviously present within the company mainly in the design department, who expresses their work by means of the design-oriented product structure, but during the transformation of information from the design department to the shop floor, both logic and information are deformed, mutilated or lost (see Engström and Medbo, 1993; Medbo, 1994).

Note that the different engineering documents within Scania and Volvo did not, and still do not, possess a coherent stringent vocabulary. Thus it was, and still is, extremely difficult to cross-read or get hold of the total mass of information available, which in fact has proved necessary in both the Volvo automobile and truck companies as well as at Scania.

Example 4:
> The photographs in Figure 4, from the Volvo Uddevalla plant design, show the components in a Volvo 740-model. Each photograph corresponds to 1/8 of an automobile, approximately 1 hour's assembly work. The suggested assembly system design comprised work station systems with three operators resulting in 20 minutes' cycle time, a division of labour suggested by the manufacturing engineers at Volvo since it was assumed to be the maximum economically viable cycle time (leading to a learning time of 4 – 8 weeks to achieve a work pace of 115 MTM).
>
> These photographs proved to be valuable when formulating and communicating the work structuring principles and methods as described in this chapter, and by placing them beside each other, they helped illuminating the assembly work in a plant where 1/8 of an automobile was assembled in eight separate assembly workshops in series as is schematically shown below in Figure 4.

---

[5] These instructions contain, among other things, illustrations of the detailed assembly work.

630   T. Engström et al.

This made it evident that such a plant would require either large intermediate buffers between each assembly workshop or a constant shifting of operators according to time differences between individual products and product variants; consider for example the components needed for an air conditioner added to the components shown in the photographs. This way the suggested division of labour would imply extra space requirements and system losses as well as a degradation of the product perception by basing it on only 1/8 of an automobile.

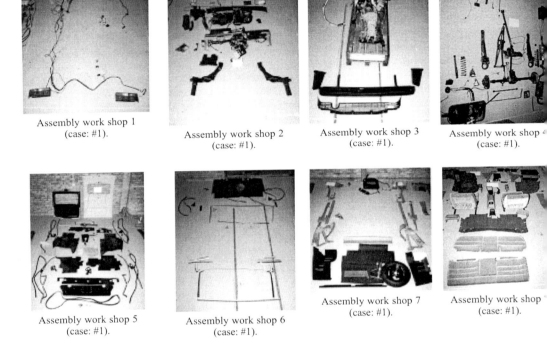

Figure 3. A disassembled automobile where the components are positioned on the floor according to their position in the automobile body.

4. PREPARING FOR DISASSEMBLY (WORK STRUCTURING PHASE B)

This work structuring phase mainly includes:

B.1. creating small cards describing the detailed assembly work (see Figure 4).
B.2. creating labels comprising information from the design-oriented product structure, (see Figure 5). These labels also contain information concerning the components suggested to belong to other types or groupings or information as to whether the same materials control code is used for one or more components fitted in different positions on the vehicle.[6]
B.3. excluding all the small cards that are not assembly-relevant due to the scope of and restrictions on the design procedure (e.g. excluding punching the identification number on the vehicle, automatic gluing of the windshield, work performed in the testing workshop after assembly, and other tasks which are performed outside the work station systems).
B.4. grouping the small cards according to a suggested assembly sequence into suggested work modules, i.e. different levels of the detailed assembly sequence, depending on specific shop floor preconditions, forming the suggested intra-group work pattern.

During the design procedure of the Autonova plant, the work was supported by having the design-oriented product structure available on line, as was also the case for the Volvo truck plant as well as for Scania. An *analysis data base* composed of different data files including information such as size of materials containers, weight, suppliers, etc. was created for the Autonova plant design. This was not possible to accomplish during the early Volvo Uddevalla plant design ten years earlier.

Thus it was possible to conveniently document the successive results from the disassembly, as well as to transfer the results for the total design of an assembly system, regarding among other things, materials requirement, space utilisation for stored materials, etc. In fact, in the Autonova case the design procedure, as well as the starting-up of this plant, was based on this specific analysis data base.

Example 5:
During the design of the Autonova plant 1 000 small cards of approximately 14 x 10 centimetres were used, containing illustrations of detailed work tasks forming a work

---

[6] There are at least three explanations for the time-consuming work to create a database related to the Volvo design-oriented product based on disassembly, i.e. to designate the correct physical component to the 'right administrative position' in the design-oriented product structure. This is a reversed process to the ongoing work in a running plant to use the bills-of-materials (derived from the design-oriented product structure) to trigger the materials to be delivered. One reason is that in the original Volvo design-oriented product structure the smallest identifiable unit (material control code) is equal to the materials address along the traditional assembly line. Thus identical components could be fitted at different positions on the product and it is time-consuming to identify and designate the components correctly. Another reason is that the information systems within Volvo are not designed for this type of analysis. They have in fact been developed over the years into a complex conglomerate of information systems suited when steering a complex, constantly changing organisation where work moves around between manufacturing facilities, sub-assembly and final assembly stations, etc. Finally, the information is constantly changing and the different files containing information are not synchronised. Therefore, each existing local assembly plant transforms the design-oriented product structure according to the specific assembly system design and the product variants manufactured (see e.g. Engström and Medbo, 1993).

operation and product variant codification. Each card included name of the work operation and the assembly time for the work operation required for two selected product variants.

These small cards were derived from the Volvo process and control instructions, which contained illustrations of the work, name of the detailed work operations, etc. The original document was extensive and included 1 – 10 pages of front illustrations as well as standardised forms of 1 – 45 pages, depending on work tasks and product variants comprising assembly sequence and materials required (defined by component number, component name, variant codification, tools, torque, control instructions, quality demand, etc.).

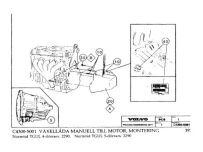
Small cards used at Autonova.

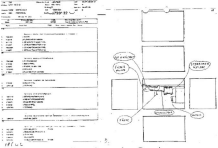
Small cards used at Scania.

Figure 4. Small cards used for guiding the disassembly at the Autonova plant design in 1996 and for the restructuring of the information system at Scania in 1998.

The method used to create these cards requires shrinking of the illustrations by means of a photocopy machine and transferring of the name of the work operation and the design-oriented product structure codes. The cards were also complemented with a sequence number, in this case from 1 to 997, first in order to be able to find the Volvo process and control instruction, i.e. the original card (some were prone to get lost), and second since the result of the procedure might need a future update due to product design change orders.

The total procedure was complicated by product design change orders which necessitated keeping various generations of information 'in the air' simultaneously. This fact called for synchronising the content on the small cards regarding the physical components of the existing automobiles disassembled, as well as the analysis data base by means of the Volvo design-oriented product structure.

On the other hand, for the restructuring of the information system at the Scania diesel engine plant, it was first a matter of understanding an existing assembly system and a number of interrelated information systems in form of Scania's design-oriented product structure, the information system which generates assembly instructions and the work orders for each diesel engine, i.e. the specification of the components required for individual engines. This, accordingly, led to the construction of small cards, of approximately 14.5 x 21 centimetres, one for each of the 43 work stations along the existing assembly line.

Each card contained six views of the engine (the components assembled at a specific work station were sketched for each view) as well as shrunk information from the work orders of the respective work stations etc. for a specific product variant. By this procedure all available information along the existing assembly line for the engine to be disassembled was gathered on these cards. This information was first checked and modified by collecting small components put into plastic bags fetched from the assembly line as well as by interviewing the operators.

The disassembly and understanding of the information systems was facilitated by the complement of folders for each work station that contained assembly instruction, engineering drawings, work orders for other product variants etc.

During the disassembly of the diesel engine, guided by the small cards, cross-reference of the removed components with the small components fetched from the assembly line, and marking of the appropriate documents contained in the binders resulted in the understanding of both the physical product, the existing assembly system and the information systems. This procedure underlined e.g. various anomalies of the present information system.

94 82 10
bandklamma 45.
3730-1001
2
2.0

Label used in 1987 during disassembly used for the design of the Volvo Uddevalla plant (case #1).[7]

GOLVMATTA,BLA A 1 001
V FRAM
1310002 8552010 011 1
V-STYRD GLES LUG BLA MAT
GOLVBEKLÄDNAD
golvmatta främre V, V-styrd, lågnivå

Label used in 1989 during disassembly used for the design of the Volvo Torslanda plant (case #2).[8]

PLUGG,SVART
FIXERING AV BOTTEN-
^SVALLARKABLAGE
LEDNING,BOTTENSVALLARE
16st 3760 010 515 1264326 011100
flera åtgångar
V:2 C3760-5001 deladat LB014
80
Mont.fas: Delfas: Delad med:

Label used in 1995 during disassembly used for the design of the Autonova plant (case #4).[9]

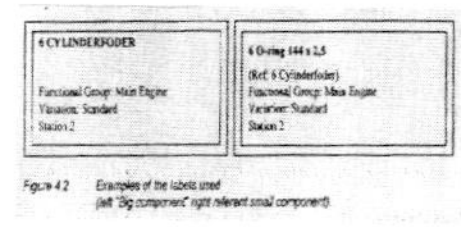

Label used in 1998 for disassembly used for the restructuring of the information systems at the Scania assembly line (case #5).

Figure 5. Example of different labels containing product information used for disassembly.

[7] These labels were used for our first work structuring. i.e. the disassembly was aimed at identifying an assembly-oriented product structure, as well as the suggested detailed assembly sequence for the first automobiles assembled in the Volvo Uddevalla plant. In this case a delimited number of specific product variants were first considered.

[8] These labels were used for disassembly comparing three specifically different product variants (denoted "A1", "B1" and "C1" on these labels) during a period when no formal product- and process information regarding parallel flow assembly system existed within Volvo. The Uddevalla plant was being designed using inferior information support since this system was under development and the responsibility for the information quality was not defined.

[9] These labels were used to guide the design, for all product variants, of the detailed intra-group work pattern, materials feeding techniques, analysis data base, etc., as well as to guide the unpacking and sorting of components delivered for the first 40 product variants manufactured in Autonova. The unpacking refined the analysis database still further. Note that in this case the assembly-oriented product structure was already regonised. During the actual disassembly, a procedure started to designate the right material control code to the correct assembly position on the automobile. This was performed by letting the components with the same material control code but different assembly positions have as many labels as number of positions chosen, resulting in adapted material control codes through splitting of the original codes.

See Figure 5 for example of labels containing product information used during disassembly generated from the analysis date base. These labels complemented the small cards shown in Figure 4, by having all information necessary to relate the label to the design-oriented product structure. This meant information such as e.g. manufacturing engineering data, materials feeding information and in some cases even selected data from other plants as reference.

Consequently it becomes possible to decompose e.g. one single product using its components as representative of all product variants. In practice the component, e.g. one seat, was removed from the automobile body according to the small card, and the labels for all variants of seats were placed on or fitted to this specific seat. Thus, as has been the case in the Autonova plant design, the product information in the form of labels and small cards could be up-to-date, while the product decomposed could even be old.

5. DISASSEMBLY AND CHECKING THROUGH ASSEMBLY (WORK STRUCTURING PHASE C):

This work structuring phase mainly includes:

C.1. Successively disassembling the product guided by the suggested assembly sequence and the preliminary work modules represented by the grouping of the small cards. The cards are positioned on tables, sometimes divided by wooden laths, in order to overview the component and allowing work modules to be redefined by regrouping small cards.

C.2. Successively positioning the components on the floor dividing suggested work modules by the wooden laths, including positioning the correct labels on or beside the respective component including correcting the analysis data base as the work goes on. Thus giving input to for example the design of the materials feeding technique.

C.3. Rechecking the disassembly and the analysis data base by guiding selected expertise through the disassembled components.

C.4. Final rechecking by assembling the product.

The small cards ought to be positioned on tables near the corresponding components. Any questions and assumptions must be noted on the small cards or on white boards as the work goes on in order to systematically decompose the product. This makes it possible to have extra personnel assist in guiding the work structuring phases, thus speeding up the work, or as has been the case in the Autonova plant design, to let operators who were going to be responsible for specific work on the product perform the disassembly.

During this process each component was marked with the labels (see Figure 5). The authors also sewed all small components together with the appropriate illustrations onto 21 large (220 x 120 centimetres) white sheets of paper[10]. This allowed the practitioners and reserchers to

---

[10] These paper sheets were also used by the new operators to learn the assembly work in the training workshop, as well as for the initial identification of the small components suited for packing in plastic bags, a unique materials feeding technique especially developed for the Volvo Uddevalla plant (Johansson and Johansson, 1990).

acquire in-depth learning of the product, as well as to establish the interrelationship between the components and the design-oriented product structure. This work was also facilitated by using illustrations reduced in size from those provided by the central Volvo product and process department (see Figure 4), and by the construction of specially designed illustrations as a contextual visual aid. These illustrations contained several interrelated levels and used a standardised outline for normalising the illustrations. The vehicle is viewed diagonally from behind, as if entering an automobile on the driver's side (see Engström, Hedin and Medbo, 1992) and example 6.

Example 6:
   Disassembly of a Volvo truck in the experimental workshop in 1988 – 89. During this work a complete truck was disassembled. The components were either placed on the shop floor or, as was the case with all leads, as illustrated in Figure 6, positioned and fixed at a number of wood mock-up frames. Thus a number of systematic, normalised illustrations were created according to similar principles as used for automobiles during the Volvo Uddevalla plant design. This work was the starting point of the introduction of the parallel product flow, long cycle time assembly system at the Volvo Tuve plant, in 1991.

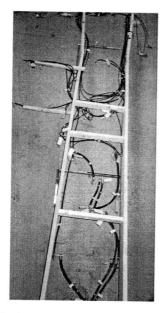
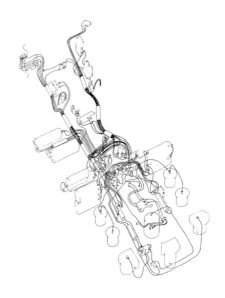

Wood mock-up of truck frame and leads.   Leads normalised against a transparent outline.

Figure 6. To the left there is a mock-up of a truck frame for fixation of leads removed during the disassembly. Four different mock-ups were required, one for each system of leads and to the right there is an illustration of leads normalised against the transparent outline. The 'transparent' components (air hoses for suspension, various tanks, etc.) shown only in form of contours are expressing that these components are omitted.

Finally, in the Autonova case, the disassembled products are also assembled and the analysis data base was used in order to mirror other product variants which were assembled to verify the hierarchical product structuring scheme and a corresponding assembly work structuring scheme. The principle of mirroring product information was performed for the design of the Autonova plant where the analysis data base was first developed and refined during the manufacturing of the Volvo 850-model and later mirrored in and used for the manufacturing of the C70-model. Thereby the manufacturing of the new model was extensively facilitated, since a preliminary assembly-oriented product structure was available before the materials for the new product arrived at the plant according to the generic characteristics of automotive vehicles, e.g. the plus/minus relationships (Engström, 1993).

Note that the purpose of the design procedure could differ. For example, in the Volvo Uddevalla case it was initially a question of finding a logical grouping of the components, which in fact proved possible by some thinking and moving around of the components of a disassembled automobile on the shop floor of the experimental workshop.

As a result of this procedure, and by the aid of photos of the components taken in the experimental workshop, it was possible to recognise an assembly-oriented product structure. The main groups in this structure were for the automobile: (0) Doors; (1) Leads for electrics, air and water; (2) Drive line, (3) Sealing and decor and (4) Interior. The first group being sub-assembly work, while the other four were work on the automobile body. These main groups of components imply not only a general classification applicable to automobiles but also, importantly, a classification based on five generic characteristics always present in all vehicles (see also example 1).

The detailed assembly work is then derived from the assembly-oriented product structure according to specific levels, where the highest level, depending on the specific assembly system designs, is work modules. Accordingly, an overall taxonomy, in the form of the assembly-oriented product structure was first stipulated, and different levels of the detailed assembly sequence, depending on specific shop floor preconditions (work group size, competence overlap within the group, ergonomic preferences, etc.) are later derived from this classification. This procedure allows, among other things, an implicit defined interrelation between materials and tools, work descriptions and other types of production documents in accordance with the work hypothesis for design of parallel product flow, long cycle time assembly systems.

In the case of the design of the Autonova plant ten years later, these main groups were known, as well as the general work structuring principles and methods. Thereby the work was primarily concentrated on a search for intra-group work patterns based on work modules as well as a search for the design of materials feeding techniques and focused on the creation of an analysis database to support the design and running-in of the plant as well. The analysis data base was also used for the unpacking and sorting of the materials for the first automobiles.

6. CONSIDERING THE EFFECT OF PRODUCT VARIANTS (WORK STRUCTURING PHASE D)

This work structuring phase mainly includes:

D.1. Detailing the assembly information gained according to variance introduced by different product variants, that are not necessary to disassemble since they could be grasped intellectually by the analysis data base and work performed in phase A. Thereby it became possible to generalise the hierarchical product structuring scheme and the hierarchical assembly structuring scheme to include all product variants through the identification of the so-called *variant tracks* (see Figure 6).

D.2. Identification of variant tracks corresponding to characteristics more or less obvious due to the choice of the assembly-oriented product structure. These tracks correspond to the need for e.g. overlapping competence between operators within or between work station systems, i.e. these tracks may or may not call for extra work, as is evident from Table 2.[11]

D.3. Grasping the differences in assembly work stipulated by a decomposed reference product variant. This could be either by rough estimations or by taking advantage of the existing product and manufacturing process data.

Table 2
Time spread in per cent of assembly work on the automobile body in comparison to a disassembled reference product variant according to the design procedure performed for the Volvo Uddevalla plant design in 1987.

	Reference variant [%]:	Variant 1 [%]:	Variant 2 [%]:	Variant 3 [%]:	Variant 4 [%]:	Variant 5 [%)]	Variant 6 [%]:
Type of variant:*	B230FS, 4D,	B230FS, 4D, ABS-brakes	B230FS, 4D, Aircondition	B230FS, 4D, Sunroof	B230FS. 4D	B230FT, 4D	B230FT, 4D, ABS-brakes Aircondition Sunroof
Operator 1:	25	0	12	0	0	0	17
Operator 2:	25	0	0	20	14	0	31
Operator 3:	25	0	1	16	0	0	21
Operator 4:	25	2	0	16	0	3	20
Operator 5:	20**	3	13	0	1	1	55

* Extremely brief description B230FS = 2.3 litre suction engine with injection, B230FT = 2.3 litre turbo engine with injection. 4D = four doors and 5D = five doors.
** This operator performed only sub-assembly work corresponding to 20% extra in relation to the 100% work on the automobile body.

---

[11] In e.g. the Uddevalla case information from the central Volvo product and process department, information in the form of motion-and-time studies, specifying assembly times for specific detailed work tasks, was available for all product variants. Thus we did not e.g. need to disassemble more than two product variants in the case of Uddevalla and one in the case of the Autonova plant design.

In this case it was an assumption of a distribution of the assembly work into four equal parts of work on the automobile body (i.e. 25%) which of course would not be possible to achieve in practice. The time given in Table 2, stipulating additional time for six product variants in relation to the reference variant, was used for the design of different intra-group work patterns, thereby assuming an ideal balancing of the work for the reference variant. The estimation of relevant assembly times for different product variants was a tedious work based on work mapping in the training workshop at the Volvo Uddevalla plant (see Engström and Medbo, 1994) by coding every component according to the product variant codification, as well as by composing and decomposing detailed work tasks organised according to the design-oriented product structure.

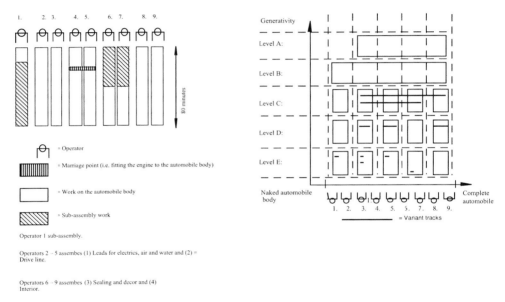

Intra-group work pattern at the Volvo Uddevalla plant.   Variant tracks at the Volvo Uddevalla plant.

Figure 7. A schematic outline of how the product variation during assembly is described with the aid of the assembly-oriented product structure. To the left there is the intra-group work pattern for one complete automobile comprised out of work modules, while to the right the variant tracks organised according to generativity are shown, i.e. the product characteristics forms so-called variant tracks through the assembly-oriented product structure. On the basis of this generative distribution the characteristics are grouped on levels from A to E, where level A represents 'obvious' (e.g. 4 or 5 doors, sunroof or not, etc.) and level E 'unpredictable' characteristics from the operator's point of view.

7. CONCLUSIONS

To summarise, the results from the work structuring phases briefly described above are a hierarchical product structuring scheme and a corresponding hierarchical assembly work structuring scheme. These schemes are necessary e.g. as inputs when calculating production capacity considering assembly time constraints. Continued assembly system design is discussed elsewhere, e.g. in Engström and Medbo (1994); the complete design procedure involves design of intra-group work patterns, the detailed work station system layout, the product flow pattern, the materials feeding techniques, definitions of sub-assembly work tasks suitable for integration into the work station system, etc.

The work structuring phases in the design procedure reported above underlines that both an overview and detailed information are required to mould the work to achieve the structural congruence needed for efficient and flexible parallel product flow, long cycle time assembly systems. This requires a combined design procedure amalgamating 'analogous' physical products and an assembly-oriented product structure gained through reforming of existing product and manufacturing process data. Thus it is possible to interrelate the 'shop floor reality' of existing product and manufacturing process data to the future 'shop floor reality', i.e. the assembly systems not yet designed.

From the authors' point of view the cases reported here are not pure examples of participative design procedures as has sometimes been the international perception of unorthodox Swedish assembly system design (Grotingen, Gustavsen and Héthy, 1989). Instead some of these design procedures include an engineering approach, supported by established knowledge from social science (e.g. Karlsson, 1978; Nilsson, 1981, 1995). Accordingly, the true core of the cases reported have been internationally misunderstood as a human relations approach, dimmed by terms like participation and humanisation.

Generally speaking, the experience from and involvement in the design of the Volvo Uddevalla plant and the Autonova plant ten years later, emphasises the importance of transferring design procedures and design experiences between large industrial development projects. In the cases reported above, due to various circumstances, this responsibility has gradually become the role of the researchers; see e.g. Ellegård (1989) and Granath (1991). This has in some respects meant avoiding reliance solely on consultants or internal expertise in these matters, which has influenced large organisations not to fabricate carbon copies of other plants. How much of this knowledge that has in fact been assimilated by the organisations is however debatable.

The experiences reported are consistent with the concept of agile manufacturing which emphasises methodologies of implementation of advanced manufacturing systems and concepts (see for example Kidd, 1994). This can be interpreted as stressing the technical dimensions of the socio-technical design paradigm, which for long time has influenced Scandinavian manufacturing systems designs (van Eijnatten, 1993) and specifically the assembly systems designs within the Swedish automotive industry, including the cases reported here; a design paradigm which has been criticised for neglecting operationalising of technical dimensions (Linder, 1990). This discourse may, however, basically be due to an assumed contradiction between social and technical sciences. Being an engineering scientist is often an advantage, though, since the manufacturing industry can, more or less directly, access and apply the research results into engineering designs.

However, if the research, as was case for the Volvo Uddevalla plant design, included not only practitioners and engineering scientists, but also researchers in architecture, as well as social scientists within the areas of pedagogy, human geography and psychology, it proved feasible to satisfy various critical restrictions, resulting in a substantial broadening of the set of designs of assembly systems.

REFERENCES:

Bozer, Y. A. and McGinnis, L. F. 1992. Kitting versus line stocking: A conceptual framework and a descriptive model. International Journal of Production Economics, Vol. 28, pp. 1 – 19.

DeJong, J. R., 1957. The Effects of Increasing Skill on Cycle Time and its Consequences for Time Standards. Ergonomics, Vol. 1, pp. 51 – 60.

Ellegård, K., 1989. Akrobatik i tidens väv – En dokumentation av projekteringen av Volvo´s bilfabrik i Uddevalla. Chorus 1989:2 Kulturgeografiska Institutionen, Göteborgs Universitet, Göteborg. (In Swedish).

Engström, T., Hedin, H.and Medbo, L., 1992. Design Analysis by means of Axonometric Hand-Drawn Illustrations. In: Proc. The International Product Development Management Conference on New Approach to Development and Engineering, pp. 147 – 157.

Engström T. and Medbo L., 1994. Intra–group Work Patterns in Final Assembly of Motor Vehicles. Published in International Journal of Operations & Production Management, Vol. 14, No. 3, pp. 101 – 113.

Engström T., 1991. Future Assembly Work – Natural Grouping. Design for Everyone. Y. Queinnec, and F. Daniellou (Eds.), In: Proc. The 11th Congress of the International Ergonomic Association, Taylor & Francis Ltd, London, Vol. 2, pp. 1 317 – 1 319

Engström, T. and Medbo, L., 1992. Preconditions for Long Cycle Time Assembly and Its Management – Some Findings. International Journal of Operations & Production Management, Vol. 12, No. 7/8, pp. 134 – 146.

Engström, T. and Medbo, L., 1993. Naturally Grouped Assembly Work and New Product Structures. International Journal of Technology Management, Vol. 7, No. 4/5, pp. 302 – 313.

Engström, T., Jonsson, D. and Medbo, L., 1997. Successive Assembly System Design Based on Disassembly of Products. In: Proc. H. Mueller, J.-G. Persson, and K. R. Lumsden (Eds.), The Creation of Prosperity, Business and Work Opportunities Through Technology Management, Proceedings of the Sixth International Conference on Management of Technology – MOT97, Vol. 1, The Swedish Society of Mechanical Engineers, Naval Architects and Aeronautical Engineers, Göteborg, Vol. 2, pp. 264 – 273.

Engström, T., Jonsson, D. and Medbo, L., 1999. Developments in Assembly System Design: The Volvo Experience. In: Y. Lung, J. J.Chanaron, T. Fujimoto and D. Raff (Eds.), Coping With Variety: Flexible Production Systems for Product Variety in the Automobile Industry, Ashgate, Aldershot. (In press).

Goldman, S. L., Nagel, R. N. and Preiss, K., 1995. Agile Competitors and Virtual Organizations – Strategies for Enriching the Customer. Van Nostral Reinhold, International Thomson Publishing Inc., New York.

Granath, J. Å., 1991. Architecture, Technology and Human Factors – Design in a Socio-Technical Context. Industrial Architecture and Planning, Chalmers University of Technology, Göteborg. (Ph.D. thesis).

Grotingen, P., Gustavsen, B. and Héthy L. 1989. New Forms of Work Organization in Europe. Transaction Publishers, New Brunswick, New Jersey.

Johansson, M. I. and Johansson, B. 1990. High Automated Kitting System for Small Parts – A Case Study from the Volvo Uddevalla Plant. In: Proc. The 23rd International Symposium on Automotive Technology & Automation, Automotive Automation Limited, Vienna, pp. 75 – 82.

Karlsson, U., 1978. Alternativa produktionssystem till lineproduktion Department of Sociology, Göteborg University, Göteborg (Ph.D. thesis in Swedish).

Kidd, P. T. And Krawowski, W. 1994. Advances in Agile Manufacturing – Integration technolgy, organization and people. IOS Press Amsterdam.

Kidd, P. T., 1994. Agile Manufacturing Forging New Frontiers. Addison-Wesley Publishing Company, Wokingham.

Lindér, J. O., 1990. Värdering av flexibel produktionsorganisation utifrån sociotekniska principer. Institutionen för industriell organisation. Chalmers Tekniska Högskola, Göteborg. (Ph.D. thesis in Swedish).

Marton, F., 1970. Structural Dynamics of Learning, Gothenburg Studies in Educational Sciences 5. Almqvist & Wiksell Publishers, Stockholm. (Ph.D. thesis).

Marton, F., 1981. Phenomenography – Describing the world around us. Instructional Science, Vol. 10, pp. 177 – 200.

Medbo, L., 1994. Product and Process Descriptions Supporting Assembly in Long Cycle-Time Assembly. Department of Transportation and Logistics, Chalmers University of Technology, Göteborg. (Licentiate thesis).

Medbo, L., 1999. Materials Supply and Product Descriptions for Assembly Systems – Design and Operation. Department of Transportation and Logistics, Chalmers University of Technology, Göteborg. (Ph.D. thesis).

Nilsson, L. (1981). Yrkesutbildning i nutidshistoriskt perspektiv. Pedagogiska Institutionen. Göteborgs Universitet, Göteborg. (Ph.D. thesis in Swedish).

Nilsson, L. (1995). The Uddevalla Plant – Why did it succeed with a holistic approach and why did it come to an end?. In: Å. Sandberg (Ed.), Enriching production – Perspectives on Volvo's Uddevalla plant as an alternative to lean production, Avebury, Aldershot, pp. 75 – 86.

Portolomeos, A. and Schoonderwal, P., 1998. Restructuring the Information Technology and Communication Systems at Scania Engine Plant. School of Technology Management and Economics, Chalmers University of Technology, Göteborg. (MOP-thesis).

van Eijnatten, F. M., 1993. The Paradigm that Changed the Work Place. The Swedish Center for Working Life, Stockholm, Van Gorcum, Assen.

Wild, R., 1975b. On the Selection of Mass Production Systems. International Journal of Production Research, Vol. 13, No. 5, pp. 443 – 461

Part VI

STRATEGIC APPROACH FOR AGILE MANUFACTURING

Reegineering and Agile Manufacturing Development

G. Doumeingts[a,b], H. Kromm[a], Y. Ducq[a], and S. Kleinhans[b]

[a] Laboratoire LAP/GRAI, Université Bordeaux I,
351 cours de la libération, 33405 Talence cedex, France – Tel : +33 5 56 84 65 30 – fax : +33 5 56 84 66 44 - Email : doumeingts@lap.u-bordeaux.fr

[b] Société GRAISOFT SA
Parc d'activités Favard, 16 cours du Général de Gaulle, 33170 Gradignan, France

1. INTRODUCTION

In the economic environment of today, many enterprises need to adapt their organisation in order to reach their strategic objectives. These objectives traduce the necessary performances for the improvement of competitiveness. This competitiveness implies the following characteristics:
- to be able to define the global performance for the enterprise and the action plan to achieve them, by the definition of a Business Plan,
- to re-engineer the enterprise in order to improve the performances on several points of view: cost, quality, time, flexibility, reactivity...
- to choose and to implement rapidly and efficiently I.T. solution,
- to evaluate continuously the performances by implementing an efficient, coherent and adequate Performance Indicator System,
- to be able to compare the enterprise practices against the best industrial practices by Benchmarking,
- to deploy the quality procedures,
- to collect and to manage the knowledge of enterprise.

All these characteristics are today included in the concept of the Agile Manufacturing Enterprise. This new organisation gives the manufacturers the ability to react fast to sudden, unpredictable change in customer demand for its products and services and make a profit.

First, we demonstrate here how the GRAI Methodology, based on the GRAI Model, allows to improve the competitiveness of the enterprise by supporting all the previous techniques. And more especially, how GRAI Methodology could support Agility concept in the enterprise.

Second, this chapter presents a case study to explain how GRAI Integrated Method (GIM) one of the method of the GRAI methodology was used to improve the global performances of a management structure.

2. MODELLING TECHNIQUES AND AGILE MANUFACTURING

2.1 Definition of Enterprise Modelling

The Enterprise Modelling is a research domain which appeared in the nineties, following the development of the BPR (Business Process Re-engineering) techniques.

We underline that the enterprise modelling techniques started in the eighties particularly in the domain of manufacturing and Production Management : simulation, production management models, information systems modelling. A good example was given by the US Air Force Project ICAM (Integrated Computer Aided Manufacturing) at the beginning of eighties which among the results, proposed IDEF 0, 1, 2 (Icam DEFinition).

Today, the Enterprise Modelling is the representation of a part or of the set of enterprise activities at a global and a detailed level, using activities, processes, in order to understand its running. In this definition, it is necessary to take in account not only the technical aspects but also the economic, social and human ones. The Enterprise Modelling Techniques allow to describe the running of the enterprise in terms of : objectives, structure, functionalities, evolution, relationships with environment (customers and suppliers).

Enterprise Modelling techniques can support the modelling of various new concepts as extended enterprise, virtual enterprise or agile manufacturing.

Based on the definition of Agile Manufacturing, we propose to use GRAI Methodology as a support to model this type of system.

2.2 The Agile Manufacturing System

The term of agile manufacturing emerged from a study of US Manufacturing conducted by a team of seniors industrial practitioners and academic researchers. One of the main remarks given by the final document is the following one:

"Agility requires integrated flexible technologies of production with the skill base of knowledgeable work and with flexible management structures that stimulates the co-operative initiatives between firms".

In that way, the agile manufacturing brings innovative products to market very quickly, responds quickly to market and customer demands for new products and product features. It has also installed a re-programmable, re-configurable and continuously changeable production system which is capable of operating economically with very small lot sizes.

An important aspect of agile manufacturing is the ability to develop some strategic relationships with customers and suppliers. It seems so that the Agility concept is associated to the integration of the supply chain.

The agile manufacturing appears so as a new form of production structure which ensures a continuous adaptation in an external environment of constant and unpredictable change. The manager of a such structure has to make it evolves in perpetual way in order to maintain its competitiveness in achieving the strategic performances he received.

R. N. Nagel [1] defines four main strategic principles:
1. Agile competition will be based on ability to thrive on change and uncertainty. Therefore use an entrepreneurial organisational strategy.
2. In an agile environment organisations sell skills, knowledge and information over time. Therefore invest to increase the strategic impact of people and information on the bottom line.
3. In an agile organisation co-operation enhances competitive capability. Therefore use the virtual company model inside and outside as a dynamic organisational strategy.

4. In an agile world customers either: pay a fee for skills materials and modest profit for products; or pay a percentage of the perceived value (to them) for solutions. Therefore adopt a value based strategy to configure your products and services into solutions for your costumers.

The GRAI methodology which the characteristics appears as suitable for the modelling of Agile Manufacturing, is described here after.

2.3 The GRAI Methodology

GRAI methodology [2] uses several concepts focused on the life cycle of the Production System design or Enterprise Model. These system life cycle phases are User Requirements definition, Modelling/Analysis, Design, Development, Implementation, Operating and Evolution.

Today, GRAI Methodology is covering all the phases of the system life cycle as shown figure 1. It was not the case in the past. This figure as well presents the abstraction levels concepts.

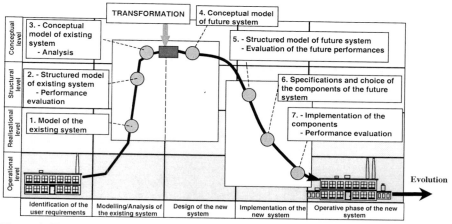

Figure 1: The various phases of the GRAI Methodology

Because of its main characteristics, the definition of the enterprise modelling techniques could be apply to the GRAI Model, and by extension, to the GRAI Methodology

One originality of GRAI Methodology is to allow to process various improvements actions as BPR, Re-engineering, Choice and Implementation of Software Package (ERP type) or advanced IT applications (workflow), Definition and Implementation of Performance Indicators, Benchmarking, by using always the same model, the same formalisms and the same generic structured approach.

An another originality is that the methodology is usable for industrial companies as for services.

The GRAI Methodology is a methodology based on:
- A GRAI reference Model and a GRAI Control model
- Some formalisms to describe the concepts of the GRAI Model,
- A structured approach to implement the methodology

2.3.1 The GRAI Reference Model

At the conceptual level, the GRAI Model is a recursive structure (figure 2) which is composed of three systems :

The physical system transforms raw materials, or components, into output Products, therefore creating a flow of materials through the physical means (equipment, machines,...) organised according to various layouts.

The decision system (DS) is split up according to two axis. On a vertical axis, the DS is decomposed according to the type of decisions : strategic, tactical, operational. On an horizontal axis, the criteria of decomposition is functional : in the figure 2, the main functions of an enterprise are represented. Conceptually, a Decision Centre (DC) is defined as the cross of a column and a decision level. This presentation is not further detailed but the complete DS take in account two types of decisions: periodic decisions and event driven decisions.
It is important to remark that the horizontal layer gives the Business Process View.
The information system contains all information needed by the running of the system: it must be structured in an hierarchical way according to the structure of the decision system.

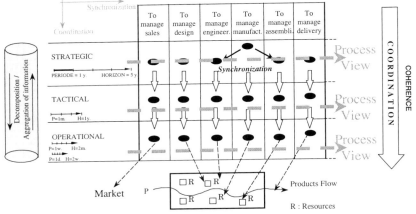

Figure 2: The global GRAI conceptual reference model

2.3.1.1 The GRAI Control Model

The GRAI Reference Model gives the generic structure of the Model. It is necessary to define now the GRAI Control Model which allows to control precisely the physical system.

The Physical Model mainly consists of the Product Flow and the Resources. We define an activity as the transformation of a product by (one or several) resources.

It is necessary to synchronise the availability of products to be transformed and the Resources (see Figure 3).

The criteria for the co-ordination is a temporal one. A couple of temporal characteristics defines each level: "Horizon" (the interval of time over which the decision extends, i.e. remains valid) and "Period" (the interval of time after which we reconsider the set of decisions). In such a structure, the horizon is a sliding horizon.

The objective of the GRAI Model is to control the Physical System in an optimal way. To do that, it is necessary to control the flow of products (to manage the products) and to synchronise it, in the time, with the availability of the resources (to manage the resources). This synchronisation is performed by a third function : "to plan". So, the three basic functions to control the physical system are: "To manage the products", "To manage the resources", and

"To plan". So, the basic GRAI model to control the physical system can be represented by Figure 3.

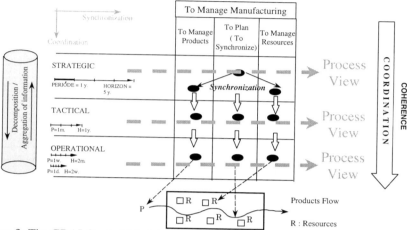

Figure 3: The GRAI Control Model

Thus, to get theoretically the complete GRAI GRID, we have to combine the Global GRAI Conceptual Reference Model and the GRAI Control Model.

2.3.1.2 GRAI Micro Model

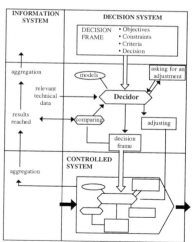

Figure 4: A GRAI Micro Model

A decisional centre corresponds by definition to the intersection of a function and a level (detailed model). At any level of the structure, it can be decomposed by using the same decomposition criteria: a physical component, a decisional component and an information component (Figure 4): that is the principle of a recursive structure because we have the same structure than for the global model.

The physical component is the view of the physical system at the level of decision considered (a production manager does not see the physical system as a foreman does).

The decisional part is made of one decision maker (could be a PLC or a person) and of various elements helping him to make the decision. A decision Centre is describes using the GRAI Net model.

2.3.2 Modelling formalisms

The modelling formalisms concern the physical system, the decision system and the information system. For the physical system, in a first phase we describe the activities on a static point of view. Actigrams and S/R (Stock/Resources) are used. For the dynamic behaviour, we transform these models to be able to use Simulation.

For the Decision System we use the GRAI grid at the global level and the GRAI nets at the detailed level. It enables the identification of decision makers, responsibility and authority, linking decision makers and decision making.

The formalism used to describe the Information System is the Entity / Relationship modelling (figure 8). It describes the data structure in coherence with the Decision System.

We had one view for the three sub system : the functional view which use the Actigrams formalism. This last view ensures the integration of the others views through a common model.

At last, a formalism is used for the modelling of the processes of the enterprise called Extended Actigrams (see Figure 5).

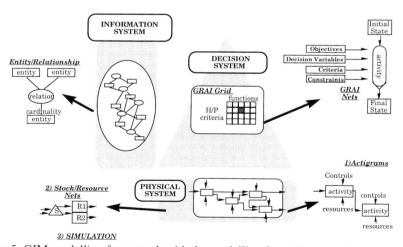

Figure 5: GIM modelling framework with the modelling formalisms used

Validation between models

The whole desired models will be obtained only after having applied validation rules between models. Its aim is to impose a global coherence to the study. Among various elements, data can be noticed to be a common concept for the various models. Data are vector of communication between sub-systems, they are thus the validating element. So, the basic point of the validation between models will be realised by imposing data coherence.

2.3.3 The GRAI Methodology generic Structured Approach

2.3.3.1 Basic principles of the structured approach

The first part of the structured approach is the determination of the user requirements. According to the user requirements of the future system, the GRAI Methodology provides specifications in terms of: organisation, information technology and manufacturing technology, which will allow to build this new system. The structured approach has four main phases : modelling, analysis, design and implementation.

1. Modelling phase : Elaboration of the AS IS Models : this phase allows elaborate the model of the existing system in order to understand better its running and to detect the points to improve.

2. Analysis phase : this phase is starting from the various models of the previous phase. Based on the rules deduced from the GRAI Model and from other models, we determine the strong points of the AS IS models and the points to improve for the future system.

3. Design phase : Elaboration of the TO BE Models : the TO BE Model is design based on the analysis phase results but also based on the objectives given to the system.

The objectives of the TO BE Model is to give all the characteristics of the future system in order to choose a solution.

In such a phase, we give the conceptual model but also the technical specifications to choose the most adapted solution.

4. Implementation phase : the goal of the implementation phase is to acquire the various components of the solution and to implement them according to the chosen organisation.

2.3.3.2 GRAI Methodology actors

We have seen before that GRAI Methodology is strictly structured. In terms of participants, it requires the following groups shown figure 6:

• a project board composed by the management of concerned system. The goals of this group are to give the objectives of the study, and to orient the study according to the results of the main phases.
• a synthesis group composed of the main future users and main deciders of the concerned system. This group has to give their knowledge, to follow the development of the study, and to check the results of the various stages.
• an analysis group composed of : an analyst (several if necessary), whose job is, in particular, to collect all the data needed for the design (in reality, an analyst as meant in GRAI terms, is a true sensor).
• the interviewed group which has to provide the information required.

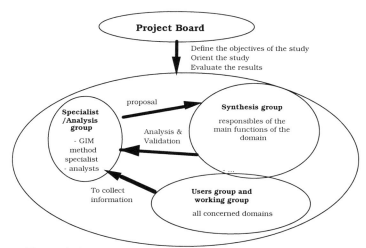

Figure 6: Groups involved in the GIM method application

2.4 The Methodological Tree

This part will present the various methods developed on the basis of the concepts of the GRAI Model and using the GRAI structured approach.

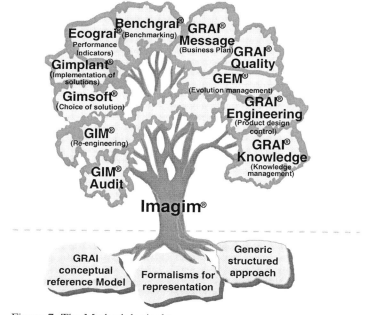

Figure 7: The Methodological tree

There are eleven different methods presented on the following tree (Figure 7):
- GIM Audit for rapid analysis of Enterprise using GIM formalisms (5 days)
- GIM (GRAI Integrated Method) for re-engineering problems
- GIMSOFT to perform specifications, and to choose packages (e.g. ERP...)
- GIMPLANT to implement chosen specific packages (e.g. ERP...)
- ECOGRAI to define and to implement a Performance Indicator System
- BENCHGRAI to perform benchmarking studies
- GRAI Message for the elaboration of Business Plan
- GRAI Quality: to collect the procedures and the operation modes and linking them to the concerned Business Processes
- GEM (GRAI Evolution Method) to manage the enterprise evolution.
- GRAI engineering: to control the product design activities
- GRAI Knowledge: allows to manage the set of knowledge collected when performing of the various modelling modules.

It is quite sure that GIM, GEM and GRAI Message ensure to introduce Agility in manufacturing.

3. HOW TO USE GRAI METHODOLOGY TO ACHIEVE AGILITY

3.1 GRAI Methodology and Agility

As it is noticed before, the GRAI Model is covering the whole life cycle of the production system. The most recent works [3] shown that we are now able to manage the evolution of the enterprise in order to achieve strategic objectives.

To do that, one needs to take in account the actual state of the system (AS IS) on one hand, and on a other hand, in the frame of the strategic goals (Should Be), the model of the expected system (Next Step) could be build. The manager always measures the performance statement of his activities, and he is so capable to correct the trajectory towards the target. He could manage many little projects in the same time and in coherency with his global objectives (figure 8).

Figure 8: GRAI evolution Method principles

This approach is quite close to the agile management system that ensures to adapt quickly to positive changes in order to meet customers needs. An Agile management system does a proactive management as GEM does.

This approach is the first stage toward a good evolution. In fact, it is an Agility point of view applying to the enterprise Modelling Techniques.

The second stage to implement is the integration of the different concepts attached to Agility notion in the frame of the GRAI Model.

3.2 Integration of Agility in The GRAI Model

The only way to improve the competitiveness of the enterprise by using several techniques in a coherent way is to use a Reference Model which describes the concepts and gives the structure of the Enterprise, some formalisms to express them and a structured approach to implement it.

One of the main interests of the GRAI Model is to be generic enough in order to be usable for any kind of Enterprises (discrete, process,...) and recently for services.

The GRAI Model is the key relationship between the various methods of the GRAI Methodology.

As we mentioned above, an Agile manufacturing should be able to [4]:
 1. analyse the existing problems
 2. propose alternative solutions in management
 3. be actively involved in the implementation of the adopted solution
 4. participate in continuous improvement

Therefore, GRAI Methodology can support the design and the implementation of an Agile manufacturing structure. Because, on one hand, GEM ensures to manage actively the continuous improvement of the enterprise by the control of several main design projects. And on an other hand, GRAI Model achieves the modelling, the analysys and the design of adapted solutions for the existing system.

3.3 A Global Approach for Agility Achievement

It is now possible to improve the agility of the enterprise through an integrated approach that proposes to managers some tools to model, to analyse the enterprise and to solve his problems, and a global method (GEM) that ensures to improve the performances in a coherent way [6]. This global approach is represented with the following figure (figure 9):

For each step of the evolution of figure 9, the models become more and more refined, and the objectives more detailed. In the same time, the enterprise becomes more proactive and could better anticipate the market needs or the changes of the environment.

The existence of these models contribute strongly to the agile structure improvement by ensuring to develop and to facilitate the management of the skills and knowledge of the people within the company.

Moreover, GEM provides the integration of design processes and manufacturing processes.

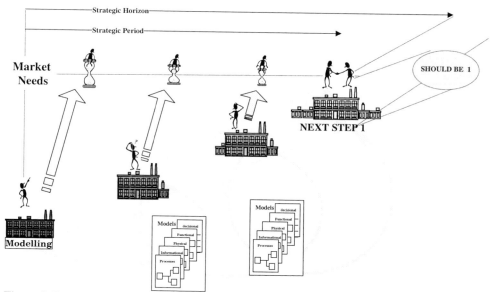

Figure 9: Integrated Method for Agility implementation

4. APPLICATION OF GRAI METHODOLOGY TO DESIGN A BUSINESS UNIT

4.1 Presentation of the enterprise

The case study presented in this paper concerned an International Company which has the leadership on its market.

Among several problems, this enterprise has to improve many points to face up a very strong concurrency from Asian Countries. More especially, it has to reduce very strongly its delivery lead time.

The Chairman and chief officer of the group said:

"*During the quarter, far-reaching changes were made in all three business groups* **to create new global organisation solutions, distinct from products and services**. *These new organisations will allow us more effective local deployment of the full range of the enterprise services.*" This point of view is now very current in the management. It concerns the expecting of a new organisation closer to the market needs and in fact closer to the environment. This is typically the Agile Manufacturing Concept.

So, the global industrial organisation will be based on centres of excellence where each production system will be oriented on one product line in a manufacturing unit. This last point participate to the modularization of the products on one hand, and to the knowledge improvement on the other hand.

4.2 Products

Currently, the enterprise proposes:
1. 1600 products dispatched on 80 families for one application domain,
2. 700 products dispatched on 120 families for the other application domain,

These products can have various uses. They are also included in two great categories: Assemble To Order and Engineering to Order. A result of the project was a precise identification of the whole products which are manufactured by the enterprise.

The results of the enterprise in terms of quantities are the following ones:

1- The first application Domain:
- 45% different products dispatched in 50% families represent 80% of the deliveries.
- 25% of these deliveries are concerning only one product
- Trade products represent an important part of the whole deliveries.
- The last year, data analysis showed a particular decrease of the deliveries due to a decrease of one major product.

2- The second application domain
- 400 products dispatched in 90 families represent 80% of the deliveries
- Assemble To Order products compose the majority of the deliveries

4.3 Objectives of the Study

The global objective assigned to the study is double:
1. to model and to analyse the results of the modelling phase. Then to design the solutions to implement in order to improve the global performances.
2. to perform a specifications book for the implementation of a new version of the existing ERP package.

The performance improvement should ensure the enterprise flexibility in a first time and moreover agility through the separation of the Commercial activity and the Manufacturing activity. The company can adapt to the system customer requirements through the commercial structure without disturb the Production structures. In fact the Business Unit must behave as filter between external changes and production system.

The evolution of existing ERP software is becoming essential for the enterprise because the various entities of the company have developed many specific software applications that generate a lot of disturbances. So the current one was not appropriate to the enterprise development and moreover don not offer all the necessary useful functions for the design and implementation of the Business Unit.

In order to evaluate the results, the commercial top management define four performance indicators:
1. Lead time Reliability
2. Conformity of products
3. Lead time Flexibility
4. Stock immobilisation lead time

These indicators should be measurable by the future structure and the design choices need to take into account these expected performances.

Global objectives of the study

The design of the Business Unit will constitute the main lever to reach the following enterprise objectives.

In fact, the most important problem of the new organisation is that the delivery lead time can increase due to the breaking off of the logistic chain.

Global Enterprise objectives
to increase the enterprise competitiveness
to simplify procedures associated to sells, production, purchase...
to separate manufacturing and commercial activities at a world level
Detailed project objectives
to design the Business Unit
to reengineer a manufacturing system and a commercial unit
to facilitate the evolution of the ERP

Actors involved in the study

For each study sites, three different groups were involved during the two phases of GIM. The composition of each group is described below for the Business Unit Design :

Group name	Composition
Project Board	Project manager, Head of Division
Synthesis Group	Project manager, Sales manager, Purchase manager, Business Unit responsible, Planning manager.
Specialists/analyst Group	One project engineer of the agent

The very difficult context of re-engineering led to many perturbations that made the groups changed during the study. This change had increased the global lead time of the project.

4.4 Planning of the Modelling/Analysis Phase

The planning of the modelling phase is included in the global planning of the study below (figure 10).

The study began by a Project Board meeting in order to define the objectives of the study.

Then five Synthesis Group meetings were organised in order to perform and to valid the various models.

These models were elaborated based on information collected by interviews between the Synthesis Group meetings.

The duration of a Project Board meeting was about one hour and a half, for a Synthesis Group meeting about three hours and for an interview about two hours. An additional difficulty was the very high flexibility of the majority of the managers. This flexibility led people to be often moved which implied to change twice the synthesis group.

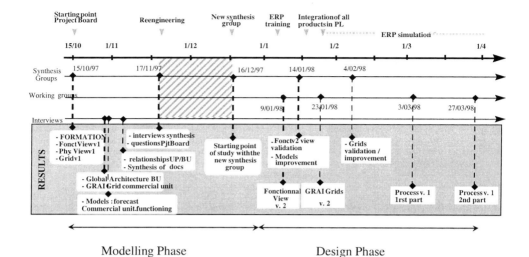

Figure 10: Global planning of the study

4.5 Presentation of the Modelling Phase Results

The objective of this phase was to perform the modelling of the enterprise running. We deduced the analysis results from the modelling results. It is important to notice that this study differ strongly from classical ones because there is no existing system. The main base of work for people involved was also the knowledge they have.

The modelling phase achieved the following results :
- The complete new organisation through:
    - The whole flows identification.
    - The Business Unit position identification.
    - A product typology.
- A representation of the relationships between the Business Unit and the other division actors through:
    - The modelling and the analysis of a manufacturing system.
    - The modelling and the analysis of a business unit

These relationships have been translate into GRAI models with GRAI grids. This representation shows the complexity of the decisional and physical flows.
The main change of the whole structure of the enterprise is represented below (figure 11).

In fact, one Business Unit will have many Production units all over the world as partners. Whereas, to consolidate the order assessment, the Business Unit needs to have all the Production Capacities at the same time. Due to the timetable, the forecast is becoming very complicated. This is one of the several problems that the study needs to solve (Figure 12).

An other problem is the physical aspect of the Business Unit:
1. is it a virtual one that use trucks subcontractors ?
2. is it a real one with a physical plants ?

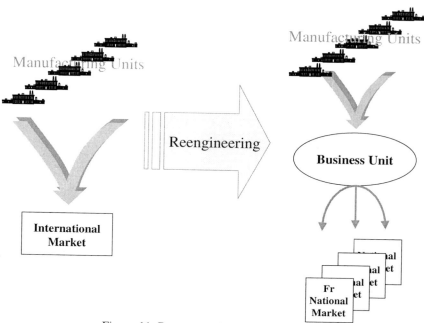

Figure 11: Representation of the main structural changes

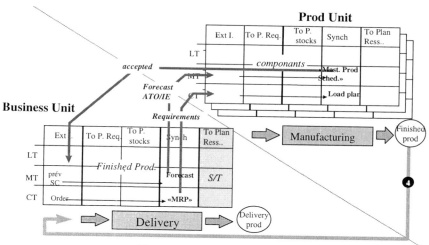

Figure 12: key relationships between Units

The first step to perform in order to respect the top management objectives is the development of a common typology of products based on APICS one (figure 13). It is necessary to identify precisely the products which pass in transit in the Business Unit and the others ones.

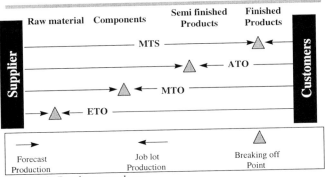

Figure 13: Product typology

1. MTS: Make to Stock products represent the current flow of products.
2. ATO: Assemble to Order products represent both the products made in the enterprise and the products made in some external enterprises.
3. MTO (Make to Order products) and ETO (Engineering to Order) have been excluded of the study because they result of working group composed by the customers, the marketing **and** the manufacturing system. So the Business Unit will not intervene in the supply chain of these products.

4.6 Presentation of the Analysis Phase Results: Design Proposals

The Business unit design differs from classical GIM applications because there was no existing system. So the first phase of modelling /analysis ensures the of modelling a global new organisation.

The main result of the modelling phase is an architecture of the new organisation and the global objectives to achieve. This architecture was refined with several models of processes and decisional organisation.

This phase was finally the most important phase of the study. The various Business Processes were designed in a coherent way [5], in order to reach the strategic objectives.

This part of the methodology is very useful to look for an efficient way for the design.

4.6.1 Design phase results: Decisional Architecture

First, three different architectures were designed for the three product families identified before. These architectures are modelled with GRAI Grids. An example of GRAI grid for the "Make to Stock" products is given bellow (Figure 14).

For the different GRAI grids, the whole relationships between Business Unit and production centres were defined. The main objective was also to maintain the delivery lead time during the implementation phase.

The different decision centres were detailed with the use of GRAI nets. They ensured to explain accurately the procedures to perform the concerned activities.

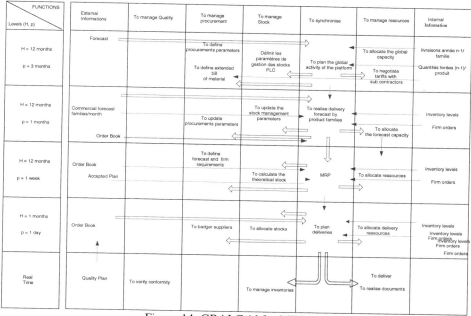

Figure 14: GRAI Grid for MTS Products

4.6.2 Design phase results: main business processes

To complete the decisional view, decision makers needed to identify the main business processes. The eight business processes which were designed are the following:

1. To manage forecasts
2. To manage orders
3. To manage requirements
4. To manage inputs
5. To mange deliveries
6. To manage inventories
7. To define items
8. To define the whole Business Unit management parameters

Each business process was modelled with extended actigram formalisms as presented in Figure 15.

A procedure is linked with each process and for each level of detail. The procedures explain the links, the supports, the actors, and the different ways of setting up. These documents are the most important of the study because they are the main level for evolution management. All processes were designed to respect the main objective: the Business Unit must not increase the lead time of the logistic chain. In a first time, it had to respect existing lead times, and in a second time, it had to reduce them.

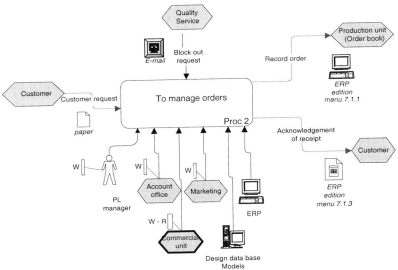

Figure 15: Process model example

5. CONCLUSION

The case presented here explains how the enterprise modelling techniques and especially GRAI Integrated Method could be used to implement strategic aspects in the enterprise. Implementation of an ERP or a Performance Indicators System, Business Process reengineering, reengineering of a decisional system... are some of these aspects.

The Business Unit is typically a strategic consideration in order to improve the lead time, the flexibility or the reactivity: in one word the Agility.

The main results of this study ascertain that Modelling Techniques are quite essential to improve strategic performances. The multiple application domains of GRAI Methodology allow to support in an efficient way the reengineering of a company toward Agile Manufacturing Type.

The Global approach developed here were followed as showed in Figure 16. The figure 16 shows that the first modelling allowed to have a global view of the market needs. The elaboration of the global organisation allowed to begin to answer to the market needs and to begin to design the logistic platform and the agencies. The precise models allowed to improve the market need answer and to precise the organisation of agencies and logistical platform. Finally, the improvement of the models allowed to obtain the final architecture for the agencies and for the logistical platform, this architecture answering precisely to the market needs.

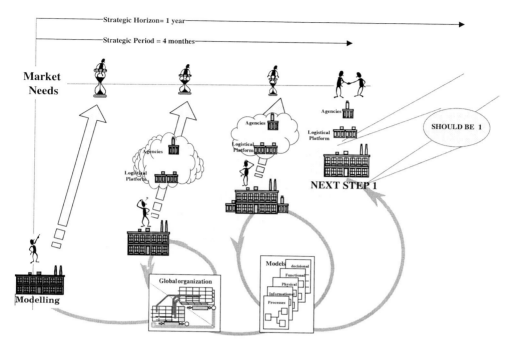

Figure 16: Global approach for agility implementation

REFERENCES

[1] Doumeingts G, Vallespir B. and Chen D, "GRAI Grid Decisional Modelling" In Handbook on Architecture of Information Systems - Edited by P. Bernus, K. Mertins, G. Schmith, International Handbook on Information Systems, Springer Verlag, pp 313-337 (1998)
[2] R.G.Nagel, "Introduction and overview of agility" AMEF Second Annual Conference- 15-18 December. (1992)
[3] Malhéné N, "Gestion du processus d'évolution des systèmes industriels : Conduite et méthode", PhD - University Bordeaux 1 - January 2000
[4] C.J. Backhouse, N.D. Burns, "Agile value chains for manufacturing- implications for performance measurement". International Journal of Agile Manufacturing Systems- volume 1 Number 2, (1999)
[5] Ducq Y, "Contribution à une méthodologie d'analyse de la cohérence des systèmes de production dans le cadre du modèle GRAI", PhD - University Bordeaux 1 - February 1999
[6] Ducq Y, kromm H and Vallespir B, "Integration of design and operating performances for manufacturing systems" in proceedings of 11th International Working Seminar on Production Economics, Igls-Innsbrück, Austria, 21-25 February, (2000).

Corporate Knowledge Management in Agile Manufacturing

Dr. Michael Thie and Dr. Dragan Stokic

ATB, Institute for applied Systems Technology Bremen GmbH,
Wiener Str. 1, 28359 Bremen, Germany

1. INTRODUCTION

A relevance of an appropriate management of Corporate Knowledge is rapidly increasing in modern agile manufacturing. The knowledge of a modern enterprise represents often its highest value. Due to dynamic changes of requirements at a current market, the products have to be often changed. Therefore, to ensure maintenance of its leading edge at the market, a company must ensure preservation and maintenance of its know-how. In other words, the management of Corporate Knowledge is often the most critical issue for a survival of an enterprise. On the other hand, an efficient communication within an enterprise cannot be reduced to only exchanging data, but must take into account the exchange of knowledge. For example, an effective management of Corporate Knowledge is necessary for time and cost-efficient continuous process improvement in flexible manufacturing companies, being an imperative to maintain their competitiveness and increase market shares.

An appropriate management of corporate knowledge for different tasks in an enterprise leads to a number of benefits in terms of an increase of efficiency and quality, reduction of wastes, an increased flexibility, etc. Specifically, effective knowledge management for process improvements tasks lead to number of benefits such as reduction of efforts/costs for searching for the potential reasons of problems and for process improvement planning, training of new employees in process improvement related tasks, introduction of new products/product models/variants, planning and the execution of changes in processes (especially in the run-up phases of the re-designed process), introduction of new (similar) processes in the enterprise related to preventive elimination of problems, as well as reduction of risks when changes in processes/products are done, e.g. potential wastes and re-work in run-up phase, reduction of down times of machines and workers etc.

On top of that, the application of the appropriate methods and tools for management of knowledge improve the possibilities to involve all employees in identifying the potential problems, searching for reasons of potential problems and planning of measures, as well as for better balancing of work of employees, easier transfer of knowledge from one area to another, or from one plant to another etc.

The management of knowledge (including the representation, capturing, accumulation and transfer of such knowledge in working environments) is highly complex, especially in large enterprises, such as the automotive and aerospace industry. However, in modern networked small companies effective sharing of knowledge is a critical issue, as well. Sharing of knowledge among different areas, departments, plants, different players in a virtual company etc. requires application of advanced methodologies and ICT (information and communication technology) tools. However, knowledge sharing problems are related to a number fundamental problems such as ontology problems, correlation of different types of knowledge, treatment of experience-based knowledge etc. This is the reason why the corporate knowledge in practice is often managed in heuristic ways based upon insufficiently systematic methodologies. The opportunities offered by IST are often inadequately used.

The chapter includes an overview of the basic requirements upon effective management of corporate knowledge in agile manufacturing, as well as of different approaches and tools to satisfy these requirements. Since management of knowledge for process improvement is an excellent example to demonstrate the relevance and potential benefits of corporate knowledge management systems for agile manufacturing, a more detailed insight in the advanced methodology to manage corporate knowledge for process improvement is provided, as well as a description of a technology basis to support this methodology. Two examples, case studies, present practical application of the methodology. The social/human aspects being crucial elements of the management of corporate knowledge are discussed as well.

2. BASIC REQUIREMENTS ON KNOWLEDGE MANAGEMENT

2.1. Generic requirements

An effective management of knowledge within a large or a small enterprise is of decisional importance for a number of tasks, both those of strategical relevance such as definition of company strategical objectives, definition of marketing strategy, design of new products, design/planning of processes etc., as well as normal (every day) tasks such as planning of production, monitoring of processes and quality, execution of manufacturing operations etc.

There are a number of generic requirements upon an efficient knowledge management system, which are common for management of knowledge for different tasks within an enterprise. Here are listed some of them [1]:

- Management of knowledge must be highly reliable and robust, ensuring high availability of knowledge in appropriate form and 'quantity' and with adequate level of detail (abstraction).
- Corporate knowledge is often experience-based which arises a problem of a representation and re-use of such knowledge, i.e. the requirement is to define appropriate way to handle (i.e. capture, formulate) such experienced based knowledge.
- Knowledge within an enterprise is normally distributed among a number of employees and an efficient management system has to use a common 'language' acceptable for different people who have to share the knowledge (ontology problems). Even if two groups of experts use the same representations for their ontologies and have the same conceptualization of the problem addresses, they still may use different vocabularies, and may have different views of the domain, which leads to partially overlapping ontologies. Therefore, the knowledge management system should enable mapping between ontologies, which in general case is a very difficult problem.
- An effective corporate knowledge management must ensure the integration of knowledge over different departments in an enterprise or over different plants (in large companies) or among a number of networked companies.
- A management of corporate knowledge must include a correlation of different 'types' of knowledge e.g. knowledge on processes, on products, quality etc.

- Knowledge gathered in different life cycle sectors of a product or a process (e.g. conceptual phase, prototype phase, 0-series production, serial production etc.) has to be made reusable.

For example, a management of knowledge for process improvements tasks is connected to solving numerous problems listed above: such knowledge is often strongly experience-based, and the results of the attempts to replace it by formalized procedures and tools (e.g. statistical methods and tools) are very limited (especially when taking into account a high complexity of products and the high flexibility of modern production processes), the knowledge on process improvement must be correlated/integrated with the process and product knowledge from different areas in the company and provided by a number of people with varying expertise and know-how (e.g. shop-floor employees, employees from product/process design) etc. On the top of that, a management of knowledge for process improvement imposes a number of requirements originating from the specific characteristics of the knowledge related to process improvements. This is valid for management of knowledge for each specific task in a company (e.g. management of knowledge for product design has a number of specificities, often strongly depending on the type of products, which have to be taken into account when dealing with this knowledge).

2.2. Requirements on Knowledge Management for Process Improvements

In order to present in more detail the requirements upon an effective corporate knowledge system, we shall consider a management of knowledge for process improvement. Starting from the system control point of view, process improvements in manufacturing enterprises are normally performed in two main ways (see Figure 1):

- In the feed-forward path when the process/products are designed and/or re-designed, the measures/actions are planed on how to optimize the processes in reference to different criteria (costs, efficiency, quality etc.) and/or to prevent potential problems occurring and to ensure the efficient run-up of the (re-) designed process (these tasks are normally performed by product and process design/planning departments).

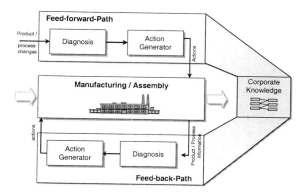

Figure 1: General structure of process improvement

- In the feed-back path, the processes/products are monitored and if problems/deviations from an optimal (defined) state are identified, the measures/actions have to be planned in order to eliminate these problems (these tasks are normally performed by process – shop-floor control departments).

In both paths two basic steps/tasks have to be carried out:
- Identification of the (potential) problems based on the monitoring of the processes/products or when changes in the process/products are planned, and identification of the reasons of the (potential) problems (diagnostic task).
- Planning of improvement/actions to eliminate the (potential) reasons of the problems (action generator task).

To efficiently support the execution of these process improvement steps and enable the effective sharing of knowledge the following approaches have to be applied:
- As indicated in Figure 1, the full benefits of the knowledge available can be achieved if the two main tasks for process improvement for both the feed-forward and feed-back paths share the same knowledge (e.g. knowledge on the problems that occurred within a run-up phase after a product/process re-design, and collected within the feed-back loop can be re-used to bring improvements for the planning of the run-up phase after a next re-design).
- A utilization and correlation of product and process related knowledge to identify the problems and define appropriate actions is needed (e.g. observation of deviations of product features can be used to identify the (potential) problems in processes).
- A high amount of knowledge/information is available in different ICT systems in a company and have to be collected and provided in appropriate form for the process improvements tasks. Therefore, what is needed is an effective combination of processing/integration of information/knowledge from different sources/systems in the companies, communication to enable collection and usage of knowledge distributed in networked organization and multi-media systems for the capturing and presentation of knowledge.

In order to enable sharing and re-use of knowledge for process improvements, appropriate methods and ICT tools are needed. These methods & tools have to enable handling of the different tasks in both paths (see Figure 1), i.e. the effective management of corporate knowledge needed to support the two main PI steps (diagnosis, action generator). Therefore, basically two methods and tools methods and tools are needed:
1. A method and tool for capturing and utilization of knowledge for identification of the reasons of the problems (diagnostic task).
2. A method and tool for capturing and utilization of knowledge for planning of actions to eliminate the reasons of problems and improve processes (action generator task).

Special benefits from these methods & tools can be achieved in the feed-forward path, based upon knowledge gathered in the feed-back path. This means that in the feed-back loop, the tools have to support the operators to identify the problems in the processes, based upon monitoring (measuring) of different process parameters (e.g. machine temperature, pressure etc.) and product features (e.g. product quality relevant features), as well as to search for the reasons for problems and disturbances and to define actions to eliminate these reasons (e.g. exchange the tools, re-programming of machine etc.). The knowledge gathered can be then re-used for the feed-forward path, e.g. once the product is changed again, or (similar) machine/tool is replaced etc. The tools will provide the operator with a list of potential problems which can occur in this situation, support him to identify the reasons of these problems and what actions to take to ensure that these problems do not re-occur (e.g. check machine parameters, apply specific tool for product change, adjust transport system etc.).

Similarly, requirements upon systems for management of knowledge related to other tasks in an enterprise can be developed taking into account specific aspects of the tasks considered and knowledge needed.

3. METHODS AND TOOLS FOR MANAGEMENT OF CORPORATE KNOWLEDGE IN AGILE MANUFACTURING

3.1. Methods for management of knowledge

There are a number of generic methods for management of corporate knowledge in agile manufacturing. These are knowledge modeling methods, knowledge based systems (e.g. rule based systems, probabilistic rule based systems etc.), case based reasoning methods, as well as causal graphs and qualitative reasoning methods based on imprecise, uncertain and fuzzy knowledge, multivariable data analysis to perform clustering, correlations, data fusion/aggregation and to enable the selection/classification of information etc. [2, 3]. Most of the methods are still not able to solve the above mentioned complex problems such as those related to the mapping of ontologies etc. However, considerable advances have been achieved in methods for knowledge elicitation from experts to acquire first (informal) models of the problems (see e.g.[4, 5]).

Effectiveness of different methods can be best studied depending on application of area of the corporate knowledge. Knowledge related to design of products often requires very specific management methods depending on specific features of the products. Similar is valid for management of knowledge for other tasks within agile manufacturing. For example, possible approaches for management of knowledge needed for detecting problems in the processes, identification of the reasons of problems and planning of process improvement activities include several key approaches:

- Statistical Approaches (methods for the monitoring of processes, e.g. Statistical Process Control, Control Charts etc.),
- System & Model Based Approaches (e.g. simulation, process parameter comparison etc.) to analyze, understand and identify the behavior of a system.
- System & Knowledge Based Approaches (Methodology for ontology building such as KADS, Problem solving methods, Rule based Systems, Case based reasoning, Fuzzy Control, FMEA, Ishikawa method, etc.),
- Learning Approaches (e.g. neural networks etc.).

In order to be efficiently applied in complex agile manufacturing processes these approaches have to fulfil different criteria such as appropriateness for application on the shop-floor, minimal efforts for creating the evaluation programs, high reliability of the results, robustness etc. The evaluation of the approaches shows that each approach has certain advantages but also some disadvantages w.r.t. these criteria and generic/specific requirements stated in the previous section. Figure 2 provides an overview of advantages and disadvantages of these approaches for identification of problems in manufacturing and assembly lines and for searching of reasons of problems as well as for planning activities.

For example, statistical approaches require no modeling efforts and are relatively easy to use, but the problems related to the statistical significance of insufficient sample size (note that the sample size of process and product features measurements are often very restricted even in mass production, due to often changes and high flexibility of agile manufacturing; especially in the run-up phases batches are very small), and the interpretation aspects are critical [6-9]. On the other hand, system and model based approaches have good interpretation possibilities and no need for a high sample size, but they require high modeling efforts and are sensitive to changes in processes. It can be concluded that there is no single approach, which fulfils all user requirements. However, a combination of different approaches, is likely to enable an efficient management of corporate knowledge, as presented in Section 4.

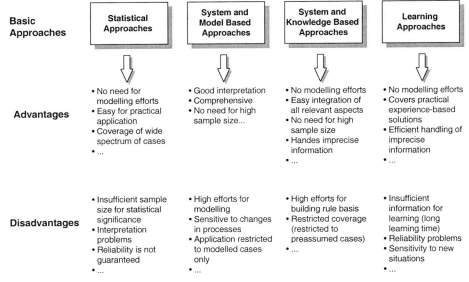

Figure 2: Evaluation of considered approaches

3.2. Tools for management of knowledge

There are a number of relatively simple to use tools for management of well encapsulated and formalized knowledge (e.g. documentation management systems, configuration management tools, intelligent manuals etc.) which, however, do not enable to effective capturing and sharing of knowledge, and, which therefore, can not fulfil requirements for management of knowledge for different (especially more complex) tasks within a company. On the other hand, more complex tools enabling to perform formulation, accumulation and re-use of knowledge are often not very practicable due to a number of inadequacies, such as ontology problems, difficulties in representation of experience-based knowledge, complex usage (e.g. complex modeling procedures), etc. Advances have been achieved in certain domains, e.g. agent-based approaches in software engineering domain, but the available tools still not applicable for management of corporate knowledge for many complex tasks.

A number of generic tools for a management of knowledge available at the market (e.g. shells for a development of rule-based systems, generic case base reasoning tools, generic modeling and simulation tools, generic statistical tools etc.) require a lot of efforts to be made applicable for management of knowledge needed for specific tasks. Tools aimed at specific tasks (e.g. tools for quality assurance, simulation of specific automation systems etc.) often are not applicable for a wide spectrum of tasks (which in turn restricts possibilities for knowledge sharing), or are limited to specific industrial sectors and/or company types.

A protection of corporate knowledge captured within information technology tools is one of the critical problem as well (e.g. problem of protection of knowledge in Internet and Intranet etc.).

The technology basis including different standards such languages for knowledge management e.g. KQML (Knowledge Query and Manipulation Language) V2 for inter-agent communication, ISO 99's Conceptual Graphs and KIF (Knowledge Interchange Format) for content description and processing, leading transport protocols and Web interfaces (HTML,

XML), object models for component development (COM, JavaBeans) and exchanging (CORBA, DCOM, RMI) is promising to enable a generation of a number of tools for management of corporate knowledge in various domains in a very near future.

3.3. State-of-the-art in practice

Due to a lack of appropriate methods and tools, the management of knowledge for the detection of the potential problems, identification of the reasons of the (potential) problems, and planning measures for process improvement are currently carried out in most companies without a specific systematic method. Similar conclusion can be made for management of knowledge for many tasks in manufacturing companies. Knowledge/information available within different systems in the company is difficult to re-use by employees, when needed. A lack of appropriate approaches and tools for management corporate knowledge leads to a number of inefficiencies such as: (a) the same (similar) problems are often solved several times and at different processes and/or plants ('re-inventing the wheel'), (b) inappropriate re-use of knowledge from different areas and over different life-cycle sectors of a product, (c) only a very low number of highly experienced experts are able to efficiently contribute to solving problems, creating difficulties in job rotation, balancing of work and involving young employees, (d) efficiency of the execution of the many tasks is inappropriate (e.g. too high efforts for planning of measures, too high efforts & costs for monitoring of processes etc.).

3.4. Research challenges

One may conclude that in order to enable an effective management of corporate knowledge, several fundamental research problems still have to be solved:
1. One of the most challenging problem is how to establish a unique, practicable knowledge management system which can be applied in the different tasks (e.g. in process improvement tasks in both feed-forward and feed-back paths), and by this enable re-use of knowledge. Investigation of new knowledge representation approaches, taking into account ontology integration aspects, have to enable the development of a knowledge management system applicable in different tasks.
2. Innovative application-oriented approaches for the effective utilization of corporate knowledge about relationships relevant for different tasks have to be developed enabling the integration of process and product-related knowledge, which in turn requires solving the problems of correlation between different product- and process-models & knowledge.
3. Further research on the advanced knowledge engineering methods is needed to establish efficient ways for the re-use of knowledge over 'similar' processes and products, different product life-cycle sectors etc.
4. Research activities on the identification of new forms for a representation of the experienced-based corporate knowledge are requested.
5. Development of approaches for 'learning from experience', i.e. efficient incorporation of the learning capabilities in knowledge management tools is a challenging issue as well.

4. ADVANCED MANAGEMENT AND INTEGRATION OF CORPORATE KNOWLEDGE FOR PROCESS IMPROVEMENT

As explained above, the process improvements tasks in an enterprise are an excellent example on how an effective management of corporate knowledge may bring considerable benefits. On the other hand, due to complexity and generic character of management of knowledge for process improvements, solutions for this problem offer good insight in possible approaches for management of corporate knowledge for a number (especially more complex) tasks. Therefore, this section focuses upon approaches for management of knowledge for process improvements.

4.1. Business cases for re-use of knowledge on process improvement

The process improvements can be effectively supported by an efficient knowledge management system, assuming that such system enables to capture, accumulate and re-use information and knowledge from human resources and overall production technology in the manufacturing plants, as well as enables to integrate the knowledge/information included in different systems in a manufacturing company. It can be identified that there are at least three areas (business cases) where the efficient sharing of knowledge for process improvement may considerably increase the effectiveness of the improvements of the manufacturing processes:

a) *Sharing of knowledge gathered within different product life cycle sectors,* i.e. sharing of knowledge from prototype manufacturing, so-called 0-serie production, serial production etc.: for example, the knowledge applied for tuning/maintenance/programming the processes (machines, tools) or eliminating the quality or other process problems and disturbances within the 0-serie production, can be re-used in serial production to ensure an efficient run-up, or when products or processes are changed (e.g. material change). The typical problem is the transfer of knowledge from the product prototype phase to the latter sectors of the product life cycle. Especially in the prototype life-cycle sector there is a need for a strong coupling of knowledge in the feed-back and feed-forward paths, as well as for knowledge/information exchange among different systems (e.g. CAD, product management system, process planning and control systems etc.). The knowledge accumulated in this prototype phase has to be re-used for the run-up of the 0-series of the product (part). For example, the welding process of the car body requires to accumulate a lot of knowledge, which can be re-used for the run-up phases of the latter life cycle sectors of the product (i.e. in 0-series and in serial production). On the other hand, the knowledge from the run-up phases in the serial production can be re-used for the run-up of the prototypes of the similar products.

b) *Sharing of knowledge on different 'similar' products and processes:* e.g. the knowledge for tuning/improvement of the operation of machines/processes, can be re-used for similar processes (e.g. over different milling machines, or over different similar assembly tasks) or when producing 'similar' products (e.g. 10 and 12 cylinder engine blocks, similar product models etc.). In many manufacturing processes a number of problems reoccur with very low frequency. For example, problems in certain processes in manufacturing and assembly of aircrafts sometimes reoccur with a sampling interval of a year or more, and certain processes related to product prototypes (aircrafts) are performed every 5-10 years etc. Normally larger changes at the aircrafts are done each 6-12 months, after producing certain amount of same products, but a number of products often have to be customized for different customers. In addition, the products and parts are often produced in very small batches. Therefore, a number of same products/parts produced are often very low, which means that it is not possible to manage the knowledge on process improvement using classical statistical approaches. The knowledge on each problem is often 'singular', and what is needed is a system to efficiently store and manage such knowledge and enable its re-use over different (similar) processes and products. This is especially critical in the complex assembly processes (e.g. fuselage assembly) where knowledge from different areas and systems are needed to perform changes with minimum efforts and wastes. This means that the knowledge has to be structured in a way that it can be re-used for similar products (e.g. similar fuselage models) and processes (e.g.

similar manufacturing or assembly operations). The methods and tools have to support the storing of knowledge from different processes and products and transfer it into a form to make it available for other employees (even) after a longer time period (e.g. when a new prototype of an aircraft has to be produced) in different departments of the company.
c) *Knowledge sharing over different plants* ('virtual' enterprise): e.g. knowledge gathered on one manufacturing line in one plant can be re-used on other lines of the same company or of suppliers or other members of a 'virtual enterprise'. Production of different variants of the same products within different plants is a typical example where sharing of knowledge on process improvement may play crucial role. Production normally starts with a basic version of a product to be split up into different variants. Implementation of the methods and tools for management of knowledge can help to improve the process in all phases of the facility's life cycle. They can save knowledge and experience gained during the run-up of the basic version to be re-used for the run-ups of the following variants and in different plants. Therefore, it can be enabled to share knowledge gathered over different plants. Obviously, the knowledge from the later product life-cycle sectors can be re-used for the earlier life-cycle sectors of the following variants. A clear benefit w.r.t. a reduction of time and efforts for the introduction of similar processes in other plants related to the preventive elimination of problems can be achieved.

4.2. Basic methodology

As indicated in Section 2, in order to enable a sharing of knowledge for process improvement in (at least) these three areas, two methods and tools are needed:
1. Knowledge-management method & tool for the diagnostic task which supports the user to identify (potential) problems when the process/product are (re-)designed and specifically in the run-up phase of a process change, as well as supports the identification of the (potential) reasons of problems (diagnostic tool).
2. Method & tool for management of knowledge for action generation task guaranteeing that for the identified reasons of problems, appropriate measures can be defined (planning tool).

The basic methodology for the management of knowledge on process improvement within these methods & tools is a combination of:
- knowledge based systems to capture the experts experience and know-how (transferred into a set of rules),
- statistical approach to support diagnostics (taking into account the problem of small sample size),
- modeling of processes and products involved in order to enable efficient knowledge structuring,
- learning approach to update knowledge and parameters and structure of models,
- fuzzy approaches to represent vague (experienced) based knowledge,
- problem-solving approaches.

The knowledge to be used within these tools can be structured in two main categories:
1. so-called 'a priori' knowledge, e.g. set of 'a priori' rules (where under a rule should be understood a set of complex interconnections using fuzzy approaches) representing generic (but rather static) knowledge on processes, products, and their relations, and which are defined in advance and independent of the actual execution of the processes,
2. so-called 'additional knowledge' which introduces dynamic knowledge/information on an actual/ specific situation (e.g. knowledge on the actual duration of processes, the actual tool used to manufacture a specific part, the actual characteristic of a part at which a

problem is identified etc.); therefore, additional knowledge covers the actual dynamic behavior of the processes or parts and changes in the processes or parts; additional knowledge may also include the results of statistical analysis; the additional knowledge needed for a concrete application is strongly application dependent.

The *method & tool for diagnostics* is organized in the following way:

1. In the feed-back path the results of measurement of different process parameters (e.g. machine parameters, machine throughput) and products parameters (e.g. product quality parameters) are analyzed to identify critical deviations/problems in the process or product. The problems are grouped/classified in such a way that each problem in a group has similar classes of causes. 'A priori' knowledge is used to identify a problem and to assign it to one or more predefined problem classes. 'Additional knowledge' on the actual processes or products (e.g. in the form of statistical values) is then used to modify the assessment of the problem to problem classes. In a case when a problem is identified directly by a user and the problem description is available, the problem will be classified by the user into the predefined problem classes.
2. Once the problem class for an occurred problem is known the problem cause has to be identified. Application of the 'a priori' rules which establish a relation (probability) between the problem class and the problem causes allows for the identification of possible problem causes. Additional information/knowledge on the actual processes or products are used to increase or decrease the probability of the problem causes. In some cases there is a need to get additional measurements in order to decide which of the potential problem causes is actually 'responsible' for the problem that occurred (i.e. to increase the observability of the system).
3. In the feed-forward path the method and tool acts similarly: once the desired change(s) in the processes/products are defined, the 'a priori' rules are applied to define the probability of different problems which may occur. The results of application of this 'a priori' knowledge are then modified based upon the additional knowledge/information in order to obtain a more accurate set of potential problems.

The *method & tool for planning* is organized in both feed-back and feed-forward paths in similar ways: Once the actual or potential problem cause is identified, actions have to be defined as to how this cause can be removed in order to prevent its appearance or repetition. The actions (additional measurements, actions to improve the process) have to be defined based on knowledge w.r.t. possible measures in each specific case. The definition of measures is similar to the identification of problem causes. The relation between problem causes and measures is defined in 'a priori' rules, i.e. these rules define which of the possible actions may be appropriate for removing a specific cause. However, in order to cover each specific situation, additional knowledge on processes and products has to be used, i.e. the rules for the handling of 'additional knowledge' have to be applied in order to change the 'relation' between a problem cause and possible measures. To support identification of 'optimal' measures problem-solving approaches are also applied.

4.3. Approaches for knowledge management

The basic background modules of both the tools represent knowledge based systems which include a set of 'a priori' rules, as well as a set of rules for handling 'additional knowledge' (for changing the probability of different causes). The following approaches are applied to solve some basic problems for effective definition and management of this knowledge basis:

1. The basic assumption is that the *knowledge from two process improvement paths can be shared*, i.e. the knowledge base used for modules in both paths can be the same. This

means that it is necessary to apply the same forms of knowledge that are captured and represented in both paths and enable application of the same knowledge in both directions: the knowledge on a relation between a specific problem and its possible cause(s), which are often related to certain changes in the processes, used (or defined) within a process monitoring in a feedback path has to be re-used in a feed-forward path to identify possible problems when (similar) changes in the processes are foreseen. Integration of different ontologies is very difficult even if ontologies can be translated into an interchange format like KIF or defined into the context of Ontolingua, since the agreement on common terms and the relationship between terms is not sufficient. The above explained combination of modeling, knowledge and statistical approaches offers good possibilities to establish a simple practical solution. If the sets of processes, problem causes, problems, process/product changes etc. are systematically well structured to enable a 'one-to-one' mapping from one set to another, then the knowledge (rules) which is interconnecting these sets is applicable in both directions. A problem is to ensure that whenever a new rule is identified in a feed-back path (e.g. a new relation between a change in a process and a problem which appears) it can be re-used in the future planning of a 'similar' change (i.e. in the feed-forward path).

2. *The proposed approach assumes an effective utilization of corporate knowledge about relationships relevant for process improvement tasks*, i.e. correlation/integration of process and product-related knowledge, often included in different legacy systems (e.g. in PPC, process monitoring systems, CAD). This requires an open system which can integrate and use knowledge/information from different systems, which in turn requires solving the specific problems of correlation between different process and product models. For example, a process model in a PPC system may not be identical to the one to be used by the knowledge management tool since the models in these tools are primarily oriented to adequate structuring of knowledge on process improvement, while in the PPC systems the models are used for scheduling, sequencing and other planning and control tasks. Therefore, a correspondence between these models must be established. Here, the standardization aspects (i.e. standardization of the interfaces between different systems which incorporate the required knowledge) are of essential importance.

3. *The advanced knowledge engineering methods* are applied to support re-use of knowledge over 'similar' processes and products, over different product life-cycle sectors and over different plants. For example, the grouping principles are applied for identification of 'similar' process and product features. Fuzzy logic is useful for the fine tuning of the method: it may be used to describe the notion of similarity as well as the relevance of particular information. The tools have to ensure the re-use of knowledge within the three above identified business areas (e.g. knowledge accumulated in the prototype-phase can be re-used in the latter life cycle sectors).

4. *The experienced-based corporate knowledge* is often not possible to express in a deterministic form. For the integration of this knowledge into the rules there is a strong need to support vague terms that may appear in the premises (i.e. 'the manufacturing area has recently been redesigned', 'the part was in storage for a long time' etc.). An efficient way to handle such sentences is to model them via fuzzy sets [10]. Fuzzy sets enable the quantitative modeling of simple sentences (e.g. 'design change is recent') consisting of linguistic variables (e.g. 'design change'), primary terms (e.g. 'recent'), and optional linguistic hedges (e.g. 'very', 'not'). To provide a basis for the model, each variable must be assigned a physical quantity (e.g. time from the moment when the design is changed) and its range of change. This approach enables that majority of the terms which the user wants to put into the knowledge base in a 'vague' form can be represented by such fuzzy sets.

5. The tools have intrinsic meta-knowledge that will allow for the deriving of reasons for problems/action plans for process improvement and/or guide the user on the basis of history information (e.g. on the basis of similarity of the new error to recently detected and resolved problems). This means that the tools include learning capabilities and by this support *'learning from experience'*. The applied approaches are:
   - Automatic changes of rules/adding of new rules based on user reasoning: if the user decides that the reason for the problem is different from the proposal of the tool and gives 'explanation' of his reasoning, a new rule can be automatically generated.
   - Statistics on the identified reasons of problems may 'automatically' change the relation (probability) between the problem and possible problem causes.
   - Automatic comparison of the relevant information in order to support reasoning on the causes of problems. When a problem in one class appears again, the system compares actual (related) information with the stored information on the previously detected/identified problems in the same class. Based on this, the system may directly support the user to draw conclusions for the reasons of the current problem. However, the problems making difficulties in application of this approach are: What relevant information should be stored? How to correlate information, e.g. how to treat 'small' differences in information (which is again the problem of 'similarity' as explained above). Generic solutions for these problems are still missing.
6. *Procedures for planning of actions* (e.g. planning of optimal quality checking procedures) are based upon the *problem–solving approaches* and an integrated corporate knowledge. The method is applied to facilitate development of the decision trees which can act as problem solvers to a spectrum of applications related to process improvements tasks. This approach is especially appropriate for process improvements action generation, since it enables comparing different possibilities through the decision tree (at various levels of abstraction) and by the step-wise identification of the optimal action plan. This approach is integrated with the information/knowledge needed for action planing (which range from deterministic information such as costs and time needed for different actions, or re-programming actions which could be carried out to correct processes, up to 'vague' knowledge on experience in performing different actions), and with the 'learning from experience' approaches enabling learning 'automatically' from 'history'.

These approaches for management of knowledge for process improvements are exhibiting a number of advantages over other approaches. As presented in Section 6, practical experience with such approaches shows that they can be effectively applied in a wide spectrum of process improvements tasks in modern manufacturing companies. It is very obvious, that such or similar approaches can be applied for management of knowledge for other (complex) tasks in agile manufacturing.

5. TECHNOLOGY BASIS FOR MANAGEMENT OF CORPORATE KNOWLEDGE

Different architectures are available for integration of corporate knowledge. Here, we shall briefly present an example of an architecture applied for management of corporate knowledge for process improvement, specifically for management of knowledge related to quality problems. Two tools QUPROC (QUETA Tool for Identification of QUality Critical PROCesses) and QUPLAN (QUETA Tool for Quality Related Action PLANning) have been developed for management of the knowledge in the scope of the European project QUETA (QUality Engineering Tools for Assembly and Small Batches Manufacturing) [1-3], applying approaches described in Section 4. The tools are generic, i.e. they can be applied for different companies and for different business cases. In order to enable their easy integration in different environments the following architecture is applied.

The architecture is based on the so called three-tier software architecture, meaning separating functionality for presentation (user interface), application logic and stored information -see Figure 3:

- Presentation layer - this is the User Interface layer. The screens are using information both from components and from data stored in legacy system. A.
- Component layer - including core functionality of the tools QUPROC and QUPLAN. All data is stored in the Data Store Layer (both internal and external data).
- Data Store layer - all data needed for components and presentation layer are implemented here. All accesses to stored data for defined objects are implemented by IDL and use of adapters.

All communication between the Layers is based upon Interface Objects using IDL (Interface Definition Language). Each application has its own User Interface, i.e. the choice of GUI is specific for every company (business case) depending on computer environment

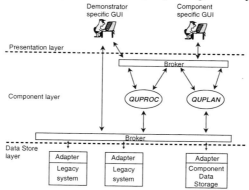

Figure 3: Architecture for integration of corporate knowledge (an example)

and tools that is normally used for GUI-development. One example is that it is possible to use Internet browsers as the demonstrator GUI. It is essential that no data is stored in the Component Layer in order to get a computer environment independent solution.

Technologies for knowledge integration: To implement the above described architecture for the knowledge integration different technologies for integration of distributed systems are commercially available,. The two main software infrastructures are CORBA (Common Object Request Broker Architecture) and DCOM (Distributed Component Request Broker). Besides different philosophies from technical perspective the main difference is that DCOM can mainly be used in the Windows-area whereas CORBA can be use on more or less all platforms and operating systems. Since a lot of legacy systems in industrial companies are running under other operating systems then Windows, CORBA is used for the implementation of the presented architecture.

6. CASE STUDIES

The following sections present 2 case studies, in order to show the application of the methods and tools for corporate knowledge management for process improvement. The case studies concentrate on quality related problems, but similar requirements are valid for other process improvements.

6.1. Case Study 1: Complex Assembly in Aerospace Industry

The quality related problems in assembly, in general, are characterized by the fact that identification of the reasons for problems and quality critical processes, is more complex than in other manufacturing processes. The reasons for the quality related problems, which appear in assembly include: problems/errors in part manufacturing (which include design related problems, storage aspects etc.) and problems/errors in the assembly processes themselves. For example, if two or more parts cannot be mated (assembled together) the reasons for this may be either improper manufacturing/design of each of the parts, or the way (process) the assembly is carried out. The elastic features of parts always play an important role in the part assembly process. Therefore, when identifying the quality critical processes, based on the problems (errors) detected in assembly, one is confronted with the situation that the reasons for a problem could be numerous and might happen in any previous production step. The main characteristics of assembly processes in the aerospace industry are: (a)The probability of a problem occurrence in assembly is higher than in manufacturing processes, (b) the identification of critical processes requires the correlation of a large amount of process/quality information, (c) the number of produced products is very small (small batch assembly – as explained in Section 4.1), (d) assembled parts are often produced in small batch manufacturing.

Such a situation implies that in the aerospace industry the identification of the critical processes is strongly reliant on the experience and know-how of the experts. The above mentioned knowledge based methods and tools QUPROC and QUPLAN are used to support the quality improvement of a final assembly of aircraft's. The problem addressed in this application area is the assembly of pipes for the hydraulic systems of the aircrafts [11].

In the past several problems occurred w.r.t. the final assembly of pipes. Pipes could not be assembled and extensive actions were necessary to solve the problems. This caused high additional efforts and costs to identify the causes of the quality problems and to define actions to improve the process and product quality. However, in order to identify the true cause and to define appropriate actions, comprehensive knowledge along the entire production process about the process- and product-relation and their relation to errors and error causes in a structured form is necessary. Based on the CAD drawings, work orders and other documents from the design & planning area, the pipes are assembled. If it is not possible to assemble a part as required, a solution is sought by the operator or in more complex problems by production or quality engineers to solve the actual problem. These errors will be documented in a quality documentation system which is available at all assembly stations (Figure 4). From time to time the error documentation is analyzed to identify critical errors, i.e. reoccurring errors or important errors (e.g. those asking for high costs to be solved). For these errors the identification of error causes has to be carried out.

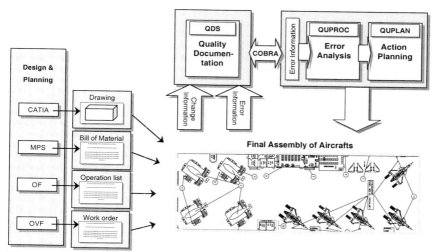

Figure 4: Integration of tools for process improvement in aircraft assembly [11]

Based on the error type the analysis is started in a specific assembly area (final or fuselage assembly, pipe production etc.). First the critical process has to be identified. Necessary information to evaluate the knowledge base is collected from other systems or an operator. In a similar way the error cause and actions to improve the process are identified.

The tools QUPROC and QUPLAN require a large amount of information which are normally available in other legacy systems of the company. To get the required information automatically, the tools are integrated with these legacy systems as described in Section 41.5. The following classes of information are needed from different systems (e.g. PPC, QDS, BDE etc.) in order to use the tools in the considered application area ('final assembly'):

- Process information (information on processes within the production of aircraft's),
- Product information (information on parts, subassemblies or the aircraft),
- Error information (information on the error and on the faulty product),
- Change information (information on product and process changes).

The systems which have to be integrated are using different hardware and operating systems. To support the integration, CORBA is used to get data from different legacy systems and provide the functionality in a web browser at different places in the company.

Example: To better understand the benefits of such a method for process improvement the usage of the tools will be described through an example. The starting point for an analysis is the description of an occurred error. In the example considered, a pipe can not be assembled. The following processes could be responsible for this error:

- Design & Planning: The given documents (drawings, working instructions) for the pipe or related parts have an error.
- Production of Pipes: The pipe did not reach the target, since an error occurred in the bending of the pipe.
- Fuselage assembly: The fuselage did not reach the target, since the fuselage assembly was not executed in the right way.
- Final assembly: The error occurred in final assembly since other parts were not assembled as defined in the working instructions.

Accordingly QUPROC calculates based on knowledge of the error/process-relations the 'a priori'-probabilities for the critical processes (Figure 41.5).

Since it is not possible to identify the critical process with only the 'a priori'-probabilities, additional information is used in modification rules to update the probabilities for the concrete error situation. For the actual error the following information is relevant:
- Verification if the given documents for the faulty or related part were changed recently.
- Verification if the faulty pipe was inspected after the bending process.
- Verification if small changes were made to a related part in final assembly, since this could have influences upon the pipe assembly.

The necessary information can be found in the different legacy systems of the company. Using the implemented interfaces and internal functionality QUPROC is able to evaluate the necessary information automatically. Based on the application of the rule base, the probabilities for the critical processes calculated that the responsible process can be identified with a sufficient probability (Figure 5).

Once the critical process 'Production of Pipes' is identified, the list of possible error causes could be drastically reduced. In a similar way the error cause will be identified and the action to improve the process will be defined.

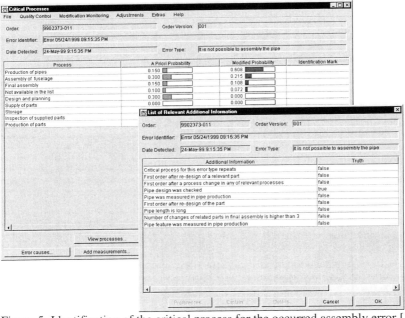

Figure 5: Identification of the critical process for the occurred assembly error [11]

The presented analysis of an occurred error in final assembly shows that for such complex cases a lot of different information and knowledge is necessary. The benefits achieved using the tools QUPROC and QUPLAN for process improvement in the final assembly of aircraft's include reduction of errors in final assembly by identification of the true causes for errors, reduction of efforts and time for the identification of reasons for errors by the availability of error knowledge, less dependency on employees with high experience by enabling less experienced employees to analyze errors, reduction of efforts and time to reach a stable assembly process after changes, preventive avoidance of errors etc.

6.2. Case Study 2: Small Batch Manufacturing in Automotive Industry

The production process in the automotive industry is characterized by the manufacturing of a large number of complex final products. Manufacturing of parts for cars includes, mainly, high volume manufacturing and a certain amount of small batch manufacturing. Due to the high customization of cars, and the increasing number of variants/models, there is also a clear trend towards increasing small batch manufacturing in the automotive industry. The manufacturing of parts in such batches is often done within an FMS (flexible manufacturing system). Generally speaking, in the automotive industry parts are manufactured in small batches in preparation for series production, in the phase of the expiration of a production series, for after series production (e.g. spare parts), in the testing phase (i.e. manufacturing of test parts), for manufacturing of special tools and operational equipment, etc. Due to the high complexity of the end-product as well as due to the high variety of products and often changes of products/introduction of new product/models in FMS, changes in the production processes and design are very frequent and, therefore, transition phases occur very often in the considered processes. The most FMS have a variety of parts with a batch size between 20 and 500. The tools QUPROC and QUPLAN are applied in such an FMS in the automotive industry.

This FMS is comprised of 4 machining centers, 1 coordinate measuring machine, 1 robot for part machining and handling, 1 storage rack with 80 part or tool pallet bays and 3 load transfer stations. The single devices are connected by a rail-guided vehicle (RGV). The production program does not allow for the use of classical methods of Statistic Process Control (SPC). For an effective use of classical methods, a critical number of about 1000 parts per order with a typical number of 125 parts for the basic process analysis is needed. These numbers are not reached in most cases at the pilot site. Therefore specific quality control mechanisms for small batch manufacturing were required. Figure 41.6 presents the integration of the tools QUPROC and QUPLAN in the process chain of the FMS. The information system AiP includes information serving for generation of NC programs and the measurement program. The results of measurement are introduced in the QUPROC tool which enables to group features of the same product (e.g. radius of different holes at the same engine cylinder block made by the same drilling machine/tool) or of different products (e.g. 10 and 12 cylinder engine blocks) and by this get a bugger sample for which statistical analysis can be applied. The grouping is also performed based on knowledge (rules) integrated in the tool. The tools QUPROC and QUPLAN for corporate knowledge management within the FMS are used in the same way as in the first study case, i.e. in final assembly. The tools are successfully applied to enable the grouping of products features, the interpretation of the statistical results and effective identification of quality critical processes in small-batch FMS and by this, reduce the efforts and time required by manual procedures for the analysis of the true causes of errors, reduce inspection activities, support the control of the production process, establish feedback from the quality control results to the job scheduling of the FMS and reduce quality problems by applying preventive quality assurance measures.

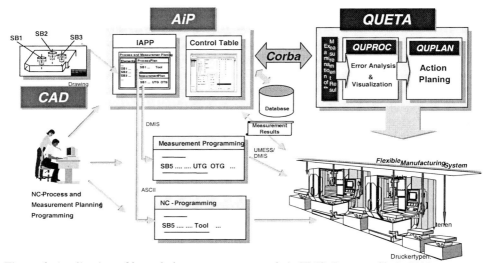

Figure 6: Application of knowledge management tools in FMS: Process Chain [11]

7. HUMAN AND SOCIAL IMPLICATIONS OF MANAGEMENT OF CORPORATE KNOWLEDGE

Implications on working conditions: An effective management of corporate knowledge enables new ways of executing of a number of tasks in manufacturing companies. For example tasks related to process improvements in the manufacturing companies, specifically process monitoring and set-up on process changes, can be executed in a considerably more efficient way once the knowledge, available in an enterprise, is provided to the employees. This may considerably reduce the stress which employees normally experience both within the run-up phases after product/process changes and in keeping production running at the different life-cycle phases of the product. Management of knowledge has specific influence on ways of co-operation among employees within the companies, e.g. between the shop-floor and planning areas, contributing to a removal of departmental barriers. Management of knowledge in modern companies opens different promotion prospects for the employees where judgement of the employees is based on how they contribute, re-use and share the overall knowledge in the company, and by this bring a new, improvement- and team-oriented culture into the companies. Provision of improved learning and training conditions for the employees through accumulated corporate knowledge, contribute to the quality of working conditions especially for shop-floor people making their jobs much more challenging and shifting job skill requirements from a lower to higher level. By enabling the re-use of knowledge from similar products and processes, or knowledge gathered over different product life cycle sectors and among different plants, an increase in the safety of the working conditions can be guaranteed, e.g. in the run-up phases which are often related to considerable risks.

Implication on labor market: Appropriate management of knowledge may strongly contribute to strengthening the labor market and the development of the individual's skills by enabling easier learning from experience for the employees which in turn enables: (a) easier circulation of employees, specifically within large companies (job-rotation over different subsidiaries), (b) higher mobility of workers on the entire market, etc. Provision of methods and tools for accumulating corporate knowledge makes employment of people easier, especially for first time employees, women and people in less developed countries due to reduced efforts for training and the involvement of inexperienced employees in different complex tasks. Enabling easier installation of new plants based on the knowledge accumulated in appropriate management systems reduce the reluctance of the large companies to invest in new manufacturing lines in different countries.

Ethical issues: An essential objective of an efficient management of corporate knowledge is deeply ethically founded: to reduce differences between people and offer equal opportunities to employees to contribute to their companies and the quality of their lives and working conditions based on a sharing of knowledge. Establishing of ICT based systems for management of corporate knowledge can be seen as an extraordinary example of how advanced information technology can be used to establish radically new ethical relations among people, increase tolerance and ways of co-operation within companies. Of course, since sharing of knowledge are strongly related to employees and their work, the systems must include all measures to protect information on employees which will be included in the system and which may have any negative consequences upon them, and to ensure adequate privacy of data. On the other hand, one of the most critical issue is a protection of knowledge of an enterprise, arising a number of ethical and legislative problems.

8. CONCLUSIONS

The chapter provides a brief insight in main problems and requirements regarding management of corporate knowledge, being one of the most challenging and critical issue in agile manufacturing. The possible approaches for management of corporate knowledge are briefly overviewed, and a methodology for management of knowledge concerning process improvement in manufacturing companies is presented in a more detail, as an example of how the problems related to management of corporate knowledge can be approached.

As already stressed out, a number of problems in this area still has to be solved, in order to enable efficient management of knowledge based on usage of modern information and telecommunication technology. The problems related specifically to management of knowledge of networked companies (often small companies) are requiring urgent solutions: modern information and technology offers excellent opportunities for efficient internetworking of companies, but the problems of efficient management of common knowledge within such a networked (virtual) enterprise is still to be solved.

REFERENCES

1. Wangberg, K.-G. et al.: IT based Quality Engineering Tools for Assembly and Small Batches Manufacturing, in: Changing the Ways We Work - Shaping the ICT-solutions for the Next Century (Eds. N. Martensson et al.), ISO Press, Amsterdam, Berlin, Oxford, Tokyo, Washington, 1998, pp. 141-153.
2. Thie, M; Stokic, D.: Knowledge Based Methods and Tools for TQM in Small Batch Flexible Manufacturing and Complex Assembly, in: Intelligent Systems for Manufacturing: Multi-Agent Systems and Virtual Organizations, (Eds. Luis M. Camarinha-Matos et al.) Kluwer Academic Publishers, Boston, Dordrecht, London, 1998, pp. 459-468.

3. Thie, M.; Stokic, D.: Quality Assurance in Assembly and Small Batch Manufacturing using Knowledge Based Approaches, in: Life Cycle Approaches to Production Systems: Management, Control, Supervison, ASI Conference, Bremen, 1998, pp. 109-115.
4. J. Angele, D. Fensel, D. Landes, and R. Studer: Developing Knowledge-Based Systems with MIKE, Journal of Automated Software Engineering, Vol. 5(4), October 1998, pp. 389-418.
5. T.R. Gruber, A Translation Approach to Portable Ontology Specifications. Knowledge Acquisition, Vol. 6., No. 2., 1993, pp. 199-221.
6. Al-Salti, M., and Statham, A., A review of the Literature on the use of SPC in Batch Production, Quality and Reliability Eng. Intern., Vol. 10., 1994, pp. 49-61.
7. Bothe, D.: Statistical Process Control For Short Production Runs, International Quality Institute, Northville, Michigan, 1989.
8. Dale B.G.; Shaw P.: Statistical process control in PCB manufacture: what are the lessons?, IEE Proceedings-A, Vol. 139, No. 4, July 1992.
9. A. Wald, 'Sequential Analysis", Dover Publications, INC., New York, 1947.
10. Angstenberger, J.; Lieven, K.; Weber, R.; Zimmermann, H.-J.: Fuzzy Technologien – Prinzipien, Werkzeuge, Potentiale, VDI-Verlag, Düsseldorf, 1993.
11. Field-Test and Evaluation: Results of testing and evaluation of tools, Public Report of the ESPRIT project QUETA, Bremen, 1999.

Agile Manufacturing Strategic Options

Dr. Vicky Manthou and Dr. Maro Vlachopoulou

University of Macedonia, Department of Applied Mathematics, 156 Egnatia str., 540 06, Thessaloniki, Greece

In the years leading to the 21st century the radical environmental changes, in particular the customers expectations, the evolution of information technologies and communications, the globalization of the markets, and the nature of competition are forcing enterprises to rethink business strategic options. There is a shift from geographically limited closed enterprises to global open enterprises. Agile manufacturing strategy is the key for the alignment of the enterprise systems to changing business needs in order to achieve competitive performance. It aims at accomplishing manufacturing flexibility and responsiveness to the new market needs. Agility is an enterprise-wide response to an increasingly competitive and changing business environment, based on four fundamental principles: Enrich the customer, master change and uncertainty, leverage human and information resources, co-operate to compete (Goldman et al, 1995; Metes et al, 1997).

1. MANUFACTURING STRATEGIC OPTIONS AND THEIR IMPLICATIONS

Manufacturing strategy term encompasses everything it takes to satisfy customers, from determining which products they will buy, to deciding how to produce them, and planning for their delivery. There has been considerable attention over the years on contemporary concepts of manufacturing strategy. Traditional manufacturing, production planning and control, just in time management, and total quality management initiatives have all been adopted and adapted, as appropriate, to suite the needs of manufacturing companies. Furthermore, new forms of supply chain management, where long collaborative partnerships between customers, suppliers and producers have been developed. Terms such as world class manufacturing, and lean manufacturing have been used to differentiate these forms of manufacture from traditional forms of production. More recently, however, the concept of agile manufacturing has emerged whereby organizational leanness must be matched with the flexibility to respond to dynamic market conditions and competitive. Companies now move from being manufacturing-driven to customer drive (Maskell, 1994; Maskell & Assoc., 1999).

Table 1 summarizes the main strategic aspects according to the evolution of the manufacturing systems. In particular, customer and product policy, the organizational process, the cooperation and competitiveness, the information technology and communication and the human resources are examined.

1. Traditional manufacturing. A company adopts mass production as its primary philosophy, faces complicated systems that don't fit together, has long production cycle time, and high inventory. The lack of business control, strategic vision and flexibility are the main drawbacks of these systems. One of the standard problems of traditional manufacturing is that

Table 1
Evolution of manufacturing strategic options-systems

Systems→ Aspects ↓	Traditional manufacturing	PPC	World class/lean manufacturing	Agile
Customer/ product policy	-Mass and manufacturing driven production -Long production cycle time	-Improved customer service	-Reduced time to market -Lead time reduction -Fixed term production cycles -Short cycle times -Zero defects	-Customer prosperity -mass customization -small quantities of highly specialized products -Globalization
Organizational process	-Inflexibility -High inventory -Less business control -Departmentalisation -Lack of strategy	-Greater flexibility -Reduced inventory -Reduced production cost -MRPII-ERP	-Processes under control -TQM-Quality Control -JIT -Zero inventories -Productivity improvement	-Management of change and uncertainty -Multiple organizational structures -BPR -Rapid decision making
Cooperation/ Competitiveness	-Cost/Price relationship	-Dependability -Reliability	-Value added chain -Continuous improvement	-Competitiveness through cooperation -Virtual organizations -Networks
Information technology - Communication	-Isolated computer systems -Record inaccuracy	-Interface of manufacturing information with marketing and finance	-Integration -Simplicity -Flexibility -Accessibility -Openness	-Open sharing of information
Human Resources	-No employee involvement	-Coordination among different departments within the firm -better communication	-Empowered employee -Cross-trained employees	-Employee responsibility, adaptability and innovation -Empowered teams -Reward, recognition

there is no communication between the employees of the different departments, resulting in disintegration between the business functions because there is very little planning and vision from management. Competition is based on cutting production cost in order to achieve price differentiation.

2. *Production Planning and Control.* When a company moves into gaining manufacturing control aims at improving customer service and flexibility, reliability and delivery speed. Material Requirements Planning (MRP), Manufacturing Resource Planning (MRPII) and Enterprise Resource Planning (ERP) systems allow the companies to manage inventory, sales order entry, production costs, business and sales planning, and improve flexibility, control and communication. The philosophy behind this is to use computer systems and other mechanisms to control the business.

3. *World class/lean manufacturing.* Bringing a company to world class status Total Quality Management methods are implemented to bring manufacturing processes under control and create continuous improvement. Just in Time manufacturing techniques involve cellular manufacturing, quick change overs, and small batch sizes, managing low inventories, and cutting down waste. Lean production aims to optimize performance of the production system against a standard of perfection to meet unique customer requirements. The strategy is to identify value-added activities and functions according to the customer's needs, to organize production as a continuous flow, to design concurrently product and process and to apply production control throughout the life of the project. Close cooperation with suppliers and empowerment of the work force are also key characteristics of the lean organization.

4. *Agile.* The focus of agility is to provide the strategic direction, the methodology and the capabilities to help companies to be competitive in an environment where change and unpredictability are the order of the day. Agile manufacturers must recognize customer prosperity, virtual teaming, leveraging of technology, process redesign. Agility requires customer response with high quality, low cost, and innovative products and services, significant management skills and highly skilled and trained employees, and decision making at a local level. In an agile environment information is the primary enable resource.

2 AGILE MANUFACTURING

Agility in competitive performance is the ability of a business to prosper in rapidly changing, fragmented global market place, establishing the processes that allow an enterprise to master change so as to be able to offer high quality, high performance, customer-configured products and services. Agility is not simply responding to changes, but having the capabilities and processes to successfully respond to constant and unpredictable change. Agility is built upon the company's foundation of world class or lean manufacturing methods, coupled with an organization that is physically, technologically and managerial established for rapid and constant unpredictable change (Vernadat, 1999; Maskell, 1999).

Agile competition (manufacturing) is an umbrella term that embraces a wealth of strategic options. These options include:
- Innovative alliances between suppliers, customers, and other companies even competitors in the pursuit of value added service.
- Management through leadership, motivation, support and trust.
- Rapid decision making not in centralized management but at functional level.
- Exploitation of powerful concepts of information and communication technologies.

- Integrated and flexible technologies of production.
- Distribution authority supported by information technology.
- Customer integrated processes for designing, manufacturing, marketing, and supporting all products and services.
- Mass customization versus mass production.
- Competition of the basis of customer perceived value and cooperative partnerships.

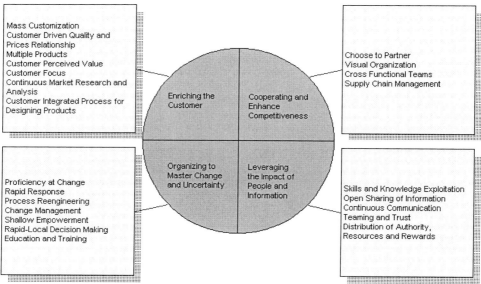

Figure 1. Strategic Dimensions of Agile Competition

Agile manufacturing concepts are the key to future competitiveness, but many of these concepts are still in the development state. There are no clearly established roadmaps to achieve agility. However, four basic strategic dimensions have been identified: Enriching the customer, cooperating and enhance competitiveness, organizing to master change and uncertainty and leveraging the impact of people and information (Goldman et al, 1995; DeVor et al, 1997). These dimensions will provide a useful context for examining agility requirements (Figure 1).

1. Enriching the customer. The examination of the value added to the customer by the use of the company's products and services is the main concern for an agile approach. A better understanding of customer's needs is a prerequisite for meeting changing customer needs on an ongoing basis. Customers feel enriched by products or services that they perceive as solutions to their individual requirements. Thus, products and services can be priced based on their value as solutions, not on their cost to produce. An unlimited variety of products, information, and extra services or technical support add value to the customers. The design process must be integrated with the manufacturing process. Having the customer participate in the design of the product can significantly enhance it. This close cooperation allows for the development of service-rich products that can evolve over time, as the customer and the company work closely together. This leads to the development of long term relationships.

2. *Cooperating and enhance competitiveness.* An agile company cooperates internally and with other companies, even competitors in order to bring highly specific customized products to market, no matter where resources and customers are located. Cooperation is necessary within an organization as a means of synchronizing people and organizational units. Agile enterprises will concentrate on their core competencies, and subcontract the remaining functions to other firms. A network of subcontractors will be sustained on short term for the length of an individual project. In this sense, organizations become virtual, and anything that is not a core competency becomes a candidate for outsourcing. (Baker, 1996) Cooperation of this kind requires respect, trust, training and openness. Inter- and intra-organizational relationships are dynamic and partnership changes over time. The beginning of the information age has made possible the ability to create widely diverse virtual corporations that can quickly and effectively address the needs of the customers and the market place. Internet technology is an appropriate telecommunication infrastructure for supporting this kind of virtual cooperation (McGaughey, 1999). Internal cooperation through cross functional teams and empowerment of local work force and external cooperation for managing the supply chain as a strategy of choice enhances the competition.

3. *Organizing to master change and uncertainty*. The only confident prediction that can be made about change is that it will continue and accelerate. Agility must develop its human, technical, and managerial capabilities and related processes to take advantage of change. Agile enterprise engineering increases organization's effectiveness by enabling it to identify and manage strategically critical tasks and processes. Process engineering becomes a strategically driven activity focused on day to day removing constraints. Training, practice, and experience at change management are essential, as reorganization becomes routine. Significant management skills, wide distribution of expertise and authority, and local decision-making to address local customers are the challenges of agility. The traditional management style of command and control must be replaced by leadership, motivation, and trust.

4. *Leveraging the impact of people and information.* Agility requires an entrepreneurial culture that "leverages the impact of people and information on operations" (DeVor, 1997). Rigid internal and external boundaries and structures should not suffocate organizations. Enterprises must learn to share information rather than control it. The agile company is characterized by a sharing of needed information across the organization rather than within a traditional hierarchical structure. Products are delivered to the marketplace faster and pinpointed to the customer's requirements. The availability of information and the means to transform it and transmit it determines how efficiently and effectively an organization will respond to changes in business processes and practices. New information systems technology and the wide access to Internet allows direct communication for internal and external constituents. It enhances an organization's ability to disseminate the company's vision and strategy, to educate and train people, to provide accurate data and information, and contributes substantially to the development of a more knowledgeable workforce. Educated and skilled workers organized into teams need empowerment within the boundaries of business principles to enrich the customer. An agile company by nurturing an entrepreneurial spirit, provides distribution of authority, continuous communication, teaming and trust, open sharing information systems and value based compensation.

3 VIRTUAL ENTERPRISE

The concept of Virtual Enterprise (VE) challenges the way manufacturing systems are planned and managed. It is a temporary alliance of enterprises that come together to share skills and resources in order to better respond to business opportunities and whose cooperation is supported by computer networks (Browne & Zhang, 1999).

Table 2
Virtual enterprise

Principles	Capabilities	Drawbacks
• Link complementary core competencies • Address short term objective • Depends on sophisticated Information and Communication Technology • Develop world class solutions for customers • Access to a wide range of specialized resources • Turn change into market opportunity • Share infrastructure and risk	• Independence of individual members • Reshape the virtual enterprise/member's change according to project • Flexibility in the use of people and technology • Short term, high return collaboration • Adaptability to diverse customer needs • Total integrated supply chain • Participation of small and medium size manufacturers • Reduction of operational cost	• Intensive information sharing • Telematics and information technology infrastructure • Contingency Risk and Security Risk • Network Organizational Structure

A VE consists of a number geographically widely distributed facilities and yet closed linked conceptually in terms of dependencies and material, information and knowledge flows, which come together in order to design and manufacture a product. In particular the flows relating to the physical manufacturing operations are only one part of a complex flow of information and knowledge concerned with underlying requirements, the logic of design decisions, the availability of inventory, contractual obligations related to customers, suppliers, employees and government agencies. The need for forming a VE by these facilities arises from their complementary capabilities in the product development process, that could be designed as a concurrent engineering model. The different companies that participate in virtual organization provide precisely the capabilities needed for each individual customer project. This creates significant flexibility and efficiency in the use of people and technology in terms of value-added to customers. It also gives the chance to small and medium size manufacturers to participate in the virtual enterprise while getting access of the resources of a large organization. The agile manufacturing strategic option of a virtual enterprise, is to put together the specific skills and competencies of its component organizations needed for the specific project without carrying the overhead for capabilities not needed. Each company within the web provides the service it does best to produce a final product with speed, efficiency, quality and without waste. An agile enterprise offers an integrated value added supply chain with flexibility, high quality and productivity. The collaboration in a virtual enterprise reduces operational cost, saves time and money while expanding market opportunities and maximizes past investment by reuse of existing applications (Table 2).

The heterogeneity of the many difference sub-systems involved makes the total management of an agile enterprise a complex problem. The flexible integration of a range of disparate IT applications is a key requirement for modern global enterprises. One major requirement for the formation of a virtual enterprise is a transparent interface mechanism for configuring and reconfiguring the units. The communication protocols of such an interface should be standard, easy to implement and flexible in order to enable the units accessing needed information through an existing data communication network.

Given the global nature of collaborative work, tool-sets are required that include integration of Internet and Intranet technologies with engineering and design repositories, document management, and software tools for computer-supported collaborative work (CSCW), workflow management (WFM), and project management for reliable and secure information platforms. Other major drawbacks to the virtual enterprise according to Menagh (1996) are the isolation between business partners, managers and staff, the contingency risk and security risk with a workforce that is spread out, having staff "employed" at different times, in different places and under different conditions. In the virtual enterprise there is a shifting set of short term relationships connected by computer networks versus a physical entity with a stable mission and location. There is a need for a comprehensive and unified theory for the coordination and control of globally distributed inter-organizational work, as well as a contingency theory, that covers the cultural and structural assumptions of the virtual enterprise.

4. LEAN PRODUCTION

Lean production aims at reducing all necessary human, material and financial inputs to the production process (Womack, 1991). Control mechanisms exist to bring the processes under control. It organizes production as a continuous flow, eliminates departmentalisation and batch processing leading to a short lead-time, high quality and low cost. Lean practice seeks to refine operations so unnecessary cost is eliminated from the creation, production, storage or distribution of goods and services. Lean manufacturing systems operate on a continuous improvement philosophy based on the removal of waste from the system in order to maximize the potential of price, quality, on-time delivery, and availability of required quantity (Hill, 1993). The lean system has enabled managers to eliminate increase of unit cost and to reduce lead or cycle time. It gave manufacturers greater flexibility in managing the production process and new ways to enhance the exploitation of labor.

Lean production was developed by Toyota, to eliminate waste. (Womack, 1996). Toyota is commonly accepted to be the architect of lean production. Their system was developed in the 1950s and 1960s and did not make extensive use of Information and Communication Technology. As others began to follow the lean production model, however, sophisticated EDI systems were developed to streamline the planning and the ordering of components and sub-systems, and to establish closer communication links with suppliers and dealers. Thus, the roots of electronic commerce in lean production were as an assisting technique for the just-in-time system.

Lean production stresses simplicity of methods and empowerment of the shop floor personnel. Simple methods allow personnel to use effectively the system and always to strive for perfection. Many Japanese pioneers in a lean production system simplify production processes.

The key principles of a lean system providing value added activities are (James-Moore, 1997):

1. Customers communication. Customers require improved products or services. Knowledge about customers profile and needs determines value added activities and functions. The strategy is to identify value-added activities and functions according to the customers need and eliminate anything that does not add value. In this sense lean systems provide strong communication between customer, product, processes and supplier.

2. Quality management. Quality is approached via quality management, not quality control. The lean system designs products to be quality robust and continually improves product quality. Suppliers are evaluated and performance measurements can be established in all aspects of the value stream. Close cooperation with supplier on quality and design issues during the design stage, ensures reliability, quality and service to be built into the product. Quality is the major determinant of supplier relationships. Employee participation in quality enhancement is widespread. Continuous quality improvement is seen as a mean to reduce cost as well as win orders.

3. Improvement of processes and flow. With each movement of material, value is added by streamlining material flow. The entire set of actions required to bring a product from its raw materials to the customer is examined and improved. Just In Time techniques are used to minimize work-in-process inventory and to ensure consistent quality. Uninterrupted flow results in short lead-time, high quality and low cost.

4. Empowerment of workers. A firm that empowers its workers gives them the authority to solve problems on their own. That means that they have the authority to stop production to solve emerging problems, instead of first waiting for top management guidance. The objective of worker's empowerment is having workers involved in problem solving at the shop floor level. Employees are cross-trained and can perform a variety of jobs. Few hierarchical levels (flat hierarchical structure) enhance open communication and mutual trust and team work. Outstanding staffing is attained with effective recruiting, selection, training, and retention.

Lean production remained limited by its focus on short, fixed term production cycles hindering the achievement of even greater flexibility and capacity utilization. The lean production system is no longer considered good enough for an enterprise to maintain its competitive position. It cannot enable a corporation to optimally exploit unanticipated customer opportunities by being able to quickly adapt or change over the production process. Lean systems cannot deal with frequent fluctuations in demand. Lean Production is highly vulnerable to production disruptions and it cannot deal very effectively with the things that cannot be controlled, in an environment where change and unpredictability are the order of the day. It can only correct the processes, so the processes are under control.

Agility embraces lean. It reduces unnecessary costs more effectively by shifting emphasis away from cost cutting to increased throughput, market segmentation and rapid creation of new services and products. Agility combines lean with high throughput rather that cost cutting. Agility builds upon the lean system's principles in terms of streamlining the work process and continuously improving quality, while overcoming its limitations in order to make everything in a production process dynamic and reconfigurable. Consequently, tactics such as continuous improvement, Just in Time and Total Quality Management are viewed as means to agility. Unlike lean production, the agile enterprise is based on short-term relationships.

5 TOTAL QUALITY MANAGEMENT

Total Quality Management is a strategic approach to business improvement. It focuses on process and product or service innovation to improve continually quality. Quality of a product or service is the customer's perception of the degree to which it meets his expectations. The principles common to TQM include an emphasis on customer focus, continuous improvement and teamwork. According to Zahedi (1995) "Total quality management, as a framework, is a collection of ideas, concepts, and tools, all designed to promote quality throughout an organization in all its functions...The word total underlies the all-encompassing nature of quality, and the word management removes quality from its purely technical scope and generalizes it to include organizational and behavioral components of the organization as well." Merlyn and Parkinson (1994) defined TQM as "the integration of quality management methods, concepts, and beliefs into the culture of the organization to bring about continuous improvement". The main problem in accomplishing this is how to manage change in organizational behavior since attitude and culture changes are often required.

Based on a comprehensive review of the quality management literature, various factors that are essential for successful adoption of Total Quality Management in an organization have been identified: (Saraph et al, 1989; Flynn, 1995; Powell, 1995; Gunasekaran, 1999)

- *Top management commitment*. Top management actively supports and communicates the quality program.
- *Customer involvement*. The contact between the organization and the customer, before, during, and after the sale to determine product requirements and set standards is managed with quality.
- *Supplier partnership*. The suppliers require strict quality specifications and adoption of quality principles.
- *Employee involvement*. Training of all employees (including management) in quality principles and in problem-solving skills and teamwork approaches. Employee interaction with customers and suppliers as well as in planning and design phases. Empower people to make decisions through leadership and coaching.
- *Benchmarking*. Identify best practices of other organizations and use them as a standard.
- *Organization's quality culture*. Make sure that everybody in the firm knows what quality means to the firm and how it can be maintained. Monitoring and controlling the quality is necessary.
- *Process improvement*. Reduce cycle time for order processing, product/service development, and delivery. Plan to reduce defects continuously.
- Total Quality Management seeks continuous improvement of quality, productivity, flexibility, timeliness, and customer responsiveness. Companies who use TQM are committed to (Pegels, 1995):

- Even better, more appealing, less variable quality of the product or service.
- Even quicker, less variable response-from design and development through supplier and sales channels, offices and plants all the way to the final user.
- Even greater flexibility in adjusting to customers' shifting volume and mix requirement.
- Even lower cost through quality improvement, rework reduction, and non-value-adding waste elimination.

Agility implementation widely employs Total Quality Management, as a potential source of sustainable competitive advantage.

6. BUSINESS PROCESS REENGINEERING

Business Process reengineering (BPR) is an approach to defining necessary and desirable changes in how business should be carried out. The new processes are likely to require any number of changes: different roles and responsibilities for individuals, new communication patterns, use of new tools, different team arrangements and possibly whole new team-oriented approaches to planning and managing work. Reengineering means rethinking and redesigning the above processes by which added value is created. In BPR an effort is made to enlarge the scale of improvement. Information Technology provides the infrastructure and tools, which fundamentally change organizations, but management provides the strategic business vision that transforms technology into competitive advantage.

Table 3
Differences between TQM and BPR

	Improvement	Innovation
Level of Change	Incremental	Radical
Starting Point	Existing Process	Clean Slate
Frequency of Change	One-time/Continuous	One-time
Time Required	Short	Long
Participation	Bottom-up	Top-Down
Typical Scope	Narrow, within functions	Broad, cross-functional
Risk	Moderate	High
Primary Enabler	Statistical Control	Information Technology
Type of Change	Cultural	Cultural/Structural

Source: Davenport (1993), p. 11

Michael Hammer and James Champy (1993), define business process reengineering as "the fundamental rethinking and radical redesign of business processes to bring about dramatic improvements in critical, contemporary measures of performance, such as cost, quality, service, and speed." BPR can also be defined as "the critical analysis and radical redesign of existing business processes to achieve breakthrough improvements in performance measures." (Grover et al, 1995) The common element is that the change occurs across the whole process. A focus is placed on company-wide processes rather than individual functions or departments. Processes should be targeted where change will lead to competitive advantage.

There are many misconceptions as to the nature of business process reengineering. Reengineering precedes reorganization. It focuses on rethinking work from the ground up, eliminating work that is not necessary and finding better, more effective ways of doing work. Reengineering is not reduction of people to achieve short term cost decrease. TQM and BPR share a cross-functional orientation. Davenport (1993) notes that quality specialists tend to focus on incremental change and gradual improvement of processes, while proponents of reengineering often seek radical redesign and drastic improvement of processes. Differences between the two are provided by Davenport (1993) (Table 3)

Business process reengineering is essentially value engineering applied to the system to bring forth, sustain, and retire the product, with an emphasis on information flow. Analyzing the business process, non value added functions are identified, eliminated and a new process is developed. Reengineering is not simply about making an organization work efficiently. It is about creating value for the customer defined as high quality, low cost, and increased response time (Grover et al, 1995).

Organizations adopting business process reengineering benefit through improvements of customer response and reduced costs. More specifically these benefits are:

Increase speed of change and responsiveness to customer's requirements. There are savings in direct engineering time for field investigation, design, drafting and preparation of specifications and in management of the engineering process. Organizations can quickly respond to customers inquires on the schedule or status of a project, because customer service and marketing acquire on line access to work management and design information with multiple access keys.

Elimination of non value added rules and controls. Actions that are unnecessarily increase complexity without providing value should be identified and eliminated.

Cost reduction. BPR principles are used to lower the overall management and production cost while at the same time improve quality and customer service.

Better coordination of information, technology and human resources. Improvements of cross functional relationships increase speed and efficiency. Sharing of information results in less uncertainty in work planning.

While many companies have reported impressive gains, many others have failed to achieve the major improvements they sought through reengineering projects. (Keen, 1991; Bashein et al, 1994; Cardarelli et al, 1998)

Obstacles that can result in failure for enterprise reengineering projects include:

Lack of sustained management commitment and leadership. Direction in the form of strategic vision must come from the top management. On going management support is essential because reengineering changes the way the organization does business and can generate opposition from internal affected individuals.

Lack of strategic context. King (1994) views the primary reason of BPR failure as overemphasis on the tactical aspects and the strategic dimensions being compromised. He notes that most failures of reengineering are attributable to the process being viewed and applied at tactical, rather than strategic, levels.

Unrealistic scope and expectations. A short-term focus and expectations concerning the difficulty of BPR can lead to quick solutions that do not solve the fundamental business problems. Misperceptions and misunderstandings about BPR are common. Expectations of senior management may not be realistic.

Resistance to change. Employees are oriented to the business culture, trained to do a job, gain experience, succeed and grow within the environment. BPR is a radical change. The new processes are likely to require any number of changes such as different roles and responsibilities for individuals, new communication patters, use of new tools, different team arrangements and possibly whole new team-oriented approaches to planning and managing work. If change has not been viewed as logical, necessary and an ongoing part of the operation then resistance will appear.

There are a wide variety of strategic dimensions to process reengineering (Davenport, 1990, King, 1994) These dimensions include:

Developing the Business Vision and Process Objectives. Cost reduction, time reduction, quality improvement, and employee empowerment are a series of business objectives driven by the organization's vision that can provide valuable guidance to reengineering teams.

Identification of non added value processes. Identify processes that conflict with the business vision and don't add value. Choose the processes according to their importance, feasibility and dysfunction building blocks of operational capabilities that translate into value, setting business apart from the competition.

Identifying New Product and Market Opportunities. A valuable potential by product of the reengineering effort is the identification of new business opportunities.

Identifying IT Levers. New information technology is being implemented to support the work of the teams. Awareness of IT capabilities can and should influence process design.

Developing a Human Resources Strategy. Pursuing technological potential without exploiting human potential will not yield the outcomes sought from BPR effort. The purpose of BPR should be to increase the value of people whether through process or policy redesign, training, or access to information.

A new approach to BPR is required to take advantage of change. Process engineering becomes a strategically driven exercise focused on removing constraints. It is an integral part of the day-to-day life of the organization.

7. JUST-IN-TIME

Just in Time manufacturing is a systems approach to developing and operating a manufacturing system, based on continuing elimination of waste and consistent improvement in productivity (Maskell, 1989; Wallace, 1990) It has been part of the Japanese manufacturing industry approach for several decades. The key point in understanding JIT is that JIT is a continuously goal oriented process to eliminate waste and improve productivity. The perfect environment for JIT is the repetitive production environment. If demand is not predictable and product variety cannot be constrained then JIT solutions are not suitable.

The fundamental concepts to JIT production system are:

Elimination of waste: Production of only the minimum necessary units in the smallest possible quantities (lot sizes) at the latest possible time.

Employee participation: Employees need to be trained as to their role in this operational philosophy in order to gain cooperation, and acceptance.

Integrated systems: Successful implementation depends on integration of manufacturing, quality, materials and supply systems with the Human Resource system.

Using a JIT system the following benefits can be obtained:

- Inventory reduction (raw, work-in-progress, finished products)
- Increase inventory turn overs
- Lead time reduction
- Quality improvement
- Reduction in working capital and cash flow improvement
- Better labour and equipment utilisation
- Increase in worker motivation
- Team work increase
- Reduction of batch sizes
- Higher productivity
- Improved customer service

Since each manufacturing process is different, it is up to the individual company to determine the final application of JIT. The goal of a JIT approach is to develop a system that allows a manufacturer to have only the materials, equipment, and people on hand required to do the job. To achieve this goal several aspects and implementation elements have to be taken into consideration. (Millar, 1990; Abraham et al, 1990; Zhu et al, 1994)

Human aspects. Total employee improvement can be obtained with quality circles, education, work groups, and job enrichment. Job enrichment provides rob-rotation, flexible workforce and cross training. Cross training is necessary when workers are encouraged to operate multiple machines. "The biggest mistakes a company can make when implementing JIT is to ignore interpersonal skills, education and training" (Adair-Heeley, 1989). Training should concentrate on technical and environmental aspects of JIT. JIT implementation should be initiated from the top, with full support of all managerial level (Evans, 1990). During the JIT implementation process a group of experts from different functional areas within an organization must work together as a team. As JIT implementation involves a significant organizational change, employee commitment is crucial to its success.

Quality aspects. Quality within JIT manufacturing must be inherent in the process. Several approaches such as quality at the source (JIDOKA), error proofing (POKA YOKE), and statistical Process Control can be used. Every individual is responsible for the quality of products or components he produces. If anything goes wrong, the operator has the right to stop the production line. While the entire line is stopped, the other workers will do their machine maintenance work. JIT also emphasizes quality starting at the source of supply by demanding quality certificates from vendors.

Technical aspects. Several techniques are used for the JIT system. The JIT technique is a pull system based on not producing units until they are needed.

Integration-Communication aspects. JIT integration can be found in manufacturing, quality, materials supply with the human resource system. Planning across functional boundaries is essential. All possible channels of communication must be strengthened when JIT implementation takes place.(Johnston, 1989; Helms, 1990)

JIT implementation will not have maximum affect unless technical solutions are integrated with people solutions. It is not strategies and plans that make things happen, it is people. Agile enterprises can improve their responsiveness to customers by implementing Just In Time systems to reduce setup, batch sizes, and work in process inventory.

8. COMPUTER INTEGRATED MANUFACTURING

Computer Integrated Manufacturing (CIM) can be considered as an advanced business philosophy that unifies a company's administration, engineering and manufacturing. The information technology plays a central role for planning and controlling the manufacturing process. It uses computers and communication networks to transform automated manufacturing systems into interconnected systems that cooperate across all organizational functions. CIM requires a new management perspective and careful planning of each technical element in conjunction with training. The goal of CIM is to remove all the barriers between all the functions within an operation, to encourage marketing, order entry, accounting, design, manufacturing, quality control, shipping and all the other departments to work closely together throughout the process. It provides information by linking each operation task by computer, giving decision makers access to needed information. Tasks can be performed in parallel, not in sequence. Real CIM potential lies in creating a network of people and activities to accelerate decision making, minimize waste, and speed up response to customers while producing a high quality product. CIM must be thought of as a strategic policy within a company. Commitment is required at all levels of the company. It can be costly, and can require changes in policies that may be difficult for those accustomed to the old methods to accept.

Some of the benefits of computer integrated manufacturing systems are:

- Cost reduction. Information handling is the way to reduce manufacturing time. Improved accuracy and time savings can translate into reduced costs and process time for operation. Better use of capital resources through work automation results in higher productivity and lower cost. The automation of the entire production process shifts management's emphasis from supervising people to supervising machines.
- Quality improvements. CIM supports customer satisfaction resulting from the elimination of waste from the design, engineering and production cycle.
- Greater production control. Company's efficiency increases through work simplification and automation, better production schedules planning and better balancing of production workload to production capacity.
- Faster responsiveness to the market. Improved product development cycles, high levels of human and capital resource productivity, improved quality, and short delivery time, lead CIM users to a rapid response to the market place.
- Reduced Inventory. Reduced investment in production inventories and facilities through work simplification, and just in time inventory policies.
- Small lot manufacturing. CIM is based on small lot sizes and offers greater variety of products.

In the past years several surveys have attempted to investigate the major barriers to CIM (Shank & Govindarajan, 1992; Zammuto & O'Conner, 1992). They include:

Managers attitude. Managers view CIM as a technology than as a concept. Successful implementation of CIM means optimization of the entire process instead of individual production processes. Lack of understanding the technology and suitable infrastructures, contributes to managers failure to appreciate CIM.

Top management commitment. CIM installation must start from the top with a commitment to provide the necessary time, money and other resources needed to make the changes that CIM requires.

Integration. One of the strongest means to implement CIM is integration, which has to be established consistently at several levels at the same time (i.e. people's behaviour and organization, product and manufacturing processes, material and information flows).

Organizational structure. The existing structure of the organization must be altered to facilitate cooperation between manufacturing, accounting, marketing, engineering, and information systems department.

Cost. Many companies are experiencing difficulties in developing cost patterns to define specific objectives and justify CIM cost.

A company adopting CIM must take into consideration the strategy and compatibility of CIM with the overall goals of the firm. While CIM can be costly to implement, difficult to transition, and requires a total commitment the benefits are seen in increased quality, cost reductions, and faster work flow. Successful adoption of CIM gives the company a competitive weapon in the global market.

9. FLEXIBLE MANUFACTURING SYSTEMS

A flexible manufacturing system (FMS) is an integrated computer-controlled complex of numerically controlled machine tools, automated material and tool handling and automated measuring and testing equipment that, with a minimum of manual intervention and short change over time, can process any product belonging to certain families of product its stated capability and to a predetermined schedule. It is a component technology of CIM.

Flexible Manufacturing Systems (FMS) react quickly to product and design changes. They include a number of work stations, an automated material-handling system, and system supervisory computer control. FMS enable firms to produce specialized designs for smaller market segments, increase capability to produce a greater variety of products while at the same time reduce delivery lead times and inventory requirements (Sethi, 1990). FMS enable manufacturing to build volume across products to achieve economies of scope and ensure more consistent product quality (Table 3)

Control optimization, material flow efficiency, setup efficiency, and data flow efficiency can be achieved by using Flexible Manufacturing Systems. Economical machining is achieved by:

- Exploiting the flexibility and productivity of numerically controlled machine tools for the production of smaller and medium sized lots
- Utilising costly production equipment more effectively through the reduction or elimination of setups
- Changing parts, tools, and machining programs automatically

The emphasis on long production runs to exploit economies of scale must be replaced by thinking of how to employ wide production runs to exploit economies of scope. "Increased competition means more volatile markets, shorter life cycles, and more sophisticated buyers, which have all contributed to flexibility's emergence as a new strategic imperative" (Suarez et al, 1995).

Agility is often confused with some of the above business reforms. Its implementation widely employs the above strategies and practices as means to develop a company-wide strategy of sustained global competitiveness in the face of change. An integrated set of disciplines for building or changing the manufacturing system of an organization is often interchangeably used that transform work processes and organizations in ways that range from incremental to radical.

Table 3
Advantages of FMS (over conventional manufacturing)

Low Inventories
Short Lead times
High Quality
Low Production Cost
Consistent Delivery Performance
High Equipment Utilization
Good Flexibility

A significant determinant of the likelihood of success or failure of any particular strategy is the ability to match the technique to the desired outcome and the organizational environment. No one methodology or approach is right for every organization and every instance and there are documented successes and failures for every transformational strategy. Information technology provides the infrastructure and tools, which fundamentally change organizations, but management provides the strategic business vision that transforms technology into competitive advantage.

REFERENCES

1. Abraham, Y., Holt, T., and Kathawala, Y. (1990), Just-in-time: Supplier-side Strategic Implications, *Industrial Management and Data Systems*, 3, pp. 12-17
2. Adair-Heeley, C.B. and Garwood, R.D., (1989), Helping Teams Be the Best They can Be: the Message in the Milk Bottle, *Production and Inventory Management Review and APICS News*, 9, 7, pp 22-5
3. Baker, J. (1996), Less lean but considerably more agile, *Financial Times*, p. 17
4. Bashein, B., Markus, L, and Riley P. (1994), Preconditions for BPR success and how to Prevent Failures, Information Systems Management, spring, pp 7-13
5. Browne, J. & Zhang, J. (1999), Extended and Virtual Enterprises - Similarities and Differences, *International Journal of Agile Management Systems*, 1, 1 pp 30-36
6. Cardarelli, D.P., Agarwal, R., and Tanniru, M. (1998), Organizational Pitfalls of Reengineering, Information Systems Management, spring, pp 34-39
7. Davenport, T.H. (1993), *Process Innovation*, Harvard Business School Press, Boston, MA
8. Davenport, T.H., & Short, J.E. (1990), *The new industrial engineering: Information Technology and Business Process Redesign*, Sloan Management Review, 31, 4 pp. 11-27
9. DeVor, R., Graves, R., and Mills, J.J. (1997), Agile manufacturing research: accomplishments and opportunities, *IIE Transactions*, 29, pp. 813-23
10. Evans, J.R. Anderson, D.R. Sweeney, D.J. and Williams, T.A. (1990), *Applied Production and Operations Management*, 3rd ed., West Publishing Company, St Paul, MN
11. Flynn, B., Schroeder, R. and Sakakibara, S. (1995), The impact of quality management practices on performance and competitive advantage, *Decision Sciences*, 26, pp 659-691

12. Goldman, S., Nagel, R. and Preiss, K. (1995), *Agile Competitors and Virtual Organizations*, van Nostrand Reinhold, New York, NY,
13. Grover, V., Jeong, S.R. Kettinger, W.J. and Teng, J.T.C. (1995), The Implementation of Business Process Reengineering, *Journal of Management Information Systems*, 12, 1, pp. 109-144.
14. Gunasekaran, A. (1999), Enablers of total quality management implementation in manufacturing: a case study, *Total Quality Management*, 10, 7 pp 987-996
15. Hammer, M. & Champy, J. (1993), *Reengineering the Corporation: A Manifesto for Business Revolution*, Harper Collins, New York
16. Helms, M.M., Thibodoux, G.M., Haynes, P.J. and Pauley, P., (1990), Meeting the Human Resource Challenges of JIT through Management Development, *Journal of Management Development*, 9, 3, pp. 28-34
17. Hill, T. (1993), *Manufacturing Strategy*, 2nd ed., Macmillan Press Ltd., London
18. James-Moore, S.M. and Gibbons, A., (1997), Is lean manufacture universally relevant? An investigative methodology, *International Journal of Operation and Production Management*, 17, 9
19. Johnston, S.K., (1989), JIT: Maximizing Its Success Potential, *Production and Inventory Management Journal*, 30, 1, pp. 82-6
20. Keen, P. (1991), *Shaping the Future: Business Design through Information Technology*, Harvard Business School Press, Cambridge, MA
21. King, W.R. (1994), Process Reengineering: The Strategic Dimensions, *Information Systems Management,* 11, 2, pp. 71-73
22. Maskell, B. (1989), *Just-In-Time--Implementing The New Strategy,* Hitchcock Publishing, Carol Stream, Illinois
23. Maskell, B. (1994), *Software and the Agile Manufacturer-Computer Systems and World Class Manufacturing*, Productivity Press, Portland, Oregon
24. Maskell, B. Associates Inc., (1999), *The Journey to Agility*, www.maskell.com/4box.htm
25. Maskell, B., (1999), An Introduction To Agile Manufacturing, www.maskell.com/
26. McGaughey, R.E. (1999), Internet technology: contributing to agility in the twenty-first century, *International Journal of Agile Management Systems*, 1, 1, pp 7-13
27. Menagh, M (1996), *Virtues and Vices of the Virtual Corporation*, http://careers.competerworld.com/111395.html
28. Merlyn, V. and Parkinson, J. (1994), *Development, Effectiveness: Strategies for IS Organizations, Transition*, John Wiley and Sons, New York, NY
29. Metes, G., Gundry, J. and Bradish, P. (1997), *Agile Networking: Competing Through the Internet and Intranets,* Prentice-Hall PTR, Upper Saddle River New Jersey
30. Millar, I. (1990), Total Just-in-Time, *Industrial Management and Data Systems*, 2, pp1-10
31. Pegels, C.C. (1995), Total Quality Management: A Survey of Its Important AspectsBoyd and Fraser Publishing Co, , Danvers, MA
32. Powell, T.C. (1995), Total quality management competitive advantage: a review and empirical study, *Strategic Management Journal*, 16, pp. 15-37
33. Saraph, J., Benson, G. and Schroeder, R. (1989), An instrument for measuring the critical factors of quality management, *Decision Sciences*, 20, pp. 810-829
34. Sethi, A.K. and Sethi, P.S. (1990), Flexibility in Manufacturing: A Survey, *International Journal of Flexible Manufacturing Systems*, 2, pp 289-328

35. Shank, J.K. and Govindarajan, V. (1992), Strategic cost analysis of technological investments, *Sloan Management Review*, Autumn, pp. 39-51
36. Suarez, F.F., Cusumano, M.A., Fine, C.H. (1995), An Empirical Study of Flexibility in Manufacturing, *Sloan Management Review*, fall, pp. 25-32
37. Vernadat, F.B. (1999), Research Agenda for Agile Manufacturing, *International Journal of Agile Management Systems*, 1, 1, pp. 37-40
38. Wallace, T.F. (1990), MRPII and JIT Work together in Plan and Practice, *Automation,*. 37, 3, pp 40-42
39. Womack, J., Jones, D., and Roos, D. (1991), *The Machine that changed the World: The Story of Lean production*, Harper Perennial, New York, NY
40. Womack, J.P. and Jones, D.T. (1996), Lean Thinking, Simon and Schuster, New York, NY.
41. Zahedi, F. (1995), *Quality Information Systems*, Boyd and Fraser, Danvers, MA
42. Zammuto, R.F. and O'Conner, E.J. (1992), Gaining advanced manufacturing technologies benefits: the role of organizational design and culture, *Academy of Management Review*, October, pp. 710-28
43. Zhu, Z., Meredith, P.H., and Makboonprasith, S., (1994), Defining Critical Elements in JIT Implementation: A Survey, *Industrial Management and Data Systems,*. 5, pp 3-9

Virtual Enterprise Engineering in Support of Distributed and Agile Manufacture

R. H. Weston, R. Harrison and A. A. West

MSI Research Institute, Loughborough University, Leicestershire LE11 3TU, UK.

1. PHYSICALLY DISTRIBUTED ENGINEERING AND MANUFACTURE

To enable operation on a global scale Manufacturing Enterprises (MEs) are composed of consortia of companies, where partners have complementary roles and responsibilities. This enables appropriate access to physically distributed consumer markets and to the resources (e.g. materials, components, competences, capacities, finance and so forth) needed to serve those markets in timely and effective ways. The explicit description of common business, engineering, manufacturing and logistical processes can facilitate co-ordination of partner interactions even though process resourcing needs to be physically distributed. This is exemplified below, with reference to a global ME designed to operate in the automotive sector.

Automotive products are sophisticated but mature, with well-proven engineering methods and high levels of reliability. However, customer desires and requirements dictate a need for product customisation and distribution in high volumes globally. It follows that having the most effective manufacturing plant and best product designers is no guarantee of sector leadership or even long term competitive behaviour (Harbers 1996). It is also necessary to form influential and effective global partnerships, centred on the distributed interworking of people, computers and plant. This can ensure that the interchange of material (including components, sub-assemblies, assemblies and vehicles), information and knowledge leads to the right volume of custom products of suitable cost and quality, on time and at the right locations around the world.

Ford and Mazda are aggressively pursuing strategies to rationalise the global engineering of both power-trains and vehicles. Their latest collaborative engine programme is to produce a new generation of 4 and 5 cylinder in-line engines between 1.8 and 2.3 litre capacity. This so called I4/I5 programme has partners located around the world with the following centres of expertise: Engine Product Engineering in Hiroshima; Business Office in Germany; Manufacturing Engineering in the UK; and Manufacturing Plants Phase 1 in Mexico, Phase 2 in North America, Phase 3 in Hiroshima, Phase 4 in Europe and Phase 5 a second plant in North America. The collective production capacity of all plants when completed will be over two million engines a year in almost any mix of types. The distribution of the builders of automotive production machinery working on the I4/I5 programme is roughly: 70% in Europe, 25% in North America and 5% in other locations. To co-ordinate partner interactions on this development programme principle goals have been defined and adopted with respect to *common engineering, common manufacturing process* and *full cross shipment capacity between all manufacturing plants.*

Financial and organisational considerations are driving leaders in the automotive sector to become increasingly dependent on distributed partnerships (Harbers 1996; Anon 1997). For example, cost reduction (at least in the short term) may be realised if the size of permanently employed engineering teams is reduced, hence a need to outsource many (possibly most) systems (including production machinery) design and implementation activities. A natural consequence of this is that a single engineering team might be responsible for projects throughout the world.

Global engineering projects such as the I4/I5 programme are extremely complex and costly. There is no obvious 'right way' of engineering things but inappropriate engineering decisions could have disastrous repercussions. There follows an implicit need to be able to develop common models that co-ordinate and support the interworking and decision making of distributed engineering teams in such a way that multiple viewpoints of team members are considered, with proper account taken of the business, cultural and technical issues involved. This complex decision making is made even harder to optimise because of its temporal nature. Change will be constantly occurring, e.g. with respect to: the composition and behaviour of partner organisations; new customer desires and needs; competitor actions; product choice and performance; the influence of financial markets around the globe; the availability of new methods, processes and technology; and the required location and volumes of products and sub-products. The impact on engineering decision making of some of these changes will be fairly predictable, but the effect of others may be impossible to predict. As by definition the scope of any global engineering project will be large, it will not be practical to cater for all future eventualities. What is more, even if prediction were possible, the inclusion of redundant capabilities into IT and machine systems on the basis that they will be needed later can lead to loss of performance and/or further inflation of costs. It can be argued therefore that a global consortium might most effectively invest in engineering decisions that lead to systems that have an inherent ability to cope with change more effectively than competitor enterprises. In so doing a global consortium should pursue a dual strategy of (i) building systems that are inherently capable of being changed and (ii) developing and deploying methods and tools that enable it to realise rapid and effective change.

2. NEED FOR MODELS IN SUPPORT OF DISTRIBUTED ENGINEERING

With reference to the I4/I5 global engineering programme, this section illustrates the need for models that represent reality from a number of complementary perspectives such that they can structure and support interactivity between distributed engineering partners.

2.1. Illustrative Need for Interaction between Engineering Partners

Figure 1 typifies current automotive product development processes. These processes span a development cycle of three to four years and include: product design (from concept to production), production machinery design and build, and the installation and commissioning of production plant (Anon 1994). Manufacturing related activities alone can amount to over 5,000 project line items for each phase of an engine programme (such as I4/I5). Hence the complete production development process is extremely complex, with a very large number of simultaneous product and process engineering activities ongoing, many of which involve complex and unpredictable interaction. Yet further complexity arises as these engineering activities are resourced by distributed partners who will have individual foci of concern, levels of skill and expertise, be subject to conflicting project, parent company and personal goals, and so forth. Nevertheless those activities must be focused on the production of the first complete product off the production line at month zero (see Figure 1).

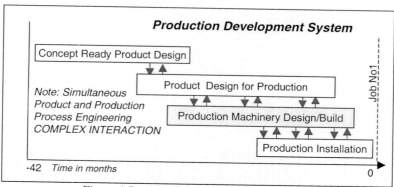

Figure 1 Production development process

To support and integrate activities of engineering partners, largely ad hoc integration methods and mechanisms are used in current programmes such as I4/I5. The situation is illustrated by Figure 2 for the production machinery design and build process. Machine and control system implementation is based around use of a fragmented set of specialist tools and methods that include various CAD packages, ladder logic programming and structured design, diagnostic and coding methods (Michel 1990; Anon 1998; Price 1994; Ford 1994). As ad hoc integration methods are deployed to co-ordinate the fragmented use of heterogeneous tools, it follows that there is no commonly maintained or understood representation or visualisation of production machines through their lifetime. Nor is there any overall computer executable model capable of supporting 'what if' analysis of machine designs, machine behaviours or of alternative ways of engineering or changing production machines (such as in response to an unpredicted product change). As a consequence current production machinery design and build processes used in the automotive sector are themselves inflexible, in the sense that little support is provided for the reconfiguration and reuse of production machinery as products (and related business, social and technical factors) change in an unpredictable way. The present engineering process used to create a new machine is largely paper-based. Processing requirements are translated into timing diagrams to describe the behaviour of the various machine elements. These are then interpreted by specialist members of the design team to produce the first cut machine design and its associated control logic. The physical machine design (2D and 3D CAD based) and the control system design activities (e.g. PLC control logic, I/O system configuration) remain isolated from one another only merging at the implementation phase (when actual machine testing occurs) at a very late stage in the machine build.

To summarise, an automotive company can compete on the basis of its ability to cope with complexity and uncertainty in (i) the required operation and composition of its production machines and (ii) the engineering methods, activities and personnel required to make, design and realise modifications to production machines.

However, the global teams responsible for programmes like I4/I5 face major challenges in co-ordinating their attempts to deal with flexibility issues which they have difficulty in resolving in an organised and timely way because: (1) changes occur continuously, often late in a project; (2) communication and co-ordination problems arise due to cultural, language, geographical and time differences; (3) frequently skill and knowledge mismatch and shortages occur; (4) lack of effective software support tools and difficulty in a single project to commit to new tool development; (5) technology (e.g. video conferencing, the web and electronic data exchange) is deployed but on an ad hoc basis.

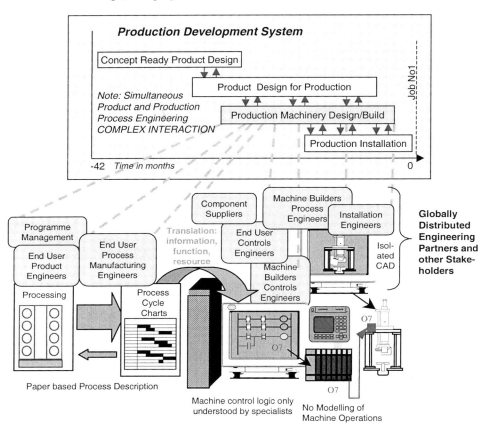

Figure 2 A view of current engineering processes and partners

2.2. A Distributed Virtual Engineering Environment

To meet key aspects of engineering partner interaction requirements related to the I4/I5 a distributed virtual engineering environment is under development at Loughborough University, this environment being known as COMPANION (COmmon Model for PArtNers in automatION) (see Figure 3). Focus is on enabling global partners to co-operate throughout the life-cycle of component-based production machinery. The engineering processes supported by the environment are being evaluated to determine factors that would inhibit a broad industrial take up of a virtual engineering approach. The COMPANION environment is designed to support teams of distributed engineers (e.g. customer [product, process and controls], machine builder [process, controls and installation] and component suppliers) and programme managers throughout the machine lifecycle (e.g. initial design, build and evaluation, mid-life product change and reuse).

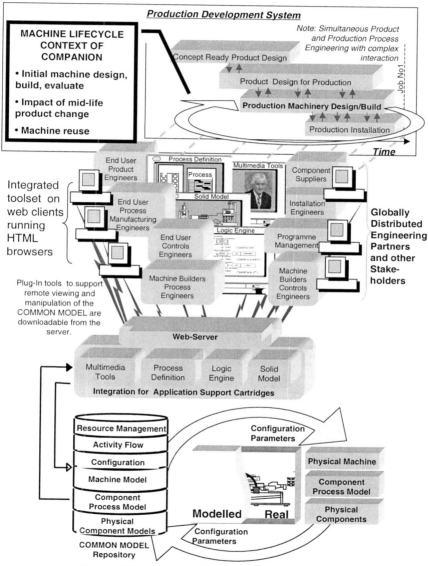

Figure 3 The distributed engineering environment

Concepts that underpin the development of the COMPANION environment include:

(1) Partner interaction requirements are defined and flexibly configured based on the reuse of multi-perspective enterprise models, thereby promoting the reuse of similar concepts in other application domains and providing improved means of handling uncertainty in any given application instance.
(2) Use of Internet (e.g. plug in tools based around web clients running HTML browsers) and database technologies to facilitate partner interactions by communicating common and consistent representations of various processes, designs and components, making these models globally visible and understood by relevant parties.
(3) Making change visible (to all relevant parties) as and when it occurs.
(4) Handling complexity and reducing necessary levels of skill and knowledge for system design and implementation.
(5) Enabling remote use of the virtual engineering capability via suitable application tools (e.g. process definition, component configuration, logic analysis, machine visualisation and Internet service access).
(6) Creating an effective distributed toolset with benefit and value that is clearly understood in business terms.

The impact in timesaving through more effective initial machine design, build and evaluation is expected to be considerable. A target of 50% has been set for saving in the commissioning time to 'ramp up' to full production capability. Currently the time involved is typically 5 months and a 50% reduction will save multiple tens of million $US on a engine programme. Additional value should arise when implementing a 'mid-life' product change and thereby enabling more effective and timely machine reuse. A target of halving the production downtime when making a 'mid-life' product change could potentially make savings of approaching ten million $US. The value of accurately modelling machines to enable more effective operation, maintenance and reuse would potentially enable savings of many multiple tens of million $US per phase of a typical engine programme over the life-time of production machinery.

3. NEXT GENERATION ENTERPRISE ENGINEERING REQUIREMENTS

I4/I5 distributed engineering requirements represent but one example of a vast number of largely unique distributed engineering needs currently facing company partnerships operating globally. This section identifies common facets of that need and thereby a requirements pull on next generation Enterprise Engineering (EE) methods, concepts and tools.

3.1. Status Review

Distributed MEs are complex and unique organisms that must be *change capable*, i.e. by sufficiently well designed that it can do the range of jobs in hand (i.e. be *programmable*) yet be capable of meeting unforeseen requirements and conditions of certain types. It also follows that if a distributed ME is to continue to thrive, it must renew (i.e. *recompose* and *reconfigure*) itself rapidly and effectively, as new opportunities and threats arise. It follows that a distributed ME should deploy suitable EE and Enterprise Integration (EI) methods and tools to ensure sufficient *change capability rate*. Adequate change capability and change capability rate is required at various levels, namely at business system, manufacturing system, resource system and resource levels. Inflexibility with system blocks or organisation structures at any one of these levels can impede responses at other levels.

It follows that some form of higher level of intelligence is required to observe, model, analyse, predict, plan, control and manage distributed MEs during their lifetime. As noted earlier, this is the role of Manufacturing Business Systems (MBEs) and thence teams of people who co-operate on EE and EI projects. In the case of a globally distributed ME formed from company partnerships, EE and EI requirements will be particularly complex, unique and uncertain (when compared with the case of a single company ME operating locally). It is not surprising therefore that current EE practice (as typified by the use of consultancy methods and frameworks) leads to deficient MEs in the following respects.

(1) Current generation distributed MEs are insufficiently change capable because:

(i) Rather than being uniquely engineered, certain aspects of MEs conform to semi-generic and idealised models of requirements and solutions developed by a number of different third parties. Typically these semi-generic models are embedded into systems engineering methods (e.g. as used by an external consultant) and software tools (e.g. into ERP packages). This requires different (even competing) manufacturing organisations to use similar processes, systems and information models to realise products and services.

(ii) Even individual company MBSs comprise teams of personnel, who deploy a mismatch of methods and tools over various timeframes. They have different (even conflicting) purpose, often with unknown relationships one with another. Consequently MB sub-systems seldom operate holistically, compete for resources and mutually constrain each other's achievable states. To minimise mismatch problems, MBSs often impose a fairly rigid framework but this can overly constrain the kind of creative and innovative behaviours needed to cope with uncertainty. MEs get around some of these problems and attain new states simply because people take innovative short cuts, etc., but seldom will it be known whether their actions represent a net plus for the ME and normally explicit information about the short cut will not be retained for future reuse.

(iii) Generally as ME size is increased such as by forming distributed partnerships, the MBS role will increase in complexity. If decomposition techniques are used to alleviate complexity problems then new integration problems arise. Any inflexible glue between parts will introduce flexibility constraints. For example, the introduction of additional layers of hierarchy is known to dampen response times.

(2) Factors constraining change capability rate in distributed MEs include:

(a) Current generation enterprise systems are inherently difficult to break into and change. Current techniques used to achieve IT system interoperation (between software applications, software tools and supporting information systems) tend to leave existing elements largely unaltered. Rather than risk modifying existing, often poorly understood elements they tend to introduce new front ends and glue that constrains unwanted achievable states. But typically this will overly constrain subsequent system recomposition and reconfiguration and leave a legacy of unwanted problems and poorly documented designs.

(b) Current methods and techniques used to change enterprise processes and systems are deficient and very time consuming to apply. Seldom is a sufficiently holistic view taken of system requirements change. Consequently current practice is typically characterised by spending significant sums of money (multiple millions of pounds per project) and taking a long time to make enterprise state changes of limited scope. Even during the

project, new requirements for state change typically emerge. This may even make an original state change obsolete before it becomes operational.
(3) Current EE practice does not exploit well the human intellectual capital available to distributed MEs. Nor does it account people as assets with knowledge and skills that can be exploited. There are many reasons for this, including pressure to think short-term and be pragmatic. However, a root cause is that exploiting human capital to induce state change is not a simple matter, particularly where changes can impact on those contributing to the change.
(4) Current EE practice does not naturally generate common or holistic understandings of required enterprise states, behaviours and ways of moving between different states and behaviours. Indeed, seldom are key change processes well known and documented. It is even less likely that any MB subsystem will be used to record the knowledge used, decisions made and the resultant impact of a change process, even where such a process is carried out repetitively.

It follows that there is significant elbowroom for improving EE and EI concepts, methods and tools, notwithstanding the fact that there cannot be any ideal MBS that will serve all needs of different distributed MEs.

3.2. Implied Requirements for Next Generation EE and EI

General requirements of next generation EE and EI approaches can be deduced with reference to current problems associated with composing and configuring distributed enterprise systems. Namely they should possess means of:
(1) Realising *flexible and scalable connections* between enterprise building blocks (both human and technical). In general, interoperation between blocks will be required at all levels of MEs and the blocks will be physically distributed.
(2) Gaining ready access to *explicit definitions of enterprise building blocks* in terms of:
   (a) 'how' building blocks should be integrated together (i.e. how they should be physically and logically connected, the interaction and information protocols they require, their physical location, etc.) so that they can interoperate with each other.
   (b) 'what' building blocks are capable of (individually and collectively) in terms of their functional capabilities, capacity to do things and likely behaviours when doing things.
(3) Developing a *holistic definition of what enterprise activities need to be done, and in what order and by when*.
(4) *Matching capabilities* [of (2b)] *to requirements* [of (3)] rapidly and effectively.
(5) *Implementing decisions made* [during (4)] into actual recompositions and/or reconfigurations of enterprise systems, thereby as required changing the way enterprise building blocks function and interoperate over specified timeframes.

4. VIRTUAL AND REAL ENTERPRISE BUILDING BLOCKS

This section differentiates classes and types of system building block. Discussion is directed towards exemplifying how representations of virtual (or modelled) enterprise processes, systems and components can be used by distributed engineering teams, thereby meeting requirements identified in section 3.2 (and particularly requirements (2), (3) and (4)).

4.1. Classes of Enterprise System

Before deciding how we might model enterprise systems, as virtual systems, it is necessary to consider the nature of current systems used industrially.

Normally the competitive delivery of services and products will require systematic (rather than ad hoc) approaches to dealing with complexity and uncertainty. As discussed earlier, this will necessitate use of a problem decomposition technique. Proper problem decomposition should identify a set of manageable sub-problems for which solution fragments (capable of operating in an integrated and systematic manner with other solution fragments) can be generated. Because the problem space will be subject to unpredictable change, the ideal problem decomposition may change, as might the idealised design of solution fragments and the manner in which these solution fragments will need to be implemented and integrated to produce an operational whole.

The philosopher will observe that the systematic application of a decomposition technique presupposes that the complex problem space is sufficiently well understood that an effective set of problem segments can be defined and tackled. Seldom will this be the case in any distributed ME. It follows that any decomposition of something as complex as a distributed ME must be sub-optimal, as must the design and subsequent interoperation of its sub-systems. Nonetheless only by breaking down a problem space into manageable and "known" classes of sub-problem it becomes practical to address individual problem segments with the aim of generating near-ideal sub-solutions. Practical experience has shown that certain approaches to problem decomposition fare much better than others and generate solution fragments with desirable qualities; such as with an inherent capability to accommodate certain types of unpredictable change (Skyttner 1996, Weston 1999(a)). Therefore that the proper use of a good decomposition of enterprise engineering problems will in general be preferable (in terms of improved enterprise response times, efficiency and quality levels) to an *ad hoc* one. However the extent to which this is true is likely to depend upon the complexity of the problem being addressed, the previous experience of the enterprise engineering team involved and on characteristic problems associated with previous *ad hoc* problem solving techniques deployed. Also we may conclude that in general as the problem focus is narrowed the solutions developed will tend closer towards ideal ones, when they are viewed within their local focus of concern. However when these solutions are viewed with a wider (enterprise wide) focus of concern they may be valued as being of lower quality (i.e. if they fit overall requirements less well). This emphasises the need for the holistic design of solutions.

Industry has deployed many forms of enterprise system to bring order where there was chaos. Current systems used by companies to structure and support *What*, *How* and *Do* activities may take the form of undocumented or documented procedures that individuals or teams of people follow. They may be informed by wall charts, spreadsheets, plant mimics, databases of information etc. They may be supported by computerised information networks and databases, requiring strict sign-off procedures. They may be transaction processing based, e.g. in support of manufacturing control and planning. They may take the form of paper or electronic tickets (e.g. a form of Kanban) which control process flows. They may be based on the use of business rules supported by message interchange, which may or may not be enforced by a computer system. Computer modelling tools can also be used in various ways to provide graphical means and models that help to structure and support the design of products, processes, plant layouts and machines. Computer applications which embed some model of activities and resources (such as workflow, project management tools and ERP systems) are also used to structure and support the planning, scheduling, dispatching and monitoring of operations, tasks, jobs and material flows and the allocation and utilisation of resources.

In general however, existing systems used industrially have been developed and deployed in a fragmented way and do not conform to any "big picture" of enterprise needs or solutions. This situation remains despite standards development under programs like AMRF (Jones and Saleh 1990) and NCS (Albus et al 1987). We may infer therefore, that their design has not been developed in a 'holistic' way and that essentially they represent an *ad hoc*, localised decomposition of enterprise requirements. Despite this fact current systems take a variety of forms that have evolved to a state where individually they may function well. However invariably their fragmented and heterogeneous design causes major difficulty when developing large scale computer systems (Barber and Weston 1998). Consequently often both informal and formal systems are deployed in tandem (each consuming resources) that may be both paper and computer based. Even in a single enterprise seldom are different systems capable of even rudimentary interoperation without significant human effort. Normally there will be an overlap in the scope of systems typically resulting in the replication of information and models; sub-optimal business processes; and duplicate use of resources. Such deficiencies are likely to be most troublesome in situations where process chains cross system boundaries.

Not surprisingly in view of the fragmented origin of current generation enterprise systems there is no generally agreed classification of them. However Figures 4 to 6 show three alternative system classifications that reflect their different scope and foci of concern. Figure 4 shows a hierarchical classification of manufacturing systems in common industrial use. The overall scope of this particular set of systems is focused primarily on 'Do' and 'How' activities (ISO 14258 1998). Also their design is based on the presumption that the products, services and business processes of an enterprise have been previously defined. Despite further and progressive narrowing of the focus of concern (in respect to ERP, MES, LCS, SCADA and Equipment Systems in general) each class of system in Figure 4 should organise and support interworking between human and technical resources so that they tackle complex problems in their domain of concern in a systematic and effective manner. In so doing their

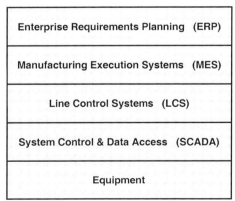

Figure 4 Common System Classes Deployed Industrially to Systemise the Operation of Manufacturing Processes

efforts should contribute to a broader purpose of the enterprise. It should also be noted at this point that because of the levels of uncertainty and complexity involved seldom can any of these system classes be fully automatic, i.e. in general each will require the involvement of human resources and their ability to function intuitively.

On considering common characteristic properties of current generation industrial systems it is widely reported that each class of system in Figure 4: (1) is time consuming (order of 6 months to 3 years) and costly to specify and install; (2) embeds some generalised model of requirements which often does not closely match specific requirements of its host enterprise; and (3) is very time consuming and costly to develop if changes required are outside the original system scope.

Furthermore if more than one class of these systems needs to be deployed in an enterprise:
   (a) often they do not interoperate effectively without further significant developmental effort (particularly if the systems used are supplied by different vendors)
   (b) often it is impractical to develop large scale systems which operate effectively in alignment with enterprise-wide requirements and even where it is practical often duplicative resources are expended in ensuring that the systems work harmoniously.

This is no more than can be expected as in general current generation industrial systems will not have been designed and developed with a common "big picture" in mind, therefore their underlying characteristic properties cannot be expressed in common terms.

Figure 5 shows a wider scope decomposition of enterprise systems. This positions the system classes of Figure 4 with respect to other common classes of business process found in different types and sizes of manufacturing enterprise (Berztiss 1996; Weston 1999(a)). There are many proprietary examples of these complementary systems that in general also suffer from deficiencies (1), (2), (3), (a) and (b) albeit that their relative impact on the profitability of the business may be quite different.

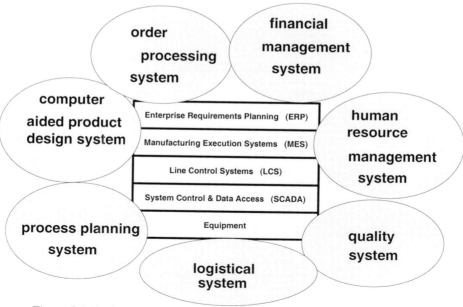

Figure 5 A Business Process Oriented Classification of Enterprise Systems

Figure 6 shows a third theoretical system classification. This has been developed to reflect the need to systemise the use of human and technical resources as they carry out W, H and D activities (ISO 14258, 1998). In this third classification the purpose of each class is as follows.

Business Systems – structure and support multiparty involvement in defining *What* an enterprise should do, helping to reassess the purposes of an enterprise in the light of changing environmental circumstances. They should develop holistic views about the vision, strategy, business objectives, business rules and general operating principles of an enterprise.

Process Systems – structure and support multiparty involvement in defining *How* the *What* can be achieved in an enterprise-wide and essentially implementation independent way. These systems can focus on developing process oriented models to specify requirements of enterprise systems and resources in terms of value adding activities.

Operational Systems – transform enterprise-wide descriptions of *How* the enterprise should operate into more detailed, implementation dependent descriptions of *How* specific groups of resources should interoperate to *Do* the activities required. Therefore these systems should determine ways of organising, planning, scheduling, co-ordinating and monitoring various groupings of human and technical resources so that they can achieve designated *Do* activities effectively and efficiently.

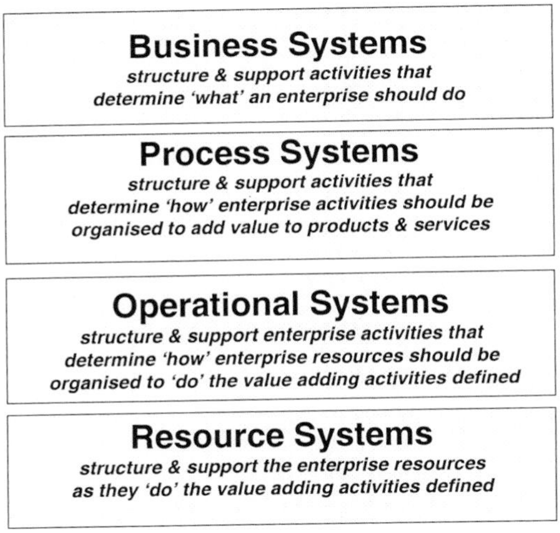

Business Systems
structure & support activities that determine 'what' an enterprise should do

Process Systems
structure & support activities that determine 'how' enterprise activities should be organised to add value to products & services

Operational Systems
structure & support enterprise activities that determine 'how' enterprise resources should be organised to 'do' the value adding activities defined

Resource Systems
structure & support the enterprise resources as they 'do' the value adding activities defined

Figure 6 A System Classification with respect to W, H and D Activities

Resource Systems – control and support the operation of groupings of resources so that they *Do* specified sets of enterprise activities in the right order, on time and hence realise products and services in a competitive manner.

Clearly enterprises physically require only one set of systems not three. However, there can be many more than three separate representations (or views) of one physical system set, where each representation is developed to suit a different purpose. The key point, however, is that if alternative views are to be used by different members of a systems engineering team they should remain consistent even though there will be a need to develop systems (as requirements change) and customise them (so that they can be used in different types of enterprise). By developing multiple and consistent views of enterprise systems it should also prove possible to flesh out a "big picture" of (A) system requirements, and (B) system solutions that can be expressed in common terms and facilitate interoperation in alignment with global and local requirements. The availability of such a "big picture" should lead to very significant benefit for end user enterprises (in terms of better systems at lower cost) and for vendors of enterprise systems and their components (in terms of increased reuse of their systems).

The foregoing illustrates a need to develop:
(i) holistic models of enterprise systems that characterise common enterprise needs and provides unified views of known solutions;
(ii) improved means of specifying and designing a specific set of enterprise systems for a given enterprise, by making reference to generic model fragments defined via (i);
(iii) improved means of realising a physical set of enterprise systems that conform to specified requirements and designs under (ii) and can interoperate in a holistic yet reconfigurable way.

4.2. 'Real' and 'Virtual' Building Blocks of Enterprise Systems

During the lifetime of any distributed ME there will be a set of operational instances of enterprise system types described above. Collectively they will structure and support the way in which available human and technical resources observe, model, analyse, predict, plan, renew, control, operate (to directly add value to products and services) and manage the enterprise. Each system instance will have a lifetime. Normally this will be distinct from the lifetime of other enterprise systems and that of its host ME.

Consider common life phases of a typical system. According to CIMOSA, system life phases include: 'requirements specification', 'conceptual design', 'detailed design and implementation description', 'system implementation and build', and 'system operation and maintenance'. This does not presuppose that the lifetime of systems corresponds to a sequence of steps from specification through to operation and maintenance. However lifecycle constraints inherent in systems engineering methods commonly used by industry tend to make this the norm if not the rule. Contemporary consultancy based systems engineering methods commonly used industrially are fundamentally deficient. They do not provide a coherent basis for representing multi-perspective representations of systems during each life phase that can readily be reused by members of engineering teams to innovatively customise, develop, extend and integrate enterprise systems.

New systems engineering methods are emerging that have an inherent capability to support iteration between life phases as required. The fundamental basis on which the new methods function is that consistent sets of idealised models (of (1) real system building blocks and (2) alternative ways of structuring system blocks into real systems) can be usefully deployed to facilitate life cycle engineering. These idealised models can be referred to as virtual models as their use can facilitate development of models of systems within a virtual engineering environment comprising a suitable set of modelling tools. Thereby the capabilities and behaviours of alternative system compositions can be studied before it is necessary to commit to costly and time-consuming effort in building and changing real systems. Essentially this reuse of virtual models corresponds to the reuse of semantic information and can result in rapid prototyping and commissioning of flexibly integrated enterprise systems (Weston et al 1995; Weston et al 1998; Weston 1998).

Importantly any given virtual model should not seek for all life phases to model everything that is important about a real system and its building blocks. This is impractical. Even if it were possible to model all aspects of interest to engineering partners with sufficient accuracy such a model is likely to be overly complex and therefore difficult to capture and reuse. Rather it is better to (1) build and use suitable sets of more primitive virtual models (with well defined relationships between modelled primitives) that can adequately represent reality from common perspectives of concern, and (2) to use well defined modelling constructs to represent each aspect of concern coded up by a given set of virtual models. Furthermore, generally it will not be practical to automate (i.e. remove humans from) the processes of specifying, conceptually designing, detailing the design, building, operating and maintaining enterprise systems. Hence the virtual models and virtual engineering environment should be designed to structure and support (i.e. semi-automate) common roles and life cycle activities carried out by engineering teams comprising human (and possibly technical) change actors.

Public domain engineering methodologies and frameworks typified by CIMOSA (Kosanke 1992), GRAI-GIM (Vallespir et al 1991), IEM (Spur et al 1996), UML (Quatrani 1998) and Model-Driven CIM (Weston 1995) define and operationalise the use of a backbone of enterprise modelling constructs. Collectively the modelling constructs they provide can structure and support the development and use of multi-perspective virtual models of enterprise systems. Typical perspectives include function, activity, process, behaviour, control, information, resource, organisation and cost aspects of enterprise systems. Suppose therefore that the role of a member of an enterprise modelling team can be supported by a capability to analyse possible dynamic behaviours of alternative compositions of system building blocks. Such a team member will (probably unwittingly) use the backbone of constructs via use of some graphical front-end to an analytic modelling facility provided by a virtual engineering environment. This will enable the capture and development of formal models of system requirements (e.g. in terms of a sequence of required system states and data access needs) as well as formal models of the functional capabilities and behaviours of alternative system compositions. Thereby it becomes practical to seek a close match between capabilities and requirements before any real system is constructed. Additionally other engineering team members, possibly with software generation and system implementation responsibilities could use the virtual models so developed. By such means it is practical and effective to structure and semi-automate various software generation activities, such as by automatically generating supporting information structures, facilitating the programming of real system blocks, and so forth.

4.3. System Modules and Components

For many decades end user manufacturing companies and their system component suppliers have understood that significant benefit (essentially manifest in increased *change-capability* and *change-capability rate*) can be gained by producing modular machines and systems. Practical experience and industrial development over many decades has identified common building blocks (such as transducers, drive systems, gantry units, software drivers and interpolation algorithms) that can be reused in different applications. Similarly, human resources (i.e. people) are deployed in various roles as a part of a bigger system (of human and technical resources). As such, people (with different skills, knowledge, capacities and behavioural characteristics) can be classified as reusable building blocks or modules of enterprise systems. It follows that technical and human resources (i.e. 'blocks of functionality') can be 'organised' or 'configured'[1] into systems. Also it follows that there may be opportunities to reuse these 'blocks of functionality' in different systems, provided that their capabilities, capacities and qualities can be mapped adequately well onto requirements in any given target application area. This has given rise to common use of the term 'component', particularly by vendors of IT systems, to describe *reusable* and *configurable system* elements. However, when choosing a so-called component, questions that should be asked include: Can the 'components' in question be readily used in the target application or will significant systems engineering effort and time be involved? Can these 'components' be readily reconfigured as requirements change? Can they be readily integrated with components from other vendors?

Unfortunately use of the term 'component' is often abused as many proprietary system building blocks are referred to as system components, yet their functional capabilities are not well defined and consequently they cannot readily be integrated into a larger scale system.

The notion that software can be componentised has grown as a prominent topic in computer science since the mid 1970s. As software systems grew in size structural descriptions were essential, and sometimes mandated. The seminal paper of De Remer and Kron (1976) on 'programming in the large' gave the notion of two programming levels, namely via a job control language and a sequential programming language. This led to the Module Interconnection Language (MIL). Another seminal paper by McIllroy (1976) heralded the emergence of structured methods (mainly in the USA) and data-centred methods (mainly in Europe), an example of which was the DRACO software reuse project of Freeman (1987).

During the 80s software architecture gained currency. Software began to be described by high level representations of software subsystems together with interconnection and interaction definitions. A burst of activity in the mid 80s led to notions of software integrated circuits by Cox (1986). Also around this time the intimate link between software architecture and software component reuse was recognised.

Since the early 90s the study of patterns has gained much prominence. Seminal work of Gamma, Helm, Johnson and Vlissides (1995) categorised commonly occurring problems in a context that designers and analysts have encountered. Similar ideas of reusing patterns of experience are now being applied widely in human computer interaction, education and business application domains.

By the late 80s it became generally understood that software component reuse only works well in a narrow domain and where the components reused have a proven track record of working well in that domain. Commercially exploited work of Grady Booch (1987) emphasised this point. Also it became evident that if less than say 100 components are required in a domain, complex search approaches will be superfluous. These observations point out the importance of good domain analysis using for example techniques such as those pioneered by Schlumberger on 'technology notebooks' (Guilermo Arango, 1993). They also emphasise the importance of considering business and social issues governing reuse in a specific domain, such as human resistance and difficulties associated with reuse and lack of incentives so to do.

Increasing numbers and varieties of software applications and tools are being deployed in MEs to structure and support the interworking of humans. Consequently it is a natural step to reuse the techniques developed to reuse software components to underpin the reuse of 'higher level' enterprise components (e.g. companies, departments, teams, people, software

---

[1] Configuration: the term 'configuration' is commonly used in technical circles, to imply that structural relationships are established between various elements of a system.

applications, software tools, manufacturing machines and building blocks of software applications, software tools and manufacturing machines). However, as identified in section 3.2 2(a) and 2(b), to facilitate systems engineering this will require explicit models of enterprise components, i.e. sets of virtual models that provide idealised representations of enterprise components from different perspectives of concern. As intimated in section 4.2 and explained further in section 8, the use of virtual models of well defined system components within a distributed engineering environment must ultimately lead to the development and widespread industrial use of new generations of change-capable enterprise system.

5. COMPUTATIONAL INFRASTRUCTURES – A KEY ENABLER FOR BUILDING FLEXIBLE DISTRIBUTED SYSTEMS

We have seen how enterprise modelling concepts, methods, frameworks and tools provide enablers of distributed engineering environments. Such an environment can be deployed in an organised way by members of systems engineering teams to specify and design enterprise systems comprising suitable sets of well defined virtual components. Through simulation and multi-perspective analysis, virtual system compositions and behaviours can be analysed rapidly and effectively before new systems are build or existing systems are changed. However, the timeliness and efficiency benefits associated with virtual systems design will be compromised unless virtual design specifications can be manifest rapidly and effectively as real enterprise systems comprising a suitable composition of real system building blocks. Therefore a suitable way of transforming from virtual system designs to real system solutions is required.

Essentially this requirement implies a need to have:
(a) sufficiently close correspondence between explicit descriptions used to model virtual components (at all ME levels) and the various views of real components those models represent.
(b) sufficiently close correspondence between virtual and real architectural constructs used to connect virtual components and real building blocks respectively.
(c) means of realising flexible and scalable connections between virtual components and real building blocks (human and technical) at all ME levels, so that virtual, hybrid and real change capable systems can be rapidly and effectively built.

The remainder of this section will explain how computational infrastructures, such as the Internet naturally provide concepts and mechanisms that can facilitate requirements (a) through (c), that are in effect a restatement of requirements expressed in section 3(2) and 3(1).

5.1 Purposes of a Computational Infrastructure

The term 'infrastructure' is used in many walks of life. Invariably a developed country will have a good infrastructure inasmuch that its transport (e.g. road, rail, air, river, sea), power (e.g. nuclear, electrical, hydro) and information (e.g. radio, TV, computer) networks fit their purpose well. This can provide a firm basis on which inhabitants of that country can conduct their individual and collective work and pleasure activities in an effective, enjoyable and (as required) competitive way. There are many types of infrastructure used by the personnel in MEs to help them carry out strategic (what), tactical (how) and operational (doing) activities. Such an infrastructure may take the form of complex, domain specific functions and services (such as that provided by finance, human resource or engineering departments) or more general purpose services and utilities (such as the provision of factory air, electrical power or a computer network).

Like other infrastructures, logically we should expect a computational infrastructure to underpin (i.e. structure and support) the activities of users[2] thereby improving their capability and capacity to carry out individual and collaborative work. In general, users will have various roles, jobs, tasks, responsibilities, commitments, etc., which need to be organised so that they realise products and services on time and with competitive cost and quality. The people, software applications, software tools and manufacturing machines involved will be referred to here as *system actors,* as their organised groupings will be systems. It follows that the interoperation of system actors and systems must be enabled and targeted. This is by no means a trivial problem as in a single ME there may be hundreds of system actors who will have both individual and collective goals and objectives.

When using an infrastructure to achieve interworking between system actors, it is evident from Figure 7 that the 'connection complexity' grows linearly with increase in the number of system actors (as illustrated by Figure 7). Hence the use of an infrastructure can prove highly beneficial, by enabling simpler and inherently scalable systems to be built. Whereas if direct connection is made between system actors the connection complexity grows in a square law fashion, i.e. the maximum number of connections required will be:

$$\sum_{p=1}^{p=n} (p-1)$$

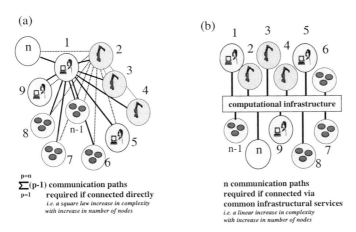

Figure 7 Illustrative Use of a Computational Infrastructure

A second important outcome arising from *using a computational infrastructure* is that it leads to improved change capability and change capability rate, i.e. the magnitude of changes made can be greater as can the rate at which changes are made. This arises because system actors only need to know how to use infrastructure services and do not need to know about the capabilities, interaction requirements and behaviours of other system actors, as is typically the case if direct connection is made. By simplifying the nature of the interfaces to system

---

[2] Users: in this context the users of a computational infrastructure will be both human and technical system actors operating in a manufacturing enterprise.

actors (a) new connections to system actors can be made readily and quickly and old connections easily removed, (b) it becomes easier to standardise interfaces to system actors, and (c) it much simplifies the development of virtual models of system actors. However, knowledge about the functional capabilities of system actors does need to be dealt with somewhere (i.e. either within the infrastructure itself, by some part of the system itself, or by change actors (with responsibility for realising system changes) or by methods and tools they use to engineer the system during its lifetime). Also knowledge about how system actors must interact and use infrastructure services (so that system-wide goals and objectives can be achieved) must also be maintained. We will return later to the issue of where this systems integration knowledge might best reside, however if systems integration knowledge is lost a likely consequence will be inflexibility.

A third important outcome arising from the *use of a computational infrastructure*, is that its use *can much simplify issues concerned with physically distributing system actors*, i.e. physically locating them at different parts of a single plant, or into different plants located around the globe. This third outcome arises primarily from the inherent separation of systems integration knowledge from connection (or interface) definitions related to system actors. It is becoming common practice now to retain location knowledge about system actors within some form of distribution service provided by computational infrastructures. We will also return to this issue later but this property means that systems built on an infrastructure can have much increased change-capability and change-capability rate when compared with systems built with direct connections between building blocks.

It follows that the use of a computational infrastructure will facilitate the development of *flexible and scalable* connections between ME building blocks, thereby underpinning physical aspects of (re)composing enterprise systems. However, ME building blocks must have an associated computational capability, so that they can utilise infrastructure services. Consequently people will require a supporting computer system capable of accessing the infrastructure services required, as will manufacturing machines. Provided this precursor requirement is met, building blocks can function as part of a bigger system that has few geographical restrictions.

In practice, however, any infrastructure will have a finite bandwidth. Inevitably this will place restrictions on the number of users (who can utilise infrastructure services at any instance in time) and/or on the quality of service (QoS) provided. Also we must bear in mind that a user requiring multimedia connectivity will in general consume more bandwidth than one requiring a simple digital data connection.

5.2 Use of Computer Networks and Integration Services – Building Blocks of Computational Infrastructures

Over the last two decades Local Area (Computer) Networks (or LANs) have become everyday tools used in support of many facets of our lives. LANs can be characterised by their topology, protocol and medium. Technically a LAN becomes an infrastructure when integration services are built on top of it, i.e. when a so-called middle layer of software is introduced between the network and its users, as illustrated by Figure 8. In fact commonly this middleware comprises a number of layers of software that use the low-level capability of LANs, to transfer data, messages, commands and information between remote computers. Thereby system actors can utilise and infrastructure service in such a way that they don't need to understand details of networks or how to communicate with them. Also users can work in designated groupings as part of a wider system (or subsystem) whilst being located around the globe.

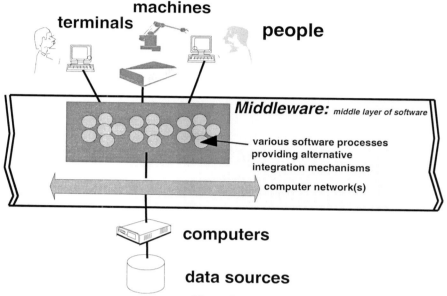

Figure 8

Key elements of the Internet, which provides a computational infrastructure that traverses the globe, are shown in Figure 9. Clearly there are performance limitations on its use but the Internet provides means of 'connecting' very large numbers of ME system actors concerned for example with product design, development, marketing, sales, manufacture and delivery. Its use is allowing so-called 'virtual enterprises' to operate in a distributed, yet effective way.

Where performance requirements dictate (e.g. in terms of guaranteed QoS and hence data-rates, connection times, etc.) companies can have their own LANs and middleware services (sometimes called Intranets) to connect system actors within the locality of a plant. Intranets and Internets used in office environments can be based on the use of similar topology (e.g. a bus network), protocol (e.g. Ethernet and TCP-IP) and media (e.g. twisted pair, co-axial and fibre-optic cabling and satellite data links). However, alternative types of LAN and middleware services (such as Fieldbuses) will generally be required on the shop floor as they guarantee (via the use of so called deterministic protocol) connection times and data-rates between system actors, which in the case of a shop floor will typically be plant sensors, actuators, machine controllers and so forth. Common examples of Fieldbuses used within MEs, and within products made by MEs include: 'Profibus', 'FIP', 'CAN', Control Bus, Interbus and Circos. Some Fieldbuses are designed with process control applications in mind (and therefore provide power down the line, and provide low cost network connections. Others are intended mainly for use in discrete manufacturing environments, and may be used to connect machine controllers to actuators, therefore their low level and middleware services guarantee QoS in terms of message transfer and device synchronisation.

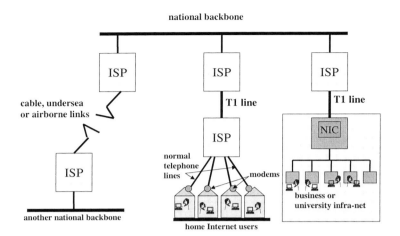

Figure 9 Illustration of Internet Topology

5.3 Common Systems Integration Services Required from a Computational Infrastructure

From the foregoing discussion we may deduce general requirements of a computational infrastructure suitable for use by manufacturing businesses. Namely it should provide a set of common computational services that can be deployed by various (human and technical) users (i.e. system actors) who may even be physically located around the globe, in such a way that as required the services provide a suitable *structure* and *technology* for individual and collective interworking. Bearing in mind the nature of enterprise systems its human and technical users will require: (1) *connection services,* that allow use of local computers to receive and transmit various packets of data, (2) *distribution services,* that allow data packets to be transmitted between remotely located computers and their users in a manner that is independent of the physical location of computers, (3) *interaction services,* that allow user actions to be synchronised such as via exchange of special packets of data called messages, (4) *information sharing services,* that allow users to access and update data of common concern in a meaningful way, (5) *presentation services,* that allow users to input and output data in a form they require (e.g. as floating point numbers or as graphical symbols) and (6) *system configuration and management services,* that enable system structure to be defined and used to organise user interactions and thereby their use of technology. Furthermore it will require some organised means of dealing with security of data issues (such as when moving sensitive data between companies) and operational failure of the infrastructure (particularly where time-critical interaction is required between system components). Figure 10 illustrates some of these common infrastructure requirements.

Also evident from the nature of enterprise systems and the requirement for businesses to evolve their structures and behaviours over time is that some users will need to be designated roles as change actors. Once again these users could be technical (e.g. software processes) or human actors who may have multiple roles and hence multiple responsibilities for carrying out *what, how* and *do* types of enterprise activity.

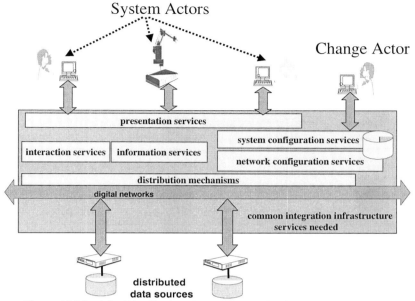

Figure 10 Need for Distribution, Interaction and Information Services
to Facilitate Systems Integration

6. THE INFORMATION INTEGRATION PARADIGM

Various forms of the information integration paradigm have been widely used industrially. The basic idea is that interoperation between system actors can be structured by making reference to a data structure that describes relationships between common data entities used by system actors in a given system. This system integration approach is one currently advocated by leading ERP vendors. One reason for this is that typically the scope of ERP systems is broad in comparison to other proprietary enterprise systems, as it can provide support of activities in financial, sales, engineering, manufacturing, warehousing and logistic domains of MEs. Hence major ERP vendors recommend that their customer businesses achieve enterprise-wide integration by building upon the use of the data structure underpinning the operation of their products. Of course this would suit the ERP vendor as it will lock the end user into that particular ERP product. However, this practice is reported to constrain business process design and business process operation in any single enterprise. This is the case because ideally manufacturers need a unique set of business processes to compete on a long term basis and do not want to be driven into a mode of operation that is common to that of competitors who also use the same ERP product and hence the same primary data structure. In general end users will know better (than vendors) what they want from a data structure. Unfortunately the end user might not know a better way of getting what they want. Clearly a single vendor product (such as an ERP package) will not provide full coverage and support for all enterprise activities, hence the end user business must at the very least develop some of their own complementary data structures, or have these provided by another product from a different vendor. Indeed, in general we have seen that an ME will need to use a range of software application packages (such as CAD/CAM, CAPP, Data Acquisition, Line Controllers, SCADA, and Material Handling systems as well as ERP) and invariably these will come from alternative vendors. Because each package will have its own underlying data structure there inevitably arises problems of incompatibility (and hence systems integration) problems. Hence an end user will not really solve its integration problems by adopting the use of a single product data structure. Nonetheless many big companies are doing this kind of thing. Small companies may not foresee any other viable option at present, particularly in situations where conformance to such a data structure might increase the probability of being able to partner with a big company.

Figure 11 shows three alternative ways of adopting the use of a common data structure within a system, which typically will be embedded into the operation of one of the subsystems (such as an ERP package), where the choice of primary data structure will normally default to the software application having the widest coverage of system issues. Clearly option 2 is better than option 1 from a change-capability point of view as by deploying common and scalable infrastructure services new system actors (e.g. new salesmen, new engineering systems, etc.) can be added into the overall system with relative ease provided that the activities they perform can conform to use of the common data structure. Whereas if direct connection is made to the primary subsystem, system change is likely to be more costly and time consuming to achieve.

Unfortunately however even the option 2 version of the information integration paradigm will result in limited 'change-capability' and 'change-capability rate' in installed systems. Typical problems found are as follows. In general the information integration paradigm does not inherently place restrictions on the order and use of data entities. However, all distributed systems require mechanisms to control data access and updates, otherwise data may become corrupt. The temporal nature of data will not (in general) be encoded by the types of data structure used to underpin current ME vendor products. Hence potential concurrency problems will arise, that may be particularly difficult to solve in cases where MEs generate many products and services and deploy many concurrently operating processes during their realisation. Another major problem arises when the primary data structure used to integrate a system is embedded into the application code of one of the system actors (e.g. an ERP system). Under such conditions it may be necessary to invoke functionality of the system actors (e.g. it might be necessary to 'explode' a bill of materials) simply to access or update a particular data entity. The need to invoke functionality, simply to access data, can act as a major performance bottleneck and can significantly reduce the inherent 'change-capability' of enterprise systems. In this respect option 3 (of Figure 11) is preferable to options 1 and 2, where the primary data structure is managed by a separate database management system. However, even here there needs to be some system entity included to maintain the integrity of the database by appropriately controlling the order and use of data items.

The philosopher will observe that essentially the integration paradigm is a bottom-up, vendor (and hence IT) centred view of the world and does not match ME needs overly well. However currently, in the IT system domain, vendors are the 'pipers' and 'pipers call the tune'.

Virtual Enterprise Engineering in Support of Distributed and Agile Manufacture 725

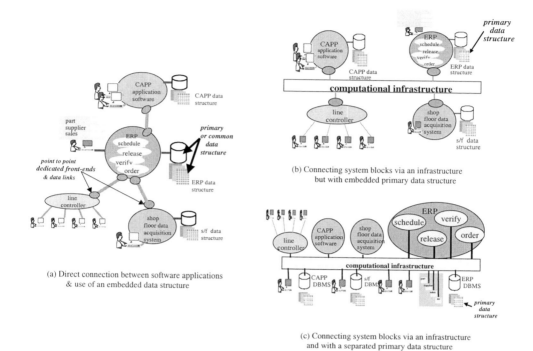

Figure 11 Illustrating the Use of Alternative Information Integration Approaches

7. THE PROCESS INTEGRATION PARADIGM

This alternate emerging integration paradigm (to that of the information integration paradigm) is based on the reuse of models of processes. Here process descriptions are used to determine how system actors should interoperate and how they need to share information entities. Subsequently the process descriptions are used to drive the runtime interoperation of system actors.

Figure 12 shows an example process model to which relationships describing interactions, information requirements and knowledge requirements can be attached. Generally this requires use of a modelling tool such as the ProcessWise workbench (Weston et al 2000), ARIS (Scheer 1992) or SEW-OSA (Aguiar and Weston 1993). Having modelled the processes required, multiple instances of these processes (corresponding for example to different projects, products, services, etc.) could be instantiated and run in a computer executable form. This might for example involve the use of graphical modelling primitives (provided by a modelling tool) to build models and the automatic transformation of these models into equivalent and executable Petri-Nets (such as via the use of facilities provided by a workflow management system). Indeed use of the process integration paradigm is fundamental to that of many current generation workflow management tools that function to control the flow of documents and commands/messages between system actors.

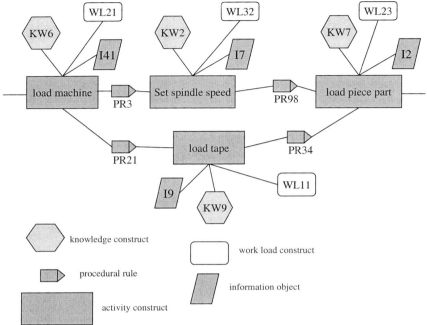

Figure 12 Example Process Model with Knowledge, Information and Workload Constructs Defined

Figure 13 shows how in principle this might be achieved. Clearly this approach is actually based on the combined use of the process integration paradigm, the information integration paradigm and a computational infrastructure. It should also be evident that the process integration paradigm takes a top-down, user-centred view of requirements and translates these into computer software processes. These processes can subsequently execute as some form of virtual model of the system structure during system runtime, thereby maintaining modelled relationships between system entities. Essentially the approach can naturally lead to 'change-capable' systems as new activity based relationships can be coded into a system by making changes to a graphically represented process model. Note, having defined a new graphical model the modelling tool can automatically generate the executable code needed to structure and/or control system operation.

The system engineering approach depicted by Figure 13 is potentially very powerful and can lead to the generation of custom built data structures that closely match unique business process requirements of MEs. However, in practice major systems integration difficulties still remain. Some important reasons for this are outlined in the following.

The approach illustrated by Figure 13 has no real notion about system actors and how they should be assigned roles, responsibilities, commitments etc. related to the process model. In practice today these actors tend to come in big lumps such as a person or team, an ERP or CAPP package, a robot or NC machine. Hence the system designer needs to know about what the various possible system actors can do (i.e. their capabilities, capacities and behavioural qualities) and what they need in the way of specific commands, messages, information and knowledge to do various roles they can be assigned. If no well-structured mechanisms are provided to support such issues they will be handled in an ad hoc way and this will restrict the 'change-capability rate' of system solutions developed. Also in practice may system actors (such as an ERP system) have their own way of doing things and have their own data structures that may conflict with those defined by the process model.

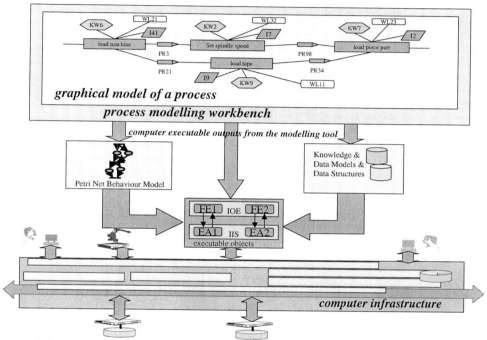

Figure 13 System Integration Approach Based on Use of Process Models

Another key problem of the top-down process-oriented approach is that in practice exceptions occur that alter process flows in an unpredictable way. Hence an effective way of programming in exceptions must be found and this raises some difficult conceptual issues when many concurrently executing processes requiring intuitive human activities and interactivity need to be modelled. Indeed this kind of problem is currently limiting the practicability (and hence benefits gained) from the use of workflow management tools.

8. THE COMPONENT-BASED SYSTEMS ENGINEERING PARADIGM

The component-based system-engineering paradigm essentially represents corresponds to leading edge systems development approaches that combine a use of the process-oriented integration paradigm (illustrated by Figure 13) with a component-oriented way of describing and using system actors.

8.1 Component Interoperation Using a Computational Infrastructure

Figure 14 illustrates conceptually how next generation, reusable components of enterprise systems could interoperate using the integration services of an infrastructure.

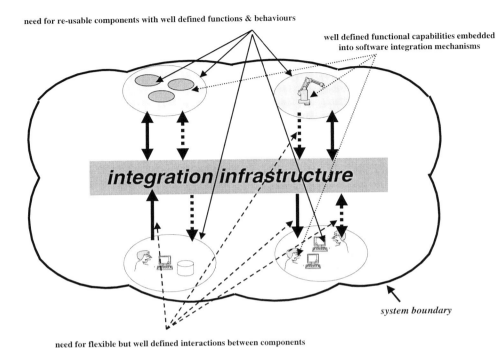

Figure 14 Conceptual Representation of Next Generation Enterprise Components Interoperating via an Integrating Infrastructure

These *components* will require an embedded integration capability that enables them to register and use infrastructure services. From the viewpoint of being able readily to integrate them into wide-scale systems, the system *components* must conform to an explicit representation or virtual model of their functional capabilities, interaction requirements and possible behaviours. The virtual models so defined will be used by system designers (or even automatically by software system functions) to select and configure the use of appropriate groupings of *components* so that they collectively provide the functional capabilities, interaction requirements and behaviours needed from the system as a whole (Weston 1999(b)). To enable this to happen in an effective way, system configuration and management tools will be required which lend structure to component interactions so that they achieve designated purposes of the system. This concept does not presuppose that *components* will be co-located. Indeed theoretically they may be distributed globally (or even across the galaxy) provided that the computational infrastructure provides the services required to enable the remotely located *components* to interoperate in the manner required.

From this discussion it is evident that the computational infrastructure should enable distributed components to connect to and use its services and thereby enable each component connected to interact with other distributed components. In general this will require services that facilitate things like data transfer, data storage, data sharing, message transfer and co-ordination of activities and distribution and navigation services that facilitate location transparency.

8.2 Use of Model-Based Integration Structure

Foregoing discussion about the need to explicitly define, realise and change organisation *structures* and *technology* for interworking highlights a need also to explicitly represent the "integration structure" (see Figure 15) of systems. The role of this structure, which can also take the form of a computer executable virtual model, will be to organise and control component interoperation in both virtual and real environments so that their collective behaviours can be targeted at specific system-wide goals. Typically this 'structure' will comprise a software architecture that operates as part of a wider enterprise structure designed to facilitate interworking between system actors and change actors as they carry out W, H and D enterprise activities. *Integration structure* can take the form of an organisational structure or framework that defines static relationships between components. Alternatively *integration structures* might take the form of a control hierarchy that actively uses functions to control the behaviour of component groups, based for example upon decisions about the way in which one or more components are behaving or could behave under new conditions.

It follows that *integration structures* should restrain and focus system behaviour, whereas *integration infrastructures* should enable and support system behaviour.

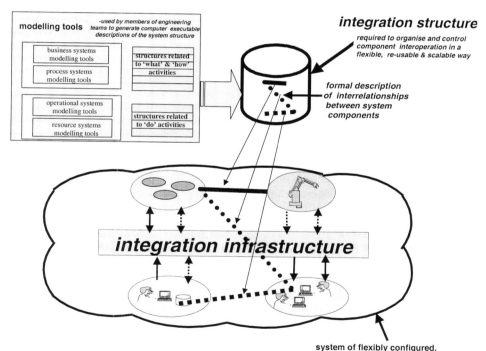

Figure 15 Roles of Integration Structure, Integration Infrastructure and Component Models

It is important to bear in mind that the way that any given *integration infrastructure* and *integration structure* is realised can have a significant impact on characteristic properties of resultant systems, e.g. in terms of their "flexibility", "performance" and "ease of use". Theoretically various options exist. However, to improve the change capability rate of a system it will be preferable to separate out *integration infrastructure* and *integration structure* elements in a system from each other and from specific system functionality and data structures embedded into a given selection of system components. This can also have very beneficial results in terms of system scalability, reuse and ease of development and can enable component, infrastructure and structure developments to be asynchronous. However in practice contemporary IT systems used industrially are typically implemented as a tangled web of interconnected infrastructure, structure and application specific elements. Previously this may have been because of a lack of understanding of the issues involved and associated disbenefits, because of technological constraints or simply pragmatism. Particularly in the case of wide scale systems however, the result has been high cost systems, long lead-times and solutions that may only fit their purpose acceptably well for a small fraction of their lifetime.

To enable flexible systems to be rapidly configured, developed (in terms of functional capabilities and behaviours) and extended (in terms of scale) during their lifetime it is evident that the integration structures used must enable components to be flexibly configured. Whereas different choices of integration mechanism and services are likely to be required to allow various types and implementations of enterprise component to interact in different ways.

A direct consequence of using a *component oriented approach to the life-cycle engineering of systems in which virtual models are used to structure the interworking of a selection of reusable components* is that have an inherent capability to accommodate change through 'system reaction' as discussed earlier. This is a distinctly different approach from that of supporting largely anticipated system change by embedding redundant functional capabilities (i.e. programmability) into a system or into one or more of its building blocks; such as that implicit in contemporary systems engineering approaches based on systems programming or parameterisation to cope with planned or anticipated change. Whereas use of a virtual model-driven, component-based engineering paradigm will in general be preferable where systems may or may not be subsequent to known classes of change during their life-time as it will facilitate 'system reaction' and 'system pro-action' and not incur unnecessary financial cost arising from the inclusion of unused functionality. However this will only be true if the use of well defined system decompositions does not impose an unacceptable performance overhead and is adequately supported by suitable integration mechanisms and tools.

Interestingly use of a virtual model-driven systems engineering approach can embed a 'change ethic' into system components as well as into configured systems. The two alternatives are akin to notions of *evolvable components* and *evolvable* (or readily maintainable) *architectures* being currently researched by the software engineering community (Kawalek and Greenwood, 1998; Clements, 1997). A set of *basic* (or non-readily evolvable) *components* could be configured via use of an *evolvable architecture* (based on the use of a suitable *infrastructure* and model driven *system structure*) to produce a larger grained *evolvable component* and so on. The means of evolving (a component or architecture) could range from being achieved completely automatically (at one extreme, such as via the use of software agent technology), to it being achieved (at the other extreme) manually and in an ad hoc way (Weston 1999(c)). Whereas (in between these extremes) evolution could be achieved via a model-driven structure where the model is captured and developed by members of systems engineering teams, i.e. in a hybrid fashion by deploying an enterprise modelling framework to organise inputs from a team of system architects, system designers, system implementers, system managers and maintenance engineers.

9. CONCLUSIONS – STEPS TOWARDS NEW EE AND EI PARADIGMS

It follows that more scientific and effective EE and EI approaches will become available to industry over the next decade. Furthermore we can expect these advances to lead towards a realisation of 'The Opportunity Concept' illustrated by Figure 16.

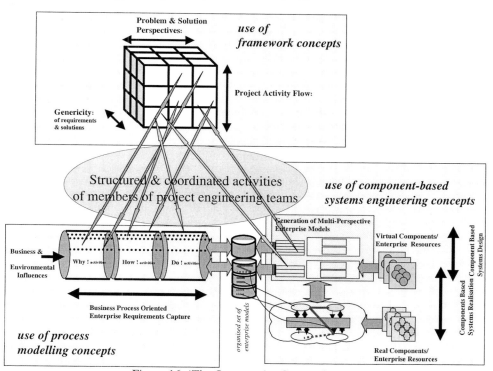

Figure 16 'The Opportunity Concept'

Such an approach to EE and EI should facilitate the process-oriented capture of system requirements within a framework of system engineering concepts. These requirements will be expressed in terms of the W, H and D activities (and their interrelationships) needed to renew and realise the purposes of an enterprise and should be defined holistically by enterprise personnel with the necessary knowledge and skills. The function of the framework concepts will be to structure system development into a more effective whole (while not overly constraining) the contributions made by members of project engineering teams, i.e. it will function to unify their different perceptions and experiences about the processes involved. Whereas the process orientation will allow team members to focus their innovation on the definition and development of sets of value-added and managerial activities that can be resourced in an effective manner. As illustrated by Figure 16 a combined role of the framework and process modelling concepts will be to target requirements capture on the design and configuration of large-scale enterprise systems from components.

The integration structure governing the interoperation of various sub-systems of enterprise components will be defined and realised using a virtual engineering environment (i.e.

organised grouping of enterprise modelling and model execution tools). Thereby the reuse of semantic information will be facilitated and system recomposition and reconfiguration enabled as the enterprise needs to respond to unforeseen requirements and conditions.

Level in ME	Source of Change Capability	Source of Change Capability Rate
Enterprise (multiple, concurrent W, H, D processes)	• reactivity of processes • reconfigurability of enterprise	• ease of developing and changing process compositions and configurations.
Individual Processes (realised by systems of sub-systems)	• reactivity of system compositions • reconfigurability of processes	• ease of developing and changing sub-system compositions and configurations.
Sub-Systems (organised groupings of components and hence resources)	• reactivity of sub-system compositions • reprogrammability of sub-systems	• ease of developing and changing sub-system compositions and configurations.
Individual enterprise component resources (e.g. people, machines, software processes)	• reactivity of component compositions • reprogrammability of components	• ease of developing and changing sub-system compositions and configurations.
Modular building blocks of enterprise components (e.g. machines and software resources	• reactivity of component compositions • reprogrammability of modular blocks	• through new modular block selection.

Table 1 Sources of Inherent Flexibility Arising from the Opportunity Concept

Table 1 indicates how the opportunity concept would lead to improved 'change-capability' and 'change-capability rate'. Assumptions made in the construction of this table include:
(1) Enterprise modelling tools are used to (a) determine suitable component compositions and configurations that match models of W, H and D process requirements (based for example on system simulation and performance analysis); (b) automatically generate the process co-ordination and data structure code needed to structure system operation; (c) runtime system exceptions can be handled either by pre-programming responses of components, by rapidly and effectively modifying system compositions (through 'system reaction') and/or including components capable of adaptation or by recomposing/reconfiguring the system; (d) that suitable infrastructure services are available to support necessary component interoperation.
(2) System components are available at an appropriate level of granularity and with the required capabilities, capacities and behavioural qualities are available to carry out the roles and responsibilities assigned to them.

REFERENCES

1. Aguiar, M.W.C. and Weston, R.H., "CIM-OSA and stochastic time Petri nets for behavioural modelling and model handling in CIM systems design and building" Proc. I.MechE. 1993. Vol 207. Part B: Journal of Engineering Manufacture. pp 147-158.
2. Albus, J., Barbara, A. and Nagel, R. (1981) NASA/NBS Standard Reference Model for Telerobot Control System Architecture (NASREM). Technical report, National Bureau of Standards, Technical Note 1235, 1987.
3. Anon., Project Analysis Data: In Line and Diesel Engine Operations, Ford Motor Company, Dunton UK, 1994.
4. Anon, "Next Generation Machines and Practices", Report to the automotive industry published by Giddings & Lewis, 1997.
5. Anon., "IEC Standard for Programmable Controllers, 1131 Pt 3: Programming Languages", Working Group 65A (Secretariat 67), 1998.
6. Arango G., Shoen E, Pettengill R. Design as Evolution and Reuse. Advances in Software Reuse. Selected Papers from the Second International Workshop on Software Reusability, March, 1993, Lucca, Italy, IEEE Computer Society Press, pp 9-18.
7. Barber M.I., Weston, R.H., "Scoping Study on Business Process Reengineering: Towards Successful IT Application", *Int. Journal of Production Research*, 1998, Vol.36, No.3, pp575-601.
8. Berztiss, A, Software Methods for Business Reengineering, Springer Verlag, New York, 1996. ISBN 0-3878-945563-9.
9. Booch G. Software Components with Ada, Structures, Tools and Subsystems, Benjamin/Cummings Publishing Company, Menlo Park, 1987.
10. Clements, P.E., 1997, "Requirements for Software Systems Which Enable the Flexible Enterprise", Managing Enterprise-Stakeholders, Engineering, Logistics & Achievement (ME-SELA '97), pp433 --438. Mechanical Engineering Publications Limited
11. Cox B. Object-Oriented Programming – An Evolutionary Approach, Addison-Wesley, Reading, 1986.
12. De Remer F, Kron H. Programming in the Large Versus Programming in the Small. In: IEEE Transactions on Software Engineering, June 1976, pp 312-327.
13. Ford Motor Company, General Motors Corporation, Chrysler Corporation, Requirements of Open, Modular Architecture Controllers, Version 1.1, 1994.
14. Freeman P. Conceptual Analysis of the Draco Approach to Constructing Software Systems. IEEE Transactions on Software Engineering, 1987 and included in IEEE Tutorial: Software Reusability, 1987.
15. Gamma E, Helm R, Johnson R, Vlissides J. Design Patterns – Elements of Reusable Object-Oriented Software, Addison-Wesley, Reading, Massachusetts, 1995.
16. Harbers W O, "Ford Automation Strategies and Needs", Automation Research Corporation: Automation Strategies Forum, Boston, Massachusetts, June 1996.
17. ISO (1998) "Industrial automation systems – concepts and rules for enterprise models", British Standards Institute, BS ISO 14258: 1998, BSI, Chiswick, London.
18. Jones, A. and Saleh, A. (1990), "A Multi-level/ Multi-layer Architecture for Intelligent Shop floor Control", *Int. Journal of Computer Integrated Manufacture*, 3, 1 pp60-70.
19. Kawalek, P. and Greenwood, M., 1998, Modelling in the Context of Systems, http://www.cs.man.ac.uk/ipg/sebpc.html

20. Kosanke, K., (1992), "CIMOSA – A European development for Enterprise Integration". In Enterprise Integration Modelling (C. Petrie, Ed.), MIT Press, Cambridge, Ma., pp. 180-188.
21. McIllroy M. Mass-Produced Software Components in Software Engineering Concepts and Techniques, Petrocelli/Charter, Belgium, 1976, pp 88-98.
22. Michel, G., Programmable Logic Controllers: Architecture and Applications, John Wiley & Sons, Chichester, England, 1990.
23. Price A.D. "Quality Methods in Software Control: the need for STEPS", Ford Motor Company (UK) Ltd, 1994.
24. Quatrani T, Visual Modelling with Rational Rose and UML. Addison Wesley. 1998. ISBN 0-201-301016.3.
25. Scheer, A. W., (1992), "Architecture of Integrated Information Systems", Springer-Verlag, Berlin, ISBN 3-540-55131-X.
26. Skyttner, L., (1996), "General Systems Theory" MacMillan Press, London. ISBN 0-333-61833-5.
27. Spur, G., Mertins, K. and Jochem, R., (1996). Integrated Enterprise Modelling, Beuth Verlag, Berlin.
28. Vallespir, B., Chen, D., Zanettin, M & Doumeingts, G., "Definition of a CIM Architecture within the ESPRIT Project 'IMPACS'". Computer Applications in Production Engineering: Integration Aspects, G. Doumeingts, J. Browne and M. Tomljaprovich (Eds.), Elsevier, Amsterdam, pp731-738, 1991.
29. Weston, R. H., Edwards, J. M. and Hodgson, A., 1995, "Model Driven CIM: A framework and toolset for the design and management of open CIM systems" January 1995, Final Report to EPSRC/CDP.
30. Weston, R. H., 1998, "A comparison of the capabilities of software tools designed to support the rapid prototyping of flexible and extendible manufacturing systems" *Int. Journal of Production Research*, 1998, Vol. 36, No. 2, 291-312.
31. Weston, R. H., Edwards, J. M. and Hodgson, A., 1998, "Enabling methods, tools and utilities for enterprise integration: Core project", July 1998, Final Report to EPSRC/CDP.
32. Weston, R.H., 1999(a) "A model-driven, component-based approach to reconfiguring manufacturing software systems", *Int. Journal of Operations and Production Management*, Special issue on Responsiveness in Manufacturing, Vol.19, No.8, 1999.
33. Weston, R. H., 1999(b), "Manufacturing Business Systems: Requirements and Current Practice" Manufacturing Business Systems module course notes.
34. Weston, R.H., 1999(c), "Reconfigurable, Component-Based Systems and the Role of Enterprise Engineering Concepts", Special Issue of *Computers in Industry* on CIMOSA. Ed. F. Vernadat and K. Kosanke. Volume 40, Number 2-3, November 1999.
35. Weston R.H., Clements P E, Shorter D N, Carrott A J, Hodgson A and West A A, (2000), On the Explicit Modelling of Systems of Human Resources, Special Issue of *Int. Journal of Production Research* on Modelling, Specification and Analysis of Manufacturing Systems, (Eds.) Francois Vernadat and Xiaolan Xie.

Putting the pieces together using standards

Ricardo Jardim-Gonçalves[a] and Adolfo Steiger-Garção[a]

[a]UNINOVA – CRI, Centro de Robótica Inteligente, Departamento de Engenharia Electrotécnica da Faculdade de Ciências e Tecnologia da Universidade Nova de Lisboa, Quinta da Torre, P2825-114 Caparica, Portugal

While integration of data between networked heterogeneous systems is currently an active area of research, the exact role that this issue has to play in the Agile Manufacturing (AM) domain and the specific benefits it has to offer, has not been well established so far. *Putting the pieces together using standards* intends to contribute in this field and addresses technical aspects related with the analysis, integration and validation of data and required mechanisms to connect Applications to Information Systems (IS) for AM environments. The integrating methods, architectures, models and toolkits presented in this chapter result from the work developed and in progress for industrial and research international projects (e.g., funStep and ECOS ESPRIT projects).

1. AGILE MANUFACTURING AND DATA INTEGRATION

Nowadays Information Society organisations are being confronted with new opportunities and challenges in the workspace and the market place. In a global networked economy, organisations will try constantly to shift the boundaries of their operations and collaborations to competitively exploit new business models and markets. In this scenario, industrial organisations are starting to seek Agile Manufacturing (AM) Environments to enable them to catch the resultant from these emergent new opportunities. But, to make it possible, specialised and dedicated Information Systems (IS) are required since they play a vital role [1].

The typical problem found in companies interested to join such environments, and slowing down the creation of AM Environments, is directly related with their previous investments in equipment and software. Usually focused to solve local and particular problems, it causes information segmentation and makes impracticable the functional integration with third parties, due to the incompatibility in data access and through data format representation. Even if conformance in data format and access is achieved and verified, in general reliable interoperability of information semantics is not [2].

At this moment the use of Internet is most of the time for direct access by humans, where mainly graphical information is exchanged (e.g., HTML) and not structured data. To use Internet as the network integrator channel for AM, enabling computers to communicate automatically among them, is an issue that has been worked out by the scientific community [3][4][5]. The lack of structured data models and tools able to be immediately adopted in

Internet-based environments is one of the main gaps identified. Utilisation of standards like STEP and XML seem to be able to open new possibilities in this area.

As a wide perspective, IS frameworks for AM should cover a horizontal approach (i.e., non-industrial sector oriented), based on a consolidation of proposals in complementary manufacturing sectors, where most of them presented and delivered results in a vertical perspective (i.e., industrial sector oriented), supported by parametrical models to be adjusted to the required scenarios of participating organisations [6].

Due to the inherent particularities of large size companies and small ones (SMEs), usually the same approach cannot be considered as a common one, and adaptations of the general framework's methods and tools must be examined and determined in order to fulfil the particular requirements of each kind of organisation [7].

A basic requirement to achieve the necessary flexibility in AM environments is through software [8]. Applications must be prepared with suitable mechanisms and interfaces easily adaptable for fast and reliable plug-and-play. Although software is typically more malleable than hardware, continuous changes and updates can gradually degrade it [9]. Therefore, a strategy for integration of software applications together with architectures to interface in AM environments, should be considered towards an open platform.

Data modelling, sharing and exchange, reuse of models, automatic code generators and software libraries, together with the possibility to incorporate expertise and knowledge representation, is a challenge to face when working in environments supported by heterogeneous platforms and concepts. To include mechanisms for conformance testing will faster assure interoperability among applications

The use of effective and *de facto* standards to represent data, knowledge and services has shown to be very important on the consolidation and generalisation of IS in this scope, helping interoperability between systems and the sharing of information, like product or business data between the participant parties. Real examples ready to be used in AM environments are: 1- the OMG standard [10], defining interfaces for services of a Product Data Management (PDM) system in a distributed and object oriented environment; 2- the STEP standard [11], defining the representation of product data to be managed by the PDM system, or; 3- the XML [12], for structured data exchange using internet.

Thus, the integration and mix of different standards and *de facto* standards become the basis to implement a complete and harmonised environment, although its aims, scope, suitability for the purpose and the possibilities for integration of different standards has to be explained and clarified for those aiming to adopt them to avoid misuse (e.g., in terms of data representation, UN/EDIFACT, STEP and XML: how can they interact, its benefits, advantages and drawback).

In the same way, software technologies should be integrated in an open and scalable architecture to support normative methods and mechanisms for system's interoperability (e.g., CORBA/IIOP, DCOM, RPC, RMI), code mobility (e.g., using JAVA) and information modelling (e.g., DTD, EXPRESS), exchange (e.g., STEP#21, XML), share and access (e.g., ODBC, SDAI, DOM).

2. TECHNOLOGIES FOR AGILE MANUFACTURING

Information compatibility and interoperability between systems is a requirement to achieve AM. The ability to efficiently share and exchange data, among the heterogeneous parties willing to be plugged to an integrated systems, is decisive to achieve a collaborative integrated world. Compliance in syntax and semantics play fundamental roles to make possible a complete and reliable information exchange (Figure 1). In this context, conformity on the data format exchanged and interoperability of data contents should be guarantied.

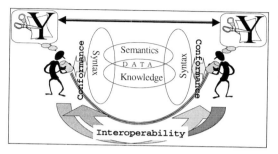

Figure 1 – Framework for compliant information exchange

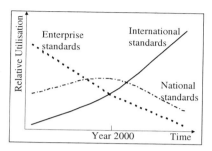

Figure 2 – Trends regarding International National and Enterprise standards

An essential issue to achieve this goal is the adoption and implementation of international standardised methodologies and technologies for data exchange, including those *de facto*, as a natural evolution from national and enterprises's (i.e., proprietary) ones. The recent technologies and methods available today are suitable to give an important help to support and assist this aim. Based on standards and standards *de facto*, these technologies will open the system's doors to the world, bringing plug-and-play mechanisms and facilitators to the parties.

Figure 2 depicts the trends in this evolution, based on a study done by PDES, Inc[1][13].

2.1 Data Exchange and Data Share

Data exchange and data sharing between systems can be effective only if both systems support compatible data models.

Systems based in standard methodologies and modelling can guarantee, under the scope of its business, better information flow and connection with third parties interested to deal with them [14]. This will not only give an automatic flow of information inside companies, but also a better capacity to external communication, as long as its parties adopt the same normalised methods and application protocols.

Data exchange and data share are differentiated by the fact that in the first case the owner of the data keeps internally a master copy of the data and give an instance of the exchange model to the requester, without any care concerning data synchronisation in respect to the original data, i.e., the requester receives the data and assumes it as its proprietor. In the second case, data is centralised and controlled in such way that all systems have access to the same data, and a policy for transactions is well defined to assure the communication and consistence of the data [11].

In both cases, to have standards as implementation methods for such environments gives the opportunity to have more systems able to immediately plug in it and join the consortium,

---

[1] PDES, Inc. is a corporation started in late 80s with the specific goal of accelerate developing and implementing ISO 10303 (STEP). Today, it incorporates several major U.S. and European companies and the American government agency NIST.

thought a standard is a system independent, well-documented and stable International agreement, thus attractive to be adopted as a confident investment.

In the advent of an explosion of standards for data model representation, special attention should be deserved in the possibility for overlapping of concepts covered by several of them. When the same concept is in the scope of different standard models, and consequently is described in different ways, harmonisation covering the different levels of specialisation and interpretation from those models has to be considered and a mapping strategy defined. If not, it takes the risk of inconsistency and difficulties in interoperability and reuse between those standards.

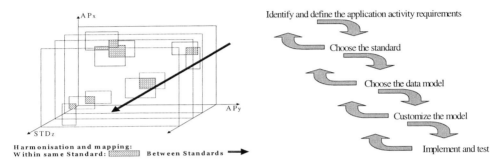

Figure 3. Overlap between APs within a standard and between standard data models.

Figure 4. Method to select a specific standard and a particular AP.

Figure 3 depicts this problem of overlapping and harmonisation between Application Protocols and Standards. The plan APx-APy defines the universe of APs for a specific standard. For the same standard, the overlapping between different APs is identified as dashed area in the plan APx-APy. Between standards, the same issue is identified in the intersection of APx-APy plan across axis STDz.

The method to select a specific standard and a particular Application Protocol (AP) can follow the steps described in Figure 4. Feedback should be considered to resume the selection process.

1. *Identifying and defining the application activity requirements.* Depict a clear understanding of what information is to be exchanged, the required frequency for exchange, inputs and outputs, mechanisms for control and the parts related with the activity. IDEF0 methodology is an example that can give a valuable help in this stage.
2. *Choosing the standard.* Deal with the trading parties and get an agreement on what standard to use and what mapping to adopt, if needed. Some examples are: STEP, for Product data representation, Edifact (mostly in Europe), X12 (mostly in USA) or XML for business data exchange.
3. *Choosing the data model.* Could be a STEP AP, an Edifact or X12 message or a harmonized model.
4. *Customizing the model.* A standard model in a specific context should be a superset of the needs. In this case an Implementation Convention (IC) should be agreed between trading

parties and the not used parts should be identified and not considered. Otherwise, some extensions can be required and added to the model after definition of a previous IC.
5. *Implementing and testing.*

2.2 Data Modelling and Representation

A conceptual model is an abstraction for computational realisation of a world of entities (e.g., physical, concept, relationship, method, constructor, fact, rule). It is through the realisation of such model that information can flow appropriately between the parties in an AM environment. The conceptual data model is thus the basic enabler for information transfer.

Heterogeneous distributed parties plugged in an integrated platform, will rely on a *global harmonised data model* to support the interoperability of its applications and services. This considers the control and management activities of a specific consortium agreement for a certain activity and in a point of time.

Several standards and *de facto* standards exist for data modelling. Next sections will present some of them though suitable for use in AM environments.

2.2.1. STEP– STandard for the Exchange of Product model data

STEP - STandard for Exchange Product model data [11][15], is an ISO (International Organisation for Standarisation) / TC 184 (Technical Committee: Industrial automation systems and integration) / SC4 (Subcommittee: Industrial data) international standard officially identified as ISO10303. STEP aims to provide a set of mechanisms capable of representing models and data related with the product during its life cycle [16][17].

It publishes a proposal for a methodology for development, implementation and validation of an open architecture for exchange and share of product data, together with a set of public data models as Application Protocols (APs) [18].

During the last years this community has been working in several definitions of APs for some of the recognised main production system areas, as are the automotive, aircraft, electrical/electronics, shipbuilding, oil and gas and building and construction.

The creation of a global model to support large-scale company requirements is at this moment one challenge to be address by the international scientific community. STEP has been presented as a viable alternative to the current state of multiple, incomplete, overlapped and proprietary data formats, seeking solid and reliable data exchange between partners using heterogeneous systems [19] [20].

STEP is mainly contributing for worldwide open systems networking communication of product data for neutral data exchange between heterogeneous systems, both in-house and with third parties, long-term archiving, system-independent architecture, flexible migration policies, and also contribution to paperless and life-cycle maintenance support.

The complete architecture of STEP can be found at *STEP On A Page* [18]. STEP is organised in series of Parts, each one devoted to a specific function on the global STEP's architecture:

- Overview – Part1
- Description methods – Part11
- Data specification –Application Protocols (Parts 200s), Application Interpreted Constructs (Parts 500s), Integrated resources (Application resources-Parts 100s, Generic Resources-Parts 41-99), Modules (Parts 1000s)
- Implementation methods – Physical data file (Part21), Data access methods (Parts 22-29)
- Conformance testing – General concepts (Part31), Requirements for Test Labs and Clients Test Methods for File and data access methods (Parts 32-35), Abstract Test Suites (Parts 300s)

2.2.2 XML - eXtensible Markup Language

The eXtensible Markup Language (XML) is a meta-language describing a standard format devoted for structured data interchange over the Internet. XML has being developed by the World Wide Web Consortium (W3C) and results in a flexible extension of HTML (Hypertext Markup Language), which has being universally adopted as the standard for data web browsing over Internet with a pre-defined fixed data structure using tags [12].

Specifically created for flexible data presentation using a tag-based mechanism, XML gives the possibility to have customised data models for structured data interchange over Internet using a human-readable format, with the possibility of definition of new tags.

For validation, each XML data document can have a pointer to its data structure as a Document Type Definition (DTD). The DTD provides the metadata description corresponding to the data format in the exchange file. Thus, after parsed the XML data can be search out and interpreted by applications via its API called Document Object Model (DOM) [21].

Most Internet browsers are supporting now XML or plan to do it very soon. This means that XML-based solutions in an Open Platform environment can reduce set-up costs, enable prompt integration with any browser-enabled system, making possible immediate plug-in and data exchange with a wide automatic human interfacing facilities.

On the other hand, XML can enable product data exchange as an alternative to the STEP neutral format Part#21, as an encoding of the STEP schema instance, and a new part of STEP, its Part#28, is in this moment being worked out by this community.

Not formerly created to handle very large files, as it is the typical case for example for the exchange of geometric data, XML emerges as an appropriate choice for exchanging business or product data in all other cases

The possibility to have STEP embedded in XML, integrating all conformance and interoperability testing for instance, as a reference link for product description in a business data exchange message described in XML, will bring new opportunities for integration of heterogeneous environments. The same could happen with EDI in terms of its messages representation, where XML has the possibility to enable easier Web-based EDI solutions [22][23].

XML does not provide a standard mechanism for displaying the data contents of an XML data file, as there is in HTML. This separation between data definition and data presentation makes easier the use of XML for different purposes. Anyway *stylesheets* for XML can be defined for XML data presentation, and there are several implemented interfaces already defined for these languages in some Web-browsers (e.g., CSS and XSL).

The XML Stylesheet Language Transformations (XSLT) is a programming language devoted to transform XML data files [24]. An example of that are transformations from XML to HTML to enable immediate visualisation of its data by any of the popular Web-browser. Other examples are the use of XSLT to map XML to another standard or to run rules for verification, validation and report on its data.

2.2.3 EDI - Electronic Data Interchange

The acronym EDI, Electronic Data Interchange, is usually related with the use of general electronic transfer of business data from one independent computer system to another, using a standard data format as defined by the ANSI X12 [25] standard (mainly in the USA) or by the UN/EDIFACT (United Nations/Electronic Data Interchange for Administration, Commerce and Transport) [26] standards.

STEP and EDI are viewed as complementary standards, one addressing product data the other business data, and already there were some projects launched aiming to interconnect and harmonise them, willing do it without any modification on any of these standards [27].

2.3 Distributed Object Oriented Data Technologies

Object-orientation is founded in the premises that objects could model the world. One object is realised as one instance of an *Entity* or *Class of objects*, which describes everything about the object, e.g., its structure, functionalities and methods for data access. In short, it represents the Abstract Data Types (ADT) containing object's data structure and functionalities.

Its methods represent the interface for the objects and through mechanisms of inheritance, objects can be specialised extending its features, providing reusability. This new class, usually called inherited-class or sub-class, inherits the structure and behaviour of the original class, providing a complete extension based on reusability.

Objects behave as a black box where exchange and assignment of data is done via messages using its methods. Therefore, messages define the interface to the object and everything an object can do is described by its message interface.

In classical programming languages (e.g., PASCAL, C), code and data have been developed separately and are not formally connected, where functions can operate on more than one type of structure, and more than one function can operate on the same structure. Although not having automatic object-oriented facilities behind them, it is possible to use in such languages the object-oriented paradigm.

Services and interfaces to support the distribution of objects have been established to supply a common platform for heterogeneous environments. Next sub-sections present some examples.

2.3.1 CORBA/POA/IIOP

Common Object Request Broker Architecture (CORBA) is a standard specification developed by the Object Management Group (OMG) consortium to support communication of objects on one machine to other objects running on same or different machines in a distributed and heterogeneous environment [28].

OMG is one of the biggest open consortia in the world and its members are mainly software vendors, software developers and users interested in object-oriented standards for integration of applications in such environments [10]. Nowadays, OMG specifications have been used in sectors like Manufacturing, Health care and Electronic commerce.

CORBA is represented by a set of libraries that implement broker functions and invocation conventions over a network, and a compiler for Interface Definition Language (IDL), the

CORBA's language to define interfaces that generates interface specific bindings of the communication libraries to specific programming languages.

An IDL specification defines the Application Programming Interfaces (API), describing in a standard way the specific mechanisms for access and manage of distributed objects. Therefore, the distributed system can be constructed in an independent manner, since object's interfaces are well defined and described in a normalised way.

OMG defines CORBA consisting of four groups of entities [29]: the Object Request Broker (ORB), the CORBAservice, the CORBAfacility and the CORBAdomain. The ORB is a specification for distributed objects communication, and the remainder are functionalities. These four groups of entities provide the required mechanisms to put objects communicating with each other in distributed heterogeneous platforms together with ready-to-use functionalities and proving different levels of abstraction and covering different domain areas (e.g., accounting, user interface).

The CORBA's Basic Object Adapter (BOA) is a standard specifying how an object is activated in a CORBA environment, and acts as the communication manager for the objects with the ORB. With new version 3.0 of CORBA, BOA is being replaced with the Portable Object Adapter (POA), giving solution for some problems existent in BOA like portability across ORB implementations [30].

CORBA 2.0 introduced the Internet Inter-ORB Protocol (IIOP), allowing objects developed for an ORB to communicate with objects developed by other ORB over TCP/IP [31]. This was an important issue to bring interoperability to CORBA environments towards open platforms.

2.3.2. PDM enablers

Product Data Management (PDM) Schema is a harmonised STEP data model developed with the aim to provide a consistent mapping from a set of Application Protocols addressing PDM issues [32]. PDM enablers are an example of an API described in IDL working on top of the PDM Schema. These API intends to cover the most important services within engineering activities. Examples of such services are Document, Product Structure, Configuration, Manufacturing management and Import/Export mechanisms of STEP exchange files.

The specification of PDM enablers releases a set of standard services for CORBA-enabled environments. Accordingly, any system that adopts this API under a specified standard network interface can cooperate immediately with this environment and use it, taking the advantage of such services.

PDM enablers are a good example of complementary work based in two standard technologies, where the add-on provided by a standard layer on the top of standard-based data representation brings a standard way for the access and exchange of data realised as services for distributed and heterogeneous environments.

2.3.3. C++

The programming language C++ is a general-purpose language [33] that emerges as an extension of the language C to support the object-oriented paradigm. Consequently C code is compatible and can be incorporated into C++ programs.

C++ is an executable language, thus platform-dependent, and uses compile-time (i.e., early) binding of its data structures. Although this characteristic allows generating high run-time efficiency and small executable code size, this means that classes of objects are specified at compile time and its structure and behaviour will be fixed during run-time.

C++ is at the present an extremely popular programming language, and now most of new C compilers are actually C/C++ compilers.

2.3.4. JAVA/Java Beans

Java is an object-oriented language started to be developed by Sun Microsystems about one decade ago, with the objective to be a portable programming language enabled to directly run on Web browsers and computers with direct access to Internet [34].

Java syntax is very similar to that C++. Its main extensions are the garbage collection facilities, a feature that automatically manages the allocating and de-allocating application's memory, and to be an interpreted language running on a virtual machine, the Java Virtual Machine (JVM).

This means that Java programs generate run-time portable code, the *bytecode*, and can run without recompilation on any platform that supports the JVM. This platform-independent characteristic allows that a program written once can run anywhere over Internet without recompilation, facilitating and stimulating the mobility of objects and functionalities over the net.

Due to its suitability to be used over Internet environments and with Web-browsers, Java has gather more and more enthusiasts, becoming now a promising programming standard for Internet and Intranet environments, and have been supported by most of the popular web-browsers.

A JavaBean is a component consisting of one or more Java classes, which makes public its structure and mechanisms [35]. It provides a common interface for creating and working with components using Java. Similar to other component technologies, JavaBeans are designed to have an easy interface and to be reusable in a distributed and heterogeneous environment.

To maintain a repository of such components ready for use allows software to be developed at a higher level, reducing significantly the time and effort required to develop applications and facilitating the development of new JavaBean components.

The possibility to have JAVA interoperating with other standards, e.g., STEP, is an important issue to stimulate the usage of such standard over Internet. An example of it is to have a JAVA application or component for product visualisation over Internet with data represented in STEP neutral format. To be used in these cases, a JAVA binding for STEP SDAI has been developed [36], and first releases start appearing in the market.

2.3.5. Sockets, RPC and RMI

Sockets, Remote Procedure Calls (RPC) and Remote Method Invocation (RMI), are technologies devoted to support the exchange of data in a network.

Socket is a mechanism letting you send and receive data in datagram format messages over a network RPC allows remote data exchange based on procedure calls in a non-object oriented paradigm. RMI extends the RPC concept supporting the object-oriented paradigm [29]. All these three technologies represent in the presented order a growing level of abstraction in terms of data exchange support for distributed heterogeneous environments.

Available to be implement under several programming languages, sockets are at lowest level with better performance but less functionality. RPC and RMI are at a higher level of abstraction for remote procedure and method invocation. Although RMI is only available for JAVA, it provides some more useful mechanisms than RPC, like Java serialization facilities to send and manage objects sent as arguments and return values over the Internet.

RMI also provides a registry service maintaining the names of the remote objects, allowing that remote objects be used as local ones, and implements a distributed mechanism for garbage collector to keep track of remote objects.

2.3.6. ActiveX/DCOM/ATL

ActiveX is the name that Microsoft has given to its object-oriented program technologies and tools, which the Component Object Model (COM) is the key one. When enabled to be used in a distributed network, the COM is known as Distributed Component Object Model (DCOM). DCOM is a technology developed by Microsoft for its platforms with similar CORBA objectives, although at present only Windows and Macintosh systems support ActiveX [37].

An ActiveX environment is formed by ActiveX components, running as independent and autonomous programs that can be executed anywhere over an ActiveX-enabled network. Each component is known as an ActiveX control and works like a Java applet. An ActiveX control can be created in any programming language that recognizes Microsoft's COM, and be reused by any other applications in the same computer or in others in a distributed network. Any application program that uses the COM program interfaces can run an ActiveX control and stimulate its reuse, like in objects for WEB pages in Internet applications.

The Active Template Library (ATL), formerly called ActiveX Template Library, is a Microsoft program library to be used in developments of ActiveX program components in C++ [38]. ATL is designed to automate most of the work associated with the development of COM and support the creation of fast and memory-efficient objects.

ATL intends to improve the implementation of COM components, substituting the usage of Microsoft Foundation Classes (MFC), since MFC produces large code for each component, slowing down the process of its download via Internet.

2.3.7. CGI/Servlets

In a Web-based environment, when a user using a browser requests to the server a Web page, typically the server sends back the requested page without any additional processing. However, when the user would like to get data after some processing in the server side, typically the Web server passes the client's input to an application program running in the server side that processes the input data and send back the correspondingly response.
The Common Gateway Interface (CGI) is part of the Web's HTTP protocol, and it is a standard way for a Web server to pass a request from a client to an application program and send back its response through the Internet to the user [39].
When a Java Virtual Machine is running in the server, such applications can be implemented in the server side using the Java programming language, and are thus called Java servlets [29].

3. SUPPORT FOR ADOPTION OF STANDARDS FOR DATA EXCHANGE

Nowadays the ability to share and exchange data quickly and efficiently among different computing environments is decisive though in agile environments. The representation of such data in digital format is the basic technology to support that, and the use of standards to represent it, is a requirement to achieve the intents [40].

One of the main problems usually found not contributing for an easy adoption of standards for data exchange, is the complexity and effort required to develop the interfaces to adopt it. This is truer for large environments where models for data exchange are very often updated and flexibility in its interfaces is required. Therefore, to be possible to keep all parties plugged in the integrated environment updated, interfaces should be bring to date faster in order to take the full advantage of the integrated system [41]. Figure 5 depicts the general scenarios for integration of two Applications.

In order to give some support for the development and update of these interfaces for standard-based data exchange, standards include Implementation Methods (IM) together with standardised data access interfaces (SDAI).

Besides that, after developing the interfaces, validation procedures should be performed to assure conformity of the application with the adopted standard. Also, to assure a complete interoperability with this application with all other parties integrated in the global system, verification and validation should be performed with all parties in order to assure a reliable exchange of information conforming with the standard, as syntactically as semantically.

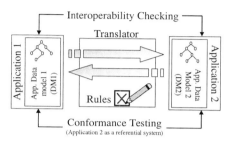
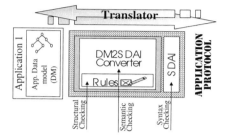

Figure 5 –Integration of two Applications using a translator.

Figure 6 – SDAI and the developments of translators adopting a Standard.

3.1 Interfacing with Standards and Standards "De Facto"

For one application to adopt a Standard for Data Exchange, it should implement the interface that bridges the Application's data represented usually in proprietary format, translating such proprietary data format to the Standard's.

In order to make easier the development of these translators, an Application Programming Interface (API) is normally defined by the Standard, providing a description of the required basic mechanisms to support the implementation of the translator for data described by the Standard model.

Defined as a Standard, this API works as the normative layer establishing the communication between the data objects themselves and those objects implemented by the application, and is very often called Standard Data Access Interface (SDAI). Figure 6 depicts the role of SDAI in the development of a translator between an Application internal data format and a standard Application Protocol.

To those interested to adopt a standard, the importance to have this APIs tested and ready to be used and linked by applications is enormous, since it will facilitate its adoption in a convenient level of abstraction without the requirements in terms of details of implementation for those developing the translators for import and export of data.

For Application1 to adopt a standard Application Protocol (AP), it needs to develop a translator for data import and export. Such translator can be developed on the top of a SDAI that provides the basic mechanisms to access and manage the data in Standard format. Therefore, the required development is thus the Application Data Model to SDAI converter (DM2SDAI), which links with the SDAI to have direct access to the Application1's object data [42].

Another important feature a SDAI should provide is the possibility to support early and late bindings of the Application Protocol for the programming language. Early binding means that SDAI has a static structure dedicated exclusively to manage the specific AP for what it was developed. Late binding is based in a general implementation of SDAI assisted by a dictionary of data ready to support any application model, and managed in run-time.

While the early binding approach is static, although with an easier and faster interface for implementation purposes, the late binding is very flexible allowing run-time model changes and updates.

Examples of these APIs are the STEP's SDAI and the XML's DOM.

3.1.1. STEP's Standard Data Access Interface (SDAI)

The ISO10303 STEP describes its Standard Data Access Interface (SDAI) in its part #22, specifying its functionalities in a general and neutral way independently of a programming language [43].

Others parts of this standard (e.g., Part#23, Part#24), provides a description of SDAI binding for specific programming languages (e.g., C, C++), that when implemented are ready to be linked with the Application using such programming language.

This API enables the management of the data structures for the standard model, and provides the functionalities to handle and instantiate its objects. Objects and data dictionary are managed in an SDAI-repository, at attribute and entity level.

The export of data in standard format is done using specific procedures, and reflects the data stored in the SDAI's repository. The import of data using the SDAI specific functions allows reading data in standard format and populating the SDAI's repository to be accessed afterwards through SDAI mechanisms.

3.1.2. XML's Document Object Model (DOM)

The Document Object Model (DOM) is an API to access data represented in XML format [21]. This API understands the XML data described as a tree-based representation, and defines the mechanisms required to navigate across such tree in width and depth. They enable access and handle of its elements and attribute values as tree data nodes, allowing insert and delete of such nodes, and the conversion of the tree structure back into XML data format.

The root of the tree is the XML document. Each root's child represents the top-level instances of XML data Elements. Each Element can have related attributes and other Elements as children nodes, representing the data content and its sub-elements, which may have also children, and so on.

The DOM is thus a useful tool to handle and manage XML data format files, viewing its structure mapped as a DOM tree. Modification or production of new XML documents as output, or construction of a new DOM tree by beginning and after convert it to XML, can be done using this API. This mechanisms provide a very flexible way of access and produce XML data format output, usually easier than simply writing or reading directly to a file in that format.

3.2. Validation, Conformance and Interoperability Testing

Conformance and interoperability testing are procedures that should be performed to validate and assure the quality of the global integrated system, as a monitoring procedure or when a new party is plugged in. This issue is even more sensitive when different standards (e.g., Application Protocols) are concerned, and semantics and harmonisation of concepts and structures have to be realised.

Validation is directly related with the fact that erroneous assumptions in the early phase of development (e.g., data modelling) will cause correction work in later stages and consequently worst quality. To guarantee the syntactical correctness of a data model with a standard is not the main problem because it comprehends the complete formal description of its methods and grammar, and parsers are usually available.

The main difficulty is typically related when dealing with semantics. The more semantics are included the more complex the conceptual data model is. Thus, standard models have been including little formal semantics to avoid not realistic implementations due to the complexity of its representation. Together with difficulties related with the extension of the model not giving easily a complete overview of the global model, this causes too much freedom in the model interpretation by the user.

From the architecture of Figure 5, the Syntactical validation is performed by SDAI, and provided as a library ready to be used and adopted. The DM2SDAI implementer will assure the structural validation (e.g., check of array bounds, attribute value out of type range) and verification of semantics.

4. PUTTING THE PIECES TOGETHER...

In the new and emerging Agile Manufacturing environments, where multiple heterogeneous applications should cooperate in a non-permanent style, an open architecture for integration of information where applications are able to be integrated in a fast way is a requirement to realistically enable the plug of such application during the time of the join contract [44].

The information exchange should be based in an harmonised set of standard data models, i.e. Application Protocols, covering the purpose of the contract, and to be adopted by the parties attracted to join such AM environment. Is through a platform adopting data structured according to this integrated data model that information will flow within AM environment, and the "pieces will be putted together" [45]

4.1. Information Model and System's Architecture

The information model plays a key role in an integrated environment. The usage of standards to describe it will open the platform since any other application can have access to a standard and know how to adopt its applications and prepare to be ready to join the AM environment. This model should describe the AM consortium requirements and to assure that these requirements are fulfilled by the standards adopted.

A mixed of different standards (Figure 7) can be adopted if this is the best way to cover all the requirements (e.g., STEP and XML). The mix of such standards should be assembled in a consistent manner, assuring interoperability and conformance between them.

Methods and tools to implement the model must exist already tested and ready to use, as well mechanisms and tools to validate the coming applications adopting the standard model, assuring that applications are reliable, semantically compatible, and compliant to the model specifications.

Tests to verify compliance of applications willing to join the AM environment should be performed to assure compatibility in terms of data exchange among parties and reliability of the global integrated environment.

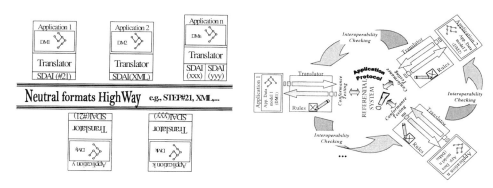

Figure 7 – Typical environment mixing Neutral Formats for Data Exchange.

Figure 8 – Environment for Conformance Testing and Interoperability Checking.

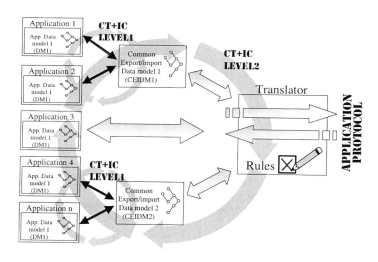

Figure 9 – Hierarchical or multi-level Integration of Applications

These tests should be performed at two levels.
- First, there is an Application certification resulting from Conformance Testing (CT) procedures against a Referential System (RS), verifying compliance of such system in terms of structure and semantics of data exchange.
- Second, Interoperability Checking (IC) between all applications joining that environment. This is normally an arduous task implying availability from all other parties to run the

complete testing process. Although seen as a complement from the CT procedures, this task is important to better assure a reliable integrated environment among all parties.

Figure 8 shows the typical environment for Conformance Testing and Interoperability Checking.

4.2 Linking the Parties

To link in a correct manner the parties willing to join an open environment is the central aspect for the success of an AM environment. A fundamental piece is the development of the translators enabling the applications to adopt the standards supported by this environment.

Though that some of the applications constituting such environment can be already interoperable, two levels of integration should be considered (Figure 9):

1- The first level consists using already existent import/export facilities by some applications for a specific data format, and integrate those applications using it. Afterwards, to develop a translator to convert such format to the AM environment's standard one.

2- For those application not already supporting common import/export facilities, they should develop a translator to convert directly from its internal data model to the standard.

In this scenario, also two levels of Conformance Testing and Interoperability Checking should be considered. One at level of the already existent import/export mechanisms, usually already previously tested. The other, at the level of the new developed translators, this now to be tested.

4.2.1 Architecture of the translator

As depicted above in Figure 6, the translator is the entity responsible to convert the data represented in a Data Model format (e.g., Application DM) to another data model format (e.g., Application Protocol). When translating to a Standard model (e.g., STEP), very often the standard already provides the Application Programming Interface for data access, known as Standard Data Access Interface (SDAI). Therefore, the development required is the DM2SDAI module.

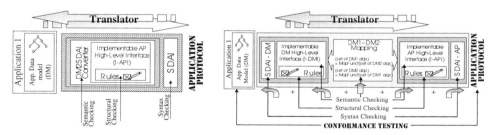

Figure 10 – Architecture of translator using I-API   Figure 11 – Global Architecture of the translator

The DM2SDAI module must be able to access the Application data in DM format and convert it to SDAI commands for AP neutral data format representation, and vice-versa. Also, the DM2SDAI should handle and interpret the AP model rules, for validation and compliance of the data exchange. Beside data conversion, the DM2SDAI module should also assure the structure and the semantics of the converted data.

The architecture of DM2SDAI should consider the existence of a high-level interface on the top of SDAI (I-API, Implementable AP High-Level Interface), that implements the AP data structure based on the AP model declaration in a programmable way. The programming language used should be one compatible with the Application and SDAI. This I-API works as an Abstract Data Model (ADM) for the AP, mapping directly with the AP and providing the structure and functionalities required for the data handle and validation of rules.

Using I-API for export of data, the translator's developer needs to instantiate ADM's objects from the correspondent class of objects represented in the AP model, and then to call the respective methods for data access and conversion.

YFor import, it needs to ask for a specific object, or set of objects, represented in the AP neutral format, to be uploaded and represented as instantiated objects of the I-API ADM.

The validation process is done calling the respective rule checker on the instantiated objects of the ADM structure. (Figure 10)

Detailing the architecture of the translator towards its global overview, on the Application side there is exactly a similar high-level interface on the specific data access interface for the data represented in the Application's Data Model format. The construction and procedures of such I-API are exactly the same as it is for the AP model, where the High-level interface now represents the ADM for the Application data model.

Therefore, to link the Application ADM to the AP ADM it is needed to implement mapping procedures. This mapping establishes the transformation rules between one ADM to the other, in a bi-directional way. To achieve it, the mapping policy should be defined, for instance as a table, and during the conversion process a set of selected objects represented in ADM1 format are transformed through the respective mapping function to a set of ADM2 objects. In the reverse, the same is valid.

The mapping task is one of the most importants for the translator, since it reflects the conversion of the data structures, where the data semantics and structural rules must be conserved during the transformations process. Figure 11 shows the global architecture of the translator.

As a summary, the translator role is to convert the data represented in one format to another. It should be developed in such programming language prepared to link directly to the Application requiring it. The SDAI modules are the low-level interface with the native (i.e., Application) data model, providing the required mechanisms for an easier handling and management of the model data. SDAI is the first level front-end with the native data structures.

On top of SDAI, a high-level interface (I-API and I-DMI) is developed on both sides as an Abstract Data Model of the Application models. This second level interface reflects the application model described in a declarative format using a programming language, plus the mechanisms required to import, export and validate the data from the native format to this intermediate one. Due to its higher level, it becomes more suitable for programming and manipulation for further integration with the required mapping procedures.

The abstraction of the second level is important since it puts the data to be exchanged ready for mapping, after a previous translation if needed, from the native format to the intermediate one.

The mapping module bridges the two second levels, performing the data transformation between the ADMs.

This translator's architecture also allows a multi-level policy for conformance testing. Verification of conformance starts at SDAI level, mainly doing syntax checking with the native data structure. When compliant, conformance is verified on the second level for I-API and I-DMI, in this case already including structural and semantic validation. When conforming, testing procedures will act on the Mapping module, for complementary semantic compliance verification. At the end, a final test should be performed at Applications level for conclusive conformity and certification of the translator. Figure 11 and Figure 12 depict these levels of Conformance Testing.

The integration of two Applications can follow a same approach (as depicted in Figure 8) or if willed by both parties, developing a unique translator for direct link as depicted in Figure 13. Interoperability checking should also be done for data exchange compliance.

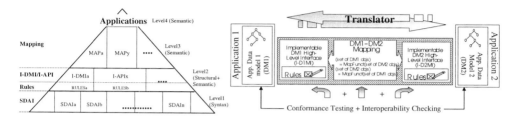

Figure 12 – Levels of Conformance Testing

Figure 13 – Integration of two Applications using the proposed translator.

4.2.2. Automatisms

The integration of Applications in an Open Platform requires the development of translators to enable such Application to adopt the common data format (i.e., Neutral format) defined for data exchange. These translators, as explained in the previous sections, are very often not easy to develop. Main reasons for such difficulty are the complexity of the standards for data representation being adopted, and the amount of software required to develop the translators for data conversion.

To provide automatically parts of the translator can accelerate the development of converters, reducing the amount of software required to develop, and reducing the number of errors associated with the production of such software.

The architecture proposed in Figure 11 shows a mirrored system, where 5 main modules are identified, i.e., SDAI-DM, I-DMI, DM1-DM2 Mapping, I-API and SDAI-AP.

The SDAI-AP comes usually in toolkits available in the market ready to support the low level interface with the import/export mechanisms for standard data model adoption. Such portion of software comes already tested and certified, assuring a correct implementation of the standard through its import/export mechanisms.

The I-API module is the piece of software intended to implement the Abstract Data Model from the declarative Application Protocol model. It represents in a programming language the structure of the AP plus the set of methods and mechanisms to populate its data structure and validate semantics. I-API is then the programmatic High-level interface corresponding directly to the AP model description. It will represent an image of the AP data objects as instances of the ADM. The production of I-API modules can be generated automatically using an automatic code generator [46][47][48] (let's call it *Genesis*) that, reading the AP meta-data dictionary described using a standard-based Data Model Language (e.g., ISO10303 EXPRESS), creates automatically the correspondent ADM plus a set of pre-defined access methods and functionalities (e.g., for import/export or rules validation)(Figure 14, Figure 15).

Having a process to certify the I-API generator (i.e., *Genesis*), assuring the code generated is able to produce data compliant with the AP, will guarantee that anyone using the generated ADM will produce/consume compliant AP data. Using such code will reduce implementation errors and speed up the certification process of the complete translator.

On the Application side, there is also a SDAI (i.e., SDAI-DM) for low-level interface with the internal Application data structure, provided by the Application developer.

The I-DMI module should be also generated automatically using the *Genesis* tool, giving for its input the description of the Application DM using a Standard-based Data Model Language.

The mapping module is one of the most sensitive and intense tasks to perform to establish the conversion process. Having a process to describe such mapping using a specific language (e.g., tabular form with associated transformation functions, or EXPRESS-X), enables the possibility to have a tool (i.e., MapData) generating automatically part of the code for mapping procedures required for the Mapping module. Having a way to certificate MapTool, will assure the Conformance of the translator as explained for the *Genesis* case.

By means of this approach using certified automatic generators of software to assist the development of translators will accelerate drastically its development and assure a faster way to have them certified and compliant through an AP.

Since all other translator's modules are already partially generated automatically and certified, therefore the main effort in the development of the translator will be focused on the SDAI-DM that should be developed by the Application developer, devoted to provide the low-level interface to access its internal data.

To develop it and certify it is thus the main effort required.

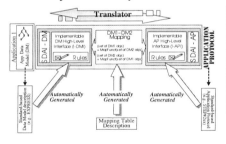

Figure 14 – Role of Automatic Code Generators for development of translators

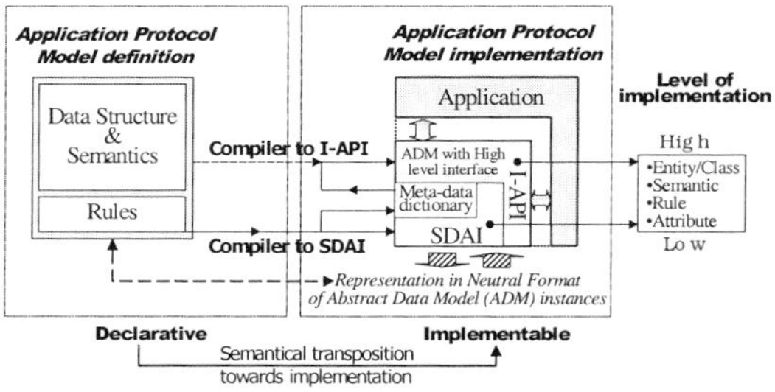

Figure 15 – The general architecture of Genesis tool

5. CASE STUDIES

The work presented in this chapter results mainly from the bibliographic compilation, research and developments done by the authors contributing for the results of the European projects presented in summary below. Details are available from their public reports and deliverables.

5.1. CIMTOFI Project

The basic aim of the BRITE/EURAM CIMTOFI (CIM sysTems with improved capabilities fOr Furniture Industry) project was to reduce the "Idea-to-Production" cycle for furniture product development, by integration of tools selected in the market [49]. Tools and methods developed and implemented in this project intended to contribute for the introduction of the CIM philosophy in a step-by-step way, not only in furniture industry, but also in other industrial sectors. The tools integrated were modelled using EXPRESS, and the integration tasks supported by a Standard-based Integration Platform called SIP.

5.2. ROADROBOT Project

The main goal of the ESPRIT III RoadRobot (Operator assisted Mobile Road Robot for Heavy Duty Civil Engineering Applications) project was the development of a generic architecture to be used to design and develop various modules of an automated road construction site [50][51]. A road paver and an excavator working on the field were used to test the integrated system functionality. RoadRobot's information models were described in EXPRESS and the information integration platform was based in such developed for CIMTOFI project.

5.3. FUNSTEP Project

The ESPRIT IV funStep (Development of a STEP-based Environment for the Manufacturer-Customer Integration in the Furniture Industry) project aims to develop a general integrated STEP based environment for the manufacturer-customer integration and for the support of the management information flows in the factory, applied to the furniture industry [7][52][53][54]. The project have been developed a STEP based specification of information models regarding product development, customer orders processing, process planning, and customer's project management. While the modelling work developed during the CIMTOFI project addressed the manufacturer process (internal to the factory) of the furniture industry, the funStep project addresses the business one (external) where several applications have been adapted and integrated using a STEP based integration platform.

5.4 ECOS Project

The ESPRIT IV ECOS (Lite e-commerce operative scaleable solution for SMEs) is a 30 months project started in late 1998 [6]. ECOS aims to develop a set of services and software tools to implement standard-based data exchange to assist in electronic business activities among industrial collaborators in the furniture sector chain, i.e. material suppliers, furniture manufacturers and commerce. This project has considered existing standardisation initiatives like ISO/STEP for product data modelling, and UN/Edifact and XML for business related data exchange. The available technology at the SMEs and the existent internet-related technology including data security and privacy have been used and adapted when required. ECOS has a vertical approach (furniture sector oriented), although its architecture based on that developed during the previous presented projects, can be applied to other sectors, resulting so in a set of services and software tools to be exploited regardless industrial sector.

6. CONCLUSIONS AND FUTURE RESEARCH DIRECTIONS

This chapter presents technologies able to support the integration of information for AM environments, adapting and integrating methods, architectures, platforms, models and toolkits resulting from pilots and demonstrators from industrial and research international projects (e.g., funStep and ECOS ESPRIT projects).

At the present time several technologies are suitable to help the design and construction of open platforms for AM environments. Some of these techniques and methods seem appropriate to set-up and construct open platforms to assist standard-based data exchange and data share on distributed and heterogeneous environments.

One of the major tasks when integrating applications in such open platforms is the development of translators to convert the application's internal data format to those formats adopted by the platform. To have methods and tools ready to be used to automate the development of these translators is an important challenge to consider. Its use saves in development time and guarantees compliance with the standard models adopted in a more rapidly way, enhancing the automatic incorporation of one application with all others already integrated in the environment.

Demonstrative pilots have been developed during last years intending to demonstrate the aptness of use of such technologies for these environments, although further effort have still to be done to reach a referential architecture for real AM environments.

Clustering of organisations and industrial interest groups (e.g., funStep Interest Group [55]) supported by industrial association and research and development centres (e.g., International STEP Centres [56]), plays a fundamental role to drive and accelerate the creation of such environments in the industry.

6.1 Future Research Directions

Future research directions includes:

- Development of integrated meta-protocols with direct reusing of existent APs, including its Abstract Test Suites and Conformance Testing methods, seeking a global enhanced reusability and effectiveness in the production of standard data models.
- Improvement of the automatic development of application's translators.
- Design and development of dedicated software on the top of solid and already proved platforms, like Relational Data Bases with Object Oriented layer on the top, as the basis for the design and set-up of recognised open platforms for AM environments.
- Support for reliability and operability of the AM environments, through Quality assurance of Modelling and Interchange of information through Conformance testing and Interoperability checking, for virtual standard-based integrated platforms and environments.
- Study, analysis and statement of cases.

Acknowledgements

The authors of this chapter would like to thank, without exception, all FCT/UNL and UNINOVA colleagues and the partners working with them in the CIMTOFI, RoadRobot, funStep and ECOS projects' consortium and, in particular, to acknowledge the European Commission by the financed budget and its support and trust in their ideas and developments.

REFERENCES

1. Camarinha-Matos, L. and Afsarmanesh, H., 1999, *Tendencies and general requirements for virtual enterprises*, IFIP Working Conference on Infrastructures for virtual enterprises, Porto 1999, Chapter 2, Kluwer Academic Publishers, ISBN 0-7923-8639-6.

2. Jardim-Gonçalves, R., et. al., 1997, *Implementation of computer integrated manufacturing systems using SIP: CIM case studies using a STEP approach*, International Journal of Computer Integrated Manufacturing special issue on Design and Implementation of CIM systems, vol. 10, no. 1-4, pp172-180, Taylor & Francis, ISSN 0951-192X/97
3. Jardim-Gonçalves, R., Sousa, P., Pimentão, J.P., Steiger-Garção, A., *Furniture commerce electronically assisted by way of standard-based integrated environment – the ESPRIT funStep project proposal*, 1999, pp. 129-136, ICE'99 - 5th International Conference on Concurrent Enterprising, The Hague, Netherlands, CCE-DMEOM, UK, ISBN 0-9519759-8-6.
4. Osório, L., Antunes, C., Barata, M., 1999, *The PRODNET communication infrastructure*, IFIP Working Conference on Infrastructures for virtual enterprises, Porto 1999, Chapter 11, Kluwer Academic Publishers, ISBN 0-7923-8639-6.
5. Ducroux, F., *The IT integration supporting the extended enterprise*, 1999, pp. 137-145, ICE'99 - 5th International Conference on Concurrent Enterprising, The Hague, Netherlands, CCE-DMEOM, UK, ISBN 0-9519759-8-6.
6. http://www.funstep.org/funstep/ecos/ecos.htm
7. http://www.funstep.org/funstep/
8. Kim, Yoohwan, Podgurski, A., *A flexible software architecture for Agile Manufacturing*, 1997, IEEE International Conference on Robotics and Automation, Albuquerque, New Mexico, USA.
9. Quinn, Roger and Causy, Greg, *An agile manufacturing workcell design*, IIE Transactions on Design and manufacturing, Special Focused Issue, 1995.
10. http://www.omg.com/
11. ISO TC184/SC4, *IS - ISO 10303, Part1 - Overview and Fundamentals Principles*, 1994
12. http://www.xml.com/pub/98/10/guide0.html
13. http://pdesinc.scra.org/
14. Clements, P., *Standard support for the virtual enterprise*, International Conference on Enterprise Integration Modeling Technology – ICEIMT'97, 1997, Torino, Italy. http://www.mel.nist.gov/workshop/iceimt97/pap-cle2/stdspt2.htm
15. *Introducing STEP – The foundation for product data exchange in the aerospace and defence sectors*, Government of Canada, 1999, ISBN 0-662-64382-8.
16. ISO TC184/SC4, *IS - ISO 10303, Part11 – EXPRESS Language reference manual*
17. ISO TC184/SC4, *IS - ISO 10303, Part21 – Clear text encoding of the exchange structure*
18. http://www.mel.nist.gov/sc5/soap/
19. Starzyk, D., *STEP and OMG Product Data Management specifications – A guide for decision makers*, OMG Document mfg/99-10-04 and PDES Inc. Document MG001.04.00, 1999. http://pdesinc.aticorp.org/whatsnew/step_omg_guide10_99.pdf
20. Hars, A., *Better control with PDM*, Byte Magazine, February 1998.
21. http://www.w3.org/TR/REC-DOM-Level-1
22. Rawlins, M., *Future EDI-An overview of emerging technologies which might replace X12 and Edifact*, 1998, The Journal of Electronic Commerce, Spring, ISSN 972-783-9573. http://www.metronet.com/~rawlins/edifutur.doc
23. Raman, D., *XML/EDI Cyber assisted business in practice*, 1999, TIE Holding NV, ISBN 90-805233-2-1
24. http://www.jclark.com/xml/xt.html
25. http://www.x12.org/international/EntryPoint.htm
26. http://www.x12.org/international/edif.htm

27. Phelps, T., *Electronic Commerce and Product Development–Improving the Work of Supply Chains EDI Forum*, v 10, n 1, 1997. http://www.erim.org/cec/papers/ecandpd.html
28. Mowbray, T. and Ruh, W., *Inside CORBA-Distributed object standards and applications*, 1997, Addison Wesley Longman, ISBN 0-201-89540-4.
29. McCarty, B. and Cassady-Dorion, L., *Java distributed objects*, 1998, Sams, ISBN 0-672-31537-8
30. *ORB Portability Joint Submission (Final), Part 1 of 2, orbos/ 97-05-15*, document submitted to OMG, 1997. http://www.cs.wustl.edu/~schmidt/CORBA-docs/ORBOS-97-05-15.PDF
31. Vogel, A. and Duddy, K., *Java Programming with CORBA*, 1997, John Wiley & Sons, ISBN 0-471-17986-8
32. *PDM Enablers-Joint Proposal to the OMG in Response to OMG Manufacturing Domain Task Force RFP 1, mfg / 98–02–02*, document submitted to OMG, 1998, ftp://ftp.omg.org/pub/docs/mfg/98-02-02.pdf
33. Ellis, M., Stroustrup, B., *The annotated C++ reference manual*, 1991, Addison Wesley, ISBN 0-201-51459-1
34. Eckel, B., *Thinking in Java*, 1998, Prentice Hall, ISBN 0-13-659723-8
35. Sankar, K. et al., *JAVA 1.2 class libraries-Unleashed*, 1998, SAMS, ISBN 0-7897-1292-x
36. ISO TC184/SC4, *NWI/CD - ISO 10303, Part27 – Java programming language binding to the standard data access interface with Internet/Intranet extensions*, 1999
37. http://www.microsoft.com/com/
38. Bates, J., *Creating lightweight components with ATL*, 1999, Sams, ISBN 0-672-31535-1
39. Felton, M., *CGI: Internet Programming in C++ and C*, 1997, Prentice Hall, ISBN 0137123582
40. Zarli, A., Poyet, P., *A framework for distributed information management in the virtual enterprise: the VEGA project*, The PRODNET communication infrastructure, IFIP Working Conference on Infrastructures for virtual enterprises, Porto 1999, Chapter 19, Kluwer Academic Publishers, ISBN 0-7923-8639-6
41. Jardim-Gonçalves, R., Sousa,P., Pimentão, J.P., Steiger-Garção, A., *Integrating manufacturing systems using ISO 10303 (STEP): An overview of UNINOVA projects*, 1999, International Journal of Computer Applications in Technology, pp. 39-45, Vol.12, No 1, ISSN 0-952-8091
42. Sousa, P., Jardim-Gonçalves, R., Pimentão, J.P., Steiger-Garção, A., *Seeking intelligent product development - an integrator environment based on STEP*, accepted for publication in Journal of Intelligent Manufacturing, on the special issue on "Artificial Intelligent and Expert Systems in Product Development.
43. ISO TC184/SC4, "IS - ISO 10303, *Part22 – Standard Data Access Interface*
44. Song, L., Nagi, R., *Design and Implementation of a Virtual Information System for Agile Manufacturing*, IIE Transactions on Design and Manufacturing, special issue on Agile Manufacturing, 1997, Vol. 29(10), pp. 839-857
45. West, Matthew, *Integration of industrial data for exchange access and sharing: Aproposed architecture*, ISO TC184/SC4/WG10/N195, December 1998
46. Jardim-Gonçalves, R., Sousa, P., Pimentão, J.P., Steiger-Garção, A., *Integration of furniture manufacturing systems - the funStep project approach*, Eight International Manufacturing Conference - IMCC'98, 1998, Singapore, ISBN 981-04-0209-0.
47. Sousa, P., Pimentão, J.P., Jardim-Gonçalves, R., Steiger-Garção, A., *Towards compatibility between product data libraries using the Genesis' environment*, 13th

International Conference on Systems for Automation of Engineering and research (SAER'99), 1999, September , Varna-Bulgaria
48. Sousa, P., Pimentão, J.P., Jardim-Gonçalves, R., Steiger-Garção, *ISO 10303 Application Interfaces supported by Genesis environment – The funStep ESPRIT project experience*, 3rd IMACS International Multiconference on Circuits, Systems, Communications and Computers (CSCC'99/IEEE), 1999, Athens-Greece
49. Jardim-Gonçalves, R., Barata, M., Steiger-Garção, A., *CIMTOFI Project - Brite / EURAM Project BE-3653, Deliverables and reports*, 1990/94.
50. Pimentao,J.P.; Azinhal,R.; Goncalves,T.; Steiger-Garcao,A., *RoadRobot project (ESPRIT III- 6660) - Deliverables and reports*, 1993-1994
51. Pimentão, J.P., Jardim-Gonçalves, et al., *The RoadRobot project - from theory to practice*, 1996, Lisbon, Basys'96: Balanced Auomation Systems II - Implementation challenges for anthropocentric manufacturing", pp.126-133, Chapman & Hall, London 1996, ISBN 0-412-78890-X.
52. Jardim-Gonçalves, R., Sousa, P., Pimentão, J.P., Steiger-Garção, *Furniture commerce electronically assisted by way of a standard-based integrated environment - the ESPRIT funStep project proposal*, ICE'99 – International Conference on Concurrent Enterprising, Hague, Netherlands, 1999, ISBN 0 9519759 8 6
53. Jardim-Gonçalves, R., Sousa, P., Pimentão, J.P., Steiger-Garção, Borras, M., Gresa, I., *An integrated architecture to promote furniture business. The funStep project and established industrial initiatives*, Conference in Product Data Technology Europe - PDT99, Stavangar, Norway, 1999, QMS, UK, ISBN 1 901782 03 4
54. Jardim-Gonçalves, R., Sousa, P., Pimentão, J.P., Steiger-Garção, *ESPRIT #22056: the funStep project Integration of product and business data for furniture industry*, 1998, ECPPM'98 – Product and Process modelling in the Building Industry, BRE, ISBN 1 86081 249 X
55. http://www.funstep.org/funstep/ig/ig.htm
56. http://isc.aticorp.org/

Enterprise Integration and Management

R. H. Weston and A. Hodgson

MSI Research Institute, Loughborough University, Leicestershire LE11 3TU, UK.

1. ENTERPRISE INTEGRATION

Earlier chapters have explained how a Manufacturing Enterprise (ME) exists to realise products and services for customers, whilst satisfying requirements of stakeholders. MEs comprise unique, complex and changing groupings of personnel whose activities are supported by technology. Because of inherent complexity, personnel who make decisions and take actions effecting ME behaviours require systematic ways of (a) focusing individual concerns and (b) integrating the concerns of all personnel into a coherent whole. The systematic focusing of concerns is known as Problem Decomposition (PD). Properly applied PD approaches enable groups of people to tackle complex problems, breaking them down to enable the development of well understood and easily managed sub-solutions to different aspects of that problem. However new problems emerge as a consequence of the decomposition, namely it is necessary to (1) ensure that due account is taken of all necessary concerns so that sufficient sub-solutions are developed that collectively have a capability to adequately solve the original problem, and (2) integrate the operation of developed sub-solutions so that collectively they achieve a satisfactory problem solution and the resultant whole system meets its intended purposes.

Several PD concepts, methods and techniques have been developed to achieve (1) and are embedded into various concepts, methods and tools used to achieve Enterprise Engineering (EE). Subsequent sections of this chapter will explain what is meant by EE. When an EE approach is selected this will also largely determine the manner in which PD will be achieved. The EE approach and its underpinning PDs should adequately match the scope and nature of the whole problem. Also explained is how various Enterprise Integration (EI) concepts, methods and techniques have been developed to achieve (2). This will illustrate how EI and PD processes (like conventional mathematical integration and differentiation) are inversely related.

Collectively, ME personnel have numerous and various concerns. Often, a given concern is expressed as a particular instance of a common viewpoint, e.g. from a function, activity, behaviour, process, information, organisation, resource, time, financial, quality, user, supplier or customer point of view.

Problem simplification will result from limiting the viewpoints taken about a complex problem. Such a course of action equates to the application of a PD of some kind. A natural by-product of the simplification should be improved representation, visualisation, communication and general deployment of information pertaining to a selected viewpoint and its related concern. In the context of EE, often this information can usefully be captured in the form of computer-executable models designed to support strategic, tactical and operational decision-making and action-taking on a continuing basis.

1.1 Illustrative Use of Process Oriented Viewpoint

During the lifetime of a ME and of the various products and services it generates, a finite set of human and technical resources must be utilised to realise the types of activity classified by ISO and listed in Table 1. Particular instances of these activities will need to be carried out in a logically ordered fashion. Such a logically ordered activity set is referred to as an enterprise process (or simply a process). Enterprise processes (also called activity flows) exist at various levels of granularity. Large-grained processes (such as so-called business processes) typically traverse conventional organisational boundaries in MEs (e.g. teams, sections, departments, business units and companies) and can usefully represent a flow of activities that can be assessed in business or value-added terms. Also, large-grained activities typically comprise an organised set of smaller-grained processes that in turn can be expressed as some time-ordered set of common tasks or unit operations, and relationships between tasks and operations.

	"What" activities	"How" activities	"Do" activities
Plan and build phase (e.g. before sell/buy title transfer)	• Develop goals • Define strategy • Define product needs	• Develop requirements • Define concept • Design product • Plan to produce product • Plan to support product	• Procure parts • Produce product • Test product • Ship product
Use and operate phase (e.g. after sell/buy title transfer)	• Define support needs • Define use	• Define use requirements • Define support requirements	• Use product • Support product
Dispose & recycle phase (e.g. after product is no longer useful)	• Define recycle/ dispose needs	• Define recycle/ dispose requirements	• Recycle product • Dispose product

Table 1 Example *what*, *how* and *do* enterprise activities (after ISO 14258:1998)

On a continuing basis MEs should develop competitive behaviours so that they thrive within complex markets. Hence MEs require strategic assessment processes (so-called 'what' processes, ISO 14258) to decide what an enterprise should do in the medium-to-long term. Having decided what to do, 'how' enterprise processes are required to determine how the 'what' can be achieved. Generally this will require the specification and development of new products and the specification and implementation of changes to product-realising processes to enable new and old products to be produced in the correct quantities, at the right place and on time. Subsequently it is necessary to determine the frequency and time frames over which direct value-adding enterprise processes (i.e. 'doing' processes) must be invoked based on analysis of: product and service requirements of customers; the capabilities and capacities of existing human and technical resources; the availability of materials, components and equipment from suppliers; and so forth.

At any given time in a typical ME, many instances of enterprise processes will be concurrently active. Hence there often arises a need for human and technical resources to have multiple roles in and responsibilities for 'what', 'how' and 'do' enterprise processes, including the co-ordination and management of interactions between executing processes. These processes can be described by computer-executable models encoding temporal relationships between enterprise activities, necessary information, knowledge and workload requirements to carry out activities and groups of activities, and so forth [CIMOSA Association 1996; GERAM 1999]. This allows the capabilities of systems of human and technical resources to be modelled analytically and matched to groups of activities. In this way, roles and responsibilities can be systematically assigned to human and technical resources [Weston et al 2000].

This discussion exemplifies use of a process-oriented view of EE. Over the last decade, process-orientated modelling has become popular in industry and academia, primarily because it naturally provides a backbone modelling concept capable of coding system dynamics. Generally this backbone will be some model of temporally-related activities to which other multi-perspective models can be attached, such as descriptions of information, knowledge, workload and function requirements necessary to achieve activities, as well as descriptions of designated roles, responsibilities and commitments of allocated resources (i.e. people and technology that possess capabilities that are needed to realise the activities concerned. Further, if semantic information (defining enterprise activities and their associated modelling views) can be adequately captured in the form of computer-executable models, it can be developed and reused to facilitate process and system change, both within a given ME as requirements change unpredictably and in other similar MEs.

1.2 Illustrative Use of Resource and Object-Oriented Viewpoints

Resources are required to execute the processes of an enterprise, thereby physically realising conceptual requirements defined by process descriptions. Vernadat [1996] classified resource types as follows: input items (parts, products, raw material, documents, etc.); human resources; technological resources (tools, machines, devices, software packages, etc.); information resources (data and knowledge); financial resources; energy resources; and time.

The focus here is on resources with capabilities to realise functional requirements of processes, i.e. on human and technical resources that can be deployed as enterprise agents or actors to carry out enterprise activities.

The enterprise modelling community has recognised that there should be a clear separation (or decoupling) between descriptions of processes and resources. This is because they are distinct types of entity: the former being conceptual, the latter real, the former specifying requirements, the latter offering a potential solution with capabilities to realise those requirements. It follows that in general the processes and resources in an enterprise should be managed separately, even though their life cycles will necessarily be interwoven.

A given process may be realised by alternative sets of resources involving different levels of automation. However one set may outperform other sets in a given situation. For example, new metal-cutting operations may be uniquely defined in conceptual terms, when seeking to induce process change in a ME. In general, however, the operations defined can be performed by different combinations of people, simple tools and computer-controlled tools, each combination having potential advantage in alternative metal-cutting scenarios.

Human and technical resources in an enterprise normally have to behave and operate in systematic ways, as part of bigger, organised sets of resources, i.e. they form part of one or more enterprise resource systems. Figure 1 shows a simple representation of a ME and its primary enterprise systems. In a given ME there may be multiple instances of these enterprise systems, as might be required for instance to resource a number of manufacturing plants distributed around the globe. Therefore at an abstract level we can characterise a ME as follows:

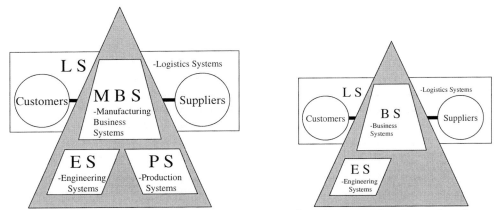

Figure 1. Simple models of manufacturing enterprises and service enterprises focused on the functionality of enterprise systems

A set of human and technical resources organised into business, engineering, logistical and manufacturing systems that interoperate in a structured way to realise concurrently the various processes of an enterprise. Thereby the resources of the ME (1) add value to materials, components and systems received from suppliers and deliver profit to its stakeholders (2) plan and control the repetitive resourcing of processes in such a manner that the ME supplies products and services to its customers and (3) develop and review enterprise processes and resources in order to renew and realise strategic purposes of the enterprise as it changes over time.

Resources are modelled from various viewpoints. For example, the Integrated Enterprise Modelling (IEM) approach [Spur et al 1996] models resources as instances of a generic object class. By so doing it can define attributes of resources, such as their functional capabilities and their availability over time. The CIMOSA approach to resource modelling [CIMOSA Association 1996] is slightly different in that it provides two essential constructs to model *resource requirements* and *resource objects*. The former construct is used to express resource requirements from the viewpoint of enterprise activities (these being elemental building blocks of processes), and the latter is used to express functional capabilities available to instances of resource classes. By matching *resource requirements* to *capabilities of resource objects*, CIMOSA provides a flexible (and hence readily-changeable way) of managing couplings between process models and models of systems of resources. The approach to the Modelling of Human Resources (MHR) reported by Weston et al [2000] extends the use of CIMOSA resource modelling concepts to match information, knowledge, and workload requirements of processes to relevant capabilities of groups of human resources. Thereby the MHR approach provides methods and mechanisms to decide how to allocate roles and responsibilities for enterprise operations to stereotypical humans.

2. ENTERPRISE MANAGEMENT

The Manufacturing Business System (MBS) of a ME (refer again to Figure 1) will normally be the dominant enterprise system with ongoing responsibility for the operation and performance of the ME as a whole. Consequently, responsibility for enterprise management is largely attributable to a MBS. This will typically involve (1) developing strategy, (2) developing, planning, controlling and observing tactical decision-making, and (3) financially controlling, observing and analysing the performance of 'Doing' processes. Therefore a typical MBS must resolve business and related financial concerns on various time frames and will be the primary ME interface to the outside world. This set of responsibilities will be further complicated where MEs operate globally in partnership with other companies, each with their own business, engineering, production and logistical systems (refer again to Figure 1). This in turn requires a MBS to have access to adequate 'internal knowledge' (about the ME) and 'external knowledge' (of other enterprises [stakeholders, partners and competitors], financial and product markets, of political and environmental policies, local communities and so forth). From the law of requisite variety (Ashby 1964) we can deduce that a MBS needs to be highly complex and sophisticated because it must possess greater functional capabilities than the sum of the entities (i.e. rest of the ME) it has responsibility for.

To achieve its purposes, a typical MBS should bring to bear the collective human know-how available to an enterprise on strategic and tactical matters (i.e. to resource and underpin 'How' and 'What' processes concerned with innovation and renewal activities), as well as to oversee and observe operational matters, i.e. to direct value-adding activities, (or 'Doing' processes) that achieve product and service realisation for customers. To cope with the high levels of complexity involved, generally a MBS will be decomposed into a number of Manufacturing Business subsystems (MB subsystems) that operate in a loosely-coupled manner. Generally MB subsystems will themselves comprise appropriate groupings of enterprise personnel (i.e. human resources) supported by suitable technological resources, who deliver results within specified timeframes and in a manner defined by an organisation structure that will have been pre-designed (possibly mainly) to target 'doing' processes but also to facilitate innovation and renewal. Proprietary software applications (such as those found in an enterprise requirements planning (ERP) IT product) can be used to lend structure and partially reuse intellectual capital available to an enterprise. Indeed ERP (and predecessor MRP and MRPII) systems are commonly deployed by MEs as a primary MB subsystems. However, a present-day ERP system only provides partial coverage of the set of MBS capability requirements. Also, such a sub-system can only provide a structure and instrument to facilitate inter-working but cannot itself generate a competitive response to external or internal requirements. Generally, it is appropriate groupings of enterprise personnel who will have the required knowledge and skills, which can be focused and enabled systematically by a proprietary IT product such as ERP. Similarly other MB subsystems can be deployed, based on the use of various management consultancy methods, enterprise engineering frameworks and knowledge and information management tools. These methods, frameworks and supporting software tools can harmonise efforts of enterprise personnel, such as those involved in large multi-disciplinary projects, by bringing together the concerns, knowledge and skills of individuals in a team to achieve goals in timely and cost-effective ways. However, as in the case of ERP systems they provide but a structure and an instrument for interworking. They do not automate ME re-engineering, nor do they automate the management of ongoing ME operations.

Based on the foregoing, Table 2 summarises common enterprise management functions and activities, while Figure 2 illustrates conceptually how common types of MB subsystems are used. Collectively these subsystems must perform all required roles of a MBS and will comprise various multi-disciplinary teams of people, with sectional purposes and concerns, whose actions are structured and supported by technical resources (i.e. IT and machine systems). The table and figure are merely illustrative and are not intended to be complete. The type of ME (e.g. whether it is an engineer-to-order (ETO), make-to-order (MTO), assemble-to-order (ATO), make-to-stock (MTS), etc. type of business), the complexity, variability and volatility of the products it makes and the services it offers, and the working culture in a given ME will greatly influence the distribution of roles and responsibilities (amongst enterprise systems and their component human and technical resources), as well as the priorities placed on and the frequencies with which enterprise systems need to enact their roles.

Function Type	Example MBS Activities
Observe...	Market requirements, competitor strengths and stakeholder needs.
	Political, economic, social and technical conditions.
	Business patterns and trends.
	(Measure) performance of W (what), H (how) and D (do) processes in ME.
Model...	Markets, patterns and structures.
	Alternative W, H and D processes and operating policies.
	Partnership structures, capability distributions.
	Effects of uncertainty and risk.
Analyse...	Strategies for renewal.
	Strategies and tactics for innovation.
	Impact of alternative W, H and D processes and operating policies.
Predict...	Importance of alternative change projects.
	Required order, component, stock and asset levels.
	Human resource requirements.
Plan...	The formation of new partnerships.
	New system and capital equipment projects.
	New product introduction scenarios.
	Process and quality improvement programmes.
	Knowledge capitalisation activities.
Control...	Resource allocation and distribution.
	Interactions between W, H and D processes.
	New system and capital equipment programmes.
Manage...	Financial budgets.
	Activity, product and information flows.
	Human resource recruitment and training.

Table 2 Typical enterprise management functions and activities

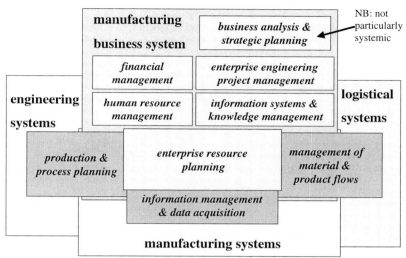
Figure 2. Some key MB sub-systems

3. ENTERPRISE FLEXIBILITY AND IMPLICATIONS FOR ENTERPRISE MANAGEMENT AND ENTERPRISE INTEGRATION

In the context of manufacturing, Buzacott (1982) defined flexibility as 'the ability of a system to process a wide variety of parts or assemblies, without intervention from outside to change the system'. He differentiated 'job flexibility' (i.e. the ability of the system to cope with different jobs to be processed by a system) from 'machine flexibility' (i.e. the ability of the system to cope with changes and disturbances at machines and workstations). Zalenovic (1982) made similar observations identifying two types of production system capability in terms of their 'design adequacy' (which is similar to Buzacott's 'machine flexibility') and 'adaptation flexibility' (which relates to the time it takes to change between jobs). Garrett (1986) pointed out that a flexible manufacturing system should be able to cope with 'external disturbances' (e.g. fluctuation in customer demand, product prices, product mix, action of competitors) and 'internal disturbances' (such as equipment breakdowns, variable task times, queuing delays, rejects, reworks). Slack (1983) also defined two types of flexibility, namely the range of states a system can adopt and the time (or cost) of moving from one state to another. Essentially, Slack's observations are consistent, albeit not coincident with earlier and more general observations of Mandelbaum (1978) regarding the capabilities of most physical systems, namely that such systems can be categorised with respect to their 'state flexibility' (i.e. a capability to continue functioning despite changes in the environment) and 'action flexibility' (i.e. a capability to take new actions).

Many authors have pointed to penalties arising from the provision of flexibility. Like the main body of flexibility literature, early discussion on this topic centred on the 'cost' of flexible, stand-alone machines. Stereotypical here is the view that 'state flexibility' (as defined by Mandelbaum) found in a typical industrial robot can much increase its range of potential applications (at least from a vendor perspective). However, because it will have a sub-optimally designed mechanical system and a sophisticated computer control system, from the user perspective it will typically perform less well (e.g. in terms of number of operations per minute, in terms of placement accuracy and in terms of costly support services) than a special-purpose machine designed specifically for the job in hand (Weston 1989).

Based on an analysis of the literature and previous experimental findings of the authors and their fellow researchers, a set of flexibility concepts is defined below that will be used in this and the next chapter. These concepts are centred on the 'change capability' and 'change capability rate' of enterprise systems as follows.

Change Capability: This is defined as the ability of a system to reach a range of states, and essentially is a extension of notions described by Slack [1983] (as the range of viable states of a system) and by Mandelbaum [1978] (as 'state flexibility'). The extended definition separates out three change capability classes and relates them to change scenario types in which they can most beneficially be applied, as follows:

(a) *Programmability:* The ability to program system behaviour and/or composition so that a system can reach a range of well-known states, thereby providing means of coping with change of a predictable nature.

(b) *Reactivity:* The ability to react to changes of an unpredictable nature by modifying system behaviour or composition.

(c) *Pro-activity:* The ability to prepare for modification of system behaviour or composition, such as by predicting and anticipating change requirements in uncertain environments.

Change Capability Rate: This is defined as the ability of a system to change at a given rate with a specified unit of engineering resource. This concept is a re-representation of Slack's 'time or cost of change' and Zalenovic's 'adaption flexibility' [Zalenovic 1982]. It is assumed that 'change capability rate' will depend on:

(1) *The structure and mechanisms used to integrate the various parts of a system,* as these can enable or constrain subsequent change to the composition and behaviours of component-based systems.

(2) *The change processes and supporting technology that are available to modify composition and behaviours.*

(3) *Properties of systems building blocks* (such as their grain size, explicit knowledge of their functional and interfacing capabilities, etc.) that will determine the ease with which they can be integrated into bigger systems.

Subsequent sections of this and the following chapter explain the impact of using appropriate EI and EM approaches on (1), (2) and (3). We will see that the size (or granularity) of enterprise building blocks is of important concern, as is the ability of building blocks to bind together (non-exclusively) in new and apparently seamless configurations which achieve new system behaviours while preserving needed system qualities, either by prescribing the extent of adaptability of components, or perhaps constraining their behaviour by mechanisms (e.g. mediating services) which enforce globally-desirable system behaviour. Recent flexibility literature has focused on the difficulties (and hence the costs) of flexibility in large systems comprising people, software tools, software applications and manufacturing machines. In general it is evident that the larger the system, the more difficult it will be to identify all states required and to cater for all eventualities in advance. Indeed, when enterprise requirements and/or environmental conditions are uncertain it may be preferable to be positioned to add new functional capabilities into a system (by using an appropriate set of systems engineering and systems integration approaches and by supporting the application of these approaches with well-proven methods and tools) rather than seek to initially include redundant, programmable functionality.

It is assumed that when designing and building an enterprise system a balance point should be struck where:

i) sufficient redundant capabilities are incorporated into an initial system composition so that various alternative behaviours can be readily programmed to enable the system to reach a predictable range of states, possibly on a frequent basis.

ii) the system is so structured, and integration mechanisms so provided, that the system composition can be modified rapidly and effectively, such as by invoking a change process capable of adding new system capabilities, thereby enabling the system to react in an effective and timely manner in response to known classes of change as they occur unpredictably and possibly infrequently;

iii) the system is decommissioned and replaced when changes in requirements or environmental conditions are outside the scope of the structure, integration mechanisms and change processes referred to under (ii)

In general, teams of human and technical 'change actors' will modify the composition and behaviours of enterprise systems, through system programming, system reaction and system pro-action. These 'change actors' may be external to or component parts of, the system being modified. Chapter 4 explains that significant improvements in design adequacy and change capability can accrue through systematically composing large-scale enterprise systems from well-proven, well-defined and change-capable lower-level system blocks.

4. BUSINESS ANALYSIS AND BUSINESS SOLUTION STRUCTURE

It follows that the activities of human and technical resources in any ME should be co-ordinated and supported by appropriate methods and tools so that all personnel responsible for ME development and operation continue to refresh and access a shared view of what the business needs. This can be facilitated by explicitly defining common (current and future) requirements. Table 3 shows possible elements of a shared view, which will be termed a generalised business solution structure (Weston 1999), and lists examples of methods, frameworks and tools that can be used to specify elements of that structure. Subsequent sections will explain how such methods, frameworks and tools can be used. However it is important to emphasise that the meta model (i.e. model of models) specified by Table 3 is no more than a generalised guide. A specific shared view (such as a specific instance of the meta model and its component models) must be tailored to individual enterprise needs before it can be usefully applied.

An output of business analysis and strategic planning will be a strategic plan. This plan will typically include a portfolio of prioritised 'change projects' to advance ME performance. The plan might identify necessary modifications to enterprise processes (and their relationships) or suggest changing the way processes are realised by systems of human and technical resources. Examples of types of change project are illustrated by Table 4. In general these projects will require significant investment, complex project management and will seek to make a significant impact on one or more enterprise systems (possibly including the MBS itself). Because these change projects can have a major impact on the future health of MEs, it is vital that they are carried out in a timely and cost-effective manner by teams comprising enterprise personnel who have the necessary multi-disciplinary knowledge and skills, and their external advisors (such as consultants with large project management and systems integration skills, and IT suppliers of selected software applications and software tools).

View of the Solution	Output Model or Specification	Interpretation / Example	Example method, framework or tool
Need to generate a mission or vision statement.	A specification of the business purpose of the enterprise and its systems.	E.g. to be the highest quality electronics supplier to the European computer market.	Methods: Idons for thinking, Scenario Analysis Representation: CIMOSA objectives
Need to analyse and define extreme influences, market forces, customer requirements, supplier capabilities and competitor abilities.	A definition of customer requirements and metrics, and their causal effects on the business processes within the enterprise.	E.g. customers primarily require high quality (reproducibility) assemblies at the lowest price. This dictates concentration on production process control.	Methods: Porters five forces, External SWOT analysis, Market segmentation, Buyer Value Segmentation, PEST Analysis, Representation: Portfolio Analysis
Need to understand the capabilities and knowledge held within the enterprise that can be exploited in relation to the capabilities of competitors.	A description of the capabilities of human and technical resources available with respect to the processes under consideration.	E.g. the enterprise holds particular expertise in rapid fault diagnosis systems.	Methods: Porter's Value chain, GAP analysis, Internal SWOT, SSM Representation: CIMOSA Resource constructs, ARIS constructs
Need to understand the mix of business processes and rules that govern the way the business satisfies its requirements and mission.	A definition of activity flows for each business process at an appropriate level of granularity that fulfils the business process requirements and goals.	E.g. the enterprise must recruit personnel able to communicate effectively and quickly to people in all areas of the organisation.	Methods: Process Modelling, Object Oriented Design, Systems Thinking, Beer's Viable System model, Representation: Stella, ProcessWise, ARIS constructs, GRAI-GIM
What resources (people, machines, information and finances) are required to support the overall business structure?	A description of the resources required (people, machines, information and finances) to support each activity in each business process.	E.g. the enterprise must be prepared to finance ongoing training of personnel for communication and troubleshooting skills.	Methods: Beer's Viable system model, process modelling Representation: Stella, CIM-OSA resource allocation
What organisational structures are required to support W, H & D activities? (roles, relationships, etc.)?	Which particular organisational entities will be responsible for each business activity and how will this be co-ordinated (organisational infrastructure)?	E.g. the enterprise should foster team-based structures to ensure rapid response and ownership of problems and resolutions.	Methods: Beers Viable System Model, SSM, Systems Thinking Representation: Stella, GRAI-GIM, PRIMA, PACT
What are the general information requirements in support of W, H & D activities?	What information is required to support the business processes and how is it to be controlled and co-ordinated?	E.g. the enterprise should provide systems that make available information regarding previous production problems and resolutions.	Methods: IDEF, O-O Analysis Representation: SEMATECH, PACT, SHADE, POSC, ARIS, IEM, CIMOSA.
Need to define timescales and deliverables for the implementation of business change.	Need to define a plan of activities and resource configuration to implement change (people, timescales, etc.).	E.g. need to recruit trainers within 2 months and develop statistical process monitoring within 6 months.	Methods: Project management tools, Constraint theory Representation: PERA, PRIMA

Table 3 Business analysis issues leading to a business solution structure (Weston et al 1998)

- Develop new partnership with supplier.
- Geographically relocate part of a business process.
- Introduce new business rules.
- Introduce a new approach to managing and assessing the impact of change.
- Re-engineer a sales process
- Re-engineer a production process.
- Change organisation structures and job/task allocation procedures.
- Introduce a new product type.
- Introduce a new engineering system.
- Introduce new quality procedures.
- Implement a new redundancy policy.
- Introduce new terms and conditions of service.
- Introduce new training programmes.

Table 4 Examples of Change Projects

Each change project will typically be markedly different from other change projects in a portfolio and will require a different project team. However, normally the project will have common life phases, as follows: conceptually specify requirements; conceptualise alternative 'candidate' (systems) solutions; financially justify necessary investments; detail and select between alternative solutions; implement new systems or make modifications to existing systems, and all that entails in terms of seeking to modify working practices, company cultures, and so on.

Over the last two decades, generalised approaches to large-scale project engineering have emerged. There had been two main sources of that development, namely: (1) by management consultancies, systems integrators and large end-users who have developed their own (i.e. proprietary) flavours of methods and frameworks to achieve alternative types of project engineering and (2) by public domain consortia, who primarily have sought to generalise and develop methods and frameworks from source (1) into the form of public domain specifications of reference architectures and methodologies (that define concepts to structure and facilitate enterprise engineering). To distinguish between them, approaches from source (1) will be referred to in the following as 'consultancy methods', whilst approaches from source (2) will be termed 'public domain methodologies'. However, in general both facilitate use of decomposition techniques (via an architectural framework) and both recommend use of a flow of project activities (i.e. a method or methodology) at some level of detail and prescription. Consultancy methods have been used often and successfully, but in general their specification is closed and is not linked to a formal approach to enterprise modelling. Although public domain methodologies have been used less frequently than consultancy methods, where their use is based on a modelling framework, the project results can be reused as an enterprise asset. Also their formal specification is open to end-users and vendors of process and product systems and this can facilitate improvements in interoperation between multi-vendor enterprise systems and subsystems.

5. EXAMPLE 'CONSULTANCY METHOD'

Space constraints dictate a need to briefly describe but one consultancy method. The choice here is the OPENframework methodology, developed originally by a consultancy group within the company International Computers Limited but it has been used by over 400 practitioners world-wide. The methodology is a set of well-documented and method-supported precepts and guidelines to ensure that business process re-engineering and systems integration projects are managed effectively and that they give rise to solutions that are aligned with business needs. The methodology is also used to assess customer needs, to define criteria for success and to structure design, development and delivery processes. Whether developing a standalone product or a full business process re-engineering solution, the principles remain the same.

A key precept of OPENframework is that it is necessary to provide separately (but to maintain 'connections' between) means of addressing *business*, *social* and *technical* issues. It is assumed that with respect to each of these issue sets, problem decomposition (PD) is necessary to cope with (a) the levels of complexity involved, (b) the need to represent the different perspectives and timeframes of alternative groups of 'change actors' and (c) the need to promote the reuse of alternative enterprise engineering concepts, methods, tools and knowledge (such as in the form of enterprise models). However, we know that the adoption of PD techniques raises coherence and co-ordination issues related to reintegrating constituent perspectives and solutions. Therefore there is a need to maintain adequate 'connections' between the conceptualisation of business processes and systems and their manifestations as real groupings of operations and resources (including IT systems). Without well-defined connections, business process improvements will be constrained by difficulties in the cost-justification of proposals.

From a business perspective, OPENframework's checklist of concepts ensures a proper appraisal of existing and proposed enterprise processes and also a proper appraisal of the human and technical resource qualities required to satisfy customer and market requirements. It also flags up the need to consider the impact of the environment, i.e. political, economic, social and technical (PEST) issues, to facilitate competitor analysis and consider the potential role of trading partnerships. The result of this business analysis should be a description of a suitable *enterprise configuration* (which can be termed a business architecture). This description should document in some form a strategy, a set of business processes, and their qualities, organisational structures and systems, necessary assets (including resources) and infrastructural needs. A key test of the power of the consultancy method concerns its ability to maintain links (possibly informal cause-and-effect relationships) between different business issues. Such links are essential in order to determine suitable responses (i.e. defined changes in enterprise configuration) to changing requirements, trading conditions, etc.

From a social perspective, OPENframework structures the way in which knowledge about the culture of an enterprise is (or could be) used to realise business process requirements. Its checklist of concepts should help to determine business rules and organisation structures that promote a suitable balance between control and innovation. This should lead naturally to the assignment of work groups and human roles. Again, it will be important to maintain linkages between different aspects of the *social system* so designed, and cause-and-effect relationships with business issues. As for business analysis, use of the method should elicit views from personnel with a stake in requirements and solutions, via structured interviews, the use of work groups, etc.

From a technical perspective, OPENframework defines what computer tools and systems are required to support the social system in realising business processes. It can also help to decide what activities could be automated. In so doing, it should help to unify requirements and constraints of users, system managers, software applications, system developers and service providers. The outcome is typically a description of a technical architecture that defines links between business process requirements, application systems and components, and infrastructure support needs. In turn this can define requirements for user interfaces, distributed application services, information management, and a suitable choice of hardware platforms. Again, OPENframework maintains casual links to social and business issues.

OPENframework maintains linkages between business, social and technical issues by attributing *qualities* and *capabilities* to different problem perspectives. By maintaining the consistency of these quality and capability attributes across process and system boundaries, a mutual coupling mechanism is created to help to co-ordinate the innovative contributions of different actors. Typically, these qualities would include potential for change, performance, usability, availability and security. Figure 3 represents an abstraction of the OPENframework consultancy method, which reflects requirements identified in the foregoing description.

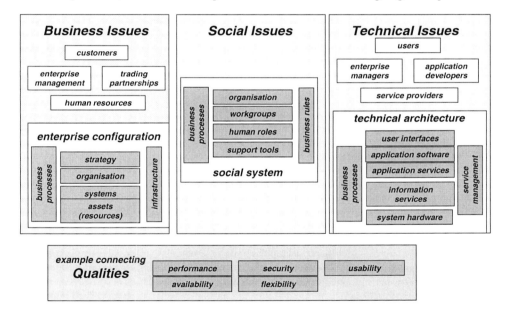

Figure 3. Consultancy framework abstraction

As indicated earlier, one 'consultancy method' may be better than another in terms of its scope or depth of coverage in a given area. Also, one approach may be informal, and another semi-formal with a formalism based on the use of one or more enterprise modelling tools. Choice and coverage of enterprise modelling tool(s) will often reflect the nature of the problems commonly tackled by a consultant/integrator.

A primary benefit of formal modelling is that the models created can be reused as an enterprise asset. This knowledge can be used to appraise (e.g. based on simulation) alternative business scenarios, to analyse and enhance business processes, to facilitate change to organisational structures, to appraise use of alternate types of resource (e.g. to decide if an operation should be automated or not) and so on. Furthermore, formal models of the consultancy method itself can lend structure that can help to co-ordinate and support detailed downstream system design and implementation life phases. Similarly model-based structures can also be used to help to co-ordinate run-time decision-making and action-taking.

Despite the evident benefits of formal modelling the authors are unaware of any wide-scope consultancy method that is comprehensively supported by modelling tools.

To summarise, consultancy methods will continue to be used very successfully in co-ordinating human activities required on large enterprise change projects. This is because their guidelines and checklists bring to bear the expertise of key personnel on complex problems, solutions to which can realise new strategic directions. However, generally their use of modelling is limited, ad-hoc and fragmented. The main consequence of this is that valuable knowledge and information is lost and cannot be reused in subsequent projects. Hence all enterprise engineering projects based on the use of current generation consultancy methods are one-offs and are high cost projects as a result. Furthermore, typically their timeframe is long (of the order of nine months) so that for many MEs it is time to change again, before the last set of major changes has been actioned and become operational.

6. USES OF A PUBLIC DOMAIN ENTERPRISE ENGINEERING METHODOLOGY AND FRAMEWORK

Most public domain methodologies can be distinguished from consultancy methods in that they have a 'backbone' modelling structure and an associated set of modelling concepts. The modelling backbone allows modelling concepts to be operationalised in a structured manner. This can produce analytic, reusable models of an enterprise from various viewpoints, which represent the differing concerns of the 'change actors' involved. Well-known public domain methodologies that have a backbone modelling structure include CIMOSA, GRAI-GIM (Vallespir et al 1991), IEM (Spur et al 1996) and ARIS (Scheer et al 1994). These methodologies support three main classes of decomposition, namely:

Decomposition Class I: Dimension View This class of decomposition is concerned with representing different problem perspectives, i.e. supporting focused concerns (or views) of 'change actors'. An introduction to the need for this decomposition class was given in Section 2 of this chapter. Ideally, sufficient modelling concepts should be provided here to enable appropriate foci of concern about business, social (including human and cultural) and technical issues to be represented at different levels of abstraction. However, current generation public domain methodologies provide better modelling concept coverage for technical issues than for business and social issues. However, the modelling concepts that they make available allow problem and solution representations (or models) to be captured and reused, and considered in relationship to each other so that any conflicts can be resolved. Typical models captured by CIMOSA, GRAI-GIM and ARIS encode function, process, behaviour, control, information, organisation, resource, economic and other viewpoints on system requirements, conceptual and detailed system designs, implementation descriptions and interoperating systems. The methodologies also provide other concepts to allow mappings to be established between different viewpoints. This is necessary to enable holistic models to be developed, as a consistent and structured set of model fragments of concern to a team of system designers, builders, developers, etc. The structural arrangements between concepts also need to be explicitly maintained to ensure that sub-solutions can be developed independently (e.g. with different local objectives in mind and on different timeframes) yet be consistent with each other so that they can be reintegrated into an effective and flexible whole solution.

Decomposition Class II: Dimension Life Phase To varying degrees GRAI-GIM, CIMOSA, IEM and ARIS also provide concepts that lend 'project structure' to enterprise engineering activities. This is necessary to ensure that the various parties (change actors) involved carry out engineering activities in an agreed order and timeframe. In general, consultancy methods are strong with respect to this class of decomposition but less strong with regard to Class I and Class III. Class II decompositions may be of prime concern with respect to enabling the reuse of solution fragments and models that have previously been captured, by placing constraints on flows of enterprise engineering activities. Therefore in one scenario it may be necessary to ensure that the requirements of a system are defined and verified before a conceptual solution is conceived and detailed, whereas in another scenario it may be more appropriate to enable the use of some iterative combinations of top-down and bottom-up methods (sometimes called middle-up-down methods) and their supporting concepts.

Decomposition Class III: Dimension Genericity A third set of concepts provided by CIMOSA, ARIS, IEM and GRAI-GIM relates to the generality of representations (or models) of enterprise requirements and system solutions. Generic models and generic modelling concepts will have universal application but as a consequence will be very abstract and contain little detailed knowledge or information that can be reused. An example generic model is an *Ithink* archetype, whereas generic modelling concepts include *Ithink* 'stock', 'flows' and 'converters'. Partial (sometimes called Domain) models and partial modelling constructs are designed for use in a particular enterprise domain. Semi-generic models of a sales process, a human resource management system or an ERP system are examples of partial models. Reference to a semi-generic model of ERP data structures can be used as the basis of a dialogue between an ERP vendor and user, when vendors sell and users select such a MBS sub-system. Whereas particular models are used uniquely, e.g. to describe specific enterprise processes, specific enterprise systems, specific vendor products, etc. Several commercial modelling tools that complement either CIMOSA, GRAI-GIM or ARIS modelling concepts can be used to support human users as they 'particularise' and 'generalise' models, i.e. move between different levels of abstraction, detail and generalised use.

Figure 4 is essentially a re-representation of the so-called CIMOSA cube, which indicates the orthogonal nature of the three classes of decomposition described. Models that reside within different segments of this cube can be used to generate suitable sets of models for different partners in a global enterprise engineering team.

In order to develop world standards on enterprise modelling that can underpin large change projects (and hence re-engineer enterprises), an international IFAC/IFIP working group has been active for nearly a decade. This so-called Task Force has defined GERAM (a Generic Reference Architecture and Methodology). GERAM seeks to unify definitions used by researchers world-wide in respect of methods, models and tools needed to conceptualise and build the integrated enterprise [Bernus and Nemes 1995]. GERAM is generic in that it applies to most, potentially all, types of enterprise. The coverage of the GERAM framework spans products, enterprises, enterprise integration and strategic enterprise management, with the emphasis being on the middle two. Currently GERAM is being developed by consolidating aspects of previously-analysed architectures, such as the Purdue Enterprise

Reference Architecture (PERA) [Williams 1992], the GRAI Integration Methodology, CIMOSA and TOVE [Fox 1992]. In this context it is also important to reference complementary European standardisation initiatives and particularly CEN TC310, WG1's ENV40 003. ENV 40 003 is now an agreed European Community framework for modelling operational processes in an enterprise [Shorter 1994].

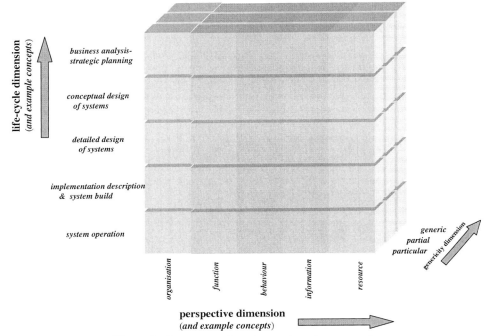

Figure 4. Common enterprise engineering concepts

Many common conclusions have been drawn from GERAM and ENV40 003 activities. They have highlighted important advantages of using models to facilitate systems analysis (using model-based *simulation*), system construction (by deploying *emulation* models, *auto-generating* fragments of system *code* and selecting pre-existing resources and components from *partial models of components*) and system execution (during which *control models*, which may be embedded in system components, support the flexible interoperation of enterprise systems). The industrial use of these concepts has demonstrated significant benefits in terms of *creating better-designed systems more rapidly* and *in facilitating subsequent system change and enhancement*. This approach to maintaining and reusing semantic information across design and control environments (i.e. along the forward path of the life-phase dimension) has been termed *model enactment* (Weston 1996). As depicted in Figure 5, by maintaining explicit mappings between system models and entities of a physical system, much-improved life-cycle engineering support can be facilitated. However, as yet it has only proved practical to reuse a limited amount of semantic information in such a way. Nonetheless, where it has proved to be practical (such as in the domain of workflow systems) model-driven solutions can be deployed to 'conceptualise holistically, whilst implementing incrementally towards well-defined business objectives and goals' (Barber and Weston 1998).

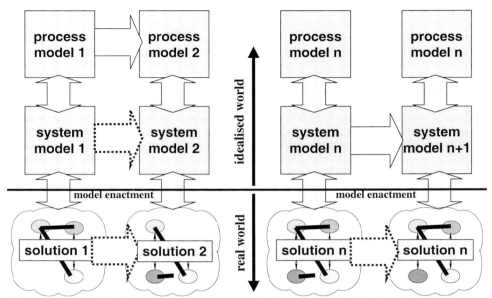

Figure 5. Illustrating how model enactment can lead to rapid and effective change to system compositions and configurations, i.e. it can facilitate action flexibility.

It is evident therefore that benefit can accrue if the models themselves are *living things*. As such, models will have a life-cycle that should be supported by appropriate modelling tools. The tools should support *model capture* (i.e. formation of the models), *validation* (in terms of their completeness, consistency and correctness), *their particularisation and generalisation* (from generic to particular versions and vice versa), *their enactment* (in terms of supporting simulation, emulation and execution) and *their maintenance, ongoing development and reuse* (i.e. their change, release and enhancement).

7. CAPABILITIES OF CURRENT GENERATION ENTERPRISE MODELLING TOOLS

There are in excess of a hundred commercial vendors of general-purpose enterprise modelling tools, with an estimated annual market of US$79 billion (Goransen, ICEIMT 92 and 97). Also claimed is a growing reliance by industry on the use of such tools. Indeed the research literature reports on many and varied applications of these tools such as in support of: capabilising and reusing knowledge; business strategy development; analysing market and environmental conditions; business process re-engineering; the management of large enterprise engineering projects; tactical and operational decision making; IT system design and engineering; and so forth. In general, however, the modelling capability provided by any single tool will be limited. Hence, industry is required to select appropriate modelling technology for each job in hand, based on an understanding of the type and form of the models a given tool can produce, alternative ways in which the models can be captured and developed, and so on. Consequently at any point in time various groupings of people in a single company may be using a number of different tools and for a variety of purposes that may or may not be linked. Neither the experience of producing these models, nor the models themselves, inform or facilitate the subsequent development of models for different purposes.

Weston et al (1995) conducted a review of existing and emerging EI methods, techniques and tools. Since that date the initial classification has been updated and developed regularly as the MSI Research Institute's researchers have evaluated and advanced the use of various methods and tools. The initial evaluation and classification work positioned the capabilities of existing methods and tools with respect to:
(a) *the scope of coverage provided*, in terms of the extent of coverage of the life-cycle engineering cube depicted in Figure 4,
(b) *the completeness and usability of the models produced*, in respect of the information content captured which can be used to semi-automate the generation of elements of a physical system, as depicted by Figure 6.

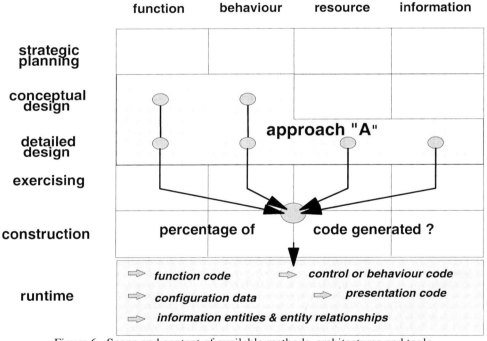

Figure 6. Scope and content of available methods, architectures and tools.

Most existing enterprise modelling methods and tools naturally fall into one of two camps. One of these camps corresponds to methods and tools conceived originally to *support the design of manufacturing enterprises* (and hence are candidate tools with capability to underpin the operation of various MBS sub-systems) whereas in the second camp are methods and tools originally conceived to *facilitate software engineering*. Example residents at the first camp are: visioning and scenario-generation tools; market and competitor analysis tools; financial analysis and accounting tools; risk analysis methods and tools; forecasting and sales analysis methods and tools; organisational design methods; business process analysis methods and tools; enterprise integration methods and frameworks; simulation tools for workflow management and factory layout; project planning tools; and general purpose manufacturing systems modelling and diagramming tools. Figure 7 positions the 'centre of gravity' of coverage of a selection of these methods and tools, some of which have over time migrated the scope of their life-cycle engineering coverage downstream to provide limited model enactment capabilities (typically providing a level of support for applications systems development and system configuration). Generally speaking however, as indicated by Figure 8 (which positions a small selection of methods and tools within the CIMOSA cube) the main focus of support (and prime functionality) of typical methods and tools from this camp align with strategic planning and conceptual design life phases and with organisation, function, control and information perspectives of the comparison cube. Hence they are each candidate elements of MBS (and hence Enterprise Management systems).

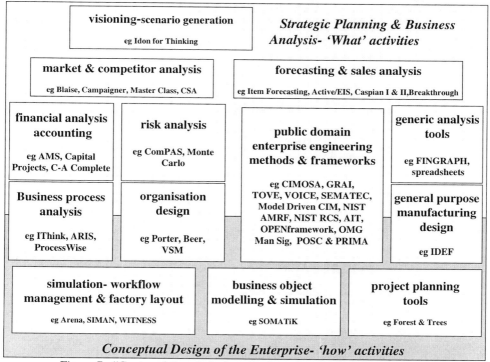

Figure 7. "Centre of Gravity" of a selection of methods and tools.

Conversely, example residents at the software engineering camp include: structured application design and construction methods; formal modelling methods; custom-designed Computer Aided Software Engineering (CASE) tools and second generation, general purpose CASE tools; various application development tools; and a host of infrastructural services, human interface system components, and application service interface tools.

coverage of typical enterprise engineering frameworks & tools

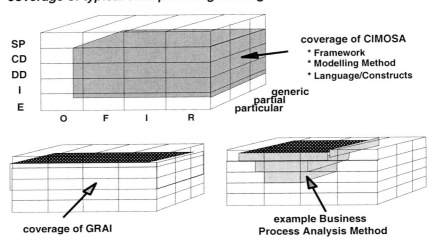

coverage of typical software engineering methods

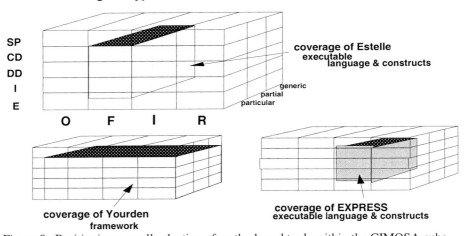

Figure 8. Positioning a small selection of methods and tools within the CIMOSA cube

Figure 9 provides a classification of a small fraction of the software engineering methods and tools in common use. In general, methods and tools that belong to the software engineering camp possess an inherent capability to enact models and thereby can be used to produce various fragments of executable software systems, which, if deployed appropriately, can collectively facilitate systems integration in a flexible and scalable way. However, as illustrated by Figure 9, seldom will a method and tool with a software engineering origin have coverage which extends upwards beyond conceptual design on the life-phase dimension and normally their attention will be focused sharply on software aspects of systems engineering.

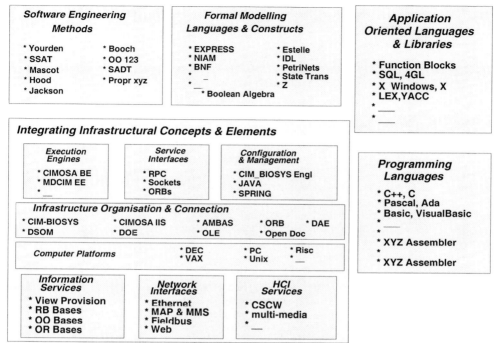

Figure 9. A classification of software engineering methods and technologies

There will typically be considerable overlap of coverage of individual software engineering tools and significant differences in their underlying concepts and the ways in which they are implemented. Nonetheless there are many ongoing exciting technical developments within this camp, particularly with respect to the bringing together of: infrastructure technology, application development tools, groupware, distributed object technology, multi-database management techniques, semantic integration, enterprise ontologies, object-relational technology and object-oriented system and software design techniques. The anticipated joint deployment of some of these technologies is already giving rise to notions of 'component warehousing' and the expected birth of application design and development tools. These tools will facilitate the configuration of systems from re-usable software components, involving their selection from a warehouse, and the semi-automatic generation of code to 'glue' the components together. This should lead on to a new generation of more flexible and effective IT systems that require reduced engineering effort and much shorter lead-times. Furthermore these developments are likely to be fuelled by 'partnering' amongst major IT systems suppliers.

Notwithstanding the ongoing extension in coverage of many research prototypes and commercially-available enterprise modelling methods and tools, as yet any single one of them provides only very limited coverage of the enterprise engineering problem domain, in terms of scope and completeness. Stated more positively however, we may conclude from this, and the fact that major benefits can be realised from deploying a single tool, that orders of magnitude increase in benefits may be possible following a unification in the use of a significant number of enterprise modelling methods and tools. Indeed this possibility is attracting considerable interest world-wide. However, efforts aimed at achieving tool integration will be complicated by the need to unify multi-platform, multi-vendor tools in a manner that enables users to select and deploy a personalised toolset. Additionally the methodological use of a toolset should be controlled in a systematic manner that can be customised by the working practice in a host company.

8. PERFORMANCE OF CURRENT GENERATION APPROACHES TO ENTERPRISE INTEGRATION AND MANAGEMENT

A recent survey of business process re-engineering projects in the UK found that current-generation enterprise systems are insufficiently responsive to changing requirements and conditions faced by manufacturing companies (Barber and Weston 1998). The time frame of large scale systems engineering projects is of the order of six months to four years and, typically, multi-partner developments expend budgets of around US$10M or greater. Enterprise systems of wide scope typically have multiple purposes and their life cycle engineering will require the resolution of multi-disciplinary, uncertain and competing concerns. Hence, there is a requirement to co-ordinate the involvement of teams of people. Typically there will be a number of EE projects ongoing in MEs, and there may be shared membership of project teams. However, normally an individual project team will be assigned responsibility and resources to detail organisation design changes and to implement their particular chunk of change. With current practice, seldom will an implementation team have available to them a holistic understanding of strategic and operation benefits and disbenefits that can arise from the changes they specify and for which they have responsibility. Therefore it is unlikely that they will be able to reason in an ongoing way about whether the objectives of their project should be re-targeted during its lifetime. It follows that current approaches to large scale organisation design and change in industry are essentially based on a policy of "specify desirable changes to organisational structure and technology, then launch a series of largely unrelated, medium-term, open-ended implementation projects to realise change as and when it is politically and financially viable to do so".

Despite evident limitations, conventional approaches to organisation (re)design and engineering change can work well under essentially steady-state conditions, as collectively and over a period of time the people concerned evolve and tune the organisational structures that bind their collective activities (e.g. through common learning they develop the organisation culture and improve their practices) and on a long term basis they develop suitable technology to underpin their interworking. However this is not the case for organisations that require frequent, large-scale change (Hammer and Champy 1993, Johnson and Kaplan 1987, Hamel 1993). Firstly, conventional design and implementation approaches lead to fairly rigid structures and technologies that cannot be changed sufficiently, rapidly, or effectively (Goldman et al 1995, Barber and Weston 1998). The same structures and technologies deliberately imposed to organise and constrain unwanted behaviours can place severe constraints on new behaviours. Consequently the ability to respond to a new set of requirements or to assimilate the use of new technologies within products and product-realising processes may be compromised (Peters 1988, Goldman et al 1995). Secondly, conventional approaches to organisation design are more akin to an 'art' than a 'science'. Seldom will the structures and technology deployed be explicitly described in a form that promotes reasoning about them (Warnecke 1993, Weston 1998). Generally, therefore, it will not be possible to scientifically determine how any given organisation should be developed to facilitate an improved response to new business opportunities and threats.

9. CONCLUSIONS

Public domain EE methods and frameworks and EI technology promise to provide industry with improved means of handling complexity and uncertainty thereby enabling the development of large-scale systems that are change capable. Primarily this is because they provide a semi-formal, yet customisable structure for PD and EI that can target and co-ordinate the efforts of engineering teams and can facilitate the reuse of semantic information coded by multi-perspective models of the enterprise. It has also become practical and effective to build EE environments (comprising an appropriate selection of proprietary enterprise modelling tools) that can facilitate the shared development and reuse of multi-perspective models. Such an environment is described in the next chapter. Potentially therefore, emerging EE and EI methods, concepts and tools can drive down the costs and lead-times of large-scale systems engineering projects and (possibly even more importantly) can facilitate engineering change on a continuing basis even where future problems (and their associated requirements, solutions and systems) are uncertain.

However, to-date this potential has been realised in very few MEs while the very large majority of companies continue to deploy conventional EE approaches, based for example on the one-off use of a consulting methodology. Even the successes thus far have generally involved significant up-front research and development effort. However, as explained in the next chapter, exemplar public domain EE and EI approaches that have been developed for use in important ME domains can be expected to provide the basis for next generation practice in many industries.

REFERENCES

1. Ashby, R., (1964), "An Introduction to Cybernetics", Chapman & Hall, London.
2. Barber M.I., Weston, R.H., "Scoping Study on Business Process Reengineering: Towards Successful IT Application", *Int. J. of Prod. Res.*, 1998, Vol.36, No.3, pp575-601.
3. Bernus, P. & Nemes, L., 1995, "A Framework to Define a Generic Enterprise Reference Architecture and Methodology (GERAM)". Div. Rep. No. MTM 366 CSIRO Div of Manuf Tech, Preston, Victoria 3072 Australia.
4. Buzacott J A. The Fundamental Principles of Flexibility in Manufacturing Systems, Proceedings of 1st International Conference on FMS, Brighton, UK, 20-22 October 1982, pp 13-22.
5. CIMOSA Association (1996), CIMOSA Technical Baseline. CIMOSA Association e.V., Stockholmer Str.7. D-70734 Boeblingen, Germany, April.
6. Fox, MS., "The TOVE Project, towards a common sense model of the Enterprise". Enterprise Integration, C. Petrie (Ed), Cambridge, MA, MIT Press (1992).
7. Garrett, S.E., "Strategy First: a case in FMS justification", in K. E. Stecke and R Suri (Eds.), Proceeding of the second ORSA/TIMS Conference on FMS; Operations Research Models and Application, Amsterdam: Elsevier Science Publishers, 1986, pp. 17-29.
8. GERAM 1999: Generalised Enterprise Reference Architecture and Methodology, Version 1.6.3 (March 1999) IFIP–IFAC Task Force. http://www.cit.gu.edu.au/~bernus/taskforce/ geram/versions/geram1-6-3/v1.6.3.html

9. Goldman, S.L., Nagel R.N. and Preiss, K., 1995, "Agile Competitors and Virtual Organisations", Van Nostrand Reinhold Pub., New York. ISBN 0-442-01903-3.
10. Hamel, G., (1993), Strategy as Stretch and Leverage, Harvard Business Review, 71, pp75-84.
11. Hammer, M., and Champy, J., 1993, Reengineering the Corporation. *New York: Harper Business*.
12. Johnson, T.H. and Kaplan, R.L., (1987), "Relevance Lost", Harvard Business School Press, Boston.
13. Mandelbaum M. 1978, Flexibility in Decision Making: An Exploration and Unification. PhD Thesis, Dept. of Industrial Engineering, University of Toronto, Ontario, 1978.
14. Peters, T., (1988), "Thriving on Chaos", Macmillan, London.
15. Shorter, D. "An Evaluation of CIM modelling constructs. Evaluation report of constructs for views according to ENV 40 003". Computer in Industry, 24(1994) 159-236.
16. Scheer, A.W. and Kruse, C., (1994), "ARIS – Framework and Toolset. A Comprehensive Business Process Reengineering Methodology." Proc. 4th Int. Conf. on Automation, Robotics & Computer Vision (ICARCV 94), Singapore 8-10 Nov., pp327-331.
17. Slack, N., "Flexibility as a manufacturing objective", International Journal of Operations and Production Management, Vol. 3, No. 3, 1983, pp. 4-13.
18. Spur, G., Mertins, K. and Jochem, R., (1996). Integrated Enterprise Modelling, Beuth Verlag, Berlin.
19. Vallespir, B., Chen, D., Zanettin, M & Doumeingts, G., "Definition of a CIM Architecture within the ESPRIT Project 'IMPACS'". Computer Applications in Production Engineering: Integration Aspects, G. Doumeingts, J. Browne and M. Tomljaprovich (Eds.), Elsevier, Amsterdam, pp731-738, 1991.
20. Vernadat, F., "CIMOSA - A European Development for Enterprise Integration, Part 2: Enterprise Modelling". Enterprise Integration Modelling, C. Petrie (Ed.). The MIT Press, Cambridge, MA, pp189-204, 1992.
21. Vernadat, F.B., *Enterprise Modelling and Integration: Principles and Applications*, 1996 (Chapman & Hall, London), ISBN 0 412 60550 3.
22. Warnecke, H. J., (1993), "The Fractal Company – A Revolution in Corporate Culture", Springer-Verlag, New York, ISBN 3-540-56537.
23. Weston R H., Universal Machine Control System Primitives for Modular Distributed Manipulator Systems. *International Journal of Production Research*, 1989, Vol. 27, No. 3, pp 395-410.
24. Weston R.H., (1996), "'General Issues Involved in Enacting Enterprise Models'", Proceedings of Process Modelling for Enterprise Integration with CIMOSA, March 1996, CIMOSA Association, Boblingen, Germany, pp 5.1-5.15.
25. Weston, R. H., Edwards, J. M. and Hodgson, A., 1995, "Model Driven CIM: A framework and toolset for the design and management of open CIM systems" January 1995, Final Report to EPSRC/CDP.
26. Weston, R.H., 1999, "A model-driven, component-based approach to reconfiguring manufacturing software systems", Int. J. of Operations and Prod. Management, Special issue on Responsiveness in Manufacturing, Vol.19, No.8, 1999.
27. Weston, R.H., Gascoigne, J. D. and Gilders, P. J. (1998), "First Stage Development of the Draft Mandate Generated at Boulder, February 1998", ISO TC 184/SC5/WG1 and the IFAC/IFIP Task Force on Enterprise Architecture, Paris, April 1998.

28. Weston, R. H., 1998, "A comparison of the capabilities of software tools designed to support the rapid prototyping of flexible and extendible manufacturing systems" Int. J. Prod. Res., 1998, Vol. 36, No. 2, 291-312.
29. Weston R.H., Clements P E, Shorter D N, Carrott A J, Hodgson A and West A A, 2000, "On the Explicit Modelling of Systems of Human Resources", to be published in *International Journal of Production Research* Special Issue on Modelling, Specification and Analysis of Manufacturing Systems, Ed. Francois Vernadat and Xiaolan Xie.
30. Williams, T.J., "The Purdue Enterprise Reference Architecture", Instrument Society of America, Research Triangle Park, NC (1992).
31. Zalenovic D M., Flexibility – A Condition for Effective Production Systems, *International Journal of Production Research*, Vol. 20, No. 3, 1982, pp 319-337.

Gaining Agility Through Supply Chain Management

Tareq Suleman and Mohamed Zairi
Management Center, University of Bradford, Bradford, United Kingdom.

1. INTRODUCTION

The 1990's has seen the emergence of a new management concept – Supply Chain Management (SCM). Cynics would say, so what? – Management fads have come and gone like any other fashion, its just another acronym that will soon loose its appeal. But will it? SCM has been around for a very long time – if not in its present guise. For decades, companies have constantly tried to achieve competitive advantage through inventory reduction, improved customer-supplier relationships, lean manufacturing, etc. SCM which was introduced in the late 1980's – early 1990's, is the concept that has brought together these techniques towards building an integrated supply chain, which will bring benefits to both suppliers and customers such as: cycle-time reduction, improved customer satisfaction, reduced inventory, etc.

Through the following fictional transcript of an annual vendor meeting the two sides of SCM can be shown very clearly: the blunt meaning and the euphemistic translation (Thomas, 1998). Although hypothetical, the above passage could be true of any organisation – with the different attitudes towards the supply chain.

This project seeks to investigate fully the new business management concept of SCM – to show that it has the foundation and resilience like JIT and TQM, of not becoming another expensive business failure. In essence, SCM looks to integrate the whole process of moving goods, to the ultimate end-user – the consumer, from raw materials to finished goods.

The project's objectives were to:
- study the evolution and development of SCM
- document best practice SCM organisations
- document Critical Success Factors (CSF's) for effective SCM
- develop using CSF's, an Audit tool for benchmarking the applications of SCM
- test and validate Audit tool by highly recognised organisations
- finally, develop a best practice approach towards SCM implementation

2. SUPPLY CHAIN MANAGEMENT

The business world has observed great changes in the latter part of the 20th Century, with major progress and changes in the global business environment:
- Political – there has been a stabilisation of many global economies due to political events around the world, e.g. the fall of the Soviet Union, the re-unification of East and West Germany, etc... Markets have opened up and companies have become 'global' to gain competitive advantage and therefore, not only survive but become world leaders.
- Ecological – New legislation by governments to protect the environment and an increasingly aware and active publics concern over health issues and the quality of life, has also been a major factor.
- Social – Consumers habits and expectations are changing rapidly as they seek to improve their standard of living.
- Technological – This, by far, has had the greatest effect on many businesses with the advent of more complex and advanced computer systems, the Internet, EDI, etc... Technology offers the freedom to organisations to be more creative in bypassing the barriers that come along in the process of conducting business.

"Welcome to our third annual Admiral Auto suppliers meeting," said Carl Krass, director of supplier relations. "Let me now introduce our president, Skip Stalker. Because Mr. Stalker is brutally honest, I will interpret his remarks for you today."

"I'll get right to the point," said Stalker. "We posted lousy numbers last year, so I expect everyone here to cut their prices."

"Mr Stalker said good morning," said Krass. "Today, I ask for your help in developing creative solutions that will improve our shareholder value."

"A 5% price reduction is mandatory," said Stalker.

"Mr Stalker said that he was proud of our relationship with you," said Krass. "Yet Admiral Auto faces tremendous competition, both domestically and offshore. To remain competitive, we plan to introduce a program of Supply Chain Management."

"I'll be blunt: if you don't agree, we can look for other suppliers," said Stalker. "Mr. Stalker said that partnerships would be a critical component of our long-term strategy," said Krass. "We want you to grow as a valued member of the Admiral family."

"Remember, scores of foreign suppliers would kill for the chance to do business with one of the world's top automakers," said Stalker. "And many of their economies are in the hopper right now, so the exchange rates have become pretty darn attractive."

"Mr. Stalker said that the Supply Chain Management will provide us with significant global opportunities," said Krass.

"We plan to reduce our supplier base, so some of you won't be here next year," said Stalker.

"Mr. Stalker said that we should work openly in an atmosphere of co-operation, not one of leverage and intimidation," said Krass. "You should keep an eye open for ways to improve our costs," said Stalker. Mr. Stalker said that Supply chain Management is a two way street," said Krass. "We obviously welcome your suggestions."

"We will reduce the number of Admiral facilities by 40% next year, said stalker. "In accordance, we will require suppliers to carry more of our inventory."

"Mr. Stalker said that one of our goals is inventory reduction," said Krass. "Through Just In Time deliveries, we can maximise our efficiencies and save money." ...

..."My entire compensation package and my summer home in the islands – is based on achieving this plan," said Stalker. "So do it!"

"Mr. Stalker said that this proposal fits well with the theme of this year's conference, building partnerships for 2000," said Krass. "Now, I will be happy to answer any questions."

(Thomas, 1998)

To stay competitive and become world class organisations, businesses have had to improve their processes and operations internally to keep pace with the fast moving changes (Handfield and Nichols, 1999). The origins of many popular management concepts can be traced back over the past few decades. In the 1960's, businesses were in the early stages of developing comprehensive marketing strategies, the purpose of which was to meet and exceed customer needs and build 'customer loyalty' along the way. This continued through the 1970's with organisations realising that to support these marketing strategies they had to excel in other areas of their business. Investments and new ideas in R&D, engineering and manufacturing, ensured customers needs were translated into the required specification, at the right quality, but above all, at a fair price. The 1980s, was a prosperous decade and resulted in an increased demand for all types of products – both new and old. Customers needs were ever changing and organisations had to become flexible and reactive to meet this new demand. To cope, new products had to be introduced, as did new automation and processes. Management concepts such as MRPII, FMS, JIT, TQM, etc… were being implemented by many organisations so as to become leaner and more competitive.

In the 1990's, many businesses were heading towards world class manufacturing status – but the changes that led to this were all internally driven. Managers realised that their businesses alone could not meet customer needs 100% - the material, information and service input of their suppliers had a major impact on them. Also, no longer was it enough for a product to be manufactured to a high degree of quality, but it should also be available in the right quantity, at the right price, at the right time and at the right place. Together, with the recent re-awakening of the old concept of Logistics, this led to an increase in attention on not only the supply base, but also the organisations overall strategies from sourcing and procurement to warehousing and customer service.

Organisations have realised that they have to step out of their 'cocoons' and manage not only themselves, but also the complex network of all 'upstream' firms that provide the inputs and downstream firms that are responsible for product delivery or the after sales service to the end consumer. A new era was born – the concept of Supply Chain Management (SCM).

2.1 SCM Definition

The supply chain, by definition, encompasses all activities associated with moving goods from raw material stage through to end consumers. It includes sourcing and procurement, production scheduling, order processing, inventory management, transportation, warehousing, and customer service. More importantly, it encompasses the information systems used to monitor these activities (Palevich, 1999). Supply Chain Management (SCM) is then the integration of all these key business activities through improved relationships at all levels of the supply chain (internally, suppliers and customers), to achieve competitive advantage and hence, a world class organisation.

A major misconception about SCM, to this day, is that some people still refer to it as Logistics. Logistics is a key component of SCM and is associated with moving the product through the supply chain to the consumer. The definition given by the Council of Logistics (Palevich, 1999) is the process of planning, implementing and controlling the efficient, effective flow and storage of goods, services and related information from point of origin to point of consumption, to meet the customer requirements. In other words, quite simply - the process of getting the right product, to the right customer, at the right time and at the right price.

2.2 SCM Concept

The definition of SCM can be further illustrated if we consider the example of an individual firm and both its upstream suppliers and downstream distribution channel (see Fig.2.1). The diagram in Fig.2.1 is a very simplified example of a supply chain and could be that of a small manufacturing firm (the 'focal' firm). The shaded box represents the focal firms full supply chain, while the bi-directional arrows represents both product and material flow as well as accompanying information flows within the supply chain. The number of feedback loops makes this an 'integrated' system. In the upstream supplier network, S1

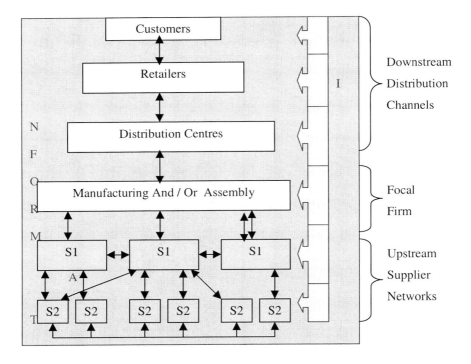

Fig. 2.1: The Supply Chain (Adapted from Handfield and Nichols,1999)

represents the '1st Tier Suppliers', while S2 denotes '2nd Tier Suppliers'. The supply chain encompasses, sourcing and procurement, production scheduling, order processing, inventory management, transportation, warehousing, customer services as well as the management of information systems. Most organisations are simultaneously members of a number of supply chains. In each of these chains, they could be offering or receiving a number of products, raw materials, services, etc, ... A supply chain can be complicated and long - as in the case of the aerospace or automotive industries, or simple and short – as in the case of the local family bakery.

The supply chain from a focal organisational viewpoint encompasses its own *internal operations*, its *upstream supplier networks* and *downstream distribution channels* (Handfield and Nichols, 1999).

2.2.1 Internal Operations

The major activity that takes place within the focal firm is the transformation of inputs supplied by its network of suppliers into inputs (see Fig.2.2). The transformation could involve the manufacture of parts or components or the assembly of manufactured and/or bought-in parts/sub-assemblies. Key areas here are: order processing and production scheduling.
- Order processing translates customer requirements into actual orders. In many organisations this area also involves detailed customer interaction to build 'partnership relationships', so there is no mix up in areas such as product specification, delivery lead-times and after-sales service.
- Production scheduling converts actual orders into production tasks for the shopfloor. Systems such as MRPII and JIT can be used in such areas to assist in scheduling, WIP management, shopfloor control, etc...

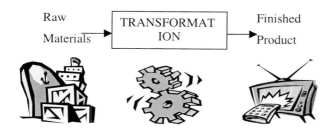

Fig. 2.2: Internal Operations

2.2.2 Upstream Supplier Networks

The management of upstream external suppliers is a major part of SCM. The upstream supplier networks consist of firms that supply inputs (raw materials, parts, assemblies, etc...) to the focal firm. These inputs can be direct from internal divisions of the company or indirect from external suppliers.

The firm that supplies directly to the focal organisation is called a '1st Tier Supplier'. It too, though has its own set of suppliers called '2nd Tier suppliers'. This list can go on, as supply chains are basically a series of linked supplier-customer relationships (see Fig.2.3) until we come to the ultimate supplier – the planet Earth. The Earth provides the essential raw materials for many world organisations, e.g. wood, iron core, coal, petroleum, etc...

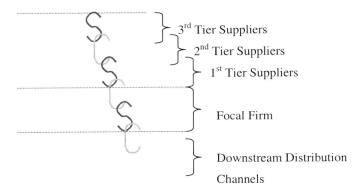

Fig.2.3: Supplier-Customer Relationships

Organisations employ a host of managers (purchasing, materials, supply, etc...) to ensure that inputs such as raw materials arrive in the right quantities, at the right place and the right time. Tasks for such managers can be sourcing, negotiation of contracts, vendor rating, planning, forecasting and relationship management – both internally with functional departments, and externally with suppliers.

2.2.3 Downstream Distribution Channels

The firms external downstream consists of all distribution channels and processes of the supply chain that help the product reach its ultimate destination – the end consumer. This involves inventory management, warehousing, distribution networks, retailers, after sales service, etc... Activities such as after sales and maintenance services ensures continual customer satisfaction and helps the firm not only manage its forward supply chain, but also its backward supply chain.

The length of internal supply chains and the subsequent delivery channels can vary quite enormously, depending upon the type of product. In the aerospace industry for example, the internal supply chain is fairly long, but its distribution channel is quite short - where the finished product is delivered directly to the customer. In the case of sweets or chocolates, the internal supply chain is short but the distribution channel is fairly long and extensive. From point of manufacture the packaged sweets/chocolates are sent to distributors, these are then sent to retailers (supermarkets, grocers, etc...) and only then does the consumer purchase them for consumption.

2.3 SCM's Advantages

In most industries, the supply chain represents 70-75% of the operating budget expense. SCM reduces operating costs, improves asset productivity and reduces order cycle time – all leading to increased customer satisfaction and increased company profits.

If a company is not effectively managing its supply chain, then not only could it be loosing vital competitive advantage, but also falling behind the competition, who are vigorously implementing a total integrated supply chain.

The following two studies carried out by leading exponents of the SCM concept should be motivation enough for any organisation, however small, to start the implementation process (Quinn, 1998):
1. A study by the research and consulting firm of Pittiglio, Rabin, Todd and McGrath found that companies considered to be best-practice organisations in moving product to market enjoyed a 45% supply chain cost advantage over their median competitors. Their order-cycle time was half that of the competition; their inventory days of supply were 50% less. These leading companies met their promised delivery dates 17% more often than the competition.
2. A study by management consultants, A.T.Kearney of Chicago, found that inefficiencies in the supply chain could waste as much as 25% of an organisation's operating costs.

2.4 SCM and The Environment

As discussed earlier, the business world has observed great changes in the latter part of the 20th Century, not least in the environmental/ecological areas. New legislation by governments to protect the environment and an increasingly aware and active public has made this a critical issue for many businesses – no longer can they ignore it and hope that it will go away.

In the 1990's, companies are integrating their supply chain processes to lower costs and better service their customers. So how do businesses balance the need for environmental awareness on one side and the need for competitive advantage on the other? Many organisations have brought these issues to the fore by including environmental issues in their strategic planning. Focal organisations are meeting members of their supply chain (suppliers and purchasers) to ensure that they meet and exceed the environmental expectations of not only their governments, but also their customers.

A new area of SCM that has emerged over the past few years has been that of Recycling. Organisations are now collecting and recycling products that have reached the end of their Product Life Cycles (PLC) and are no longer required by the customer. Many firms have now extended their downstream distribution channels to include recycling of not only the product but also its packaging. The aim for many organisations is to become 'Eco-friendly' and eventually, recycle all used products into new or return materials that cannot be used for recycling back to the Earth, without harming the environment.

In a recent study (Walton et al, 1998), a number of supply chain Environmentally Friendly Practices (EFP) were identified:
1. EFP in Product Design and Purchased Materials: Product design and purchasing personnel should work together to influence environmental improvements in their own and their suppliers products. This can be achieved by substituting or changing material specifications for purchased products, and avoiding the use of hazardous or Environmental Protection Agency (EPA)-regulated materials where possible, to reflect the environmental agenda for the company.
2. EFP in Product Design Processes: Product design processes must consider the lifecycle of all materials used in the product, including 'cradle to grave' considerations. This can be accomplished by promoting a dialogue between designers and material experts and the use of tools such as life cycle analysis, Quality Functions Deployment (QFD), etc.
3. EFP in Supplier Process Improvements: Purchasing managers must proactively influence suppliers processes, since liability for non-compliance to environmental regulations extends to all supply chain members. This can be accomplished by understanding core

supplier processes and materials, and the environmental regulations associated with these processes and materials. High-level support of these activities is critical for success.
4. EFP in Supplier Evaluation: The methods used and the criteria emphasized for supplier evaluation must reflect the strategy direction of the buying company's environmental initiatives. This is accomplished by first selecting criteria, which focuses on meeting government regulations, followed by proactive criteria focused on process improvements.
5. EFP in Inbound Logistics Processes: Suppliers must help buying companies change inbound logistics processes to reduce waste (e.g., packaging) which, in turn, can yield an operational advantage (e.g., cost and ease of assembly). This is accomplished by initiatives such as reusable containers for material delivery.

The concluding message provided in the study is (Walton et al, 1998): companies must proactively manage supply chain environmental initiatives and seek higher benchmarks rather than simple compliance with government regulations. Proactive strategies must be supported with adequate resources and cannot just be given 'lip service'. A number of cross-functional and inter-organizational processes must be addressed, including product design, supplier processes, supplier evaluation systems and inbound logistics. Proactive environmental strategies also have another benefit: cost and waste reduction objectives can be achieved.

3. SCM TOOLS

Along with the *internal operations*, *upstream supplier networks* and *downstream distribution channels* discussed in chapter two, the following SCM tools (Handfield & Nichols, 1999), are a further aspect of an integrated supply chain and provides a firm foundation on which an organization can build upon:
        Information Systems Management
        Inventory and Time Management
        Relationship Management

The integrated supply chain diagram introduced in Fig.2.1 will be used to assist in describing how critical these three tools are on the whole supply chain.

3.1 Information Systems Management

The pace of technological change has had a profound effect on all aspects of modern life. The immense degree of transition created by the continuous technological advancements has radically altered society and especially so in the communications field. The various types of technologies have made it possible to make infinite amounts of information available to almost anyone; the only equipment needed being a personal computer.

The rapid development in computers and associated hardware and software (e.g. laptop's, Internet, EDI,...) has allowed many organisations to become world class businesses by building linkages and speeding up information transfer/sharing with other businesses in their supply chain. This in turn has led to major cost savings and in many, a 'paperless' organisation. The paperless organisation concept refers to businesses embracing new ideas in the field of information technology – E-commerce and SCM software are two examples that will be discussed later. Organisations that have implemented the paperless organisation concept 100% have seen major cost savings and improvements in the quality and flow of information:
- There is no longer a need for paper purchase orders / invoices, goods received notes or the manual 'matching' process carried out in the accounts department.

- Companies have developed systems whereby finished goods products and their associated codes are stored and coded into a database. This allows stock levels to be monitored more easily and orders placed automatically.
- With the advances in IT, real-time communication has become a potent reality. Organisations can now communicate on-line with all members of their supply chain – from 1st and 2nd tier suppliers to distributors and end consumers.

The improvements outlined above can not only be seen in firms functional departments, but also their transactions with other members of the supply chain – because everything is done electronically with little or no human intervention.

The supply chain encompasses an organisation's upstream as well as downstream networks, therefore any information systems implementation should be an integrated approach, whereby information can be passed or shared by all members (see Fig2.1). The IT infrastructure of each member of the supply chain should be in a position to support this communication effort, as well as its databases and operating systems.

The sharing of information in a supply chain is critical to allow other tools of SCM such as JIT and Relationship management to work effectively and efficiently:
- JIT – information sharing allows JIT deliveries to occur between suppliers and customers, reduce inventory and hence cut costs.
- Relationship Management – information sharing allows two-way communication between a focal firm and its suppliers and customers. Changes in consumer demand can be managed, payment of goods can be quicker, etc...

For an effective supply chain, managers have to first see what type of information is required and check that it is accurate. Distorted information at one level can have a snowball effect throughout the whole supply chain. An example of this could be a rather erratic forecast driving demand through the supply chain and building unnecessary levels of stock or giving extended lead times due to stock outs. Secondly, how this information can be used and shared across their supply chain through the use of emerging technologies.

As mentioned earlier, information systems in the supply chain are critical both internally within a focal organisation and externally with both upstream suppliers and downstream distribution channels. Much has been written about internal organisational information systems and the benefits that can be derived from its implementation. Today, however, to take the supply chain concept further, an equal amount of importance is being placed on external systems: Inter-Organisational Information Systems (IOIS), as stated in (Handfield and Nichols, 1999). IOIS's are described by (Bakos, 1991) as 'systems based on information technologies that cross-organisational boundaries'.

In its simplest form, an IOIS consists of a database of information that can be used by two or more members of a supply chain. The integrated system would be electronically linked between organisations to help speed up some of the trivial or repetitive tasks (order status, product lead-times, payment information, ...) that usually goes on between them. The level of information sharing between organisations may differ for many reasons (information security, type of relationship between the parties, ...), but many are implementing some type of integrated IOIS, as it is an on-line – real time system that reduces cost and brings about major productivity savings throughout the supply chain(see Fig.3.1).

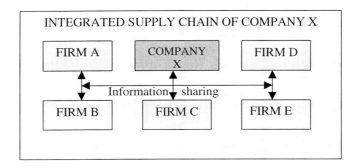

Fig. 3.1: Fully Integrated IOIS System

The world of IT is changing at such a pace that today's fad could be tomorrows dud. This is the case for SCM; more sophisticated and cheaper hardware and software are being developed to support the importance of information in the supply chain.

3.1.1 Electronic Tools and Techniques

A prime example of 'information freedom' is the 'Internet' or the 'Information Super Highway', as it is sometimes called.

The Internet gives instant access to thousands of users locally and also millions of users globally, with thousands joining the global network each month (Long and Long, 1998). Potential customers can quickly find various companies through the various search engines available through the Internet without even leaving the comfort of their own desks. Many businesses have embraced the Internet as a tool to help them achieve a competitive advantage. The following are the major uses that the Internet can offer to a business: to establish a presence both nationally and internationally, serving customers more efficiently, making business information available to all and Internet Marketing and Advertising.

From a business point of view, the most common uses of the Internet are the World Wide Web (www), Electronic Mail (E-mail) and Electronic Commerce (E-Commerce). Generally, there are numerous applications of the Internet. Some of these include Information gathering, marketing and advertising, discussion groups, on-line shopping, etc. The information available on the Internet is endless (Collin, 1997).

When people talk about using the Internet they normally refer to the use of the **World Wide Web**. The WWW, or Web, as it is also known is an application of the Internet, which is a graphical and an easy to use system that offers an enormous amount of information to be accessed. Any firm that wishes to have a presence on the WWW will have to obtain a Web Site, which stores the various pages that can be viewed. The WWW is the part of the Internet, which is primarily used for information gathering from around the world. This is the area of the Internet, which has pages and pages of information together with visual images and sound. Many organisations have 'home pages' on the web, which can be accessed by anyone and these basically act like advertisements for the firm. Although money can be saved by reduced printing costs of brochures, for Web pages to be any use at all, they must be kept up to date. However, the advantage being that while a brochure will soon go out of date, a well sustained Web site will not.

Using **E-mail** through the Internet allows the exchange of messages with various organisations through out the world. Once initial set-up costs for connection to the Internet have been paid for, there is usually no charge for sending or receiving E-mail. The use of E-mail is probably the most popular feature of the Internet because it is fast, easy to use, inexpensive, saves paper and usually the one that draws people to the Internet in the first place (Randall, 1995). E-mail is cheaper than using the telephone or faxing; the scenario of engaged lines is also avoided. In order to increase business efficiency, E-mail messages can be easily and accurately prioritised. This can avoid the interruption of unimportant calls.

There are mainly two forms of **Electronic Commerce**:
(1) Where businesses can provide a service for customers to buy products over the Internet and;
(2) Electronic tools such as Electronic Data Interchange (EDI) where businesses can interact with other organisations such as their suppliers and exchange information such as purchase orders, invoices and eventual payment. The process of sending orders by phone, post or fax can be tedious, time consuming and costly. When information from one end is received at the other end, it has to be retyped into another system where, as with EDI, the information is already in the system and orders for example are ready to be processed. Human intervention is therefore minimised, thereby reducing the risks of mistakes. Other forms of electronic commerce are electronic funds transfers, electronic publishing, etc. (Palevich, 1999) shows an example of some of these tools being used.

According to experts (Anonymous-Accountancy, 1998), the main problem which is holding the Internet back as a complete business medium is the lack of a determined and protected standard for participating in financial transactions. For this reason, many people may browse the Internet and enquire about certain products but do not actually purchase them through the Internet. However, many industry analysts believe that with continuous developments, electronic commerce will become routine in the future for all businesses. Further developments of the Internet are Intranets. These are networks similar in design and technology to the Internet but are specific to only one organisation. Intranets are being developed by organisations to improve their internal information systems and then being extended to join forces with suppliers and customers in their own supply chains.

Other relevant technologies that have emerged recently to assist in supply chain management are bar coding/scanning of products, data warehousing and warehouse management systems (Palevich, 1999).

3.1.2 Supply chain Software

There are currently hundreds of supply chain or related software available on the market, and the feeling is that this area will see continued growth due to the popularity of the SCM concept. According to (Hicks, 1997) there are four types of supply chain software:
1. Enterprise Resource Planning (ERP) Software: Formerly known as MRP, ERP is based on the concepts of MRP which are, however, integrated with other functions of the organisation that are not manufacturing/planning, but are somehow related. ERP systems represent the software infrastructure that facilitates the flow of information between all functions in a company (manufacturing, logistics, finance, human resources, etc.) ERP systems can be visualised as huge database applications for storing transaction data.

When an order is received by a sales department it is entered onto a computer. This order then generates an internal order to the warehousing department (if it is in stock) or the manufacturing department (if it is not in stock) so it can be produced. As a result, specific parts may have to be re-ordered from suppliers. Each step, in turn, will have accounting implications (debit the accounts receivable account, purchase supplies, etc.)

ERP deals with the toughest problem of all – organising and executing the millions of transactions that operating a complex business entails. The systems use very little or no modelling and problem solving techniques – basically it is a data processing system, that sees the supply chain as database management problem, not a mathematical modelling and optimisation problem. Recently, however, some ERP systems are being implemented with problem modelling/solving capabilities.

2. Supply Chain Management (SCM) Software: Unlike ERP, SCM software doesn't try to track down all business transactions. The software focuses primarily on matching up supply and demand of businesses, rather than tracking individual transactions themselves. Most of these applications are based on sophisticated forecasting methods for demand planning, a production planning and scheduling module for supply planning, and some analysis tools for examining supply-demand match-ups.

The best SCM tools turn on their ability to gather and synthesise actual point-of-sale data and turn it into useful demand data, while pulling together actual supply information (inventory-on-hand information, production schedule status, etc.) relevant to decision making. Some of these tools have a strong slant towards the transaction side, some towards the modelling and optimisation side, but all of them aim to bridge the gap between real time data collection and monitoring, and off-line analysis.

SCM software has been growing rapidly over the years, as businesses who have implemented ERP in the past, have come to realise that it is no longer enough to know what is going on – they also need to know what to do about it. Hence, companies often use ERP, for their day-to-day running and use SCM systems to make decisions about improving their week to week supply chain operations.

3. Constraint Based Optimisation Tools: This is a heuristic, or rules-based approach to finding solutions to very complex deterministic models. This approach closely integrates the building of a system model with the rules used to find solutions for it. Often this process can provide pretty good solutions to huge problems while it's still early enough to do something about it.

With optimisation tools, operations research and mathematical techniques are used to model and solve specific operations issues within a business. Focusing on areas such as transportation resource optimisation, vehicle routing and scheduling, inventory allocation and manufacturing schedule optimisation, these tools do not claim to be total solution. Rather, they are decision-making tools to be used in addition to software required to run a business.

4. Analysis Tools: They are used largely for analysis and understanding of system dynamics, or as strategic design tools. Many old-traditional vendors have failed to adopt optimisation capabilities, and while they play a role in academic research and high-level planning, their lack of focus on defining a relevant role in real-world decision making has severely hurt their future prospects.

3.2 Inventory & Time Management

The second tool that has to be effectively utilised by organisations is that of inventory and time (cycle) management.

Products are evolving, markets are changing and technology is advancing rapidly. The business world is changing rapidly for both customers and organisations. Customers of today are more and more demanding, they want high quality products and services – manufactured and delivered to the highest standards. Organisations, therefore, have to respond to these needs, otherwise the competition is out there – ready to pounce and lead customers away. Businesses have realised that they need to be more responsive to the needs of their customers and offer shorter cycle times in both manufacturing and delivery. In other words, become 'lean, mean business machines' – its not jut good enough to survive, they have to become world leaders. Supply chain mistakes, therefore, cannot be tolerated because as mentioned earlier, customers can turn way to the competition.

An effective supply chain that utilises the tools of inventory and cycle management will be in a position in any situation to deliver top quality products, quickly and accurately. These companies have implemented business wide philosophies of Quick Response (QR) and flexibility into their day-to-day culture (Handfield & Nichols, 1999). Basically, Handfield and Nichols describe these companies as paying attention to time – be it lead-time reduction, time compression, QR, etc , these time based methods are an effective means by which a company can grow, make large profits and control operational problems of overheads and inventory costs.

Inventory has always been a problem for many organisations, but with an integrated supply chain, inventory can decrease throughout the chain. Instead of holding large stockpiles of products – time based firms cut their costs by managing their businesses so that inventory/materials flow smoothly with little or no delay between members of the supply chain, in the required quantities, right destination, at the right time and at the required specification.

3.2.1 Logistics

As mentioned previously in chapter two, logistics is an integral part of SCM and is associated with moving inventory (materials or goods) smoothly through a supply chain (see Fig.2.1). The definition given by the council of logistics management is – 'Logistics is the process of planning, implementing and controlling the efficient, effective flow and storage of goods, services and related information from point of origin to point of consumption to meet the customer requirements'(Palevich, 1999).

Logistics includes the following main areas:
- Warehousing
- Inventory Management
- Transportation

Engler and Natalie (1997) state that according to the US department of commerce, nearly 60% of all Fortune 500 companies logistics costs are spent on transporting products from manufacturers to distribution centres or retailers. Furthermore, transportation costs chew up 2% to 8% of a companies sales…therefore, for a multibillion dollar organisation, shaving even 1% off these costs can add up. So huge is its impact on SCM, many organisations are implementing logistics or finding new ways to enhance logistics capabilities throughout their supply chains – with world class SCM organisations such as Wal-Mart leading the way.

Two major areas that are impacting on logistics are the role of IT and third-party logistics vendors. Many companies prefer to keep 100% control of their logistic processes, but evidence (Engler and Natalie, 1997) has shown that outsourcing and building partnerships with third party logistics providers is becoming more and more popular. The KPMG survey (Engler and Natalie, 1997), of 360 logistics professionals has shown that almost 40% of respondents planned to rely more heavily on third party providers in the closing year.

Logistics is one area where paperwork never seems to end, from purchase orders to routes and maps. The survey by KPMG, further indicated that whilst most of the respondents believed that, IT would eventually have a major impact on logistics, only 21% said that such systems are currently well-integrated with the logistics processes within their organisations. Typical IT tools that can alleviate much of the many problems associated with logistics range from warehouse management systems and bar-coding and scanning to transport routing systems.

3.2.2 Cycle Time

As mentioned earlier in the chapter, organisations that pay attention to time and deliver customer requirements in the right quantity, at the right time and at the right place, will gain competitive advantage over their rivals. The reduction of the time required to deliver a product to a consumer is a major cycle time problem for many businesses. Within an integrated supply chain, the overall cycle time reduction exercise, can be broken down into its components throughout the supply chain. The three parts that make up the supply chain – upstream suppliers, focal organisation and downstream distribution channels, can all look at ways to reduce their own internal cycle-times (see Fig 2.1).

Areas for cycle time improvement are:
- Planning / Forecasting Cycles
- Procurement / Purchasing Cycle
- Manufacturing Cycle (batching, lean manufacture, machine changeovers,...)
- Inbound Logistics (transportation, goods in, warehousing, inspection, ...)
- Customer service (order processing, ...)
- Outbound logistics (transportation, warehousing, documentation, ...)
- Customer Service (returns, warranties, ...)

Much has been written about methods/techniques used to reduce cycle times within specific organisations. As (Handfield and Nichols, 1999) state, there is no one 'right-way' to do this, instead they propose a method based on the process-improvement approach developed by (Harrington, 1991) which is focused on cycle time performance:

1. Establish a Cycle Time Reduction Team (CTRT) - team members should represent all functions of the organisation that have a direct impact on the process being investigated.
2. Develop an understanding of the given supply chain process and current cycle time performance. CTRT teams in 'workshops', develop an understanding of the current process and its associated cycle time performance characteristics.
3. Once point two has been completed, identify opportunities for cycle time reduction.
4. Develop and implement recommendations for cycle time reduction.

3.2.3 Pull-Push Strategies

A major area that has a great bearing on cycle times is that of pull and push strategies. Although a pull strategy (JIT) is synonymous with SCM, push strategies cannot be discarded completely – because a pull strategy that works for one organisation might not work for another. That organisation might then decide to use a hybrid system that will be 'fit for purpose'. (Tompkins, 1998) explains the different types of strategies very simply and accurately:

<u>Push Strategy:</u> In a push environment, a product goes directly from the palletizer at the manufacturing facility to the distribution facilities. There's absolutely no room for storage. Product is made and shipped to the distribution centres based on forecasts of demand. The advantages of this system are:
1. There's no requirement for warehouses at the manufacturing sites
2. Customer satisfaction is high because plenty of stock is available and delivery time is short
3. Product is shipped from the factory in full truckloads, saving costs and improving scheduling and handling.

However, the Push strategy has the following disadvantages:
1. Forecasts are inaccurate, safety stock is required at different points and therefore larger distribution centres
2. The larger the distribution centres are, the more safety stock is required – thus increasing costs and decreasing the ability to forecast.

<u>Pull strategy:</u> Here a product is not shipped until something – a retail sale, for example, triggers an order for replenishment. Advantages for this system are that:
1. There is no finished goods or manufacturing inventory
2. There is no concern about forecast accuracy because of the pull system
3. Pulling allows links in the chain to be cut, as deliveries are made direct to retailers.

There are, however, two minor disadvantages to the pull system:
1. Small lots may have to be manufactured, which may cause inefficient manufacturing
2. The assumption that all customer orders can be fulfilled may not be possible for a number of reasons.

<u>Hybrid System:</u> A system of this kind may allow an organisation to push its popular products and pull the slow movers. If a product is being made that everybody uses in quantity- the push system is used. The forecast doesn't have to be that accurate and if a stockout does occur, the organisation can always catch up. The slow moving products can be left back in a central location and pulled from that. The advantage of a hybrid system is that a smaller warehouse can be used at manufacturing and the higher volume products move in truckload quantities to the distribution centres. The advantages of a pull over a push system are:
- Reduced stock transfers
- Reduced safety stock
- Direct customer shipments
- Improved customer service

3.3 Relationship Management

Relationship management is the key tool that links all members of a particular supply chain. How strong or weak it is, will dictate the strength of the supply chain and its effectiveness (see Fig2.1).

Effective relationship management in a supply chain will breed mutual trust and confidence through all members, by breaking down not only functional barriers internally, within an organisation, but also creating vital links with 1st and 2nd tier suppliers as well as downstream distribution channels.

People are an important part of the SCM concept and for that matter, relationship management. All organisations within a supply chain must ensure that all their employees from top down understand the SCM concept and the importance of supplier-customer relationships. Training and awareness workshops can facilitate this. An important factor in all this is that while the internal training workshops will educate people within an organisation, a fully integrated approach would require cross-organisational workshops. The starting point for these workshops would be the focal organisation initially understanding it own internal processes and ensuring there was cross-functional integration. Only then would meetings be set up with suppliers and customers, so as to understand their processes, share information on SCM and ultimately build trust and confidence.

For an effective supply chain, the adversarial relationships of the past, where there was no mutual trust between parties must be put aside for a more open and two-way relationship – where both parties can benefit. Much has been written about the different types of relationships that exist between suppliers and customers and the different expectations they have of each other. (Groves and Valsamakis, 1998) state from their research that there are basically three generic models for supplier-customer relationships, they are adversarial, semi-adversarial and the partnership model. Section C in the Audit tool chapter 5.0 is based around this model. (Moody, 1992) stated that there were 24 characteristics in the customer/supplier relationship (see Fig. 3.2). These characteristics were used by the Association for Manufacturing Excellence (AME), who completed a survey on industry executives, to describe 'best customer'. Early supplier involvement, mutual trust and involvement in product design were rated as most important, while, surprisingly characteristics that were not rated highly were negotiation and award process, technology sharing and training and education.

Choosing a supplier on price alone will not guarantee 100% customer satisfaction when it comes to product specification, deliver times, etc... In firms where SCM has been implemented or is about to be implemented, suppliers are chosen on a long-term basis, so that an effective partnership can be built up with clear goals and objectives.

To build mutual trust between two parties, an effective vendor rating or performance-measurement system between suppliers and customers can be set up, to measure against different parameters laid down in the negotiations, e.g. delivery lead-times, price, specifications, etc... A word of warning when using performance-measurement systems, is that it should not be solely based on price alone, as this could breed mistrust and break down a crucial link in the supply chain (Groves and Valsamakis, 1998).

The pre-requisite for effective relationships both internally and externally within a supply chain is communication. Communication within a supply chain must be multi-directional. When used effectively, it will virtually eliminate misunderstanding between two parties and lead to much more involvement in the common task by all members of the supply chain.

Effective communication can lead to better decisions, increased trust and commitment and therefore, speedier implementation of ideas and plans.

Early supplier involvement	Business growth potential
Negotiation and award process	Profitability
Schedule sharing	Supplier evaluation
Length of planning horizon	Purchasing & other contact personnel professionalism
Schedule stability	Site visits
Mutual trust	Training and education
Technical support	Crisis management/response
Technology sharing	Preferred status
Involvement in product development	Partnership for growth program
Quality initiatives	Management commitment to partnerships
Response to cost reduction ideas	Purchasing commitment to partnerships
Identification of cost drivers	

Fig. 3.2: Customer/Supplier Relationship Areas

4. SCM IMPLEMENTATION

'Thinking is easy, acting is difficult, and to put ones thoughts into action is the most difficult thing in the world.' Goethe (Supply chain council web page, 1999)

In the attempt to achieve competitive advantage and world class performance, many organisations fail to design their Supply Chains thoroughly and effectively. Instead, most concentrate on only one or two of the supply chain building blocks and, therefore, fail to realise the full potential and benefits that can be achieved.
The Critical Sources Factors (CSF's) identified and used in the Audit tool have been used to develop a best practice approach to Supply Chain Management Implementation.

Figure 4.1 shows the six pillars of SCM- the bedrock principles an organisation will have to attain to achieve a fully integrated Supply Chain Management Program.

4.1 Commitment and Leadership

Commitment must come from the top, but like all things the Supply Chain exercise will fail if it does not have the wholehearted support of every employee within the organisation (Palevich, 1999).

If change is imposed or forced from the top, employees can very easily get disillusioned and loose trust. What is needed is a clear vision and mission statement from the top that can be broken down through the organisation as plans and tasks, so employees can become empowered and participate with little or no encouragement. Leadership is, therefore, essential in this environment, from the CEO down to the line manager. That is because an effective supply chain management program requires major transformations, changes and collaborations within the focal organisation itself and outside supply chain members. There could be difficult times for the focal organisation with changes requiring time and persistence, therefore, leadership and commitment from the top is essential in showing the way.

The mission statement must be aligned with the overall business strategy (Keifer and Novack, 1999). The key to supply chain strategy should be the level of service the company intends to provide it's different customer segments, so as to compete successfully.

Many successful supply chain organisations appoint one executive with overall responsibility of the supply chain program to oversee its implementation and then it's effective day-to-day operation.

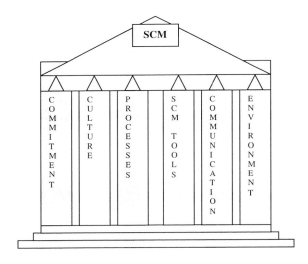

Fig. 4.1: Six pillars of SCM

4.2 Culture and Infrastructure

For an effective implementation, leaders of organisations need to build commitment and confidence, through developing a culture that supports change (Oakland, 1998). Everybody from the top down needs to understand why their organisation exists, what it does, how it does it and who does what?

People cannot get motivated if they are blind to the business and its surroundings. A blame-free culture of empowerment with freedom to make decisions is required. Jobs need to be designed so that they are challenging and departmental barriers need to be torn down.

People are a major factor for any management concept implementation and none more so than SCM. Everybody needs to be involved; their needs recognised, training offered both in house and outside the organisation and relationships need to be fostered through teamwork (Edwards, 1998). Cross-functional integration allows company wide skills and abilities to be used to manage core processes.

The infrastructure of the organisation and its supply chain members need to be understood fully (Anonymous-Occupational Health and Safety, 1998). There could be a need for the re-organisation of physical networks of plants and distribution facilities, to best meet overall business goals. Certain manufacturing and logistical activities could be outsourced to third party vendors as a result.

The supply chain network and its processes, both internally and externally need to be examined as part of the reorganisation. Only the critical members of the 'focal' firms supply chain need to be recognised – too much detailed investigation will lead to a network that is highly complex, making process management a virtual impossibility. To assist the investigation, process maps and flow charts can be used to plot the supply chain members (Upstream 1st and 2nd tier suppliers, the 'focal' organisation and downstream distribution channels) and their related processes, to identify areas for improvement or expansion. The exercise should initially start within the focal organisation and its internal supply chain – only when these are understood should attention move to external members.

4.3 Processes

There are a series of customer/supplier links that not only exist between the focal organisation and its supply chain members, but also that runs through the organisation itself. The strength of these links depends on the supply chain processes, both internally and externally within an organisation (Oakland, 1998).

A process can be seen as the transformation of a set of inputs – (raw materials) into a set of outputs (finished goods) (see Fig.2.2). Processes exist at every supplier – customer link – without efficient processes, links can be weak and mistakes are inevitable, e.g.:

a) External Organisational Process: Purchasing/Procurement department.

The purchasing/procurement department is the customer who buys a raw material from a supplier. The supply chain processes that are required to ensure that a strong supplier/customer link exist are good sourcing, negotiation, quality and specification of goods, vendor certification, inbound logistics, documentation, etc....

b) Internal Organisational Process: Warehousing and Manufacturing.

As stated earlier, for a fully integrated SCM approach to doing business – there should be full functional integration rather than a 'them and us' culture. If the warehousing department is considered the supplier and manufacturing the customer, the processes required to ensure that a strong internal supplier – customer link is maintained are good documentation, internal transportation, loading/unloading activities, etc....

There are many such processes that exist both internally and externally within the organisation. They constitute the first building blocks that have to be implemented efficiently if a totally integrated SCM is the goal. The main supply chain processes that exist are shown below:

- Forecasting (materials and finished goods)

- Planning and Scheduling (Materials and finished goods)
- Manufacturing Process
- Customer Services (Order processing, returns, warranties, ...)
- Purchasing and Procurement
- Recycling
- Logistics to include:
  1. Warehousing: (Materials receipt/inspection, warehouse management, documentation)
  2. Transportation: (Inbound/Outbound)
  3. Inventory management

Best in class approaches have to be adopted for all core processes e.g.: Quality: TQM and manufacturing/distribution: JIT philosophy. Cross-functional teams that break down departmental barriers need to be set up to work on process designs and improvements (Kiefer and Novack, 1999). These teams would be initially set up to improve internal processes and then move on to work with supply chain partners to build linkages in areas such as:
- Joint management of Inbound and Outbound Logistics
- Integrating Suppliers and Customers into the design process

Benchmarking the organisations core processes using an internally developed Audit tool or externally against other supply chain exponents or even using the SCOR model, is critical if an organisation is to gain full advantage and learn from other peoples experiences. Benchmarking external supply chain partners gives an insight to the 'focal' organisation of the processes within their suppliers and customers. Working with their partners, experiences can be shared, guidance sought/given and specifications agreed upon, to avoid problems later on (Quinn, 1998).

4.4 SCM Tools

The tools of information systems management, Inventory & Time management and Relationship management are critical tools for an organisations supply chain and its processes:

Information Systems Management:
The IT structure of the organisation must be in a position to support SCM implementation. Also, an IT strategy must be developed that is aligned with and supports key business processes – what level of data would be required at strategic, operational and functional levels to support this? (Keifer and Novack, 1999)

The organisation must be in a position to capitalise on advancements in IT – in areas of communication (WWW, E-mail, etc.) and E-commerce (EDI, Electronic Publishing, etc.).

Software systems (ERP, SCM, etc.) for demand forecasting, distribution planning, production planning and scheduling and decision-making must be implemented and fully integrated.

The aim for any organisation aiming to reach the heights of a fully integrated system is a 'paperless' organisation with effective flow of information through advancements in IT.

Inventory & Time Management:
The second tool that has to be effectively utilised by organisations is that of Inventory and Time (cycle) management. Organisations need to be in a position where they can deliver top

quality products, quickly and accurately – to do this they need to pay attention to time (Hanfield and Nichols, 1999). Cycle-time reduction techniques need to be implemented so inventory of all types, flows smoothly through a supply chain - Logistics. For a Logistics implementation, the areas of warehousing, inventory management and transportation requires effective management so that full integration occurs not only between themselves, but also with manufacturing related activities (forecasting, planning, procurement, scheduling, etc.)

Cycle-time reduction techniques can be used to reduce cycle times/lead times in all supply chain activities, from the 'focal' organisation through to the upstream suppliers and downstream distribution channels.

The decision to use Pull, Push or a Hybrid system can have a major bearing on cycle-times. The Pull strategy (JIT) is synonymous with SCM and is usually the preferred strategy for many organisations.

Relationship Management:

This is the key tool that links al members of a focal organisation and its supply chain partners. (Groves and Valsamakis, 1998) describes external supplier-customer relationships as either adversarial, semi-adversarial or partnerships. The basis of a focal organisations sourcing decisions, role of Research & Development (R&D) in both parties, quality management, information flow management and performance measurement, will be different in each of the different types of relationships.

In an integrated supply chain, the partnership model is the ideal, where strategic alliances are established with key suppliers and customers. Rather than having a host of suppliers, because of a lack of trust and confidence in any one supplier, in strategic alliances only one or two suppliers are chosen - closer, longer-term ties, are built in areas such as:
- Closer integration of R&D in areas of product, manufacturing and processes development
- Quality of products and processes
- Exchange of information and data at all levels (forecasts, growth plan, etc.)
- Performance measurement internally and supplier/customer satisfaction level assessment.

4.5 Communication

Effective communication is the fifth pillar of SCM and one of the most important ones – for without it the whole idea and message of an integrated supply chain could get lost.

Internal as well as external communications within an organisation and its supply chain are essential between employees and functional departments (Tompkins, 1998).

Communication is the primary tool that can be used by people to express their ideas and values. Senior management can use communication for effective change management and the goal of creating a SCM culture.

A communication strategy needs to be implemented, so as to point out the benefits SCM can bring to all employees as well as the organisation as a whole. This used effectively can stop resentment and opposition building towards change.

Various communication tools (Oakland, 1998), from poster campaigns and workshops to exhibitions and newsletters, can be used to project an organisations missions and objectives, to its employees as well as its supply chain partners.

4.6 Environment

New legislation by governments to protect the environment and an increasingly aware public has made this a critical area for many businesses. Environmental management should be an integral part of the SCM program – for many it is a case of do or die.

Environmental issues should not be something an organisation does if it has the spare time, but should be included in the strategic thinking and planning of the supply chain. Focal organisations should ensure that they are prepared internally, as well as working with suppliers on environmental practices such as (Walton et al, 1998):

- Environmental Friendly Practices (EFP), in product design and purchased materials.
- EFP practices in product design processes.
- EFP in supplier process improvements.
- EFP in supplier evaluation
- EFP in Inbound Logistical processes; recycling of packaging

5. CONCLUSION

Today, we've entered a new era - its called Supply Chain Management (SCM). The supply chain of any organisation (focal) embraces its internal operations, like forecasting, planning, scheduling, purchasing, logistics and customer services, as well as the external upstream supplier networks (1st and 2nd tier suppliers) and downstream distribution channels (distributors, retailers, consumers, etc.). All of these supply chain partners are 'bonded' together through the effective use of SCM tools such as information technology, inventory and time management and relationship management to ensure the effective flow of goods, from raw material stage through to end consumers.

To gain competitive advantage, businesses are turning their attentions to their supply chains to become world-class organisations and reap benefits not only for themselves, but also for their partners. Studies carried out by external consultancies (Quinn, 1998), backed by findings from this project, suggest that many organisations are acutely aware of the benefits that can be achieved through an integrated supply chain – with most citing cost reduction and increased customer satisfaction as their number one priority. Best practice organisations enjoy (Quinn, 1998):

- 45% supply chain cost advantage over their median suppliers,
- cycle-times half that of the competition
- meeting promised delivery dates 17% more often than the competition.

Environmental issues have been at the forefront in recent years. With strong government regulations and an ever-vigilant public, organisations have had to integrate environmental issues into their strategic supply chain thinking - to achieve a 'greener' supply chain (Walton et al).

BIBLIOGRAPHY

1. Bakos, Y. (1991), Information Links and Electronic Marketplaces: The role of inter-organisational information systems in vertical markets, <u>Journal of Management Information systems</u>, Fall, pp. 15-34
2. Collin, S. (1997), <u>Doing business on the Internet</u>, London: Kogan Page Ltd.

3   Edwards, N. (1998), Preventative Medicine, Supply Management, Vol. 3, No. 18, Sept, pp. 30-32.
4   Engler and Natalie (1997), The supply chains most neglected link, Software Magazine, Vol. 17, No. 2, Feb, pp. 72-79.
5   Groves, G. and Valsamakis, V. (1998), Supplier-Customer relationships and company performance, International Journal of Logistics Management, Vol. 9, No. 2, pp. 51-64.
6   Handfield, R. B. and Nichols, E. L. (1999), Introduction to Supply Chain Management, 1st Edition, New Jersey: Prentice-Hall Inc.
7   Harrington, J. H. (1991), Business Process Improvement: The Breakthrough strategy for Total Quality, Productivity and Competitiveness, New York: Mcgraw-Hill.
8   Hicks, D.A. (1997), The managers guide to supply chain and logistics problem-solving tools and techniques, IIE Solutions, Vol.29, No.10, Oct., pp.24-29.
9   Kiefer, A.W. and Novack, R.A. (1999), An empirical analysis of warehouse measurement systems in the context of supply chain implementation, Transportation Journal, Vol.38, No.3, Spring, pp. 18-27.
10 Long, L. and Long, N. (1998), Computers, USA: Prentice-Hall.
11 Moody, P.E. (1992), Customer-Supplier integration: Why being an excellent customer counts, Business Horizons, July-Aug, pp.52-57.
12 Oakland, J.S. (1998), Total Quality Management: The route to improving performance, 2nd edition, Oxford: Butterworth-Heinemann.
13 Palevich, R.F. (1999), Supply chain management, Hospital Material Management Quarterly, Vol.20, No.3, Feb, pp.54-63.
14 Quinn, F.J. (1998), Building a world-class supply chain, Logistics Management and Distribution Report, Vol.37, No.6, Jun, pp.38-44.
15 Randall, N. (1995), Teach yourself the Internet, USA: Sams Publishing.
16 Thomas, J. (1998), The two faces of supply chain management, Logistics Management and Distribution Report, Vol.37, No.3, Mar, pp. 103-104.
17 Tompkins, J.A. (1998), Time to rise above supply chain management, Transport and Distribution, Vol.39, No.10, Oct, pp.16-18.
18 Walton, S.E. and others (1998), The green supply chain: integrating suppliers into environmental management processes, International Journal of Purchasing and Materials Management, Vol.34, No.2, Spring, pp.2-11.
19 Anonymous, (1998), Accountancy, Mar, p. 42.
20 Anonymous, (1998), Benchmarking the best, Occupational Health and Safety, Mar, pp. 3-4.
21 Supply chain council and supply chain operations reference (SCOR) Model overview, (1999), URL: http://www.supply-chain.org/html/slidel.cfm

AUTHOR INDEX

A

Ahmed, P.K.	157
Altinel, I.K.	265

B

Bajgoric, N.	265, 397
Banerjee, P.	535
Bennett, David	483
Bessant, John	113
Birgoren, B.	265
Brennan, Robert W.	499

C

Chan, Michael F.S.	601
Chandra, Charu	437
Chung, Walter W.C.	601
Correa, Henrique Luiz	3

D

Doumeingts, G.	645
Dowlatshahi, Shad	417
Draman, M.	265
Ducq, Y.	645

F

Francis, David	113, 193

E

Engström, T.	621

G

Ginn, David	157
Gunasekaran, A.	25

H

Harrison, R.	703
Hodgson, A.	759
Hua, Z.-S.	535

I

Ilyasov, B.	559
Ismagilova, L.	559

J

Jain, Neelesh K.	515
Jain, Vijay K.	515
Jardim-Gonçalves, Ricardo	735
Johansen, John	53
Jonsson, D.	621

K

Katayama, Hiroshi	483
Kincses, Zoltan	337
Kleinhans, S.	645
Knowles, David	113
Kromm, H.	645

L

Lau, Henry C.W.	205

M

Manthou, Vicky	685
Mavris, Dimitri N.	95
McGaughey, Ronald E.	25, 279
Medbo, L.	621
Meredith, Sandra	113
Mezgár, István	337
Mohanty, R.P.	131

N

Nahm, A.	229

P

Pawar, Kulwant S.	175
Pego Guerra, M.A.	317
Putnik, G.D.	73

R

Riis, Jens O.	53

S

Sarkis, J.	359
Schrage, Daniel P.	95
Sharifi, Sudi	175
Smirnov, Alexander V.	437
Song, S.-J.	583
Steiger-Garção, Adolfo	735
Stokic, Dragan	665
Subba Rao, S.	229
Suleman, Tareq	785
Szczerbicki, E.	247

T

Talluri, S.	359
Thie, Michael	665
Towill, Denis R.	377

U

Unal, A.T.	265

V

Valeeva, R.	559
Vernadat, F.B.	461
Vlachopoulou, Maro	685

W

Wang, Kesheng	297
West, A.A.	703
Weston, R.H.	703, 759
Wolstencroft, V.	25
Wong, Eric T.T.	205

Z

Zairi, Mohamed	157, 785
Zhang, W.J.	317